Fundamentals of Geomorphology

The new fourth edition of *Fundamentals of Geomorphology* continues to provide a comprehensive introduction to the subject by discussing the latest developments in the field, as well as covering the basics of Earth surface forms and processes.

The revised edition has an improved logically cohesive structure, added recent material on Quaternary environments and landscapes, landscape evolution and tectonics, as well as updated information in fast-changing areas such as the application of dating techniques, digital terrain modelling, historical contingency, preglacial landforms, neocatastrophism, and biogeomorphology. The book begins with a consideration of the nature of geomorphology, process and form, history, and geomorphic systems, and moves on to discuss:

- Endogenic processes: structural landforms associated with plate tectonics and those associated with volcanoes, impact craters, and folds, faults, and joints.
- Exogenic processes: landforms resulting from, or influenced by, the exogenic agencies of weathering, running water, flowing ice and meltwater, ground ice and frost, the wind, and the sea; landforms developed on limestone; and long-term geomorphology, a discussion of ancient landforms, including palaeosurfaces, stagnant landscape features, and evolutionary aspects of landscape change.

Featuring over 400 illustrations, diagrams, and tables, *Fundamentals of Geomorphology* provides a stimulating and innovative perspective on the key topics and debates within the field of geomorphology. Written in an accessible and lively manner, and providing guides to further reading, chapter summaries, and an extensive glossary of key terms, this is an indispensable undergraduate level textbook for students of physical geography.

Richard John Huggett is a retired Reader in Physical Geography at the University of Manchester, UK.

ROUTLEDGE FUNDAMENTALS OF PHYSICAL GEOGRAPHY SERIES

Series Editor: John Gerrard

This new series of focused, introductory textbooks presents comprehensive, up-to-date introductions to the fundamental concepts, natural processes and human/environmental impacts within each of the core physical geography sub-disciplines. Uniformly designed, each volume contains student-friendly features: plentiful illustrations, boxed case studies, key concepts and summaries, further reading guides and a glossary.

For a full list of titles in this series, please visit www.routledge.com/Routledge-Fundamentals-of-Physical-Geography/book-series/RFPG

Fundamentals of Soils
John Gerrard

Fundamentals of Biogeography, Second edition
Richard John Huggett

Fundamentals of Hydrology, Second edition
Tim Davie

Fundamentals of Geomorphology, Fourth edition
Richard John Huggett

Fundamentals of Geomorphology

Fourth Edition

Richard John Huggett

Routledge
Taylor & Francis Group
LONDON AND NEW YORK

For additional information and resources please visit the *Fundamentals of Geomorphology* Companion Website at www.routledge.com/cw/huggett

Fourth edition published 2017
by Routledge
2 Park Square, Milton Park, Abingdon, Oxon, OX14 4RN

and by Routledge
711 Third Avenue, New York, NY 10017

Routledge is an imprint of the Taylor & Francis Group, an informa business

First edition published by Routledge 2002
Third edition published by Routledge 2011

British Library Cataloguing in Publication Data
A catalogue record for this book is available from the British Library

Library of Congress Cataloging in Publication Data
Names: Huggett, Richard J.
Title: Fundamentals of geomorphology / Richard John Huggett.
Description: Fourth edition. | Milton Park, Abingdon, Oxon ; New York, NY : Routledge, 2017. |
Includes bibliographical references and index.
Identifiers: LCCN 2016025168| ISBN 9781138940642 (hardback : alk. paper) | ISBN 9781138940659 (pbk. : alk. paper) | ISBN 9781315674179 (ebook)
Subjects: LCSH: Geomorphology.
Classification: LCC GB401.5 .H845 2017 | DDC 551.41--dc23
LC record available at https://lccn.loc.gov/2016025168

ISBN: 978-1-138-94064-2 (hbk)
ISBN: 978-1-138-94065-9 (pbk)
ISBN: 978-1-315-67417-9 (ebk)

Typeset in Minion and Univers
by Florence Production Ltd, Stoodleigh, Devon, UK
Printed by Bell & Bain Ltd, Glasgow

MIX
Paper from responsible sources
FSC® C007785
www.fsc.org

for my family

CONTENTS

FIGURES

PLATES

TABLES

BOXES

SERIES EDITOR'S PREFACE

We are presently living in a time of unparalleled change and when concern for the environment has never been greater. Global warming and climate change, possible rising sea levels, deforestation, desertification, and widespread soil erosion are just some of the issues of current concern. Although it is the role of human activity in such issues that is of most concern, this activity affects the operation of the natural processes that occur within the physical environment. Most of these processes and their effects are taught and researched within the academic discipline of physical geography. A knowledge and understanding of physical geography, and all it entails, is vitally important.

It is the aim of this *Fundamentals of Physical Geography Series* to provide, in five volumes, the fundamental nature of the physical processes that act on or just above the surface of the Earth. The volumes cover climatology, geomorphology, biogeography, hydrology, and soils; they are treated in sufficient breadth and depth to provide the coverage expected in a *Fundamentals* series. Each volume leads into the topic by outlining the approach adopted. This is important because there may be several ways of approaching individual topics. Although each volume is complete in itself, there are many explicit and implicit references to the topics covered in the other volumes. Thus, the volumes together provide a comprehensive insight into the totality that is Physical Geography.

The flexibility provided by separate volumes has been designed to meet the demand created by the variety of courses currently operating in higher education institutions. The advent of modular courses has meant that physical geography is now rarely taught, in its entirety, in an 'all-embracing' course but is generally split into its main components. This is also the case with many Advanced Level syllabuses. Thus students and teachers are being frustrated increasingly by lack of suitable books and are having to recommend texts of which only a small part might be relevant to their needs. Such texts also tend to lack the detail required. It is the aim of this series to provide individual volumes of sufficient breadth and depth to fulfil new demands. The volumes should also be of use to sixth-form teachers where modular syllabuses are also becoming common.

Each volume has been written by higher-education teachers with a wealth of experience in all aspects of the topics they cover and a proven ability in presenting information in a lively and interesting way. Each volume provides a comprehensive coverage of the subject matter using clear text divided into easily accessible sections and subsections. Tables, figures, and photographs are used where appropriate as well as boxed case studies and summary notes. References to important previous studies and results are included but are used sparingly to avoid overloading the text. Suggestions for further reading are also provided. The main target readership is introductory-level undergraduate students of physical geography or environmental science, but there will be much of interest to students from other disciplines, and it is also hoped that sixth-form teachers will be able to use the information that is provided in each volume.

John Gerrard

AUTHOR'S PREFACE TO FOURTH EDITION

The subject of geomorphology evolves all the time. It is important to keep textbooks in line with the latest developments in the subject, which is a compelling reason for bringing out a fourth edition five years after the third. It is also inevitable that, with hindsight, parts of the third edition are improvable, partly by updating and partly by adding more discussion of such fast-growing areas within the field as the application of digital terrain models, biogeomorphology, subsea geomorphology, long-term geomorphology, and human activity and geomorphology, including geoconservation and geoheritage. I also decided to include photographs of influential geomorphologists, where possible 'in the field', so often the place of inspiration for their geomorphological theories. Major structural changes were made to the book between the second and third editions. These changes greatly improved the sequence in which topics were covered and gave the book a firmer and more logical cohesion, so much so that I felt no further changes of that nature were needed for this edition.

I'm delighted to report that the last five years or so have seen the publication of several new books on geomorphology, including Andrew Goudie and Heather Viles's *Landscapes and Geomorphology: A Very Short Introduction* (2010), Adrian Harvey's *Introducing Geomorphology: A Guide to Landforms and Processes* (2012), Ken Gregory and John Lewin's *The Basics of Geomorphology: Key Concepts* (2014), *The Sage Handbook of Geomorphology* (Gregory and Goudie 2011), and the gargantuan fourteen volume *Treatise on Geomorphology* (Shroder 2013). I'm also pleased to say that all these laudable books complement mine: the first two are short and sweet summaries of the subject, the Gregory and Lewin book is an admirable outline of geomorphological concepts, and the last two are pitched at an advanced level.

Once more, I should like to thank many people who have made the completion of this book possible: Nick Scarle for revising some of the third-edition diagrams and for drawing the new ones. Andrew Mould for persuading me to pen a new edition. Sarah Wilkes and Elge Zigaite at Routledge, for their help with the preparation of the book. Stéphane Bonnet, Fabio De Blasio, Stefan Doerr, Karin Ebert, Cynthia Ebinger, Derek C. Ford, Neil Glasser, Stefan Grab, Adrian Hall, Mike Hambrey, Kate Holden, Karna Lidmar-Bergström, David Knighton, Marli Miller, Dave Montgomery, Phil Murphy, Alexei Rudoy, Paul Sanborn, Nick Scarle, Steve Scott, Andy Short, Wayne Stephenson, Wilf Theakstone, Dave Thomas, Heather Viles, Tony Waltham, Jeff Warburton, Clive Westlake, and Ray Womack, for letting me re-use their photographs; and for Stephen Bagnold, Edwin Baynes, Amos Frumkin, Cliff Hupp, Serge Lerouil, Waite Osterkamp, Lucas Plan, Ugo Sauro, Nick Scarle, Abi Stone, Alan Strahler, Laura Tîrlă, Tony Waltham, and Peter Wilson supplying me with fresh ones. And, as always, my wife and family for sharing the highs and lows of writing a book.

Richard John Huggett
Poynton, March 2016

AUTHOR'S PREFACE TO THIRD EDITION

The third edition of *Fundamentals of Geomorphology* comes hot on the heels of the second edition. Anonymous reviewers of the second edition suggested that some rearrangement of material might be beneficial, and I have taken most of their suggestions on board. The key changes are: splitting the first chapter into three sections, the first explaining the nature of geomorphology, the second outlining ideas about land-surface process and form, and the third introducing concepts about the history of the land surface; losing the chapter on geomorphic material and process, its components being dished out among relevant chapters elsewhere (i.e. reverting to the arrangement in the first edition – I've never known a set of reviewers so unanimous on a point as this one!); integrating much of the material in the two history chapters (14 and 15 in the second edition) at appropriate points in other chapters to reflect the importance of history in all aspects of geomorphology and to provide a better way of integrating process and historical ideas and studies; placing the karst chapter towards the end of the book; and revamping the final chapter dealing with landscape evolution as a whole. I trust that these adjustments will aid understanding. In the text, the use of **bold** type indicates that a concept or phenomenon is important in the context of the book; within Glossary definitions, it indicates that a term has its own entry.

Once again, I should like to thank many people who have made the completion of this book possible: Nick Scarle for revising some of the second-edition diagrams, for drawing the many new ones, and for colouring all of them. Andrew Mould for persuading me to pen a new edition. Stefan Doerr, Derek C. Ford, Neil Glasser, Stefan Grab, Adrian Hall, Mike Hambrey, Kate Holden, Karna Lidmar-Bergström, David Knighton, Phil Murphy, Alexei Rudoy, Nick Scarle, Wayne Stephenson, Wilf Theakstone, Dave Thomas, Heather Viles, Tony Waltham, Jeff Warburton, Clive Westlake, and Jamie Woodward for letting me re-use their photographs; and Stéphane Bonnet, Fabio De Blasio, Karin Ebert, Marli Miller, Dave Montgomery, Paul Sanborn, Steve Scott, Andy Short, Tony Waltham, and Ray Womack for supplying me with fresh ones. And, as always, my wife and family for sharing the highs and lows of writing a book.

Richard John Huggett
Poynton, June 2010

AUTHOR'S PREFACE TO SECOND EDITION

The first edition of *Fundamentals of Geomorphology* was published in 2003. I was delighted that it was well received and that I was asked to write a second edition. Anonymous reviewers of the first edition suggested that some rearrangement of material might be beneficial, and I have taken most of their suggestions on board. Cliff Ollier also kindly provided me with many ideas for improvements. The key changes are new chapters on geomorphic materials and processes and on hillslopes, the reorganizing of the tectonic and structural chapters into large-scale and small-scale landforms, and the splitting of the single history chapter into a chapter dealing with Quaternary landforms and a chapter dealing with ancient landforms. I have also taken the opportunity to update some information and examples.

Once again, I should like to thank many people who have made the completion of this book possible: Nick Scarle for revising some of the first-edition diagrams and for drawing the many new ones. Andrew Mould for persuading me to pen a new edition. George A. Brook, Stefan Doerr, Derek C. Ford, Mike Hambrey, Kate Holden, Karna Lidmar-Bergström, David Knighton, Phil Murphy, Alexei Rudoy, Nick Scarle, Wayne Stephenson, Wilf Theakstone, Dave Thomas, Heather Viles, Tony Waltham, Jeff Warburton, and Clive Westlake for letting me re-use their photographs; and Neil Glasser, Stefan Grab, Adrian Hall, Heather Viles, Tony Waltham, and Jamie Woodward for supplying me with fresh ones. And, as always, my wife and family for sharing the ups and downs of writing a book.

Richard John Huggett
Poynton, October 2006

AUTHOR'S PREFACE TO FIRST EDITION

Geomorphology has always been a favourite subject of mine. For the first twelve years of my life I lived in north London, and I recall playing by urban rivers and in disused quarries. During the cricket season, Saturday and Sunday afternoons would be spent exploring the landscape around the grounds where my father was playing cricket. H. W. ('Masher') Martin, the head of geography and geology at Hertford Grammar School, whose 'digressions' during classes were tremendously educational, aroused my first formal interest in landforms. The sixth-form field-trips to the Forest of Dean and the Lake District were unforgettable. While at University College London, I was lucky enough to come under the tutelage of Eric H. Brown, Claudio Vita-Finzi, Andrew Warren, and Ron Cooke, to whom I am indebted for a remarkable six years as an undergraduate and postgraduate. Since arriving at Manchester, I have taught several courses with large geomorphological components but have seen myself very much as a physical geographer with a dislike of disciplinary boundaries and the fashion for overspecialization. Nonetheless, I thought that writing a new, student-friendly geomorphological text would pose an interesting challenge and, with *Fundamentals of Biogeography*, make a useful accompaniment to my more academic works.

In writing *Fundamentals of Geomorphology*, I have tried to combine process geomorphology, which has dominated the subject for the last several decades, with the less fashionable but fast-resurging historical geomorphology. Few would question the astounding achievements of process studies, but plate-tectonics theory and a reliable calendar of events have given historical studies a huge boost. I also feel that too many books get far too bogged down in process equations: there is a grandeur in the diversity of physical forms found at the Earth's surface and a wonderment to be had in seeing them. So, while explaining geomorphic processes and not shying away from equations, I have tried to capture the richness of landform types and the pleasure to be had in trying to understand how they form. I also discuss the interactions between landforms, geomorphic processes, and humans, which, it seems to me, are an important aspect of geomorphology today.

The book is quadripartite. Part I introduces landforms and landscapes, studying the nature of geomorphology and outlining the geomorphic system. It then divides the material into three parts: structure, form and process, and history. William Morris Davis established the logic of this scheme a century ago. The argument is that any landform depends upon the structure of the rocks – including their composition and structural attitude – that it is formed in or on, the processes acting upon it, and the time over which it has been evolving. Part II looks at tectonic and structural landforms. Part III investigates process and form, with chapters on weathering and related landforms, karst landscapes, fluvial landscapes, glacial landscapes, periglacial landscapes, aeolian landscapes, and coastal landscapes. Each of these chapters, excepting the one on weathering, considers the environments in which the landscapes occur, the processes involved in their formation, the landforms they contain, and how they affect, and are affected by, humans. Part IV examines the role of history in understanding landscapes and landform evolution, examining some great achievements of modern historical geomorphology.

There are several people to whom I wish to say 'thanks': Nick Scarle, for drawing all the diagrams and handling the photographic material. Six anonymous reviewers, for the thoughtful and perceptive comments on an embarrassingly rough draft of the work that led to several major improvements, particularly in the overall structure; any remaining shortcomings and omissions are of course down to me. A small army of colleagues, identified individually on the plate captions, for kindly providing me with slides. Clive Agnew and the other staff at Manchester, for friendship and assistance, and in particular Kate Richardson for making several invaluable suggestions about the structure and content of Chapter 1. As always, Derek Davenport, for discussing all manner of things. And, finally, my wife and family, who understand the ups and downs of book-writing and give unbounded support.

Richard John Huggett
Poynton, March 2002

ACKNOWLEDGEMENTS

The author and publisher would like to thank the following for granting permission to reproduce material in this work:

The copyright of photographs remains held by the individuals who kindly supplied them (please see photograph captions for individual names); Figure 1.5 after Figure 3.10 from S. A. Schumm (1991) *To Interpret the Earth: Ten Ways to Be Wrong* (Cambridge: Cambridge University Press), reproduced by permission of Cambridge University Press; Figure 2.12 after Figure 3 from *Earth and Planetary Science Letters* 264, J. J. Roering, J. T. Perron, and J. W. Kirchner, 'Functional relationships between denudation and hillslope form and relief', pp. 245–58 copyright © 2007, with permission from Elsevier; Figure 3.1 after Figure 3 from Claudio Vita-Finzi (1969) *The Mediterranean Valleys: Geological Changes in Historical Times* (Cambridge: Cambridge University Press), reproduced by permission of Claudio Vita-Finzi; Figure 4.9 reprinted from Figure 15.1 in Lóczy and Sütő (2011) with permission from the author; Figure 4.10 reprinted from *Earth Surface Processes and Landforms* 35, P. K. Haff, 'Hillslopes, rivers, plows, and trucks: mass transport on Earth's surface by natural and technological processes', pp. 1157–66, copyright © 2010, with permission from Elsevier; Figure 6.9 reprinted from *Earth-Science Reviews* 69, J. W. Cole, D. M. Milner, and K. D. Spinks, 'Calderas and caldera structures: a review', pp. 1–26, copyright © 2005, with permission from Elsevier; Figure 6.18 after Figure 4.9 from M. A. Summerfield (1991) *Global Geomorphology: An Introduction to the Study of Landforms* (Harlow, Essex: Longman), © M. A. Summerfield, reprinted by permission of Taylor and Francis; Figure 7.1 after Figures 3.3 and 3.5 from G. Taylor and R. A. Eggleton (2001) *Regolith Geology and Geomorphology* (Chichester: John Wiley & Sons), Copyright © 2001, reproduced by permission of John Wiley & Sons Limited; Figures 7.5, 7.6, 9.15, and 9.20 after Figures 11.11, 11.18, 16.2, and 16.7 from C. R. Twidale and E. M. Campbell (2005) *Australian Landforms: Structure, Process and Time* (Kenthurst: Rosenberg Publishing), reproduced by permission of C. R. Twidale; Figure 9.9 reprinted from *Earth Surface Processes and Landforms* 39, P. Carling, J. Jansen, and L. Meshkova (2014) 'Multichannel rivers: their definition and classification', pp. 26–37, copyright © 2014, with permission from Elsevier; Figure 9.13 after Figure 14.1 from F. Ahnert (1998) *Introduction to Geomorphology* (London: Arnold), reproduced by permission of Verlag Eugen Ulmer, Stuttgart (the original German language publishers); Figure 9.30 after Figure 2 from *Quaternary International* 79, J. Rose, B. S. P. Moorlock, and R. J. O. Hamblin, 'Pre-Anglian fluvial and coastal deposits in Eastern England: lithostratigraphy and palaeoenvironments', pp. 5–22, copyright © 2005, with permission from Elsevier; Figure 9.35 after Figure 4 from *Catena* 49, J. Warburton, M. Danks, and D. Wishart, 'Stability of an upland gravel-bed stream,

Swinhope Burn, Northern England', pp. 309–29, copyright © 2002, with permission from Elsevier; Figure 10.4 after Figure 2 from *Geomorphology* 71, N. F. Glasser, K. M. Jansson, S. Harrison, and A. Rivera, 'Geomorphological evidence for variations of the North Patagonian Icefield during the Holocene', pp. 263–77, copyright © 2005, with permission from Elsevier; Figure 10.11 slightly adapted from Figure 6.9 in A. S. Trenhaile (1998) *Geomorphology: A Canadian Perspective* (Toronto: Oxford University Press), © Oxford University Press, Canada, reprinted by permission of the publisher; Figure 10.15 after Figure 1 from Denis Mercier (2008) 'Paraglacial and paraperiglacial landsystems: concepts, temporal scales and spatial distribution', *Géomorphologie: Relief, Processus, Environnement* 4, 223–33, reprinted with permission; Figure 13.9 slightly adapted from Figures 5.6, 5.17, and 5.21 from A. D. Short and C. D. Woodroffe (2009) *The Coast of Australia*, Cambridge University Press, Melbourne, reproduced by permission of the author, © Andy Short; Figure 14.1 after 'Plan of Poole's Cavern' from D. G. Allsop (1992) *Visitor's Guide to Poole's Cavern* (Buxton, Derbyshire: Buxton and District Civic Association), after a survey by P. Deakin and the Eldon Pothole Club, reproduced by permission of Poole's Cavern and Country Park; Figures 14.6 and 14.9 after figures in D. C. Ford and P. W. Williams (2007) *Karst Geomorphology and Hydrology*, revised edition, with permission of John Wiley & Sons, Chichester; Figures 15.2 and 15.3 after Figures 10 and 14 from K. Ebert, *Cenozoic Landscape Evolution in Northern Sweden: Geomorphological Interpretation within a GIS-Framework*, PhD dissertation no. 19, Department of Physical Geography and Quaternary Geology, Stockholm University, Sweden (2009b), ISSN: 1653-7211 and ISBN: 978-91-7155-904-3, reproduced with permission from Karin Ebert; Figure 15.9 after Figure 10 from D. K. C. Jones (1999) 'Evolving models of the Tertiary evolutionary geomorphology of southern England, with special reference to the Chalklands' in B. J. Smith, W. B. Whalley, and P. A. Warke (eds) *Uplift, Erosion and Stability: Perspectives on Long-term Landscape Development* (Geological Society, London, Special Publication 162), pp. 1–23, reproduced by permission of the Geological Society, London; Figure 15.12 after Figure 16 from P. Japsen, T. Bidstrup, and K. Lidmar-Bergström (2002) 'Neogene uplift and erosion of southern Scandinavia induced by the rise of the South Swedish Dome' in A. G. Doré, J. A. Cartwright, M. S. Stoker, J. P. Turner, and N. White (eds) *Exhumation of the North Atlantic Margin: Timing, Mechanisms and Implications for Petroleum Exploration* (Geological Society, London, Special Publication 196), pp. 183–207, reproduced by permission of the Geological Society, London; Figure 15.13 and 15.14 after Figures 5 and 7 from *Geomorphology* 233, P. Bessin, F. Guillocheau, C. Robin, J.-M. Schroëtter, and H. Bauer, 'Planation surfaces of the Armorican Massif (western France): denudation chronology of a Mesozoic land surface twice exhumed in response to relative crustal movements between Iberia and Eurasia', pp. 75–91, copyright © 2015, with permission from Elsevier.

Every effort has been made to contact copyright holders for their permission to reprint material in this book. The publishers would be grateful to hear from any copyright holder who is not here acknowledged and will undertake to rectify any errors or omissions in future editions of this book.

PART I

INTRODUCING LANDFORMS AND LANDSCAPES

CHAPTER ONE

WHAT IS GEOMORPHOLOGY?

1

Geomorphology is the study of landforms and the processes that create them. This chapter covers:

- The nature of geomorphology
- Historical approaches
- Process approaches
- Other approaches
- Geomorphological 'isms'

The word geomorphology derives from three Greek words: γεω (the Earth), μορφη (form), and λογος (discourse). Geomorphology is therefore 'a discourse on Earth forms'. The term was coined sometime in the 1870s and 1880s to describe the morphology of the Earth's surface (e.g. de Margerie 1886, 315), was originally defined as 'the genetic study of topographic forms' (McGee 1888, 547), and was used in popular parlance by 1896. Despite the modern acquisition of its name, geomorphology is a venerable discipline (Box 1.1). Today, geomorphology is the study of Earth's physical land-surface features, its landforms – rivers, hills, plains, beaches, sand dunes, and myriad others. Some workers include submarine landforms within the scope of geomorphology; and some would add the landforms of other terrestrial-type planets and satellites in the Solar System – Mars, the Moon, Venus, and so on.

Landforms are conspicuous features of the Earth and occur everywhere. They range in size from molehills to mountains to major tectonic plates, and their 'lifespans' range from days to millennia to aeons (Figure 1.1). David Mark and Barry Smith (2004) listed twenty-six broad landform types, many of which have included subtypes: basin, catchment, cave, cirque, cliff, continent, crater, cut, earth surface, floodplain, gap, ground, ground surface, moraine, mount, mount range, peak, pinnacle, plain, plateau, ridge, ridge line, terrace, trough, and valley. Geomorphology investigates landforms such as those and the processes that fashion them.

Form, process, and the interrelationships between them are central to understanding the origin and development of landforms. In geomorphology, form or morphology has three facets – **constitution** (chemical and physical properties described by material property variables),

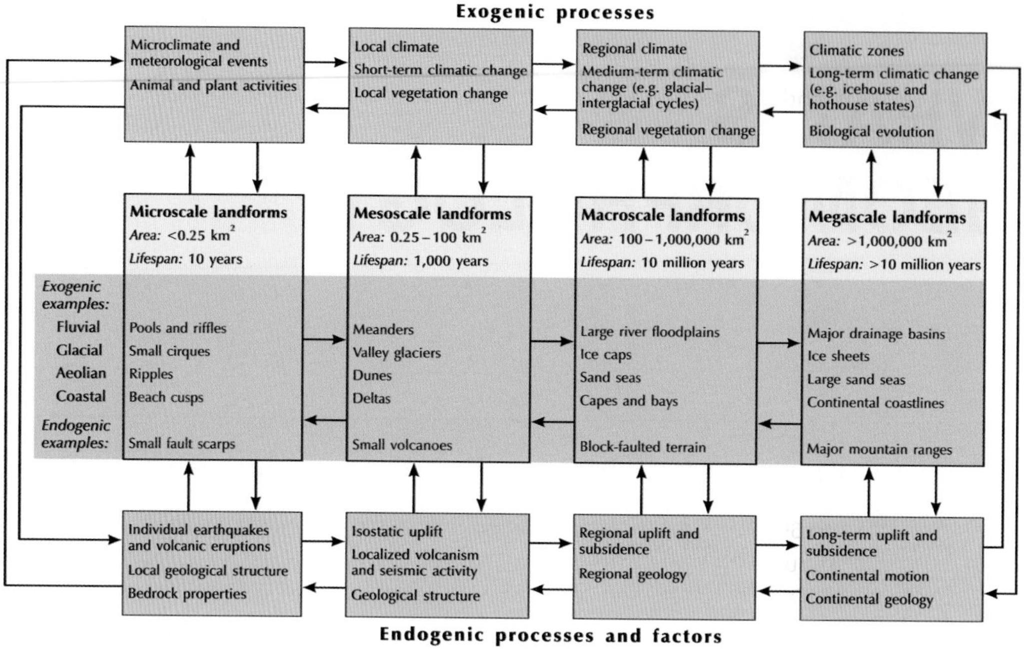

Figure 1.1 Landforms at different scales and their interactions with exogenic (external) and endogenic (internal) processes.

configuration (size and form described by geometry variables), and **mass flow** (rates of flow described by such mass-flow variables as discharge, precipitation rate, and evaporation rate) (Figure 1.2; Strahler 1980). These form variables contrast with **dynamic variables** (chemical and mechanical properties representing the expenditure of energy and the doing of work) associated with geomorphic processes; they include power, energy flux, force, stress, and momentum. Take the case of a beach (Figure 1.3). Constitutional properties include the degree of sorting of grains, mean diameter of grains, grain shape, and moisture content of the beach; configurational properties include such measures of beach geometry as slope angle, beach profile form, and water depth; mass-flow variables include rates of erosion, transport, and deposition.

Dynamic variables include drag stresses set up by water currents associated with waves (and modulated by tides), possibly by channelled water flowing over the beach, and by wind, and also include forces created by burrowing animals and humans digging beach material.

Geomorphic processes are the multifarious chemical and physical means by which the Earth's surface undergoes modification. They are driven by geological forces emanating from inside the Earth (**endogenic** or **endogene processes**), by forces originating at or near the Earth's surface and in the atmosphere (**exogenic** or **exogene processes**), and by forces coming from outside the Earth (**extraterrestrial processes**, such as asteroid impacts). They include processes of transformation and transfer associated with weathering, gravity, water, wind, and ice. Mutual interactions between form and process are the core of geomorphic investigation – form affects process and process affects form. In a wider setting, atmospheric processes, ecological processes, and

Box 1.1 THE ORIGIN OF GEOMORPHOLOGY

Ancient Greek and Roman philosophers wondered how mountains and other surface features in the natural landscape had formed. Aristotle, Herodotus, Seneca, Strabo, Xenophanes, and many others discoursed on topics such as the origin of river valleys and deltas, and the presence of seashells in mountains. Xenophanes of Colophon (c. 580–480 BC) speculated that, as seashells are found on the tops of mountains, the surface of the Earth must have risen and fallen. Herodotus (c. 484–420 BC) thought that the lower part of Egypt was a former marine bay, reputedly saying 'Egypt is the gift of the river', referring to the year-by-year accumulation of river-borne silt in the Nile delta region. Aristotle (384–322 BC) conjectured that land and sea change places, with areas that are now dry land once being sea and areas that are now sea once being dry land. Strabo (64/63 BC–AD 23?) observed that the land rises and falls, and suggested that the size of a river delta depends on the nature of its catchment, the largest deltas being found where the catchment areas are large and the surface rocks within it are weak. Lucius Annaeus Seneca (4 BC–AD 65) appears to have appreciated that rivers possess the power to erode their valleys. About a millennium later, the illustrious Arab scholar Ibn-Sina, also known as Avicenna (980–1037), who translated Aristotle, propounded the view that some mountains are produced by differential erosion, running water and wind hollowing out softer rocks. During the Renaissance, many scholars debated Earth history. Leonardo da Vinci (1452–1519) believed that changes in the levels of land and sea explained the presence of fossil marine shells in mountains. He also opined that valleys were cut by streams and that streams carried material from one place and deposited it elsewhere. In the eighteenth century, Giovanni Targioni-Tozzetti (1712–84) recognized evidence of stream erosion. He argued that rivers and floods resulting from the bursting of barrier lakes excavated the valleys of the Arno, Val di Chaina, and Ombrosa in Italy, and suggested that the irregular courses of streams relate to the differences in the rocks in which they cut, a process now called differential erosion. Jean-Étienne Guettard (1715–86) argued that streams destroy mountains and the sediment produced in the process builds floodplains before being carried to the sea. He also pointed to the efficacy of marine erosion, noting the rapid destruction of chalk cliffs in northern France by the sea, and the fact that the mountains of the Auvergne were extinct volcanoes. Horace-Bénédict de Saussure (1740–99) contended that valleys were produced by the streams that flow within them, and that glaciers may erode rocks. From these early ideas on the origin of landforms arose modern geomorphology. (See Chorley *et al.* 1964 and Kennedy 2005 for details on the development of the subject.)

geological processes influence, and in turn are influenced by, geomorphic process–form interactions (Figure 1.2).

The nature of the mutual connection between the Earth's surface processes and the Earth's surface forms has lain at the heart of geomorphological discourse. The language in which geomorphologists have expressed these connections has altered with changing cultural, social, and scientific contexts. In very broad terms, a

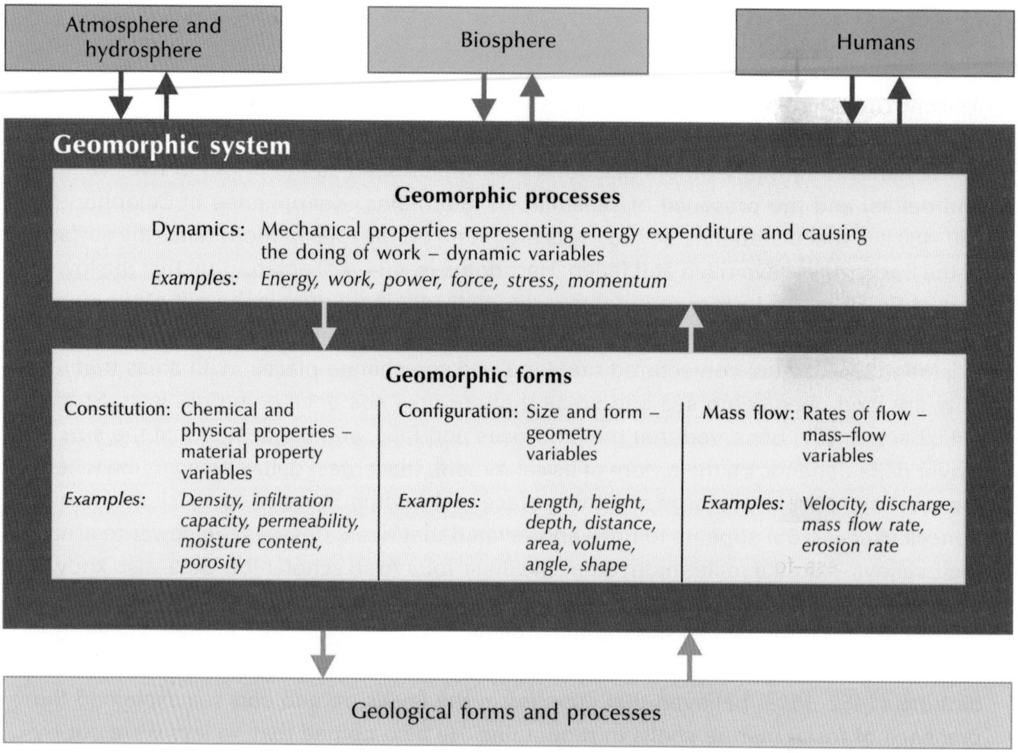

Figure 1.2 Process–form interactions – the core of geomorphology.

qualitative approach begun by Classical thinkers and traceable through to the mid-twentieth century preceded a **quantitative approach**. Early writers pondered the origin of Earth's surface features, linking the forms they saw, such as mountains, to assumed processes, such as catastrophic floods. An excellent example is the work of Nicolaus Steno (alias Niels Steensen, 1638–86). While carrying out his duties as court physician to Grand Duke Ferdinand II at Florence, Steno explored the Tuscan landscape and devised a six-stage sequence of events to explain the current plains and hills (Steno 1916 edn) (Figure 1.4). The first true geomorphologists, such as William Morris Davis and Grove Karl Gilbert, also tried to infer how the landforms they saw in the field were fashioned by geomorphic processes.

Currently, there are at least four approaches used by geomorphologists in studying landforms (Slaymaker 2009; see also Baker and Twidale 1991):

1 A process–response (process–form) or functional approach that builds upon chemistry and physics and utilizes a systems methodology.
2 A landform evolution approach that has its roots in historical geological science (geohistory) and that explores the important historical dimension of many landforms.
3 An approach that focuses on characterizing landforms and landform systems and that stems from geographical spatial science.
4 An environmentally sensitive approach to landforms, systems of landforms, and land-

Figure 1.3 Process–form interactions for a beach. (*Photograph by Andy Short*)

scapes at regional to global scales. This approach includes such topics and geoconservation and geoheritage.

This book will not look specifically at the third approach, although it will mention it in passing. Interested readers should read the fascinating paper by Jozef Minár and Ian S. Evans (2008).

The process and historical approaches dominate modern geomorphology (Summerfield 2005), with the former predominating, at least in Anglo-American and Japanese geomorphology. They have come to be called **surface process geomorphology**, or simply **process geomorphology**, and **historical geomorphology** (e.g. Chorley 1978; Embleton and Thornes 1979), although the tag 'historical geomorphology' is not commonly used. Historical geomorphology tends to focus around histories or trajectories of landscape evolution and adopts a sequential, chronological view;

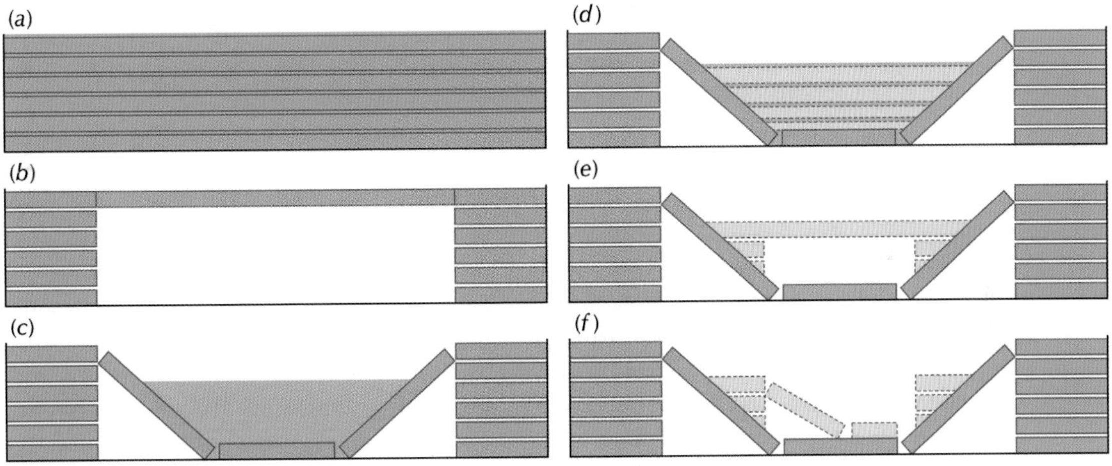

Figure 1.4 Steno's six-stage landscape history of the Tuscan region. First, just after Creation, the region was covered by a 'watery fluid', out of which inorganic sediments precipitated to form horizontal, homogeneous strata. Second, the newly formed strata emerged from their watery covering to form a single, continuous plain of dry land, beneath which the force of fires or water ate out huge caverns. Third, some of the caverns might have collapsed to produce valleys, into which rushed the waters of the Flood. Fourth, new strata of heterogeneous materials containing fossils were deposited in the sea, which now stood at higher level than it had prior to the Flood and occupied the valleys. Fifth, the new strata emerged when the Flood waters receded to form a huge plain, and were then undermined by a second generation of caverns. Finally, the new strata collapsed into the cavities eaten out beneath them to produce the Earth's present topography in the region. *Source*: Adapted from Steno (1916 edn)

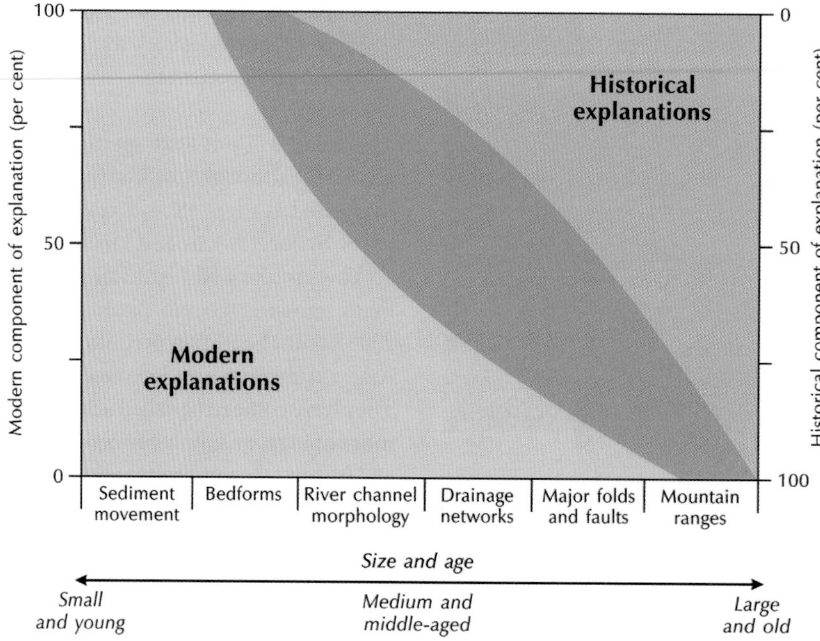

Figure 1.5 The components of historical explanation needed to account for geomorphic events of increasing size and age. The top right of the diagram contains purely historical explanations, while the bottom left contains purely modern explanations. The two explanations overlap in the middle zone, the top curve showing the maximum extent of modern explanations and the lower curve showing the maximum extent of historical explanations. *Source*: After Schumm (1985b, 1991, 53)

process geomorphology tends to focus around the mechanics of geomorphic processes and process–response relationships (how geomorphic systems respond to disturbances). Largely, historical geomorphology and process geomorphology are complementary and go hand-in-hand, so that historical geomorphologists consider process in their explanations of landform evolution while process geomorphologists may need to appreciate the history of the landforms they investigate. Nonetheless, either a process or an historical approach has tended to dominate the field at particular times. Process studies have enjoyed hegemony for some three or four decades, but sidelined historical studies are making a strong comeback.

George Gaylord Simpson (1963), an American palaeontologist, captured the nature of historical and process approaches in his distinction between 'immanence' (processes that may always occur under the right historical conditions – weathering, erosion, deposition, and so on) and 'configuration' (the state or succession of states created by the interaction of immanent process with historical circumstances). The contrast is between a 'what happens' approach (**timeless** knowledge – immanence) and a 'what happened' approach (**timebound** knowledge – configuration). In simple terms, geomorphologists may study geomorphic systems in action today, but such studies are necessarily short term, lasting for a few years or decades and principally investigate immanent properties. Yet geomorphic systems have histories that go back centuries, millennia, or millions of years. Using the results of short-term studies to predict how geomorphic systems will change over long periods is difficult owing to environmental changes and the occurrence of singular events (configuration in Simpson's parlance) such as bouts of uplift and the breakup of landmasses. Stanley A. Schumm (1991; see also Schumm and Lichty 1965) tried to resolve this problem, and in doing so established some links between process studies and historical studies. He argued that, as the size and age of a landform increase, so present conditions can explain fewer of its properties and geomorphologists must infer more about its past. Figure 1.5 summarizes his idea. Evidently, such small-scale landforms and processes as sediment movement and river bedforms are explicable with recent historical information. River channel morphology may have a considerable historical component, as when rivers flow on alluvial plain surfaces that events during the Pleistocene determined. Explanations for large-scale landforms, such as structurally controlled drainage networks and mountain ranges, require mainly historical information. A corollary of this idea is that the older and bigger a landform, the less accurate will be predictions and postdictions about it based upon present conditions. It also shows that an understanding of landforms requires a variable mix of process geomorphology and historical geomorphology; and that the two subjects should work together rather than stand in polar opposition.

HISTORICAL GEOMORPHOLOGY

All landforms have a history. Such landforms as ripples on beaches and in riverbeds and terracettes on hillslopes tend to be short-lived, so that their history will pass unrecorded unless burial by sediments ensures their survival in the stratigraphic (rock) record. For this reason, geomorphologists with a prime interest in long-term changes usually deal with relatively more persistent landforms at scales ranging from coastal features, landslides, and river terraces, through plains and plateaux, to regional and continental drainage systems. Nonetheless, ripple marks and other small-scale sedimentary features that do manage to survive can provide clues to past processes and events.

Historical geomorphology is the study of landform evolution or changes in landforms over medium and long timescales, usually timescales well beyond the span of an individual human's experience – centuries, millennia, millions, and hundreds of millions of years. It brings in the historical dimension of the subject with all its attendant assumptions and methods, and relies mainly on the form of the land surface and on the sedimentary record for its databases.

The foundations of historical geomorphology

Traditionally, historical geomorphologists strove to work out landscape history by mapping morphological (form) and sedimentary features. Their golden rule was the dictum that '**the present is the key to the past**'. This was a warrant to assume that the effects of geomorphic processes seen in action today may be legitimately used to infer the causes of assumed landscape changes in the past. Before reliable dating techniques were available, such studies were difficult and largely educated guesswork. However, the brilliant successes of early historical geomorphologists should not be overlooked.

William Morris Davis

The '**geographical cycle**', expounded by **William Morris Davis** (Plate 1.1), was the first modern theory of landscape evolution (e.g. Davis 1889, 1899, 1909). It assumed that uplift takes place quickly. Geomorphic processes, without further complications from tectonic movements, then gradually wear down the raw topography. Furthermore, slopes within landscapes decline through time – maximum slope angles slowly lessen (though few field studies have substantiated this claim). So topography is reduced, little by little, to an extensive flat region close to baselevel – a **peneplain**. The peneplain may contain occasional hills,

Plate 1.1 William Morris Davis.
(Photograph courtesy of Waite Osterkamp)

called **monadnocks** after Mount Monadnock in New Hampshire, USA, which are local erosional remnants, standing conspicuously above the general level. The reduction process creates a time sequence of landforms that progress through the stages of **youth**, **maturity**, and **old age**. However, these terms, borrowed from biology, are misleading and much censured (e.g. Ollier 1967; Ollier and Pain 1996, 204–5). The 'geographical cycle' was designed to account for the development of humid temperate landforms produced by prolonged wearing down of uplifted rocks offering uniform resistance to erosion. It was extended to other landforms, including arid landscapes, glacial landscapes, periglacial landscapes, to landforms produced by shore processes, and to karst landscapes.

William Morris Davis's **'geographical cycle'**– in which landscapes are seen to evolve through stages of youth, maturity, and old age – must be regarded as a classic work, even if it has been superseded (Figure 1.6). Its appeal seems to have lain in its theoretical tenor and in its simplicity (Chorley 1965). It had an all-pervasive influence on geomorphological thought and spawned the once highly influential field of **denudation chronology**.

Eduard Brückner and Albrecht Penck

Other early historical geomorphologists used geologically young sediments to interpret Pleistocene events. **Eduard Brückner** and **Albrecht Penck**'s work on glacial effects on the Bavarian Alps and their forelands provided the first insights into the effects of the Pleistocene ice ages on relief (Penck and Brückner 1901–9; Plate 1.2). Their classic river-terrace sequence gave names to the main glacial stages – Donau, Gunz, Mindel, Riss, and Würm – and sired Quaternary geomorphology (see Appendix for the divisions of geological time).

Modern historical geomorphology

Historical geomorphology has developed since Davis's time, and geomorphologists no longer squeeze the interpretation of longer-term changes of landscapes into the straitjacket of the geographical cycle. They rely now on various chronological analyses, particularly those based on stratigraphical studies of Quaternary sediments, and upon a much fuller appreciation of geomorphic and tectonic processes (e.g. Brown 1980). Observed stratigraphical relationships furnish **relative chronologies** (events placed in order of occurrence but without accurately fixed dates); **absolute chronologies** derive from sequences dated using historical records, radiocarbon analysis, dendrochronology, luminescence, palaeomagnetism, and so forth (p. 66). Historical studies tend to fall into two groups: Quaternary geomorphology and long-term geomorphology.

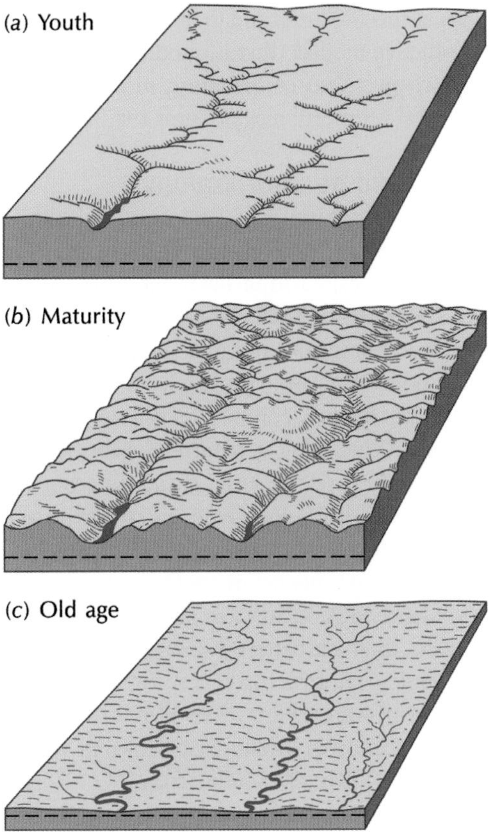

Plate 1.2 Eduard Brückner and Albrecht Penck in the summer of 1893 near Flims (Vorderrhein, Graubünden, Switzerland) during joint fieldwork for *Die Alpen im Eiszeitalter* (*The Alps in the Ice Age*). (*Photograph is the frontispiece from* Klima-Geomorphologie *by Julius Büdel (1977), courtesy of Gebrüder Borntraeger* (www.schweizerbart.de))

Figure 1.6 William Morris Davis's idealized 'geographical cycle' in which a landscape evolves through 'life-stages' to produce a peneplain. (a) Youth: a few 'consequent' streams (p. 237), V-shaped valley cross-sections, limited floodplain formation, large areas of poorly drained terrain between streams with lakes and marshes, waterfalls and rapids common where streams cross more resistant beds, stream divides broad and ill-defined, some meanders on the original surface. (b) Maturity: well-integrated drainage system, some streams exploiting lines of weak rocks, master streams have attained grade (p. 234), waterfalls, rapids, lakes, and marshes largely eliminated, floodplains common on valley floors and bearing meandering rivers, valley no wider than the width of meander belts, relief (difference in elevation between highest and lowest points) is at a maximum, hillslopes and valley sides dominate the landscape. (c) Old age: trunk streams more important again, very broad and gently sloping valleys, floodplains extensive and carrying rivers with broadly meandering courses, valleys much wider than the width of meander belts, areas between streams reduced in height and stream divides not so sharp as in the maturity stage, lakes, swamps, and marshes lie on the floodplains, mass-wasting dominates fluvial processes, stream adjustments to rock types now vague, extensive areas lie at or near the base level of erosion. *Source*: Adapted from Holmes (1965, 473)

Quaternary geomorphology

The environmental vicissitudes of the last couple of million years have wrought substantial adjustments in many landforms and landscapes. In particular, climatic swings from glacial to interglacial conditions altered geomorphic process rates and process regimes in landscapes. These alterations drove some landscapes into disequilibrium, causing geomorphic activity to increase for a while or possibly to stop. This was especially true with a change in process regime as the landscape was automatically in disequilibrium with the new processes. The disequilibrium conditions produced a phase of intense activity, involving the reshaping of hillslopes, the reworking of regoliths, and the changing of sediment stores in valley bottoms.

For many years, geomorphologists working on Quaternary timescales lacked a cogent theoretical base for explaining the links between climatic forcing and geomorphic change, and adopted a rather spongy paradigm involving the concepts of thresholds, feedbacks, complex response, and episodic activity (Chorley *et al.* 1984, 1–42). That unsatisfactory situation has changed, with climatic changes induced by changes in the frequency and magnitude of solar radiation receipt – orbital forcing (p. 280) – providing in part the missing theoretical base against which to assess the complex dynamics of landform systems. The discovery was that landscape changes over periods of 1,000 to 100,000 years display consistent patterns largely forced by the interplay of climatic changes, sea-level changes, uplift, and subsidence. Orbital forcing has affected geomorphic processes before the Quaternary, as well (e.g. Sewall *et al.* 2015).

Originally, most Quaternary geomorphologists concerned themselves with local and regional changes, usually confining their enquiries to Holocene and Late Pleistocene, so to roughly the last 18,000 years of the 2.6-million-year-long Quaternary. Since the 1950s, as their knowledge of the last 18,000 years grew, Quaternary geomorphologists started applying this knowledge to earlier times. In doing so, they collaborated with other Earth scientists to produce palaeogeographical reconstructions of particular areas at specific times and to build postdictive or retrodictive models (that is, models that predict in retrospect), so contributing to a revival of historical geomorphology (Nunn 1987).

Long-term geomorphology

Studies of landforms and landscapes older than the Quaternary, or even late Quaternary, have come to be called **long-term geomorphology** (e.g. Ollier 1992). They include investigations of Cenozoic, Mesozoic, and in some cases Palaeozoic landforms. Davis's geographical cycle was in some ways the progenitor of long-term geomorphology. Later, other geomorphologists became interested in baselevel surfaces and the school of denudation chronology emerged studying the historical development of landscapes by denudation, usually at times before the Quaternary, using as evidence erosion surfaces and their mantling deposits, drainage patterns, stream long-profiles, and geological structures. Key figures in this endeavour were Sydney W. Wooldridge and David L. Linton in Britain, Eric Brown in Wales, and Lester C. King in South Africa.

Baselevel surfaces still engage the attention of geomorphologists. Indeed, since about 1990, the field of long-term geomorphology has experienced a spectacular instauration. The reasons for this lie in the stimulation provided by the plate tectonics revolution and its rebuilding of the links between tectonics and topography, in the development of numerical models that investigate the links between tectonic processes and surface processes, and in major breakthroughs in analytical and geochronological (absolute dating) techniques (Bishop 2007). The latest numerical models of landscape evolution routinely combine bedrock river processes and slope processes; they tend to focus on high-elevation passive continental margins and convergent zones; and they regularly include the effects of rock flexure (bending and folding) and isostasy (the reestablishment of gravitational equilibrium in the lithosphere following, for example, the melting of an ice sheet or the deposition of sediment). Radiogenic dating methods, such as apatite fission-track analysis (p. 70), allow the determination of rates of rock uplift and exhumation by denudation from relatively shallow crustal depths (up to about 4 km). Despite this, long-term geomorphology still depends on landform analysis and relative dating, as most absolute dating methods fail for the timescales of interest. It is not an easy task to set an accurate age to long-term development landforms, and in many cases, later processes alter or destroy them. The old landforms surviving in today's landscapes are, in the main, large-scale features that erosion or deposition might have modified before or during the Quaternary.

Plate 1.3 Grove Karl Gilbert in Colorado, USA, 1894. (*Photograph courtesy of National Academy of Sciences*)

PROCESS GEOMORPHOLOGY

The history of process geomorphology

Process geomorphology is the study of the processes responsible for landform development. In the modern era, the first process geomorphologist, carrying on the tradition started by Leonardo da Vinci (p. 5), was **Grove Karl Gilbert** (Plate 1.3). In his treatise on the Henry Mountains of Utah, USA, Gilbert discussed the mechanics of fluvial processes (Gilbert 1877), and later he investigated the transport of debris by running water (Gilbert 1914). Up to about 1950, important contributors to process geomorphology included **Ralph Alger Bagnold** (Plate 1.4), who considered the physics of blown sand and desert dunes (p. 337), and Filip Hjulström (p. 218), who investigated fluvial processes.

After 1950, several 'big players' emerged who set process geomorphology moving apace. Arthur N. Strahler was instrumental in establishing process geomorphology, his 1952 paper called 'Dynamic basis of geomorphology' being a landmark publication (Plate 1.5). He proposed a 'system of geomorphology grounded in basic principles of mechanics and fluid dynamics' that he hoped would

> enable geomorphic processes to be treated as manifestations of various types of shear stresses, both gravitational and molecular, acting upon any type of earth material to produce varieties of strain, or failure, which we recognize as the manifold processes of weathering, erosion, transportation and deposition
> (Strahler 1952, 923)

In fact, the research of Strahler and his students, and that of Luna B. Leopold (Plate 1.6) and M. Gordon Wolman in fluvial geomorphology (e.g. Leopold *et al.* 1964), was largely empirical, involving a statistical treatment of form variables (such as width, depth, and meander wavelength) and surrogates for variables that controlled them (such as river discharge) (see Lane and Richards 1997). The challenge of characterizing the geomorphic processes themselves was eventually taken up by William E. H. Culling (1960, 1963, 1965) and Michael J. Kirkby (1971). It was not until the 1980s that geomorphologists,

Plate 1.4 Ralph Ager Bagnold, pioneer of desert exploration. (a) On an expedition to Libya during the 1930s. He was the founder and first commander of the British Army's Long Range Desert Group during World War II. (b) As a world-renowned scientist. (*Photographs courtesy of Stephen Bagnold*)

in particular William E. Dietrich and his colleagues in the Universities of Washington and Berkeley, USA (e.g. Dietrich and Smith 1983), developed Strahler's vision of a truly dynamic geomorphology (see Lane and Richards 1997). There is no doubt that Strahler's groundbreaking ideas spawned a generation of Anglo-American geomorphologists who researched the small-scale erosion, transport, and deposition of sediments in a mechanistic and fluid dynamic framework (cf. Martin and Church 2004). Moreover, modern modelling studies of the long-term evolution of entire landscapes using landscape evolution models (LEMs) represent a culmination of this work (pp. 196–99; see Coulthard and Van de Wiel 2012; Tucker and Hancock 2010; Chen *et al.* 2014).

Another line of process geomorphology considered ideas about stability in landscapes. Stanley A. Schumm (Plate 1.7), a fluvial geomorphologist, refined notions of landscape stability to include **thresholds** and dynamically **metastable states** and made an important contribution to the understanding of timescales (p. 36). Stanley W. Trimble worked on historical and modern **sediment budgets** in small catchments (e.g. Trimble 1983). Richard J. Chorley (Plate 1.8) brought process geomorphology to the UK and demonstrated the power of a **systems approach** to the subject.

The legacy of process geomorphology

Process geomorphologists have done their subject at least three great services. First, they have built up a database of process rates in various parts of the globe. Second, they have built increasingly

Plate 1.5 Arthur N. Strahler in Marblehead, Massachusetts, USA around 1990. (*Photograph courtesy of Alan Strahler*)

Plate 1.6 Luna B. Leopold while Chief Hydrologist of Water Resources Division, United States Geological Survey, taking notes by the Middle Fork Salmon River, Idaho, USA. (*Photograph by William Emmett, USGS*)

Plate 1.7 Stan Schumm near Corner Brook, Newfoundland, July, 1996. (*Photograph courtesy of Waite Osterkamp*)

Plate 1.8 Richard J. Chorley, Cambridge geomorphologist. (*Photograph by William Compton, Creative Commons CC BY-SA 3.0*)

refined models for predicting the short-term (and in some cases long-term) changes in landforms. Third, they have generated some enormously powerful ideas about stability and instability in geomorphic systems (see pp. 37–47).

Measuring geomorphic processes

Some geomorphic processes have a long record of measurement. The oldest year-by-year record is the flood levels of the River Nile in Lower Egypt. Yearly readings at Cairo are available from the time of Muhammad, and some stone-inscribed records date from the first dynasty of the pharaohs, around 3100 BC. The amount of sediment annually carried down the Mississippi River was gauged during the 1840s, and the rates of modern denudation in some of the world's major rivers were estimated in the 1860s. The first efforts to measure weathering rates were made in the late nineteenth century. Measurements of the dissolved load of rivers enabled estimates of chemical denudation rates to be made in the first half of the twentieth century, and patchy efforts were made to widen the range of processes measured in the field. But it was the quantitative revolution in geomorphology, started in the 1940s, that was largely responsible for the measuring of process rates in different environments.

Since about 1950, the attempts to quantify geomorphic processes in the field have grown fast. An early example is the work of Anders Rapp (1960), who tried to quantify all the processes active in a subarctic environment and assess their comparative significance. His studies enabled him to conclude that the most powerful agent of removal from the Karkevagge drainage basin was running water bearing material in solution. An increasing number of hillslopes and drainage basins have been instrumented, that is, had measuring devices installed to record a range of geomorphic processes. The instruments used on hillslopes and in geomorphology generally are explained in several books (e.g. Goudie 1994). Interestingly, some of the instrumented catchments established in the 1960s have recently received unexpected attention from scientists studying global warming, because records lasting decades in climatically sensitive areas – high latitudes and high altitudes – are invaluable. However, after half a century of intensive field measurements, some areas, including Europe and North America, still have better coverage than other areas. And field measurement programmes should ideally be ongoing and work on as fine a resolution as practicable, because rates measured at a particular place may vary through time and may not be representative of nearby places.

Modelling geomorphic processes

Since the 1960s and 1970s, geomorphologists have tended to direct process studies towards the construction of models for predicting short-term changes in landforms, that is, changes happening over human timescales. Such models have drawn heavily on soil engineering, for example in the case of slope stability, and hydraulic engineering in the cases of flow and sediment entrainment and deposition in rivers. Nonetheless, some geomorphologists, including Michael J. Kirkby and Jonathan D. Phillips, have carved out a niche for themselves in the modelling department. These groundbreaking endeavours led to the modelling of long-term landscape evolution, which now lies at the forefront of geomorphic research. The spur to these advances in landscape modelling was huge advances in computational technology, coupled with the establishment of a set of process equations designated 'geomorphic transport laws' (Dietrich *et al.* 2003). As Yvonne Martin and Michael Church (2004, 334) put it, 'The modelling of landscape evolution has been made quantitatively feasible by the advent of high speed computers that permit the effects of multiple processes to be integrated together over complex topographic surfaces and extended periods of time'. Figure 1.7 shows the output from a hillslope evolution model; landscape evolution models will be discussed in Chapter 8.

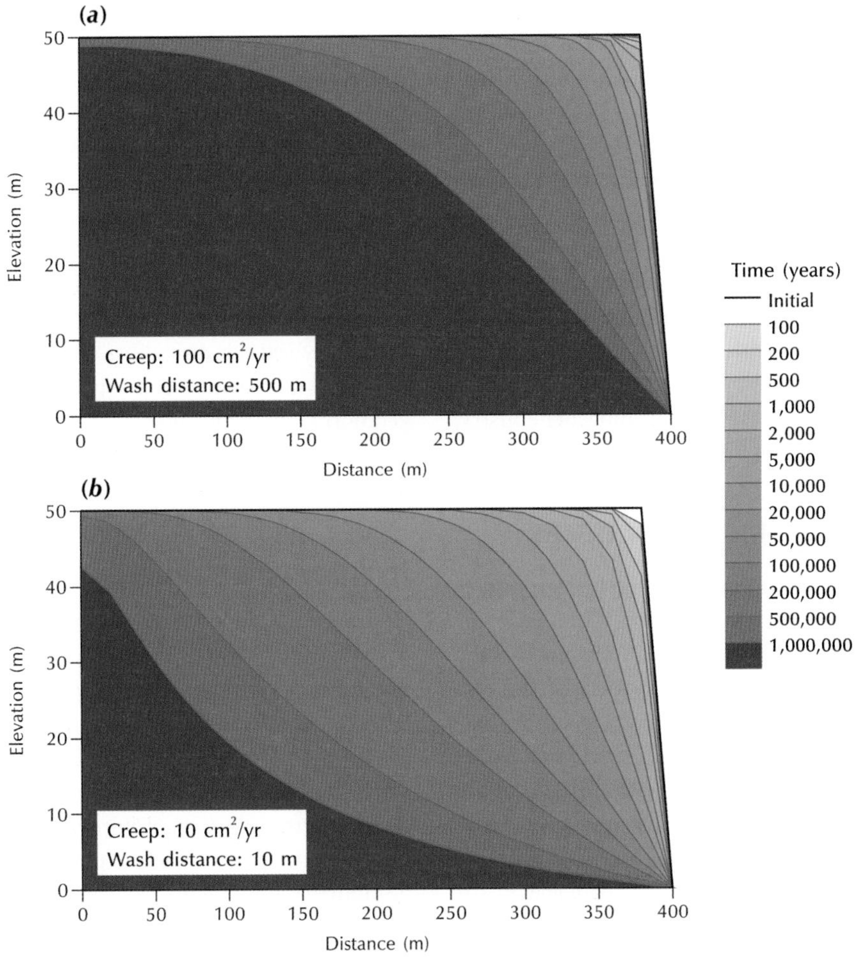

Figure 1.7 Example of a geomorphic model: the predicted evolution of a scarp bounding a plateau according to assumptions made about slope processes using a numerical model of hillslope evolution built by Mike Kirkby. (a) Slope evolution with creep processes running at 100 cm^2/year and no wash processes. (b) Slope evolution with wash process dominating.

Process studies and global environmental change

With the current craze for taking a global view, process geomorphology has found natural links with other Earth and life sciences. Main thrusts of research investigate (1) energy and mass fluxes and (2) the response of landforms to climate, hydrology, tectonics, and land use (Slaymaker 2000b, 5). The focus on mass and energy fluxes explores the short-term links between land-surface systems and climate forged through the storages and movements of energy, water, biogeochemicals, and sediments. Longer-term and broader-scale interconnections between landforms and climate, water budgets, vegetation cover, tectonics, and human activity are a focus for process geomorphologists who take a historical perspective and investigate the causes and effects of changing processes regimes during the Quaternary.

The developments in geomorphology partly parallel developments in the new field of **biogeoscience**. This rapidly evolving interdisciplinary subject investigates the interactions between the biological, chemical, and physical processes in life (the biosphere) with the atmosphere, hydrosphere, pedosphere, and geosphere (the solid Earth). It has its own journal – *Biogeosciences* – that started in 2001. Moreover, the American Geophysical Union now has a biogeoscience section that focuses upon biogeochemistry, biophysics, and planetary ecosystems.

OTHER GEOMORPHOLOGIES

A rich mix of process and historical studies dominate much modern geomorphological enquiry. However, in the past, other types of study have emphasized different aspects of the discipline. For example, structural geomorphologists, who were once a very influential group, argued that underlying geological structures are the key to understanding the origin of many landforms.

Today, a wide range of geomorphologies exists, all with some process and historical elements. Many of these arose over the last couple of decades and they are part of a wider reintegration of Earth and life sciences, the constituent disciplines within which had become increasingly specialized since the start of the nineteenth century, a time when Natural History was an all-inclusive and unified discipline. The reintegration perhaps began with Arthur George Tansley's (1935) influential idea of ecosystems, which viewed life and life-support systems (air, water, and soil) as integrated entities, and which later found unexpected support in the notion of Earth System Science (see p. 79) that takes an even broader perspective of the Earth and its geological and cosmic settings. The result is what might be styled a 'new natural history' that considers the interdependencies between the life, air, water, rocks, and landforms and their links with processes occurring deep within the Earth and processes occurring in the Solar System and beyond.

The perceived need to reintegrate subjects has led to the evolution of some of the 'new' geomorphologies – hydrogeomorphology, biogeomorphology (including phytogeomorphology and zoogeomorphology), applied geomorphology, and anthropogeomorphology (Figure 1.8). As well as these hybrid sub-disciplines, tectonic geomorphology, submarine geomorphology, climatic geomorphology, and planetary geomorphology are significant elements of modern geomorphological research.

Hydrogeomorphology

Hydrogeomorphology explores the links between hydrology and geomorphology. Water delivered to, and moving through, hillslopes, rivers, and landscapes affects geomorphic processes and geomorphic forms. Equally, geomorphic forms drive the spatial and temporal distribution of shallow groundwater. Hydrogeomorphology combines these two aspects to produce 'an interdisciplinary science that focuses on the interaction and linkage of hydrologic processes with landforms or earth materials and the interaction of geomorphic processes with surface and subsurface water in temporal and spatial dimensions' (Sidle and Onda 2004, 598).

Research into hydrogeomorphology is proving helpful in identifying hazards and understanding the impacts of land use and climatic change. Japanese geologists have published a number of outstanding works examining the role of surface and subsurface flow regimes and flowpaths on fluvial erosion and mass wasting, and ecologists have found the concept of hydrogeomorphology useful to describe the linked water and geomorphic conditions that define habitats in wetlands, rivers, and other environments.

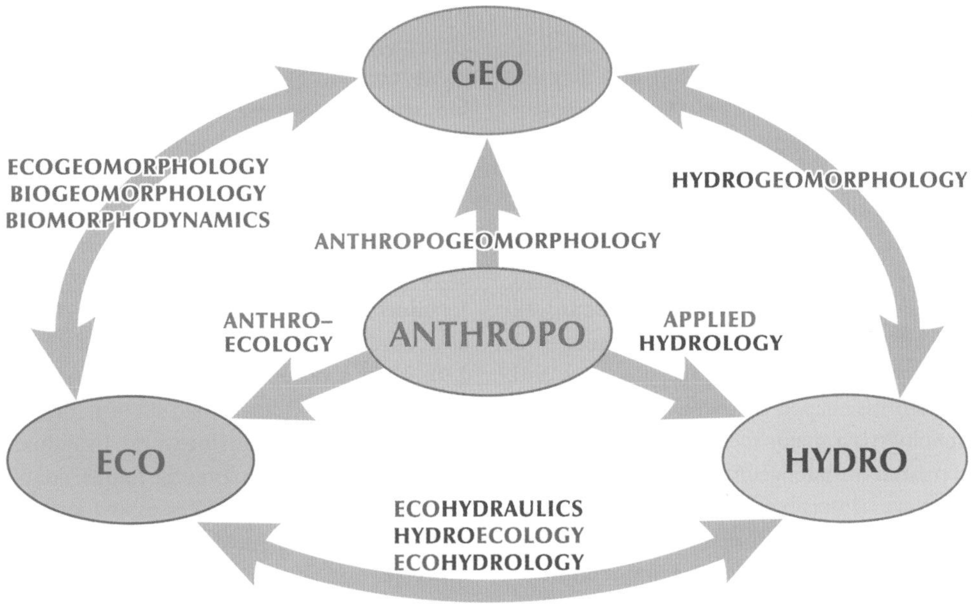

Figure 1.8 Some 'new' geomorphologies resulting from the reintegration of specialized disciplines. Joseph Wheaton and his colleagues call it the 'Eco-Geo-Hydro Mess' that results from interdisciplinary scientists at the interface of the physical and biological sciences struggling to come up with names for their disciplines. The seven combinations of geo-, hydro-, and eco- roots are invented terms sprinkled throughout the literature for the novel disciplines. The anthropo- root combines with geo- to give anthropogeomorphology. The term anthropohydrology appears not to exist (fortunately), with applied hydrology being its nearest equivalent. The word anthropo-ecology was coined by Ferdinand W. Haasis in 1935, but it is rarely used and is probably best avoided. However, Erle C. Ellis (2015) used anthroecology to couple 'ecology with society across evolutionary time through integrative scientific frameworks including anthroecosystems, anthrosequences, anthrobiogeography, anthroecological guilds and anthroecological succession' (Ellis 2015). *Source*: Partly adapted from Wheaton *et al.* (2011)

Biogeomorphology and ecogeomorphology

Plants, animals, and other living organisms (especially microorganisms) influence land form and geomorphic processes. In turn, land form and geomorphic processes influence plants, animals, and other living organisms. These connections between geomorphology and the living world have long been studied. Charles Darwin (1881) is usually credited as the first person to investigate such a connection. In his study of *The Formation of Vegetable Mould, Through the Action of Worms, with Observations on Their Habits* (Darwin 1881), he explained the process by which worms ingest soil at depth and deposit it on the surface as faecal castings (Darwin 1881). Since Darwin, other studies, mainly by ecologists, have demonstrated strong interactions between life and land form. But, it was not until the twenty-first century that **biogeomorphology** and **ecogeomorphology**, which expressly explore these interactions, began to grow rapidly (see Fei *et al.* 2014).

Biogeomorphology and ecogeomorphology (and the related biomorphodynamics, phytogeomorphology, and zoogeomorphology) are attempts to integrate geomorphology with life and ecological sciences. However, the terms are not easy to define, as researchers use them in different ways (Naylor *et al.* 2002), and they did not gain popularity until the 1990s (Osterkamp and Hupp 2010). Heather Viles coined the term **biogeomorphology** in 1988 to describe the important connection between life and landforms. She suggested it recognized the influence of landforms on the distribution and development of animals, plants, and microorganisms, as well as the influence of those animals, plants, and microorganisms on geomorphic processes and the landforms shaped by them. Today, biogeomorphology is a major focus of geomorphological research, with biogeomorphologists probing the two-way interplay between organisms on the one hand and landforms and land-forming processes on the other. **Zoogeomorphology** (Butler 1992, 1995) examines animals as geomorphic agents (looking at the role of beavers and burrowing animals, for instance), while **phytogeomorphology** is the study of relationships between plants, landforms, and geomorphic processes (Howard and Mitchell 1985).

Ecogeomorphology and biogeomorphology are very similar in scope, but they arose from somewhat different research interests. Both are part of a larger set of terms that researchers use to describe studies at the interfaces between hydrology, geomorphology, and biology or ecology (or both) (Figure 1.8). Biogeomorphology tends to focus on changes in chemical and physical weathering brought about by biological feedbacks, whereas ecogeomorphology generally focuses on changes in erosion and deposition processes (Wheaton *et al.* 2011). However, the usage of the terms is inconsistent. But at root, both biogeomorphology and ecogeomorphology are chiefly concerned with the bidirectional influences of biota and landscapes upon each other.

Biomorphodynamics is a subset of ecogeomorphology. Brad Murray *et al.* (2008) defined it as the biological–physical feedback between animals and plants and physical geomorphic processes. They reason that plants and animals influence sediment transport and land-form (morphodynamic) change, and that in turn, sediment transport and land-form change influence biological development. An example is the dynamic interactions between vegetation and river flow and sediment transport that can determine river channel patterns.

Evolutionary ecogeomorphology is an extension of biogeomorphology. It was proposed by Johannes Steiger and Dov Corenblit (2012; see also Corenblit *et al.* 2008, 2011, 2015) (Figure 1.9). They actually suggested the name evolutionary geomorphology, but that term omits the key role allotted to the living world. Evolutionary ecogeomorphology explores the mutual coupling between Earth-surface processes and landforms on the one hand and ecological and evolutionary processes on the other.

Applied geomorphology

Applied geomorphology started life as the application of geomorphological knowledge and techniques to problems facing society. The first book on the subject was James Alfred Steers's (1971) *Applied Coastal Geomorphology*. Since then, the scope of applied geomorphology has expanded to include the study of humans as geomorphic agents, and in particular their deliberate or unintentional impacts on geomorphic forms and processes. This type of applied geomorphology is anthropogeomorphology.

Traditional applied geomorphology, as opposed to anthropogeomorphology, contributes answers to the problems arising from the unrelenting growth of the human population and the continuing development of the Earth's surface. It uses geomorphological knowledge and techniques, particularly mapping skills, to aid environmental management with such issues as resource evaluation, the impact of development projects, and the prediction of natural hazards. Applications include coastal erosion and beach management (e.g. Bird 1996; Viles and Spencer 1996), soil erosion, the weathering of buildings, landslide

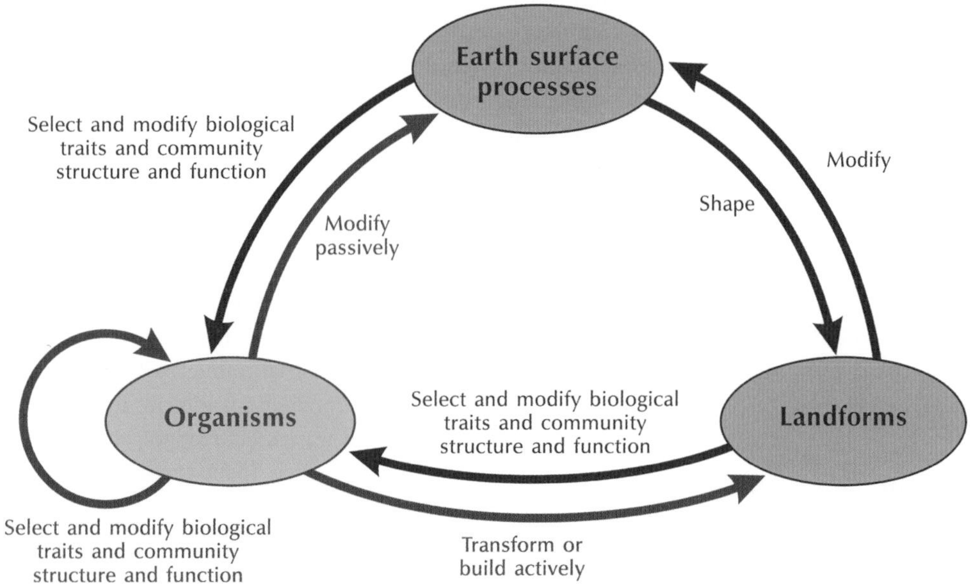

Figure 1.9 Feedback model of reciprocal interactions and adjustments between organisms and Earth surface processes and landforms. The model combines geomorphic, ecological, and evolutionary processes within an eco-evolutionary framework. Blue components and links represent physicochemical elements and controls. The green component and links represent living organisms and biological controls. Earth surface processes, landforms and living organisms co-adjust according to feedback mechanisms between them. *Source*: Adapted from Steiger and Corenblit (2012)

protection, river management and river channel restoration (e.g. Brookes and Shields 1996), and the planning and design of landfill sites (e.g. Gray 1993). Other applied issues are many and varied. Some of them are the subject of books. *Geomorphology in Environmental Planning* (Hooke 1988), for example, considered the interaction between geomorphology and public policies, with contributions on rural land-use and soil erosion, urban land-use, slope management, river management, coastal management, and policy formulation. *Geomorphology in Environmental Management* (Cooke 1990), as its title suggests, looked at the role played by geomorphology in management aspects of the environment. *Geomorphology and Land Management in a Changing Environment* (McGregor and Thompson 1995) focused upon problems of managing land against a background of environmental change.

The impending environmental changes associated with global warming have brought into a sharp focus three aspects of applied geomorphology (Slaymaker 2000b) and illustrate the value of geomorphological expertise. First, applied geomorphologists are ideally placed to work on the mitigation of natural hazards of geomorphic origin, which may well increase in magnitude and frequency during the twenty-first century and beyond. Landslides and debris flows may become more common, soil erosion may become more severe and the sediment load of some rivers increase, some beaches and cliffs may erode faster, coastal lowlands may become submerged, and frozen ground in the tundra environments may thaw. Applied geomorphologists can address all these potentially damaging changes. Second, a worrying aspect of global warming is its effect on natural resources – water, vegetation, crops, and so on. Applied geomorphologists, equipped with such techniques as terrain mapping, remote sensing, and geographical information systems, can contribute to environmental management programmes (see Downs and Booth 2011). Third, applied geomorphologists are able to translate the predictions of global and regional temperature rises into predictions of critical boundary changes, such as the poleward shift of the permafrost line and the tree-line, which can then guide decisions about tailoring economic activity to minimize the effects of global environmental change.

Anthropogeomorphology

Anthropogeomorphology is the study of human interactions with geomorphic forms and processes, especially weathering, erosion, transport, and deposition. Berl Golomb and Herbert M. Eder (1964) invented the rather cumbersome term to signify the direct human creation of landforms and human modification of natural biophysical processes. In its broadest sense, **anthropogeomorphology** is 'the study of all geomorphologically based interactions between humans and landscape' (Goudie and Viles 2010, 70).

An important recent development partly falling under the umbrella of anthropogeomorphology concerns **geodiversity** and its allied concepts – geoheritage, geosites, geoconservation, geomorphosites, and geoethics (Box 1.2). Aspects of these concepts, particularly geoconservation, will be picked up later in the book.

Tectonic geomorphology

This studies the interaction between tectonic and geomorphic processes in regions where the Earth's crust actively deforms. Advances in the measurement of rates and in the understanding of the physical basis of tectonic and geomorphic processes have revitalized it as a field of enquiry. It is a stimulating and highly integrative field that uses techniques and data drawn from studies of geomorphology, seismology, geochronology, structure, geodesy, and Quaternary climate change (e.g. Burbank and Anderson 2001).

Submarine geomorphology

This deals with the form, origin, and development of features of the sea floor. Submarine or subsea landforms cover about 71 per cent of the Earth's surface, but they are mostly less well studied than their terrestrial counterparts are. In shallow marine environments, landforms include ripples, dunes, sand waves, sand ridges, shorelines, and subsurface channels. In the continental slope transition zone are submarine canyons and gullies, inter-canyon areas, intraslope basins, and slump and slide scars. The deep marine environment contains varied landforms, including trench and basin plains, trench fans, sediment wedges, abyssal plains, distributary channels, and submarine canyons.

Planetary geomorphology

This is the study of landforms on planets and large moons with a solid crust, for example Venus, Mars, and some moons of Jupiter and Saturn. It is a thriving branch of geomorphology (e.g. Howard 1978; Carr 2012; Baker 1981, 2015; Baker *et al.* 2015). Surface processes on other planets and their satellites depend materially on their mean distance from the Sun, which dictates the annual receipt of solar energy, on their rotational period, and on the nature of the planetary atmosphere. Observed processes include weathering, aeolian activity, fluvial activity, glacial activity, and mass movements (Plate 1.9).

Climatic geomorphology

The chief climatic geomorphology exponents are French and German. Their arguments rest on the not universally accepted observation that each

Box 1.2 GEODIVERSITY AND GEOHERITAGE, GEOSITES, GEOPARKS, GEOCONSERVATION, GEOMORPHOSITES, AND GEOETHICS

Geodiversity

The origin of the word **geodiversity** (geological diversity) is difficult to pin down (Gray 2013, 9). Chris Sharples (1993) used it to describe the diversity of minerals, rocks, fossils, soils, landforms, and such associated processes as erosion and landsliding that constitute Earth's topography, landscapes, and underlying geological structures. In this, he followed the pattern set by the word biodiversity, a contraction of biological diversity, first used in 1985 and swiftly adopted by scientists and the public after 1992 when 150 nation states signed the Convention on Biological Diversity (CBD) at the United Nations Conference on the Environment and Development in Rio de Janeiro (the 'Earth Summit'). And geodiversity followed hot on the heels of the word pedodiversity, invented by Alex McBratney in 1992 to describe the variety of soil bodies in a region.

Current usage is by no means agreed, but geodiversity now includes a diversity of other diversities – lithodiversity, landform diversity, pedodiversity, hydrodiversity, and climodiversity (see Phillips 1999, 144; Ibáñez *et al.* 2013, 25). In short, geodiversity is the abiotic complement of biodiversity. **Topodiversity** (Phillips 1999, 144) and **geomorphodiversity** (Panizza 2009) are alternative words for landform diversity, and some researchers opt to separate climodiversity (climate diversity) from geodiversity (e.g. Beggs 2013).

Geodiversity today is defined as 'the variability of Earth's surface materials, landforms, and physical processes' (Hjort *et al.* 2015). Materials include rocks, soils, and water; landforms mountains, glaciers, and lakes; and processes soil formation, coastal erosion, and sediment transport. Geodiversity is now widely recognized for its scientific value and for the substantial knowledge benefits it provides for society, including records of past climate changes, the evolution of life, and understanding how Earth systems operate (Gray 2013; Gray *et al.* 2013). And over the last decade, its wider values and its relationships with landscape and biodiversity conservation, economic development, climate change adaptation, sustainable management of land and water, historical and cultural heritage, and people's health and well-being have become appreciated (Table 1.1) (e.g. Gordon *et al.* 2012; Hansom 2012; Gray 2013; Crofts and Gordon 2015). Geodiversity may not have quite the same cachet as biodiversity, but it is just as important a concept and focus for conservation efforts.

Geoheritage

Early discussion of what is now termed 'geological heritage' or 'geoheritage' concentrated on 'geological monuments' and 'significant geological features'. For instance, Eric Robinson (1989) talked about issues of Earth Science conservation, with specific mention of rock sections that are the standards for world stratigraphy (type sections or stratotypes), 'geological monuments', and educational field sites. He pointed to work done in the Republic of Ireland, where a policy for geological conservation was forming that built upon the identification of important sites started the early 1970s as part of a National Heritage Inventory. When results of the survey were published, of the 1,059 sites listed, thirty-nine were geological and nineteen geomorphological (An Foras Forbartha 1981).

continued . . .

Table 1.1 Summary of the principal values and benefits of geodiversity

Key values of geodiversity	Geodiversity provides
Maintaining life on Earth	The basic materials and cycling of water and biogeochemicals to enable food and fibre production, provide clean air and water, construction materials, and energy sources
Underpinning biodiversity and landscape	The physical basis or stage (including natural processes) that supports most terrestrial, aquatic, and marine ecosystems and species and the physical basis of the character of valued landscapes (both rural and urban) and seascapes
Economic development	Resources and assets for many aspects of economic development, including geotourism-based activities and has a profound influence on terrestrial and marine habitats, wildlife, and the use of land and water
Knowledge of Earth history, materials, and processes	Resources for scientific research and education and the knowledge base to help society adapt to climate change and to predict and mitigate natural hazards (including erosion, flooding, and slope failure)
Cultural inspiration	A powerful influence on cultural heritage through inspiration for art, sculpture, music, poetry, and literature, and on the character of the built environment through the use of different building stones
Recreation and health	A resource for a variety of recreation and outdoor activities and thus benefits for people's health and well-being

Source: Adapted from Gordon *et al.* (2012) and Hjort *et al.* (2015)

The first mention of geological heritage per se, from which the word geoheritage derives, was at the First International Symposium on the Conservation of our Geological Heritage at Digne, France in 1991 (Anon 1991). The European Working Group on Earth Science Conservation and the Conférence Permanente des Réserves Naturelles (CPRN) organized this to provide people involved in the conservation of 'our geological heritage' an opportunity to advance their work through participation in discussions on the four topics. The topics were (1) the present legal framework national and international for Earth science conservation; (2) the way and means by which the geological heritage is conserved; (3) national initiatives in Earth science conservation; and (4) the role of education and the media in the promotion of Earth science conservation. The first specific use of geoheritage as a word was at the Malvern International Conference in 1993 (O'Halloran *et al.* 1994).

However, these early forays into geoheritage did not address issues of intrinsic values and ecological sustainability, and so nature conservation programmes largely ignored them, or at best treated them as minor issues, because of their perceived lack of relevance to central issues in land and ecosystem management. However, by the mid 1990s, the scope of geoheritage had expanded to include the protection of dynamic geological and geomorphic processes and geodiversity (Sharples 1995).

Today, **geoheritage** encompasses 'global, national, state-wide, and local features of geology, at all scales that are intrinsically important sites or culturally important sites offering information or insights into the evolution of the Earth; or into the history of science, or that can be used for research, teaching, or reference' (Brocx and Semeniuk 2007, 53).

Geoconservation, geosites, and geoparks

Geoconservation deals with the conservation of non-living parts of the natural environment – geological features, landforms, and soils (geodiversity) – for heritage, science, or education (Sharples 2002). The conservation of geodiversity is now a firmly established area of research and often forms part of broader conservation efforts (see Crofts and Gordon 2015). Geoconservation involves identifying and trying to conserve geological, landform, and soil features, assemblages, systems, and processes for their intrinsic, ecological, or heritage values. Geoheritage involves concrete examples of elements of geodiversity deemed important to humans for purposes other than resource exploitation; in other words, elements with conservation significance (Sharples 2003) and with significance as sacred places (Kiernan 2015).

Geosites are places identified as worthy of geoheritage and geoconservation. They include classic outcrops or a landscape such as the Giant's Causeway. They also include engineered sites of economic importance such as museums, structures featuring striking building stones, or sites of significance to the history of geology or a nation's industrial heritage. In 2014, the Geological Society of London launched a 100 Great Geosites voted for by the public, who nominated some 400 sites that divided into ten categories (Figure 1.10).

A **geomorphosite**, a word suggested by Mario Panizza in 2001, is a geosite of a geomorphological nature; that is, a landform to which some value is attributable. Certainly, geomorphologists increasingly contribute to the UNESCO World Heritage Convention, in particular by identifying landforms and processes as superlative geomorphic phenomena and arguing why geomorphology is significant within itself, to other natural sciences, and in a wider cultural setting (see Migoń 2014).

Geoparks are areas with outstanding geoheritage that are set up primarily to encourage **geotourism** and to support local economic development (Crofts and Gordon 2015). They are not protected areas in themselves, although they may fall wholly or partly within protected areas. Nor are they rigorously identified and classified as a comprehensive global network. Many are based on community-led, voluntary initiatives and others on top-down designation. Nevertheless, the Global Network of National Geoparks or Global Geoparks Network (GGN), assisted by UNESCO, provides an international framework to conserve and enhance the value of the Earth's heritage, its landscapes, and geological formations. In 2015, the network comprised 120 national geoparks worldwide that, as from November that year, were designated as UNESCO Sites.

Many geoparks promote understanding of geological hazards, including volcanoes, earthquakes, and tsunamis. They also highlight the sustainable use and need for natural resources while encouraging respect for the environment and the integrity of the landscape. Several countries have their own National Geoparks, such as the Kanawinka Geopark in Australia, which may or may not belong to The Global Geoparks Network or to the European Geoparks Network.

Geoethics

This new branch of ethics, begun in the early 1990s but not firmly established until 2011 (Peppoloni and Di Capua 2012), considers the ethics of human interactions with the physical world. The Constitution of the International Association for Promoting Geoethics (IAGP) states that

continued . . .

Landscape

Pen y Fan, Brecon Beacons

Malham Cove, Yorkshire Dales

The Burren, County Clare

Industrial and Economic Importance

Dinorwig Power Station, Snowdonia, Wales

Channel Tunnel, England/France

The Ironbridge Gorge, Shropshire

Historical and Scientific Importance

Trotternish, Isle of Skye

Ludlow, Shropshire

Siccar Point, Berwickshire

Educational

Lyme Regis beaches, Dorset

Natural History Museum, London

Knockan Crag, Ross-Shire

Adventurous

The Cuillin Hills, Isle of Skye

Gaping Gill, Yorkshire Dales

Staffa, Inner Hebrides

Human Habitation

Hadrian's Wall, Northumberland

Callanish Stones, Lewis

Stonehenge, Wiltshire

Coastal

Lulworth Cove, Dorset

Cliffs of Moher, County Clare

Hunstanton Cliffs, Norfolk

Outcrops

Brimham Rocks, North Yorkshire

Snowball Earth Deposits, Islay, Scotland.

Coast around Craster, Northumberland

Folding and Faulting

Millook Haven, Cornwall

The Muidhe, Lochaber, Highlands

South Stack Formation, Anglesey

Fire and Ice

Cwm Idwal, Snowdonia

Glen Roy Parallel Roads, Lochaber, Highlands

Drumlins of Clew Bay, County Mayo

Figure 1.10 Examples of geosites in the UK. The photographs illustrate the first-named example in each of the ten categories. (*All photographs by Tony Waltham Geophotos except Brimham Rocks, by Richard Huggett, and Lulworth, by Nick Scarle*)

'geoethics consists of the research and reflection on those values upon which to base appropriate behaviours and practices where human activities intersect the Geosphere' (www.geoethics.org). It deals with the ethical, social, and cultural implications of Earth sciences research and practice, and represents an 'opportunity for Geoscientists to become more conscious of their social roles and responsibilities in conducting their activities' (Peppoloni and Di Capua 2015, 1).

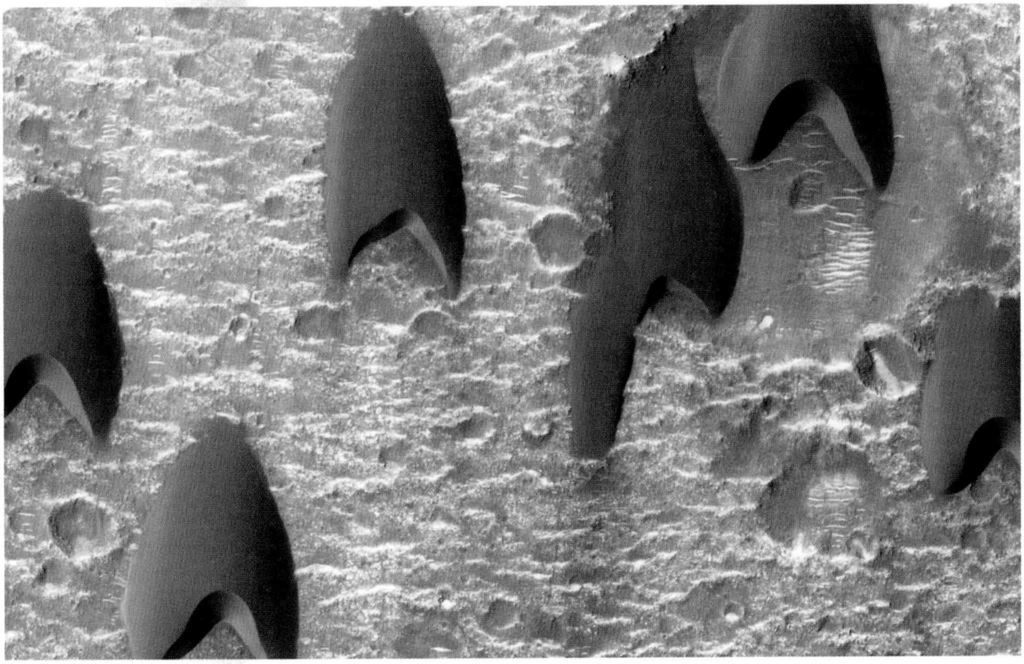

Plate 1.9 Dune field on Mars near Mawrth Vallis, indicating aeolian activity. (*Photograph by NASA/JPL-Caltech/University of Arizona*)

climatic zone (tropical, arid, temperate, for example) engenders a distinctive suite of landforms (e.g. Tricart and Cailleux 1972; Büdel 1982). Climate does strongly influence geomorphic processes, but it is doubtful that the set of geomorphic processes within each climatic zone creates characteristic landforms. The current consensus is that, owing to climatic and tectonic change, the climatic factor in landform development is more complicated than climatic geomorphologists have suggested on occasions (cf. p. 64).

The two geomorphologies

Another important aspect of twenty-first century geomorphology is a widening rift in approaches recognized by Mike Church (2010). On the one hand sits a rigorous **geophysical approach** based on the physical principles governing the breakdown and transport of materials that sees the land surface as a significant element of the Earth system. A prime example of this approach is the excellent textbook by Robert S. and Suzanne P. Anderson (2010) – *Geomorphology: The Mechanics* *and Chemistry of Landscapes*. The subtitle says it all.

On the other hand rests an approach increasingly more concerned with human social and economic values, with environmental change, with the human impact on environment, with geoconservation and with issues of social justice and equity; in short, with the multifarious aspects of anthropogeomorphology and what Peter K. Haff (2002, 2003) dubbed **neogeomorphology** and Peter Ashmore (2015) called **sociogeomorphology**. Protagonists of this approach are becoming more and more preoccupied with such issues as the broader definition of the Earth system and environmental changes within it, and with the supremacy of human agency. In this, they are less constrained by a particular methodological approach than are their geophysical approach counterparts. As Church (2010, 282) put it, 'This is a geomorphology that more readily incorporates human social and economic dimensions – as necessary – into its analyses, and that no longer treats humans as a special and somewhat exceptional agency modifying Earth's surface.' It is less easy to find a work that epitomizes this approach, but the chapter by Dénes Lóczy and László Sütő (2011) on 'Human Activity and Geomorphology' fits the bill. Both approaches have their place in geomorphology and this book will explore each of them.

GEOMORPHOLOGICAL 'ISMS': A NOTE ON METHODOLOGY

Process and historical geomorphologists alike face a problem with their methodological base. In practising their trade, all scientists, including geomorphologists, follow rules. Scientific practitioners established these rules, or guidelines. They advise scientists how to go about the business of making scientific enquiries. In other words, they are guidelines concerned with scientific methodology or procedures. The foremost guideline – the **uniformity of law** – is the premise from which all scientists work. It is the presupposition that natural laws are invariant in time and space. In simple terms, this means that, throughout Earth history, the laws of physics, chemistry, and biology have always been the same. Water has always flowed downhill, carbon dioxide has always been a greenhouse gas, and most living things have always depended upon carbon, hydrogen, and oxygen.

Three other guidelines are relevant to geomorphology. Unlike the uniformity of law, which is a universally accepted basis for scientific investigation, they are substantial claims or suppositions about how the Earth works and are open to interpretation. First, the **principle of simplicity** or, as it is commonly called in geomorphology, the **uniformity of process** states that no extra, fanciful, or unknown causes should be invoked if available processes will do the job. It is the supposition of **actualism**, the belief that past events are the outcome of processes seen in operation today; or, in Sir Archibald Geikie's (1905, 299) memorable phrase, 'the present is the key to the past'. However, the dogma of actualism is being challenged, and its flip-side – **non-actualism** – is gaining ground. Some geologists and geomorphologists are coming round to the view that the circumstances under which processes acted in the past were very different from those experienced today, and that those differences greatly influence the interpretation of past processes. So, before the evolution of land plants, and especially the grasses, the processes of weathering, erosion, and deposition would have occurred in a different context, and Palaeozoic deserts, or even Permian deserts, may not directly correspond to modern deserts.

The second substantive claim concerns the rate of Earth surface processes, two extreme views being **gradualism** and **catastrophism** (p. 46). The third substantive claim concerns the changing state of the Earth's surface, **steady-statism** arguing for a more or less constant state, or at least cyclical changes about a comparatively invariant mean state, and **directionalism** arguing in favour of directional changes.

Uniformitarianism is a widely, but too often loosely, used term in geomorphology. A common mistake is to equate uniformitarianism with actualism. Uniformitarianism was a system of assumptions about Earth history argued by Charles Lyell, the nineteenth-century geologist. Lyell articulately advocated three 'uniformities', as well as the uniformity of law: the uniformity of process (actualism), the uniformity of rate (gradualism), and the uniformity of state (steady-statism). Plainly, extended to geomorphology, uniformitarianism, as introduced by Lyell, is a set of beliefs about Earth surface processes and states. Other sets of beliefs are possible. The diametric opposite of Lyell's uniformitarian position would be a belief in the non-uniformity of process (non-actualism), the non-uniformity of rate (catastrophism), and the non-uniformity of state (directionalism). All other combinations of assumption are possible and give rise to different 'systems of Earth history' (Huggett 1997a). The various systems may be tested against field evidence. To be sure, directionalism was accepted even before Lyell's death, and nonactualism and catastrophism are discussed in geomorphological circles. In particular, geomorphology has followed the trend in the wider geosciences to embrace **neocatastrophism**, the modern version of catastrophism (e.g. Marriner *et al.* 2010). And the pressing need to understand the changes likely to occur in the Earth system over the next century and beyond has led to a new maxim: 'the present is the key to the future' (Furlani and Ninfo 2015). It is unclear just how useful a maxim that will prove to be, given that the Earth system appears to contain tipping points, that is, thresholds that once crossed lead to radically different, and possibly irreversible, system states (Brook *et al.* 2013; Lenton and Williams 2013; Williams *et al.* 2015). Partly for these reasons, Jasper Knight and Stefan Harrison (2014) questioned the applicability of uniformitarianism to geomorphology in the Anthropocene, where nonlinear system feedback is increasing, and they concluded that the present is not *the* key to the past, just a key.

SUMMARY

Geomorphology is the study of landforms. Three key elements of geomorphology are land form, geomorphic process, and land-surface history. The two complementary main brands of geomorphology are historical geomorphology and process geomorphology. Other brands include applied geomorphology, anthropogeomorphology, tectonic geomorphology, submarine geomorphology, planetary geomorphology, and climatic geomorphology. Geomorphology has engaged in methodological debates over the extent to which the present is the key to the past and the rates of Earth surface processes.

ESSAY QUESTIONS

1. To what extent are early ideas in geomorphology relevant today?
2. Explain why geomorphology encompasses a wide range of approaches.
3. To what extent is the present a key to the future and to the past?

FURTHER READING

Ahnert, F. (1998) *Introduction to Geomorphology.* London: Arnold.
A good starting text with many unusual examples.

Bloom, A. L. (2004) *Geomorphology: A Systematic Analysis of Late Cenozoic Landforms,* 3rd edn. Long Grove, Illinois: Waveland Press.
A sound text with a focus on North America.

Kennedy, B. A. (2005) *Inventing the Earth: Ideas on Landscape Development since 1740.* Oxford: Blackwell.
A good read on the relatively recent history of ideas about landscape development.

Slaymaker, O. (2009) The future of geomorphology. *Geography Compass* 3, 329–49.
Heavy on philosophy and not a smooth ride for the beginner, but worth the effort.

Strahler, A. H. and Strahler, A. N. (2006) *Introducing Physical Geography*, 4th edn. New York: John Wiley & Sons.
Comprehensive and accessible coverage of all aspects of physical geography if a general background is needed.

Summerfield, M. A. (1991) *Global Geomorphology: An Introduction to the Study of Landforms.* Harlow, Essex: Longman.
A classic after just twenty years. Includes material on the geomorphology of other planets.

Summerfield, M. A. (2005) A tale of two scales, or the two geomorphologies. *Transactions of the Institute of British Geographers*, New Series 30, 402–15.
A plainly written and thoughtful paper.

Thorn, C. E. (1988) *An Introduction to Theoretical Geomorphology*. Boston, Mass.: Unwin Hyman.
A very clear discussion of the big theoretical issues in geomorphology. Well worth a look.

CHAPTER TWO

INTRODUCING PROCESS AND FORM

2

Earth surface process and landform are key to geomorphic understanding. This chapter introduces:

- Geomorphic systems
- Geomorphic models
- Landform

GEOMORPHIC SYSTEMS

Defining systems

What is a geomorphic system?

Process geomorphologists commonly adopt a **systems approach** to their subject. To illustrate what this approach entails, take the example of a hillslope system. A hillslope extends from an interfluve crest, along a valley side, to a sloping valley floor. It is a system insofar as it consists of things (rock waste, organic matter, and so forth) arranged in a particular way. The arrangement is seemingly meaningful, rather than haphazard, because it is explicable in terms of physical processes (Figure 2.1). The 'things' of which a hillslope is composed may be described by such variables as particle size, soil moisture content, vegetation cover, and slope angle. These variables, and many others, interact to form a regular and connected whole: a hillslope, and the mantle of debris on it, records a propensity towards reciprocal adjustment among a complex set of variables. The complex set of variables includes rock type, which influences weathering rates, the geotechnical properties of the soil, and rates of infiltration; climate, which influences slope hydrology and so the routing of water over and through the hillslope mantle; tectonic activity, which may alter baselevel; and the geometry of the hillslope, which, acting mainly through slope angle and distance from the divide, influences the rates of processes such as landsliding, creep, solifluction (see p. 192), and wash. Change in any of the variables will tend to cause a readjustment of hillslope form and process.

Isolated, closed, open, and dissipative systems

Systems of all kinds are open, closed, or isolated according to how they interact, or do not interact, with their surroundings (Huggett 1985, 5–7). Traditionally, an **isolated system** is a system that is completely cut off from its surroundings and

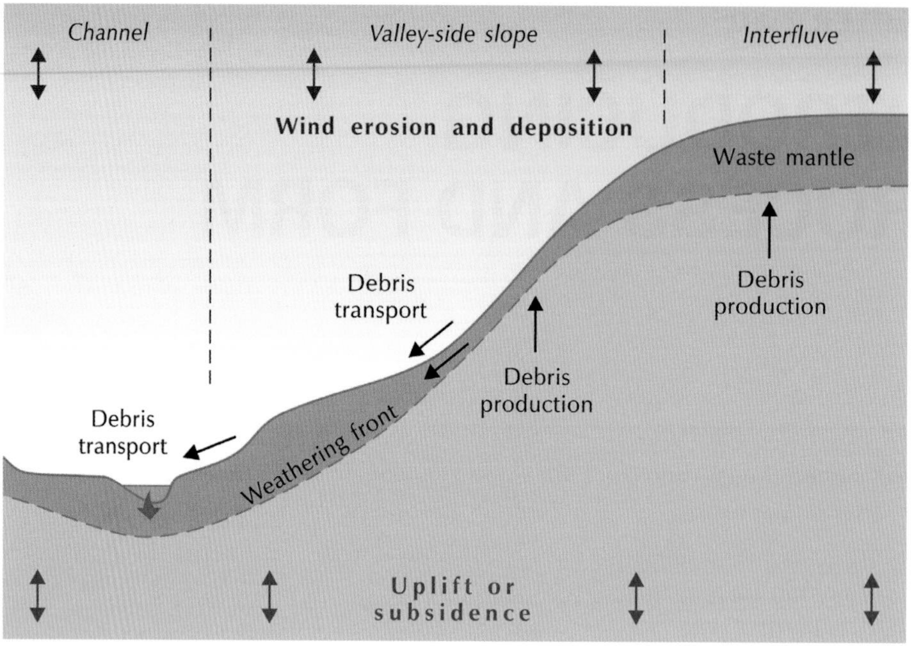

Figure 2.1 A hillslope as a system, showing storages (waste mantle), inputs (e.g. wind deposition and debris production), outputs (e.g. wind erosion), throughputs (debris transport), and units (channel, valley-side slope, interfluve).

that cannot therefore import or export matter or energy. A **closed system** has boundaries open to the passage of energy but not of matter. An **open system** has boundaries across which energy and materials may move. All geomorphic systems, including hillslopes, are open systems as they exchange energy and matter with their surroundings. They are also **dissipative systems**, which means that irreversible processes resulting in the dissipation of energy (generally in the form of friction or turbulence) govern them. Thus, to maintain itself, a geomorphic system dissipates energy from such external sources as solar energy, tectonic uplift, and precipitation.

Internal and external system variables

Any geomorphic system has **internal** and **external variables**. Take a drainage basin. Soil wetness, streamflow, and other variables lying inside the system are endogenous or internal variables. Precipitation, solar radiation, tectonic uplift, and other such variables originating outside the system and affecting drainage basin dynamics are exogenous or external variables. Interestingly, all geomorphic systems can be thought of as resulting from a basic antagonism between **endogenic** (**tectonic** and **volcanic**) processes driven by geological forces and **exogenic** (**geomorphic**) processes driven by climatic forces (Scheidegger 1979). In short, tectonic processes create land, and climatically influenced weathering and erosion destroy it. The events between the creation and the final destruction are what fascinate geomorphologists.

Classifying systems

Systems are mental constructs and defined in various ways. Two conceptions of systems are important in geomorphology: systems as process and form structures, and systems as simple and complex structures (Huggett 1985, 4–5, 17–44).

Geomorphic systems as form and process structures

Four kinds of geomorphic system may be identified: form systems, process systems, form and process systems, and control systems.

1. Form systems. **Form** or **morphological systems** are sets of form variables deemed to interrelate in a meaningful way in terms of system origin or system function. Several measurements could be made to describe the form of a hillslope system. Form elements would include measurements of anything on a hillslope that has size, shape, or physical properties. A simple characterization of hillslope form is shown in Figure 2.2a, which depicts a cliff with a talus slope at its base. All that could be learnt from this 'form system' is that the talus lies below the cliff; no causal connections between the processes linking the cliff and talus slope are inferred. Sophisticated characterizations of hillslope and land-surface forms may be made using digital terrain models.
2. Process systems. **Process systems**, which are also called **cascading** or **flow systems**, are defined as 'interconnected pathways of transport of energy or matter or both, together with such storages of energy and matter as may be required' (Strahler 1980, 10). An example is a hillslope represented as a store of materials: weathering of bedrock and wind deposition add materials to the store, and erosion by wind and fluvial erosion at the slope base removes materials from the store. The materials pass through the system and in doing so link the morphological components. In the case of the cliff and talus slope, it could be assumed that rocks and debris fall from the cliff and deliver energy and rock debris to the talus below (Figure 2.2b).
3. Form and process systems. **Process–form systems**, also styled **process–response systems**, comprise an energy-flow system linked to a form system in such a way that system processes may alter the system form and, in turn, the changed system form alters the system processes. A hillslope may be viewed in this way with slope form variables and slope process variables interacting. In the cliff-and-talus example, rock falling off the cliff builds up the talus store (Figure 2.2c). However, as the talus store increases in size, so it begins to bury the cliff face, reducing the area that supplies debris. In consequence, the rate of talus growth diminishes and the system changes at an ever-decreasing rate. The process described is an example of negative feedback, which is an important facet of many process–form systems (Box 2.1).

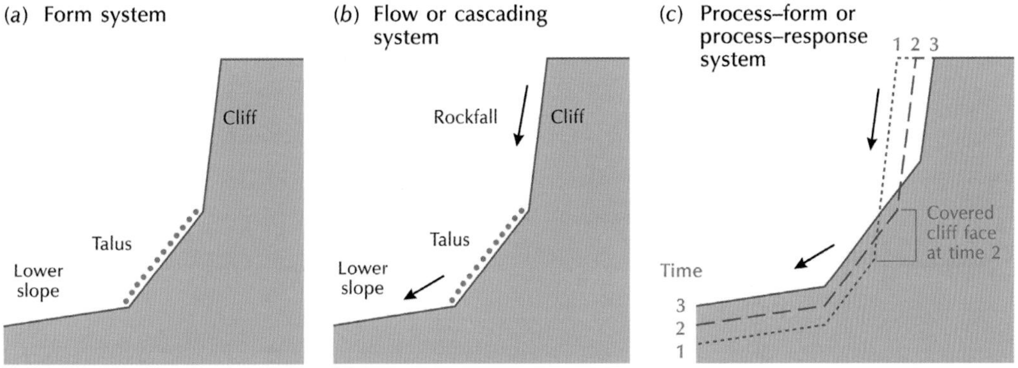

Figure 2.2 A cliff and talus slope viewed as (a) a form system, (b) a flow or cascading system, and (c) a process–form or process–response system. Details are given in the text.

4 Control systems. **Control systems** are process–form systems that interact with humans. They include managed rivers, coasts with sea defences, and some caves. Given the degree to which humans now affect the planet, it might be argued that many geomorphic systems – certainly all anthropogeomorphic systems (p. 33) – are control systems.

Geomorphic systems as simple or complex structures

Three main types of system are recognized under this heading: simple systems, complex but disorganized systems, and complex and organized systems.

1 **Simple systems**. The first two of these types have a long and illustrious history of study. Since at least the seventeenth-century revolution in science, astronomers have referred to a set of heavenly bodies connected together and acting upon each other according to certain laws as a system. The Solar System is the Sun and its planets. The Uranian system is Uranus and its moons. These structures may be thought of as simple systems.

Box 2.1 NEGATIVE AND POSITIVE FEEDBACK

Negative feedback occurs when a change in a system sets in motion a sequence of changes that eventually neutralize the effects of the original change, so stabilizing the system. An example occurs in a drainage basin system, where increased channel erosion leads to a steepening of valley-side slopes, which accelerates slope erosion, which increases stream bed-load, which reduces channel erosion (Figure 2.3a). The reduced channel erosion then stimulates a sequence of events that stabilizes the system and counteracts the effects of the original change. Some geomorphic systems also display **positive feedback** relationships characterized by an original change being magnified and the system being made unstable. An example is an eroding hillslope where a storm increases surface runoff, which increases slope erosion, which in turn causes a reduction in infiltration capacity of water, which then increases the amount of surface runoff, so promoting even more slope erosion (Figure 2.3b). In short, a 'vicious circle' is created, and the system, being unstabilized, continues changing. Feedback may occur across different spatial scales, which makes it complex (Figure 2.4).

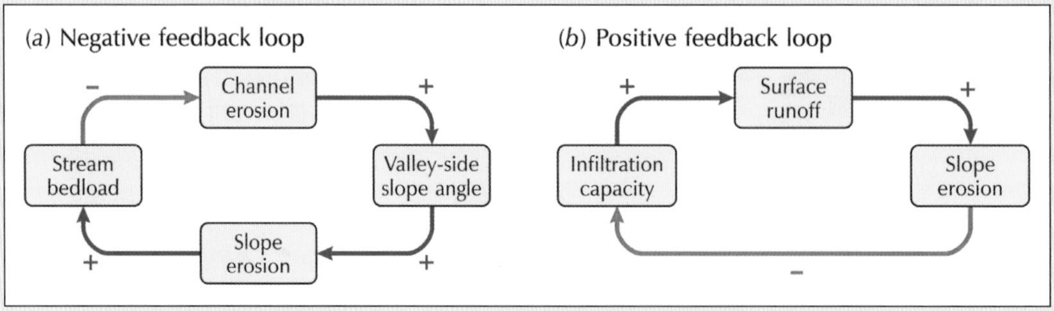

Figure 2.3 Feedback relationships in geomorphic systems. (a) Negative feedback in a valley-side slope–stream system. (b) Positive feedback in an eroding hillslope system. Details of the relationships are given in the text.

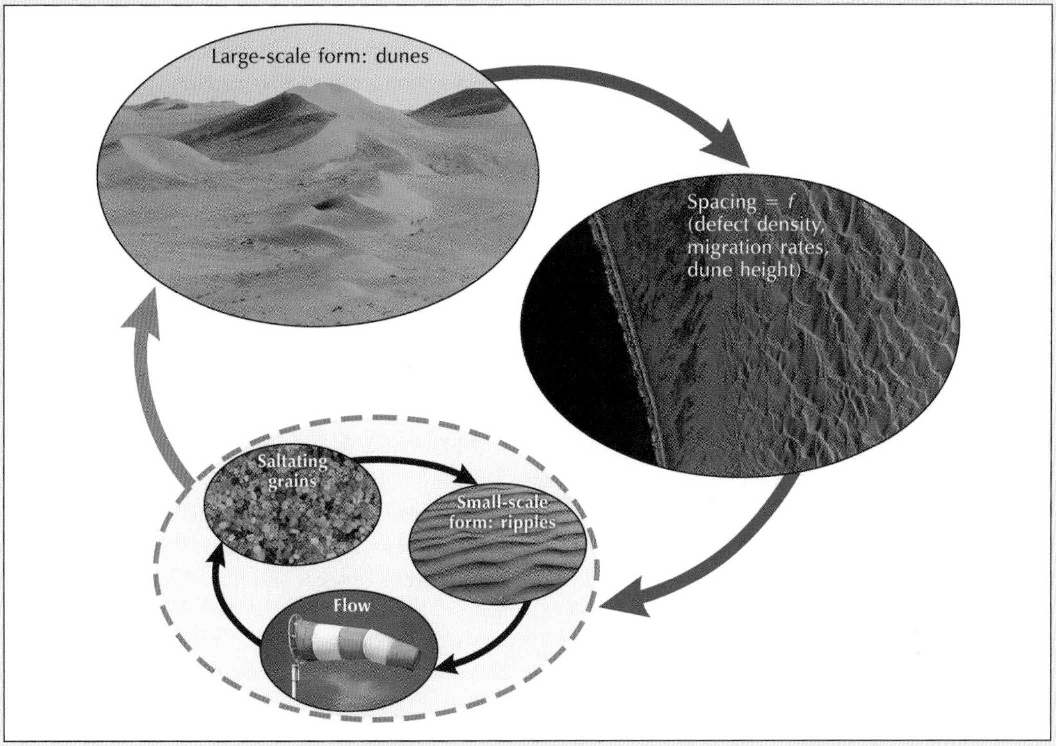

Figure 2.4 An example of complex feedback. In sandy deserts, the shape of a sand-covered bed influences the pattern of wind velocities, which influences the patterns of sediment fluxes, which then influence changes in bed shape. If this feedback loop bolsters a bed-shape perturbation, the feedback is positive. The complexity arises once a large-scale structure such as a dune emerges (*top left*), the large-scale variables connected with which (*middle right*) influence smaller-scale processes such as airflow, ripples, and saltating grains (*bottom left*). 'Defects' represent terminations of dune crestlines. *Source*: Adapted from Murray *et al.* (2014) (*Photographs of large-scale dunes and ripples in Namib Sand Sea by Abi Stone; photograph of complex dunes of Namib Sand Sea, with Conception Bay sector of the Namibian coastline showing breakers and a strand plain on the left, by NASA*)

In geomorphology, a few boulders resting on a talus slope is a simple system. The conditions needed to dislodge the boulders, and their fate after dislodgement, is predictable from mechanical laws involving forces, resistances, and equations of motion, in much the same way that the motion of the planets around the Sun can be predicted from Newtonian laws.

2 **In a complex but disorganized system**, a vast number of objects interact in a weak and haphazard way. An example is a gas in a jar. This system might comprise upward of 10^{23} molecules colliding with each other. In the same way, the countless individual particles in a hillslope mantle could be regarded as a complex but rather disorganized system. In both the gas and the hillslope mantle, the interactions are somewhat haphazard and far too numerous to study individually, so aggregate measures must be employed (see Huggett 1985, 74–7; Scheidegger 1991, 251–8).

3 In a third and later conception of systems, objects are seen to interact strongly with one another to form **systems of a complex and organized nature**. Most biological and ecological systems are of this kind. Many structures in geomorphology display high degrees of regularity and rich connections, and may be thought of as complexly organized systems. A hillslope represented as a process–form system could be placed into this category. Other examples include soils, rivers, and beaches.

System hierarchy: the scale problem

A big problem faced by geomorphologists is that, as the size of geomorphic systems increases, the explanations of their behaviour may change. Take the case of a fluvial system. The form and function of a larger-scale drainage network require a different explanation from a smaller-scale meandering river within the network, and an even smaller-scale point bar along the meander requires a different explanation again. The process could carry on down through bedforms on the point bar, to the position and nature of individual sediment grains within the bedforms (cf. Schumm 1985a; 1991, 49). A similar problem applies to the time dimension. Geomorphic systems may be studied in action today. Such studies are short term, lasting for a few years or decades. Yet geomorphic systems have a history that goes back centuries, millennia, or millions of years. Using the results of short-term studies to explain how geomorphic systems will change over long periods is beset with difficulties. Stanley A. Schumm (1985a, 1991) tried to resolve the **scale problem**, and in doing so established some links between process and historical studies (p. 9).

Table 2.1 Disturbance characteristics for selected geomorphic disturbances

Characteristics						
Frequency	**Magnitude**	**Duration extent**	**Spatial onset**	**Speed of dispersion**	**Spatial spacing**	**Temporal**
Disturbing agency						
Fire						
Frequent to rare	Low to moderate	Short	Moderate to extensive	Diffuse	Random	Rapid
Drought						
Frequent to rare	Low to moderate	Short to moderate	Extensive	Slow	Diffuse	Random to cyclical
Volcanic eruption						
Rare	Low to extreme	Short	Local to extensive	Rapid	Concentrated	Random
Eustatic sea-level change						
Rare	Moderate to extreme	Long	Extensive to global	Slow	Diffuse	Cyclical
Subsidence						
Rare to moderate	Low to high	Short to moderate	Local to moderate	Slow to rapid	Moderate to concentrated	Random
Mining						
Singular	Extreme	Short to moderate	Local	Rapid	Concentrated	Singular

Source: Adapted from Gares *et al.* (1994) and Phillips (2009)

System dynamics: stasis and change

The adoption by process geomorphologists of a systems approach has provided a common language and a theoretical basis for discussing static and changing conditions in geomorphic systems. It is helpful to explore the matter by considering how a geomorphic system responds to a **disturbance** or a change in driving force (a **perturbation**), such as a change in stream discharge. Table 2.1 shows some common perturbers of geomorphic systems and their characteristics.

Discussion of responses to disturbances in the geomorphological literature tends to revolve around the notion of **equilibrium**, which has a long and involved history. In simple terms, equilibrium is 'a condition in which some kind of balance is maintained' (Chorley and Kennedy 1971, 348), but it is a complex concept, its complexity lying in the multiplicity of equilibrium patterns and the fact that not all components of a system need be in balance at the same time for some form of equilibrium to obtain. The more recently introduced ideas of **disequilibrium** (moving towards a stable end state, but not yet there) and **non-equilibrium** (not moving towards any particular stable or steady state) add another dimension to the debate.

Equilibrium

Figure 2.5 shows eight conditions of equilibrium (a–h). **Thermodynamic equilibrium** is the tendency towards maximum entropy, as demanded by the second law of thermodynamics. In geomorphology, such a tendency would lead to a continuous and gradual reduction of energy gradients (slopes) and an attendant lessening of the rates of geomorphic processes. A featureless plain would be in a state of thermodynamic equilibrium, but virtually all landscapes are far removed from such an extreme state.

Several forms of equilibrium occur where landforms or geomorphic processes do not change and maintain static or stationary states. **Static equilibrium** is the condition where an object has forces acting upon it but it does not move because the forces balance. Examples are a boulder resting on a slope and a stream that has cut down to its base level, so preventing further entrenchment. **Stable equilibrium** is the tendency of a system to return to its original state after experiencing a small perturbation, as when a sand grain at the base of a depression is rolled a little by a gust of wind but rolls back when the wind drops. Negative feedback processes may lead to the process of restoration. **Unstable equilibrium** occurs when a small perturbation forces a system away from its old equilibrium state towards a new one. If the disturbance persists or grows, perhaps through positive feedback processes, it may lead to disequilibrium or non-equilibrium. A simple example would be a boulder perched atop a hill; a force sufficient to dislodge the boulder would lead to its rolling downslope.

In another common form of equilibrium, a geomorphic system self-maintains a constant form or steady state in the face of all but the largest perturbations. An example is a concavo-convex hillslope profile typical of humid climates with a concave lower portion and a convex upper slope, where erosion, deposition, and mass movement continue to operate, and the basic slope form stays the same. Such **steady-state equilibrium** occurs when numerous small-scale fluctuations occur about a mean stable state. The notion of steady state is perhaps the least controversial of systems concepts in physical geography. Any open system may eventually attain time-independent equilibrium state – a steady state – in which the system and its parts are unchanging, with maximum entropy and minimum free energy. In such a steady state, a system stays constant as a whole and in its parts, but material or energy continually passes through it. As a rule, steady states are irreversible. Before arriving at a steady state, the system will pass through a transient state (a sort of start-up or warm-up period). For instance, the amount of water in a lake could remain steady because gains of water (incoming river water and precipitation) balance losses through river out flow, ground-

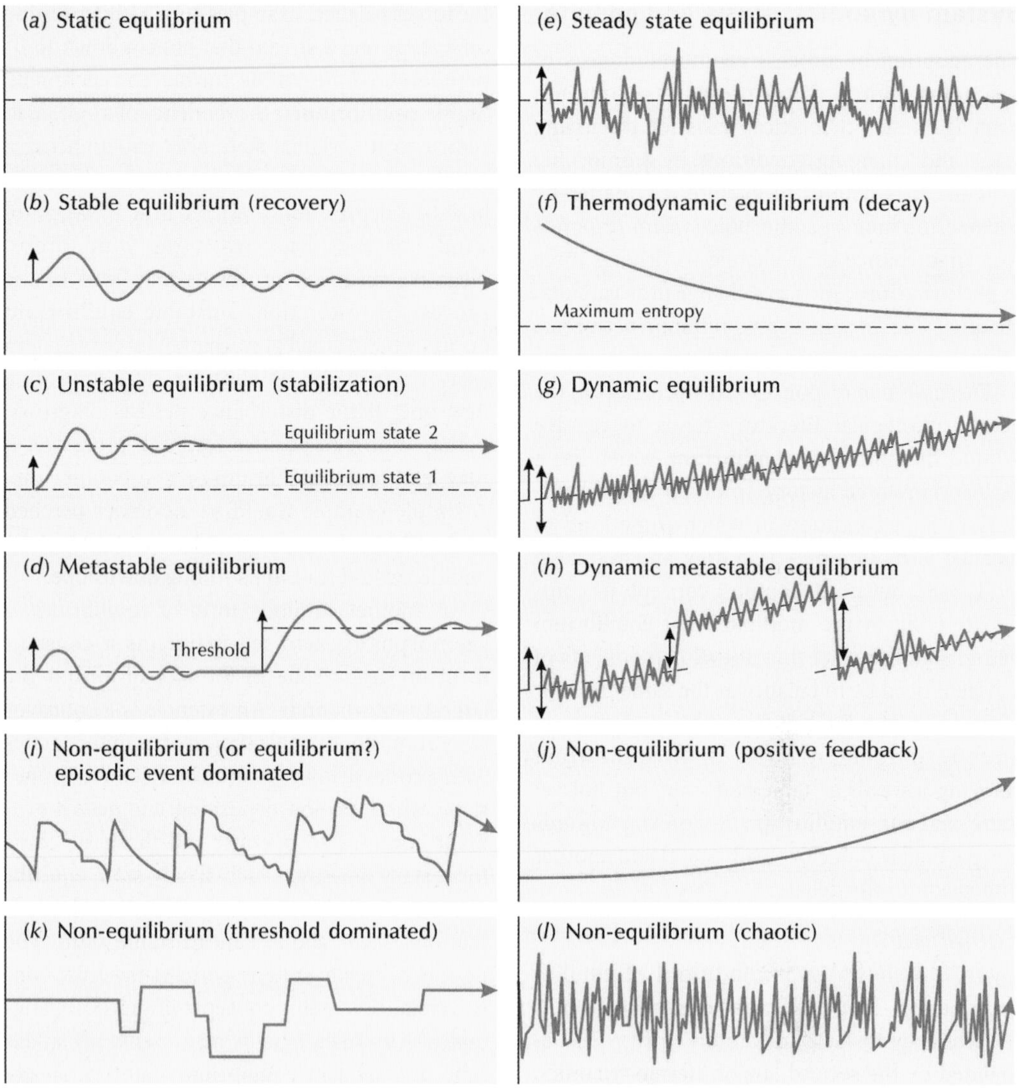

Figure 2.5 Types of equilibrium in geomorphology. *Source*: Adapted from Chorley and Kennedy (1971, 202) and Renwick (1992)

water seepage, and evaporation. If the lake started empty, then its filling up would be a transient state. **Dynamic equilibrium** is a disputatious term and discussed in Box 2.2.

From the 1960s onward, some geomorphologists began questioning simplistic notions of equilibrium and steady state. In 1965, Alan D. Howard noted that geomorphic systems might possess **thresholds** (Box 2.3) that separate two rather different system economies. Schumm (1973, 1977) introduced the notions of **metastable equilibrium** and **dynamic metastable equilibrium**, showing that thresholds within a fluvial system cause a shift in its mean state. The thresholds, which may be intrinsic or extrinsic, are not part of a change continuum, but show up as dramatic

Box 2.2 DYNAMIC EQUILIBRIUM

Chemists first used the expression **dynamic equilibrium** to mean equilibrium between a solid and a solute maintained by solutional loss from the solid and precipitation from the solution running at equal rates. The word equilibrium captured that balance and the word dynamic captured the idea that, despite the equilibrium state, changes take place. In other words, the situation is a dynamic, and not a static, equilibrium. Grove Karl Gilbert (1877) possibly first applied the term in this sense in a geomorphic context. He suggested that all streams work towards a graded condition, and attain a state of dynamic equilibrium when the net effect of the flowing water is neither the erosion of the bed nor the deposition of sediment, in which situation the landscape then reflects a balance between force and resistance. Applied to any landform, dynamic equilibrium would represent a state of balance in a changing situation. Thus, a spit may appear to be unchanging, although deposition feeds it from its landward end, and erosion consumes it at its seaward end.

John T. Hack (1960) developed Gilbert's ideas, arguing that a landscape should attain a steady state, a condition in which land-surface form does not change despite material being added by tectonic uplift and removed by a constant set of geomorphic processes (Plate 2.1). He contended that, in an erosional landscape, dynamic equilibrium prevails where all slopes, both hillslopes

Plate 2.1 John T. Hack (on left) with John Goodlett, circa 1958, on a debris fan in the Little River, Virginia, USA. (*Photograph courtesy of Cliff Hupp*)

continued . . .

and river slopes, are adjusted to each other (cf. Gilbert 1877, 123–4; Hack 1960, 81), and 'the forms and processes are in a steady state of balance and may be considered as time independent' (Hack 1960, 85). In practice, this notion of dynamic equilibrium was open to question (e.g. Ollier 1968) and difficult to apply to landscapes. Other geomorphologists have used the term dynamic equilibrium to mean 'balanced fluctuations about a constantly changing system condition which has a trajectory of unrepeated states through time' (Chorley and Kennedy 1971, 203), which is similar to Alfred J. Lotka's (1924) idea of moving equilibrium (cf. Ollier 1968, 1981, 302–4).

Currently then, dynamic equilibrium in physical geography is synonymous with a 'steady state' or with a misleading state, where the system appears to be in equilibrium but in reality is changing extremely sluggishly. Thus, the term has been a replacement for such concepts as grade (p. 234). Problems with the concept relate to the application of a microscale phenomenon in physics to macroscale geomorphic systems, and to the difficulty of separating any observed fluctuations from a theoretical underlying trend (Thorn and Welford 1994). On balance, it is perhaps better for physical geographers to abandon the notion of dynamic equilibrium, and indeed some of the other brands of equilibrium, and instead adopt the terminology of non-linear dynamics.

Box 2.3 THRESHOLDS

A threshold separates different states of a system. It marks some kind of transition in the behaviour, operation, or state of a system. Everyday examples abound. Water in a boiling kettle crosses a temperature threshold in changing from a liquid to a gas. Similarly, ice taken out of a refrigerator and placed upon a table in a room with an air temperature of 10°C will melt because a temperature threshold has been crossed. In both examples, the huge differences in state – liquid water to water vapour, and solid water to liquid water – may result from tiny changes of temperature. Many geomorphic processes operate only after the crossing of a threshold. Landslides, for instance, require a critical slope angle, all other factors being constant, before they occur. Stanley A. Schumm (1979) made a powerful distinction between **external** and **internal system thresholds**. A geomorphic system will not cross an external threshold unless forced to do so by a change in an external variable. A prime example is the response of a geomorphic system to climatic change. Climate is the external variable. If, say, runoff were to increase beyond a critical level, then the geomorphic system might suddenly respond by reorganizing itself into a new state. No change in an external variable is required for a geomorphic system to cross an internal threshold. Rather, some chance fluctuation in an internal variable within a geomorphic system may take a system across an internal threshold and lead to its reorganization. This appears to happen in some river channels where an initial disturbance by, say, overgrazing in the river catchment triggers a complex response in the river channel: a complicated pattern of erosion and deposition occurs with phases of alluviation and downcutting taking place concurrently in different parts of the channel system (see p. 41).

Plate 2.2 The terraced landscape around Douglas Creek, Wyoming, USA. The photo was taken from a bluff along the west side looking northeast. The abandoned channel that turns right near the left side of the photo is the 1961 surface. On the right side of the channel is a crescent-shaped terrace with saltcedars dated to 1937. The unvegetated, near-vertical bluff line in the centre right of the photo leads up to the 1900 surface. The valley floor steps up gently to the east to the 1882 surface. (*Photograph by Ray Womack*)

changes resulting from minor shifts in system dynamics, such as caused by a small disturbance. In metastable equilibrium, static states episodically shift when thresholds are crossed. It involves a stable equilibrium acted upon by some form of incremental change (a trigger mechanism) that drives the system over a threshold into a new equilibrium state. A stream, for instance, if forced away from a steady state, will adjust to the change, although the nature of the adjustment may vary in different parts of the stream and at different times. Douglas Creek in western Colorado, USA, was subject to overgrazing during the 'cowboy era' and, since about 1882, it has cut into its channel bed (Plate 2.2; Womack and Schumm 1977). The manner of cutting has been complex, with discontinuous episodes of down-cutting interrupted by phases of deposition, and with the erosion–deposition sequence varying from one cross-section to another. Trees have been used to date terraces at several locations. The terraces are unpaired (p. 251), which is not what would be expected from a classic case of river incision, and they are discontinuous in a downstream direction. This kind of study serves to dispel forever the simplistic cause-and-effect view of landscape evolution in which change is seen as a simple response to an altered input. It shows that landscape dynamics may involve abrupt and discontinuous behaviour involving flips between quasi-stable states as system thresholds are crossed. In dynamic metastable equilibrium, thresholds trigger episodic changes in states of dynamic equilibrium (dynamic equilibrium meaning here

Box 2.4 BIFURCATION THEORY AND GEOMORPHIC SYSTEMS

Some geomorphologists applied bifurcation theory to geomorphic systems in the late 1970s and early 1980s. They based their arguments on catastrophe theory, which is a special branch of bifurcation theory developed by René Thom (1975), and tried to use Thom's ideas (his cusp catastrophe proved a favourite) to explain certain processes at the Earth's surface. An example is John Thornes's (1983) model of sediment transport in a river, in which a three-dimensional surface defines the equilibrium value of sediment load (the response variable) in relation to stream power (control variable 1 or the splitting parameter) and the ratio of sediment sorting to sediment size (control variable 2 or the normal parameter) (Figure 2.6). For low values of the sorting–size ratio, the equilibrium surface is a single, smooth curve; beyond the point where the surface splits (point A in the diagram) and becomes complex, three equilibria points exist, the middle of which is unstable. After the splitting point, the system jumps from one of the equilibrium states to another – the transition is not smooth. The model suggests that for mixed sediments of relatively small mean size, the equilibrium sediment load is a smooth function of stream power, but that larger and better sorted sediments are entrained in a discontinuous manner, the sediment load of the rising and falling limbs of a hydrograph being very different.

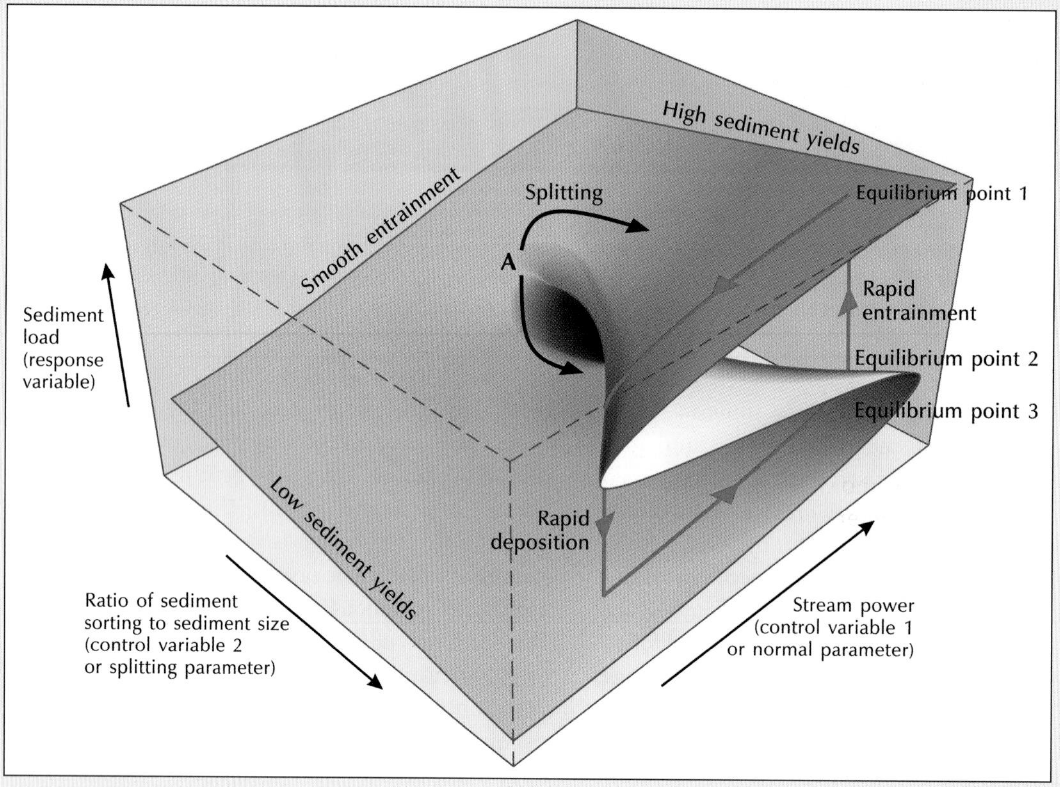

Figure 2.6 A cusp catastrophe model applied to sediment transport in a river.
Source: Adapted from Thornes (1983)

The cusp catastrophe model still has some currency, being used, for example, to explain the instability of a slip-buckling landslide, where a translational slip of rock layers leads to buckling of rock slabs near the slope base (Qin *et al.* 2001).

a trending mean state). So, dynamic metastable equilibrium is a combination of dynamic and metastable equilibria, in which large jumps across thresholds break in upon small-scale fluctuations about a moving mean. For this reason, dynamic metastable equilibrium is really a form of disequilibrium as a progressive change of the mean state occurs (Renwick 1992).

The seminal idea of **thresholds** led eventually to applications of bifurcation theory (Box 2.4) and chaos (Box 2.5) in geomorphology, which deal with non-equilibrium as well as equilibrium states (see Huggett 2007).

Box 2.5 CHAOS

Early ideas on complex dynamics and non-equilibrium within systems found a firm theoretical footing with the theory of nonlinear dynamics and chaotic systems that scientists from a range of disciplines developed, including geomorphology itself. Classical open systems research characteristically dealt with linear relationships in systems near equilibrium. A fresh direction in thought and a deeper understanding came with the discovery of deterministic chaos by Edward Lorenz in 1963. The key change was the recognition of nonlinear relationships in systems. In geomorphology, nonlinearity means that system outputs (or responses) are not proportional to system inputs (or forcings) across the full gamut of inputs (cf. Phillips 2006).

Nonlinear relationships produce rich and complex dynamics in systems far removed from equilibrium, which display periodic and chaotic behaviour. The most surprising feature of such systems is the generation of 'order out of chaos', with systems states unexpectedly moving to higher levels of organization under the driving power of internal entropy production and entropy dissipation. Systems of this kind, which dissipate energy in maintaining order in states removed from equilibrium, are **dissipative systems**. The theory of complex dynamics predicts a new order of order, an order arising out of, and poised perilously at the edge of, chaos. It is a fractal order that evolves to form a hierarchy of spatial systems whose properties are holistic and irreducible to the laws of physics and chemistry. Geomorphic examples are flat or irregular beds of sand on streambeds or in deserts that self-organize themselves into regularly spaced forms – ripples and dunes – that are rather similar in size and shape (e.g. Baas 2002; see Murray *et al.* 2009 for other examples). Conversely, some systems display the opposite tendency – that of non-self-organization – as when relief reduces to a plain. A central implication of chaotic dynamics for the natural world is that all Nature may contain fundamentally erratic, discontinuous, and inherently unpredictable elements. Nonetheless, nonlinear Nature is not all complex and chaotic. Phillips (2006, 110) astutely noted that 'Nonlinear systems are not all, or always, complex, and even those which can be chaotic are not chaotic under all circumstances. Conversely, complexity can arise due to factors other than nonlinear dynamics'.

Phillips (2006) suggested ways of detecting chaos in geomorphic systems. He argued that convergence versus divergence of a suitable system descriptor (elevation or regolith thickness, for instance) is an immensely significant indicator of stability behaviour in a geomorphic system. In landscape evolution, convergence associates with downwasting and a reduction of relief, while divergence relates to dissection and an increase of relief. More fundamentally, convergence and divergence underpin developmental, 'equilibrium' conceptual frameworks, with a monotonic

continued . . .

move to a unique endpoint (peneplain or other steady-state landform), as well as evolutionary, 'non-equilibrium' frameworks that engender historical happenstance, multiple potential pathways and end-states, and unstable states. The distinction between instability and new equilibria is critical to understanding the dynamics of actual geomorphic systems, and for a given scale of observation or investigation, it separates two conditions. On the one hand sits a new steady-state equilibrium governed by stable equilibrium dynamics that develops after a change in boundary conditions or in external forcings. On the other hand sits a persistence of the disproportionate impacts of small disturbances associated with dynamic instability in a non-equilibrium system (or a system governed by unstable equilibrium dynamics) (Phillips 2006). The distinction is critical because the establishment of a new steady-state equilibrium implies a consistent and predictable response throughout the system, predictable in the sense that the same changes in boundary conditions affecting the same system at a different place or time would produce the same outcome. In contrast, a dynamically unstable system possesses variable modes of system adjustment and inconsistent responses, with different outcomes possible for identical or similar changes or disturbances.

Non-equilibrium

Figure 2.5 also shows four types of non-equilibrium (not tending towards any particular stable or steady state), which range from a system lurching from one state to another in response to episodic threshold events, through a continuous change of state driven by positive feedback and threshold-dominated abrupt changes of state, to a fully chaotic sequence of state changes. These non-equilibrium interpretations of response in geomorphic systems come from the field of **dynamic systems theory**, which embraces the buzzwords **complexity** and **chaos**. The argument runs that steady states in the landscape may be rare because landscapes are inherently unstable. This is because any process that reinforces itself keeps the system changing through a positive feedback circuit and readily disrupts any balance obtaining in a steady state. The 'instability principle', which recognizes that, in many landscapes, accidental deviations from a 'balanced' condition tend to be self-reinforcing, formalizes this idea (Scheidegger 1983). It explains why cirques tend to grow, sinkholes increase in size, and longitudinal mountain valley profiles become stepped. The intrinsic instability of landscapes is borne out by mathematical analyses that point to the chaotic nature of much landscape change (e.g. Phillips 1999; Scheidegger 1994).

Reaction, relaxation, resistance, resilience, and recursion

In many geomorphic systems, change in system form trails behind a change in input (a disturbance). This lag is the time taken for some mechanism to react to the changed input and is called the **reaction time** (Figure 2.7a). In the case of river bedload, particles will not react to increased discharge until a critical shear stress is applied. In other words, a system response requires the crossing of a threshold. Another reason for a lag between a changed input and a change in form is that the input and the form response are separated geographically. A case in point is pyroclastic material ejected from a volcanic vent, which cannot change the elevation of the land surface surrounding the volcano until it has travelled through the atmosphere. It is common in geomorphic systems for system form to be unable to keep pace with a change in input, which delays the attainment of a new equilibrium state irrespective of any reaction-time effects (Figure 2.7b). The time taken for the system to adjust to the changed input is the **relaxation time**. Geomorphic systems may possess reaction times and relaxation times, which combine to give the system **response time**. In summary, the reaction time is the time needed for a system to start responding to a changed input and the relaxation time is the time taken for the system to complete the response.

Resistance is the ability of a geomorphic system to avoid or to lessen responses to driving forces. It has two components – strength and capacity. The 'strength' of a system is measurable as literal strength (as in shear strength), chemical or mechanical stability (as in mineral stability, hardness, cohesion), or susceptibility to modification (as in soil erodibility). These must be compared with some measure related to the magnitude of the drivers of change. For instance, the ratio of shear strength to shear stress ratio is used in assessing slope stability (p. 85). A geomorphic system may also resist changes to inputs by absorbing them, and the ability to do so depends on the system's 'capacity'. So, what happens to sediment delivered to a channel by a landslide or by soil erosion from fields will depend in part on the sediment transport capacity of the stream: if the stream has a low transport capacity, then the sediment will accumulate; if it has a high transport capacity then it will be removed. **Resilience** is the ability of a system to recover towards its state before disturbance. It is a direct function of the dynamical stability of the system. A geomorphic system in a steady state will display resilience within certain bounds. **Recursion** involves the changes in the system following a disturbance feeding back upon themselves. Recursive feedbacks may be positive, reinforcing and thus perpetuating or even accelerating the change, or negative, slowing or even negating the change (p. 34).

Magnitude and frequency

Interesting debates centre on the variations in process rates through time. The 'tame' end of this debate concerns arguments over **magnitude** and **frequency** (Box 2.6), the pertinent question here being which events perform the most geomorphic work: small and infrequent events, medium and moderately frequent events, or big but rare events? The first work on this issue concluded, albeit provisionally until further field work was carried out, that events occurring once or twice a year perform most geomorphic work (Wolman and Miller 1960). Some later work has highlighted the geomorphic significance of rare events. One study revealed that low-frequency, high-magnitude events greatly affect stream channels

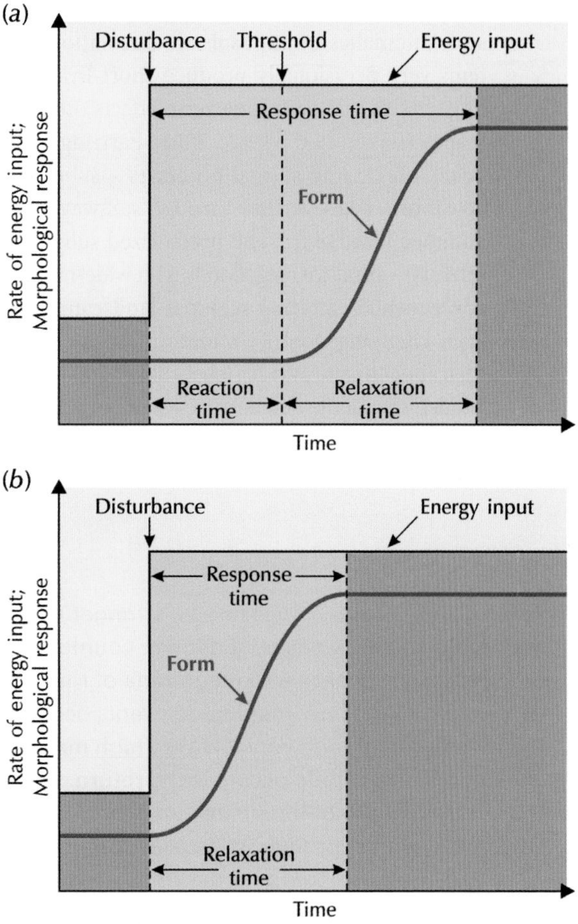

Figure 2.7 (a) Reaction time and (b) relaxation time in geomorphic systems.

(Gupta 1983). Other work suggested that large-scale anomalies in atmospheric circulation systems very occasionally produce short-lived but huge floods that have long-term effects on landscapes (Baker 1977, 1983, 2008; Partridge and Baker 1987). And several processes - subglacial volcanism, lake overflow through spillways, ice-dammed lake failure, and pressurized subglacial outburst - produce megafloods with wide-ranging consequences, altering regional landscapes and even changing ocean circulation patterns that then alter climate and sea level (Baker 2009). The 'wilder' end of the debate engages hot arguments over **gradualism** and **catastrophism** (Huggett 1989, 1997a, 2006). The crux of the gradualist–catastrophist debate is the seemingly innocuous question: have the present rates of geomorphic processes remained much the same throughout Earth surface history? Gradualists claim that process rates have been uniform in the past, not varying much beyond their present levels. Catastrophists make the counterclaim that the rates of geomorphic processes have differed in the past, and on occasions, some of them have acted with suddenness and extreme violence, pointing to the effects of massive volcanic explosions, the impacts of asteroids and comets, and the landsliding of whole mountainsides into the sea.

Box 2.6 MAGNITUDE AND FREQUENCY

As a rule of thumb, bigger floods, stronger winds, higher waves, and so forth occur less often than their smaller, weaker, and lower counterparts do. Indeed, graphs showing the relationship between the frequency and magnitude of many geomorphic processes are right skewed, which means that a lot of low-magnitude events occur in comparison with the smaller number of high-magnitude events, and very few very high magnitude events. The frequency with which an event of a specific magnitude occurs is the **return period** or **recurrence interval**, which is calculated as the average length of time between events of a given magnitude. Take the case of river floods. Observations may produce a dataset comprising the maximum discharge for each year over a period of years. To compute the **flood–frequency relationships**, the peak discharges are listed according to magnitude, with the highest discharge first. The recurrence interval is then calculated using the equation

$$T = \frac{n+1}{m}$$

where T is the recurrence interval, n is the number of years of record, and m is the magnitude of the flood (with $m = 1$ at the highest recorded discharge). Each flood is then plotted against its recurrence interval on Gumbel graph paper and the points connected to form a frequency curve. If a flood of a particular magnitude has a recurrence interval of 10 years, it would mean that there is a 1-in-10 (10 per cent) chance that a flood of this magnitude (2,435 cumecs in the Wabash River example shown in Figure 2.8) will occur in any year. It also means that, on average, one such flood will occur every 10 years. The magnitudes of 5-year, 10-year, 25-year, and 50-year floods are helpful for engineering work, flood control, and flood alleviation. The 2.33-year flood ($Q_{2.33}$) is the mean annual flood (1,473 cumecs in the example), the 2.0-year flood ($Q_{2.0}$) is the median annual flood (not shown), and the 1.58-year flood ($Q_{1.58}$) is the most probable flood (1,133 cumecs in the example).

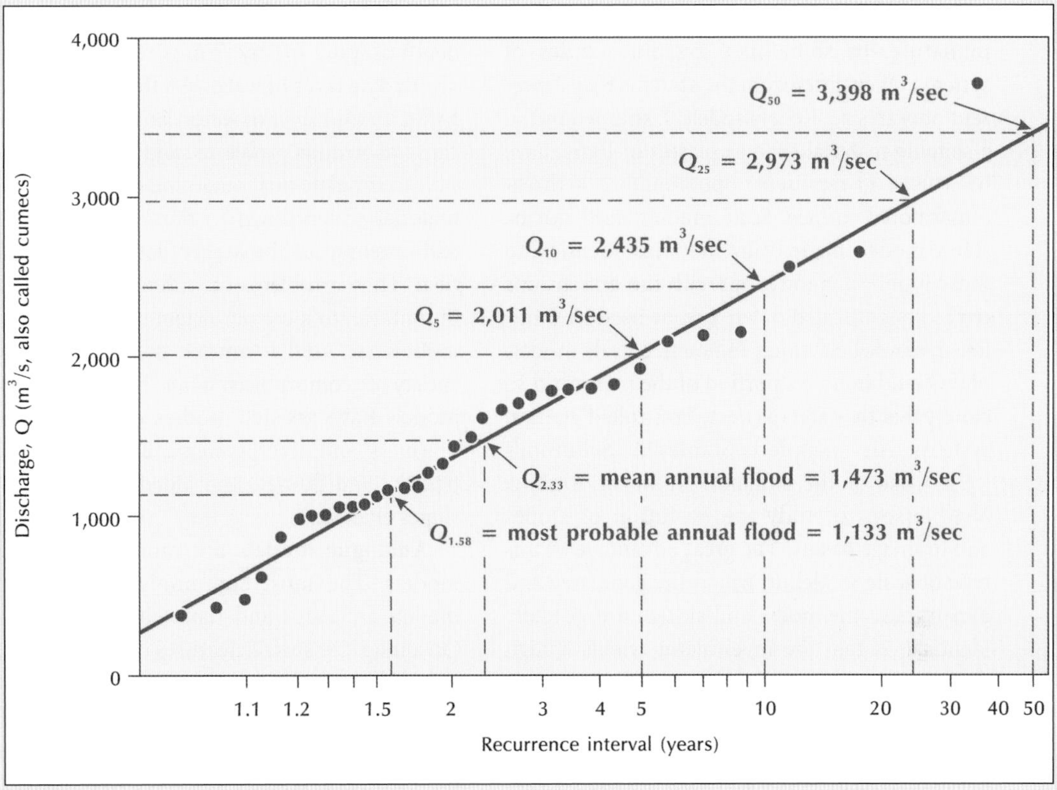

Figure 2.8 Magnitude–frequency plot of annual floods on the Wabash River, at Lafayette, Indiana, USA. See text for details. *Source*: Adapted from Dury (1969)

The dichotomy between gradualists and catastrophists polarizes the spectrum of possible rates of change. It suggests that there is either gradual and gentle change, or else abrupt and violent change. In fact, all grades between these two extremes, and combinations of gentle and violent processes, are conceivable. It seems reasonable to suggest that land-surface history has involved a combination of gentle and violent processes.

GEOMORPHIC MODELS

In trying to single out the components and interrelations of geomorphic systems, some degree of abstraction or simplification is necessary: the landscape is too rich a mix of objects and interactions to account for all components and relationships in them. The process of simplifying real landscapes to manageable proportions is **model building**. Defined in a general way, a geomorphic **model** is a simplified representation of some aspect of a real landscape that happens to interest a geomorphologist. It is an attempt to describe, analyse, simplify, or display a geomorphic system (cf. Strahler 1980).

Geomorphologists, like all scientists, build models at different levels of abstraction (Figure 2.9). The simplest level involves a change of scale. In this case, a **hardware model** represents the system (see Mosley and Zimpfer 1978). There are two chief kinds of hardware model: scale models and analogue models. **Scale** (or **iconic**) **models** are miniature, or sometimes gigantic, copies of systems. They differ from the systems they represent only in size. Relief models, fashioned out of a suitable material such as plaster of Paris, have been used to represent topography as a three-dimensional surface. Scale models need not be static: models made using materials identical to those found in Nature, but with the dimensions of the system scaled down, can be used to simulate dynamic behaviour. In practice, scale models of this kind imitate a portion of the real world so closely that they are, in effect, 'controlled' natural systems. An example is Stanley A. Schumm's (1956) use of the badlands at Perth Amboy, New Jersey, to study the evolution of slopes and drainage basins. The great advantage of this type of scale model, in which the geometry and dynamics of the model and system are virtually identical, is that the investigator wields a high degree of control over the simplified experimental conditions. Other scale models use natural materials, but the geometry of the model is dissimilar to the geometry of the system it imitates – the investigator scales down the size of the system. The process of reducing the size of a system may create a number of awkward problems associated with scaling. For instance, a model of the Severn estuary made at a scale of 1:10,000 can easily

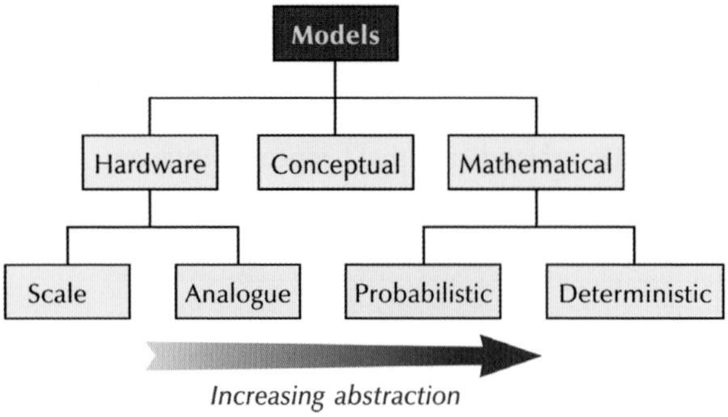

Figure 2.9 Types of model in geomorphology.
Source: After Huggett (1993, 4)

preserve geometrical and topographical relationships. However, when adding water, an actual depth of water of, say, 7 m is represented in the model by a layer of water less than 0.7 mm deep. In such a thin layer of water, surface tensions will cause enormous problems, and it will be impossible to simulate tidal range and currents. Equally, material scaled down to represent sand in the real system would be so tiny that most of it would float. These problems of scaling are usually surmountable, to a certain extent at least, and scale models are used to mimic the behaviour of a variety of geomorphic systems. For example, scale models have assisted studies of the dynamics of rivers and river systems using waterproof troughs and flumes, and aided studies of talus slopes (Plate 2.3).

Analogue models are more abstract scale models. The most commonly used analogue models are maps and remotely sensed images. On a map, the surface features of a landscape are reduced in scale and represented by symbols: rivers by lines, relief by contours, and spot heights by points, for instance. Remotely sensed images represent, at a reduced scale, certain properties of the landscape systems. Maps and remotely sensed images are, except where a series of them is available for different times, static analogue models. Dynamic analogue models may also be built. They are hardware models in which the system size is changed, and in which the materials used are analogous to, but not the same as, the natural materials of the system. The analogous materials simulate the dynamics of the real system. In a laboratory, the clay kaolin can be used in place of ice to model the behaviour of a valley glacier. Under carefully controlled conditions, many features of valley glaciers, including crevasses and step faults, develop in the clay. Difficulties arise in this kind of analogue model, not the least of which is the problem of finding a material that has mechanical properties comparable to the material in the natural system. However, they can prove a very useful tool, for example in studying long-term landscape development (Plate 2.4; see

Plate 2.3 (*right*) An analogue model simulating talus development (De Blasio and Sæter 2009). The model used a 1.5-m-long board sloping at 37.5° (a tad lower than the angle of repose of the rocky material) and bolted to a frame of aluminium and steel. Compacted angular grains were glued to the board with epoxy to increase the friction angle and avoid particle slippage against the base. Grains of basalt in five size classes (each a different colour), were dropped from a suspended plate at the top of the slope. The ratio between table length and maximum particle size was about a hundred, which agrees with the ratio of talus length to maximum boulder diameter in the field. At the start of the experiment, the grains developed a gradation along the slope similar to the gradation found on natural talus slopes, where small grains settle at the top and large grains roll downwards to the bottom section. However, after a transient period dominated by single-particle dynamics on the inert granular medium, the talus evolution was more variable than expected. Owing to the continuous shower of falling grains, the shear stress at the bottom of the upper granular layer increased, so initially producing a slow creep downslope that finally collapsed in large avalanches and homogenizing the material. (*Photographs by Fabio De Blasio*)

Bennett *et al.* 2015 and Clarke 2014, 2015 for examples).

Conceptual models are initial attempts to clarify loose thoughts about the structure and function of a geomorphic system. They often form the basis for the construction of mathematical models. **Mathematical models** translate the ideas encapsulated in a conceptual model into the formal, symbolic logic of mathematics. The language of mathematics offers a powerful tool of investigation limited only by the creativity of the human mind. Of all modes of argument, mathematics is the most rigorous. Nonetheless, the act of quantification, of translating ideas and observations into symbols and numbers, is in itself nothing unless validated by explanation and prediction. The art and science of using mathematics to study geomorphic systems is to discover expressions with explanatory and predictive powers. These powers set mathematical models apart from conceptual models. An unquantified conceptual model is simply a body of ideas and cannot be tested without further development; it is simply a body of ideas. A mathematical model, on the other hand, is testable by

Plate 2.4 An analogue model for simulating long-term landform evolution with uplift and variable rainfall rate (Bonnet and Crave 2003). The model used a paste of pure silica grains (mean grain size of 0.02 mm) mixed with water, the content of which ensured that the paste had a vertical angle of rest and that water infiltration was negligible. The paste was placed in a box with a vertically adjustable base, the movements of which were driven by a screw and a computer-controlled stepping motor. During an experimental run, uplift was simulated by raising the base of the box at a constant rate, so pushing out the paste from the top of the box. Precipitation was generated by a system of four sprinklers. These delivered water droplets with a diameter of approximately 0.01 mm, which was small enough to avoid any splash dispersion at the surface of the model. The precipitation rate could be controlled by changing the water pressure and the configuration of the sprinklers. The surface of the model was eroded by running water at its surface and grain detachment and transport occurred mainly by shear detachment through surface runoff. (*Photographs by Stéphane Bonnet*)

matching predictions against the yardstick of observation. By a continual process of mathematical model building, model testing, and model redesign, the understanding of the form and function of geomorphic systems should advance.

Three chief classes of mathematical model assist the study of geomorphic systems: stochastic models, statistical models, and deterministic models. The first two classes are both probabilistic models. **Stochastic models** have a random component built into them that describes a system, or some facet of it, based on probability. **Statistical models**, like stochastic models, have random components. In statistical models, the random components represent unpredictable fluctuations in laboratory or field data that may arise from measurement error, equation error, or the inherent variability of the objects being measured. A body of inferential statistical theory exists that determines the manner in which the data should be collected and how relationships between the data should be managed. Statistical models are, in a sense, second best to deterministic models: they can be applied only under strictly controlled conditions, suffer from a number of deficiencies, and are perhaps most profitably employed only when the 'laws' determining system form and process are poorly understood. **Deterministic models** are conceptual models expressed mathematically and containing no random components. They are derivable from physical and chemical principles without recourse to experiment. It is sound practice, therefore, to test the validity of a deterministic model by comparing its predictions with independent observations made in the field or the laboratory. Hillslope models based on the conservation of mass are examples of deterministic models (p. 197).

FORM

The two main approaches to form in geomorphology are description (field description and morphological mapping) and mathematical representation (geomorphometry).

Field description and morphological mapping

The only way fully to appreciate landforms is to go into the field and see them. Much can be learnt from the now seemingly old-fashioned techniques of field description, field sketching, and map reading and map making.

The mapping of landforms is an art (see Dackombe and Gardiner 1983, 13–20, 28–41; Evans 1994). Landforms vary enormously in shape and size. Some, such as karst depressions and volcanoes, may be represented as points. Others, such as faults and rivers, are linear features that are best depicted as lines. In other cases, areal properties may be of prime concern and suitable means of spatial representation must be employed. Morphological maps capture areal properties. **Morphological mapping** attempts to identify basic landform units in the field, on aerial photographs, or on maps. It sees the ground surface as an assemblage of landform elements. **Landform elements** are recognized as simply curved geometric surfaces lacking inflections (complicated kinks) and are considered in relation to upslope, downslope, and lateral elements. They go by a plethora of names – facets, sites, land elements, terrain components, and facies. The 'site' (Linton 1951) was an elaboration of the 'facet' (Wooldridge 1932), and involved altitude, extent, slope, curvature, ruggedness, and relation to the water table. The other terms were coined in the 1960s (see Speight 1974). Figure 2.10 shows the land surface of Longdendale in the Pennines, England, represented as a morphological map. The map combines landform elements derived from a nine-unit land-surface model (p. 201) with depictions of deep-seated mass movements and superficial mass movements. Digital elevation models lie within the ambits of landform morphometry and are dealt with below. They have greatly extended, but by no means replaced, the classic work on landform elements and their descriptors as prosecuted by the morphological mappers.

Geomorphometry

A branch of geomorphology – **geomorphometry** or **landform morphometry** – studies quantitatively the form of the land surface and landforms (see Hengl and Reuter 2009; Pike *et al.* 2009). It has two component parts – general geomorphometry and specific geomorphometry (Evans 2012). **General geomorphometry** studies the land surface form, which is continuous and covers the entire globe (both subaerial and subaqueous), as well as the surface of other planets, moons, and asteroids. **Specific geomorphometry** studies the geometric and topological characteristics of landforms, which are bounded and sometimes discontinuous segments of a land surface. There are areas of the land surface not delimited as landforms that pose as yet unanswered questions (Evans 2012).

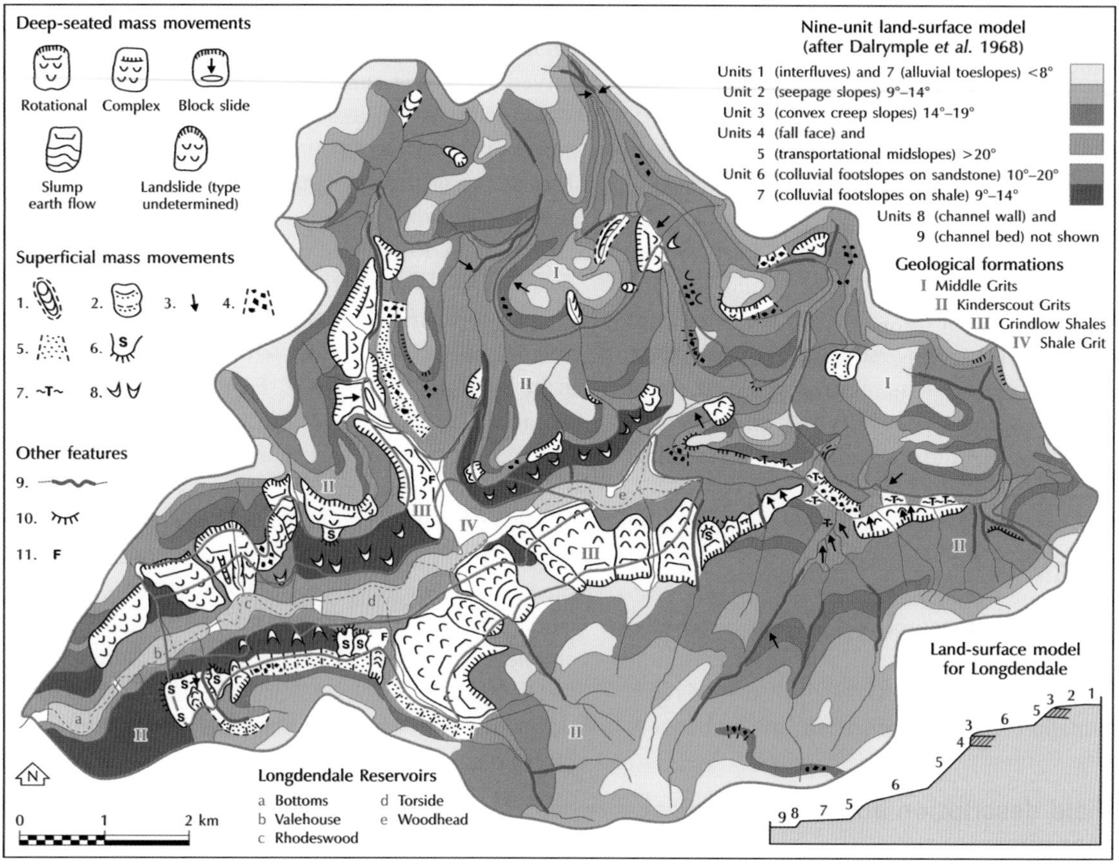

Figure 2.10 Morphological map of Longdendale, north Derbyshire, England. The map portrays units of a nine-unit land-surface model, types of mass movement, and geological formations. The superficial mass movements are: 1 Mudflow, earthflow, or peat burst; 2 Soil slump; 3 Minor soil slump; 4 Rockfall; 5 Scree; 6 Solifluction lobe; 7 Terracettes; 8 Soil creep or block creep and solifluacted material. The other features are: 9 Incised stream; 10 Rock cliff; 11 Valley-floor alluvial fan. *Source*: Adapted from Johnson (1980)

Geomorphometry in the modern era is traceable to the work of Alexander von Humboldt and Carl Ritter in the early and mid-nineteenth century (see Pike 1999). It had a strong post-war tradition in North America and the UK, and it has been 'reinvented' with the advent of remotely sensed images and **Geographical Information Systems (GIS)** software. The contributions of geomorphometry to geomorphology and cog-nate fields are legion. Geomorphometry is an important component of terrain analysis and surface modelling. Its specific applications include measuring the morphometry of continental ice surfaces, characterizing glacial troughs, mapping sea-floor terrain types, guiding missiles, assessing soil erosion, analysing wildfire propagation, and mapping ecoregions (Pike 1995, 1999). It also contributes to civil engineering, transportation, public works, oceanography, planetary science and space exploration, the mass entertain-ment business (video games and motion picture animations), and military operations (Pike *et al.* 2009).

Digital elevation models

The resurgence of geomorphometry since the 1970s is in large measure due to two developments. First is the light-speed development and use of **GIS**, which allow input, storage, and manipulation of digital data representing spatial and aspatial features of the Earth's surface. The digital representation of topography has probably attracted greater attention than that of any other surface feature. Second is the development of **Electronic Distance Measurement** (**EDM**) in surveying and, more recently, the **Global Positioning System** (**GPS**), which made the very time-consuming process of making large-scale maps much quicker and more fun.

The spatial form of surface topography is modelled in several ways. Digital representations are referred to as either **Digital Elevation Models** (**DEMs**) or **Digital Terrain Models** (**DTMs**). A DEM is 'an ordered array of numbers that represent the spatial distribution of elevations above some arbitrary datum in a landscape' (Moore *et al.* 1991, 4). DTMs are 'ordered arrays of numbers that represent the spatial distribution of terrain attributes' (Moore *et al.* 1991, 4). DEMs are, therefore, a subset of DTMs. Topographic elements of a landscape can be computed directly from a DEM (p. 203). Further details of DEMs and their applications are given in several books (e.g. Wilson and Gallant 2000; Huggett and Cheesman 2002) and a review paper by John P. Wilson (2012). Geomorphological applications are many and various, including modelling geomorphic processes, identifying remnant inselbergs in northern Sweden (p. 455), and assessing the erosional impacts of ice sheets on Baffin Island, eastern Canada (Box 2.7).

Remote sensing

Modern digital terrain representations derived from remotely sensed data greatly aid the understanding of Earth surface processes. Applications of remote sensing to geomorphology (and to the environmental sciences in general) fall into four periods. Before 1950, the initial applications of aerial photography were made. From 1950 to 1970 was a transition period from photographic applications to unconventional imagery systems (such as thermal infra-red scanners and side-looking airborne radars), and from low-altitude aircraft to satellite platforms. From 1972 to 2000, the application of multispectral scanners and radiometer data obtained from operational satellite platforms predominated. Since about 2000, a range of new remote sensing techniques has led to a proliferation of information on terrain.

Raw elevation data for DEMs are derivable from photogrammetric methods, including stereo aerial photographs, satellite imagery, and airborne laser interferometry, or from field surveys using GPS or total stations (a total station is an electronic theodolite integrated with an electronic distance meter that reads distances from the instrument to a particular point; it is usually linked to a data-logger and automated mapping software). If stereo aerial photographs and satellite images are the sources for elevation data, there will be a complete coverage of the land-scape at the resolution of the image or photographs. An advantage of using satellite images is that they are already in digital format. Airborne laser interferometry uses scanners to provide high-resolution surface measurements. An example is Light Detection And Ranging (LiDAR). Although LiDAR is a relatively young and complex technology, it provides a technique that is accurate, that is suitable for areas of rugged and difficult terrain, and that is increasingly affordable. LiDAR works by measuring the laser-pulse travel time from a transmitter to a target and back to the receiver. The laser pulse travels at the speed of light, so very accurate timing is required to obtain fine vertical resolutions. As the aircraft flies over an area, a scanning mirror directs the laser pulses back and forth across-track. The collected data is a set of points arranged across the flight-line. The combination of multiple flight-line data provides coverage for an area. An extremely useful characteristic of LiDAR is its ability to

Box 2.7 ASSESSING THE IMPACT OF ICE SHEET EROSION ON BAFFIN ISLAND USING GIS AND A DEM

Areas of low-relief, resistant, Precambrian bedrock surfaces in the Northern Hemisphere were repeatedly covered by ice sheets during the late Cenozoic. The erosional impacts of the ice are not well understood. Karin Ebert (2015) used GIS to assess the impacts of glacial erosion on a portion of Baffin Island, a tilted shield block in the eastern Canadian Arctic. She used two geomorphic parameters derived from GIS analysis – lake density and terrain ruggedness (Figure 2.11a, b) – to gauge erosional impact. The resulting patterns, as shown in Figure 2.11 c and d, conformed well to published data from other sources.

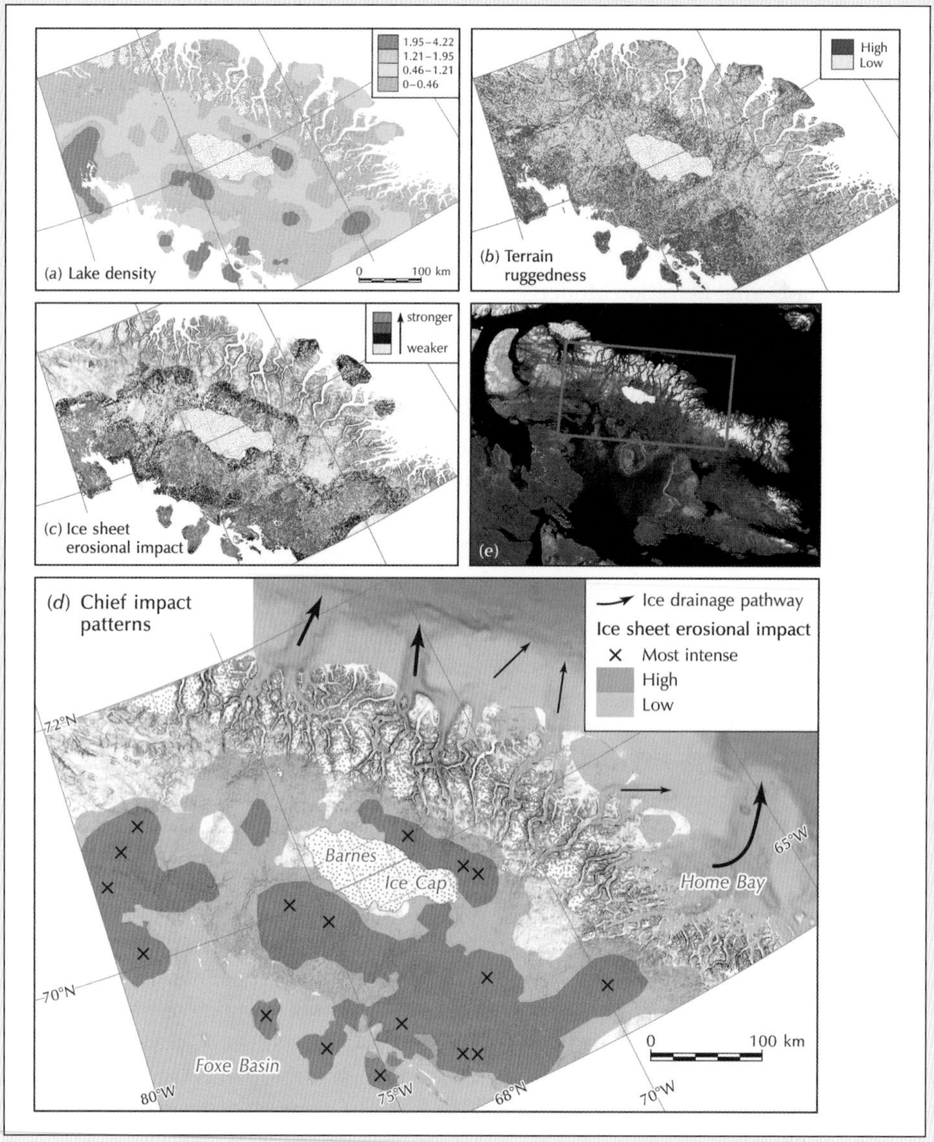

Figure 2.11 Baffin Island: geomorphic parameters and glacial erosion. (a) Lake density. (b) Terrain ruggedness. (c) Ice-sheet erosional impact the study area as a raster overlay of best results, calibrated with data from previous studies. (d) Chief erosional ice sheet impact patterns derived from lake density and terrain roughness. *Source*: After Ebert (2015)

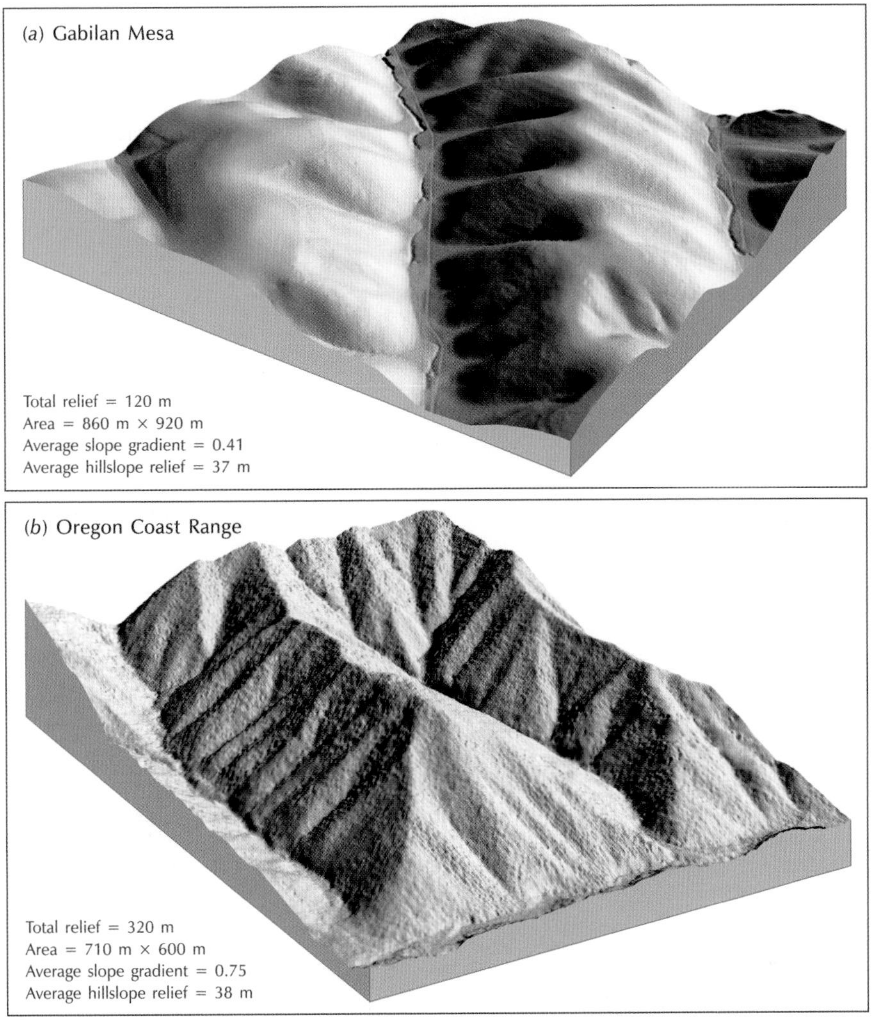

Figure 2.12 Airborne altimetry data: perspective shaded relief images of Gabilan Mesa (*top*) and Oregon Coast Range (*bottom*) study sites using high-resolution topographic data acquired via airborne laser altimetry. Steep, nearly planar slopes of the Oregon Coast Range contrast with the broad, convex Gabilan Mesa slopes. *Source*: After Roering *et al.* (2007)

penetrate the vegetation canopy and map the ground beneath.

Terrestrial Laser Scanner (TLS) and Airborne Laser Swath Mapping (ALSM) technology, using LiDAR technology, now provide high-resolution topographic data with advantages over traditional survey techniques, including the capability of producing sub-metre resolution DTMs, and high-quality land-cover information (Digital Surface Models or DSMs) over large areas (Tarolli *et al.* 2009; Tarolli 2014). New topographic data has aided geomorphic studies, including the analysis of land-surface form, landsliding, channel network structure, river morphology and bathymetry, the recognition of palaeosurfaces, and tectonics (Figure 2.12). Figure 2.13 shows the

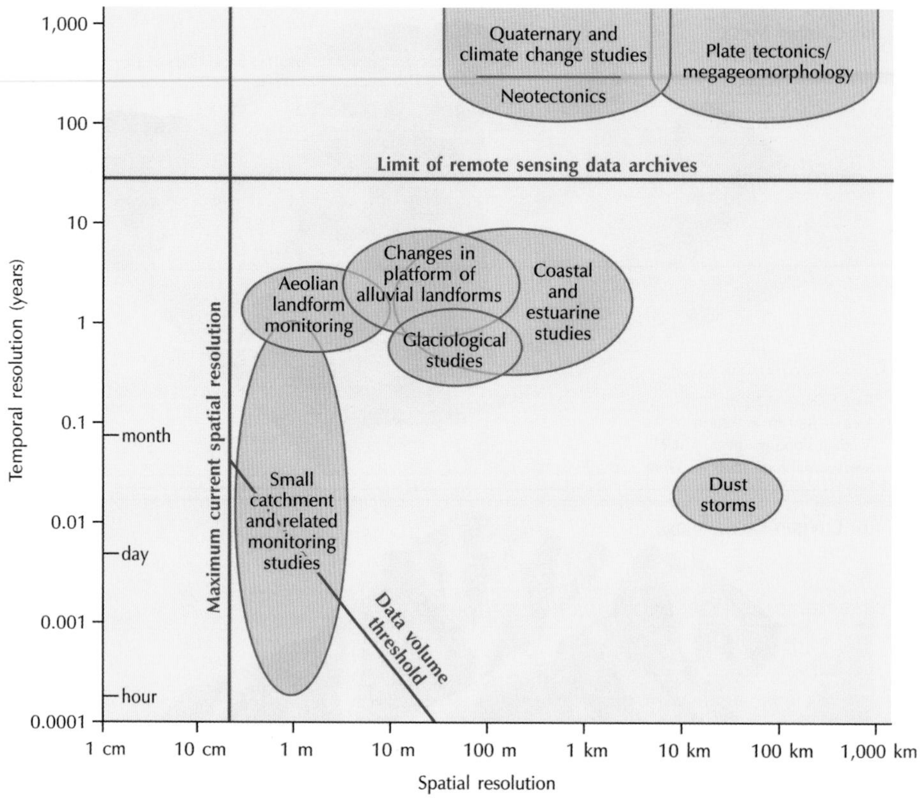

Figure 2.13 Constraints of spatial and temporal resolutions of satellite sensors on geomorphic studies. *Source*: Adapted from Millington and Townshend (1987) and Smith and Pain (2009)

current spatial and temporal resolution of satellite sensors for geomorphic studies.

SUMMARY

Geomorphologists commonly use a systems approach to their subject. Form systems, flow or cascading systems, process–form or process–response systems, and control systems are all recognized. Hugely important are ideas about stasis and change, with equilibrium and non-equilibrium views providing a focus for much debate. Non-equilibrium views grew from notions of complexity and chaos. The language of systems concepts employs such terms as negative feedback and positive feedback, reaction, relaxation, thresholds, and magnitude and frequency. Great achievements using systems-based arguments include notions of stability, instability, and thresholds in landscapes, the last two of which belie simplistic ideas on cause and effect in landscape evolution. Magnitude and frequency studies have led to unexpected results: at first, geomorphologists believed that medium-magnitude and medium-frequency events did the greatest geomorphic work, but some studies now suggest that rare events such as immense floods may have long-lasting effects on landforms. Geomorphic models are exceedingly useful tools. Scale and analogue hardware models, conceptual models, and mathematical models all play a role in the advancement of geomorphological understanding.

Geomorphic form is describable by morphological maps and, more recently, by geomorphometry. Geomorphometry today uses digital elevation models, remote sensing, and GIS and is a sophisticated discipline.

ESSAY QUESTIONS

1 Discuss the pros and cons of a 'systems approach' in geomorphology.

2 Explain the different types of equilibrium and non-equilibrium recognized in geomorphic systems.

3 Discuss the pros and cons of scale models in geomorphology.

4 To what extent have remote sensing and GIS revolutionized geomorphology?

FURTHER READING

Allen, P. A. (1997) *Earth Surface Processes*. Oxford: Blackwell Science.
An outstanding account of geomorphic processes.

Butler, D. R. (1995) *Zoogeomorphology: Animals as Geomorphic Agents*. Cambridge: Cambridge University Press.
An engaging account of the role of animals in landscape development.

Goudie, A. S. (ed.) (1994) *Geomorphological Techniques*, 2nd edn. London and New York: Routledge.
Covers the topics not covered by the present book – how geomorphologists measure form and process.

Goudie, A. (1995) *The Changing Earth: Rates of Geomorphological Processes*. Oxford and Cambridge, Mass.: Blackwell.
A good survey of spatial and temporal variations in the rates at which geomorphic processes operate.

Ritter, D. F., Kochel, R. C., and Miller, J. R. (1995) *Process Geomorphology*, 5th edn. Long Grove, Illinois: Waveland Press.
A good, well-illustrated, basic text with a fondness for North American examples.

Stoddart, D. R. (ed.) (1997) *Process and Form in Geomorphology*. London: Routledge.
This book will give the flavour of process geomorphology and more.

3

CHAPTER THREE

INTRODUCING HISTORY

The Earth's surface has a history that leaves traces in present-day landscapes and sediments. These traces make possible the partial reconstruction of long-term landscape change. This chapter introduces:

- Reconstruction of land-surface history
- Vestiges of past landscapes
- Dates and rates in land-surface history
- Unforeseen events and geomorphic systems

RECONSTRUCTING GEOMORPHIC HISTORY

The problem with measuring geomorphic processes is that, although it establishes current operative processes and their rates, it does not provide a dependable guide to processes that were in action a million years ago, ten thousand years ago, or even a hundred years ago. In trying to work out the long-term evolution of landforms and landscapes, geomorphologists have three options open to them – stratigraphic and environmental reconstruction, chronosequence studies, and numerical modelling.

Stratigraphic and environmental reconstruction

Fortunately for researchers into past landscapes, several archives of past environmental conditions exist: tree rings, lake sediments, polar ice cores, mid-latitude ice cores, coral deposits, loess, ocean cores, pollen, palaeosols, sedimentary rocks, and historical records (see Huggett 1997b, 8–21). Sedimentary deposits are an especially valuable source of information about past landscapes. In some cases, geomorphologists may apply the **principles of stratigraphy** to the deposits to establish a relative sequence of events. Colluvium for example, which builds up towards a hillslope base, is commonly deposited episodically. The result is that distinct layers are evident in a section, the upper layers being progressively younger than the lower layers. If such techniques as **radiocarbon dating** or **dendrochronology** can date these sediments, then they may provide an absolute timescale for the past activities on the hillslope, or at least the past activities that have left traces in the sedimentary record (p. 66). Recognizing the origin of the deposits may also be possible – glacial, periglacial, colluvial, or whatever. Moreover, sometimes geomorphologists use techniques of environmental reconstruction to establish the climatic and other environmental conditions at the time of sediment deposition.

To illustrate the process of stratigraphic and environmental reconstruction, take the case of the river alluvium and colluvium that fills many valleys in countries bordering the Mediterranean Sea. Claudio Vita-Finzi (1969) pioneered research into the origin of the valley fills, concluding that almost all alluvium and colluvium was laid down during two episodes of increased aggradation (times when deposition of sediment outstripped erosion). Figure 3.1 is a schematic reconstruction of the geomorphic history of a valley in Tripolitania (western Libya). The key to unlocking the history of the valleys in the area was datable archaeological material in the fluvial deposits. Vita-Finzi found three main deposits of differing ages. The oldest contains Palaeolithic implements and seems to have accumulated during the Pleistocene. Rivers cut into it between about 9,000 and 3,000 years ago. The second deposit accumulated behind dams built by Romans to store water and retain sediment. Late in the Empire, flood-waters breached or found a way around the dams and cut into the Roman alluvium.

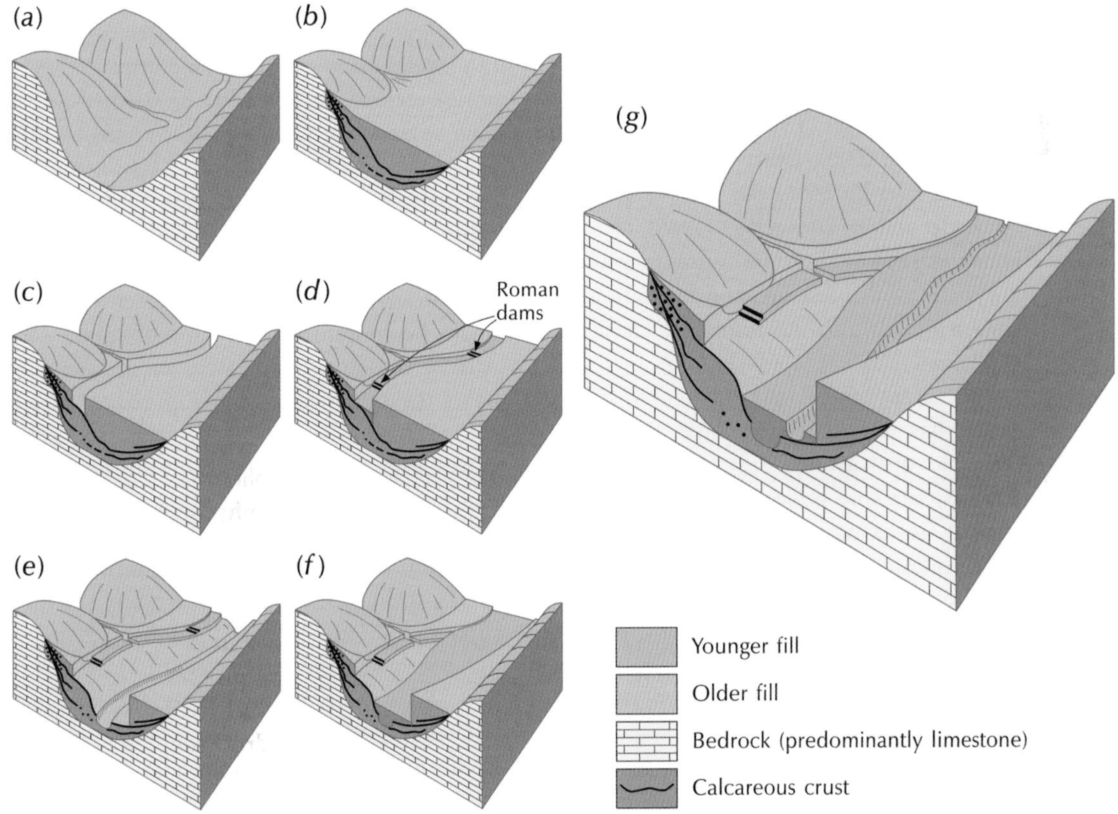

Figure 3.1 A reconstruction of the geomorphic history of a wadi in Tripolitania, western Libya. (a) Original valley. (b) Deposition of Older Fill. (c) River cut into Older Fill. (d) Roman dams impound silt. (e) Rivers cut further into Older Fill and Roman alluvium. (f) Deposition of Younger Fill. (g) Present valley and its alluvial deposits. *Source*: After Vita-Finzi (1969, 10)

Rivers built up the third deposit, within the downcut wadis. This deposit contained reworked Roman and earlier material as well as pottery and charcoal that dated it to the Medieval Period (AD 1200–1500). The deposition of this Younger Fill was followed by reduced alluviation and downcutting through the fill.

Wider examination of alluvia in Mediterranean valleys allowed Vita-Finzi to recognize an Older Fill dating from the Pleistocene and a Younger Fill dating from about AD 500–1500. The Older Fill was deposited as a substantial body of colluvium (slope wash) under a 'periglacial' regime during the last glacial stage. The Younger Fill was a product of phases of erosion during the later Roman Imperial times, through the Dark Ages, and to the Middle Ages. Vita-Finzi believed it to be the result of increased erosion associated with the climate of the Medieval Warm Period or the Little Ice Age, a view supported by John Bintliff (1976, 2002). Other geomorphologists, including Karl Butzer (1980, 2005) and Tjierd van Andel and his co-workers (1986), favoured human activity as the chief cause, pointing to post-medieval deforestation and agricultural expansion into marginal environments. The matter is still open to debate (see p. 260).

The recent global environmental change agenda has given environmental reconstruction techniques a fillip. **Past Global Changes (PAGES)** is a core project of the IGBP (International Geosphere–Biosphere Programme). It concentrates on two slices of time: (1) the last 2,000 years of Earth history, with a temporal resolution of decades, years, and even months; and (2) the last several hundred thousand years, covering glacial–interglacial cycles, in the hope of providing insights into the processes that induce global change (IGBP 1990). Examples of geomorphological contributions to environmental change over these timescales may be found in the book *Geomorphology and Global Environmental Change* (Slaymaker *et al.* 2009; see also Slaymaker 2000a).

Landform chronosequences

Another option open to the historical geomorphologist is to find a site where a set of landforms differ from place to place and where that spatial sequence of landforms may be interpreted as a time sequence. Such sequences are called **topographic chronosequences**, and the procedure is sometimes referred to as **space–time substitution** or, using a term borrowed from physics, **ergodicity**.

Charles Darwin used the chronosequence method to test his ideas on coral-reef formation. He thought that barrier reefs, fringing reefs, and atolls occurring at different places represented different evolutionary stages of island development applicable to any subsiding volcanic peak in tropical waters. William Morris Davis applied this evolutionary schema to landforms in different places and derived what he deemed was a time sequence of landform development – the geographical cycle – running from youth, through maturity, to senility. This seductively simple approach is open to misuse. The temptation is to fit the landforms into some preconceived view of landscape change, even though other sequences might be constructed. A study of south-west African landforms since Mesozoic times highlights the significance of this problem, where several styles of landscape evolution were consistent with the observed history of the region (Gilchrist *et al.* 1994). Users of the method must also be warned that not all spatial differences are temporal differences – factors other than time exert a strong influence on the form of the land surface, and landforms of the same age might differ through historical accidents. Moreover, it pays to be aware of **equifinality**, the idea that different sets of processes may produce the same landform. The converse of this idea is that landform is an unreliable guide to process. Given these consequential difficulties, it is best to treat chronosequences circumspectly.

Trustworthy topographic chronosequences are rare. The best examples normally come from manmade landscapes, though there are some landscapes in which, by quirks of history, spatial differences are translatable into time sequences. Occasionally, field conditions lead to adjacent hillslopes being progressively removed from the action of a fluvial or marine process at their bases. This has happened along a segment of the South Wales coast, in the British Isles, where cliffs have formed in Old Red Sandstone (Savigear 1952, 1956). Originally, the coast between Gilman Point and the Taff estuary was exposed to wave action. A sand spit started to grow. Wind-blown and marsh deposits accumulated between the spit and the original shoreline, causing the sea progressively to abandon the cliff base from west to east. The present cliffs are thus a topographic chronosequence: the cliffs furthest west have been subject to subaerial denudation without waves cutting their base the longest, while those to the east are progressively younger (Figure 3.2). Slope profiles along Port Hudson bluff, on the

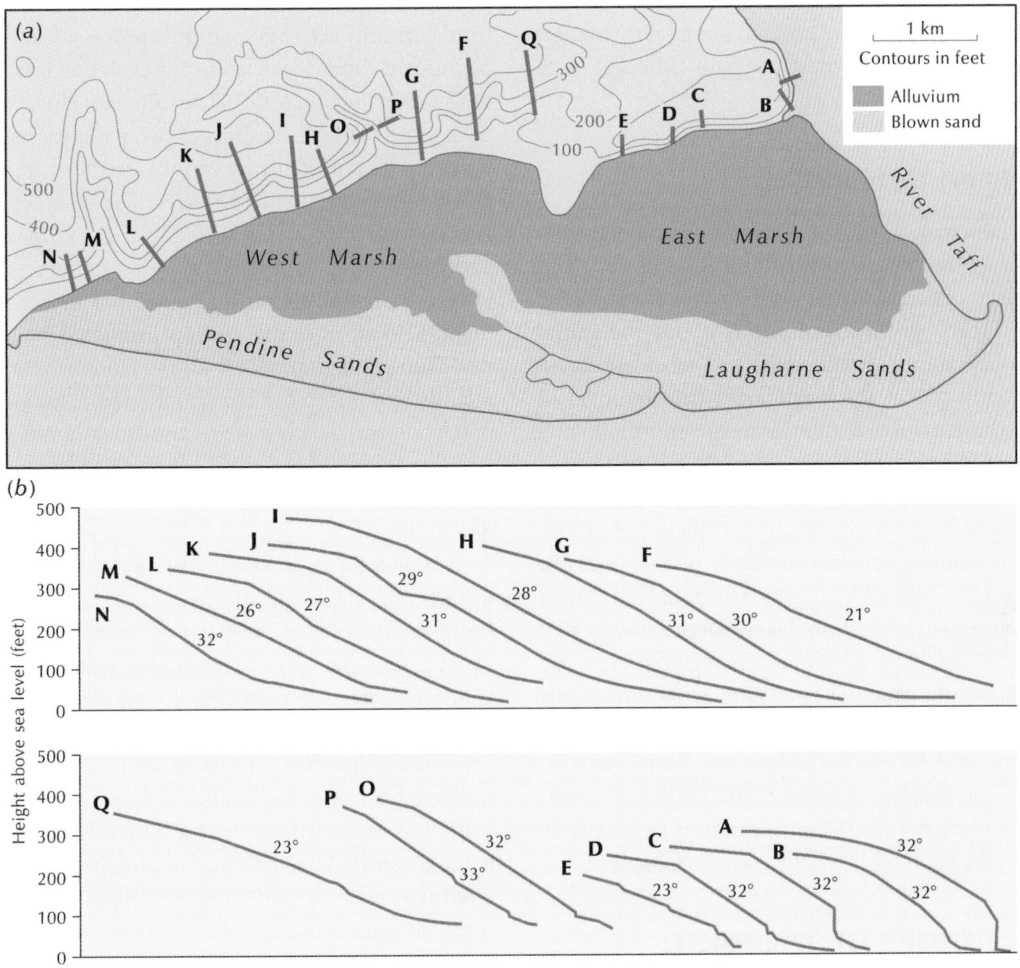

Figure 3.2 A topographic chronosequence in South Wales. (a) The coast between Gilman Point and the Taff estuary. The sand spit has grown progressively from west to east so that the cliffs to the west have been longest-protected from wave action. (b) The general form of the hillslope profiles located on Figure 3.2a. Cliff profiles become progressively older in alphabetical order, A–N. *Source*: From Huggett (1997b, 238) after Savigear (1952, 1956)

Mississippi River in Louisiana, southern USA, reveal a chronosequence (Brunsden and Kesel 1973). The Mississippi River was undercutting the entire bluff segment in 1722. Since then, the channel has shifted about 3 km downstream with a concomitant cessation of undercutting. The changing conditions at the slope bases have reduced the mean slope angle from 40° to 22°.

Recent applications of space–time substitution include the identification of thresholds in the behaviour of river systems (Fryirs *et al.* 2012) and the evolution of a submarine canyon–channel system in the passive progradational Ebro Margin lying off the east coast of Spain (Micallef *et al.* 2014).

Numerical modelling

Mathematical models of landscapes predict what happens if a particular combination of slope and river processes is allowed to run for millions of years, given assumptions about the initial topography, tectonic uplift and subsidence, and conditions at the boundaries (the removal of sediment, for example). Some geomorphologists would argue that these models are of limited worth because environmental conditions will not stay constant, or approximately constant, for millions or even hundreds of thousands of years. Nonetheless, the models do show the broad patterns of hillslope and land-surface change that occur under particular process regimes. They also enable the study of landscape evolution as part of a coupled tectonic–climatic system with the potential for feedbacks between climatically influenced surface processes and crustal deformation (see pp. 97–9). Some examples of long-term landscape models will be given in Chapter 8.

VESTIGES OF THE PAST: RELICT FEATURES

'Little of the earth's topography is older than the Tertiary and most of it no older than Pleistocene' (Thornbury 1954, 26). For many decades, this view was widely held by geomorphologists. Research over the last twenty years has revealed that a significant part of the land surface is surprisingly ancient, surviving in either relict or buried form (see Twidale 1999). These survivors from long-past climatic and environmental regimes were almost invariably created by processes no longer acting on them. Such landforms are **relicts**. Relict landforms and landscapes may endure for thousands, millions, tens of millions, or hundreds of millions of years. As Arthur L. Bloom (2002) put it, just a few very young landforms result from currently active geomorphic processes, and because the timescale of landscape evolution is far longer than the timescale of late Cenozoic climate changes, nearly all landscapes are palimpsests, written over repeatedly by various combinations of climate-determined processes. For instance, it is common for a cliff, a floodplain, a cirque, and many other landscape features to survive longer than the climatic regime that created them. Seldom does the erosion promoted by a new climatic regime renew all the landforms in a landscape. Far more commonly, remnants of past landforms are preserved. Consequently, most landscapes are a complex collection of landforms inherited from several generations of landscape development.

It is helpful to distinguish relict landforms from a non-glacial perspective and relict landforms from a glacial perspective (Ebert 2009a). From a non-glacial perspective, the term relict landform applies to many landforms worldwide (Bloom 2002). From a glacial perspective, a relict landform is one that cold-based ice (p. 284) has preserved, owing to the fact that little or no deformation takes place under ice continuously frozen to the ground (Kleman 1994). The term **preglacial landform** refers to any landform older than a specified glaciation.

Relict landforms

In some landscapes, the inherited forms were fashioned by processes similar to those now operating there, but it is common to find polygenetic landscapes in which the processes responsible for a particular landform no longer operate. The clearest and least equivocal example of this is the glacial and periglacial landforms left as a vestige of the Ice Age in mid-latitudes. Many of the glacial landforms discussed in Chapter 10 are relicts from the Pleistocene glaciations. In upland Britain, for instance, hillslopes sometimes bear ridges and channels that were fashioned by ice and meltwater during the last ice age. In the English Lake District, U-shaped valleys, roches moutonnées, striations, and so on attest to an icy past. However, not all signs of glaciation are incontrovertible. Many landforms and sediments found in glaciated regions, even those regions buried beneath deep and fast-flowing ice, have no modern analogues. Landforms with no modern analogues include drumlins, large-scale flutings, rogen moraines (p. 299), and hummocky topography. This means that drumlins are not forming at present and the processes that fashion them cannot be studied directly but can only be inferred from the size, shape, composition, and location of relict forms. Glacial landforms created by Pleistocene ice may be used as analogues for older glaciations. For instance, roches moutonnées occur in the geological record: abraded bedrock surfaces in the Neoproterozoic sequence of Mauritania contain several well-developed ones, and others have been found in the Late Palaeozoic Dwyka Tillite of South Africa (Hambrey 1994, 104).

Other polygenetic landscapes are common. In deserts, ancient river systems, old archaeological sites, fossil karst phenomena, high lake strandlines, and deep weathering profiles are relict elements that attest to past humid phases; while stabilized fossil dune fields on desert margins are relicts of more arid phases. In the humid tropics, a surprising number of landscape features are relict. Researchers working in the central Amazonian Basin (Tricart 1985) and in Sierra Leone (Thomas and Thorp 1985) have unearthed vestiges of fluvial dissection that occurred under dry conditions between about 20,000 and 12,500 years ago. In New South Wales, Australia, a relict karst cave that could not have formed under today's climate has possibly survived from the Mesozoic (Osborne and Branagan 1988).

Relict land surfaces

In tectonically stable regions, land surfaces, especially those capped by duricrusts, may persist for 100 million years or more, witness the Gondwanan and post-Gondwanan erosion surfaces in the Southern Hemisphere (King 1983). Some weathering profiles in Australia are 100 million years old or even older (Ollier 1991, 53). Remnants of a ferricrete-mantled land surface surviving from the early Mesozoic era are widespread in the Mount Lofty Ranges, Kangaroo Island, and the south Eyre Peninsula of South Australia (Twidale *et al.* 1974). Indeed, much of south-eastern Australia contains many very old topographical features (Young 1983; Bishop *et al.* 1985; Twidale and Campbell 1995). Some upland surfaces originated in the Mesozoic era and others in the early Palaeogene period; and in some areas the last major uplift and onset of canyon cutting occurred before the Oligocene epoch. In southern Nevada, early to middle Pleistocene colluvial deposits, mainly darkly varnished boulders, are common features of hillslopes formed in volcanic tuff. Their long-term survival indicates that denudation rates on resistant volcanic hillslopes in the southern Great Basin have been exceedingly low throughout Quaternary times (Whitney and Harrington 1993).

The palaeoclimatic significance of these finds has not passed unnoticed: for much of the Cenozoic era, the tropical climatic zone of the Earth extended much further polewards than it does today. Indeed, evidence from deposits in the landscape, as well as evidence in the palaeobotanical record, indicates that warm and moist conditions extended to high latitudes in the North Atlantic during the late Cretaceous and Palaeogene periods. Julius Büdel (1982) was convinced that Europe suffered extensive etchplanation during Tertiary times (p. 459). Signs of ancient saprolites and duricrusts, bauxite and laterite, and the formation and preservation of erosional landforms, including tors, inselbergs, and pediments, have been detected (Summerfield and Thomas 1987). Traces of a tropical weathering regime have been unearthed (e.g. Battiau-Queney 1996). In the British Isles, several Tertiary weathering products and associated landforms and soils have been discovered (e.g. Battiau-Queney 1984, 1987). On Anglesey, which has been a tectonically stable area since at least the Triassic period, inselbergs, such as Mynydd Bodafon, have survived several large changes of climatic regime (Battiau-Queney 1987). Karin Ebert (2009b) has recognized many inselbergs in northern Sweden formed before the Quaternary and surviving late Cenozoic glaciations (Plate 3.1). In Europe, Asia, and North America many karst landscapes are now interpreted as fossil landforms originally produced under a tropical weathering regime during Tertiary times (Büdel 1982; Bosák *et al.* 1989).

The connection between landforms and climate is the subject of considerable dispute, with protagonists being on the one hand climatic geomorphologists, who believe that different climatic zones cultivate distinct suites of landforms, and on the other hand those geomorphologists who are unconvinced by the climatic argument, at least in its most extreme form. This debate has relevance to the interpretation of relict landscape features (Box 3.1).

Box 3.1 RELICT LANDFORMS AND CLIMATIC GEOMORPHOLOGY

Climatic geomorphologists have made careers out of deciphering the generations of landforms derived from past climates. Their arguments hinge on the assumption that present climatic zones tend to foster distinctive suites of landforms (e.g. Tricart and Cailleux 1972; Büdel 1982; Bremer 1988). Such an assumption is certainly not without foundation, but many geomorphologists, particularly in English-speaking countries, have questioned it. A close connection between process regimes and process rates will be noted at several points in the book (e.g. pp. 177–9). Whether the set of geomorphic processes within each climatic zone creates characteristic landforms – whether a set of morphogenetic regions may be established – is debatable.

Climatic geomorphology has been criticized for using temperature and rainfall data, which provide too gross a picture of the relationships between rainfall, soil moisture, and runoff, and for excluding the magnitude and frequency of storms and floods, which are important in landform development. Some landforms are more climatically zonal in character than are others. Arid, nival, periglacial, and glacial landforms are quite distinct. Other morphoclimatic zones have been distinguished, but their constituent landforms are not clearly determined by climate. In all morphoclimatic regions, the effects of geological structure and etching processes are significant, even in those regions where climate exerts a strong influence on landform development (Twidale and Lageat 1994). It is likely that, for over half the world's land surface, climate is not of overarching importance in landform development. Indeed, some geomorphologists opine that landforms, and especially hillslopes, will be the same regardless of climate in all geographical and climatic zones (see Ruhe 1975).

The conclusion is that, mainly because of ongoing climatic and tectonic change, the climatic factor in landform development is not so plain and simple as climatic geomorphologists have on occasions suggested. Responses to these difficulties go in two directions – towards complexity and towards simplicity. The complexities of climate–landform relations are explored in at least two ways. One way is to attempt a fuller characterization of climate. A recent study of climatic landscape regions of the world's mountains used several pertinent criteria: the height of timberline, the number and character of altitudinal vegetational zones, the amount and seasonality of moisture available to vegetation, physiographic processes, topographic effects of frost, and the relative levels of the timberline and permafrost limit (Thompson 1990). Another way of delving into the complexity of climatic influences is to bring modern views on fluvial system dynamics to bear on the question. One such study has taken a fresh look at the notion of morphogenetic regions and the response of geomorphic systems to climatic change (Bull 1992). A simpler model of climatic influence on landforms is equally illuminating (Ollier 1988). It seems reasonable to reduce climate to three fundamental classes: humid where water dominates, arid where water is in short supply, and glacial where water is frozen (Table 3.1). Each of these 'climates' fosters certain weathering and slope processes. Deep weathering occurs where water is unfrozen. Arid and glacial landscapes bear the full brunt of climatic influences because they lack the protection afforded by vegetation in humid landscapes. Characteristic landforms do occur in each of these climatic regions, and it is usually possible to identify past tropical landscapes from clay minerals in relict weathering profiles. It seems reasonable, therefore, by making the assumption of actualism (p. 28), to use these present climate–landform associations to interpret relict features that bear the mark of particular climatic regimes. Julius Büdel (1982, 329–38), for instance, interprets the 'etchplain stairways' and polja of central Dalmatia as relicts from the late Tertiary period, when the climate was more 'tropical', being much warmer and possibly wetter. Such conditions would favour polje formation through 'double planation' (p. 459): chemical decomposition and solution of a basal weathering surface under a thick sheet of soil or sediment, the surface of which was subject to wash processes.

Table 3.1 A simple scheme relating geomorphic processes to climate

Climate	Weathering process	Weathering depth	Mass movement
Glacial	Frost (chemical effects reduced by low temperatures)	Shallow	Rock glacier Solifluction (wet) Scree slopes
Humid	Chemical	Deep	Creep landslides
Arid	Salt	Deep	Rockfalls

Source: Adapted from Ollier (1988)

Plate 3.1 Kuormakka, a remnant inselberg in northern Sweden surviving late Cenozoic glaciations. (*Photograph by Karin Ebert*)

DATING EVENTS AND LANDFORMS

A broad range of methods is now available for dating events in Earth history (Table 3.2). Some are more precise than others. Four categories are recognized: numerical-age methods, calibrated-age methods, relative age-methods, and correlated-age methods (see also Walker 2005).

Numerical-age methods produce results on a ratio (or absolute) timescale, pinpointing the times when environmental change occurred. This information is crucial to a deep appreciation of environmental change: without dates, nothing much of use can be said about rates. **Calibrated-age** methods may provide approximate numerical ages. Some of these methods are refined and enable age categories to be assigned to deposits by measuring changes since deposition in such environmental factors as soil genesis or rock weathering (see McCarroll 1991). **Relative-age methods** furnish an age sequence, simply putting events in the correct order. They assemble the 'pages of Earth history' in a numerical sequence. The Rosetta stone of relative-age methods is the principle of stratigraphic superposition. This states that, in undeformed sedimentary sequences, the lower strata are older than the upper strata. Some kind of marker must be used to match stratigraphic sequences from different places. Traditionally, fossils have been employed for this purpose. Distinctive fossils or fossil assemblages can be correlated between regions by identifying strata that were laid down contemporaneously. This was how such celebrated geologists as William ('Strata') Smith (1769–1839) first erected the stratigraphic column. Although this technique was remarkably successful in establishing the broad development of Phanerozoic sedimentary rocks, and rested on the sound principle of superposition, it is beset by problems (see Vita-Finzi 1973, 5–15). It is best used in partnership with numerical-age methods. Used conjointly,

Table 3.2 Methods for dating Quaternary and Holocene materials

Method	Age range (years)	Basis of method	Materials needed
Sidereal methods			
Dendrochronology	0–10,000	Growth-rings of live trees or correlating ring-width chronology to other trees	Trees and cultural materials (e.g. ships' timbers)
Varve chronology	0–200,000	Counting seasonal sediment layers back from the present, or correlating a past sequence with a continuous chronology	Glacial, lacustrine, marine, soil, and wetland deposits
Sclerochronology[a]	0–800	Counting annual growth bands in corals and molluscs	Marine fossiliferous deposits
Isotopic methods			
Radiocarbon	100–60,000	Radioactive decay of carbon-14 to nitrogen-14 in organic tissue or carbonates	A variety of chemical and biogenic sediments
Cosmogenic nuclides[a]	200–8,000,000[b]	Formation, accumulation, and decay of cosmogenic nuclides in rocks or soils exposed to cosmic radiation	Surfaces of landforms
Potassium–argon, argon–argon	10,000–10,000,000+	Radioactive decay of potassium-40 trapped in potassium-bearing silicate minerals during crystallization to argon-40	Non-biogenic lacustrine deposits and soils, igneous and metamorphic rocks
Uranium series	100–400,000[c]	Radioactive decay of uranium and daughter nuclides in biogenic chemical and sedimentary minerals	Chemical deposits and biogenic deposits except those in wetlands
Lead-210	<200	Radioactive decay of lead-210 to lead-206	Chemical deposits and wetland biogenic deposits
Uranium–lead, thorium–lead[a]	10,000–10,000,000+	Using normalized lead isotopes to detect small enrichments of radiogenic lead from uranium and thorium	Lava
Radiogenic methods			
Fission-track	2,000–10,000,000[d]	Accumulation of damage trails (fission tracks) from natural fission decay of trace uranium-238 in zircon, apatite, or glass	Cultural materials, igneous rocks
Luminescence	100–300,000	Accumulation of electrons in crystal lattice defects of silicate minerals resulting from natural radiation	Aeolian deposits, fluvial deposits, marine chemical and clastic deposits, cultural materials, silicic igneous rocks
Electron-spin resonance	1,000–1,000,000	Accumulation of electrical charges in crystal lattice defects in silicate minerals resulting from natural radiation	Cultural materials, terrestrial and marine fossils, igneous rocks
Chemical and biological methods			
Amino-acid racemization	500–1,000,000	Racemization of L-amino acids to D-amino acids in fossil organic material	Terrestrial and marine plant and animal remains

continued . . .

Table 3.2 *continued*

Method	Age range (years)	Basis of method	Materials needed
Obsidian hydration	100–1,000,000	Increase in thickness of hydration rind on obsidian surface	Cultural materials, fluvial gravels, glacial deposits, clastic deposits in lakes and seas, silicic igneous and pyroclastic rocks
Lichenometry[a]	20–500	Growth of lichens on freshly exposed rock surfaces	Exposed landforms supporting lichens
Geomorphic methods			
Soil-profile development	8,000–200,000	Systematic changes in soil properties owing to weathering and pedogenic processes	Soils and most landforms
Rock and mineral weathering	0–300,000	Systematic alteration of rocks and minerals owing to exposure to weathering agents	Landforms
Scarp morphology[a]	2,000–20,000	Progressive change in scarp profile (from steep and angular to gentle and rounded) resulting from surface processes	Fault scarp and other landforms with scarp-like features (e.g. terraces)
Correlation methods			
Palaeomagnetism			
Secular variation	0–10,000	Secular variation of the Earth's magnetic field recorded in magnetic minerals	Suitable cultural materials, sediments and rocks
Reversal stratigraphy	800,000–10,000,000+	Reversals of the Earth's magnetic field recorded in magnetic minerals	Suitable sediments and igneous rocks
Tephrochronology	0–10,000,000+	Recognition of individual tephra by their unique properties, and the correlation of these to a dated chronology	Pyroclastic rocks
Palaeontology			
Evolution of microtine rodents	8,000–8,000,000	Progressive evolution of microtine rodents	Terrestrial animals' remains
Marine zoogeography	30,000–300,000	Climatically induced zoogeographical range shifts of marine invertebrates	Marine fossiliferous deposits
Climatic correlations[a]	1,000–500,000	Correlation of landforms and deposits to global climate changes of known age	Most sedimentary materials and landforms

Notes:

a Experimental method.
b Depends on nuclide used (beryllium-10, aluminium-26, chlorine-36, helium-3, carbon-14).
c Depends on series (uranium-234–uranium-230, uranium-235–protactinium-231).
d Depends on material used (zircon and glass, apatite).

Source: Adapted from Sowers *et al.* (2000, 567)

CHAPTER FOUR

GEOMORPHOLOGY AND THE EARTH SYSTEM

4

The Earth's topography results from the interplay of many processes, some originating inside the Earth, some outside it, and some on it. This chapter covers:

- Grand cycles of water and rock
- The wearing away and the building up of the land surface
- Tectonics, erosion, and climate
- Humans as geomorphic agents
- Life as a geomorphic agent

THE EARTH'S SURFACE IN ACTION: MOUNTAIN UPLIFT AND GLOBAL COOLING

Over the last 40 million years, the uplift of mountains has been a very active process. During that time, the Tibetan Plateau has risen by up to 4,000 m, with at least 2,000 m in the last 10 million years. Two-thirds of the uplift of the Sierra Nevada in the USA has occurred in the past 10 million years. Similar changes have taken place (and are still taking place) in other mountainous areas of the North American west, in the Bolivian Andes, and in the New Zealand Alps. This period of active mountain building seems to link to global climatic change, in part through airflow modification and in part through weathering. Young mountains weather and erode quickly. Weathering processes remove carbon dioxide from the atmosphere by converting it to soluble carbonates. The carbonates are carried to the oceans, where they are deposited and buried. It is possible that the growth of the Himalaya scrubbed enough carbon dioxide from the atmosphere to cause a global climatic cooling that culminated in the Quaternary ice ages (Raymo and Ruddiman 1992; Ruddiman 1997). This shows how important the geomorphic system can be to environmental change.

ROCK AND WATER CYCLES

The Earth's surface – the **toposphere** – sits at the interfaces of the solid lithosphere, the gaseous atmosphere, and the watery hydrosphere. It is also the dwelling-place of many living things. Gases, liquids, and solids are exchanged between these spheres in three grand cycles, two of which – the **water** or **hydrological cycle** and the **rock cycle** – are crucial to understanding landform evolution. The third grand cycle – the **biogeochemical cycle** – is the circulation of chemical elements (carbon, oxygen, sodium, calcium, and so on) through the upper mantle, crust, and ecosphere. It is less significant to landform development than the other two cycles, although some biogeochemical cycles regulate the composition of the atmosphere, which in turn can affect weathering.

Water cycle

The **hydrosphere** – the surface and near-surface waters of the Earth – is made of **meteoric water**. The water cycle is the circulation of meteoric water through the hydrosphere, atmosphere, and upper parts of the crust. It connects with the circulation of deep-seated **juvenile water** associated with magma production and the rock cycle. Juvenile water ascends from deep rock layers through volcanoes, where it issues into the meteoric zone for the first time. On the other hand, meteoric water held in hydrous minerals and pore spaces in sediments, known as **connate water**, may be removed from the meteoric cycle at subduction sites, where it is carried deep inside the Earth.

The land phase of the water cycle is of special interest to geomorphologists. It sees water transferred from the atmosphere to the land and then from the land back to the atmosphere and to the sea. It includes a **surface drainage system** and a **subsurface drainage system**. Water flowing within these drainage systems tends to be organized within **drainage basins**, which are also called **watersheds** in the USA and **catchments** in the UK. The basin water system may be viewed as a set of water stores that receive inputs from the atmosphere and deep inflow from deep groundwater storage, that lose outputs through evaporation and streamflow and deep outflow, and that are linked by internal flows. In summary, the basin water runs like this. Precipitation entering the system is stored on the soil or rock surface, or is intercepted by vegetation and stored there, or falls directly into a stream channel. From the vegetation it runs down branches and trunks (stemflow), or drips off leaves and branches (leaf and stem drip), or it is evaporated. From the soil or rock surface, it flows over the surface (overland flow), infiltrates the soil or rock, or evaporates. Once in the rock or soil, water may move laterally down hillsides (throughflow, pipeflow, interflow) to feed rivers, or it may move downwards to recharge groundwater storage, or it may evaporate. Groundwater may rise by capillary action to top up the rock and soil water stores, or it may flow into a stream (baseflow), or may exchange water with deep storage.

Rock cycle

After the Earth had evolved a solid land surface and an atmosphere, the water cycle and plate tectonic processes combined to create the rock cycle. The **rock cycle** is the repeated creation and destruction of crustal material – rocks and minerals (Box 4.1). Volcanoes, folding, faulting, and uplift all bring igneous and other rocks, water, and gases to the base of the atmosphere and hydrosphere. Once exposed to the air and meteoric water, these rocks begin to decompose and disintegrate by the action of weathering. Gravity, wind, and water transport the weathering products to the oceans. Deposition occurs on the ocean floor. Burial of the loose sediments leads to compaction, cementation, and recrystallization, and so to the formation of sedimentary rocks. Deep burial may convert sedimentary rocks into metamorphic rocks. Other

Box 4.1 ROCKS AND MINERALS

The average composition by weight of chemical elements in the lithosphere is oxygen 47 per cent, silicon 28 per cent, aluminium 8.1 per cent, iron 5 per cent, calcium 3.6 per cent, sodium 2.8 per cent, potassium 2.6 per cent, magnesium 2.1 per cent, and the remaining eighty-three elements 0.8 per cent. These elements combine to form **minerals**. The chief minerals in the lithosphere are **feldspars** (aluminium silicates with potassium, sodium, or calcium), **quartz** (a form of silicon dioxide), **clay minerals** (complex aluminium silicates), **iron minerals** such as limonite and hematite, and **ferromagnesian minerals** (complex iron, magnesium, and calcium silicates). **Ore deposits** consist of common minerals precipitated from hot fluids. They include pyrite (iron sulphide), galena (lead sulphide), blende or sphalerite (zinc sulphide), and cinnabar (mercury sulphide).

Rocks are mixtures of crystalline forms of minerals. There are three main types: igneous, sedimentary, and metamorphic.

Igneous rocks

These form by solidification of molten rock (magma). They have varied compositions (Figure 4.1). Most igneous rocks consist of **silicate minerals**, especially those of the felsic mineral group, which comprises quartz and feldspars (potash and plagioclase). **Felsic minerals** have silicon, aluminium, potassium, calcium, and sodium as the dominant elements.

Other important mineral groups are the **micas**, amphiboles, and **pyroxenes**. All three groups contain aluminium, magnesium, iron, and potassium or calcium as major elements. Olivine is a magnesium and iron silicate. The micas, amphiboles (mainly hornblende), pyroxenes, and olivine constitute the **mafic minerals**, which are darker in colour and denser than the felsic minerals. **Felsic rocks** include diorite, tonalite, granodiorite, rhyolite, andesite, dacite, and granite. **Mafic rocks** include gabbro and basalt. **Ultramafic rocks**, which are denser still than mafic rocks, include peridotite and serpentine. Much of the lithosphere below the crust is made of peridotite. Eclogite is an ultramafic rock that forms deep in the crust, nodules of which are sometimes carried to the surface by volcanic action. At about 400 km below the surface, olivine undergoes a phase change (it fits into a more tightly packed crystal lattice whilst keeping the same chemical composition) to spinel, a denser silicate mineral. In turn, at about 670 km depth, spinel undergoes a phase change into perovskite, which is probably the chief mantle constituent and the most abundant mineral in the Earth.

Sedimentary rocks

These are layered accumulations of mineral particles derived mostly from weathering and erosion of pre-existing rocks. They are clastic, organic, or chemical in origin. **Clastic sedimentary rocks** are unconsolidated or indurated sediments (boulders, gravel, sand, silt, clay) derived from geomorphic processes. Conglomerate, breccia, sandstone, mudstone, claystone, and shale are examples. **Organic sedimentary rocks** and **mineral fuels** form from organic materials. Examples are coal, petroleum, and natural gas. **Chemical sedimentary rocks** form by chemical precipitation in oceans, seas, lakes, caves, and, less commonly, rivers. Limestone, dolomite, chert, tufa, and evaporites are examples.

continued . . .

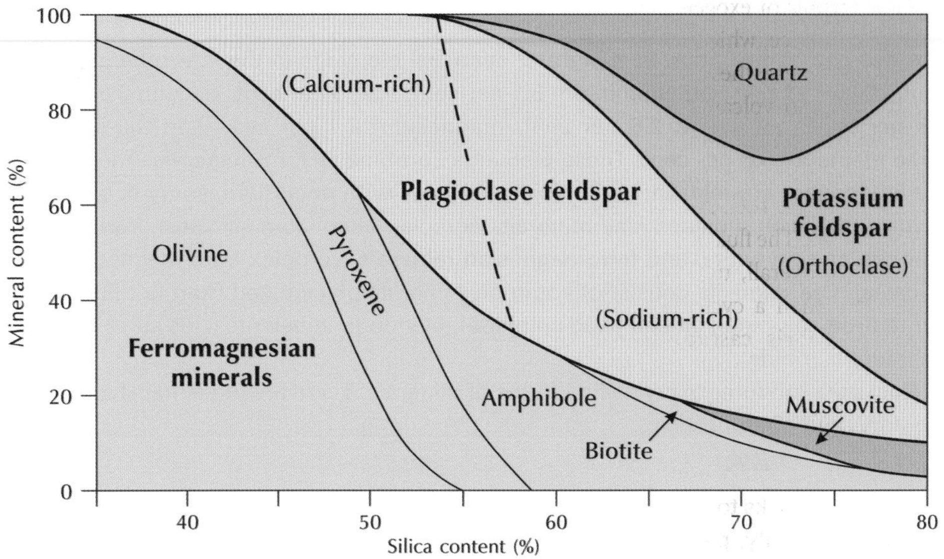

	ULTRABASIC	BASIC	INTERMEDIATE		ACID
EXTRUSIVE Fine-grained	(rare)	Basalt	Andesite	Trachyte	Rhyolite Obsidian
INTRUSIVE Medium-grained	(rare)	Dolerite	Microdiorite	Microsyenite	Microgranite
INTRUSIVE Coarse-grained	Peridotite	Gabbro	Diorite	Syenite	Granite

Figure 4.1 Igneous rocks and their component minerals. The classification is based on the silica content, which produces an ultrabasic–acid axis. The terms 'acid' and 'basic' are not meant to suggest that the rocks are acidic or alkaline in the customary sense, but merely describe their silica content.

Metamorphic rocks

These form through physical and chemical changes in igneous and sedimentary rocks. Temperatures or pressures high enough to bring about recrystallization of the component minerals cause the changes. Slate, schist, quartzite, marble, and gneiss are examples.

deep-seated processes may produce granite. If uplifted, intruded or extruded, and exposed at the land surface, the loose sediments, consolidated sediments, metamorphic rocks, and granite may join in the next round of the rock cycle.

Weathering, transport, and deposition are essential processes in the rock cycle. In conjunction with geological structures, tectonic processes, climate, and living things, they fashion landforms and landscapes. Volcanic action, folding, faulting, and uplift may all impart potential energy to the toposphere, creating the 'raw relief' on which geomorphic agents act to fashion the marvellously multifarious array of landforms found on the Earth's surface – the physical toposphere. Geomorphic or exogenic agents are wind, water, waves, and ice, which act from outside or above the toposphere; these contrast with endogenic (tectonic and volcanic) agents, which act upon the toposphere from inside the planet.

The surface phase, and particularly the land-surface phase, of the rock cycle is the domain of geomorphologists. The flux of materials across the land surface is, overall, unidirectional and is a cascade rather than a cycle. The basics of the **land-surface debris cascade** are as follows. Weathering agents move into the soil and rock along a weathering front, and in doing so, bring fresh rock into the system. Material may be added to the land surface by deposition, having been borne by wind, water, ice, or animals. All the materials in the system are subject to transformations by the complex processes of weathering. Some weathering products revert to a rock-like state by further transformations: under the right conditions, some chemicals precipitate out from solution to form hardpans and crusts. And many organisms produce resistant organic and

Box 4.2 ROCKS AND RELIEF: DIFFERENTIAL EROSION

The ability of rocks to resist the agents of denudation depends upon such factors as particle size, hardness, porosity, permeability, the degree to which particles are cemented, and mineralogy. **Particle size** determines the surface area exposed to chemical attack: gravels and sands weather slowly compared with silts and clays. The hardness, mineralogy, and degree of rock cementation influences the rate at which weathering decomposes and disintegrates them: a siliceous sandstone is more resistant to weathering than a calcareous sandstone.

Permeability is an important property in shaping weathering because it determines the rate at which water seeps into a rock body and dictates the internal surface area exposed to weathering (Table 4.1). As a rule, igneous and metamorphic rocks are resistant to weathering and erosion. They tend to form the basements of cratons, but where they are exposed at the surface or are thrust through the overlying sedimentary cover by tectonic movements they often give rise to resistant hills. English examples are the Malvern Hills in Herefordshire and Worcestershire, which have a long and narrow core of gneisses, and Charnwood Forest in the Midlands, which is formed of Precambrian volcanic and plutonic rocks. The strongest igneous and metamorphic rocks are quartzite, dolerite, gabbro, and basalt, followed by marble, granite, and gneiss. These resistant rocks tend to form relief features in landscapes. The quartz-dolerite Whin Sill of northern England is in places a prominent topographic feature (p. 134). Basalt may cap plateaux and other sedimentary hill features. Slate is a moderately strong rock, while schist is weak.

Sedimentary rocks vary greatly in their ability to resist weathering and erosion. The weakest of them are chalk and rock salt. However, the permeability of chalk compensates for its weakness and chalk resists denudation, sometimes with the help of more resistant bands within it, to form cuestas (p. 149), as in the North and South Downs of south-east England. Coal, claystone, and siltstone are weak rocks that offer little resistance to erosion and tend to form vales. An example from south-east England is the lowland developed on the thick Weald Clay. Sandstone is a moderately strong rock that may form scarps and cliffs. Whether or not it does so depends upon the nature of the sandstone and the environment in which it is found (e.g. Robinson and Williams 1994). Clay-rich or silty sandstones are often cemented weakly, and the clay reduces their permeability. In temperate European environments, they weather and are

continued . . .

eroded readily and form low relief, as is the case with the Sandgate Beds of the Lower Greenland, south-east England. In arid regions, they may produce prominent cuestas. Weakly cemented sands and sandstones that contain larger amounts of quartz often form higher ground in temperate Europe, probably because their greater porosity reduces runoff and erosion. A case in point is the Folkestone Sands of south-east England, which form a low relief feature in the northern and western margins of the Weald, though it is overshadowed by the impressive Hythe Beds cuesta. Interestingly, the Hythe Beds comprise incoherent sands over much of the Weald, but in the west and north-west they contain sandstones and chert beds, and in the north and north-east the sands are partly replaced by interbedded sandy limestones and loosely cemented sandstones. These resistant bands produce a discontinuous cuesta that is absent in the south-eastern Weald, but elsewhere rises to form splendid ramparts at Hindhead (273 m), Blackdown (280 m), and Leith Hill (294 m) that tower above the Low Weald (Jones 1981, 18). However, in general, hillslopes on the aforementioned sandstones are rarely steep and usually covered with soil. Massive and more strongly cemented sandstones and gritstones normally form steep slopes and commonly bear steep cliffs and isolated pillars. They do so throughout the world. Details of the influence of rocks upon relief will be discussed in Chapters 5 and 6.

Table 4.1 Porosities and permeabilities of rocks and sediments

Material	Representative porosity (per cent void space)	Permeability range (litres/day/m^2)
Unconsolidated		
Clay	50–60	0.0004–0.04
Silt and glacial till	20–40	0.04–400
Alluvial sands	30–40	400–400,000
Alluvial gravels	25–35	400,000–40,000,000
Indurated: sedimentary		
Shale	5–15	0.000004–0.004
Siltstone	5–20	0.0004–40
Sandstone	5–25	0.04–4,000
Conglomerate	5–25	0.04–4,000
Limestone	0.1–10	0.004–400
Indurated: igneous and metamorphic		
Volcanic (basalt)	0.001–50	0.004–40
Granite (weathered)	0.001–10	0.0004–0.4
Granite (fresh)	0.0001–1	0.000004–0.0004
Slate	0.001–1	0.000004–0.004
Schist	0.001–1	0.00004–0.04
Gneiss	0.0001–1	0.000004–0.004
Tuff	10–80	0.0004–40

Source: Adapted from Waltz (1969)

inorganic materials to shield or to support their bodies. The weathered mantle may remain in place or it may move downhill. It may creep, slide, slump, or flow downhill under the influence of gravity (mass movements), or moving water may wash or carry it downhill. In addition, the wind may erode it and take it elsewhere.

The land-surface debris cascade produces landforms. It does so partly by selectively weathering and eroding weaker rocks, a process called **differential erosion** (Box 4.2).

Biogeochemical cycles

The **biosphere** powers a global cycle of carbon, oxygen, hydrogen, nitrogen, and other mineral elements. These minerals circulate with the ecosphere and are exchanged between the ecosphere and its environment. The circulations are called **biogeochemical cycles**. The land phase of these cycles is intimately linked with water and debris movements.

Interacting cycles

The water cycle and the rock cycle interact (Figure 4.2). John Playfair was perhaps the first person to recognize this crucial interaction in the Earth system, and he was perhaps the great-grandfather of **Earth System Science** (Box 4.3). Here is how he described it in old-fashioned but most elegant language:

> We have long been accustomed to admire that beautiful contrivance in Nature, by which the water of the ocean, drawn up in vapour by the atmosphere, imparts in its descent, fertility to the earth, and becomes the great

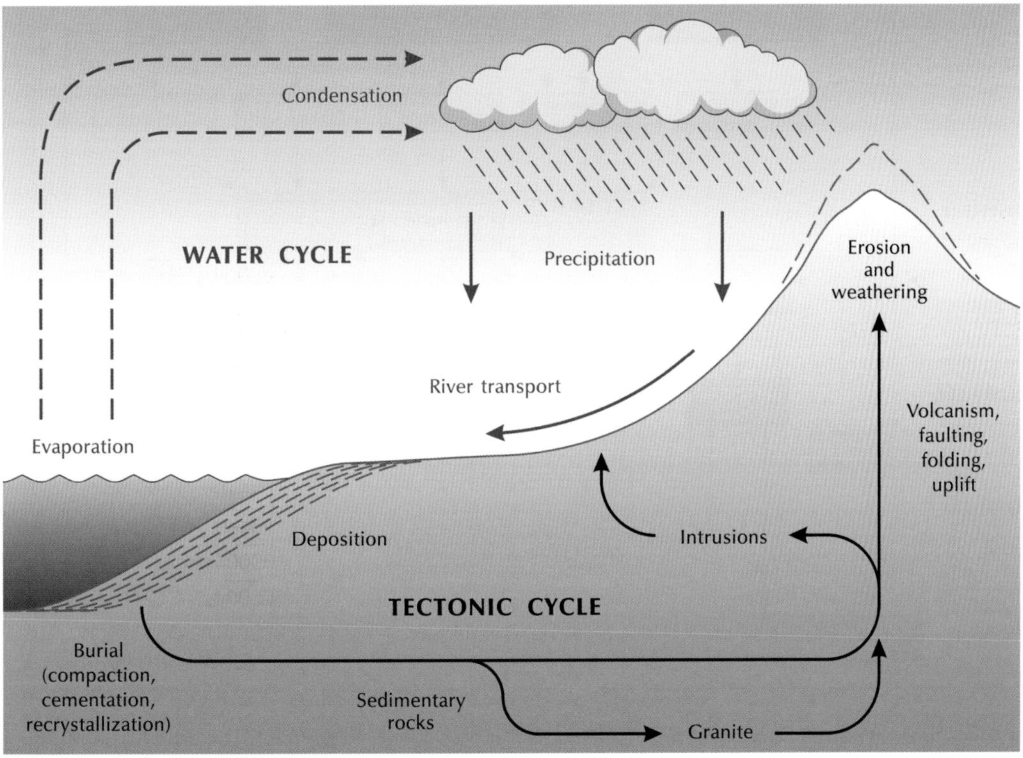

Figure 4.2 The rock cycle, the water cycle, and their interaction.

Box 4.3 EARTH SYSTEM SCIENCE

Earth system science takes the view that all the terrestrial spheres interact in a basic way: the solid Earth (lithosphere, mantle, and core), atmosphere, hydrosphere, pedosphere, and biosphere are interdependent (Figure 4.3). From a geomorphological perspective, a key suggestion of this view is that denudation processes are a major link between crustal tectonic processes and the atmosphere and hydrosphere (Beaumont *et al.* 2000). Mantle convection largely drives tectonic processes, but the denudational link with the atmosphere–hydrosphere system has a large effect. In turn, tectonic processes, acting through the climatic effects of mountain ranges, influence the atmosphere. Similarly, the Earth's climate depends upon ocean circulation patterns, which in turn are influenced by the distribution of continents and oceans, and ultimately upon long-term changes in mantle convection.

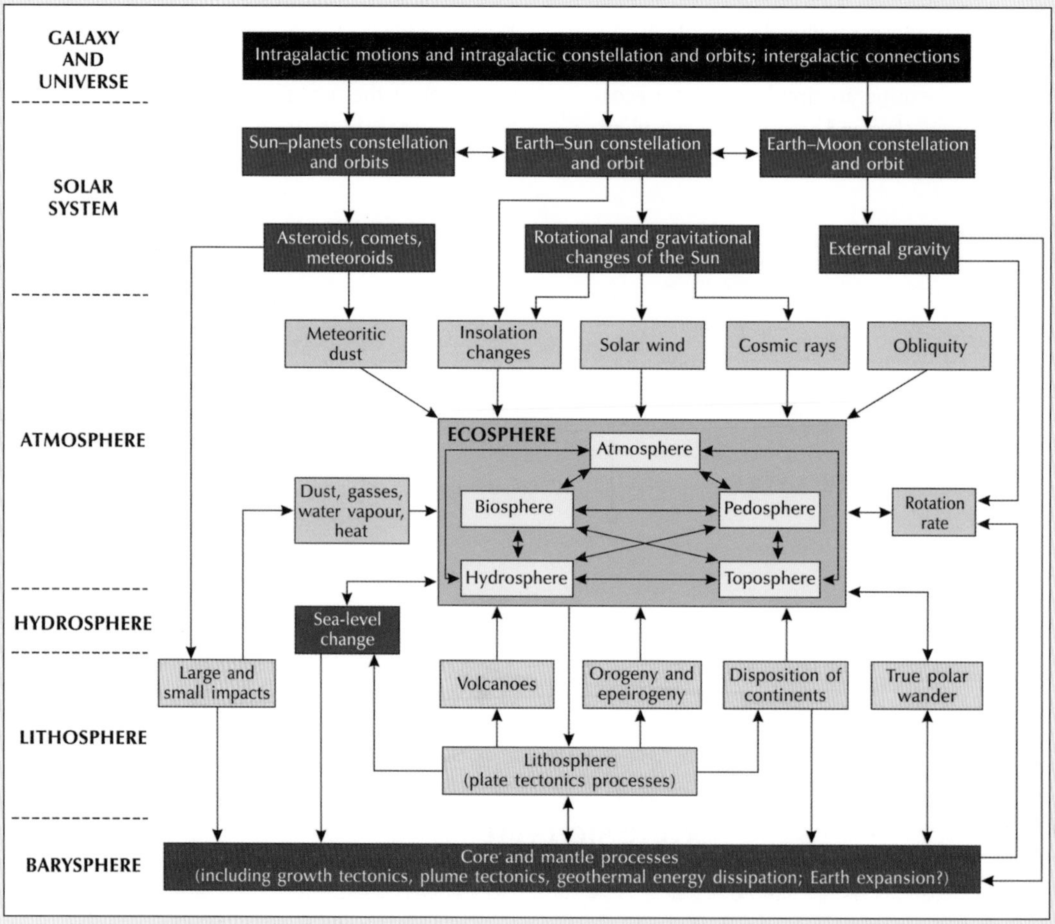

Figure 4.3 Interacting terrestrial spheres and their cosmic and geological settings. *Source*: Adapted from Huggett (1991, 1995, 1997b)

The denudational link works through weathering, the carbon cycle, and the unloading of crustal material. Growing mountains and plateaux influence chemical weathering rates. As mountains grow, atmospheric carbon dioxide combines with the fresh rocks during weathering and is carried to the sea. Global cooling during the Cenozoic era may have been instigated by the uplift of the Tibetan Plateau (p. 73). Increase in chemical weathering associated with this uplift has caused a decrease in atmospheric carbon dioxide concentrations over the last 40 million years (Raymo and Ruddiman 1992; Ruddiman 1997). The interaction of continental drift, runoff, and weathering has also affected global climates during the last 570 million years (Otto-Bliesner 1995). The removal of surface material by erosion along passive margins, as in the Western Ghats in India, causes a different effect. Unburdened by part of its surficial layers, and in conjunction with the deposition of sediment in offshore basins, the lithosphere rises by 'flexural rebound', promoting the growth of escarpments that wear back and are separated from inland plateaux that wear down (p. 125).

cause of vegetation and of life; but now we find, that this vapour not only fertilizes, but creates the soil; prepares it from the soil rock, and, after employing it in the great operations of the surface, carries it back into the regions where all its mineral characters are renewed. Thus, the circulation of moisture through the air, is a prime mover, not only in the annual succession of seasons, but in the great geological cycle, by which the waste and reproduction of entire continents is circumscribed.

(Playfair 1802, 128)

DENUDATION AND DEPOSITION

Weathering and erosion

Weathering is the decay of rocks by biological, chemical, and mechanical agents with little or no transport. It produces a mantle of rock waste. The **weathered mantle** may stay in place, or it may move down hillslopes, down rivers, and down submarine slopes. Gravity and fluid forces impel this downslope movement. The term **mass wasting** is sometimes used to describe all processes that lower the ground surface. It is also used more specifically as a synonym of **mass movement**, which is the bulk transfer of bodies of rock debris down slopes under the influence of gravity. **Erosion**, which is derived from the Latin (*erodere*, to gnaw; *erosus*, eaten away), is the sum of all destructive processes by which weathering products are picked up (entrained) and carried by transporting media – ice, water, and wind. Most geomorphologists regard **transport** as an integral part of erosion, although it could be argued, somewhat pedantically, that erosion is simply the acquisition of material by mobile agencies and does not include transport. Water is a widespread transporting agent, ice far less so. Moving air may erode and carry sediments in all subaerial environments. It is most effective where vegetation cover is scanty or absent. Winds may carry sediments up slopes and over large distances (see Simonson 1995). Dust-sized particles may travel around the globe. **Denudation**, which comes from the Latin *denudare*, meaning 'to lay bare', is the conjoint action of weathering and erosion, which processes simultaneously wear away the land surface.

Water and ice in the pedosphere (including the weathered part of exposed rocks) may be regarded as liquid and solid components of the weathered mantle. Weathered products, along with water and ice, tend to flow downhill along lines of least resistance, which typically lie at right angles to the topographic contours. The flowlines run from mountain and hill summits to sea floors. In moving down a flowline, the relative proportion of water to sediment alters. On hillslopes, there is little, if any, water to a large body of sediment. Mass movements prevail. These take place under the influence of gravity, without the aid of moving water, ice, or air. In glaciers, rivers, and seas, a large body of water bears some suspended and dissolved sediment. Movement occurs through glacial, fluvial, and marine transport.

Transport

A river in flood demonstrates sediment transport, the dirty floodwaters bearing a burden of material derived from the land surface. As well as the visible sediment, the river also carries a load of material in solution. Geomorphologists often distinguish between sediment transport, which is essentially mechanical, and solutional transport, which is essentially chemical; they also discriminate between processes involving a lot of sediment moving *en masse* – mass movement – and sediment moving as individual grains more or less dispersed in a fluid – fluid transport (cf. Statham 1977, 1). In mass movement, the weight of sediment is a key controlling factor of motion, whereas in fluid transport the action of an external fluid agency (wind or water) is the key factor. However, the distinction blurs in case of slow mass movements, which resemble flows, and in the continuous transition from dry moving material to muddy water.

Geomorphic forces

The transport of all materials, from solid particles to dissolved ions, needs a force to start and maintain motion. Such forces make boulders fall from cliffs, soils and sediment move down hillslopes, and water and ice flow along channels. For this reason, the mechanical principles controlling movement underpin the understanding of transport processes (Box 4.4).

Box 4.4 FORCE AND RESISTANCE

A body will not move unless a force is applied, and its movement will not continue without the sustained exertion of a force. Likewise, forces act on a body at rest that are in balance while the body remains stationary. For this reason, forces are immensely important in determining if the transport of sediments takes place.

A **force** is an action in a specified direction that tends to alter the state of motion of a body. An equal and opposite force called the reaction always balances it. A boulder resting on the ground exerts a vertical force on the ground due to its weight; the ground exerts a force of the same magnitude in the opposite direction on the boulder; and, if it did not do so, the boulder would sink into the ground. Forces result from the acceleration of a body. If a body is not subject to an acceleration, then it cannot exert a force in any direction. At the Earth's surface, most bodies are subject to the acceleration due to gravity and exert a force in the direction of gravity, which is approximately vertically. The magnitude of this force is generally the weight of the body in a static condition (but, if the body is moving, the force alters).

Forces have direction and magnitude. If two or more forces are acting on a body, then the magnitude and direction of a resultant force is determinable. For example, a sediment grain entrained in flowing water is subject to several forces: a vertical force pushing it vertically upwards in the flow, the force of its own weight dragging it down vertically, and the downstream force of the flowing water carrying it along the river channel. The magnitude and direction of all these forces dictate the net direction in which the grain will travel and so whether it will stay suspended or sink to the riverbed. If a single force is known, its effects in different directions (its components) can be worked out. Take the case of a boulder on a hillslope (Figure 4.4). The weight of the boulder acts vertically in the direction of gravity, but the reaction with the ground surface prevents the boulder from moving in that direction. Nonetheless, movement downslope is possible because the weight of the boulder is resolvable into two forces – a force normal to the slope, which tends to hold the boulder in place, and a force parallel to the slope, which tends to move the boulder downhill. Normal and parallel reaction forces balance these. Now, the boulder will not move unless the downslope force can overcome the resistance to movement (friction) to counter the parallel reaction force. Once the downslope force exceeds the surface resistance, the boulder will accelerate, and its reaction then involves an inertia force due to the boulder's accelerating down the slope. This means that a smaller downslope force component is required to continue the motion at constant velocity, in the same way that it is easier to pull a sledge once it is moving than it is to start it moving.

Resistance is fundamental to transport processes. Without resistance, Earth surface materials would move under the force of gravity until the landscape was all but flat. Many factors affect resistance, but none so much as friction. Friction exists between bodies and the surface over which they move. It occurs between where matter in any state (solid, liquid, gas) comes into contact, as in solids on solids, solids on fluids, fluids on fluids, and gases on solids or fluids. In a river, friction occurs at the fluid bed contact and within the water, owing to differential velocity of flow and turbulent eddies. In the case of a boulder at rest on a flat surface, if no lateral force is applied to the boulder, then the frictional resistance is zero as there is no force to resist.

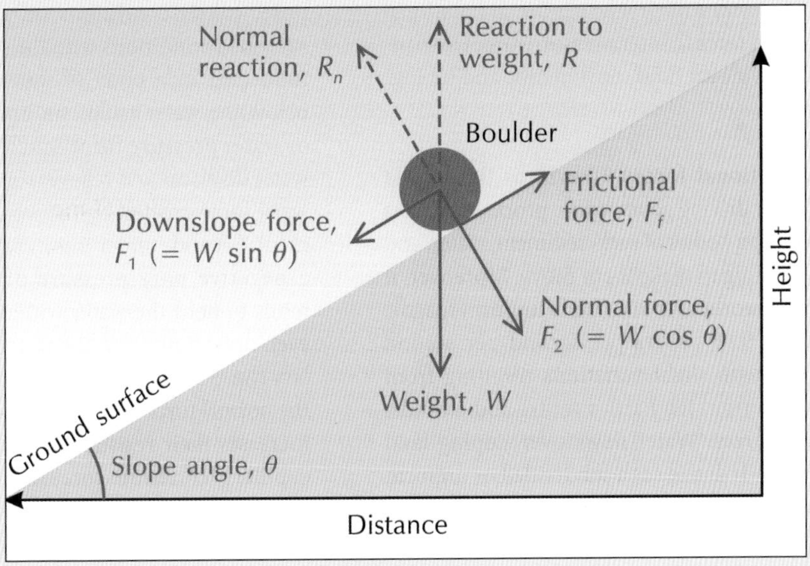

Figure 4.4 Forces acting upon a boulder lying on a hillside.

continued . . .

If a lateral force, F, is applied, then the frictional force, F_f, increases to balance the force system. At a critical value for F, the frictional resistance, generated between the boulder and the surface, will be unable to balance the applied force and the boulder will start to accelerate. For any given surface contact

$$\frac{F_{critical}}{R_n} = \text{a constant} = \mu_s.$$

As the ratio is constant, the force required to move the boulder increases in proportion with R_n (the normal reaction, which, on a flat surface, is equal to the weight of the boulder).

The forces that drive sediment movement largely derive from gravity, from climatic effects (heating and cooling, freezing and thawing, winds), and from the action of animals and plants. They may act directly, as in the case of gravity, or indirectly through such agencies as water and wind. In the first case, the force makes the sediment move, as in landslides; while, in the second case, the force makes the agency move (water for instance) and in turn the moving agency exerts a force on the sediment and tends to move it, as in sediment transport in rivers. The chief forces that act upon geomorphic materials are gravitational forces, fluid forces, water pressure forces, expansion forces, global fluid movements, and biological forces.

1 **Gravitational forces**. Gravity is the largest force for driving geomorphic processes. It acts directly on bodies of rock, sediment, water, and ice, tending to make them move. Moreover, it acts the world over at a nearly uniform magnitude of 9.81 metres per second per second (m/s^2), with slight variations resulting from distance from the Earth's centre and latitude.
2 **Fluid forces**. Water flows over sloping land surfaces. It does so as a subdivided or uniform sheet or as channel flows in streams and rivers. Water is a fluid so that it moves in the direction of any force that is applied to it, and no critical force is necessary. So water flows downhill under the influence of its own weight, which is a gravitational force. Moving the water uses only part of the downslope force, and the portion left after overcoming various resistances to flow may carry material in the flow or along the water–ground contact. The water also carries dissolved material that travels at the same velocity as the water and essentially behaves as part of the fluid itself.
3 **Water pressure forces**. Water in soil and sediment creates various forces that can affect sediment movement. The forces in saturated (all the pores filled) and unsaturated (some of the pores filled) conditions differ. First, under saturated conditions with the soil or sediment immersed in a body of water (for example, below the water table), an upward buoyancy or **water pressure force** equal to the weight of water displaces and relieves some of the downward force created by the weight of the sediment. Second, under unsaturated conditions, a **negative pore pressure** or **suction force** tends to hold the water within the pores and even draw it up from the water table by capillary rise. Such negative pore pressure increases the normal force between sediment grains and increases their resistance to movement. This capillary cohesion force keeps sandcastles from collapsing. Falling raindrops also create a force when they strike the ground. Depending on their size and terminal velocity, they may create a force strong enough to move sediment grains.

to form various minerals, including siderite, limonite (bog iron), and vivianite. **Calcium carbonate** carried in groundwater precipitates in caves and grottoes as sheets of flowstone or as stalagmites, stalactites, and columns of dripstone (p. 443). It sometimes precipitates around springs, where it encrusts plants to produce tufa or travertine (p. 437). **Evaporites** form by soluble-salt precipitation in low-lying land areas and inland seas. They include halite or rock salt (sodium chloride), gypsum (hydrated calcium sulphate), anhydrite (calcium sulphate), carnallite (hydrated chloride of potassium and magnesium), and sylvite (potassium chloride). Evaporite deposits occur where clastic additions are low and evaporation high. At present, evaporites are forming in the Arabian Gulf, in salt flats or *sabkhas*, and around the margins of inland lakes, such as Salt Lake, Utah, USA. Salt flat deposits are known in the geological record, but the massive evaporite accumulations, which include the Permian Zechstein Basin of northern Europe and the North Sea, may be deep-water deposits, at least in part.

Chemicals precipitated in soils and sediments often form hard layers called **duricrusts**. These occur as hard nodules or crusts, or simply as hard layers. The chief types are mentioned on p. 169.

Biogenic sediments

Ultimately, the chemicals in **biogenic sediments** and **mineral fuels** come from rock, water, and air. They are incorporated into organic bodies and may accumulate after the organisms die. Limestone is a common biogenic rock. The shells of organisms that extract calcium carbonate from seawater form it. **Chalk** is a fine-grained and generally friable variety of limestone. Some organisms extract a little magnesium as well as calcium to construct their shells – these produce magnesian limestones. **Dolomite** is a calcium–magnesium carbonate. Other organisms, including diatoms, radiolarians, and sponges, utilize silica. These are sources of **siliceous deposits** such as chert and flint and siliceous ooze.

The organic parts of dead organisms may accumulate to form a variety of biogenic sediments. The chief varieties are **organic muds** (consisting of finely divided plant detritus) and **peats** (called **coal** when lithified). Traditionally, organic materials are divided into sedimentary (transported) and sedentary (residual). Sedimentary organic materials are called dy, gyttja, and alluvial peat. Dy and gyttja are Swedish words that have no English equivalent. **Dy** is a gelatinous, acidic sediment formed in humic lakes and pools by the flocculation and precipitation of dissolved humic materials. **Gyttja** comprises several biologically produced sedimentary oozes. It is commonly subdivided into organic, calcareous, and siliceous types. Sedentary organic materials are peats, of which there are many types.

Sedimentary environments

The three main sedimentary environments are **terrestrial**, **shallow marine**, and **deep marine**. A single sedimentary process dominates each of these: gravity-driven flows (dry and wet) in terrestrial environments; fluid flows (tidal movements and wave-induced currents) in shallow marine environments; and suspension settling and unidirectional flow created by density currents in deep marine environments (Fraser 1989). Transition zones separate the three main sedimentary environments. The **coastal transition zone** separates the terrestrial and shallow marine environments; the **shelf-edge–upper-slope transition zone** separates the shallow and the deep marine environments.

Sediments accumulate in all terrestrial and marine environments to produce depositional landforms. As a rule, the land is a **sediment source** and the ocean is a **sediment sink**. Nonetheless, there are extensive bodies of sediments on land and many erosional features on the ocean floor. Sedimentary deposits are usually named after the processes responsible for creating them. Wind produces **aeolian deposits**, rain and rivers produce **fluvial deposits**, lakes produce **lacustrine**

deposits, ice produces **glacial deposits**, and the sea produces **marine deposits**. Some deposits have mixed provenance, as in **glaciofluvial deposits** and **glaciomarine deposits** (also spelt glacifluvial and glacimarine). On land, the most pervasive 'sedimentary body' is the **weathered mantle** or **regolith**. The thickness of the regolith depends upon the rate at which the weathering front advances into fresh bedrock and the net rate of erosional loss (the difference between sediment carried in and sediment carried out by water and wind). At sites where thick bodies of terrestrial sediments accumulate, as in some alluvial plains, the materials would normally be called sediments rather than regolith. However, regolith and thick sedimentary bodies are both the product of geomorphic processes. They are thus distinct from the underlying bedrock, which is a production of lithospheric processes.

Gravity, water, and wind transport unconsolidated weathered material in the regolith across hillslopes and down river valleys. Local accumulations form stores of sediment. Sediment stored on slopes is **talus**, **colluvium**, and **talluvium**. Talus is made of large rock fragments, colluvium of finer material, and talluvium of a fine and coarse material mix. Sediment stored in valleys is **alluvium**. It occurs in alluvial fans and in floodplains. All these slope and valley stores, except for talus, are fluvial deposits (transported by flowing water).

DENUDATION AND GLOBAL CLIMATE

Measurements of the amount of sediment annually carried down the Mississippi River were made during the 1840s, and Archibald Geikie worked out the rates of modern denudation in some of the world's major rivers in the 1860s. Measurements of the dissolved load of rivers enabled estimates of chemical denudation rates to be made in the first few decades of the twentieth century. Not until after the 'quantitative revolution' in geomorphology, which started in the 1940s, were rates of geomorphic processes measured in different environments and a global picture of denudation rates pieced together.

Mechanical denudation

Measuring denudation rates

Overall rates of **denudation** are judged from the dissolved and suspended loads of rivers, from reservoir sedimentation, and from the rates of geological sedimentation. Figure 4.7a depicts the pattern of sediment yield from the world's major drainage basins, and Figure 4.7b displays the annual discharge of sediment from the world's major rivers to the sea. It should be emphasized that these figures do not measure the total rate of soil erosion, since much sediment is eroded from upland areas and deposited on lowlands where it remains in store, so delaying for a long time its arrival at the sea (Milliman and Meade 1983). Table 4.3 shows the breakdown of chemical and mechanical denudation by continent.

Factors controlling denudation rates

The controls on mechanical denudation are so complex and the data so sketchy that it is challenging to attempt to assess the comparative roles of the variables involved. Undaunted, some researchers have tried to make sense of the available data (e.g. Fournier 1960; Strakhov 1967). Frédéric Fournier (1960), using sediment data from seventy-eight drainage basins, correlated suspended **sediment yield** with a climatic parameter, p^2/P, where p is the rainfall of the month with the highest rainfall and P is the mean annual rainfall. Although, as might be expected, sediment yields increased as rainfall increased, a better degree of explanation was found when basins were grouped into relief classes. Fournier fitted an empirical equation to the data:

$$\log E = -1.56 + 2.65 \log (p^2/P + 0.46 \log \bar{H} - \tan \theta)$$

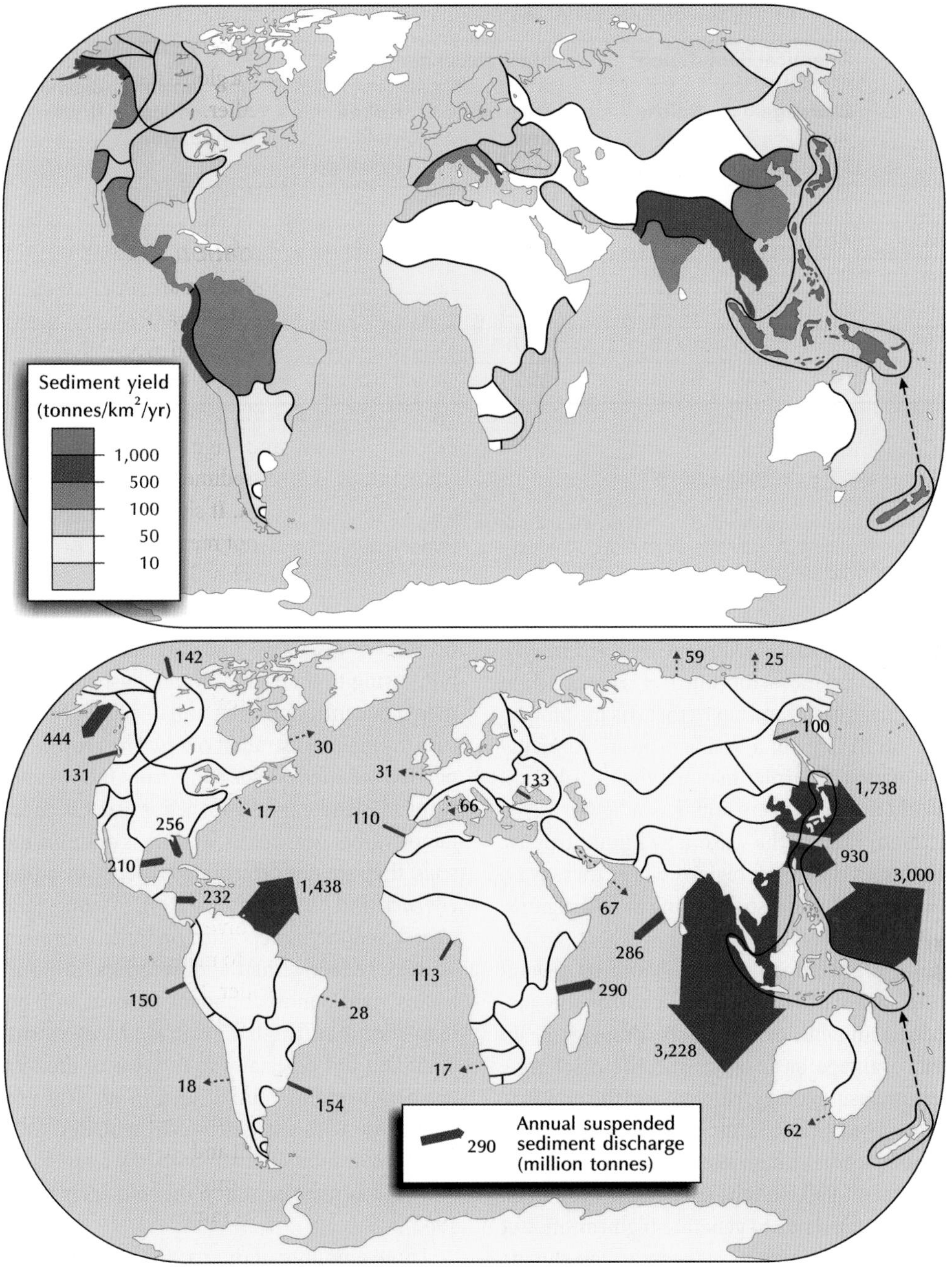

Figure 4.7 (a) Sediment yield of the world's chief drainage basins. Blank spaces indicate essentially no discharge to the ocean. (b) Annual discharge of suspended sediment from large drainage basins of the world. The width of the arrows corresponds to relative discharge. Numbers refer to average annual input in millions of tonnes. The direction of the arrows does not indicate the direction of sediment movement. *Source*: Adapted from Milliman and Meade (1983)

Table 4.3 Chemical and mechanical denudation of the continents

Continent	Chemical denudation[a]		Mechanical denudation[b]		Ratio of mechanical to chemical denudation	Specific discharge (l/s/km²)
	Drainage area (10^6 km²)	Solute yield (t/km²/yr)	Drainage area (10^6 km²)	Solute yield (t/km²/yr)		
Africa	17.55	9.12	15.34	35	3.84	6.1
North America	21.5	33.44	17.50[c]	84	2.51	8.1
South America	16.4	29.76	17.90	97	3.26	21.2
Asia	31.46	46.22	16.88	380	8.22	12.5
Europe	8.3	49.16	15.78[d]	58	1.18	9.7
Oceania	4.7	54.04	5.2	1,028[e]	19.02	16.1

Notes:

a Data from Meybeck (1979, Annex 3).
b Data from Milliman and Meade (1983, Table 4).
c Includes Central America.
d Milliman and Meade separate Europe (4.61 × 10^6 km²) and Eurasian Arctic (11.17 × 10^6 km²).
e The sediment yield for Australia is a mere 28 t/km²/yr, whereas the yield for large Pacific islands is 1,028 t/km²/yr.

Source: After Huggett (1991, 87)

where E is suspended sediment yield (t/km²/yr), p^2/P, is the climatic factor (mm), $\bar{H}$ is mean height of a drainage basin, and tan θ (theta) is the tangent of the mean slope of a drainage basin. Applying this equation, Fournier mapped the distribution of world mechanical erosion. His map portrayed maximum rates in the seasonally humid tropics, declining in equatorial regions where there is no seasonal effect, and also declining in arid regions, where total runoff is low.

John D. Milliman (1980) identified several natural factors that appear to control the suspended sediment load of rivers: drainage basin relief, drainage basin area, specific discharge, drainage basin geology, climate, and the presence of lakes. The climatic factor influences suspended sediment load through mean annual temperature, total rainfall, and the seasonality of rainfall. Heavy rainfall tends to generate high runoff, but heavy seasonal rainfall, as in the monsoon climate of southern Asia, is very efficacious in producing a big load of suspended sediment. On the other hand, in areas of high, year-round rainfall, such as the Congo basin, sediment loads are not necessarily high. In arid regions, low rainfall produces little river discharge and low sediment yields; but, owing to the lack of water, suspended sediment concentrations may still be high. This is the case for many Australian rivers. The greatest suspended sediment yields come from mountainous tropical islands, areas with active glaciers, mountainous areas near coasts, and areas draining loess soils: they are not determined directly by climate (Berner and Berner 1987, 183). As one might expect, sediments deposited on inner continental shelves reflect climatic differences in source basins: mud is most abundant off areas with high temperature and high rainfall; sand is everywhere abundant but especially so in areas of moderate temperature and rainfall and in all arid areas save those with extremely cold climates; gravel is most common off areas with low temperature; and rock is most common off cold areas (Hayes 1967).

Large amounts of quartz, in association with high ratios of silica to alumina, in river sediments indicate intense tropical weathering regimes. Work carried out on the chemistry of river sediments has revealed patterns attributable to differing weathering regimes in (1) the tropical zone and (2) the temperate and frigid zones. River sands with high quartz and high silica-to-alumina ratios occur mainly in tropical river basins of low relief, where weathering is intense enough (or has proceeded uninterrupted long enough) to eliminate any differences arising from rock type, while river sands with low quartz content but high silicate-alumina ratios occur chiefly in the basins located in temperate and frigid regions (Potter 1978). A basic distinction between tropical regions, with intense weathering regimes, and temperate and frigid regions, with less intense weathering regimes, is also brought out by the composition of the particulate load of rivers (Martin and Meybeck 1979). The tropical rivers studied had high concentrations of iron and aluminium relative to soluble elements because their particulate load was derived from soils in which soluble material had been thoroughly leached. The temperate and arctic rivers studied had lower concentrations of iron and aluminium in suspended matter relative to soluble elements because a smaller fraction of the soluble constituents had been removed. This broad pattern will almost certainly be distorted by the effects of relief and rock type. Indeed, the particulate load (p. 217) data include exceptions to the rule: some of their tropical rivers have high calcium concentrations, probably owing to the occurrence of limestone within the basin. Moreover, in explaining the generally low concentrations of calcium in sediments of tropical rivers, it should be borne in mind that carbonate rocks are more abundant in the temperate zone than in the tropical zone (cf. Figure 14.2, p. 414).

Climate and denudation

Ignoring infrequent but extreme values and correcting for the effects of relief, overall rates of denudation show a relationship with climate (Table 4.4). Valley glaciation is substantially faster than normal erosion in any climate, though not necessarily so erosion by ice sheets. The wide spread of denudation rates in polar and montane environments may reflect the large range of rainfall encountered. The lowest minimum and,

Table 4.4 Rates of denudation in climatic zones

Climate	Relief	Typical range for denudation rate (mm/millennium)	
		Minimum	Maximum
Glacial	Normal (= ice sheets) Steep (= valley glaciers)	50 1,000	200 5,000
Polar and montane	Mostly steep	10	1,000
Temperate maritime	Mostly normal	5	100
Temperate continental	Normal Steep	10 100	100 200+
Mediterranean	–	10	?
Semi-arid	Normal	100	1,000
Arid	–	10	?
Subtropical	–	10?	1,000?
Savannah	–	100	500
Tropical rain forest	Normal Steep	10 100	100 1,000
Any climate	Badlands	1,000	1,000,000

Source: Adapted from Saunders and Young (1983)

possibly, the lowest maximum rates of denudation occur in humid temperate climates, where creep rates are slow, wash is very slow owing to the dense cover of vegetation, and solution is relatively slow because of the low temperatures. Other conditions being the same, the rate of denudation in temperate continental climates is somewhat brisker. Semi-arid, savannah, and tropical landscapes all appear to denude fairly rapidly. Clearly, further long-term studies of denudational processes in all climatic zones are needed to obtain a clearer picture of the global pattern of denudation.

Chemical denudation

The controls on the rates of chemical denudation are perhaps easier to ascertain than the controls on the rates of mechanical denudation. Reliable estimates of the loss of material from continents in solution have been available for several decades (e.g. Livingstone 1963), though more recent estimates overcome some of the deficiencies in the older data sets. It is clear from the data in Table 4.3 that the amount of material removed in solution from continents is not directly related to the average specific discharge (discharge per unit area). South America has the highest specific discharge but the second lowest chemical denudation rate. Europe has a relatively low specific discharge but the second-highest chemical denudation rate. On the other hand, Africa has the lowest specific discharge and the lowest chemical denudation rate. In short, the continents show differences in resistance to being worn away that cannot be accounted for merely in terms of climatic differences.

The primary controls on chemical denudation of the continents can be elicited from data on the chemical composition of the world's major rivers (Table 4.5). The differences in solute composition of river water between continents result partly from differences of **relief** and **lithology**, and partly from **climatic differences**. Waters draining off the continents are dominated by calcium ions and bicarbonate ions. These chemical species account for the dilute waters of South America and the more concentrated waters of Europe. Dissolved silica and chlorine concentrations show no consistent relationship with total dissolved solids. The reciprocal relation between calcium ion concentrations and dissolved silica concentrations suggests a degree of control by rock type: chiefly sedimentary rocks underlie Europe and North America, whereas mainly crystalline rocks underlie Africa and South America. However, because the continents mainly consist of a heterogeneous mixture of rocks, it would be unwise to read too much into these figures and to overplay this interpretation.

Table 4.5 Average composition of river waters by continents[a] (mg/l)

Continent	SiO_2	Ca^{2+}	Mg^{2+}	Na^+	K^+	Cl^-	SO_4^{2-}	HCO_3^-	Σ_i[b]
Africa	12	5.25	2.15	3.8	1.4	3.35	3.15	26.7	45.8
North America	7.2	20.1	4.9	6.45	1.5	7	14.9	71.4	126.3
South America	10.3	6.3	1.4	3.3	1	4.1	3.5	24.4	44
Asia	11	16.6	4.3	6.6	1.55	7.6	9.7	66.2	112.5
Europe	6.8	24.2	5.2	3.15	1.05	4.65	15.1	80.1	133.5
Oceania	16.3	15	3.8	7	1.05	5.9	6.5	65.1	104.5
World	10.4	13.4	3.35	5.15	1.3	5.75	8.25	52	89.2

Notes:
a The concentrations are exoreic runoff with human inputs deducted.
b Σ_i is the sum of the other materials.

Source: Adapted from Meybeck (1979)

Many factors affect the natural chemical composition of river water: the amount and nature of rainfall and evaporation, drainage basin geology and weathering history, average temperature, relief, and biota (Berner and Berner 1987, 193). According to Ronald J. Gibbs (1970, 1973), who plotted total dissolved solids of some major rivers against the content of calcium plus sodium, there are three chief types of surface waters:

1 Waters with low total dissolved solid loads (about 10 mg/l) but large loads of dissolved calcium and sodium, such as the Matari and Negro rivers, which depend very much on the amount and composition of precipitation.
2 Waters with intermediate total dissolved solid loads (about 100–1,000 mg/l) but low to medium loads of dissolved calcium and sodium, such as the Nile and Danube rivers, which are influenced strongly by the weathering of rocks.
3 Waters with high total dissolved solid loads (about 10,000 mg/l) and high loads of dissolved calcium and sodium, which are determined primarily by evaporation and fractional crystallization and which are exemplified by the Rio Grande and Pecos rivers.

This classification has been the subject of much debate (see Berner and Berner 1987, 197–205), but it seems undeniable that climate does have a role in determining the composition of river water, a fact borne out by the origin of solutes entering the oceans. Chemical erosion is greatest in mountainous regions of humid temperate and tropical zones. Consequently, most of the dissolved ionic load going into the oceans originates from mountainous areas, while 74 per cent of silica comes from the tropical zone alone.

Further work has clarified the association between chemical weathering, mechanical weathering, lithology, and climate (Meybeck 1987). Chemical transport, measured as the sum of major ions plus dissolved silica, increases with increasing specific runoff, but the load for a given runoff depends on underlying rock type (Figure 4.8). Individual solutes show a similar pattern. Dissolved silica is interesting because, though the rate of increase with increasing specific discharge is roughly the same in all climates, the actual amount of dissolved silica increases with increasing temperature (Figure 4.8b). This situation suggests that, although lithology, distance to the ocean, and climate all affect solute concentration in rivers, transport rates, especially in the major rivers, depend first and foremost on specific river runoff (itself related to climatic factors) and then on lithology.

Regional and global patterns of denudation

Enormous variations in sediment and solute loads of rivers occur within particular regions owing to the local effects of rock type, vegetation cover, and so forth. Attempts to account for regional variations of denudation have met with more success than attempts to explain global patterns, largely because coverage of measuring stations is better and it is easier to take factors other than climate into consideration. Positive correlations between suspended sediment yields and mean annual rainfall and mean annual runoff have been established for drainage basins in all parts of the world, and simply demonstrate the fact that the more water that enters the system, the greater the erosivity. Solute loads, like suspended sediment loads, exhibit striking local variations about the global trend. The effects of **rock type** in particular become far more pronounced in smaller regions. For example, dissolved loads in Great Britain range from 10 to more than 200 t/km^2/yr, and the national pattern is influenced far more by lithology than by the amount of annual runoff (Walling and Webb 1986). Very high solute loads are associated with outcrops of soluble rocks. An exceedingly high solute load of 6,000 t/km^2/yr has been recorded in the River Cana, which drains an area of halite deposits in Amazonia; and a load of 750 t/km^2/yr has been measured in an area draining karst terrain in Papua New Guinea.

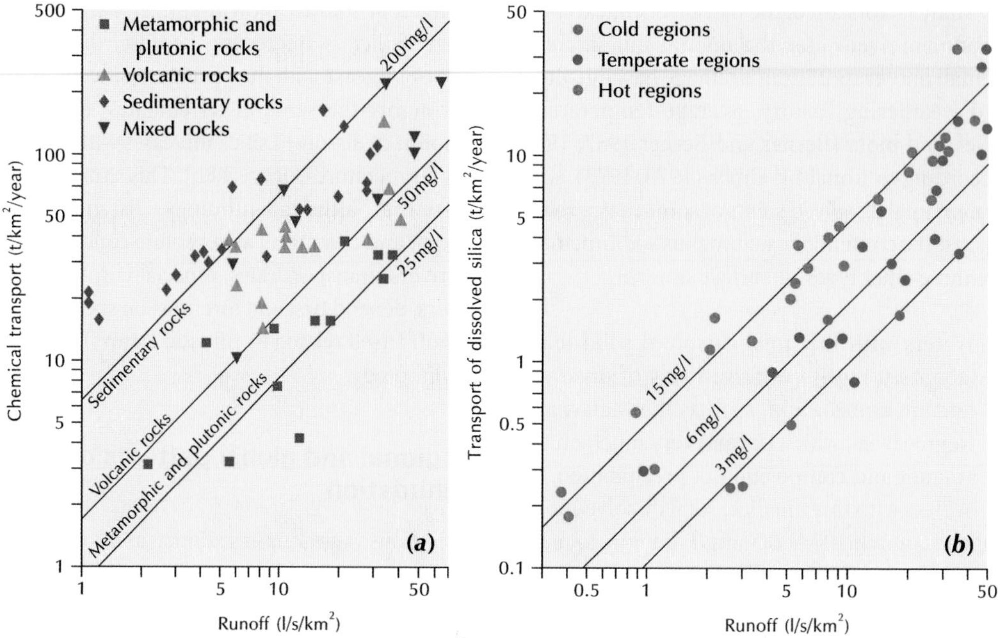

Figure 4.8 Dissolved loads in relation to runoff. (a) Chemical transport of all major ions plus dissolved silica versus runoff (specific discharge) for various major drainage basins underlain by sedimentary, volcanic, and metamorphic and plutonic rocks. (b) Evolution of the specific transport of dissolved silica for cold, temperate, and hot regions. *Source*: Adapted from Meybeck (1987)

All the general and detailed summaries of global and regional sediment yield (e.g. Fournier 1960; Jansson 1988; Milliman and Meade 1983; Summerfield and Hulton 1994) split into two camps of opinion concerning the chief determinants of erosion at large scales. Camp one sees relief as the prime factor influencing denudation rates, with climate playing a secondary role. Camp two casts climate in the leading role and relegates relief to a supporting part. Everybody seems to agree that either relief or climate, as measured by surrogates of rainfall erosivity, is the major control of erosion rates on a global scale. The problem is deciding on the relative contribution made by each factor. Jonathan D. Phillips (1990) set about the task of solving this problem by considering three questions: (1) whether indeed relief and climate are major determinants of soil loss; (2) if so, whether relief or climate is the more important determinant at the global scale; and (3) whether other factors known to influence soil loss at a local scale have a significant effect at the global scale. Phillips's results showed that slope gradient (the relief factor) is the main determinant of soil loss, explaining about 70 per cent of the maximum expected variation within global erosion rates. Climate, measured as rainfall erosivity, was less important but with relief (slope gradient) and a runoff factor accounted for 99 per cent of the maximum expected variation. The importance of a runoff factor, represented by a variable describing retention of precipitation (which is independent of climatic influences on runoff) was surprising. It was more important than the precipitation factors. Given Phillips's findings, it may pay to probe more carefully the fact that the variation in sediment yield within climatic zones is greater than the variation between climatic zones (Jansson 1988). At local scales, the influence of vegetation cover may play a critical role in dictating soil erosion rates (e.g. Thornes 1990).

Niels Hovius (1998) collated data on fourteen climatic and topographic variables used in previous studies for ninety-seven major catchments around the world. He found that none of the variables correlated well with sediment yield, which suggests that no single variable is an overriding determinant of sediment yield. However, sediment yield was successfully predicted by a combination of variables in a multiple regression equation. A five-term model explained 49 per cent of the variation in sediment yield:

$$\ln E = 3.585 - 0.416 \ln A + 4.26 \times 10^{-4} H_{max} + 0.150T + 0.095T_{range} + 0.0015R$$

where E is specific sediment yield (t/km²/yr), A is drainage area (km²), H_{max} is the maximum elevation of the catchment (m), T is the mean annual temperature (°C), T_{range} is the annual temperature range (°C), and R is the specific runoff (mm/yr). Of course, 51 per cent of the variation in sediment yield remains unexplained by the five-term model. One factor that might explain some of this unaccounted variation is the supply of erodible material, which, in geological terms, is largely determined by the uplift of rocks. Inputs of new matter by uplift should explain additional variation beyond that explained by the erosivity of materials.

A global survey of chemical and physical erosion data drew several interesting conclusions about the comparative roles of tectonics, the environment, and humans in explaining regional variations (Stallard 1995). Four chief points emerged from this study. First, in tectonically active mountain belts, carbonate and evaporite weathering dominates dissolved loads, and the erosion of poorly lithified sediment dominates solid loads. In such regions, human activities may increase physical erosion by orders of magnitude for short periods. About 1,000 m of uplift every million years is needed to sustain the observed chemical and physical erosion rates. Second, in old mountain belts, physical erosion is lower than in young mountain belts of comparable relief, perhaps because the weakest rocks have been stripped by earlier erosion. Third, on shields, chemical and physical erosion are very slow because weak rocks are little exposed owing to former erosion. And, finally, a basic distinction may be drawn between areas where soil development and sediment storage occur (terrains where erosion is limited by transport capacity) and areas of rapid erosion (terrains where erosion is limited by the production of fresh sediment by weathering).

THE GLOBAL TECTONIC AND CLIMATIC SYSTEMS

Since the 1990s, geomorphologists have come to realize that the **global tectonic system** and the **world climate system** interact in complex ways. The interactions give rise to fundamental changes in atmospheric circulation patterns, in precipitation, in climate, in the rate of uplift and denudation, in chemical weathering, and in sedimentation (Raymo and Ruddiman 1992; Small and Anderson 1995; Montgomery *et al.* 2001). The interaction of large-scale landforms, climate, and geomorphic processes occurs in at least three ways – through the direct effect of plate tectonic processes upon topography (pp. 122–9), through the direct effect of topography upon climate (and the effects of climate upon uplift), and through the indirect influence of topography upon chemical weathering rates and the concentration of atmospheric carbon dioxide.

Changes in topography, such as the uplift of mountain belts and plateaux, can influence regional climates, both by locally increasing precipitation, notably on the windward side of the barrier, and through the cooling effect of raising the ground surface to higher elevations (e.g. Ollier 2004a). Changes in topography could poten-tially have wide-ranging impacts if they interact with key components of the Earth's climatic system. In southern Africa, uplift of 1,000 m during the Neogene, especially in the eastern part of the subcontinent, would have reduced surface temperatures by roughly the same amount as during glacial episodes at high latitudes (Partridge 1998). The uplift of the Tibetan Plateau and its bordering mountains may have actively forced climatic change by intensifying the Asian monsoon (through altering surface atmospheric pressure owing to elevation increase), by creating a high-altitude barrier to airflow that affected the jet stream, and by encouraging inter-hemispherical exchange of heat (Liu and Ding 1998; Fang *et al.* 1999a, b). These forcings seem to have occurred around 800,000 years ago. However, oxygen isotope work on late Eocene and younger deposits in the centre of the plateau suggests that this area at least has stood at more than 4 km for about 35 million years (Rowley and Currie 2006).

Recent research shows that local and regional climatic changes caused by uplift may promote further uplift through a positive feedback loop involving the extrusion of crustal rocks (e.g. Molnar and England 1990; Hodges 2006). In the Himalayas, the Asian monsoon sheds prodigious amounts of rain on the southern flanks of the mountains. The rain erodes the rocks, which enables the fluid lower crust beneath Tibet to extrude towards the zone of erosion. Uplift results from the extrusion of rock and counterbalances the erosion, which reduces the land-surface elevation. Therefore, the extrusion process keeps the front range of the Himalaya steep, which encourages heavy monsoon rains, so completing the feedback loop (but see Ollier 2006 for a different view).

More recently, a study of climate and tectonic history in the eastern Himalayas showed that lateral variations in plate coupling (the degree to which tectonic plates are locked together or free to move and detectable in the plate motion record) may in some cases be a consequence of climate (Iaffaldano *et al.* 2011). It demonstrated that climate changes can affect the short-term evolution of plate motion by establishing an unequivocal link between the observed counter-clockwise rotation of the Indian plate over the last 10 million years and the increased erosion and reduced elevation along the eastern Himalayas over the same time, which was itself the result of temporal variations in monsoon intensity. In brief, the lowered gravitational potential energy in the eastern Himalayas following more vigorous erosion associated with a strengthening of the monsoon regime was a key driver of plate convergence in the region.

Carbon dioxide is a key factor in determining mean global temperatures. Over geological timescales (millions and tens of millions of years), atmospheric carbon dioxide levels depend upon the rate of carbon dioxide input through volcanism, especially that along mid-ocean ridges, and the rate of carbon dioxide withdrawal through the weathering of silicate rocks by carbonation, a process that consumes carbon dioxide. Given that carbon dioxide inputs through volcanism seem to have varied little throughout Earth history, it is fair to assume that variations in global chemical weathering rates should explain very long-term variations in the size of the atmospheric carbon dioxide pool. So what causes large changes in chemical weathering rates? Steep slopes seem to play a crucial role. This relatively new finding rests on the fact that weathering rates depend greatly on the amount of water passing through the weathering zone. Rates are highest on steep slopes with little or no weathered mantle and high runoff. In regions experiencing these conditions, erosional processes are more likely to remove weathered material, so exposing fresh bedrock to attack by percolating water. In regions of thick weathered mantle and shallow slopes, little water reaches the weathering front and little chemical weathering occurs. Interestingly, steep slopes characterize areas of active uplift, which also happen to be areas of high precipitation and runoff. In consequence, 'variations in rates of mountain building through geological time could affect overall rates of global chemical weathering and thereby global mean temperatures by altering the concentration of atmospheric CO_2'

(Summerfield 2007, 105). If chemical weathering rates increase owing to increased tectonic uplift, then CO_2 will be drawn out of the atmosphere, but there must be some overall negative feedback in the system otherwise atmospheric CO_2 would become exhausted, or would keep on increasing and cause a runaway greenhouse effect. Neither has occurred during Earth history, and the required negative feedback probably occurs through an indirect effect of temperature on chemical weathering rates. It is likely that if global temperatures increase this will speed up the hydrological cycle and increase runoff. This will, in turn, tend to increase chemical weathering rates, which will draw down atmospheric CO_2 and thereby reduce global mean temperature. It is also possible that variations in atmospheric CO_2 concentration may directly affect chemical weathering rates, and this could provide another negative feedback mechanism.

The idea that increased weathering rates associated with tectonic uplift increases erosion and removes enough carbon dioxide from the atmosphere to control climate has its dissenters. Ollier (2004a) identified what he termed 'three misconceptions' in the relationships between erosion, weathering, and carbon dioxide. First, weathering and erosion are not necessarily current processes – erosion, especially erosion in mountainous regions, may occur with little chemical alteration of rock or mineral fragments. Second, in most situations, hydrolysis and not carbonation is the chief weathering process – weathering produces clays and not carbonates. Furthermore, evidence suggests that chemical weathering rates have declined since the mid- or early Tertiary, before which time deep weathering profiles formed in broad plains. Today, deep weathering profiles form only in the humid tropics. Third, Ollier questions the accepted chronology of mountain building, which sees Tibet, the highlands of western North America, and the Andes beginning to rise about 40 million years ago, favouring instead rise over the last few million years.

HUMANS AS GEOMORPHIC AGENTS

With a few exceptions, geomorphologists generally paid lip service to the human agency in landscape change until the 1970s, when applied aspects of the subject began to take shape. From then on, studies of human impacts on the landscape have grown apace and paralleled studies of human impacts on other components of the Earth system, where current thinking in some circles suggests that humans have created a new geological epoch – the Anthropocene (Box 4.5).

Looking at human impacts on landforms and landscapes, three types of impact are distinguishable (Figure 4.9). First, human activities lead to environmental changes that in turn affect geomorphic processes. Examples of such indirect impacts are global warming and the spread of invasive species to new areas. Second, humans directly modify geomorphic processes, for instance by stripping vegetation cover and promoting soil erosion. Third, humans directly create landforms through such activities as excavation and dam building. Various aspects of these human–landscape interactions will be picked up throughout this book.

Geomorphic footprint

Over the last two centuries or so, humans have had an increasingly significant impact on the transfer of Earth materials and the modification of landforms, chiefly through agricultural practices, mining and quarrying, and the building of roads and cities. As Harrison Brown (1956, 1031) commented:

> A population of 30 billion would consume rock at a rate of about 1,500 tons per year. If we were to assume that all the land areas of the world were available for such processing, then, on the average, man [*sic*] would "eat" his way downward at a rate of 3.3 millimeters per year, or over 3 meters per millennium. This figure

Box 4.5 THE ANTHROPOCENE

Paul Crutzen made up the term Anthropocene on the spur of the moment at a conference held in 2000, when he heard someone mention the Holocene (Pearce 2007, 21). It suddenly occurred to him that the Holocene was the wrong term because the world has changed too much, so he said, 'No, we are in the Anthropocene.' Apparently, everyone was shocked, but the word Anthropocene caught on. It was first used in print the same year, as an informal term for the current geological epoch (Crutzen and Stoermer 2000; see also Crutzen 2002). Andrew Revkin had invented the similar term Anthrocene in 1992, but it failed to stick. Anthropogene is another older and rival term, suggested as a replacement for the Holocene (Nilsson 1983, 23–4), but it is not much used except by Russian geologists as a substitute for Quaternary (Gerasimov 1979). Older still is Anthropozoic Era, devised by Italian geologist Antonio Stoppani in 1873 to signify the time in Earth history when humans become a new geological force. However, a word of caution is useful here, as there are dangers in seeking precursors to new ideas, including the Anthropocene, owing in part to the very different historical contexts in which the older ideas evolved (see Hamilton and Grinevald 2015).

Anthropocene loosely means 'the age of humans', and it was chosen to highlight the central role of humankind in geology and ecology. Current thinking, at least in some circles, is that since life evolved some 3.5 billion years ago, the biosphere has passed through two stages – a microbial stage which lasted till around 650 million years ago, and a metazoan stage which lasted until very recently – and now shows signs of entering a third stage brought about by human actions that could well persist over geological timescales (Williams *et al.* 2015; Waters *et al.* 2016).

As to the onset of the Anthropocene – the date when the Holocene ended and the Anthropocene began – Crutzen and Stoermer suggested the latter part of the eighteenth century, but they recognized this moment in history was somewhat arbitrary and that rival proposals would be forthcoming. They were certainly right about that. Starting with Zalasiewicz and his colleagues in 2008, many researchers are now arguing a case for adding the Anthropocene to the geological timescale by defining it as a geological epoch in its own right, distinct from the Holocene, in recognition of humanity's unique impact on the planet. This has rekindled arguments about when the Anthropocene started.

Geological boundaries are marked by clearly recognizable and permanent signals in the rock record. To rank as a geological epoch distinct from the Holocene means finding a global signal in the rocks that will endure and be recognizable to geologists in the distant future. The question is: Have humans altered the planet to a comparable extent to past major geological events, so producing a signal in the rocks that marks the start of the Anthropocene? A thirty-eight-member Anthropocene Working Group is studying this matter, and other scientists are chipping into the debate. The chief suggestions are (1) an 'early Anthropocene' level starting in places some thousands of years ago (Edgeworth *et al.* 2015); (2) the beginning of the Industrial Revolution around 1,800 AD (Crutzen and Stoermer 2000); and (3) the 'Great Acceleration' of the mid-twentieth century (Steffen *et al.* 2015; Zalasiewicz *et al.* 2015). Current evidence suggests that the last of these has the most pronounced and globally synchronous signal.

Edgeworth and his colleagues (2015) proposed that, rather than a roughly synchronous global boundary marking it, the start of the Anthropocene should reflect the steadily growing human impacts on the planet from millennia ago to the present. They explored the transition from 'natural' soils and deposits to the material forming the archaeosphere, a term coined by Capelotti (2010) to describe the human-made portion of the geosphere. Edgeworth *et al.* (2015) called the transition Boundary A (the A standing for archaeosphere). It separates material untouched by human hands and tools from variously called 'artificial ground', 'anthropogenic ground', or 'archaeological stratigraphy'. The archaeosphere marks this boundary. The archaeosphere results from human actions – including ploughing, digging, fertilizing, bulldozing, blasting – that affect soil formation and soil properties. The boundary is in some places sharp, in other places diffuse; and it is everywhere diachronous (formed at different times in different places). Such an approach would in effect draw a line through time that was oldest in the places agriculture first developed, and become successively younger as it tracked the spread of humans and their soil-altering impacts across the globe. That boundary would encompass thousands of years, and would record the crescendo of human influence as *Homo sapiens* marched from past to present.

Towards the close of the eighteenth century, the industrial revolution accelerated as the burning of fossil fuels replaced water power. This switch in energy source produced an imprint on the Earth System recognizable worldwide and is a candidate for a start date for the Anthropocene (Crutzen and Stoermer 2000). However, while its trace will remain in geological records, the evidence of large-scale shifts in Earth System functioning before 1950 is weak (Steffen *et al.* 2015).

The Great Acceleration refers to a point around the middle of the twentieth century when the rate at which humans were altering the planet in ways that would leave indelible traces in the distant future (plastic debris, human-made metals and materials, human-made chemical signatures, the homogenization of the global biota and the extinction of many species) began accelerating. It is associated with acceleration in the growth of the human population around the same time. Accepting these arguments, a suggested date for the start of the Anthropocene is 16 July 1945, the day the first atomic bomb was detonated at Alamgordo, New Mexico. This event, and the other nuclear dentations that followed, will leave an unmistakable, long-lasting, and worldwide radioactive marker in the geological record (Zalasiewicz *et al.* 2015).

Other researchers have suggested different dates. Simon Lewis and Mark Maslin recently suggested 1610 as the start of the Anthropocene, based primarily on a dip in the atmospheric carbon dioxide concentration at that time (Lewis and Maslin 2015a, b). This, and much other work on the Anthropocene, has stirred up a very lively and stimulating debate (see, for example, Hamilton 2015). Does geomorphology have anything to contribute to this discussion? To be sure, geomorphic processes have shaped the geological record over the last few millennia, producing terrestrial and near-shore stratigraphy. For this reason, geomorphology must contribute to the debate, and the British Society for Geomorphology (BSG) has inaugurated a Fixed Term Working Group to consider the issue (Brown *et al.* 2013). Some of the issues will be picked up later in this book.

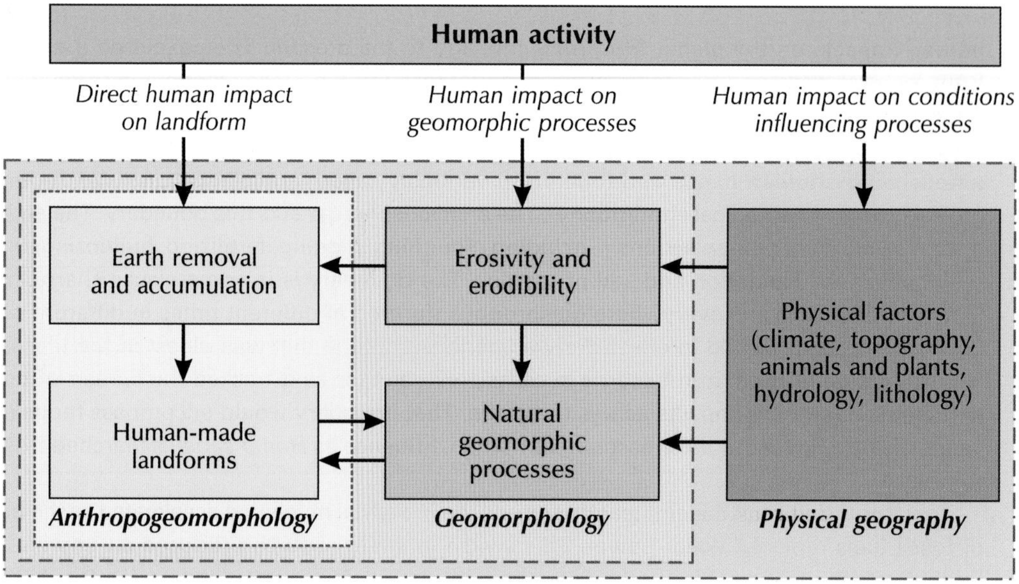

Figure 4.9 A 'human impact model', showing three ways in which human activity affects landscapes and landforms. Although this model suggests anthropogeomorphology concerns only direct human impacts, current usage of the term tends to include some indirect human impacts within its scope. *Source*: Slightly adapted with permission from Lóczy and Sütő (2011)

> gives us some idea of the denudation rates that might be approached in the centuries ahead. And it gives us an idea of the powers for denudation which lie in mankind's hands.

The 'geomorphic footprint' is a measure of the rate at which humans create new landforms and mobilize sediment (Rivas *et al.* 2006). For four study areas – one in northern Spain and three in central and eastern Argentina – new landforms were created by excavation and mining activities at a rate of 7.9 m^2 per person per year in the Spanish area and 5.93 m^2 per person per year in the Argentinean areas. The volume of sediment created by these activities was 30.4 m^3 per person per year and 6.4 m^3 per person per year for the Spanish and Argentinean areas respectively. These values convert to a sediment mobilization rate of 2.4 mm/yr for the Spanish study site and 0.8 mm/yr for the Argentinian study sites, which values exceed the rate mobilization of sediment by natural processes by an order of magnitude of two. If these figures are typical of other human-dominated areas, then Brown's denudation rates may be reached during the present century with a smaller population.

Humans have become increasingly adept at ploughing land and at excavating and moving materials in construction and mining activities. Indeed, humans are so efficient at unintentionally and deliberately moving soils and sediments that they have become the leading geomorphic agent of erosion (e.g. Hooke 2000). Placing human-induced erosion in a geological perspective demonstrates the point (Wilkinson 2005). The weathered debris stored in continental and oceanic sedimentary rocks suggest that, on average, continental surfaces have lowered through natural denudation at a rate of a few tens of metres per million years. By contrast, construction, mining, and agricultural activities presently transport sediment and rock, and lower all ice-free continental surfaces by a few hundred metres per million years. Therefore, the human species is now more important at moving sediment than all other geomorphic processes put together by an order of magnitude. This finding is supported by Peter K. Haff's (2010) more detailed study comparing technological and natural mass-action systems to see what processes move the most mass (Figure 4.10). Fluid motion of global or continental extent, as in atmospheric circulation and river flow, generate most of the world's mass-action. Technological mass-action exceeds that of all land-based geomorphic systems except rivers.

Key areas of human influence on sediment fluxes are mining and construction, agriculture, and dam building.

Mining and construction

Locally and regionally, humans transfer solid materials between the natural environment and the urban and industrial built environment. Robert Lionel Sherlock, in his book *Man as a Geological Agent: An Account of His Action on Inanimate Nature* (1922), recognized the role of human activity in geomorphic processes, and supplied many illustrations of the quantities of material involved in mining, construction, and urban development. Recent work confirms the potency of mining and construction activities in Earth surface change. In Britain, such processes as direct excavation, urban development, and waste dumping are driving landscape change: humans deliberately shift some 688 to 972 million tonnes of Earth-surface materials each year; the precise figure depends on whether or not the replacement of overburden in opencast mining is taken into account. British rivers export only 10 million tonnes of solid sediment and 40 million tonnes of solutes to the surrounding seas. The astonishing fact is that the deliberate human transfers move nearly 14 times more material than natural processes. The British land surface is changing faster than at any time since the last ice age, and perhaps faster than at any time in the last 60 million years (Douglas and Lawson 2001).

Every year humans move about 57 billion tonnes of material through mineral extraction processes. Rivers transport around 22 billion tonnes of sediment to the oceans annually, so the human cargo of sediment exceeds the river load by a factor of nearly three. Table 4.6 gives a breakdown of the figures. The data suggest that, in excavating and filling portions of the Earth's surface, humans are at present the most efficient geomorphic agent on the planet. Even where rivers, such as the Mekong, the Ganges, and the Yangtze, bear the sediment from accelerated erosion within their catchments, they still discharge a smaller mass of materials than the global production of an individual mineral commodity in a single year. Moreover, fluvial sediment discharges to the oceans from the continents are either similar in magnitude to, or smaller than, the total movement of materials for minerals production on those continents.

Mining and construction activities create new landforms. Coal mining areas are dotted with spoil tips. In Stoke-on-Trent, Staffordshire, UK, tip banks are conspicuous constructional landforms associated with coal mining in the area. They are colossal mounds of material, possibly containing more debris than that deposited by the last glaciers in the region. Increasingly, humans are building 'trash mountains' at landfill sites. The highest point in Palm Beach County, Florida, USA, is a landfill site. In England and Wales, landfill covers about 28,000 ha (a little under 0.2 per cent of the land area). Some extraction industries cause ground subsidence. Widespread subsidence has occurred because of coal mining in Ohio, USA, West Yorkshire, UK, and the southern Sydney Basin, Australia. In Cheshire, UK, salt flashes are shallow lakes or meres formed by subsidence associated with underground salt mining. In Los Angeles, USA, the extraction of petroleum from beneath the city has caused subsidence, which led to the collapse of the Baldwin Hills Reservoir dam on 14 December 1963, killing five people and destroying

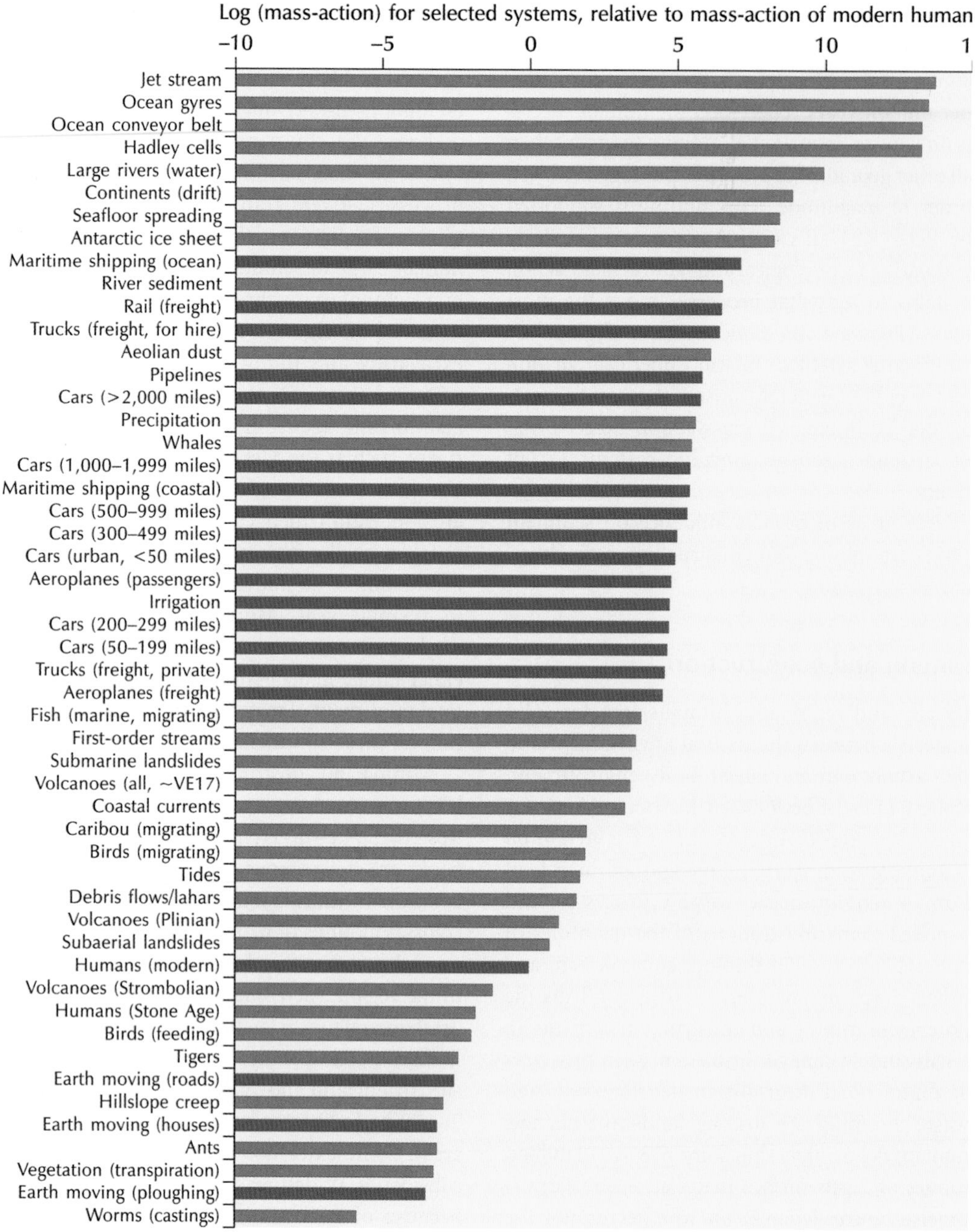

Figure 4.10 Logarithm of mass-action number, Ma, computed relative to mass-action of modern humans, for selected systems. Brown bars represent natural systems and purple bars represent technological systems. Haff (2010) defined the mass-action, *a*, of an object to be the product of its mass, *m*, times the distance it moves between changes in direction, *l*, times its average speed, *v*, between such changes: so, $a = mlv$. The mass-action number, Ma, is a metric based on mass-action to compare the mass transport effectiveness of natural and technological systems and to act as an aid to understanding the rise of technological transport as an Earth surface process. Technological systems with large mass-action tend to be discrete, self-powered units such as trucks. *Source*: After Haff (2010)

Table 4.6 Natural and mining-induced erosion rates of the continents

Continent	Natural erosion (Mt/yr)[a]	Hard coal, 1885 (Mt)	Brown coal and lignite, 1995 (Mt)	Iron ores, 1995 (Mt)	Copper ores, 1995 (Mt)
North and Central America	2,996	4,413	1,139	348	1,314
South America	2,201	180	1	712	1,337
Europe	967	3,467	6,771	587	529
Asia	17,966	8,990	1,287	1,097	679
Africa	1,789	993	–	156	286
Australia	267	944	505	457	296
Total	26,156	18,987	9,703	3,357	4,442

Note:
a Mt = megatonnes (= 1 million tonnes).

Source: Adapted from Douglas and Lawson (2001)

277 homes, and to the sinking of the bed of Long Beach Harbor by several metres, which was partly remedied by pumping salty water into the oil-bearing rocks. Similarly, water extraction from beneath Mexico City, Mexico, has produced subsidence substantial enough to damage buildings, the fall being about 2 to 8 cm per year.

Soil erosion

In transporting sediment to the oceans, rivers maintain a vital leg of the rock cycle and are a key component of the global denudation system. The amount of sediment carried down rivers is a measure of land degradation and the related reduction in the global soil resource. Many factors influence fluxes of river sediments, including reservoir construction, land clearance and land-use change, other forms of land disturbance (such as mining activity), soil and water conservation measures and sediment control programmes, and climate change. Land-clearance, most land-use change, and land disturbance cause an increase of sediment loads; soil and water conservation, sediment control programmes, and reservoir construction cause a decrease in sediment loads. A recent study provided a first assessment of current trends in the sediment loads of the world's rivers (Walling and Fang 2003). Analysis of longer-term records of annual sediment load and runoff assembled for 145 major rivers revealed that some 50 per cent of the sediment-load records contain evidence of statistically significant up-ward or downward trends, although the majority display diminishing loads. The evidence pointed to reservoir construction as probably the most important influence on land–ocean sediment fluxes, although the influence of other controls resulting in increasing sediment loads was detectable.

Dam building

The construction of dams, and other human activities, alters the amount of sediment carried by rivers to coastal environments, so affecting coastal geomorphology. Dams reduce the amount of sediment carried to coasts by about 1.4 billion tonnes per year, although soil erosion and mining and construction activities have increased it by about 2.3 billion tonnes per year (Syvitski *et al.* 2005). The increased sediment can make coastal areas less vulnerable to erosion, even if it can adversely affect coastal ecosystems. The positive and negative influences of human activities on river flow could balance each other out, but the net global result at present is that rivers carry less sediment to the coastal zone, with considerable differences on the regional level. In Indonesia, where fewer dams have meant fewer sediment-trapping reservoirs, more sediment is building up along the coastline because of human activities, chiefly deforestation. In general, Africa and Asia have seen the largest reduction in sediment to the coast. The effects of dams on rivers will be discussed in Chapter 9.

LIFE AS A GEOMORPHIC AGENT

Biogeomorphologists and ecogeomorphologists investigate the role of life in landform development. An important idea to emerge from their work is the geomorphic results of **ecosystem engineer species** (Jones 2012). There are three chief types of ecosystem engineers: **bioconstructors** (such as ants and termites, corals, mangroves, and beavers) that build new landforms; **bioeroders** (such as badgers, rabbits, and sea urchins) that wear down landforms through bioerosion; and **bioprotectors** (usually plants, including ivy and lichens, but also such animals as barnacles and zebra mussels) that shield landforms from agents of weathering and erosion, such as salts and waves (Table 4.7). Humans are bioconstructors, bioeroders, and bioprotectors rolled into one.

Bioeroders are mostly animals, but some plants cause erosion. For example, the rooting system of the Australian pine (a she-oak species and not a pine) makes the tree prone to toppling in strong winds, especially in Florida where it was introduced in 1898. The chief bioeroders are animals that burrow and animals that dig. These can displace and deposit large quantities of sediment ('spoil') at the surface in mounds or heaps. They also mix, aerate, and compact sediments while altering soil properties, which changes soil erodibility and sediment transport rates. For example, Eurasian badgers (*Meles meles*) displace soil while excavating their setts. A study in Wytham Woods, Oxfordshire, UK, revealed the rates of displacement (Coombes and Viles 2015). Individual badgers excavated between 0.19 and 4.51 m^3 per year, and the whole population displaced some 304–601 m^3 of soil during the construction of 64 setts. Over time, the setts become stabilized and persist for decades or even centuries as unique landforms.

Modern African elephants (*Loxodonta africana africana*) cause several forms of erosion (Haynes

Table 4.7 Examples of ecosystem engineers

Type	Species	Ecosystem	Location
Bioconstructor	Japanese eelgrass (*Zostera japonica*)	Mudflats	Pacific Northwest, USA and Canada
	North American beaver (*Castor canadensis*)	Cool temperate forest	North America, parts of Europe, Patagonia
Bioprotector	Swamp she-oak (*Casuarina glauca*)	Coastal	Florida, USA
	Zebra mussels (*Dreissena polymorpha*) (also bioconstructor)	Lakes and rivers	Europe and North America
Bioeroder	Australian pine (*Casuarina equisetifolia*)	Coastal	Florida, USA
	European rabbit (*Oryctolagus cuniculus*)	Grassland	Europe

Source: Based on information in Fei *et al.* (2014)

2012). They create elephant trails that cut into the ground for many kilometres on migration routes and that may be used and re-used for centuries. When digging in search of water or mineral sediments, they can excavate several cubic metres of sediments; and when wallowing, they can stir up to a cubic metre of pond sediments at each visit.

Humans are bioeroders on a massive scale. One form of human-induced bioerosion is **bombturbation** – the impact of explosive munitions upon the landscape (Hupy and Koehler 2012). Bombturbation is a disruptive force, capable of moving large bodies of sediments, and denuding landscapes to the point where small- and medium-scale changes to topography have long-term geomorphic repercussions.

A recent development in the study of ecosystem engineers is the two-way feedback between geomorphic-engineer species and the geomorphic process and landforms (Steiger and Corenblit 2012). Geomorphic-engineer species modify the dynamics of geomorphic processes and landforms, the changes in which may then feed back to populations of animals and plants and encourage natural selection of particular biological traits and adjustments in the structure and function of the ecosystem. An example comes from fluvial environments, where plants need to endure constant flows of sediment and water. Plants have evolved biological traits that help them to survive in these conditions. The traits include resistance to breakage and uprooting through flexible stems and leaves and through brittle twig bases that enable living stems to break free, as well as resilience to mechanical destruction or sediment burial through resprouting from both roots and damaged shoots and from buoyant seeds. By favouring vegetation growth at river margins, these resistance and resilience strategies help to increase the cohesiveness of the riverbank and help to retain fine sediment and nutrients within stable islands and floodplains.

An interesting suggestion is that life has a 'topographic signature' (Dietrich and Perron 2006). The argument runs that over short timescales, biotic processes mediate chemical reactions, disrupt the ground surface, expand soil, and strengthen soil by weaving a network of roots, which changes affect weathering, soil formation and erosion, slope stability, and river dynamics. Over geological timescales, biotic effects are less patent but no less significant. Animals and plants help to shape climate, and in turn, climate dictates the mechanisms and rates of erosion that constrain topographic evolution (Dietrich and Perron 2006).

SUMMARY

Three grand cycles of matter affect Earth surface processes – the water cycle (evaporation, condensation, precipitation, and runoff), the rock cycle (uplift, weathering, erosion, deposition, and lithification), and the biogeochemical cycles. Denudation encompasses weathering and erosion. Erosive agents – ice, water, and wind – pick up weathered debris, transport it, and deposit it. Transport requires forces to set material in motion and keep it moving. The chief forces that act upon geomorphic materials are gravitational forces, fluid forces, water pressure forces, expansion forces, global fluid movements, and biological forces. Eroded materials eventually come to rest. Deposition occurs in several ways to produce different classes of sediment: clastic (solid fragments), chemical (precipitated materials), or biogenic (produced by living things). Sediments accumulate in three main environments: the land surface (terrestrial sediments); around continental edges (shallow marine sediments); and on the open ocean floor (deep marine sediments). Climate partly determines denudation (weathering and erosion). In addition, geological and topographic factors affect mechanical denudation. Climate, rock type, topographic factors, and organisms influence chemical denudation. Climate, topography, and plate tectonic processes interact in complex ways. Uplift changes climates, climatic changes may increase erosion; erosion may affect the flow of crustal rocks and so influence uplift. Erosion of mountains may affect the carbon dioxide balance of the atmosphere and promote climatic change. Humans are potent geomorphic agents, currently moving more material than natural processes and making unmistakable geomorphic footprints on the land surface. Mining and construction, agricultural practices and land-use, and dam building have significant impacts upon sediment fluxes. Recent work suggests that all life, not just humans, is a potent geomorphic agent and this finding has spawned the subject of biogeomorphology and ecogeomorphology.

ESSAY QUESTIONS

1 To what extent are the Earth's grand 'cycles' interconnected?

2 Assess the relative importance of the factors that influence denudation rates.

3 Is the recognition of the Anthropocene as a new geological epoch justified?

4 How significant are humans as geomorphic agents?

FURTHER READING

Berner, E. K. and Berner, R. A. (2012) *Global Environment: Water, Air, and Geochemical Cycles*, 2nd edn. Princeton, N.J.: Princeton University Press.
An excellent introduction.

Ruddiman, W. F. (ed.) (1997) *Tectonic Uplift and Climatic Change*. New York: Plenum Press.
A detailed account of the connections between tectonics, weathering, and climate.

Westbroek, P. (1991) *Life as a Geological Force: Dynamics of the Earth*. New York: W.W. Norton.
A leisurely and winning introduction to geological and biogeochemical cycles from the perspective of Earth history.

PART II

ENDOGENIC PROCESSES

CHAPTER FIVE

PLATE TECTONICS AND ASSOCIATED STRUCTURAL LANDFORMS

5

Deep-seated geological processes and structures stamp their mark on many large landforms. This chapter looks at:

- Plate tectonic, diastrophic, and volcanic and plutonic processes
- How tectonic plates bear characteristic large-scale landforms at their active and passive margins and in their interiors
- The connections between tectonic geomorphology and large-scale landforms

SPLITTING A CONTINENT

On 14 September 2005, a 4.7-magnitude earthquake in Dabbahu, 400 km north-east of Addis Ababa, Ethiopia, was followed by moderate tremors. Between 14 September and 4 October 2005, 163 earthquakes greater than magnitude 3.9 and a small volcanic eruption (on 26 September) occurred along the 60-km Dabbahu rift segment in the Afar Depression (Figure 5.1). This volcano-seismic event marked a sudden sundering of the African and Arabian tectonic plates (Wright *et al.* 2006). It created an 8-m rift in just three weeks (Plate 5.1), a thin column of which filled up with magma forming a dyke between 2 and 9 km deep, with 2.5 km^3 of magma injection. The sudden rifting added to the long-term split that is currently tearing the north-east of Ethiopia and Eritrea from the rest of Africa and could eventually create a huge new sea. The earth movements of September 2005 are a small step in the creation of a new whole ocean that will take millions of years to complete. However, this event is unparalleled in geological investigation and it has given geologists a rare opportunity to monitor the rupture process first-hand.

continued . . .

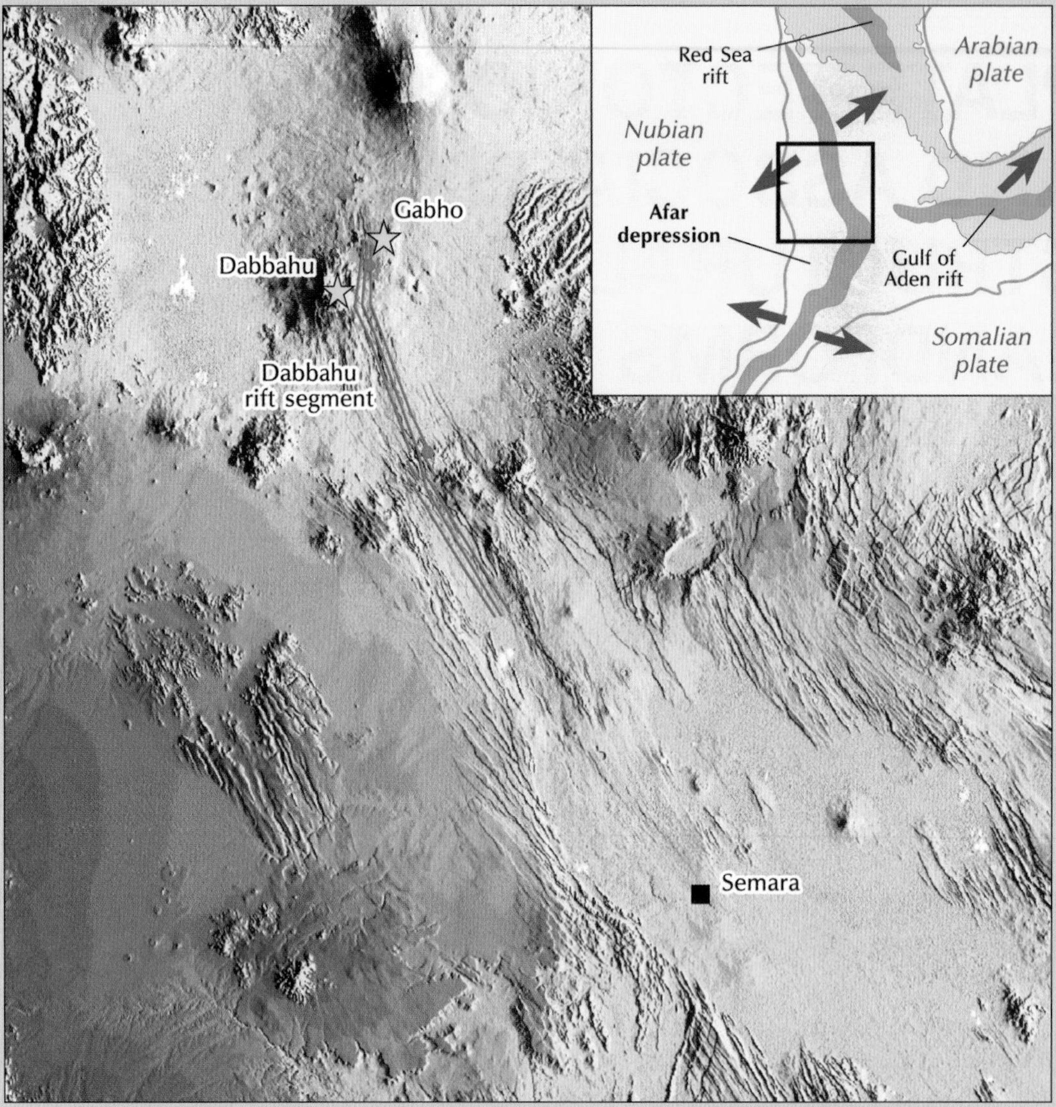

Figure 5.1 Topographic relief of the 60-km-long Dabbahu rift segment within the Afar Depression. The inset shows directions of plate divergence between the stable African (Nubian), Arabian, and Somalian plates. *Source*: Adapted from Cynthia Ebinger, Royal Holloway, University of London (Wright *et al.* 2006).

Plate 5.1 The explosive volcanic vent that opened on 26 September 2005, Dabbahu, Ethiopia, after two days of nearly continuous seismic activity. (a) The 500-m-long, 60-m-wide vent looking north. To the right lies a 200-m-wide, 4-km-long zone of open fissures and normal faults that may mark the subsurface location of the dyke. (*Photograph by Elizabeth Baker, Royal Holloway, University of London*). (b) View to the south from the north end of the vent. Notice the tunnel at the southern end. Notice also the layers of ash that built up over a period of days around the vent. The rhyolitic rocks in the foreground were blown out of the vent. (*Photograph by Julie Rowland, University of Auckland*)

TECTONICS AND LANDFORM

The ascent of internal energy originating in the Earth's core impels a complicated set of geological processes. Deep-seated processes and structures in the lithosphere (the relatively rigid and cold top 50–200 km of the solid Earth), and ultimately processes in the core and mantle, influence the shape and dynamics of the toposphere (the totality of the Earth's topography). The primary surface features of the globe are in very large measure the product of geological processes and, in particular, tectonic processes. Tectonics (from the Greek *tekton*, meaning builder or mason) involves the structures in the lithosphere, and notably with the geological forces and movements that act to create these structures. This primary tectonic influence on the toposphere expresses itself in the structure of mountain chains, volcanoes, island arcs, and other large-scale structures exposed at the Earth's surface, as well as in smaller features such as fault scarps.

Endogenic landforms may be tectonic or structural in origin (Twidale 1971, 1). **Tectonic landforms** are productions of the Earth's interior processes without the intervention of the forces of denudation. They include volcanic cones and craters, fault scarps, and mountain ranges. The influence of tectonic processes on landforms, particularly at continental and large regional scales, is the subject matter of **morphotectonics**. **Tectonic geomorphology** investigates the effects of active tectonic processes – faulting, tilting, folding, uplift, and subsidence – upon landforms. A recent development in geomorphology is the idea of '**tectonic predesign**'. Several landscape features, patently of exogenic origin, have tectonic or endogenic features stamped on them (or, literally speaking, stamped under them). Tectonic predesign arises from the tendency of erosion and other exogenic processes to follow stress patterns in the lithosphere (Hantke and Scheidegger 1999; Scheidegger 2004). The resulting landscape features are not fashioned directly by the stress fields. Rather, the exogenic processes act preferentially in conformity with the lithospheric stress (see p. 239). The conformity is either with the direction of a shear or, where there is a free surface, in the direction of a principal stress.

Few landforms are purely tectonic in origin: exogenous forces – weathering, gravity, running water, glaciers, waves, or wind – act on tectonic landforms, picking out less resistant rocks or lines of weakness, to produce **structural landforms**. An example is a volcanic plug, created when one part of a volcano is weathered and eroded more than another. A breached anticline is another example. Most textbooks on geomorphology abound with examples of structural landforms. Even in the Scottish Highlands, many present landscape features, which resulted from Tertiary etching, are closely adjusted to underlying rock types and structures (Hall 1991). Such passive influences of geological structures upon landforms are called **structural geomorphology**.

PLATE TECTONICS AND VOLCANISM

The outer shell of the solid Earth – the **lithosphere** –is not a single, unbroken shell of rock; it is a set of snugly tailored **plates** (Figure 5.2). At present there are seven large plates, all with an area over 100 million km^2. They are the African, North American, South American, Antarctic, Australian–Indian, Eurasian, and Pacific plates. Two dozen or so smaller plates have areas in the range 1–10 million km^2. They include the Nazca, Cocos, Philippine, Caribbean, Arabian, Somali, Juan de Fuca, Caroline, Bismarck, and Scotia plates, and a host of **microplates** or **platelets**. In places, as along the western edge of the American continents, continental margins coincide with plate boundaries and are **active margins**. Where continental margins lie inside plates, they are **passive margins**. The breakup of Pangaea created many passive margins, including the east coast of South America and the west coast of Africa. Passive margins are sometimes designated rifted margins where plate motion has been divergent, and sheared margins where plate motion has been

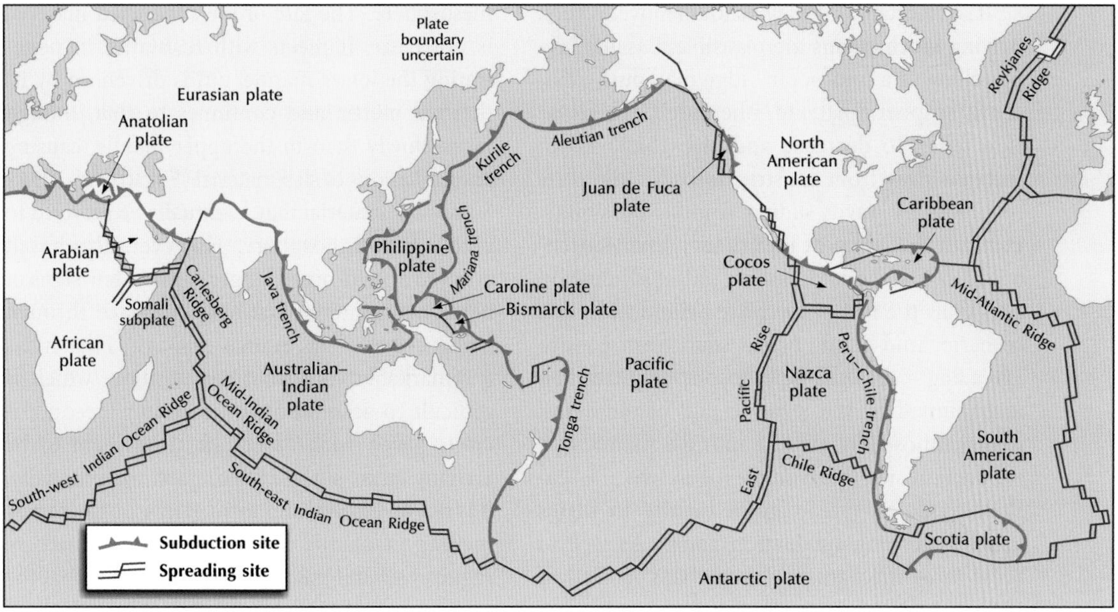

Figure 5.2 Tectonic plates, spreading sites, and subduction sites. *Source*: Adapted from Ollier (1996)

transformed, that is, where adjacent crustal blocks have moved in opposite directions. The distinction between active and passive margins is crucial to interpreting some large-scale features of the toposphere.

Earth's tectonic plates are continuously created at mid-ocean ridges and destroyed at subduction sites, and are ever on the move. Their motions explain virtually all tectonic forces that affect the lithosphere and thus the Earth's surface. Indeed, plate tectonics provides a good explanation for the primary topographic features of the Earth: the division between continents and oceans, the disposition of mountain ranges, and the placement of sedimentary basins at plate boundaries.

Plate tectonic processes

The plate tectonic model currently explains changes in the Earth's crust. This model is thought satisfactorily to explain geological structures, the distribution and variation of igneous and metamorphic activity, and sedimentary facies. In fact, it explains all major aspects of the Earth's long-term tectonic evolution (e.g. Kearey and Vine 1990). The plate tectonic model comprises two tectonic 'styles'. The first involves the oceanic plates and the second involves the continental landmasses.

Oceanic plate tectonics

The **oceanic plates** connect to the cooling and recycling system comprising the mesosphere, asthenosphere, and lithosphere beneath the ocean floors. The chief cooling mechanism is subduction. Volcanic eruptions along mid-ocean ridges produce new oceanic lithosphere. The newly formed material moves away from the ridges. In doing so, it cools, contracts, and thickens. Eventually, the oceanic lithosphere becomes denser than the underlying mantle and sinks. The sinking takes place along **subduction zones**. These are associated with earthquakes and volcanicity. Cold oceanic slabs may sink well into the mesosphere, perhaps as much as 670 km or below the surface. Indeed, subducted material may accumulate to form 'lithospheric graveyards' (Engebretson *et al.* 1992).

It is uncertain why plates should move. Several **driving mechanisms** are plausible. Basaltic lava upwelling at a mid-ocean ridge may push adjacent lithospheric plates to either side. Or, as elevation tends to decrease and slab thickness to increase away from construction sites, the plate may move by gravity sliding. Another possibility, currently thought to be the primary driving mechanism, is that the cold, sinking slab at subduction sites pulls the rest of the plate behind it. In this scenario, mid-ocean ridges stem from passive spreading – the oceanic lithosphere is stretched and thinned by the tectonic pull of older and denser lithosphere sinking into the mantle at a subduction site; this would explain why the seafloor tends to spread more rapidly in plates attached to long subduction zones. As well as these three mechanisms, or perhaps instead of them, mantle convection may be the number one motive force, though this now seems unlikely, as many spreading sites do not sit over upwelling mantle convection cells. If the mantle-convection model were correct, mid-ocean ridges should display a consistent pattern of gravity anomalies, which they do not, and would probably not develop fractures (transform faults). But, although convection is perhaps not the master driver of plate motions, it does occur. There is some disagreement about the depth of the convective cell. It could be confined to the asthenosphere, the upper mantle, or the entire mantle (upper and lower). Whole mantle convection (Davies 1977, 1992) has gained much support, although it now seems that whole mantle convection and a shallower circulation may both operate.

The lithosphere may be regarded as the cool surface layer of the **Earth's convective system** (Park 1988, 5). As part of a convective system, it cannot be considered in isolation (Figure 5.3). It gains material from the asthenosphere, itself fed by uprising material from the underlying mesosphere, at constructive plate boundaries. It migrates laterally from mid-ocean ridge axes as cool, relatively rigid, rock. Then, at destructive plate boundaries, it loses material to the asthenosphere and mesosphere. The fate of the subducted material is not clear. It meets with resistance in penetrating the lower mantle, but is driven on by its thermal inertia and continues to sink, though more slowly than in the upper mantle, causing accumulations of slab material (Fukao *et al.* 1994). Some slab material may eventually be recycled to create new lithosphere. However, the basalt erupted at mid-ocean ridges shows a few signs of being new material that has not passed through a rock cycle before (Francis 1993, 49). First, it has a remarkably consistent composition, which is difficult to account for by recycling. Second, it emits gases, such as helium, that seem to be arriving at the surface for the first time. Equally, it is not 'primitive' and formed in a single step by melting of mantle materials – its manufacture requires several stages. It is worth noting that the transformation of rock from mesosphere, through the asthenosphere, to the lithosphere chiefly entails temperature and viscosity (rheidity) changes. Material changes do occur: partial melting in the asthenosphere generates magmas that rise into the lithosphere, and volatiles enter and leave the system.

Continental plate tectonics

The **continental lithosphere** does not take part in the mantle-convection process. It is 150 km thick and consists of buoyant low-density crust (the tectosphere) and a relatively buoyant upper mantle. It therefore floats on the underlying asthenosphere. Continents break up and reassemble, but they remain floating at the surface. They move in response to lateral mantle movements, gliding serenely over the Earth's surface. In breaking up, small fragments of continent sometimes shear off; these are called **terranes**. They drift around until they meet another continent, to which they become attached (rather than being subducted) or possibly are sheared along it. As they may come from a different continent from the one they are attached to, they are called **exotic** or **suspect terranes** (p. 128). Most of the western seaboard of North America appears to

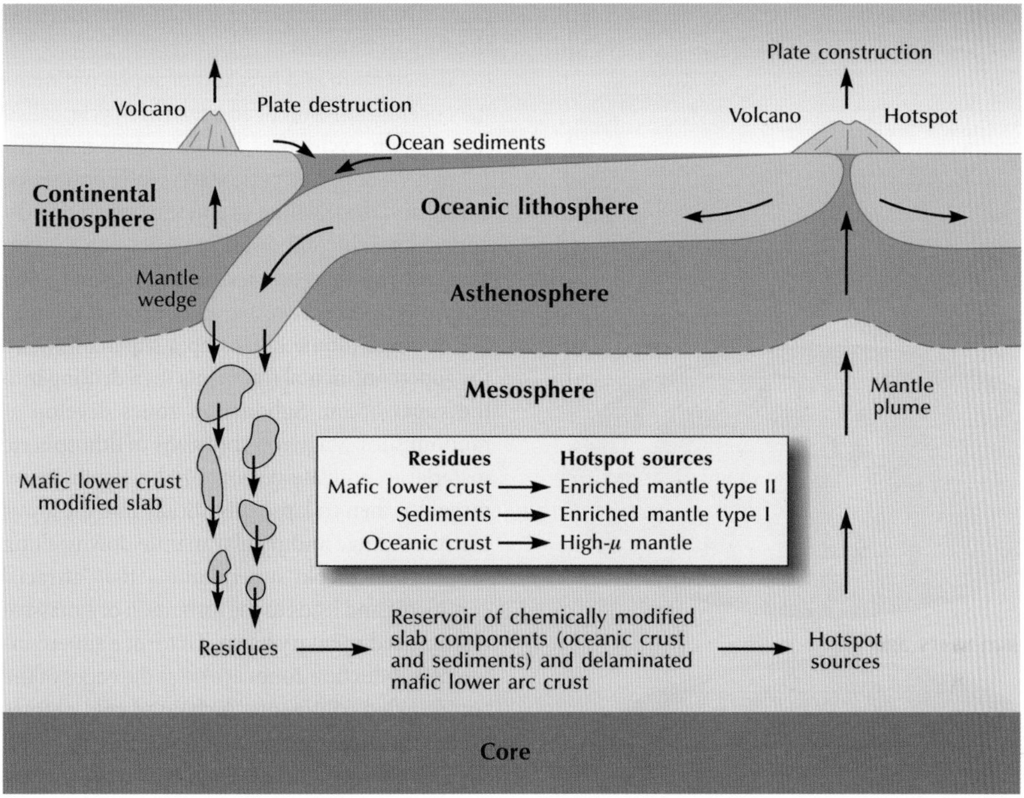

Figure 5.3 Interactions between the asthenosphere, lithosphere, and mesosphere. The oceanic lithosphere gains material from the mesosphere (via the asthenosphere) at constructive plate boundaries and hotspots and loses material to the mesosphere at destructive plate boundaries. Subduction feeds slab material (oceanic sediments derived from the denudation of continents and oceanic crust), mantle lithosphere, and mantle wedge materials to the deep mantle. These materials undergo chemical alteration and accumulate in the deep mantle until mantle plumes bear them to the surface, where they form new oceanic lithosphere. *Source*: Adapted from Tatsumi (2005)

consist of these exotic terranes. In moving, continents have a tendency to drift away from mantle hot zones, some of which they may have produced: stationary continents insulate the underlying mantle, causing it to warm. This warming may eventually lead to a large continent breaking into several smaller ones. Most continents are now sitting on, or moving towards, cold parts of the mantle. An exception is Africa, which was the core of Pangaea. Continental drift leads to collisions between continental blocks and to the overriding of oceanic lithosphere by continental lithosphere along subduction zones.

Continents are affected by, and affect, underlying mantle and adjacent plates. They are maintained against erosion (rejuvenated in a sense) by the welding of sedimentary prisms to continental margins through metamorphism, by the stacking of thrust sheets, by the sweeping up of microcontinents and island arcs at their leading edges, and by the addition of magma through intrusions and extrusions (Condie 1989). Geologists have established the relative movement of continents over the Phanerozoic aeon with a high degree of confidence, although pre-Pangaean reconstructions are less reliable than post-Pangaean

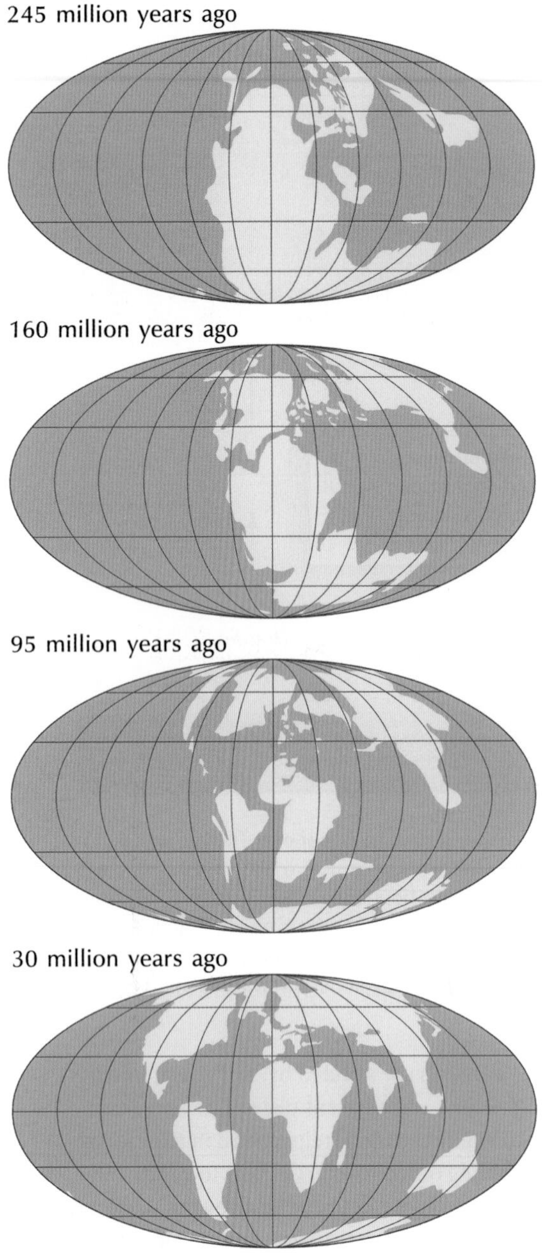

Figure 5.4 Changing arrangement of continents over the last 245 million years, showing the breakup of Pangaea, during the Early Triassic period; during the Callovian age (Middle Jurassic); during the Cenomanian age (Late Cretaceous); and during the Oligocene epoch. All maps use Mollweide's equal-area projection. *Source*: Adapted from maps in Smith *et al.* (1994)

reconstructions. Figure 5.4 charts the probable breakup of Pangaea.

The creation and breakup of supercontinents may occur as a result of an ocean life-cycle, called the **Wilson cycle** after the Canadian geologist J. Tuzo Wilson. The cycle starts with continental rifting and the opening of a new ocean and ends, approximately 800 million years later, with orogeny and then ocean closure (Wilson 1968). Plume tectonic processes may drive it (Figure 5.5). A superplume breaks up a supercontinent, the supercontinental fragments then drifting into the super-ocean. Subduction zones develop at random sites. Stagnant, cold slabs of lithospheric material accumulate at about 670 km depth. These megaliths then collapse episodically into the lower mantle. A huge and regular mantle downwelling may form – a cold superplume – that 'attracts' continents and leads to the formation of a massive cratonic sedimentary basin. To form a supercontinent, subduction zones evolve at the edges of the commingled continents. A chain of cold plumes girdles the supercontinent. The downwelling cold slabs (megaliths) squeeze out a superplume by thermally perturbing the outer core–lower mantle interface. This superplume then starts to destroy the supercontinent that created it and the cycle starts anew.

Diastrophic processes

Traditionally, tectonic (or geotectonic) forces divide into two groups: (1) diastrophic forces and (2) volcanic and plutonic forces. **Diastrophic forces** lead to the folding, faulting, uplift, and subsidence of the lithosphere. **Volcanic forces** lead to the extrusion of magma on to the Earth's surface as lava and to minor intrusions (e.g. dykes and sills) into other rocks. **Plutonic forces**, which originate deep in the Earth, produce major intrusions (plutons) and associated veins.

Diastrophic forces may deform the lithosphere through folding, faulting, uplift, and subsidence. They are responsible for some of the major features of the physical toposphere. Two categories of **diastrophism** are recognized: orogeny and epeirogeny, but these terms are a source of much confusion (Ollier and Pain 2000, 4–8). **Orogeny** literally means the genesis of mountains, and when first used it meant just that. Later, it became associated with the idea of folding, and eventually it came to mean the folding of rocks in fold belts. As mountain building is not associated with the folding of rocks, it cannot be synonymous with orogeny (Ollier 2003). **Epeirogeny** is the upheaval or depression of large areas of cratons without significant folding or fracture. The only folding associated with epeirogeny is the broadest of undulations. Epeirogeny includes **isostatic movements**, such as the rebound of land after an ice sheet has melted, and **cymatogeny**, which is the arching, and sometimes doming, of rocks with little deformation over 10–1,000 km. Some geomorphologists believe that mountains result from the erosion of areas uplifted epeirogenically (e.g. Ollier and Pain 2000, 8; Ollier 2003; see Huggett 2006, 29–30).

The relative motion of adjacent plates primarily creates the many tectonic forces in the lithosphere. Indeed, relative plate motions underlie almost all surface tectonic processes. Plate boundaries are particularly important for understanding geotectonics. They are sites of strain and associated with faulting, earthquakes, and, in some instances, mountain building (Figure 5.6). Most boundaries sit between two adjacent plates, but, in places, three plates come into contact. This happens where the North American, South American, and Eurasian plates meet (Figure 5.2). Such Y-shaped boundaries are **triple junctions**. Three plate-boundary types produce distinctive tectonic regimes:

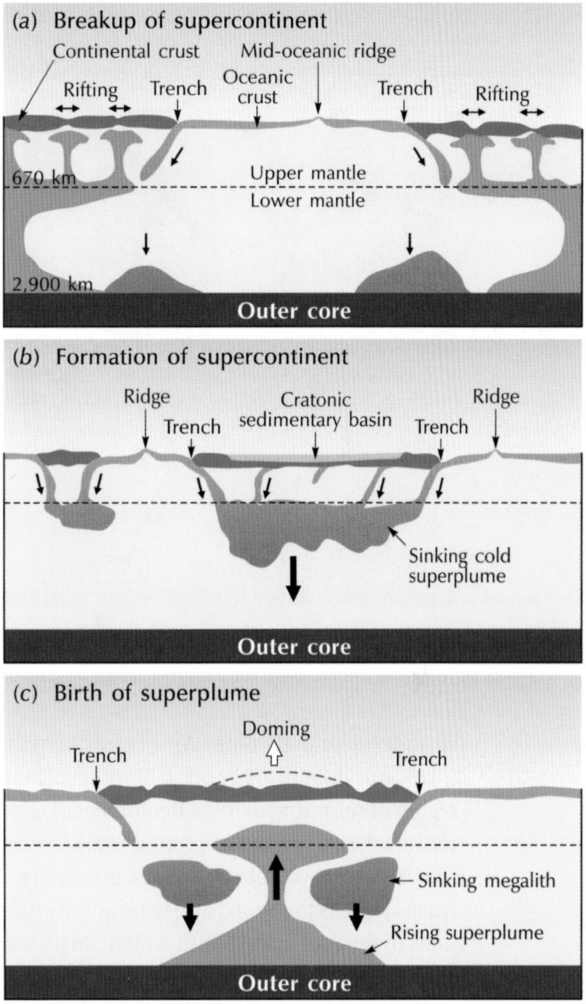

Figure 5.5 Schematic illustration of the Wilson cycle explained by plume tectonics. *Source*: Adapted from Maruyama (1994)

1 **Divergent plate boundaries** at construction sites, which lie along mid-ocean ridges, are associated with divergent tectonic regimes involving shallow, low-magnitude earthquakes. The ridge height depends primarily on the spreading rate. Incipient divergence occurs within continents, including Africa, and creates rift valleys, which are linear fault systems and, like mid-ocean ridges, are prone to shallow earthquakes and volcanism (p. 152). Volcanoes at divergent boundaries produce basalt.
2 **Convergent plate boundaries** vary according to the nature of the converging plates. Convergent tectonic regimes are equally varied; they normally lead to partial melting and the production of granite and the eruption

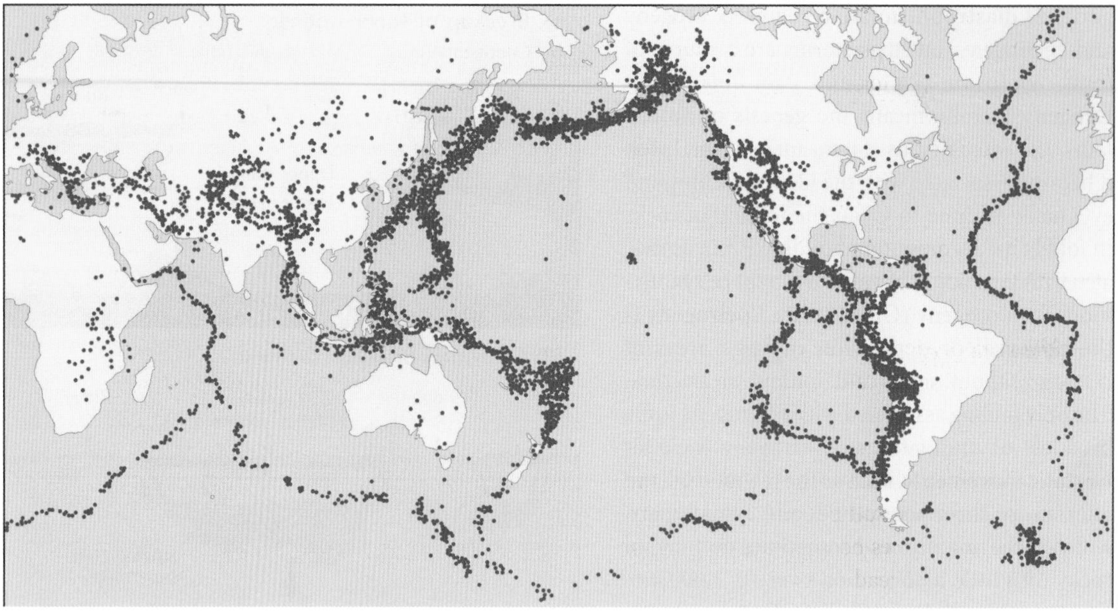

Figure 5.6 Global distribution of earthquakes. *Source*: Adapted from Ollier (1996)

of andesite and rhyolite. An oceanic trench, a volcanic island arc, and a dipping planar region of seismic activity (a Benioff zone) with earthquakes of varying magnitude mark a collision between two slabs of oceanic lithosphere. An example is the Scotia arc, lying at the junctions of the Scotia and South American plates. Subduction of oceanic lithosphere beneath continental lithosphere produces two chief features. First, it forms an oceanic trench, a dipping zone of seismic activity, and volcanicity in an **orogenic mountain belt** (or **orogen**) lying on the continental lithosphere next to the oceanic trench (as in western South America). Second, it creates **intra-oceanic arcs** of volcanic islands (as in parts of the western Pacific Ocean). In a few cases of continent–ocean collision, a slab of ocean floor has overridden rather than underridden the continent. This process, called **obduction**, has produced the Troödos Mountain region of Cyprus. Collisions of continental lithosphere result in crustal thickening and the production of a mountain belt, but little subduction. A fine example is the Himalaya, produced by India's colliding with Asia. Divergence and convergence may occur obliquely. Oblique divergence is normally accommodated by transform offsets along a mid-oceanic ridge crest, and oblique convergence by the complex microplate adjustments along plate boundaries. An example is found in the Betic cordillera, Spain, where the African and Iberian plates slipped by one another from the Jurassic to Tertiary periods.

3 **Conservative** or **transform plate boundaries** occur where adjoining plates move sideways past each other along a transform fault without any convergent or divergent motion. They are associated with strike-slip tectonic regimes and with shallow earthquakes of variable magnitude. They occur as fracture zones along mid-ocean ridges and as **strike-slip fault zones** within continental lithosphere. A prime example of the latter is the San Andreas fault system in California.

Tectonic activity also occurs within lithospheric plates, and not just at plate edges. This is called **within-plate tectonics** to distinguish it from plate-boundary tectonics.

Volcanic and plutonic processes

Volcanic forces are either intrusive or extrusive forces. **Intrusive forces** are found within the lithosphere and produce such features as batholiths, dykes, and sills. The deep-seated, major intrusions – batholiths and stocks – result from plutonic processes, while the minor, nearer-surface intrusions such as dykes and sills, which occur as independent bodies or as offshoots from plutonic intrusions, result from hypabyssal processes. **Extrusive forces** occur at the very top of the lithosphere and lead to exhalations, eruptions, and explosions of materials through volcanic vents, all of which are the result of volcanic processes.

The location of volcanoes

Most volcanoes sit at plate boundaries, along either mid-ocean ridges or subduction zones. A few, including the Cape Verde volcano group in the southern Atlantic Ocean and the Tibesti Mountains in Saharan Africa, occur within plates. These **'hot-spot' volcanoes** are surface expressions of thermal mantle plumes. Hot-spots are characterized by topographic bumps (typically 500–1,200 m high and 1,000–1,500 km wide), volcanoes, high gravity anomalies, and high heat flow. Commonly, a mantle plume stays in the same position while a plate slowly slips over it. In the ocean, this produces a chain of volcanic islands, or a **hot-spot trace**, as in the Hawaiian Islands. On continents, it produces a string of volcanoes. Such a volcanic string is found in the Snake River Plain province of North America, where a hot-spot currently sitting below Yellowstone National Park, Wyoming, has created an 80-km-wide band across 450 km of continental crust, producing prodigious quantities of basalt in the process. Even more voluminous are **continental flood basalts**. These occupy large tracts of land in far-flung places. The Siberian province covers more than 340,000 km^2. India's Deccan Traps once covered about 1,500,000 km^2; erosion has left about 500,000 km^2.

Mantle plumes

Mantle plumes appear to play a major role in plate tectonics. They may start growing at the core–mantle boundary, but the mechanisms by which they form and grow are undecided. They may involve rising plumes of liquid metal and light elements pumping latent heat outwards from the inner-core boundary by compositional convection, the outer core then supplying heat to the core–mantle boundary, whence giant silicate magma chambers pump it into the mantle, so providing a plume source. Mantle plumes may be hundreds of kilometres in diameter and rise towards the Earth's surface. A plume consists of a leading 'glob' of hot material that is followed by a 'stalk'. On approaching the lithosphere, the plume head is forced to mushroom beneath the lithosphere, spreading sideways and downwards a little. The plume temperature is 250–300°C hotter than the surrounding upper mantle, so that 10–20 per cent of the surrounding rock melts. This melted rock may then run on to the Earth's surface as flood basalt, as occurred in India during the Cretaceous period when the Deccan Traps formed.

Superplumes may develop. One appears to have done so beneath the Pacific Ocean during the middle of the Cretaceous period (Larson 1991). It rose rapidly from the core–mantle boundary about 125 million years ago. Production tailed off by 80 million years ago, but it did not stop until 50 million years later. It is possible that superplumes are caused by cold, subducted oceanic crust on both edges of a tectonic plate accumulating at the top of the lower mantle. These two cold pools of rock then sink to the hot layer just above the core, squeezing out a giant plume between them. **Plume tectonics** may be the dominant style of convection in the major part of the mantle. Two super-upwellings (the South Pacific and African superplumes) and one super-downwelling (the Asian cold plume) appear to prevail (Figure 5.7). A recent simulation showed that the southward propagation of a dominant topographic swell, which formed at least 30 million years ago in response to the African superplume, produced the current high topography of the East African Rift System (Moucha and Forte 2011). And there is evidence that the gravitational potential energy stored in the African highlands drives the extension of the EARS (Stamps *et al.* 2015).

It should be mentioned that a minority of geologists have always spoken out against plumes. However, since about the turn of the millennium the number of voices has swollen, and the validity of the plume model has emerged as a key debate in Earth science (see Foulger *et al.* 2005; Huggett 2006, 21–5).

LANDFORMS RELATED TO TECTONIC PLATES

Tectonic processes primarily determine large-scale landforms, though water, wind, and ice partly shape their detailed surface form. Geomorphologists classify large-scale landforms in many ways. One scheme rests on crustal types: continental shields, continental platforms, rift systems, and orogenic belts. It is convenient to discuss these large units under three headings – plate interiors, passive plate margins, and active plate margins.

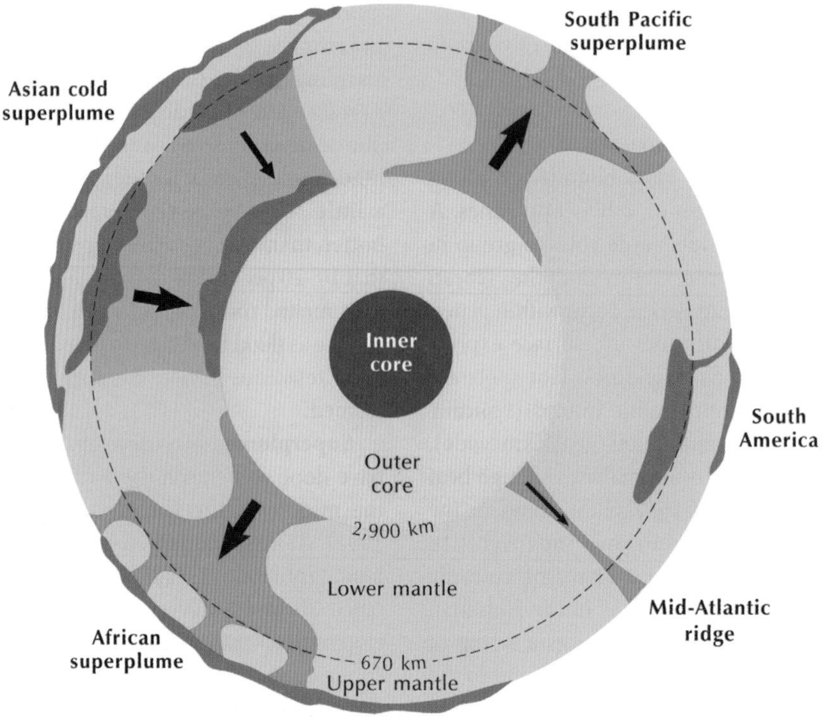

Figure 5.7 A possible grand circulation of Earth materials. Oceanic lithosphere, created at mid-ocean ridges, is subducted into the deeper mantle. It stagnates at around 670 km and accumulates for 100–400 million years. Eventually, gravitational collapse forms a cold downwelling on to the outer core, as in the Asian cold superplume, which leads to mantle upwelling occurring elsewhere, as in the South Pacific and African hot superplumes. *Source*: Adapted from Fukao *et al.* (1994)

Plate-interior landforms

Cratons are the broad, central parts of continents. They are somewhat stable continental shield areas with a basement of Precambrian rocks that are largely unaffected by orogenic forces but are subject to epeirogeny. The main large-scale landforms associated with these areas are basins, plateaux (upwarps and swells), rift valleys, and intracontinental volcanoes. Equally important landforms lie along passive continental margins, that is, margins of continents created when formerly single landmasses split in two, as happened to Africa and South America when the supercontinent Pangaea broke apart. **Intracratonic basins** may be 1,000 km or more across. Some, such as the Lake Eyre basin of Australia and the Chad and Kalahari basins of Africa, are enclosed and internally drained. Others, such as the region drained by the Congo river systems, are breached by one or more major rivers.

Some continents, and particularly Africa, possess extensive **plateaux** sitting well above the average height of continental platforms. The Ahaggar Plateau and Tibesti Plateau in North Africa are examples. These plateaux appear to have been uplifted without rifting occurring but with some volcanic activity.

Continental rifting occurs at sites where the continental crust is stretched and faulted. The rift valley running north to south along much of East Africa is probably the most famous example (p. 152), and its formation is linked with domal uplift. Volcanic activity is often associated with continental rifting. It is also associated with hot-spots.

Passive-margin landforms

Figure 5.8 shows the basic geomorphic features of **passive** or **Atlantic-type margins** with mountains (see Battiau-Queney 1991; Ollier 2004b). It seems likely that these features start as an old plain (palaeoplain) of a continental interior that breaks along a rift valley (Ollier and Pain 1997). The palaeoplain at the new continental edge, created by the rifting, experiences downwarping. Sea-floor spreading then favours the growth of a new ocean in which post-rift sediments accumulate as a wedge on the submerged palaeoplain to form a seawards-sloping basal unconformity. This is the **breakup unconformity** owing to its association with the fragmenting of a supercontinent (Ollier 2004b). Inland the palaeoplain survives as **plateaux**. Some plateaux may be depositional but most are erosion surfaces formed of uplifted palaeoplains. In areas where the sedimentary strata form folds, the uplands are bevelled cuestas and accordant, level strike ridges. The plateaux may extend over large areas or they may have suffered dissection and survive as fragments on the hardest rocks. They often retain the ancient drainage lines. **Marginal swells** are widespread asymmetrical bulges along continental edges that fall directly into the sea with steeper (2°) slopes towards the coast. They develop after the formation of plateaux and major valleys. **Great escarpments** are highly distinctive landforms of many passive margins. They are extraordinary topographic features formed in a variety of rocks (folded sedimentary rocks, granites, basalts, and metamorphic rocks) and separate the high plateaux from coastal plains. The great escarpment in southern Africa in places stands more than 1,000 m high. Great escarpments often separate soft relief on inland plateaux from highly dissected relief beyond the escarpment foot. Not all passive margins bear great escarpments, but many do (Figure 5.9). A great escarpment has even been identified in Norway, where the valleys deeply incised into the escarpment, although modified by glaciers, are still recognizable (Lidmar-Bergström *et al.* 2000). Some passive margins that lack great escarpments do possess low marginal upwarps flanked by a significant break of slope. The **Fall Line** on the eastern seaboard of North America marks an increase in stream gradient and in places forms a distinct escarpment. Below great escarpments, **rugged mountainous areas** form through the deep

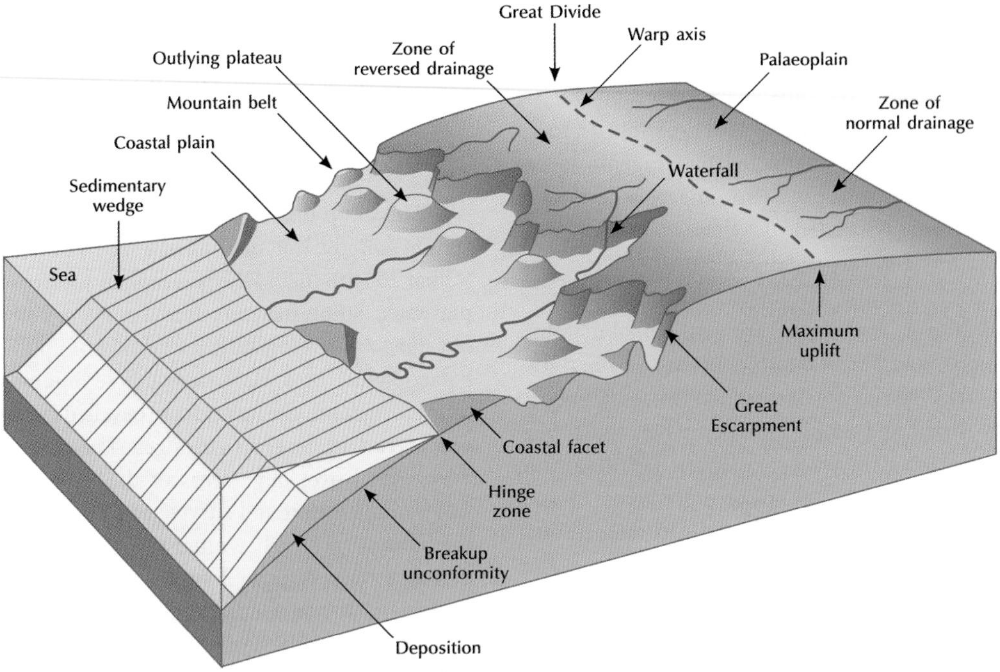

Figure 5.8 The chief morphotectonic features of a passive continent margin with mountains. *Source*: Adapted from Ollier and Pain (1997)

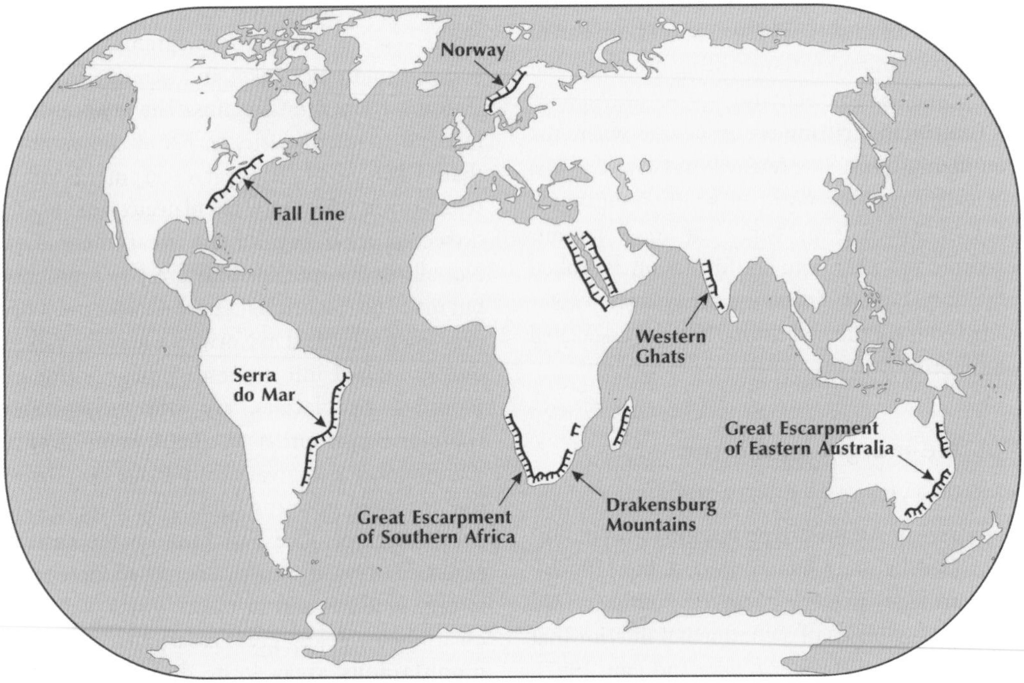

Figure 5.9 Great escarpments on passive margins. *Source*: Adapted from Summerfield (1991, 86)

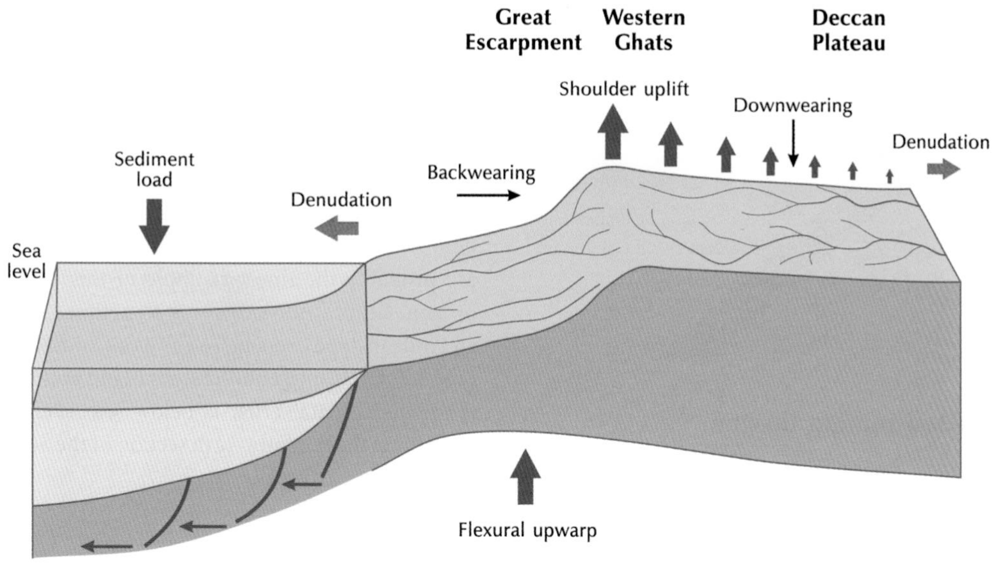

Figure 5.10 Conceptual model of passive margin denudation and shoulder uplift by flexural rebound based on the Western Ghats, India. *Source*: Adapted from Gunnell and Fleitout (2000)

dissection of old plateaux surfaces. Many of the world's large waterfalls lie where a river crosses a great escarpment, as in the Wollomombi Falls, Australia. **Lowland** or **coastal plains** lie seawards of great escarpments. They are largely the products of erosion. Offshore from the coastal plain is a wedge of sediments, at the base of which is an unconformity, sloping seawards.

Interesting questions about passive-margin landforms are starting to be answered. The Western Ghats, which fringe the west coast of peninsular India, are a great escarpment bordering the Deccan Plateau. The ridge crests stand 500–1,900 m tall and display a remarkable continuity for 1,500 km, despite structural variations. The continuity suggests a single, post-Cretaceous process of scarp recession and shoulder uplift (Gunnell and Fleitout 2000). A possible explanation involves denudation and backwearing of the margin, which promotes flexural upwarp and shoulder uplift (Figure 5.10). Shoulder uplift could also be effected by tectonic processes driven by forces inside the Earth.

Active-margin landforms

Where tectonic plates converge or slide past each other, the continental margins are said to be **active**. They may be called **Pacific-type margins** as they are common around the Pacific Ocean's rim.

The basic landforms connected with convergent margins are island arcs and orogens. Their specific form depends upon (1) what it is that is doing the converging – two continents, a continent and an island arc, or two island arcs; and (2) whether subduction of oceanic crust occurs or a collision occurs. **Subduction** is deemed to create steady-state margins in the sense that oceanic crust is subducted indefinitely while a continent or island arc resists subduction. **Collisions** are deemed to occur when the continents or island arcs crash into one other but tend to resist subduction.

Steady-state margins

Steady-state margins produce two major landforms – intra-oceanic island arcs and continental-margin orogens (Figure 5.11).

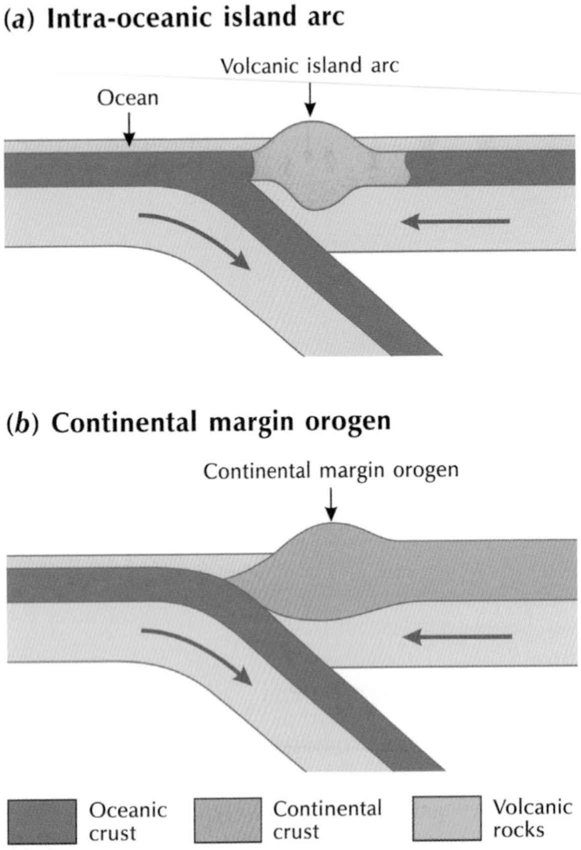

Figure 5.11 The two kinds of steady-state margins. (a) Intra-ocean island arc formed where an oceanic plate is subducted beneath another oceanic plate. These are common in the western Pacific Ocean. (b) A continental margin orogen formed where an oceanic plate is subducted beneath a continental 'plate'. An example is the Andes. *Source*: Adapted from Summerfield (1991, 57)

Intra-oceanic island arcs result from oceanic lithosphere being subducted beneath another oceanic plate. The heating of the plate that is subducted produces volcanoes and other thermal effects that build the island arc. Currently, about twenty intra-oceanic island arcs sit at subduction zones. Most of these lie in the western Pacific Ocean and include the Aleutian Arc, the Marianas Arc, the Celebes Arc, the Solomon Arc, and the Tonga Arc. The arcs build relief through the large-scale intrusion of igneous rocks and volcanic activity. A deep trench often forms ahead of the arc at the point where the oceanic lithosphere starts plunging into the mantle. The Marianas Trench, at 11,033 m the deepest known place on the Earth's surface, is an example.

Continental-margin orogens form when oceanic lithosphere is subducted beneath continental lithosphere. The Andes of South America are probably the finest example of this type of orogen. Indeed, the orogen is sometimes called an Andean-type orogen, as well as a Cordilleran-type orogen. Continental-margin island arcs form if the continental crust is below sea level. An example is the Sumatra–Java section of the Sunda Arc in the East Indies.

Collision margins

Landforms of collision margins vary according to the properties of the colliding plate boundaries. Four types of collision are possible: a continent colliding with another continent; an island arc colliding with a continent; a continent colliding with an island arc; and an island arc colliding with an island arc (Figure 5.12):

1. Continent–continent collisions create **inter-continental collision orogens**. A splendid example is the Himalaya. The collision of India with Asia produced an orogen running over 2,500 km.
2. Island arc–continent collisions occur where an island arc moves towards a subduction zone adjacent to a continent. The result is a **modified continental-margin orogen**.
3. Continent–island arc collisions occur when continents drift towards subduction zones connected with intra-oceanic island arcs. The continent resists significant subduction and a **modified passive continental margin** results. Northern New Guinea may be an example.
4. Island arc–island arc collisions are poorly understood because there are no present examples from which to work out the processes involved. However, the outcome would probably be a **compound intra-oceanic island arc**.

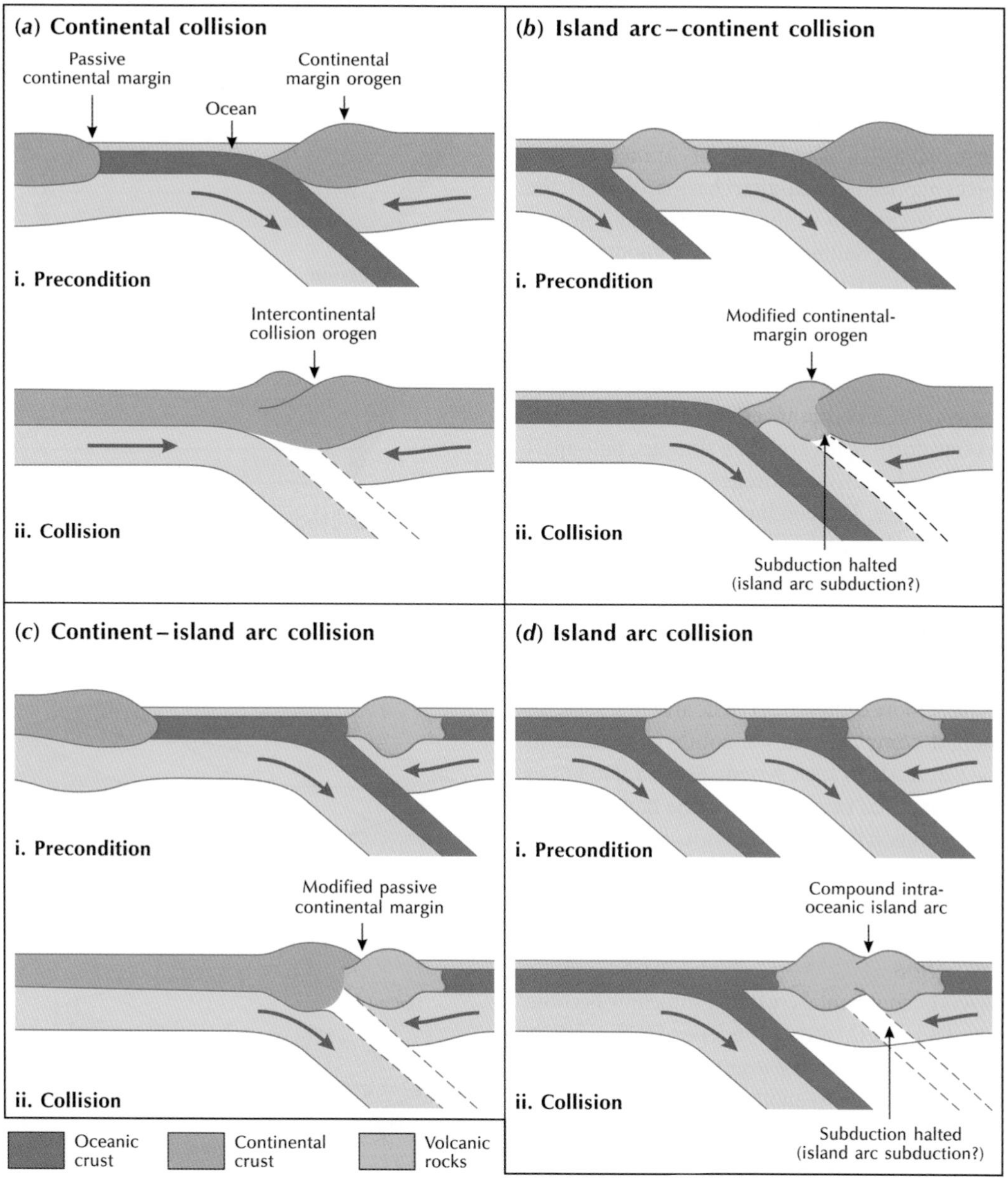

Figure 5.12 Four kinds of collisional margins. (a) Intercontinental collision orogen formed where two continental 'plates' collide. An example is the Himalaya. (b) Modified continental-margin orogen formed where an intra-oceanic island arc moves into a subduction zone bounded by continental crust. (c) Modified passive continental margin formed where a continent moves towards a subduction zone associated with an intra-oceanic island arc. (d) Compound intra-oceanic island arc formed by the collision of two intra-oceanic island arcs. *Source*: Adapted from Summerfield (1991, 59–60)

Transform margins

Rather than colliding, some plates slip by one another along **transform** or **oblique-slip faults**. Convergent and divergent forces occur at transform margins. Divergent or transtensional forces may lead to **pull-apart basins**, of which the Salton Sea trough in the southern San Andreas Fault system, California, USA, is a good example (Figure 5.13a). Convergent or transpressional forces may produce **transverse orogens**, of which the 3,000-m San Gabriel and San Bernardino Mountains (collectively called the Transverse Ranges) in California are examples (Figure 5.13b). As transform faults are often sinuous, pull-apart basins and transverse orogens may occur near to each other. The bending of originally straight faults also leads to spays and wedges of crust. Along **anastomosing faults**, movement may produce upthrust blocks and down-sagging ponds (Figure 5.14). A change in the dominant direction of stress may render all these transform margin features more complex. A classic area of transform margin complexity is the southern section of the San Andreas fault system. Some 1,000 km of movement has occurred along the fault over the last 25 million years. The individual faults branch, join, and sidestep each other, producing many areas of uplift and subsidence.

Terranes

Slivers of continental crust that somehow become detached and then travel independently of their parent body, sometimes over great distances, may eventually attach to another body of continental crust. Such wandering slivers go by several names: allochthonous terranes, displaced terranes, exotic terranes, native terranes, and suspect terranes. **Exotic** or **allochthonous terranes** originate from a continent different from that against which they now rest. **Suspect terranes** may be exotic, but their exoticism is not confirmable. **Native terranes** manifestly relate to the continental margin against which they presently sit. Over 70 per cent of the North American Cordillera is composed of displaced terranes, most of which

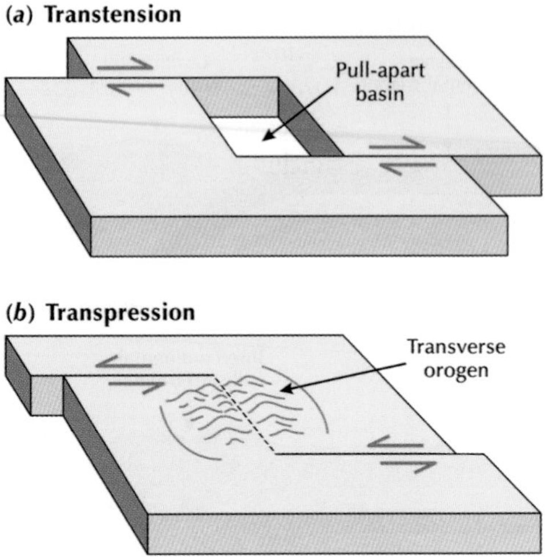

Figure 5.13 Landforms associated with oblique-slip faults. (a) Pull-apart basin formed by transtension. (b) Transverse orogen formed by transpression.

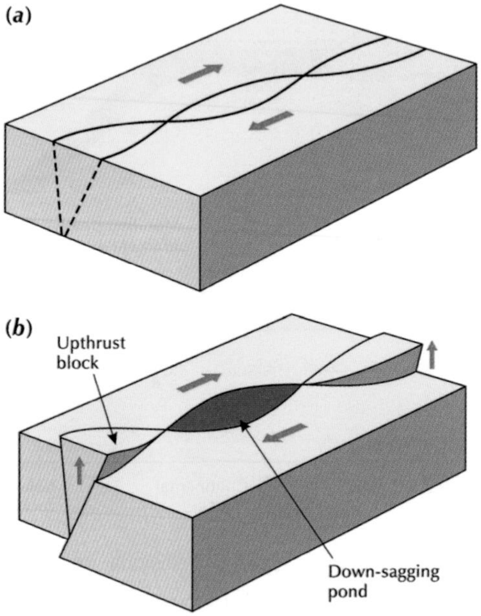

Figure 5.14 Landforms produced by anastomosing faults. (a) Anastomosing faults before movement. (b) Anastomosing faults after movement with upthrust blocks and down-sagging ponds. *Source*: Adapted from Kingma (1958)

travelled thousands of kilometres and joined the margin of the North American craton during the Mesozoic and Cenozoic eras (Coney *et al.* 1980). Many displaced terranes also occur in the Alps and the Himalayas, including Adria and Sicily in Italy (Nur and Ben-Avraham 1982).

TECTONIC GEOMORPHOLOGY AND CONTINENTAL LANDFORMS

Important interactions between endogenic factors and exogenic processes produce macroscale and megascale landforms (Figure 1.1). Plate tectonics explains some major features of the Earth's topography. An example is the striking connection between mountain belts and processes of tectonic plate convergence. However, the nature of the relationship between mountain belts (orogens) and plate tectonics is far from clear, with several questions remaining unsettled (Summerfield 2007). What factors, for example, control the elevation of orogens? Why do the world's two highest orogens – the Himalaya–Tibetan Plateau and the Andes – include large plateaux with extensive areas of internal drainage? Does denudation shape mountain belts at the large scale, and are its effects more fundamental than the minor modification of landforms that are essentially a product of tectonic processes? Since the 1990s, researchers have addressed such questions as these by treating orogens, and landscapes more generally, as products of a coupled tectonic–climatic system with the potential for feedbacks between climatically influenced surface processes and crustal deformation (Beaumont *et al.* 2000; Pinter and Brandon 1997; Willett 1999).

The **elevation of orogens** appears crucially to depend upon the crustal strength of rocks. Where crustal convergence rates are high, surface uplift soon creates (in geological terms) an elevation of around 6 to 7 km that the crustal strength of rocks cannot sustain, although individual mountain peaks may stand higher where the strength of the surrounding crust supports them. However, in most mountain belts, the effects of denudation prevent elevations from attaining this upper ceiling. As tectonic uplift occurs and elevation increases, river gradients become steeper, so raising denudation rates. The growth of topography is also likely to increase precipitation (through the orographic effect) and therefore runoff, which will also tend to enhance denudation (Summerfield and Hulton 1994). In parts of such highly active mountain ranges as the Southern Alps of New Zealand, rivers actively incise and maintain, through frequent landslides, the adjacent valley-side slopes at their threshold angle of stability. In consequence, an increase in the tectonic uplift rate produces a speedy response in denudation rate as river channels cut down and trigger landslides on adjacent slopes (Montgomery and Brandon 2002). Where changes in tectonic uplift rate are (geologically speaking) rapidly matched by adjustments in denudation rates, orogens seem to maintain a roughly **steady-state topography** (Summerfield 2007). The actual steady-state elevation is a function of climatic and lithological factors, higher overall elevations being attained where rocks are resistant and where dry climates produce little runoff. Such orogens never achieve a perfect steady state because there is always a delay in the response of topography to changing controlling variables such as climate, and especially to changing tectonic uplift rates because the resulting fall in base level must be propagated along drainage systems to the axis of the range. Work with simulation models suggests that variations in denudation rates across orogens appear to affect patterns of crustal deformation (Beaumont *et al.* 2000; Willett 1999).

For relatively simple orogens, the prevailing direction of rain-bearing winds seems significant. On the windward side of the orogen, higher runoff generated by higher precipitation totals leads to higher rates of denudation than on the drier, leeward side. As a result, crustal rocks rise more rapidly on the windward flank than on the leeward flank, so creating a patent asymmetry in depths of denudation across the orogen and producing a characteristic pattern of crustal deformation. Such modelling studies indicate that a reversal of prevailing rain-bearing winds will produce a change in topography, spatial patterns of denudation, and the form of crustal deformation (Summerfield 2007). In addition, they show that the topographic and deformational evolution of orogens results from a complex interplay between tectonic processes and geomorphic processes driven by climate.

SUMMARY

Geological processes and geological structures stamp their marks on, or in many cases under, landforms of all sizes. Plate tectonic processes dictate the gross landforms of the Earth – continents, oceans, mountain ranges, large plateaux, and so on – and many smaller land forms. Diastrophic forces fold, fault, lift up, and cast down rocks. Orogeny is a diastrophic process that builds mountains. Epeirogeny is a diastrophic process that upheaves or depresses large areas of continental cores without causing much folding or faulting. The boundaries of tectonic plates are crucial to understanding many large-scale landforms: divergent boundaries, convergent boundaries, and transform boundaries are associated with characteristic topographic features. Incipient divergent boundaries may produce rift valleys. Mature divergent boundaries on continents are associated with passive margins and great escarpments. Convergent boundaries produce volcanic arcs, oceanic trenches, and mountain belts (orogens). Transform boundaries produce fracture zones with accompanying strike-slip faults and other features. Plate tectonic processes exert an important influence upon such continental-scale landforms as mountain belts, but there is an important interplay between uplift, climate, and denudation.

ESSAY QUESTIONS

1. Discuss the idea of 'tectonic predesign'.
2. Explain the landforms associated with active margins.
3. Explain the landforms associated with passive margins.
4. Examine the factors that determine the major relief features of the Earth's surface.

FURTHER READING

Burbank, D. W. and Anderson, R. S. (2001) *Tectonic Geomorphology: A Frontier in Earth Science.* Malden, Mass.: Blackwell Science.
A detailed and insightful discussion of one of geomorphology's latest developments, but not easy for trainee geomorphologists.

Godard, A., Lagasquie, J.-J., and Lageat, Y. (2001) *Basement Regions.* Translated by Yanni Gunnell. Heidelberg: Springer.
An insight into modern French geomorphology.

Huggett, R. J. (1997b) *Environmental Change: The Evolving Ecosphere.* London: Routledge.
You may find some of the material in here of use. I did!

Summerfield, M. A. (ed.) (2000) *Geomorphology and Global Tectonics,* pp. 321–38. Chichester: John Wiley & Sons.
Not easy for the beginner, but a dip into this volume will reward the student with an enticing peep at one of geomorphology's fast-growing fields.

CHAPTER SIX

VOLCANOES, IMPACT CRATERS, FOLDS, AND FAULTS

6

The folding, faulting, and jointing of rocks creates many large and small landforms. This chapter looks at:

- How molten rocks (magma) produce volcanic landforms and landforms related to deep-seated (plutonic) processes; and how impact craters form
- How the folding of rocks produces scarps and vales and drainage patterns
- How faults and joints in rocks act as sites of weathering and produce large features such as rift valleys

GEOLOGICAL FORCES IN ACTION: THE BIRTH OF SURTSEY

On 8 November 1963, episodic volcanic eruptions began to occur 33 km south of the Icelandic mainland and 20 km south-west of the island of Heimaey (Moore 1985; Thorarinsson 1964). To begin with, the eruptions were explosive as water and magma mixed. They produced dark clouds of ash and steam that shot to a few hundred metres, and on occasions 10 km, in the air above the growing island. Base surges and fall-out of glassy tephra from the volcano built a tuff ring. On 31 January 1964, a new vent appeared 400 m to the north-west. The new vent produced a new tuff ring that protected the old vent from seawater. This encouraged the eruptions at the old vent to settle down into a gentle effusion of pillow lava and ejections of lava fountains. The lava, an alkali-olivine basalt, built up the island to the south and protected the unconsolidated tephra from wave action. After 17 May 1965, Surtsey was quiet until 19 August 1966, when activity started afresh at new vents at the older tuff ring on the east side of the island and fresh lava moved southwards. The eruptions stopped on 5 June 1967. They had lasted three-and-a-half years. Thus was the island of Surtsey created from about a cubic kilometre of ash and lava, of which only 9 per cent breached the ocean's surface. The island was named after the giant of fire in Icelandic mythology. Surtsey is about 1.5 km in diameter with an area of 2.8 km^2. Between 1967 and 1991, Surtsey subsided about 1.1 m (Moore *et al.* 1992), probably because the volcanic material compacted, the sea-floor sediments under the volcano compacted, and possibly because the lithosphere was pushed downwards by the weight of the volcano. Today, the highest point on Surtsey is 174 m above sea level.

VOLCANIC AND PLUTONIC LANDFORMS

Magma may be extruded on to the Earth's surface or intruded into country rock, which is an existing rock into which a new rock is introduced or found. Lava extruded from volcanic vents may build landforms directly. On the other hand, lava could be buried beneath sediments, be re-exposed by erosion at a later date, and then influence landform development. Intruded rocks, which must be mobile but not necessarily molten, may have a direct effect on landforms by causing doming of the surface, but otherwise they do not create landforms until they are exposed by erosion.

Volcanic forms depend very much on the composition of the lava, which affects among other things its viscosity. Basic lavas are the least viscous and flow the most readily, sometimes forming long thin flows guided by local topography. They produce effusive volcanic eruptions. Acid lavas (silica rich) are the most viscous, never flow freely, and produce explosive eruptions. In ocean basins, the magma supplying volcanoes is dominantly basaltic in composition. At divergent plate boundaries, the thin crust encourages the partial melting of magma, enabling volcanoes to form along mid-ocean ridges. In places, such as Iceland, these volcanoes rise above sea level. In the Pacific Basin, the '**andesite line**' separates a large central area of basaltic volcanoes, which includes the Hawaiian Islands, from a circum-Pacific fringe of dominantly andesite magmas, which, containing more silica than basalt, normally erupt with greater violence as explosive eruptions. The most silica-rich magmas, with compositions similar to granite, give rise to volcanoes of rhyolite or ignimbrite that also erupt explosively. The more acidic magmas are associated with subduction sites where water released from rocks in subducting plates lowers the melting temperature of the overlying mantle, producing viscous magma that rises to the surface. Examples are the volcanoes of the Andes and western North America. Hotspots, which do not necessarily occur under plate boundaries, sit atop mantle plumes and can produce pipes that vent magma to the surface, as in the Yellowstone caldera and the Hawaiian Islands, USA.

Intrusions

Intrusions form where molten and mobile igneous rocks cool and solidify without breaching the ground surface to form a volcano. They are said to be active when they force a space in rocks for themselves, and passive when they fill already existing spaces in rocks.

Batholiths and lopoliths

The larger intrusions – batholiths, lopoliths, and stocks – are roughly circular or oval in plan and have a surface exposure of over 100 km^2 (Figure 6.1). They tend to be deep-seated and are usually composed of coarse-grained plutonic rocks.

Batholiths, also called **bosses** or **plutons** (Figure 6.1a), are often granitic in composition. The granite rises to the surface over millions of years through diapirs, that is, hot plumes of rock

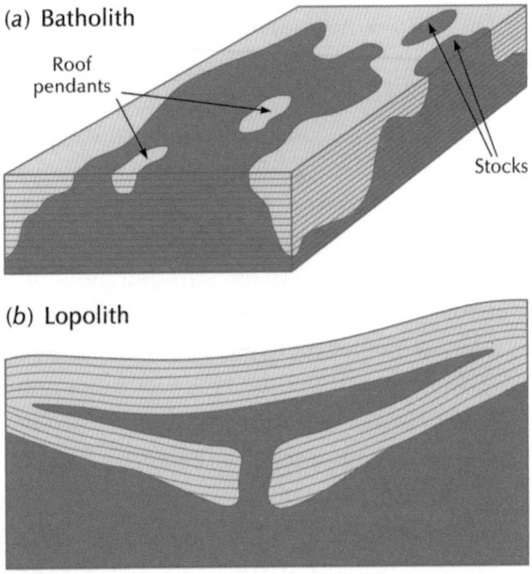

Figure 6.1 Major intrusions. (a) Batholith with stocks and roof pendants. (b) Lopolith. *Source*: Adapted from Sparks (1971, 68, 90)

ascending through cooler and denser country rock. Enormous granite batholiths often underlie and support the most elevated sections of continental margin orogens, as in the Andes. Mount Kinabalu on the island of Borneo, which at 4,101 m is the highest mountain in South-East Asia, was formed 1.5 million years ago by the intru-sion of an adamellite (granitic) pluton into the surrounding Tertiary sediments. Batholiths may cause a doming of sediments and the ground surface. This has occurred in the Wicklow Mountains, Ireland, where the Leinster granite has led to the doming of the overlying Lower Palaeozoic strata. Once erosion exposes granite batholiths, weathering penetrates the joints. The joint pattern consists initially of three sets of more or less orthogonal joints, but unloading effects pressure release in the top 100 m or so of the batholith and a secondary set of joints appears lying approximately parallel to the surface. These joints play a key role in the development of weathering landforms and drainage patterns (p. 174).

The upwards pushing of a granite pluton may produce **active gneiss domes** (Ollier and Pain 1981). These landforms occur in Papua New Guinea (e.g. Dayman dome and Goodenough dome), with ancient examples from the USA (e.g. Okanogon dome, Washington State), and many of the world's orogens. They stand 2,000–3,000 m high and are tens of kilometres across. Their formation seems to involve the metamorphosing of sediments to gneiss; the formation of granite, which starts to rise as a pluton; the arching of the gneiss by the rising pluton to form a dome of foliated gneiss; and the eruption of the dome at the ground surface, shouldering aside the bounding rocks.

Lopoliths are vast, saucer-shaped, and layered intrusions of basic rocks, typically of a gabbro-type composition (Figure 6.1b). In Tasmania, dolerite magma intruded flat Permian and Triassic sediments, lifting them as a roof. In the process, the dolerite formed several very large and shallow saucers, each cradling a raft of sediments. Lopoliths are seldom as large as batholiths. Their erosion produces a series of inward-facing scarps. The type example is the Duluth gabbro, which runs from the south-western corner of Lake Superior, Minnesota, USA, for 120 miles to the north-east, and has an estimated volume of 200,000 km^3. In South Africa, the Precambrian Bushveld Complex, originally interpreted as one huge lopolith, is a cluster of lopoliths.

Stocks or **plugs** are the largest intrusive bodies of basic rocks. They are discordant and are the solidified remains of magma chambers. One stock in Hawaii is about 20 km long and 12 km wide at the surface and is 1 km deep.

Dykes, sills, laccoliths, and other minor intrusions

Smaller intrusions exist alongside the larger forms and extrusive volcanic features (Figure 6.2a). They are classed as concordant where they run along the bedding planes of pre-existing strata, or as discordant where they cut through the bedding planes. Their form depends upon the configuration of the fractures and lines of weakness in the country rock and upon the viscosity of the intruding magma. If exposed by erosion, small intrusions can produce landforms, especially when they are composed of rock that is harder than the surrounding rock.

Dykes are discordant intrusions, characteristically 1 to 10 m wide, and commonly composed of dolerite (Figure 6.2a). They often occur in swarms. Along the coast of Arran, Scotland, a swarm of 525 dykes occurs along a 24-km section, the average dyke thickness being 3.5 m. When exposed, they form linear features that may cut across the grain of the relief. The Great Dyke of Zimbabwe is over 500 km long and averages 6–8 km wide. On occasions, dykes radiate out from a central supply point to form **cone sheets** (Figure 6.2b). **Necks** and **pipes** are the cylindrical feeders of volcanoes and appear to occur in a zone close to the ground surface. They are more common in acid igneous rocks than in basalts. They may represent the last stage of what was mainly a dyke eruption. Six dykes radiate from Ship Rock

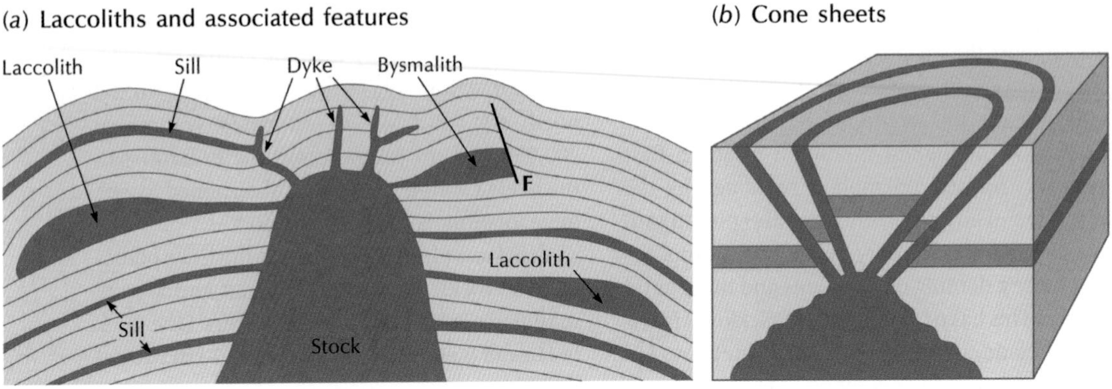

Figure 6.2 Minor intrusions. (a) Laccoliths and associated features (dykes, sills, and bysmaliths). (b) Cone sheets. *Source*: Adapted from Sparks (1971, 90, 101)

volcanic neck in New Mexico, USA (Plate 6.1). They were probably only 750 to 1,000 m below the land surface at the time of their formation.

Sills are concordant intrusions and frequently form resistant, tabular bands within sedimentary beds, although they may cross beds to spread along other bedding planes (Figure 6.2a). They may be hundreds of metres thick, as they are in Tasmania, but are normally between 10 and 30 m. Sills composed of basic rocks often have a limited extent, but they may extend for thousands of square kilometres. Dolerite sills in the Karoo sediments of South Africa underlie an area over 500,000 km^2 and constitute 15–25 per cent of the rock column in the area. In general, sills form harder members of strata into which they intrude. When eroded, they may form escarpments or ledges in plateau regions and encourage waterfalls where they cut across river courses. In addition, their jointing may add a distinctive feature to the relief, as in the quartz-dolerite Whin Sill of northern England, which was intruded into Carboniferous sediments. Inland, the Whin Sill causes waterfalls on some streams and in places is a prominent topographic feature, as where Hadrian's Wall sits upon it (Plate 6.2). Near the north Northumberland coast, it forms small scarps and crags, some of which are used as the sites of castles, for instance Lindisfarne Castle and Dunstanburgh Castle. It also affects the coastal scenery at Bamburgh and the Farne Islands. The Farne Islands are tilted slabs of Whin Sill dolerite.

Plate 6.1 Ship Rock (in the mid-distance) and one of its dykes (in the foreground), New Mexico, USA. Ship Rock is an exhumed volcanic neck from which radiate six dykes. (*Photograph by Tony Waltham Geophotos*)

Laccoliths are sills that have thickened to produce domes (Figure 6.2a). The doming arches the overlying rocks. **Bysmaliths** are laccoliths that have been faulted (Figure 6.2a). The Henry Mountains, Utah, USA, are a famous set of predominantly diorite-porphyry laccoliths and associated features that appear to spread out from central discordant stocks into mainly Mesozoic shales and sandstones. The uplift connected with the intrusion of the stocks and laccoliths has produced several peaks lying about 1,500 m above the level of the Colorado Plateau. Eroded bysmaliths and laccoliths may produce relief features. Traprain Law, a prominent hill, is a phonolite laccolith lying 32 km east of Edinburgh in Scotland. However, the adjacent trachyte laccolith at Pencraig Wood has little topographic expression.

Phacoliths are lens-shaped masses seated in anticlinal crests and synclinal troughs (Figure 6.3a). They extend along the direction of anticlinal and synclinal axes. Unlike laccoliths, which tend to be circular in plan, they are elongated. Eroded phacoliths may produce relief features. Corndon Hill, which lies east of Montgomery in Powys, Wales, is a circular phacolith made of Ordovician dolerite (Figure 6.3b).

Plate 6.2 Whin Sill, a dolerite intrusion in Northumberland, England, with Hadrian's Wall running along the top. (*Photograph by Tony Waltham Geophotos*)

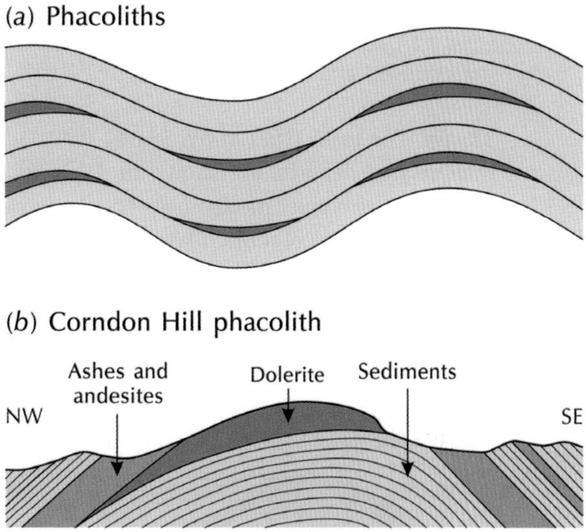

Figure 6.3 Phacoliths. (a) Occurrence in anticlinal crests and synclinal troughs. (b) Corndon Hill, near Montgomery in Wales, an eroded phacolith. *Source*: Adapted from Sparks (1971, 93, 94)

Volcanoes

Volcanoes erupt lava on to the land surface explosively and effusively. They also exhale gases. The landforms built by eruptions depend primarily upon whether rock is blown out or poured out of the volcano, and, for effusive volcanoes, upon the viscosity of the lava. Explosive or pyroclastic volcanoes blow **pyroclastic rocks** (solid fragments, loosely termed **ash** and **pumice**) out of a vent, while effusive volcanoes pour out **lava**.

Runny (low viscosity) lava spreads out over a large area, while sticky (high viscosity) lava oozes out and spreads very little. Mixed-eruption volcanoes combine explosive phases with phases of lava production. Pyroclastic rocks that fall to the ground from eruption clouds are called **tephra** (from the Greek for ashes), while both lavas and pyroclastic rocks that have a fragmented, cindery texture are called **scoria** (from the Greek for refuse).

Pyroclastic volcanoes

Explosive or **pyroclastic volcanoes** produce fragments of lava that accumulate around the volcanic vent to produce scoria mounds and other topographic forms (Figure 6.4; Plate 6.3). Pyroclastic flows and the deposits they produce are varied. **Tephra** is a term covering three types of pyroclastic material of differing grain size. Ashes are particles less than 4 mm in diameter, lapilli (from the Italian for 'little stones') are between 4 and 32 mm in diameter, and blocks are larger than

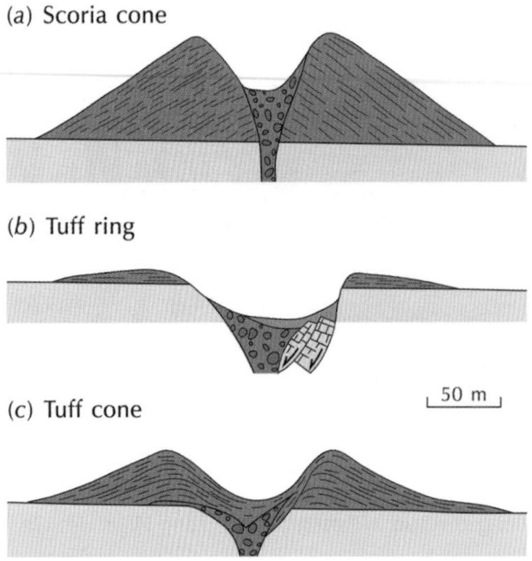

Figure 6.4 Pyroclastic volcanoes. (a) Scoria cone. (b) Tuff ring. (c) Tuff cone. *Source*: Adapted from Wohletz and Sheridan (1983)

32 mm. The main types of pyroclastic flow and their related deposits are shown in Table 6.1. Notice that two chief mechanisms trigger pyroclastic flows: (1) column collapse and (2) lava flow and dome collapse. The first of these involves the catastrophic collapse of convecting columns of erupted material that stream upwards into the atmosphere from volcanic vents. The second involves the explosive or gravitational collapse of lava flows or domes. **Pumice** contains the most vesicles (empty spaces) and blocks the least.

Plate 6.3 Cinder cone, Mono Craters, California, USA. Cinder cones are the simplest of volcanoes. They are built of cinder that falls around a vent to form a circular or oval cone, no more than 300 m or so high, usually with a bowl-shaped crater sitting at the top. (*Photograph by Kate Holden*)

Table 6.1 Pyroclastic flows and deposits

Pyroclastic flow	Pyroclastic deposit
Column collapse	
Pumice flow	Ignimbrite; pumice and ash deposit
Scoria flow	Scoria and ash deposit
Semi-vesicular andesite flow	Semi-vesicular andesite and ash deposit
Lava flow and dome collapse (explosive and gravitational)	
Block and ash flow; *nuée ardente*	Block and ash deposit

Source: Adapted from Wright *et al.* (1980)

Ignimbrites (derived from two Latin words to mean 'fire cloud rock') are deposits of pumice, which may cover large areas in volcanic regions around the world. The pumiceous pyroclastic flows that produce them may run uphill, so that ignimbrite deposits often surmount topography and fill valleys and hills alike, although valleys often contain deposits tens of metres thick known as valley pond ignimbrite, while hills bear an ignimbrite veneer up to 5 m thick. A ***nuée ardente*** is a pyroclastic flow or 'glowing avalanche' of volcanic blocks and ash derived from dense rock.

Scoria cones are mounds of scoria, seldom more than 200–300 m high, with a crater in the middle (Figure 6.4a). Young scoria scone have slopes of 33°, which is the angle of rest for loose scoria. Monogenetic volcanoes – that is, volcanoes created by a solitary eruptive episode that may last hours or years – produce them under dry conditions (i.e. there is no interaction between the lava and water). They occur as elements of scoria cone fields or as parasitic vents on the flanks of larger volcanoes. Dozens sit on the flanks of Mount Etna, Sicily. Once the eruption ceases, solidification seals off the volcanic vent and the volcano never erupts again. Monte Nuovo, near Naples, is a scoria cone that grew 130 m in a few days in 1538; San Benedicto, off the Pacific coast of Mexico, grew 300 m in 1952–3. **Scoria mounds** are like scoria cones but bear no apparent crater. An example is the Anakies, Victoria, Australia. **Nested scoria cones** occur where one scoria cone grows within another.

Maars form in a similar way to scoria cones, but in this case involving the interaction between magma and a water-bearing stratum – an aquifer. The result of this combination is explosive. In the simplest case, an explosion occurs in the phreatic or groundwater zone and blasts upwards to the surface creating a large hole in the ground. Thirty craters about a kilometre across were formed in this way in the Eifel region of Germany. These craters are now filled by lakes known as maars, which gave their name to the landforms. Some maars are the surface expression of **diatremes**, that is, vertical pipes blasted through basement rocks and that contain rock fragments of all sorts and conditions. Diatremes are common in the Swabian Alps region of Germany, where more than 300 occur within an area of 1,600 km^2. Being some 15–20 million years old, the surface expression of these particular diatremes is subdued, but some form faint depressions.

Tuff rings are produced by near-surface subterranean explosions where magma and water mix, but instead of being holes in the ground they are surface accumulations of highly fragmented basaltic scoria (Figure 6.4b). A first-rate example is Cerro Xico, which lies just 15 km from the centre of Mexico City. It formed in the basin of shallow Lake Texcoco before the Spanish drained it in the sixteenth century. **Tuff cones** are smaller and steeper versions of tuff rings (Figure 6.4c). An example is El Caldera, which lies a few kilometres from Cerro Xico.

Mixed-eruption volcanoes

As their name suggests, a mixture of lava eruptions and scoria deposits produces **mixed-eruption volcanoes**. They are built of layers of lava and scoria and are sometimes known as **stratovolcanoes** (Figure 6.5). The simplest form of strato-volcano is a simple cone, which is a scoria cone that carries on erupting. The result is a single vent at the summit and a stunningly symmetrical cone, as seen on Mount Mayon in the Philippines and Mount Fuji, Japan. Lava flows often adorn the summit regions of simple cones. Composite cones have experienced a more complex evolutionary history, despite which they

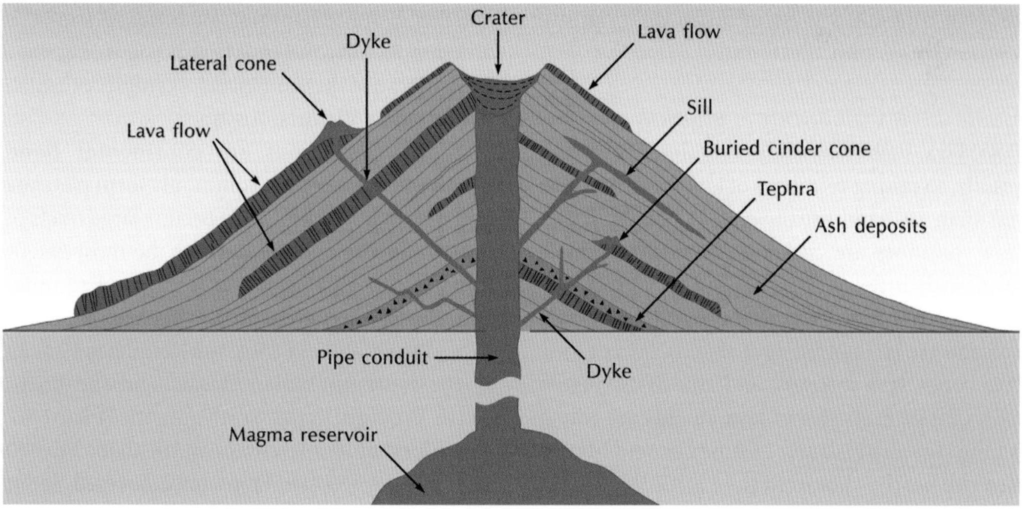

Figure 6.5 The structure of a typical stratovolcano. *Source*: Adapted from MacDonald (1972, 23)

retain a radial symmetry about a single locus of activity. In the history of Mount Vesuvius, Italy, for instance, a former cone (now Monte Somma) was demolished by the eruption of AD 79 and a younger cone grew in its place. Mount Etna is a huge composite volcano, standing 3,308 m high with several summit vents and innumerable parasitic monogenetic vents on its flanks.

Another level of complexity is found in **compound** or **multiple volcanoes**. Compound volcanoes consist, not of a single cone, but of a collection of cones intermixed with domes and craters covering large areas. Nevado Ojos del Salado, at 6,885 m the world's highest volcano, covers an area of around 70 km^2 on the frontier between Chile and Argentina, and consists of at least a dozen cones.

Volcano complexes are even more complex than compound volcanoes. They are so muddled that it is difficult to identify the source of the magma. In essence, they are associations of major and minor volcanic centres and their related lava flows and pyroclastic rocks. An example is Cordon Punta Negra, Chile, where at least twenty-five small cones with well-developed summit craters are present in an area of some 500 km^2. None of the cones is more than a few hundred metres tall and some of the older ones are almost buried beneath a jumbled mass of lavas, the origin of whose vents is difficult to trace.

Basic-lava volcanoes – shields

Basic lava, such as basalt, is very fluid. It spreads readily, so raising volcanoes of low gradient (often less than 10°) and usually convex profile. Basic-lava volcanoes are composed almost wholly of lava, with little or no addition of pyroclastic material or talus. Several types of basic-lava volcano are recognized: lava shields, lava domes, lava cones, lava mounds, and lava discs (Figure 6.6). Classic examples of **lava shields** are found on the Hawaiian Islands. Mauna Loa and Mauna Kea rise nearly 9 km from the Pacific floor. **Lava domes** are smaller than, and often occur on, lava shields. Individual peaks on Hawaii, such as

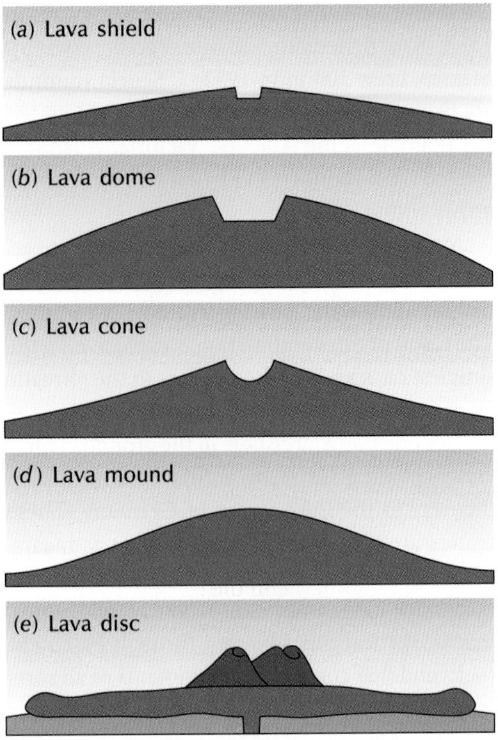

Figure 6.6 Types of basic-lava volcanoes, not drawn to scale. *Source*: Adapted from Ollier (1969, 21)

Mauna Kea, are lava domes. **Lava cones** are even smaller. Mount Hamilton, Victoria, Australia, is an example. **Lava mounds** bear no signs of craters. **Lava discs** are aberrant forms, examples of which are found in Victoria, Australia.

More voluminous are **continental flood basalts** or **traps**, which commonly form plateaux and mountain ranges. They occupy large tracts of land in far-flung places and are the most extensive terrestrial volcanic landforms. Examples include the Columbia–Snake River flood basalts in western mainland USA, the Kerguelen Plateau in the southern Indian Ocean, and the Brito-Arctic Province in the North Atlantic (Plate 6.4). The Siberian Traps covers more than 340,000 km^2. India's Deccan Traps once covered about 1,500,000 km^2; erosion has left about 500,000 million km^2.

Acid-lava volcanoes – lava domes

Acid lava, formed for instance of dacite or rhyolite or trachyte, is very viscous. It moves sluggishly and forms thick, steep-sided, dome-shaped extrusions. Volcanoes erupting acidic lava often explode, and even where extrusion takes place it is often accompanied by some explosive activity so that a low cone of ejecta surrounds the extrusions. Indeed, the extrusion commonly represents the last phase in an explosive eruptive cycle. Extrusions of acid lava take the form of various kinds of lava dome: cumulo-domes and tholoids, coulées, Peléean domes, and upheaved plugs (Figure 6.7).

Cumulo-domes are isolated low lava domes that resemble upturned bowls (Figure 6.7a). The Puy Grand Sarcoui in the Auvergne, France, the mamelons of Réunion, in the Indian Ocean, and the tortas ('cakes') of the central Andes are examples. A larger example is Lassen Peak, California,

Plate 6.4 Columnar basalt on Ulva, just west of Mull, with Ben More in the background. The basalt forms part of the Brito-Arctic Province (also called the North Atlantic Igneous Province). (*Photograph by Nick Scarle*)

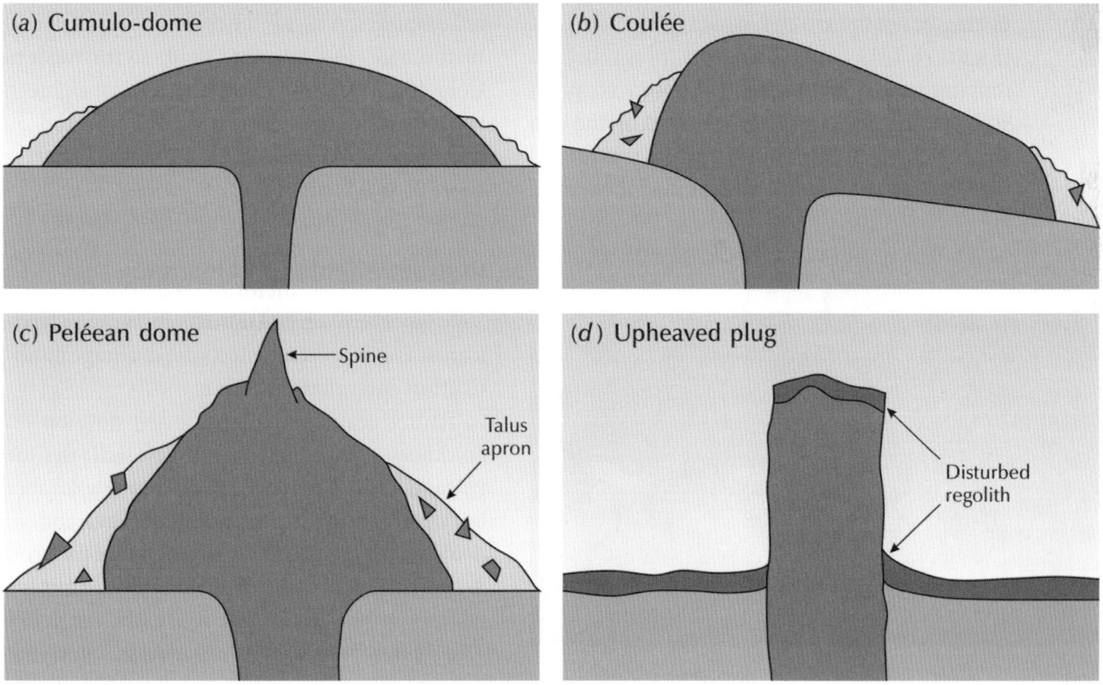

Figure 6.7 Lava domes. (a) Cumulo-dome. (b) Coulée. (c) Peléean dome. (d) Upheaved plug. Not drawn to scale.
Source: Adapted from Blake (1989)

which has a diameter of 2.5 km. **Tholoids**, although they sound like an alien race in a *Star Trek* episode, are cumulo-domes within large craters and derive their name from the Greek *tholos*, a 'domed building' (Plate 6.5). Their growth is often associated with *nuée ardente* eruptions, which wipe out towns unfortunate enough to lie in their path. A tholoid sits in the crater of Mount Egmont, New Zealand. **Coulées** are dome–lavaflow hybrids. They form where thick extrusions ooze on to steep slopes and flow downhill (Figure 6.7b). The Chao lava in northern Chile is a huge example with a lava volume of 24 km^3. **Peléean domes** (Figure 6.7c) are typified by Mont Pelée, Martinique, a lava dome that grew in the vent of the volcano after the catastrophic eruption that occurred on 8 May 1902, when a *nuée ardente* destroyed Saint Pierre. The dome is craggy, with lava spines on the top and a collar of debris around the sides. **Upheaved plugs**, also called **plug domes** or **pitons**, are produced by the most viscous of lavas. They look like a monolith poking out of the ground, which is what they are (Figure 6.7d). Some upheaved plugs bear a topping of country rock. Two upheaved plugs with country-rock cappings appeared on the Usu volcano, Japan, the first in 1910, which was named Meiji Sin-Zan or 'Roof Mountain', and the second in 1943, which was named Showa Sin-Zan or 'New Roof Mountain'.

Plate 6.5 Novarupta rhyolite tholoid formed in 1912 in the Katmai caldera, Katmai National Park, Alaska, USA. *(Photograph by Tony Waltham Geophotos)*

Calderas

Calderas are depressions in volcanic areas or over volcanic centres (Figure 6.8; Plate 6.6). They are productions of vast explosions or tectonic sinking, sometimes after an eruption (Figure 6.9). An enormous caldera formed in Yellowstone National Park, USA, some 600,000 years ago when some 1,000 km^3 of pyroclastic material was erupted leaving a depression some 70 km across. Another large caldera formed some 74,000 years ago in northern Sumatra following a massive volcanic eruption, the ash from which was deposited 2,000 km away in India. The Toba caldera is about 100 km long and 30 km wide and is now filled by Lake Toba. It is a **resurgent caldera**, which means that, after the initial subsidence amounting to about 2 km, the central floor has slowly risen again to produce Samosir Island. Large silicic calderas commonly occur in clusters or complexes. A case is the caldera complex found in the San Juan volcanic field, south-western Colorado, USA, which contains at least eighteen separate calderas between 22 million and 30 million years old. Ignimbrites from these calderas cover 25,000 km^2.

Indirect effects of volcanoes

Volcanoes have several indirect impacts on landforms. Two important effects are **drainage modification** and **relief inversion**.

Radial drainage patterns often develop on volcanoes, and the pattern may last well after the volcano has been eroded. In addition, volcanoes bury pre-existing landscapes under lava and, in doing so, may radically alter the drainage patterns. A good example is the diversion of the drainage in the central African rift valley (Figure 6.10). Five million years ago, volcanoes associated with the construction of the Virunga Mountains impounded Lake Kivu. Formerly, drainage was northward to join the Nile by way of Lake Albert

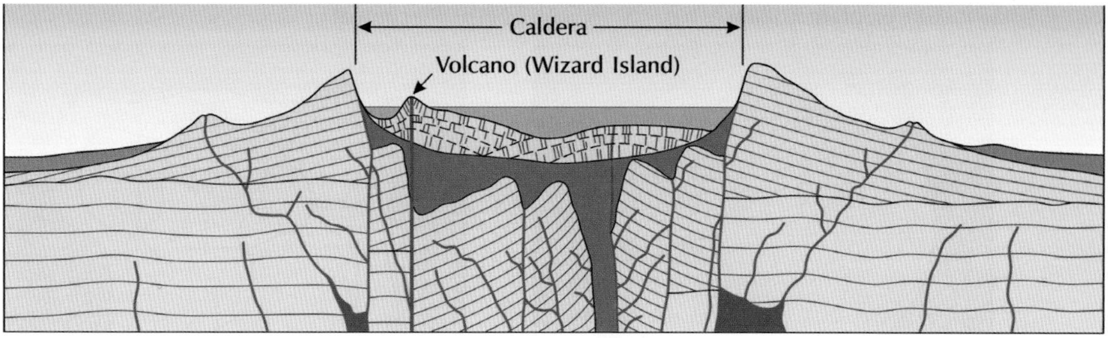

Figure 6.8 Crater Lake caldera, Oregon, USA; see Plate 6.6. *Source*: Adapted from MacDonald (1972, 301)

(Figure 6.10a). When stopped from flowing northwards by the Virunga Mountains, the waters eventually overflowed Lake Kivu and spilled southwards at the southern end of the rift through the Ruzizi River into Lake Tanganyika (Figure 6.10b). From Lake Tanganyika, the waters reached the River Congo through the River Lukuga, and so were diverted from the Mediterranean via the Nile to the Atlantic via the Congo (King 1942, 153–4).

Occasionally, lava flows set in train a sequence of events that ultimately inverts the relief – valleys become hills and hills become valleys (cf. p. 170). Lava tends to flow down established valleys. Erosion then reduces the adjacent hillside leaving the more resistant volcanic rock as a ridge between two valleys. Such inverted relief is remarkably common (Pain and Ollier 1995). On Eigg, a small Hebridean island in Scotland, a Tertiary rhyolite lava flow originally filled a river valley eroded into older basalt lavas. The rhyolite is now preserved on the Scuir of Eigg, an imposing 400-m-high and 5-km-long ridge standing well above the existing valleys.

IMPACT CRATERS

The remains of craters formed by the impact of asteroids, meteoroids, and comets scar the Earth's surface (Plate 6.7). Nearly 200 craters and geological structures discovered so far show strong signs of an impact origin (see Huggett 2006). Admittedly, impact craters are relatively rare landforms, but they are of interest.

In terms of morphology, terrestrial impact structures are either simple or complex (Figure 6.11). **Simple structures**, such as Brent crater in Ontario, Canada, are bowl-shaped (Figure 6.11a). The rim area is uplifted and, in the most recent cases, is surmounted by an overturned flap of near-surface target rocks with inverted stratigraphy. Fallout ejecta commonly lie on the overturned flap. Autochthonous target rock that is fractured and brecciated marks the base of a simple crater. A lens of shocked and unshocked allochthonous target rock partially fills the true crater. Craters with diameters larger than about

Plate 6.6 Caldera, Crater Lake, Oregon, USA, showing Wizard Island, a small cone. (*Photograph by Marli Miller*)

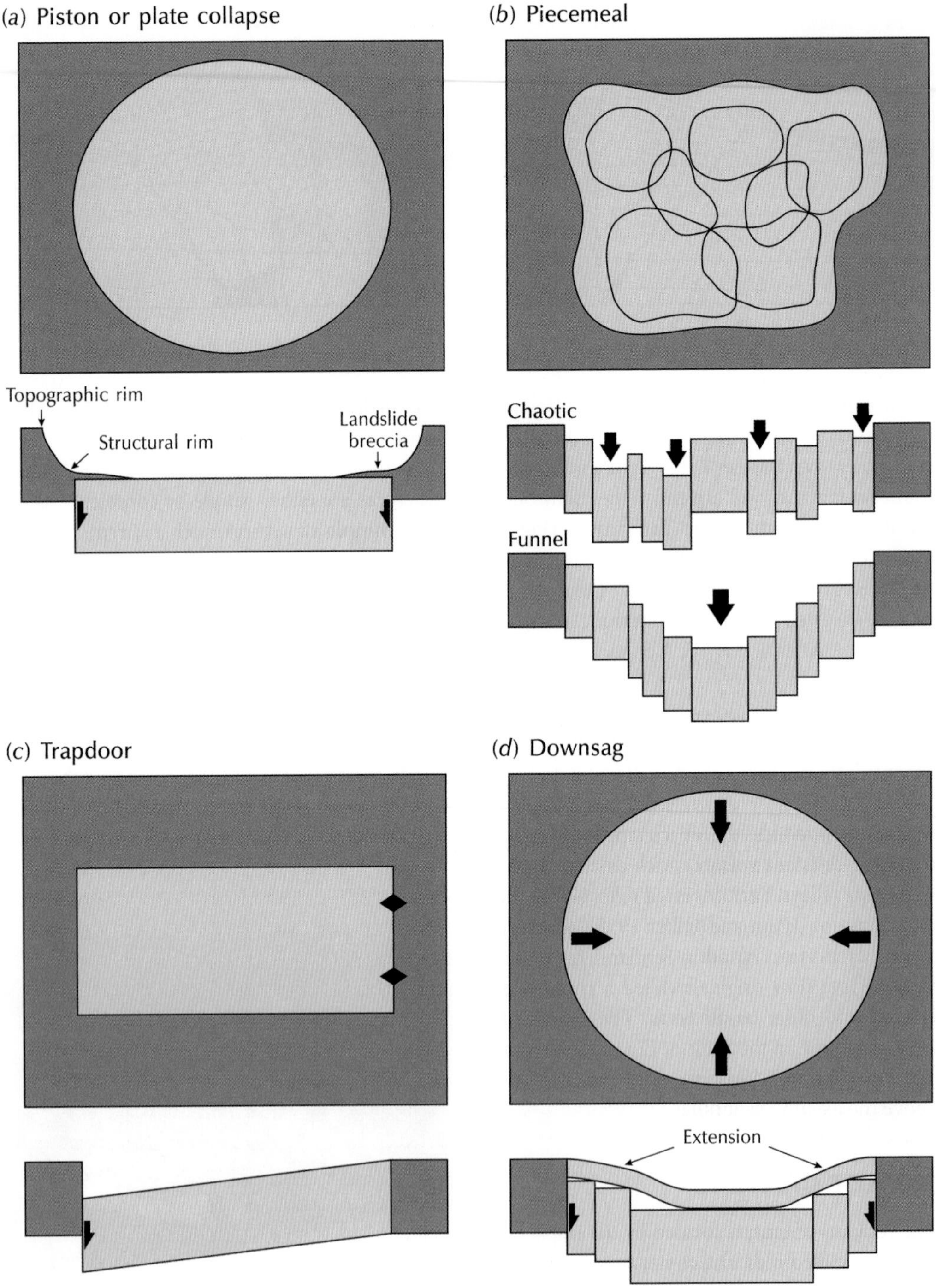

Figure 6.9 Four end-member mechanisms of caldera collapse. (a) Piston or plate collapse. (b) Piecemeal. (c) Trapdoor. (d) Downsag. *Source*: After Cole *et al.* (2005)

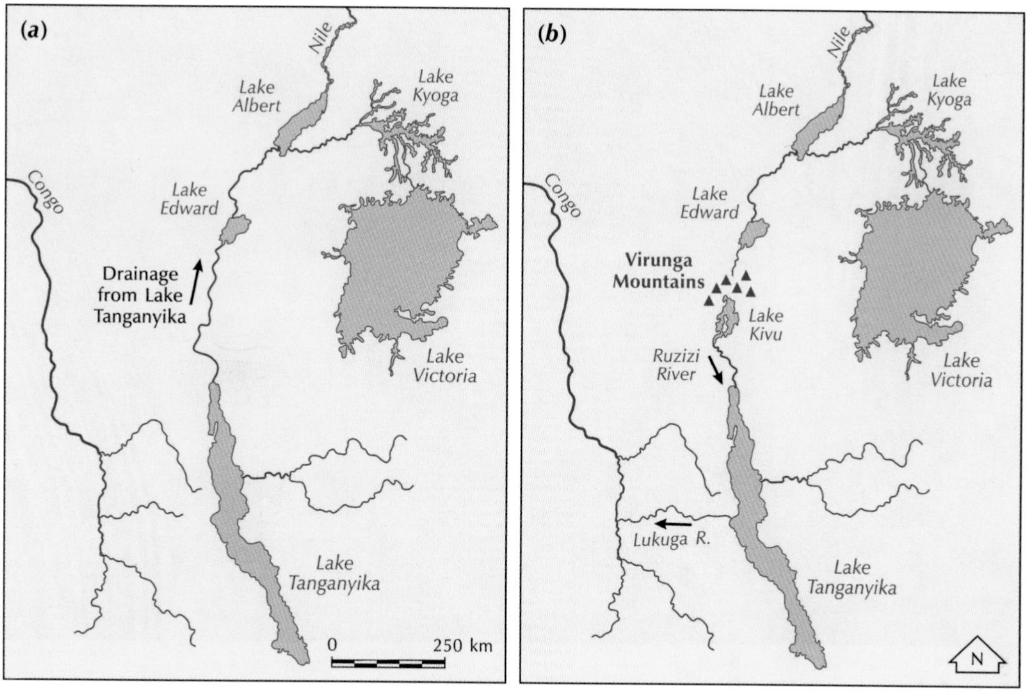

Figure 6.10 Drainage diversion by volcanoes in central Africa. (a) The Nile drainage through the Western Rift before the eruptions that built the Virunga Mountains. (b) The Nile drainage after the formation of the Virunga Mountains. *Source*: Adapted from Francis (1993, 366)

2 km in sedimentary rocks and 4 km in crystalline rocks do not have a simple bowl shape. Rather, they are **complex structures** that, in comparison with simple structures, are rather shallow (Figure 6.11b). The most recent examples, such as Clearwater Lakes in Quebec, Canada, typically have three distinct form facets. First, a structurally uplifted central area, displaying shock-metamorphic effects in the autochthonous target rocks, that may be exposed as a central peak or rings; second, an annular depression, partially filled by autochthonous breccia, or an annular sheet of so-called impact melt rocks, or a mixture of the two; and, third, a faulted rim area.

Impact craters occur on all continents. As of December 2015, 188 had been identified as impact craters from the presence of meteorite fragments, shock metamorphic features, or a combination of the two (Figure 6.12). This is a small total compared with the number identified on planets retaining portions of their earliest crust. However, impact structures are likely to be scarce on the Earth owing to the relative youthfulness and the dynamic nature of the terrestrial geosphere. Both factors serve to obscure and remove the impact record by erosion and sedimentation. Craters would have originally marked sites of all impacts. Owing to erosion, older sites are now obscure, all that remains being signs of shock metamorphism in the rocks. Thus, impacts will always leave a very long-lasting, though not indelible, signature in rocks, but the landforms (craters) they produce will gradually fade, like the face of the Cheshire Cat. The current list of known impact structures is certainly incomplete, for researchers discover about five new impact sites every year. Researchers have also found several impact structures in the seafloor.

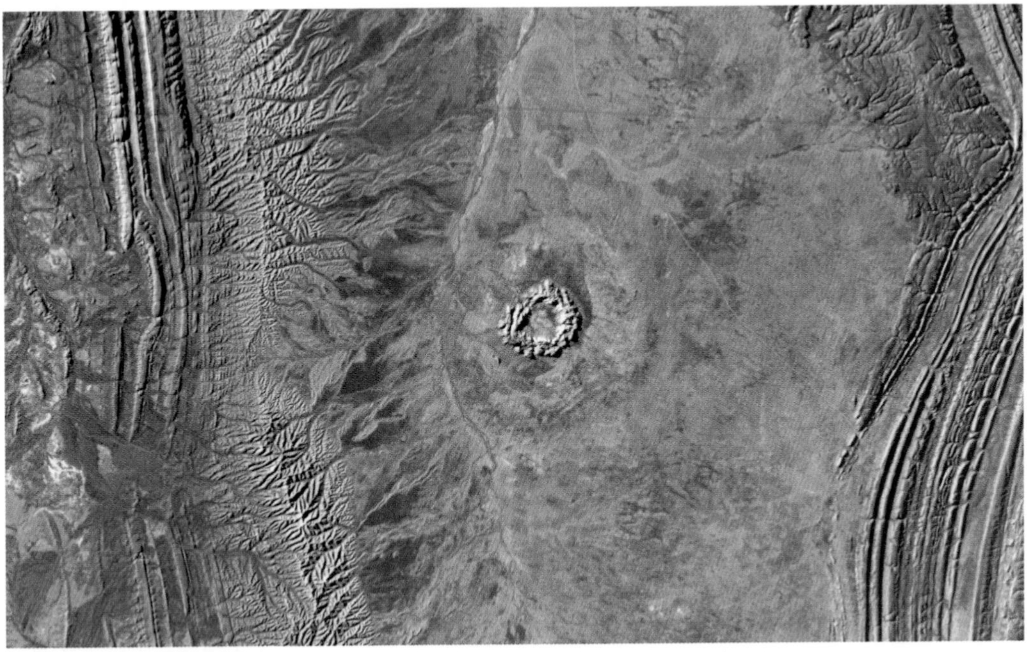

Plate 6.7 Gosses Bluff, an impact crater sandwiched between the Macdonnell Range to the north and the James Range to the south in Australia's Northern Territory, some 160 km west of Alice Springs. The impactor was probably about 1 km in diameter and crashed into the Earth about 142 million years ago. The isolated circular feature within the crater consists of a central ring of hills about 4.5 km in diameter. The greyish feature surrounding the inner ring probably marks the original boundary of the outer rim. (*Photograph courtesy of NASA*)

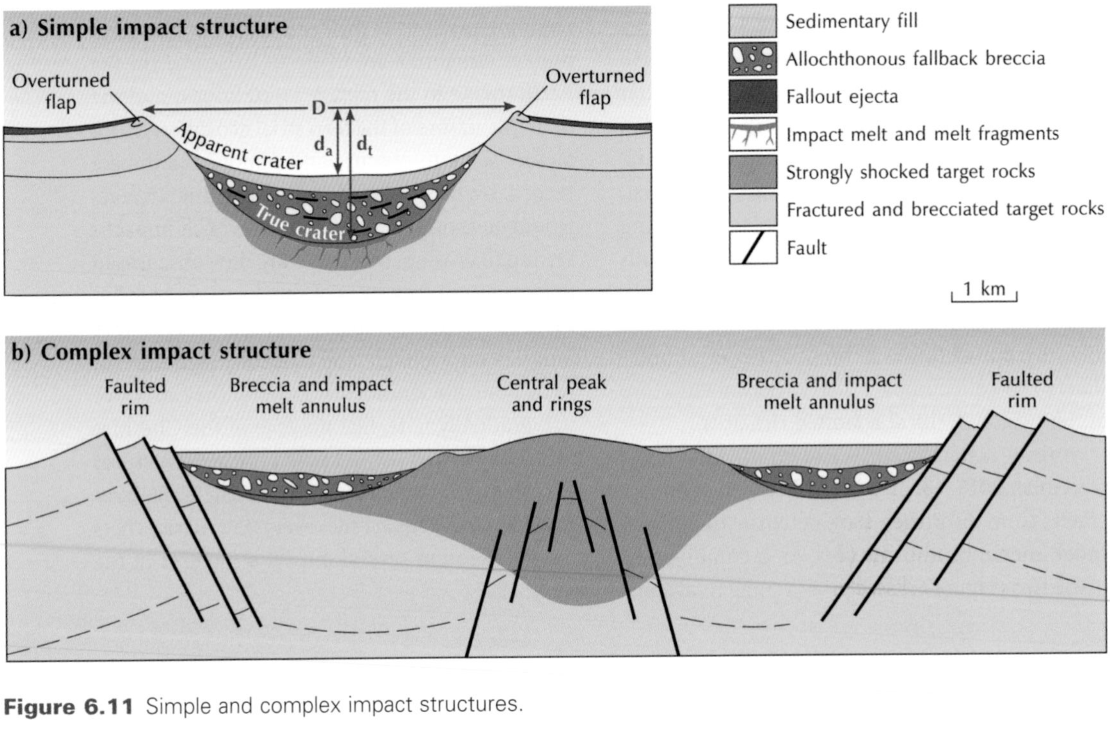

Figure 6.11 Simple and complex impact structures.

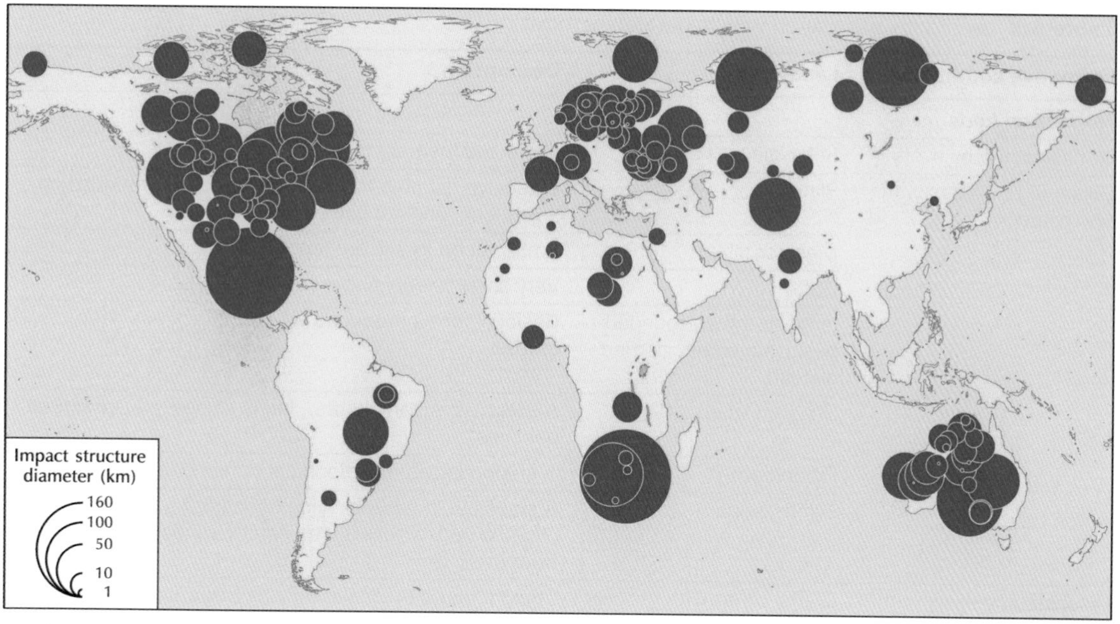

Figure 6.12 The distribution of known impact craters.

The spatial distribution of terranean impact structures reveals a concentration on the Precambrian shield areas of North America and Europe (Figure 6.12). This concentration reflects the fact that the Precambrian shields in North America and Europe have been geologically stable for a long time, and that the search for, and study of, impact craters has been conducted chiefly in those areas. It is not a reflection of the impaction process, which occurs at random over the globe.

LANDFORMS ASSOCIATED WITH FOLDS

Flat beds

Stratified rocks may stay horizontal (Plate 6.8) or they may be folded. Sedimentary rocks that remain more or less horizontal once the sea has retreated or after they have been uplifted form characteristic landforms (Table 6.2). If the beds stay flat and are not dissected by river valleys, they form large sedimentary plains (**sediplains**). Many of the flat riverine plains of the Channel Country, southwestern Queensland, Australia, are of this type. If the beds stay flat but are dissected by river valleys, they form plateaux, plains, and stepped topography (Plate 6.9). In sedimentary terrain, **plateaux** are extensive areas of low relief that sit above surrounding lower land, from which they are isolated by scarps (see Figure 6.16, p. 151). A bed of hard rock called caprock normally crowns them. A **mesa** or **table** is a small plateau, but there is no fine dividing line between a mesa and a plateau. A **butte** is a very small plateau, and a mesa becomes a butte when the maximum diameter of its flat top is less than its height above the encircling plain. When eventually the caprock is eroded away, a butte may become an isolated tower, a jagged peak, or a rounded hill, depending on the caprock thickness. In **stepped topography**, scarps display a sequence of structural benches, produced by harder beds, and steep bluffs where softer beds have been eaten away (see Plate 9.3, p. 223).

Table 6.2 Landforms associated with sedimentary rocks

Formative conditions	Landform	Description
Horizontal beds		
Not dissected by rivers	Sediplain	Large sedimentary plain
Dissected by rivers with thin caprock	Plateau	Extensive flat area formed on caprock, surrounded by lower land, and flanked by scarps
	Mesa or table	Small, steep-sided, flat-topped plateau
	Butte	Very small, steep-sided, flat-topped plateau
	Isolated tower, rounded peak, jagged hill, domed plateau	Residual forms produced when caprock has been eroded
	Stepped scarp	A scarp with many bluffs, debris slopes, and structural benches
	Ribbed scarp	A stepped scarp developed in thin-bedded strata
	Debris slope	A slope cut in bedrock lying beneath the bluff and covered with a sometimes patchy veneer of debris from it
Dissected by rivers with thick caprock	Bluffs, often with peculiar weathering patterns	Straight bluffs breached only by major rivers. Weathering patterns include elephant skin weathering, crocodile skin weathering[a], fretted surfaces, tafoni, large hollows at the bluff base
Folded beds		
Primary folds at various stages of erosion	Anticlinal hills or Jura-type relief	Folded surfaces that directly mirror the underlying geological structures
	Inverted relief	Structural lows occupy high areas (e.g. a perched syncline) and structural highs low areas (e.g. an anticlinal valley)
	Planated relief	Highly eroded folds
	Appalachian-type relief	Planated relief that is uplifted and dissected, leaving vestiges of the plains high in the relief
Differential erosion of folded sedimentary sequences	Ridge and valley topography	Terrain with ridges and valleys generally following the strike of the beds and so the pattern of folding (includes breached anticlines and domes)
	Cuesta	Ridge formed in gently dipping strata with an asymmetrical cross-section of escarpment and dip-slope
	Homoclinal ridge or strike ridge	Ridge formed in moderately dipping strata with just about asymmetrical cross-section
	Hogback	Ridge formed in steeply dipping strata with symmetrical cross-section
	Escarpment (scarp face, scarp slope)	The side of a ridge that cuts across the strata. Picks out lithological variations in the strata
	Dip-slope	The side of a ridge that accords with the dip of the strata
	Flatiron (revet crag)	A roughly triangular facet produced by regularly spaced streams eating into a dip-slope or ridge (especially a cuesta or homoclinal ridge)

Note:
a Sedimentary rocks weathered to produce the same pattern as an elephant's skin or crocodile's hide.

Source: Partly after discussion in Twidale and Campbell (1993, 187–211)

Folded beds

Anticlines are arches in strata, while **synclines** are troughs (Figure 6.13; Plate 6.10). In **recumbent anticlines**, the beds are folded over. **Isoclinal folding** occurs where a series of overfolds are arranged such that their limbs dip in the same direction. **Monoclines** are the simple folds in which beds are flexed from one level to another. An example is the Isle of Wight monocline, England, which runs from east to west across the island with Cretaceous rocks sitting at a lower level to the north than to the south. In nearly all cases, monoclines are very asymmetrical anticlines with much elongated arch and trough limbs. Anticlines, monoclines, and synclines form through shearing or tangential or lateral pressures applied to sedimentary rocks. **Domes**, which may be regarded as double anticlines, and **basins**, which may be regarded as double synclines, are formed if additional forces come from other directions. Domes are also termed **periclines**. An example is the Chaldon pericline in Dorset, England, in which rings of progressively younger rocks – Wealden Beds, Upper Greensand, and Chalk – outcrop around a core of Upper Jurassic Portland and Purbeck beds. Domed structures also form where the crust is thrust upwards, although these forms are usually simpler than those formed by more complex pressure distributions. Domes are found, too, where plugs of light material, such as salt, rise through the overlying strata as **diapirs**.

Folds may be symmetrical or asymmetrical, open or tight, simple or complex. Relief formed directly by folds is rare, but some **anticlinal hills** do exist. The 11-km-long Mount Stewart–Halcombe anticline near Wellington, New Zealand, is formed in Late Pleistocene sediments of the coastal plain. It has an even crest, the surfaces of both its flanks run parallel to the dip of the underlying beds (Box 6.1), and its arched surface replicates the fold (Ollier 1981, 59). Even anticlinal hills exposed by erosion are not that common, although many anticlinal hills in the Jura Mountains north of the European Alps remain barely breached by rivers.

Plate 6.8 Horizontal Jurassic strata seen in cliffs near Burton Bradstock, Dorset, England. These cliffs are part of the Jurassic Coast World Heritage Site; they are prone to rockfalls. (*Photograph by Nick Scarle*)

Plate 6.9 Mesas and buttes in sandstone, Monument Valley, Arizona, USA. (*Photograph by Tony Waltham Geophotos*)

The commonest landforms connected with folding are **breached anticlines** and **breached domes**. This is because, once exposed, the crest of an anticline (or the top of a dome) is subject to erosion. The strike ridges on each side tend to be archetypal dip and scarp slopes, with a typical drainage pattern, and between the streams that

Plate 6.10 Jurassic limestone strata folded into an anticline and syncline, Stair Hole, Lulworth cove, Dorset, England. (*Photograph by Nick Scarle*)

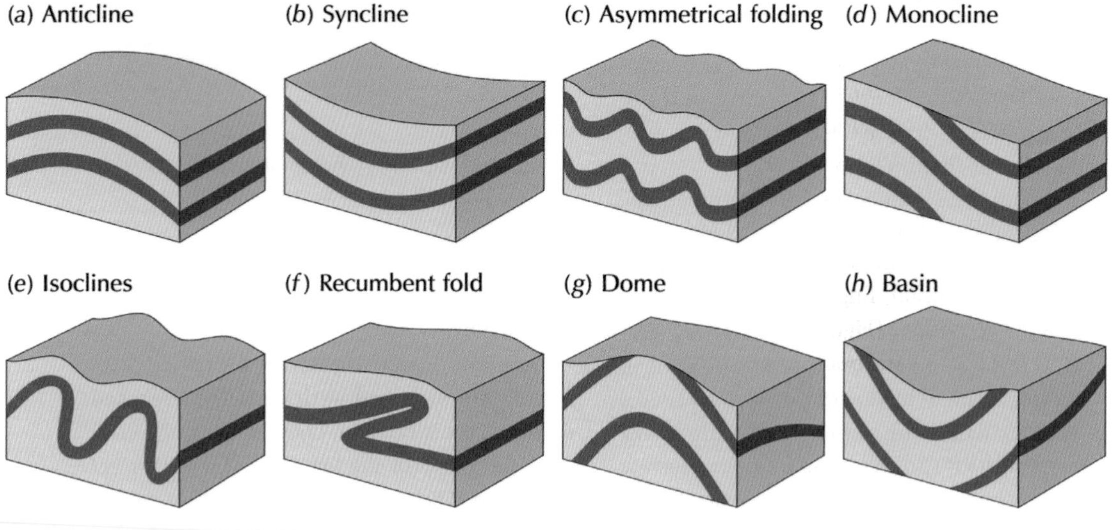

Figure 6.13 Structures formed in folded strata.

Box 6.1 DIP, STRIKE, AND PLUNGE

In tilted beds, the bedding planes are said to **dip**. The dip or true dip of a bed is given as the maximum angle between the bed and the horizontal (Figure 6.14a). The **strike** is the direction at right angles to the dip measured as an azimuth (compass direction) in the horizontal plane.

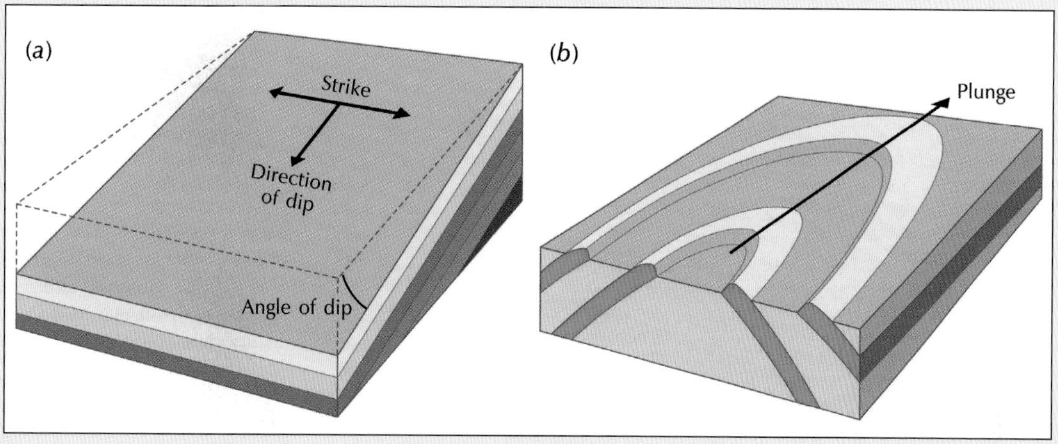

Figure 6.14 Terms relating to sedimentary structures. (a) Dip and strike. (b) Plunge.

An anticlinal axis that is tilted is said to **pitch** or **plunge** (Figure 6.14b). The angle of plunge is the angle between the anticlinal axis and a horizontal plane. Plunging anticlines can be thought of as elongated domes. Synclinal axes may also plunge.

cross the strike the dipping strata have the characteristic forms of **flatirons**, which are triangular facets with their bases parallel to the strike and their apices pointing up the dip of the rock. The strike ridges are very long where the folds are horizontal, but they form concentric rings where the folds form a dome. The scarp and vale sequence of the Kentish Weald, England, is a classic case of a breached anticline (Figure 6.15). Strike ridges may surround structural basins, with the flatirons pointing in the opposite direction.

Where strata of differing resistance are inclined over a broad area, several landforms develop according to the dip of the beds (Figure 6.16). **Cuestas** form in beds dipping gently, perhaps up to 5°. They are asymmetrical forms characterized by an **escarpment** or **scarp**, which normally forms steep slopes of cliffs, crowned by more resistant beds, and a **dip slope**, which runs along the dip of the strata. **Homoclinal ridges**, or **strike ridges**, are only just asymmetrical and develop in more steeply tilted strata with a dip between 10° and 30°. **Hogbacks** are symmetrical forms that develop where the strata dip very steeply at 40° or more. They are named after the Hog's Back, a ridge of almost vertically dipping chalk in the North Downs, England.

On a larger scale, large warps in the ground surface form **major swells** about 1,000 km across. In Africa, raised rims and major faults separate eleven basins, including the Congo basin, Sudan basin, and Karoo basin.

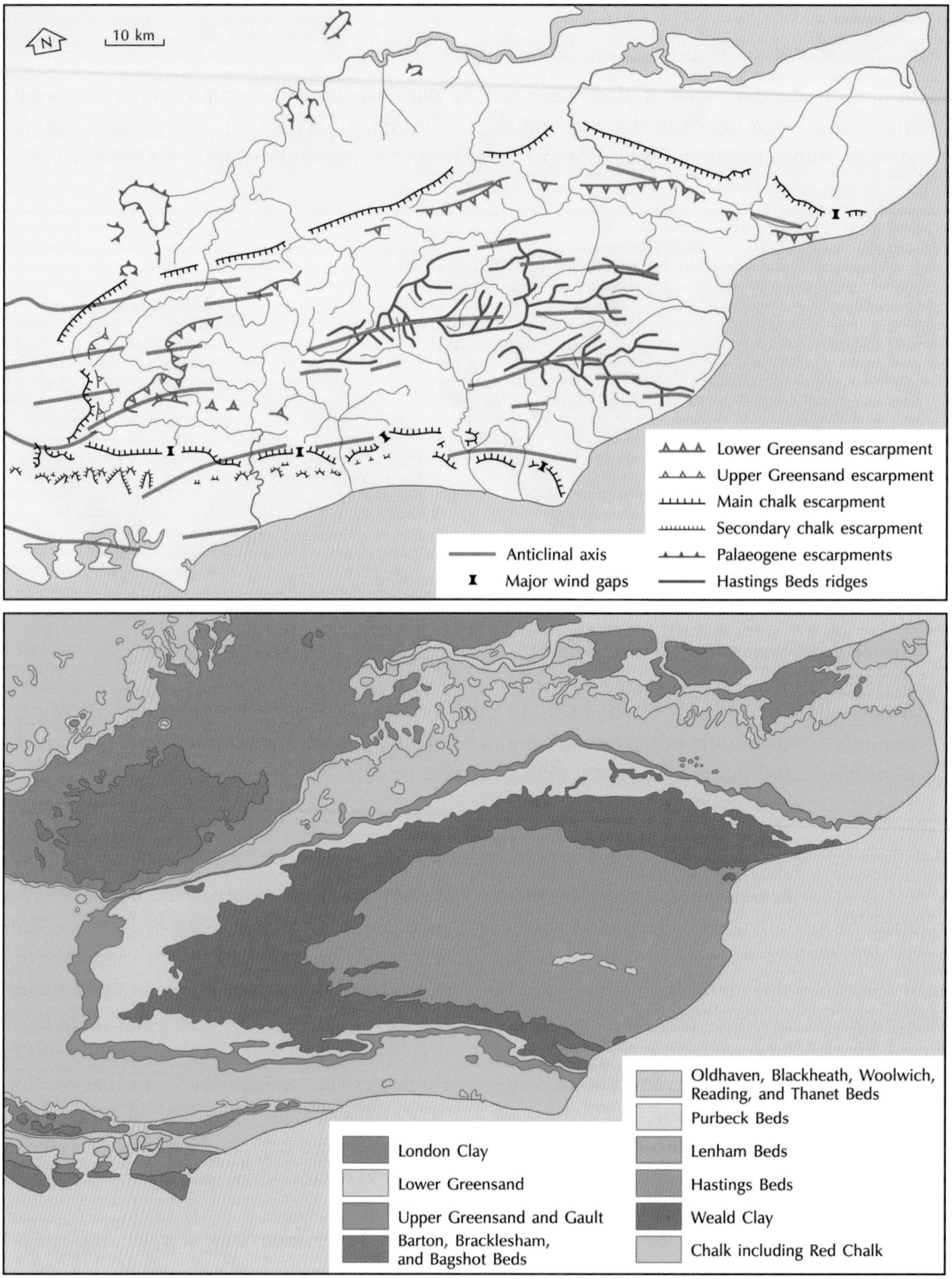

Figure 6.15 A breached anticline in south-east England. (a) Some structurally influenced topographic features. (b) Solid geology. *Source*: (a) Adapted from Jones (1981, 38)

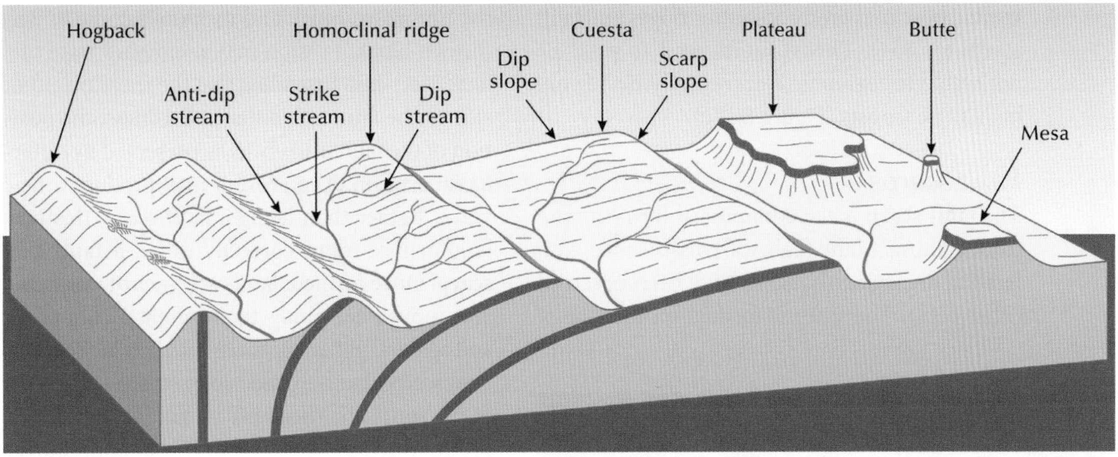

Figure 6.16 Landforms associated with dipping and horizontal strata – cuesta, homoclinal or strike ridge, hogback, butte, mesa, and plateau. The chief streams found in landscapes with dipping strata – strike streams, anti-dip streams, and dip streams – are shown. Notice that a cuesta consists of a dip slope and a steeper escarpment or scarp slope. The brown band represents a hard rock formation that caps the butte, mesa, and plateau.

LANDFORMS ASSOCIATED WITH FAULTS AND JOINTS

Faults and joints are the two major types of fracture found in rocks. A **fault** is a fracture along which movement associated with an earthquake has taken place, one side of the fault moving differentially to the other side. They are called active faults if movement was recent. Faults are commonly large-scale structures and tend to occur in fault zones rather than by themselves. A **joint** is a small-scale fracture along which no movement has taken place, or at least no differential movement. Joints arise from the cooling of igneous rocks, from drying and shrinkage in sedimentary rocks, or, in many cases, from tectonic stress. Many fractures described as joints are in fact faults along which no or minute differential movement has taken place.

Dip-slip faults

Many tectonic forms result directly from faulting. It is helpful to classify them according to the type of fault involved - dip-slip or normal faults and strike-slip faults and thrust faults (Figure 6.17). **Dip-slip faults** produce fault scarps, grabens, half-grabens, horsts, and tilted blocks. **Strike-slip**

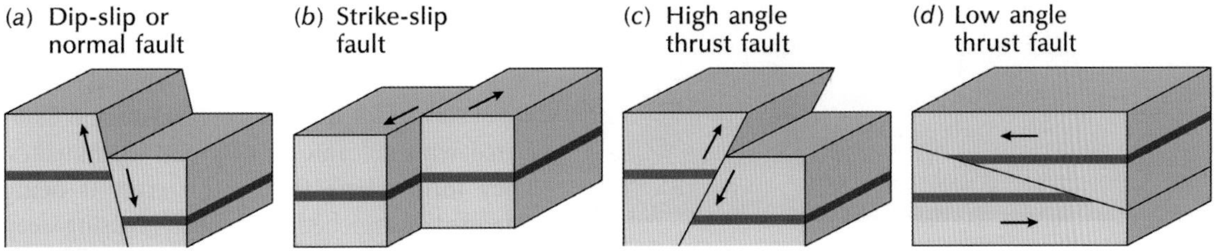

Figure 6.17 Faulted structures. (a) Normal fault. (b) Strike-slip fault. (c) High-angle reverse or thrust fault. (d) Low-angle thrust fault. *Source*: Adapted from Ahnert (1998, 233)

faults sometimes produce shutter ridges and fault scarps (p. 154). **Thrust faults** tend to produce noticeable topographic features only if they are high-angle thrusts (Figure 6.17c).

Fault scarps

The **fault scarp** is the commonest form to arise from faulting. Many fault scarps associated with faulting during earthquakes have been observed (Plate 6.11). The scarp is formed on the face of the upthrown block and overlooking the downthrown block. Erosion may remove all trace of a fault scarp but, providing that the rocks on either side of the fault line differ in hardness, the position of the fault is likely to be preserved by differential erosion. The erosion may produce a new scarp. Rather than being a fault scarp, this new landform is more correctly called a **fault-line scarp**. Once formed, faults are lines of weakness, and movement along them often occurs again and again. Uplift along faults may produce prominent scarps that are dissected by streams. The ends of the spurs are 'sliced off' along the fault line to produce triangular facets. If the fault moves repeatedly, the streams are rejuvenated to form **wineglass** or **funnel valleys** (Plate 6.12). Some fault scarps occur singly, but many occur in clusters. Individual members of fault-scarp clusters may run side by side for long distances, or they may run *en échelon* (offset but in parallel), or they may run in an intricate manner with no obvious pattern.

Plate 6.11 Fault scarp from 1959 Hebgen Lake earthquake, Montana, USA. (*Photograph by Marli Miller*)

Plate 6.12 Fault-controlled range front with wineglass valleys, Black Mountains, Death Valley, California, USA. (*Photograph by Marli Miller*)

Rift valleys, horsts, and tilt blocks

Crustal blocks are sometimes raised or lowered between roughly parallel faults without being subjected to tilting. The resulting features are called rift valleys and horsts. A **rift valley** or **graben** (after the German word for a ditch) is a long and narrow valley formed by subsidence between two parallel faults (Figure 6.18a; see also p. 140). Rift valleys are not true valleys (p. 244) and they are not all associated with linear depressions. Many rift valleys lie in zones of tension in the Earth's crust, as in the Great Rift Valley of East Africa, the Red Sea, and the Levant, which is the largest graben in the world. Grabens may be very deep, some in northern Arabia holding at least 10 km of alluvial fill. Rift valleys are commonly associated with volcanic activity and earthquakes. They form where the Earth's crust is being extended or stretched horizontally, causing steep faults to develop. Some rift valleys, such as the Rhine graben in Germany, are isolated, while others lie in graben fields and form many, nearly

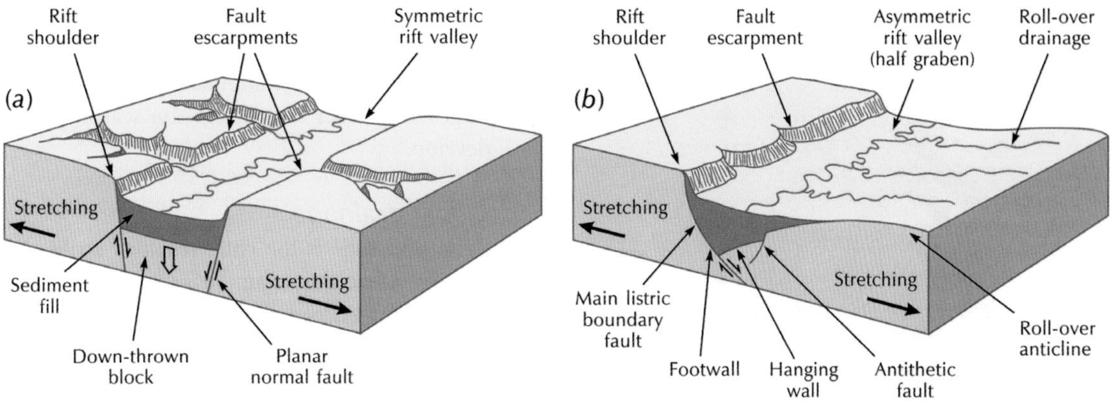

Figure 6.18 Down-faulted structures. (a) Graben. (b) Half-graben. *Source*: After Summerfield (1991, 92)

parallel structures, as in the Aegean extensional province of Greece.

A **half-graben** is bounded by a major fault only on one side (Figure 6.18b). This is called a listric (spoon-shaped) fault. The secondary or antithetic fault on the other side is normally a product of local strain on the hanging wall block. Examples are Death Valley in the Basin and Range Province of the USA, and the Menderes Valley, Turkey.

A **horst** is a long and fairly narrow upland raised by upthrust between two faults (Figure 6.19a). Examples of horsts are the Vosges Mountains, which lie west of the Rhine graben in Germany, and the Black Forest Plateau, which lies to the east of it.

Tilted or **monoclinal blocks** are formed where a section of crust between two faults is tilted (Figure 6.19b). The tilting may produce mountains and intervening basins. In the Basin and Range Province of the western USA, these are called tilt-block mountains and tilt-block basins where they are the direct result of faulting (Plate 6.13).

Dip-faults and drainage disruption

Fault scarps may disrupt drainage patterns in several ways. A fault-line lake forms where a fault scarp of sufficient size is thrown up on the downstream side of a stream. The stream is then said to be **beheaded**. Waterfalls form where the fault scarp is thrown up on the upstream side of a stream. Characteristic drainage patterns are associated with half-grabens. Back-tilted drainage occurs behind the footwall scarp related to the listric fault. Axial drainage runs along the fault axis, where lakes often form. **Roll-over drainage** develops on the roll-over section of the rift (Figure 6.18b).

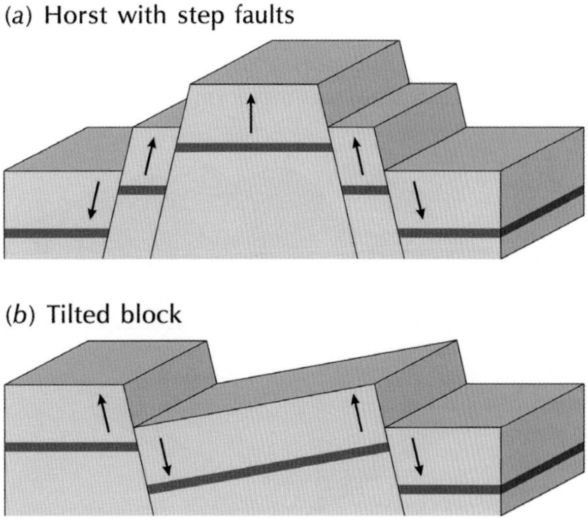

Figure 6.19 Up-faulted structures. (a) Horst. (b) Tilted block.

Plate 6.13 Tilted fault block surrounded by fans coalesced into a bajada, Eagle Mountain, south-east California, USA. (*Photograph by Marli Miller*)

Strike-slip faults

Shutter ridges and sag ponds

If movement occurs along a strike-slip fault in rugged country, the ridge crests are displaced in different directions on either side of the fault line. When movement brings ridge crests on one side of the fault opposite valleys on the other side, the valleys are 'shut off'. The ridges are therefore called **shutter ridges** (Figure 6.20).

Where tensional stresses dominate strike-slip faults, subsidence occurs and long, shallow depressions or sags may form. These are usually a few tens of metres wide and a few hundred metres long, and they may hold **sag ponds**. Where compressional stresses dominate a strike-slip fault, ridges and linear and *en échelon* scarplets may develop.

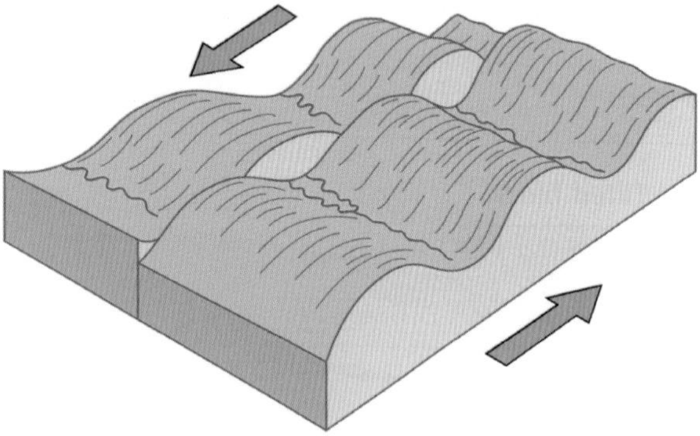

Figure 6.20 Shutter ridges along a strike-slip fault. *Source*: Adapted from Ollier (1981, 68)

Offset drainage

Offset drainage is the chief result of strike-slip faulting. The classic example is the many streams that are offset across the line of the San Andreas Fault, California, USA (Figure 6.21).

Lineaments

Any linear feature on the Earth's surface that is too precise to have arisen by chance is a **lineament**. Many lineaments are straight lines but some are curves. Faults are more or less straight lineaments, while island arcs are curved lineaments. Most lineaments are tectonic in origin. Air photography and remotely sensed images have greatly facilitated the mapping of lineaments. At times, 'the search for lineaments verges on numerology, and their alleged significance can take on almost magical properties' (Ollier 1981, 90). Several geologists believe that two sets of lineaments are basic to structural and physiographic patterns the world over – a meridional and orthogonal set, and a diagonal set. In Europe, north–south lineaments include the Pennines in England, east–west lineaments include the Hercynian axes, and diagonal lineaments include the Caledonian axes (e.g. Affleck 1970). Lineaments undoubtedly exist, but establishing worldwide sets is difficult owing to continental drift. Unless continents keep the same orientation while they are drifting, which is not the case, the lineaments formed before a particular landmass began to drift would need rotating back to their original positions. In consequence, a worldwide set of lineaments with common alignments must be fortuitous. That is not to say that there is not a worldwide system of stress and strain that could produce global patterns of lineaments, but on a planet with a mobile surface its recognition is formidable.

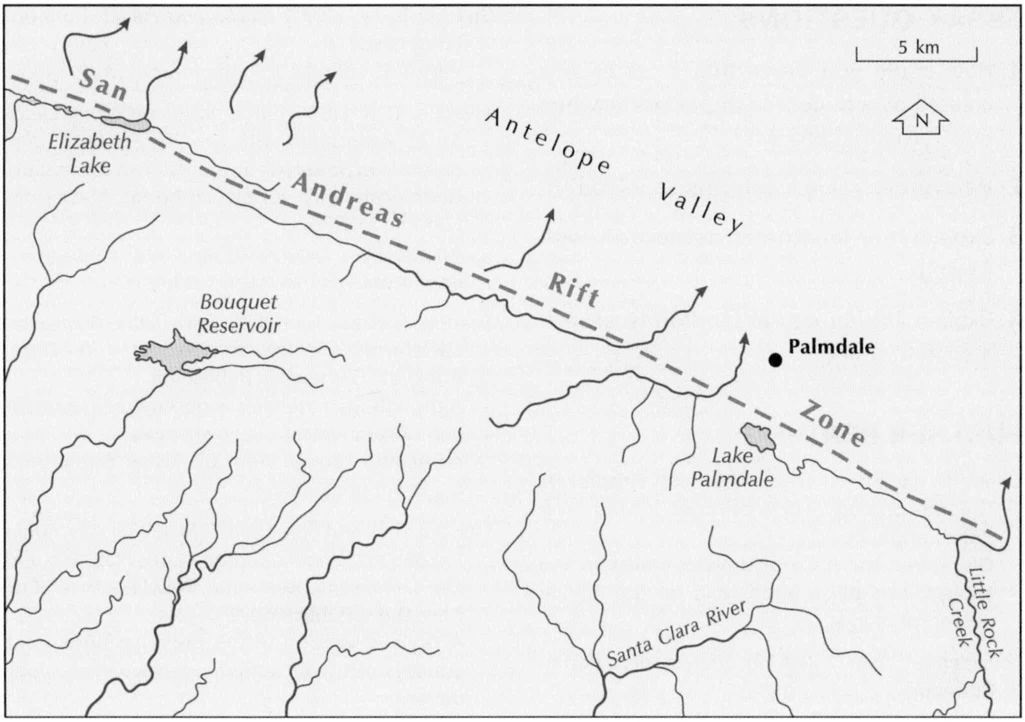

Figure 6.21 Offset drainage along the San Andreas Fault, California, USA.

SUMMARY

Plutonic and hypabyssal forces intrude molten rock (magma) into the deep and near-surface layers of the Earth respectively, while volcanic forces extrude it on to the Earth's surface. Volcanic and plutonic landforms arise from the injection of magma into rocks and the effusion and ejection of magma above the ground. Intrusions include batholiths and lopoliths, dykes and sills, laccoliths and phacoliths, all of which may express themselves in topographic features (hills, basins, domes, and so on). Extrusions and ejections produce volcanoes of various types, which are tectonic landforms. Impacts by asteroids, meteoroids, and comets pock the Earth's surface with craters and impact structures that fade with time. Flat sedimentary beds and folded sedimentary rocks produce distinctive suites of structural landforms. Flat beds tend to form plateaux, mesas, and buttes. Folded beds produce a range of landforms including anticlinal hills, cuestas, and hogbacks. Faults and joints are foci for weathering and produce large-scale landforms. Dip-slip faults may produce fault scarps, grabens, horsts, and tilted blocks. Strike-slip faults are sometimes connected with shutter ridges, sag ponds, and offset drainage.

ESSAY QUESTIONS

1 How could you prove that a crater was created by a meteorite impact rather than by a terrestrial process?

2 Why are volcanic landforms so varied?

3 Explain the landforms associated with folding.

4 Explain the structural landforms associated with rifting.

FURTHER READING

Ollier, C. D. (1981) *Tectonics and Landforms* (Geomorphology Texts 6). London and New York: Longman.
Old, given the pace of developments in the subject, but still a good read for the novice geomorphologist.

Scheidegger, A. E. (2004) *Morphotectonics*. Berlin: Springer.
Looks at the relationships between landform and tectonics. Unusual and interesting.

Sparks, B. W. (1971) *Rocks and Relief*. London: Longman.
Very old by almost any criterion, but worth a look.

Twidale, C. R. (1971) *Structural Landforms: Landforms Associated with Granitic Rocks, Faults, and Folded Strata* (An Introduction to Systematic Geomorphology, vol. 5). Cambridge, Mass., and London: MIT Press.
An excellent account of structural controls on landforms that has not dated unduly.

Twidale, C. R. and Campbell, E. M. (1993) *Australian Landforms: Structure, Process and Time*. Adelaide: Gleneagles Publishing.
Although the emphasis is on Australia, general geomorphological issues are covered. Makes a refreshing change from the usual British and North American focus.

Twidale, C. R. and Campbell, E. M. (2005) *Australian Landforms: Understanding a Low, Flat, Arid and Old Landscape*. Kenthurst, New South Wales: Rosenberg Publishing.
A revised version of the previous book, and equally useful with a stunning selection of colour plates.

PART III

EXOGENIC PROCESSES

CHAPTER SEVEN

WEATHERING AND ASSOCIATED LANDFORMS

7

The decomposition and disintegration of rock is a primary process in the tectonic cycle and landscape evolution. This chapter covers:

- Weathering processes
- Regolith and soils
- Weathering landforms
- The global pattern of leaching and weathering
- Weathering and buildings

WEATHERING IN ACTION: THE DECAY OF HISTORIC BUILDINGS

The Parthenon is a temple dedicated to the goddess Athena, built between 447 and 432 BC on the Acropolis of Athens, Greece. During its 2,500-year history, the Parthenon has suffered damage. (The Elgin Marbles, for example, now controversially displayed in the British Museum, London, once formed an outside frieze on the Parthenon). Firm evidence now suggests that continuous damage is being caused to the building by air pollution and that substantial harm has already been inflicted in this way. For example, the inward-facing carbonate stone surfaces of the columns and the column capitals bear black crusts or coatings. These damaged areas are not significantly wetted by rain or rain runoff, although acid precipitation may do some harm. The coatings seem to be caused by sulphur dioxide uptake, in the presence of moisture, on the stone surface. Once on the moist surface, the sulphur dioxide is converted to sulphuric acid, which in turn results in the formation of a layer of gypsum. Researchers are undecided about the best way of retarding and remedying this type of air pollution damage.

WEATHERING PROCESSES

Weathering is the breakdown of rocks by mechanical disintegration and chemical decomposition. Many rocks form under high temperatures and pressures deep in the Earth's crust. When exposed to the lower temperatures and pressures at the Earth's surface and brought into contact with air, water, and organisms, they start to decay. The process tends to be self-reinforcing: weathering weakens the rocks and makes them more permeable, so rendering them more vulnerable to removal by agents of erosion, and the removal of weathered products exposes more rock to weathering. Living things have an influential role in weathering, attacking rocks and minerals through various biophysical and biochemical processes, most of which are not well understood.

Weathering debris

Weathering acts upon rocks to produce solid, colloidal, and soluble materials. These materials differ in size and behaviour.

1 **Solids** range from boulders, through sand, and silt, to clay (Table 4.2). They are large, medium, and small fragments of rock subjected to disintegration and decomposition plus new materials, especially secondary clays built from the weathering products by a process called **neoformation**. At the lower end of the size range they grade into pre-colloids, colloids, and solutes.
2 **Solutes** are 'particles' less than 1 nanometre (1 nm = 0.001 micrometre) in diameter that are highly dispersed and exist in molecular solution.
3 **Colloids** are particles of organic and mineral substances that range in size from 1 to 100 nm. They normally exist in a highly dispersed state but may adopt a semi-solid form. Common colloids produced by weathering are oxides and hydroxides of silicon, aluminium, and iron. Amorphous silica and opaline silica are colloidal forms of silicon dioxide. Gibbsite and boehmite are aluminium hydroxides. Hematite is an iron oxide and goethite a hydrous iron oxide. **Pre-colloidal materials** are transitional to solids and range in size from about 100 to 1,000 nm.

Mechanical or physical weathering

Mechanical processes reduce rocks into progressively smaller fragments. The disintegration increases the surface area exposed to chemical attack. The main processes of **mechanical weathering** are unloading, frost action, thermal stress caused by heating and cooling, swelling and shrinking due to wetting and drying, and pressures exerted by salt-crystal growth. A significant ingredient in mechanical weathering is **fatigue**, which is the repeated generation of stress, by for instance heating and cooling, in a rock. The result of fatigue is that the rock will fracture at a lower stress level than a non-fatigued specimen.

Unloading

When erosion removes surface material, the confining pressure on the underlying rocks is eased. The lower pressure enables mineral grains to move further apart, creating voids, and the rock expands or dilates. In mineshafts cut in granite or other dense rocks, the pressure release can cause treacherous explosive **rockbursts**. Under natural conditions, rock dilates at right-angles to an erosional surface (valley side, rock face, or whatever). The dilation produces large or small cracks (fractures and joints) that run parallel to the surface. The dilation joints encourage rock falls and other kinds of mass movement. The small fractures and incipient joints provide lines of weakness along which individual crystals or particles may disintegrate and exfoliation may occur. **Exfoliation** is the spalling of rock sheets from the main rock body. In some rocks, such as granite, it may produce convex hills known as **exfoliation domes**. Half-Dome in Yosemite Valley, California, USA, is a classic exfoliation dome (Plates 7.1 and 7.2). In the original granodiorite intrusion, exposure to erosion leads to pressure changes that cause the dome to crack, forming shells that fall away from the mountain. Although its name suggests that half the mountain has collapsed in that manner, in fact about 80 per cent still stands. Stone Mountain, Georgia, USA, is an exfoliated inselberg.

Frost action

Water occupying the pores and interstices within a soil or rock body expands by 9 per cent upon freezing. This expansion builds up stress in the pores and fissures, causing the physical disintegration of rocks. **Frost weathering** or **frost shattering** breaks off small grains and large boulders, the boulders then being fragmented into smaller pieces. It is an important process in cold environments, where **freeze–thaw cycles** are common. Furthermore, if water-filled fissures and pores freeze rapidly at the surface, the expanding ice induces a hydrostatic or cryostatic pressure that is transmitted with equal intensity through all the interconnected hollow spaces to the still unfrozen water below. The force produced is large enough to shatter rocks, and the process is called **hydrofracturing** (Selby 1982, 16). It means that frost shattering can occur below the depth of frozen ground. **Ice segregation**, the formation of discrete bodies of ground ice in cold-environment soils, may lead to bedrock fracture (Murton *et al.* 2006).

Heating and cooling

Rocks have low thermal conductivities, which means that they are not good at conducting heat away from their surfaces. When they are heated, the outer few millimetres become much hotter than the inner portion and the outsides expand more than the insides. In addition, in rocks composed of crystals of different colours, the darker crystals warm up faster and cool down more slowly than the lighter crystals. All these thermal stresses may cause rock disintegration

Plate 7.1 Half-Dome: a classic exfoliation dome in Yosemite Valley, California, USA. The granite mountain of Half-Dome has a vertical wall 700 m high. (*Photograph by Tony Waltham Geophotos*)

Plate 7.2 Detail of exfoliation jointing in granite batholiths, Half-Dome, Yosemite Valley, California, USA. The hikers on the cable ladder give scale. (*Photograph by Tony Waltham Geophotos*)

and the formation of rock flakes, shells, and huge sheets. Repeated heating and cooling produces a fatigue effect, which enhances the **thermal weathering** or **thermoclasty**.

The production of sheets by thermal stress was once called exfoliation, but today exfoliation encompasses a wider range of processes that produce rock flakes and rock sheets of various kinds and sizes. Intense heat generated by bush fires and nuclear explosions assuredly may cause rock to flake and split. In India and Egypt, fire was for many years used as a quarrying tool. However, the everyday temperature fluctuations found even in deserts are well below the extremes achieved by local fires. Recent research points to chemical, not physical, weathering as the key to understanding rock disintegration, flaking, and splitting. In the Egyptian desert near Cairo, for instance, where rainfall is very low and temperatures very high, fallen granite columns some 3,600 years old are more weathered on their shady sides than they are on the sides exposed to the sun (Twidale and Campbell 2005, 66). Also, rock disintegration and flaking occur at depths where daily heat stresses would be negligible. Current opinion thus favours moisture, which is present even in hot deserts, as the chief agent of rock decay and rock breakdown, under both humid and arid conditions.

Wetting and drying

Some **clay minerals** (Box 7.1), including smectite and vermiculite, swell upon wetting and shrink when they dry out. Materials containing these clays, such as mudstone and shale, expand considerably on wetting, inducing microcrack formation, the widening of existing cracks, or the disintegration of the rock mass. Upon drying, the absorbed water of the expanded clays evaporates, and shrinkage cracks form. Alternate swelling and shrinking associated with wetting–drying cycles, in conjunction with the fatigue effect, leads to **wet–dry weathering**, or **slaking**, which physically disintegrates rocks.

Salt-crystal growth

In coastal and arid regions, crystals may grow in saline solutions on evaporation. Salt crystallizing within the interstices of rocks produces stresses, which widen them, and this leads to granular disintegration. This process is known as **salt weathering** or **haloclasty** (Wellman and Wilson 1965). When salt crystals formed within pores are heated, or saturated with water, they expand and exert pressure against the confining pore walls; this produces thermal stress or hydration stress respectively, both of which contribute to salt weathering.

Chemical weathering

Weathering involves a huge number of chemical reactions acting together upon many different types of rock under the full gamut of climatic conditions. Six main chemical reactions are engaged in rock decomposition: solution, hydration, oxidation and reduction, carbonation, and hydrolysis.

Solution

Mineral salts may dissolve in water, which is a very effective solvent. The process, which is called **solution** or **dissolution**, involves the dissociation of the molecules into their anions and cations and each ion becomes surrounded by water. It is a mechanical rather than a chemical process, but is normally discussed with chemical weathering as it occurs in partnership with other chemical weathering processes. Solution is readily reversed – when the solution becomes saturated some of the dissolved material precipitates. The saturation level is defined by the equilibrium solubility, that is, the amount of a substance that can dissolve in water. It is expressed as parts per million (ppm) by volume or milligrams per litre (mg/l). Once a solution is saturated, no more of the substance can dissolve. Minerals vary in their solubility. The most soluble natural minerals are chlorides of the alkali metals: rock salt or halite ($NaCl$) and potash salt (KCl). These are found only in very

Box 7.1 CLAY MINERALS

Clay minerals are hydrous silicates that contain metal cations. They are variously known as layer silicates, **phyllosilicates**, and **sheet silicates**. Their basic building-blocks are sheets of silica (Si) **tetrahedra** and oxygen (O) and hydroxyl (OH) **octahedra**. A silica tetrahedron consists of four oxygen atoms surrounding a silicon atom. Aluminium frequently, and iron less frequently, substitutes for the silicon. The tetrahedra link by sharing three corners to form a hexagon mesh pattern. An oxygen–hydroxyl octahedron consists of a combination of hydroxyl and oxygen atoms surrounding an aluminium (Al) atom. The octahedral are linked by sharing edges. The silica sheets and the octahedral sheets share atoms of oxygen, the oxygen on the fourth corner of the tetrahedrons forming part of the adjacent octahedral sheet.

Three groups of clay minerals are formed by combining the two types of sheet (Figure 7.1). The **1:1 clays** have one tetrahedral sheet combined with one flanking octahedral sheet, closely bonded by hydrogen ions (Figure 7.1a). The anions exposed at the surface of the octahedral sheets are hydroxyls. **Kaolinite** is an example, the structural formula of which is $Al_2Si_2O_5(OH)_4$. Halloysite is similar in composition to kaolinite. The **2:1 clays** have an octahedral sheet with two flanking tetrahedral sheets, which are strongly bonded by potassium ions (Figure 7.1b). An example is **illite**. A third group, the **2:2 clays**, consist of 2:1 layers with octahedral sheets between them (Figure 7.1c). An example is **smectite** (formerly called **montmorillonite**), which is similar to illite but the layers are deeper and allow water and certain organic substances to enter the lattice leading to expansion or swelling. This allows much ion exchange within the clays.

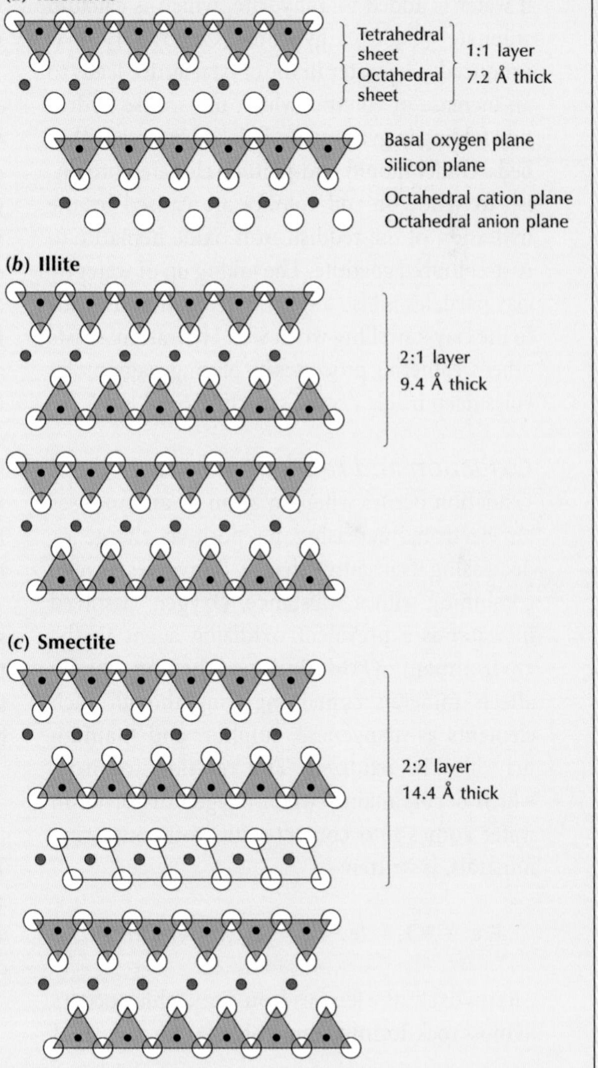

Figure 7.1 Clay mineral structure. (a) Kaolinite, a 1:1 dioctahedral layer silicate. (b) Illite, a 2:1 layer silicate, consisting of one octahedral sheet with two flanking tetrahedral sheets. (c) Smectite, a 2:2 layer silicate, consisting of 2:1 layers with octahedral sheets between. Å stands for an angstrom, a unit of length (1Å = 10^{-8} cm). *Source*: After Taylor and Eggleton (2001, 59, 61)

arid climates. Gypsum ($CaSO_4.2H_2O$) is also fairly soluble. Quartz has a very low solubility. The solubility of many minerals depends upon the number of free hydrogen ions in the water, which may be measured as the pH value (Box 7.2).

Hydration

Hydration is transitional between chemical and mechanical weathering. It occurs when minerals absorb water molecules on their edges and surfaces, or, for simple salts, in their crystal lattices, without otherwise changing the chemical composition of the original material. For instance, if water is added to anhydrite, which is calcium sulphate ($CaSO_4$), gypsum ($CaSO_4.2H_2O$) is produced. The water in the crystal lattice leads to an increase of volume, which may cause hydration folding in gypsum sandwiched between other beds. Under humid mid-latitude climates, brownish to yellowish soil colours are caused by the hydration of the reddish iron oxide hematite to rust-coloured goethite. The taking up of water by clay particles is also a form of hydration. It leads to the clay's swelling when wet. Hydration assists other weathering processes by placing water molecules deep inside crystal structures.

Oxidation and reduction

Oxidation occurs when an atom or an ion loses an electron, increasing its positive charge or decreasing its negative charge. It involves oxygen combining with a substance. Oxygen dissolved in water is a prevalent oxidizing agent in the environment. **Oxidation weathering** chiefly affects minerals containing iron, though such elements as manganese, sulphur, and titanium may also be oxidized. The reaction for iron, which occurs mainly when oxygen dissolved in water comes into contact with iron-containing minerals, is written:

$$4Fe_2 + 3O_2 + 2e \rightarrow 2Fe_2O_3 \; [e = \text{electron}]$$

Alternatively, the ferrous iron, Fe^{2+}, which occurs in most rock-forming minerals, may be converted to its ferric form, Fe^{3+}, upsetting the neutral charge of the crystal lattice, sometimes causing it to collapse and making the mineral more prone to chemical attack.

If soil or rock becomes saturated with stagnant water, it becomes oxygen-deficient and, with the aid of **anaerobic bacteria**, reduction occurs. **Reduction** is the opposite of oxidation, and the changes it promotes are called gleying. In colour, gley soil horizons are commonly a shade of grey.

The propensity for oxidation or reduction to occur is shown by the redox potential, Eh. This is measured in units of millivolts (mV), positive values registering as oxidizing potential and negative values as reducing potential (Box 7.2).

Carbonation

Carbonation is the formation of carbonates, which are the salts of carbonic acid (H_2CO_3). Carbon dioxide dissolves in natural waters to form carbonic acid. The reversible reaction combines water with carbon dioxide to form carbonic acid, which then dissociates into a hydrogen ion and a bicarbonate ion. Carbonic acid attacks minerals, forming carbonates. Carbonation dominates the weathering of calcareous rocks (limestones and dolomites) where the main mineral is calcite or calcium carbonate ($CaCO_3$). Calcite reacts with carbonic acid to form calcium hydrogen carbonate ($Ca(HCO_3)_2$) that, unlike calcite, is readily dissolved in water. This is why some limestones are so prone to solution (p. 416). The reversible reactions between carbon dioxide, water, and calcium carbonate are complex. In essence, the process may be written:

$$CaCO_3 + H_2O + CO_2 \Leftrightarrow Ca^{2+} + 2HCO_3^-$$

This formula summarizes a sequence of events starting with dissolved carbon dioxide (from the air) reacting speedily with water to produce carbonic acid, which is always in an ionic state:

$$CO_2 + H_2O \Leftrightarrow H^+ + HCO_3$$

Box 7.2 pH AND Eh

pH is a measure of the **acidity** or **alkalinity** of aqueous solutions. The term stands for the concentration of hydrogen ions in a solution, with the p standing for *Potenz* (the German word for 'power'). It is expressed as a logarithmic scale of numbers ranging from about 0 to 14 (Figure 7.2). Formulaically, pH = $-\log [H^+]$, where $[H^+]$ is the hydrogen ion concentration (in gram-equivalents per litre) in an aqueous solution. A pH of 14 corresponds to a hydrogen ion concentration of 10^{-14} gram-equivalents per litre. A pH of 7, which is neutral (neither acid nor alkaline), corresponds to a hydrogen ion concentration of 10^{-7} gram-equivalents per litre. A pH of 0 corresponds to a hydrogen ion concentration of 10^{-0} (= 1) gram-equivalents per litre. A solution with a pH greater than 7 is said to be alkaline, whereas a solution with a pH less than 7 is said to be acidic (Figure 7.2). In weathering, any precipitation with a pH below 5.6 is deemed to be acidic and referred to as **'acid rain'**.

The solubility of minerals also depends upon the **Eh** or **redox** (**reduction–oxidation**) **potential** of a solution. The redox potential measures the oxidizing or reducing characteristics of a solution. More specifically, it measures the ability of a solution to supply electrons to an oxidizing agent, or to take up electrons from a reducing agent. So redox potentials are electrical potentials or voltages. Solutions may have positive or negative redox potentials, with values ranging from about −0.6 volts to +1.4 volts. High Eh values correspond to oxidizing conditions, while low Eh values correspond to reducing conditions.

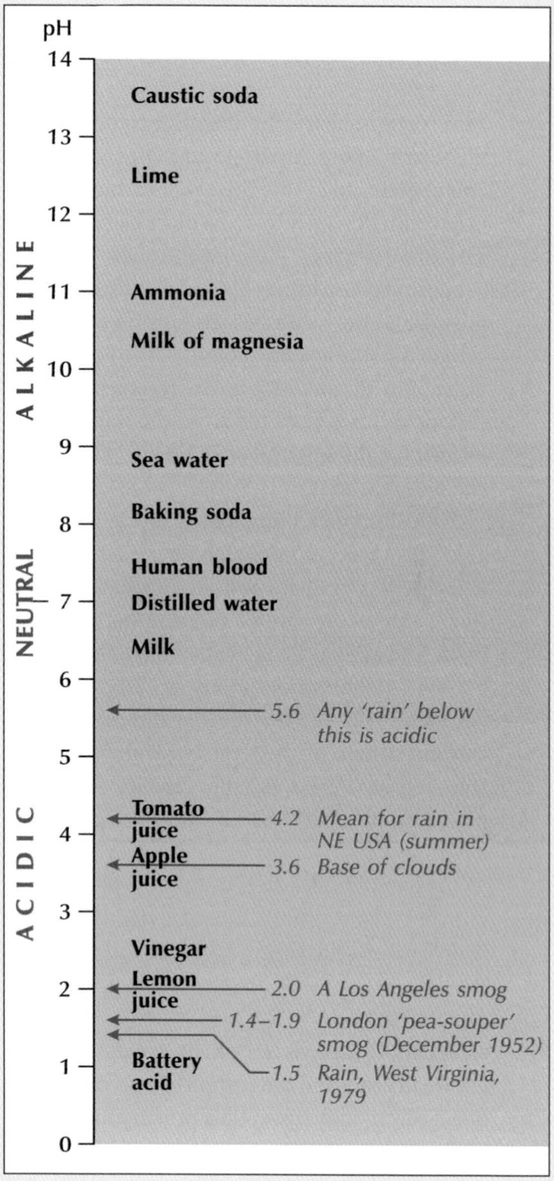

Figure 7.2 The pH scale, with the pH of assorted substances shown.

Combined, pH and Eh determine the solubility of clay minerals and other weathering products. For example, goethite, a hydrous iron oxide, forms where Eh is relatively high and pH is medium. Under high oxidizing conditions (Eh > +100 millivolts) and a moderate pH, it slowly changes to hematite.

Carbonate ions from the dissolved limestone react at once with the hydrogen ions to produce bicarbonate ions:

$$CO_3^{2-} + H^+ \Leftrightarrow HCO_3^{2-}$$

This reaction upsets the chemical equilibrium in the system, more limestone goes into solution to compensate, and more dissolved carbon dioxide reacts with the water to make more carbonic acid. The process raises the concentration by about 8 mg/l, but it also brings the carbon dioxide partial pressure of the air (a measure of the amount of carbon dioxide in a unit volume of air) and in the water into disequilibrium. In response, carbon dioxide diffuses from the air to the water, which enables further solution of limestone through the chain of reactions.

Diffusion of carbon dioxide through water is a slow process compared with the earlier reactions and sets the limit for limestone solution rates. Interestingly, the rate of reaction between carbonic acid and calcite increases with temperature, but the equilibrium solubility of carbon dioxide decreases with temperature. For this reason, high concentrations of carbonic acid may occur in cold regions, even though carbon dioxide is produced at a slow rate by organisms in such environments.

Carbonation is a step in the complex weathering of many other minerals, such as in the hydrolysis of feldspar.

Hydrolysis

Generally, hydrolysis is the main process of chemical weathering and can completely decompose or drastically modify susceptible primary minerals in rocks. In hydrolysis, water splits into **hydrogen cations** (H^+) and **hydroxyl anions** (OH^-) and reacts directly with silicate minerals in rocks and soils. The hydrogen ion is exchanged with a metal cation of the silicate minerals, commonly potassium (K^+), sodium (Na^+), calcium (Ca^{2+}), or magnesium (Mg^{2+}). The released cation then combines with the hydroxyl anion. The reaction for the hydrolysis of orthoclase, which has the chemical formula $KAlSi_3O_8$, is as follows:

$$2KAlSi_3O_8 + 2H^+ 2OH^- \rightarrow 2HAlSi_3O_8 + 2KOH$$

So the orthoclase is converted to aluminosilicic acid, $HAlSi_3O_8$, and potassium hydroxide, KOH. The aluminosilicic acid and potassium hydroxide are unstable and react further. The potassium hydroxide is carbonated to potassium carbonate, K_2CO_3, and water, H_2O:

$$2KOH + H_2CO_3 \rightarrow K_2CO_3 + 2H_2O$$

The potassium carbonate so formed is soluble in and removed by water. The aluminosilicic acid reacts with water to produce kaolinite, $Al_2Si_2O_5(OH)_4$ (a clay mineral), and silicic acid, H_4SiO_4:

$$2HAlSi_3O_8 + 9H_2O \rightarrow Al_2Si_2O_5(OH)_4 + 2H_4SiO_4$$

The silicic acid is soluble in and removed by water leaving kaolinite as a residue, a process termed **desilication** as it involves the loss of silicon. If the solution equilibrium of the silicic acid changes, then silicon dioxide (silica) may be precipitated out of the solution:

$$H_4SiO_4 \rightarrow 2H_2O + SiO_2$$

Weathering of rock by hydrolysis may be complete or partial (Pedro 1979). **Complete hydrolysis** or **allitization** produces gibbsite. **Partial hydrolysis** produces either 1:1 clays by a process called **monosiallitization**, or 2:1 and 2:2 clays through a process called **bisiallitization** (cf. p. 178).

Chelation

This is the removal of metal ions, and in particular ions of aluminium, iron, and manganese, from solids by binding with such organic acids as fulvic and humic acid to form soluble **organic matter-metal complexes**. The chelating agents are in part the decomposition products of plants and in part secretions from plant roots. Chelation encourages chemical weathering and the transfer of metals in the soil or rock.

Biological weathering

Some organisms attack rocks mechanically, or chemically, or by a combination of mechanical and chemical processes.

Plant roots, and especially tree roots, growing in bedding planes and joints have a **biomechanical effect** – as they grow, mounting pressure may lead to rock fracture. Dead lichen leaves a dark stain on rock surfaces. The dark spots absorb more thermal radiation than the surrounding lighter areas, so encouraging **thermal weathering**. A pale crust of excrement often found below birds' nests on rock walls reflects solar radiation and reduces local heating, so reducing the strength of rocks. In coastal environments, marine organisms bore into rocks and graze them (e.g. Yatsu 1988, 285–397; Spencer 1988; Trenhaile 1987, 64–82). This process is particularly effective in tropical limestones. Boring organisms include bivalve molluscs and clinoid sponges. An example is the blue mussel (*Mytilus edulis*). Grazing organisms include echinoids, chitons, and gastropods, all of which displace material from the rock surface. An example is the West Indian top shell (*Cittarium pica*), a herbivorous gastropod.

Under some conditions, bacteria, algae, fungi, and lichens may chemically alter minerals in rocks. The boring sponge (*Cliona celata*) secretes minute amounts of acid to bore into calcareous rocks. The rock minerals may be removed, leading to **biological rock erosion**. In an arid area of southern Tunisia, weathering is concentrated in topographic lows (pits and pans) where moisture is concentrated and algae bore, pluck, and etch the limestone substrate (Smith *et al.* 2000).

Humans have exposed bedrock in quarries, mines, and road and rail cuts. They have disrupted soils by detonating explosive devices, and they have sealed the soil in urban areas under a layer of concrete and tarmac. Their agriculture practices have greatly modified soil and weathering processes in many regions.

WEATHERING PRODUCTS: REGOLITH AND SOILS

There are two chief weathering environments with different types of product – weathering-limited environments and transport-limited environments. In **weathering-limited environments**, transport processes rates outstrip weathering processes rates. Consequently, any material released by weathering is removed and a regolith or soil is unable to develop. Rock composition and structure largely determine the resulting surface forms. In **transport-limited environments**, weathering rates run faster than transport rates, so that regolith or soil is able to develop. Mass movements then dominate surface forms, and forms fashioned directly by weathering are confined to the interface between regolith or soil and unweathered rock. Materials released by weathering are subject to continued weathering. This section will consider transport-limited weathering products; the next section will consider weathering-limited weathering products.

Regolith

The **weathered mantle** or **regolith** is all the weathered material lying above the unaltered or fresh bedrock (see Ehlen 2005). It may include lumps of fresh bedrock. Often the weathered mantle or crust is differentiated into visible horizons and is called a **weathering profile** (Figure 7.3). The **weathering front** is the boundary between fresh and weathered rock. The layer immediately above the weathering front is sometimes called **saprock**, which represents the first stages of weathering. Above the saprock lies **saprolite**; this is more weathered than saprock but still retains most of the structures found in the

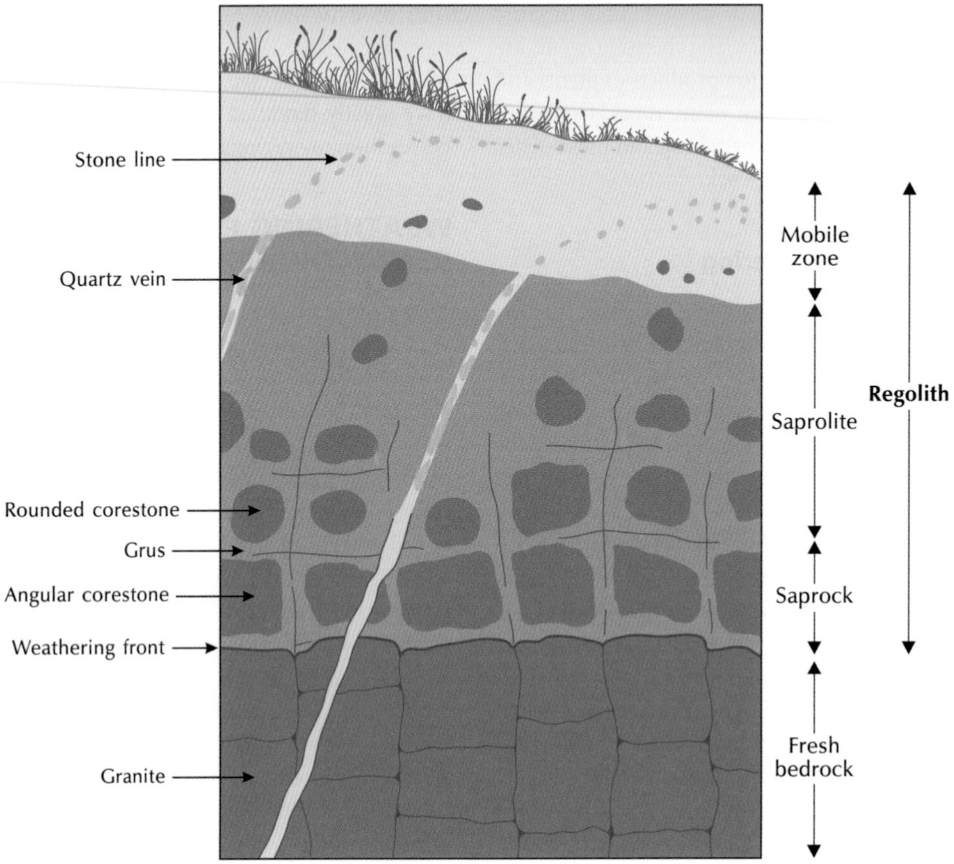

Figure 7.3 Typical weathering profile in granite. The weathering front separates fresh bedrock from the regolith. The regolith is divided into saprock, saprolite, and a mobile zone.

parent bedrock. Saprolite lies where it was formed, undisturbed by mass movements or other erosive agents. Deep weathering profiles, saprock, and saprolite are common in the tropics. No satisfactory name exists for the material lying above the saprolite, where weathering is advanced and the parent rock fabric is not distinguishable, although the terms '**mobile zone**', '**zone of lost fabric**', '**residuum**', and '**pedolith**' are all used (see Taylor and Eggleton 2001, 160).

Weathering can produce distinct mantles. The intense frost weathering of exposed bedrock, for instance, produces **blockfields**, which are also called felsenmeer, block meer, and stone fields. Blockfields are large expanses of coarse and angular rock rubble. They typically occur on plateaux in mid- and high latitudes that escaped erosion by warm-based ice during the Pleistocene, as well as polar deserts and semi-deserts. Steeper fields, up to 35°, are called blockstreams. An example is the 'stone runs' of the Falkland Islands. Some blockfields, such as those in the Cairngorms, Scotland, are relict features that predate the last advance of sheet. **Talus (scree) slopes** (Plate 7.3) and **talus cones** (Plate 7.4) are accumulations of rock fragments that fall from steep rock faces after loosening by weathering. **Debris cones** are the accumulation of material moved in debris flows.

Duricrusts and hardpans

Under some circumstances, soluble materials precipitate within or on the weathered mantle to form duricrusts, hardpans, and plinthite. **Duricrusts** are important in landform development as they act like a band of resistant rock and may cap hills. They occur as hard nodules or crusts, or simply as hard layers. The chief types are ferricrete (rich in iron), calcrete (rich in calcium carbonate), silcrete (rich in silica), alcrete (rich in aluminium), gypcrete (rich in gypsum), magnecrete (rich in magnesite), and manganocrete (rich in manganese).

Ferricrete and **alcrete** are associated with deep weathering profiles. They occur in humid to subhumid tropical environments, with alcretes favouring drier parts of such regions.

Laterite is a term used to describe weathering deposits rich in iron and aluminium.

Bauxite refers to weathering deposits rich enough in aluminium to make economic extraction worthwhile.

Silcrete, or siliceous duricrust, commonly consists of more than 95 per cent silica. It occurs in humid and arid tropical environments, and notably in central Australia and parts of northern and southern Africa and parts of Europe, sometimes in the same weathering profiles as ferricretes. In more arid regions, it is sometimes associated with calcrete.

Calcrete is composed of around 80 per cent calcium carbonate. It is mostly confined to areas where the current mean annual rainfall lies in the range 200 to 600 mm and covers a large portion of the world's semi-arid environments, perhaps underlying 13 per cent of the global land-surface area.

Gypcrete is a crust of gypsum (hydrated calcium sulphate). It occurs largely in very arid regions with a mean annual precipitation below 250 mm. It forms by gypsum crystals growing in clastic sediments, either by enclosing or by displacing the clastic particles.

Plate 7.3 Scree slopes, Oyberget, southern Norway. (*Photograph by Peter Wilson*)

Plate 7.4 Talus cone, Sierra Nevada, California, USA. (*Photograph by Marli Miller*)

Box 7.3 INVERTED RELIEF

Geomorphic processes that create resistant material in the regolith may promote **relief inversion**. Duricrusts are commonly responsible for inverting relief. Old valley bottoms with ferricrete in them resist erosion and eventually come to occupy hilltops (Figure 7.4). Even humble alluvium may suffice to cause relief inversion (Mills 1990). Floors of valleys in the Appalachian Mountains, eastern USA, become filled with large quartzite boulders, more than 1 m in diameter. These boulders protect the valley floors from further erosion by running water. Erosion then switches to sideslopes of the depressions and, eventually, ridges capped with bouldery colluvium on deep saprolite form. Indeed, the saprolite is deeper than that under many uncapped ridges.

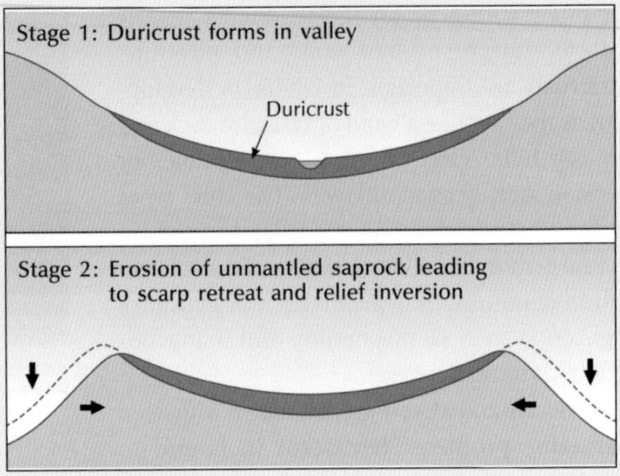

Figure 7.4 Development of inverted relief associated with duricrust formation.

Magnecrete is a rare duricrust made of magnesite (magnesium carbonate). **Manganocrete** is a duricrust with a cement of manganese-oxide minerals.

Hardpans and **plinthite** also occur. They are hard layers but, unlike duricrusts, are not enriched in a specific element.

Duricrusts are commonly harder than the materials in which they occur and more resistant to erosion. In consequence, they act as a shell of armour, protecting land surfaces from denudational agents. Duricrusts that develop in low-lying areas where surface and subsurface flows of water converge may retard valley down-cutting to such an extent that the surrounding higher regions wear down faster than the valley floor, eventually leading to **inverted relief** (Box 7.3). Where duricrusts have been broken up by prolonged erosion, fragments may persist on the surface, carrying on their protective role. The **gibber plains** of central Australia are an example of such long-lasting remnants of duricrusts and consist of silcrete boulders strewn about the land surface.

Soil

The idea of **soil** is complicated: soil, like love and home, is difficult to define (Retallack 2003). Geologists and engineers see soil as soft, unconsolidated rock. The entire profile of weathered rock and unconsolidated rock material, of whatever origin, lying above unaltered bedrock is then soil material. By this definition, soil is the same as regolith, that is, all the weathered material lying above the unaltered or fresh bedrock. It includes *in situ* weathered rock (saprolite), disturbed weathered rock (residuum), transported surficial sediments, chemical products, topsoil, and a miscellany of other products, including volcanic ash. Most pedologists regard soil as the portion of the regolith that supports plant life and where soil-forming processes dominate (e.g. Buol *et al.* 2003). This definition poses problems. Some saline soils and laterite surfaces cannot support plants – are they true soils? Is a lichen-encrusted bare rock surface a soil? Pedologists (scientists who study soils) cannot agree on these troubling issues. A possible way of dodging the problem is to define exposed hard rocks as soils (Jenny 1980, 47). This suggestion is not as daft as it might seem. Exposed rocks, like soils, are influenced by climate; like some soils, they will support little or no plant life. Pursuing this idea, soil may be defined as 'rock that has encountered the ecosphere' (Huggett 1995, 12). This definition eschews the somewhat arbitrary distinctions between soil and regolith, and between soil processes and geomorphic processes. It means that the pedosphere is the part of the lithosphere living things affect, and that 'the soil' includes sedimentary material affected by physical and chemical processes, and to far lesser degree, by biological processes. If pedologists feel unhappy with a geological definition of soil, then they can use a homegrown pedological term – solum. The **solum** is the genetic soil developed by soil-building forces (Soil Survey Staff 1999), and normally comprises the A and B horizons of a soil profile, that is, the topsoil and the subsoil.

The very strong links between soils, soil processes, geomorphology, and hydrology are seen in landscapes. Researchers have proposed several frameworks for linking pedological, hydrological, and geomorphic processes within landscapes, most them concerned with two-dimensional **catenas**. The idea of **soil–landscape systems** was an early attempt at an integrated, three-dimensional model (Huggett 1975). The argument was that dispersion of all the debris of weathering – solids, colloids, and solutes – is, in a general and fundamental way, influenced by land-surface form, and organized in three dimensions within a framework dictated by the drainage network. In moving down slopes, weathering products tend to move at right angles to land-surface contours. Flowlines of material converge and diverge according to contour curvature. The pattern of vergency influences the amounts of water, solutes, colloids, and clastic sediments held in store at different landscape positions. Naturally, the movement of weathering products alters the topography, which in turn influences the movement of the weathering products – there is f eedback between the two systems. Research into the relationships between soils and geomorphology has proved highly fruitful (e.g. Gerrard 1992; Daniels and Hammer 1992; Birkeland 1999; Schaetzl and Anderson 2005).

WEATHERING PRODUCTS: LANDFORMS

Bare rock is exposed in many landscapes. It results from the differential weathering of bedrock and the removal of weathered debris by slope processes. Two groups of weathering landforms associated with bare rock in weathering-limited environments are (1) large-scale cliffs and pillars and (2) smaller-scale rock-basins, tafoni, and honeycombs.

Cliffs and pillars

Cliffs and crags are associated with several rock types, including limestones, sandstones, and gritstones. Take the case of sandstone cliffs (Robinson and Williams 1994). These form in strongly cemented sandstones, especially on the sides of deeply incised valleys and around the edges of plateaux. Isolated pillars of rock are also common at such sites. Throughout the world, sandstone cliffs and pillars are distinctive features of sandstone terrain (see p. 453). They are eye-catching in arid areas, but tend to be concealed by vegetation in more humid regions, such as England. The cliffs formed in the Ardingly Sandstone, south-east England, are hidden by dense woodland. Many cliffs are dissected by widened vertical joints that form open clefts or passageways. In Britain, such widened joints are called **gulls** or **wents**, which are terms used by quarrymen.

On some outcrops, the passageways develop into a labyrinth through which it is possible to walk.

Many sandstone cliffs, pillars, and boulders are undercut towards their bases. In the case of boulders and pillars, the undercutting produces **mushroom**, **perched**, or **pedestal rocks**. Processes invoked to account for the undercutting include (1) the presence of softer and more effortlessly weathered bands of rock; (2) abrasion by wind-blown sand (cf. p. 338); (3) salt weathering brought about by salts raised by capillary action from soil-covered talus at the cliff base; (4) the intensified rotting of the sandstone by moisture rising from the soil or talus; and (5) subsurface weathering that occurs prior to footslope lowering.

Rock-basins, tafoni, and honeycombs

Virtually all exposed rock outcrops bear irregular surfaces that seem to result from weathering. Flutes and runnels, pits and cavernous forms are common on all rock types in all climates. They are most apparent in arid and semi-arid environments, mainly because these environments have a greater area of bare rock surfaces. They usually find their fullest development on limestone (Chapter 14) but occur on, for example, granite.

Flutes, **rills**, **runnels**, **grooves**, and **gutters**, as they are variously styled, form on many rock types in many environments. They may develop a regularly spaced pattern. Individual rills can be 5–30 cm deep and 22–100 cm wide. Their development on limestone is striking (p. 420).

Rock-basins, also called **weathering pits**, **weatherpits**, or **gnammas**, are closed, circular, or oval depressions, a few centimetres to several metres wide, formed on flat or gently sloping surfaces of limestones, granites, basalts, gneisses, and other rock types (Plate 7.5). They are commonly flat-floored and steep-sided, and no more than a metre or so deep, though some are more saucer-shaped. The steep-sided varieties may bear overhanging rims and undercut sides. Rainwater collecting in the basins may overflow to produce spillways, and some basins may contain incised spillways that lead to their being permanently drained. Rock-basins start from small depressions in which water collects after rainfall or snowmelt. The surrounding surfaces dry out, but the depression stays moist or even holds a small pool for long periods, so providing a focus for more rapid weathering. In consequence, the rock-basin expands and deepens. As rock-basins expand, they may coalesce to form compound forms. **Solution pools** (pans, solution basins, flat-bottomed pools) occur on shore platforms cut in calcareous rocks. The initiation of these various weathering cavities often involves positive feedback, as the depression tends to collect more moisture and enlarge further.

Tafoni (singular **tafone**) are large weathering features that take the form of hollows or cavities

Plate 7.5 Weathering pit on Clach Bhàn, Ben Avon, in the eastern Cairngorms, Scotland. *(Photograph by Adrian M. Hall)*

on a rock surface (Plate 7.6), the term being originally used to describe hollows excavated in granites on the island of Corsica. They tend to form in vertical or near-vertical faces of rock. They can be as little as 0.1 m to several metres in height, width, and depth, with arched-shaped entrances, concave walls, sometimes with overhanging hoods or visors, especially in case-hardened rocks (rocks with a surface made harder by the local mobilization and reprecipitation of minerals on its surface), and smooth and gently sloping, debris-strewn floors. Some tafoni cut right through boulders or slabs of rock to form rounded shafts or windows. The origins of tafoni are complex. Salt action is the process commonly invoked in tafoni formation, but researchers cannot agree whether the salts promote selective chemical attack or whether they promote physical weathering, the growing crystals prising apart grains of rock. Both processes may operate, but not all tafoni contain a significant quantity of salts. Once formed, tafoni are protected from rainwash and may become the foci for salt accumulations and further salt weathering. Parts of the rock that are less effectively case-hardened are more vulnerable to such chemical attack. Evidence also suggests that the core of boulders sometimes more readily weathers than the surface, which could aid the selective development of weathering cavities. Tafoni are common in coastal environments but are also found in arid environments. Some appear to be relict forms.

Honeycomb weathering is a term used to describe numerous small pits or **alveoli**, no more than a few centimetres wide and deep, separated by an intricate network of narrow walls and resembling a honeycomb (Plate 7.7). They are often thought of as a small-scale version of multiple tafoni. The terms **alveolar weathering**, **stone lattice**, and **stone lace** are synonyms. Honeycomb weathering is particularly evident in semi-arid and coastal environments where salts are in ready supply and wetting and drying cycles are common. A study of honeycomb weathering on the coping stones of the sea walls at Weston-super-Mare, Avon, England, suggests stages of development (Mottershead 1994). The walls were finished in 1888. The main body of the walls is made of Carboniferous limestone, which is capped by Forest of Dean stone (Lower Carboniferous Pennant sandstone). Nine weathering grades can be recognized on the coping stones (Table 7.1). The maximum reduction of the original surface is at least 110 mm, suggesting a minimum weathering rate of 1 mm/yr.

Plate 7.6 Large tafoni in granite boulder, Corsica. (*Photograph by Heather Viles*)

Table 7.1 Honeycomb weathering grades on sea walls at Weston-super-Mare, Avon, UK

Grade	Description
0	No visible weathering forms
1	Isolated circular pits
2	Pitting covers more than 50 per cent of the area
3	Honeycomb present
4	Honeycomb covers more than 50 per cent of the area
5	Honeycomb shows some wall breakdown
6	Honeycomb partially stripped
7	Honeycomb stripping covers more than 50 per cent of the area
8	Only reduced walls remain
9	Surface completely stripped

Source: Adapted from Mottershead (1994)

Joints and weathering

All rocks are fractured to some extent. A broad range of fractures exists, many of which split rock into cubic or quadrangular blocks. All **joints** are avenues of weathering and potential seats of erosion. The geomorphic significance of a set of joints depends upon many factors, including their openness, pattern and spacing, and other physical properties of the rock mass. Outcrops of resistant rocks such as granite may be reduced to plains, given time, because fractures allow water and therefore weathering to eat into the rock. If the granite has a high density of fractures, the many avenues of water penetration promote rapid rock decay that, if rivers are able to cut down and remove the weathering products, may produce a plain of low relief. This has happened on many old continental shields, as in the northern Eyre Peninsula, Australia. Even granite with a moderate density of fractures, spaced about 1 to 3 m apart, may completely decay given sufficient time, owing to water penetrating along the fractures and then into the rock blocks between the fractures through openings created by the weathering of mica and feldspar.

The weathering of granite with moderately spaced joints produces distinctive landforms (Figure 7.5). The weathering of the joint-defined blocks proceeds fastest on the block corners, at an average rate on the edges, and slowest on the faces. This differential weathering leads to the rounding of the angular blocks to produce rounded kernels or **corestones** surrounded by weathered rock. The weathered rock or **grus** is easily eroded and once removed leaves behind a cluster of rounded boulders that is typical of many granite outcrops. A similar dual process of weathering along joints and grus removal operates in other plutonic rocks such as diorite and gabbro, and less commonly in sandstone and limestone. It also occurs in rocks with different fracture patterns, such as gneisses with well-developed cleavage or foliation, but instead of producing boulders it fashions slabs known as **penitent rocks**, **monkstones**, or **tombstones** (Plate 7.8).

Another common feature of granite weathering is a bedrock platform extending from the edge of inselbergs (island mountains). These platforms appear to have formed by etching (p. 459). Inselbergs come in three varieties: **bornhardts**,

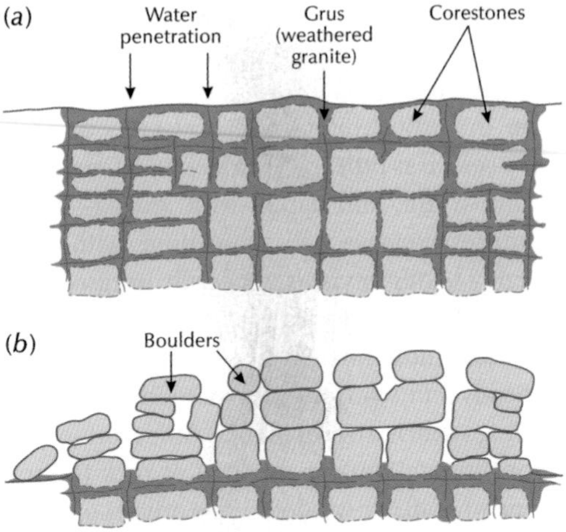

Figure 7.5 Weathering of jointed rocks in two stages. (a) Subsurface weathering occurs mainly along joints to produce corestones surrounded by grus (weathered granite). (b) The grus is eroded to leave boulders. *Source*: After Twidale and Campbell (2005, 136)

Plate 7.7 Alveoli formed in sandstone near Coos Bay, Sunset Bay State Park, Oregon, USA. (*Photograph by Marli Miller*)

Plate 7.8 Tombstone columns in basalt, Devils Postpile, California, USA. (*Photograph by Tony Waltham Geophotos*)

Plate 7.9 The Spitzkoppe (*'pointed dome'*), a granite bornhardt in Namibia. (*Photograph by Nick Scarle*)

Plate 7.10 Bornhardt granite block standing out by differential weathering, Iuiu, Minas Gerais, Brazil. (*Photograph by Tony Waltham Geophotos*)

which are dome-shaped hills (Plates 7.9 and 7.10); **nubbins** or **knolls**, which bear a scattering of blocks (Plate 7.11); and small and angular **castle koppies**. Nubbins and koppies appear to derive from bornhardts, which are deemed the basic form. Bornhardts occur in rocks with very few open joints (massive rocks), mainly granites and gneisses but also silicic volcanic rocks such as dacite, in sandstone (Uluru), and in conglomerate (e.g. the Olgas complex, also near Alice Springs, Australia); and there are equivalent forms – tower karst – that develop in limestone (p. 430). Most of them meet the adjacent plains, which are usually composed of the same rock as the inselbergs, at a sharp break of slope called the piedmont angle.

Plate 7.11 Nubbin weathering remnants in massive sandstone, Hammersley Ranges, Pilbara, Western Australia. (*Photograph by Tony Waltham Geophotos*)

One possible explanation for the formation of bornhardts invokes long-distance scarp retreat. Another plausible explanation envisages a two-stage process of deep weathering and stripping, similar to the two-stage process envisaged in the formation of granite boulders. It assumes that the fracture density of a granite massif has high and low compartments. In the first stage, etching acts more readily on the highly fractured compartment, tending to leave the less-fractured compartment dry and resistant to erosion. In the second stage, the grus in the more weathered, densely fractured compartment is eroded. This theory appears to apply to the bornhardts in or near the valley of the Salt River, south of Kellerberrin, Western Australia (Twidale *et al.* 1999). These bornhardts started as subsurface bedrock rises bulging into the base of a Cretaceous and earlier Mesozoic regolith. They were then exposed during the Early Cenozoic era as the rejuvenated Salt River and its tributaries stripped the regolith. If the two-stage theory of bornhardt formation should be accepted, then the development of nubbins and koppies from bornhardts is explained by different patterns of subsurface weathering. Nubbins form through the decay of the outer few shells of sheet structures in warm and humid climates, such as northern Australia (Figure 7.6a). Koppies probably form by the subsurface weathering of granite domes whose crests are exposed at the surface as platforms (Figure 7.6b). However, inselbergs and associated landforms in the central

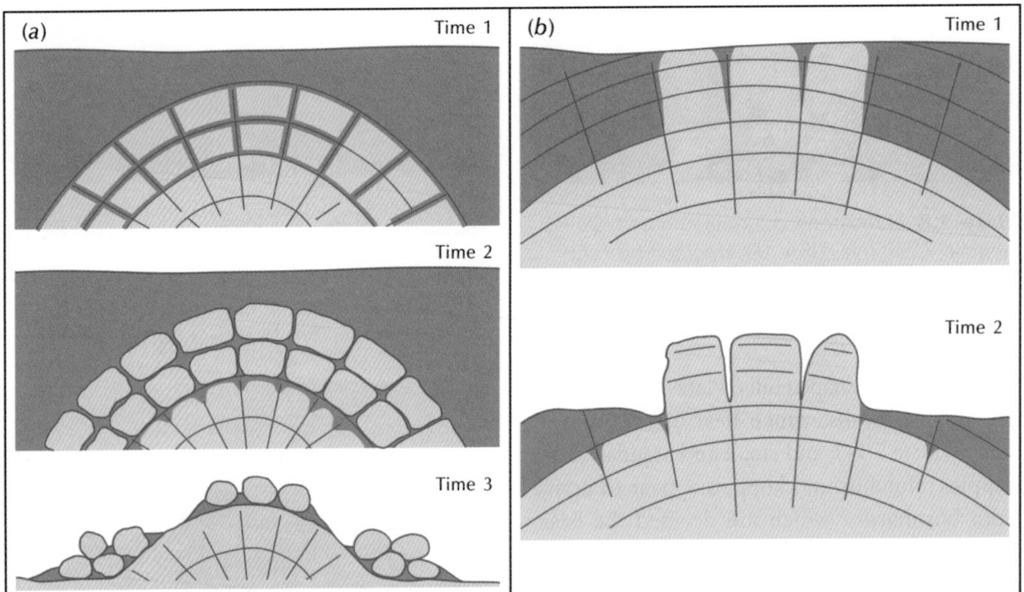

Figure 7.6 Formation of (a) nubbins and (b) castle koppies from bornhardts. *Source*: After Twidale and Campbell (2005, 137)

Namib Desert, Namibia, show no signs of deep weathering, and stripping and scarp retreat seem unlikely as formative mechanisms.

A third possibility is mantle planation (Ollier 1978). In this environment, weathering attacks any rocks protruding above the ground surface, levelling them off to create a plane surface littered with a mantle of debris. Successive bevelling episodes of mantle planation would reduce the level of the plains, leaving pockets of more durable rock as high-standing residuals with their boundaries corresponding with geological boundaries. Interestingly, therefore, three different suites of processes may produce the same suite of landforms, a case of convergent landform evolution.

Tors, which are outcrops of rock that stand out on all sides from the surrounding slopes, probably form in a similar way to bornhardts (Plate 7.12). They are common on crystalline rocks, but are known to occur on other resistant rock types, including quartzites and some sandstones. Some geomorphologists claim that deep weathering is a prerequisite for tor formation. They envisage a period of intense chemical weathering acting along joints and followed by a period when environmental conditions are conducive to the stripping of the weathered material by erosion. Other geomorphologists believe that tors can develop without deep weathering under conditions where weathering and stripping operate at the same time on rocks of differing resistance.

Plate 7.12 Granite tor, Haytor, Dartmoor, England. (*Photograph by Tony Waltham Geophotos*)

WEATHERING AND CLIMATE

Weathering processes and weathering crusts differ from place to place. These spatial differences are determined by a set of interacting factors, chiefly rock type, climate, topography, organisms, and the age of the weathered surface. Climate is a leading factor in determining chemical, mechanical, and biological weathering rates. Temperature influences the rate of weathering, but seldom the type of weathering. As a rough guide, a 10°C rise in temperature speeds chemical reactions, especially sluggish ones, and some biological reactions by a factor of two to three, a fact discovered by Jacobus Hendricus van't Hoff in 1884. The storage and movement of water in the regolith is a highly influential factor in determining weathering rates, partly integrating the influence of all other factors. Louis Peltier (1950) argued that rates of chemical and mechanical weathering are guided by temperature and rainfall conditions (Figure 7.7). The intensity of chemical weathering depends on the availability of moisture and high air temperatures. It is minimal in dry regions, because water is scarce, and in cold regions, where temperatures are low and water is scarce (because it is frozen for much or all of the year). Mechanical weathering depends upon the presence of water but is very effective where repeated freezing and thawing occurs. It is therefore minimal where temperatures are high enough to rule out freezing and where it is so cold that water seldom thaws.

Leaching regimes

Climate and the other factors determining the water budget of the regolith (and so the internal microclimate of a weathered profile) are crucial to the formation of clays by weathering and by **neoformation**. The kind of secondary clay mineral formed in the regolith depends chiefly on two things: (1) the balance between the rate of dissolution of primary minerals from rocks and

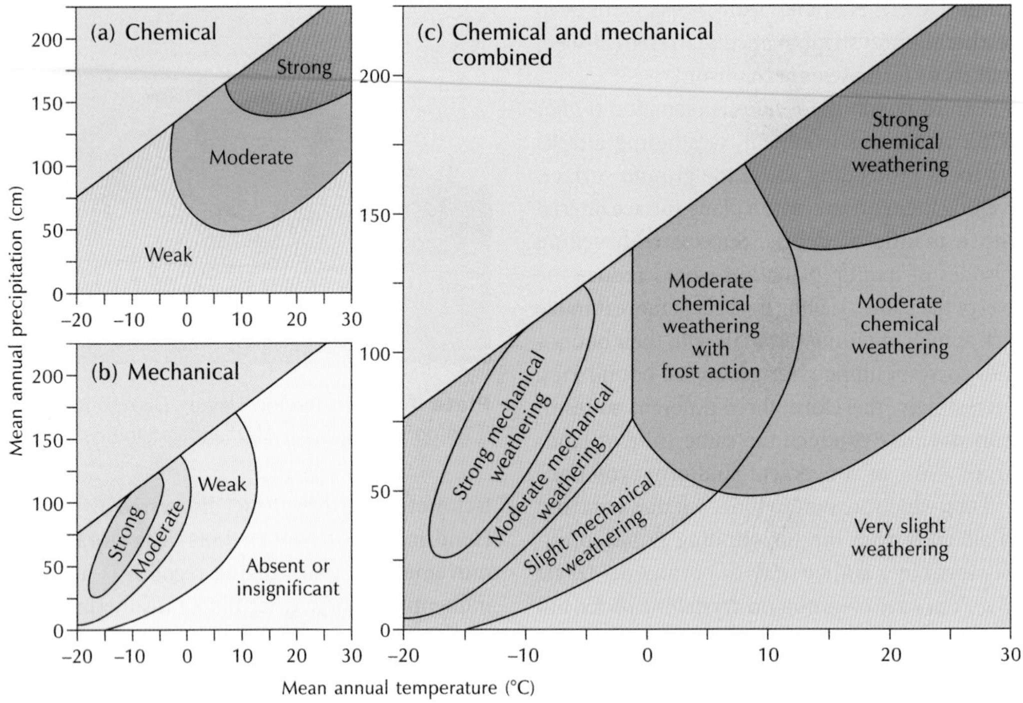

Figure 7.7 Louis Peltier's scheme relating chemical and mechanical weathering rates to temperature and rainfall. This is a classic diagram, but note that Peltier equated mechanical weathering with freeze–thaw action and overlooked thermal weathering, salt weathering, and the role played by roots, expansive clays, and rock stress relaxation. *Source*: Adapted from Peltier (1950)

the rate of flushing of solutes by water; and (2) the balance between the rate of flushing of silica, which tends to build up tetrahedral layers, and the rate of flushing of cations, which fit into the voids between the crystalline layers formed from silica. Manifestly, the leaching regime of the regolith is crucial to these balances since it determines, in large measure, the opportunity that the weathering products have to interact. Three degrees of leaching are associated with the formation of different types of secondary clay minerals – weak, moderate, and intense (e.g. Pedro 1979):

1 **Weak leaching** favours an approximate balance between silica and cations. Under these conditions the process of bisiallitization or smectization creates 2:2 clays, such as smectite, and 2:1 clays.
2 **Moderate leaching** tends to flush cations from the regolith, leaving a surplus of silica. Under these conditions, the processes of monosiallitization or kaolinization form 1:1 clays, such as kaolinite and goethite.
3 **Intense leaching** leaves very few bases unflushed from the regolith, and hydrolysis is total, whereas it is only partial in bisiallitization and monosiallitization. Under these conditions, the process of **allitization** (also termed soluviation, ferrallitization, laterization, and latosolization) produces aluminium hydroxides such as gibbsite.

Soil water charged with organic acids complicates the association of clay minerals with leaching regimes. Organic-acid-rich waters lead to cheluviation, a process associated with **podzolization**

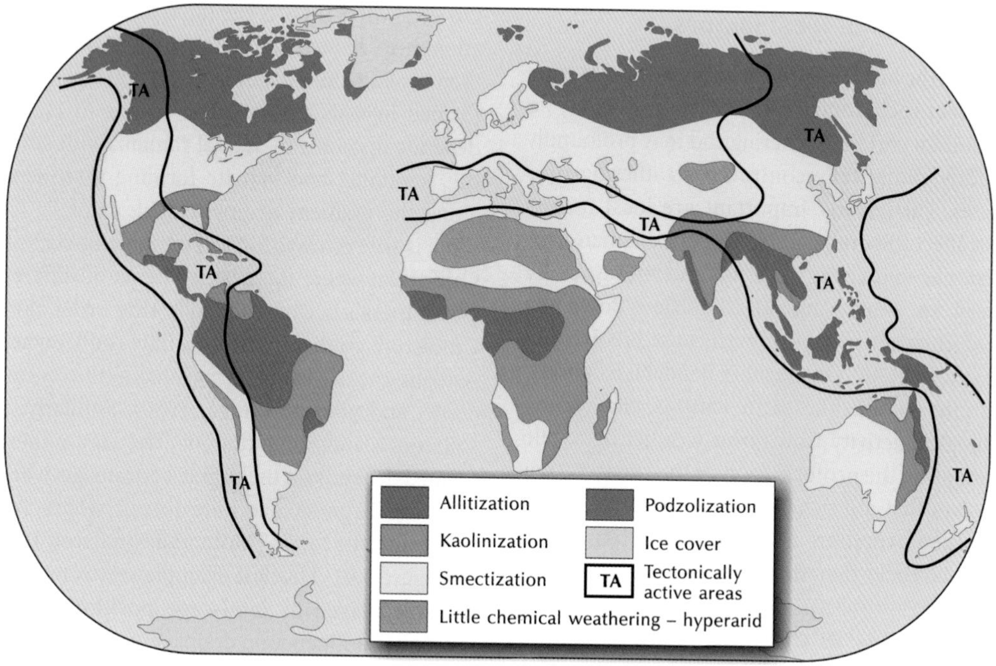

Figure 7.8 The main weathering zones of the Earth. *Source*: Adapted from Thomas (1974, 5)

in soils, which leads to aluminium compounds, alkaline earths, and alkaline cations being flushed out in preference to silica.

Weathering patterns

Given that the leaching regime of the regolith strongly influences the neoformation of clay minerals, it is not surprising that different climatic zones nurture distinct types of weathering and weathering crust. Several researchers have attempted to identify **zonal patterns in weathering** (e.g. Chernyakhovsky *et al.* 1976; Duchaufour 1982). One scheme, which extends Georges Pedro's work, recognizes six weathering zones (Figure 7.8) (Thomas 1994):

1 The **allitization zone** coincides with the intense leaching regimes of the humid tropics and is associated with the tropical rainforest of the Amazon basin, Congo basin, and South-East Asia.
2 The **kaolinization zone** accords with the seasonal leaching regime of the seasonal tropics and is associated with savannah vegetation.
3 The **smectization zone** corresponds to the subtropical and extratropical areas, where leaching is relatively weak, allowing smectite to form. It is found in many arid and semi-arid areas and in many temperate areas.
4 The **little-chemical-weathering zone** is confined to hyperarid areas in the hearts of large hot and cold deserts.
5 The **podzolization zone** conforms to the boreal climatic zone.
6 The **ice-cover zone**, where, owing to the presence of ice sheets, weathering is more or less suspended.

Within each of the first five zones, parochial variations arise owing to the effect of topography, parent rock, and other local factors. Podzolization, for example, occurs under humid tropical climates on sandy parent materials.

The effects of local factors

Within the broad weathering zones, **local factors** – parent rock, topography, vegetation – play an important part in weathering and may profoundly modify climatically controlled weathering processes. Particularly important are local factors that affect soil drainage. In temperate climates, for example, soluble organic acids and strong acidity speed up weathering rates but slow down the neoformation of clays or even cause pre-existing clays to degrade. On the other hand, high concentrations of alkaline-earth cations and strong biological activity slow down weathering, while promoting the neoformation or the conservation of clays that are richer in silica. In any climate, clay neoformation is more marked in basic volcanic rocks than in acid crystalline rocks.

Topography and drainage

The effects of local factors mean that a wider range of clay minerals occur in some climatic zones than would be the case if the climate were the sole determinant of clay formation. Take the case of tropical climates. Soils within small areas of this climatic zone may contain a range of clay minerals where two distinct leaching regimes sit side by side. On sites where high rainfall and good drainage promote fast flushing, both cations and silica are removed and gibbsite forms.

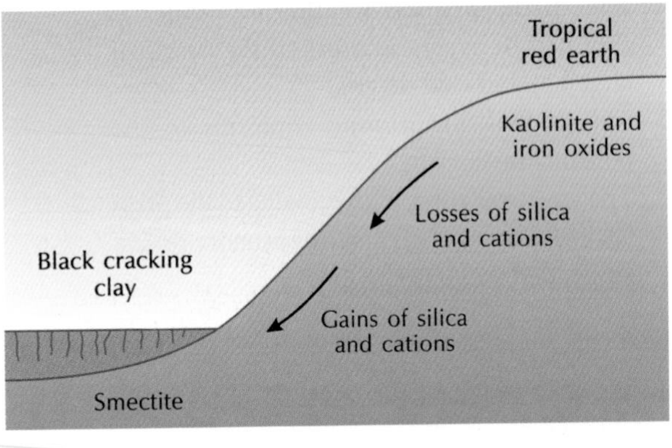

Figure 7.9 Clay types in a typical tropical toposequence.
Source: Adapted from Ollier and Pain (1996, 141)

On sites where there is less rapid flushing, but still enough to remove all cations and a little silica, then kaolinite forms. For instance, the type of clay formed in soils developed in basalts of Hawaii depends upon mean annual rainfall, with smectite, kaolinite, and bauxite forming a sequence along the gradient of low to high rainfall. The same is true of clays formed on igneous rocks in California, where the peak contents of different clay minerals occur in the following order along a moisture gradient: smectite, illite (only on acid igneous rocks), kaolinite and halloysite, vermiculite, and gibbsite (Singer 1980). Similarly, in soils on islands of Indonesia, the clay mineral formed depends on the degree of drainage: where drainage is good, kaolinite forms; where it is poor, smectite forms (Mohr and van Baren 1954; cf. Figure 7.9). This last example serves to show the role played by landscape position, acting through its influence on drainage, on clay mineral formation. Comparable effects of **topography** on clay formation in oxisols have been found in soils formed on basalt on the central plateau of Brazil (Curi and Franzmeier 1984).

Age

Time is a further factor that obscures the direct climatic impact on weathering. Ferrallitization, for example, results from prolonged leaching. Its association with the tropics is partly attributable to the antiquity of many tropical landscapes rather than to the unique properties of tropical climates. More generally, the extent of chemical weathering is correlated with the age of continental surfaces (Kronberg and Nesbitt 1981). In regions where chemical weathering has acted without interruption, even if at a variable rate, since the start of the Cenozoic era, advanced and extreme weathering products are commonly found. In some regions, glaciation, volcanism, and alluviation have reset the chemical weathering 'clock' by creating fresh rock debris. Soils less than 3 million years old, which display signs of incipient and intermediate weathering, are common in these areas. In view of these complicating factors, and the changes of climate that have occurred even during the Holocene epoch, claims that weathering crusts of recent origin (recent in the sense that they are still forming and have been subject to climatic conditions similar to present climatic conditions during their formation) are related to climate must be looked at guardedly.

WEATHERING AND HUMANS

Limestone weathers faster in urban environments than in surrounding rural areas. Archibald Geikie established this fact in his study of the weathering of gravestones in Edinburgh and its environs. Recent studies of weathering rates on marble gravestones in and around Durham, England, give rates of 2 microns per year in a rural site and 10 microns per year in an urban industrial site (Attewell and Taylor 1988).

In the last few decades, concern has been voiced over the economic and cultural costs of **historic buildings** being attacked by pollutants in cities (Plate 7.13). Geomorphologists can advise such bodies as the Cathedrals Fabric Commission in an informed way by studying urban weathering forms, measuring weathering rates, and establishing the connections between the two (e.g. Inkpen *et al.* 1994). The case of the Parthenon, Athens, was mentioned at the start of the chapter. St Paul's Cathedral in London, England, which is built of Portland limestone, is also being damaged by weathering (Plate 7.14). It has suffered considerable attack by weathering over the past few hundred years. Portland limestone is a bright white colour. Before recent cleaning, St Paul's was a sooty black. Acid rainwaters have etched out hollows where they run across the building's surface. Along these channels, bulbous gypsum precipitates have formed beneath anvils and gargoyles, and acids, particularly sulphuric acid, in rainwater have reacted with the limestone. About 0.62 microns of the limestone surface is lost each year, which represents a cumulative loss of 1.5 cm since St Paul's was built (Sharp *et al.* 1982).

Plate 7.13 Weathered balustrade on the Ashmolean Museum, Oxford, England. The balustrade has now been cleaned. (*Photograph by Heather A. Viles*)

Plate 7.14 A bust of Saint Andrew, removed from St. Paul's Cathedral because of accelerated decay, London, UK. (*Photograph by Heather A. Viles*)

Salt weathering is playing havoc with buildings of ethnic, religious, and cultural value in some parts of the world. In the towns of Khiva, Bukhara, and Samarkand, which lie in the centre of Uzbekistan's irrigated cotton belt, prime examples of Islamic architecture – including mausolea, minarets, mosques, and madrasses – are being ruined by capillary rise, a rising water table resulting from over-irrigation, and an increase in the salinity of the groundwater (Cooke 1994). The solution to these problems is that the capillary fringe and the salts connected with it must be removed from the buildings, which might be achieved by more effective water management (e.g. the installation of effective pumping wells) and the construction of damp-proof courses in selected buildings to prevent capillary rise. Building stones in coastal environments often show signs of advanced alveolar weathering owing to the crystallization of salt from sea spray.

Weathering plays an important role in releasing trace elements from rocks and soil, some of which are beneficial to humans and some injurious, usually depending on the concentrations involved in both cases. It is therefore relevant to **geomedicine**, a subject that considers the effects of trace elements or compounds in very small amounts – usually in the range of 10 to 100 parts per million (ppm) or less – on human health. For example, iodine is essential to the proper functioning of the thyroid gland. Low iodine levels lead to the enlargement of the thyroid and to the deficiency disease known as goitre. This disease is common in the northern half of the USA, probably because the soils in this area are deficient in iodine owing to low levels in bedrock and the leaching of iodine (which has soluble salts) by large volumes of meltwater associated with deglaciation. Weathering may also influence the accumulation of toxic levels of such elements as arsenic and selenium in soils and water bodies.

SUMMARY

Chemical, physical, and biological processes weather rocks. Rock weathering manufactures debris that ranges in size from coarse boulders, through sands and silt, to colloidal clays and then solutes. The chief physical or mechanical weathering processes are unloading (the removal of surface cover), frost action, alternate heating and cooling, repeated wetting and drying, and the growth of salt crystals. The chief chemical weathering processes are solution or dissolution, hydration, oxidation, carbonation, hydrolysis, and chelation. The chemical and mechanical action of animals and plants bring about biological weathering. Weathering products depend upon weathering environments. Transport-limited environments lead to the production of a weathered mantle (regolith and soil); weathering-limited environments lead to the generation of weathering landforms. The weathered mantle or regolith is all the weathered debris lying above the unweathered bedrock. Saprock and saprolite is the portion of the regolith that remains in the place that it was weathered, unmoved by mass movements and erosive agents. Geomorphic processes of mass wasting and erosion have moved the mobile upper portion of regolith, sometimes called the mobile zone, residuum, or pedolith. Weathering landforms include large-scale cliffs and pillars, and smaller-scale rock-basins, tafoni, and honeycombs. Joints have a strong influence on many weathering landforms, including those formed on granite. Characteristic forms include bornhardts and tors. Weathering processes are influenced by climate, rock type, topography and drainage, and time. Climatically controlled leaching regimes are crucial to understanding the building of new clays (neoformation) from weathering products. A distinction is made between weak leaching, which promotes the formation of 2:2 clays, moderate leaching, which encourages the formation of 1:1 clays, and intense leaching, which fosters the formation of aluminium hydroxides. The world distribution of weathering crusts mirrors the world distribution of leaching regimes. Weathering processes attack historic buildings and monuments, including the Parthenon and St Paul's Cathedral, and they can be a factor in understanding the occurrence of some human diseases.

ESSAY QUESTIONS

1 Describe the chief weathering processes.
2 Explain the chief weathering products.
3 Evaluate the relative importance of factors that affect weathering.
4 Explore the impact of weathering on human-made structures.

FURTHER READING

Goudie, A. (1995) *The Changing Earth: Rates of Geomorphological Process.* Oxford and Cambridge, Mass.: Blackwell.
A good section in here on rates of weathering.

Ollier, C. D. and Pain, C. F. (1996) *Regolith, Soils and Landforms.* Chichester: John Wiley & Sons.
An intriguing textbook on connections between geomorphology, soil, and regolith.

Pope, G. A. (2013b) Weathering in the tropics, and related extratropical processes. In: J. Shroder (Editor in Chief), G. A. Pope (ed.) *Treatise on Geomorphology, Vol. 4, Weathering and Soils Geomorphology,* pp. 179–96. San Diego, California: Academic Press.
A state-of-the-art summary.

Pope, G. A. (2013a) Weathering and sediment genesis. In: J. Shroder (Editor in Chief), G. A. Pope (ed.) *Treatise on Geomorphology, Vol. 4, Weathering and Soils Geomorphology,* pp. 284–93. San Diego, California: Academic Press.
Another excellent summary.

Taylor, G. and Eggleton, R. A. (2001) *Regolith Geology and Geomorphology.* Chichester: John Wiley & Sons.
An excellent book with a geological focus, but no worse for that.

Thomas, M. F. (1994) *Geomorphology in the Tropics: A Study of Weathering and Denudation in Low Latitudes.* Chichester: John Wiley & Sons.
A most agreeable antidote to all those geomorphological writings on middle and high latitudes.

8

CHAPTER EIGHT

HILLSLOPES

Hillslopes are an almost universal landform, occupying some 90 per cent of the land surface. This chapter will explore:

- Hillslope environments
- Hillslope transport processes and hillslope development
- The form of hillslopes
- Humans and hillslopes

HAZARDOUS HILLSLOPES

Any geomorphic process of sufficient magnitude that occurs suddenly and without warning is a danger to humans. Landslides, debris flows, rockfalls, and many other mass movements associated with hillslopes take their toll on human life. Most textbooks on geomorphology catalogue such disasters. A typical case is the Mount Huascarán debris avalanches. At 6,768 m, Mount Huascarán is Peru's highest mountain. Its peaks are snow- and ice-covered. In 1962, some 2,000,000 m^3 of ice avalanched from the mountain slopes and mixed with mud and water. The resulting debris avalanche, estimated to have had a volume of 10,000,000 m^3, rushed down the Rio Shacsha valley at 100 km/hr carrying boulders weighing up to 2,000 tonnes. It killed 4,000 people, mainly in the town of Ranrahirca. Eight years later, on 31 May 1970, an earthquake of about magnitude 7.7 on the Richter scale, whose epicentre lay 30 km off the Peruvian coast where the Nazca plate is being subducted, released another massive debris avalanche that started as a sliding mass about 1 km wide and 1.5 km long. The avalanche swept about 18 km to the village of Yungay at up to 320 km/hr, picking up glacial deposits en route where it crossed a glacial moraine. It bore boulders the size of houses. By the time it reached Yungay, it had picked up enough fine sediment and water to become a mudflow consisting of 50–100 million tonnes of water, mud, and rocks with a 1-km-wide front. Yungay and Ranrahirca were buried. Some 1,800 people died in Yungay and 17,000 in Ranrahirca.

HILLSLOPE ENVIRONMENTS

Hillslopes are ubiquitous, forming by far the greater part of the landscape. Currently, ice-free landscapes of the world are 90 per cent hillslopes and 10 per cent river channels and their floodplains. Hillslopes are an integral part of the drainage basin system, delivering water and sediment to streams. They range from flat to steep. Commonly, hillslopes form **catenas** – sequences of linked slope units running from drainage divide to valley floor. Given that climate, vegetation, lithology, and geological structure vary so much from place to place, it is not surprising that hillslope processes also vary in different settings and that hillslopes have a rich diversity of forms. Nonetheless, geomorphologists have found that many areas have a characteristic hillslope form that determines the general appearance of the terrain. Such characteristic hillslopes will have evolved to a more-or-less equilibrium state under particular constraints of rock type and climate.

Hillslopes may be bare rock surfaces, regolith and soil may cover them, or they may comprise a mix of bare rock and soil-covered areas. Hillslopes mantled with regolith or soil, perhaps with some exposures of bare rock, are probably the dominant type. They are usually designated **soil-mantled hillslopes**. However, hillslopes formed in bare rock – **rock slopes** – are common. They tend to form in three situations (Selby 1982, 152). First, rock slopes commonly form where either uplift or deep incision means that they sit at too high an elevation for debris to accumulate and bury them. Second, they often form where active processes at their bases remove debris, so preventing its accumulation. Third, they may form where the terrain is too steep or the climate is too cold or too dry for chemical weathering and vegetation to create and sustain a regolith. More generally, bare rock faces form in many environments where slope angles exceed about 45°, which is roughly the maximum angle maintained by rock debris. In the humid tropics, a regolith may form on slopes as steep as 80° on rocks such as mudstones and basalts because weathering and vegetation establishment are so speedy. Such steep regolith-covered slopes occur on Tahiti and in Papua New Guinea where, after a landslide, rock may remain bare for just a few years. Rock properties and slope processes determine the form of rock slopes. There are two extreme cases of rock properties. The first case is 'hard' rocks with a very high internal strength (the strength imparted by the internal cohesive and frictional properties of the rock). These usually fail along partings in the rock mass – joints and fractures. The second case is 'soft' rocks of lower intact strength or intense fracturing that behave more like soils. As a rule of thumb, bare rock slopes form on hard rocks. However, there are circumstances that favour the formation of bare rock slopes on soft rocks. For example, steep rock slopes may occur on mudstones and shales that lie at high elevations where the slopes are regularly undercut. Even so, such slopes denude far more rapidly than do slopes on hard rocks, and they are far more likely to develop a soil and vegetation cover (Selby 1982, 152). Some rock slopes speedily come into equilibrium with formative processes and rock properties, their form reflecting the strength of the rock units on which they have developed. Such rock slopes occur on massive and horizontally bedded rocks. On dipping and folded rocks, the form of bare rock slopes conforms to underlying geological structures.

HILLSLOPE PROCESSES

Gravity, flowing water, and temperature changes are the main forces behind hillslope processes, with the action of animals and plants being important in some situations. Weathering on hillslopes, as elsewhere, includes the *in situ* conversion of bedrock into regolith and the subsequent chemical and mechanical transformation of regolith. Several hillslope processes serve to transport regolith and other weathering products. They range from slow and continual processes to rapid and intermittent processes. Slow and continual processes fall into three categories: leaching, soil creep, and rainsplash and sheet wash.

Gravitational hillslope processes

Stress and strain in rocks, soils, and sediments

Earth materials are subject to stress and strain. A **stress** is any force that tends to move materials downslope. **Gravity** is the main force, but swelling and shrinking, expansion and contraction, ice-crystal growth, and the activities of animals and plants set up forces in a soil body. The stress of a body of soil on a slope depends largely upon the mass of the soil body, m, and the angle of slope, θ (theta):

$$\text{Stress} = m \sin \theta$$

Strain is the effect of stress upon a soil body. It may be spread uniformly throughout the body, or it may focus around joints where fracture may occur. It may affect individual particles or the entire soil column.

Materials possess an inherent resistance against downslope movement. **Friction** is a force that acts against gravity and resists movement. It depends on the roughness of the plane between the soil and the underlying material. Downslope movement of a soil body can occur only when the applied stress is large enough to overcome the maximum frictional resistance. Friction is expressed as a coefficient, μ (mu), which is equal to the angle at which sliding begins (called the **angle of plane sliding friction**). In addition to friction, cohesion between particles resists downslope movement. **Cohesion** measures the tendency of particles within the soil body to stick together. It arises through capillary suction of water in pores, compaction (which may cause small grains to interlock), chemical bonds (mainly **Van der Waals bonds**), plant root systems, and the presence of such cements as carbonates, silica, and iron oxides. Soil particles affect the mass cohesion of a soil body by tending to stick together and by generating friction between one another, which is called the **internal friction** or **shearing resistance** and is determined by particle size and shape, and the degree to which particles touch each other. The **Mohr–Coulomb** equation defines the shear stress that a body of soil on a slope can withstand before it moves:

$$\tau_s = c + \sigma \tan \varphi$$

where τ_s (tau-s) is the **shear strength** of the soil, c is soil cohesion, σ (sigma) is the **normal stress** (at right-angles to the slope), and φ (phi) is the **angle of internal friction** or **shearing resistance**. The angle φ is the angle of internal friction within the slope mass and represents the angle of contact between the particles making up the soil or unconsolidated mass and the underlying surface. It is usually greater than the slope angle, except in free-draining, cohesionless sediments. To visualize it, take a bowl of sugar and slowly tilt it: the angle of internal friction is the degree of tilt required for failure (the flow of sugar grains) to occur. All unconsolidated materials tend to fail at angles less than the slope angle upon which they rest, loosely compacted materials failing at lower angles than compacted materials. The pressure of water in the soil voids, that is, the **pore water pressure**, ξ (xi), modifies the shear strength:

$$\tau_s = c + (\sigma - \xi) \tan \varphi$$

This accounts for the common occurrence of slope failures after heavy rain, when pore water pressures are high and effective normal stresses ($\sigma - \xi$) low. On 10 and 11 January 1999, a large portion of the upper part of Beachy Head, Sussex, England, collapsed (cf. p. 364). The rockfall appears to have resulted from increased pore pressures in the chalk following a wetter than normal year in 1998 and rain falling on most days in the fortnight before the fall.

The Mohr–Coulomb equation can be used to define the shear strength of a unit of rock resting

Table 8.1 Landslides and other mass movements

Type of movement	Rock	Soil
Fall	Rockfall/icefall	Boulder/debris/silt fall
Topple	Rock block topple Rock flexural topple	Gravel/sand/silt topple
Slide	Rock rotational slide ('rock slump') Rock planar slide ('block slide') Rock wedge slide Rock compound slide Rock irregular slide ('rock collapse')	Clay/silt rotational slide ('soil slump') Clay/silt planar slide Gravel/sand/debris slide Clay/silt compound slide
Spread	Rock slope spread	Sand/silt liquefaction spread Sensitive clay spread
Flow	Rock avalanche (sturzstrom) Ice avalanche	Sand/silt/debris dry flow Sand/silt/debris flowslide Sensitive clay flowslide Debris flow Mud flow Debris flood Debris avalanche Earthflow Peat flow
Slope deformation	Mountain slope deformation Rock slope deformation	Soil slope deformation Soil creep Solifluction and gelifluction
Subsidence	Cavity collapse Settlement	

Source: Adapted from Hungr *et al.* (2014)

on a failure plane and the susceptibility of that material to landsliding, providing the effects of fractures and joints are included. Whenever the stress applied to a rock body is greater than the shear strength, the material will fail and move downslope. A scheme for defining the **intact rock strength** (the strength of rock excluding the effects of joints and fractures) has been devised. Rock mass strength may be assessed using intact rock strength and other factors (weathering, joint spacing, joint orientations, joint width, joint continuity and infill, and groundwater outflow). Combining these factors gives a rock mass strength rating ranging from very strong, through strong, moderate, and weak, to very weak (see Selby 1980).

Mass movements

Mass movements may be classified in many ways. Table 8.1 summarizes a scheme recognizing seven basic types and several subtypes, according to the chief mechanisms involved (fall, topple, slide, spread, flow, deformation, and subsidence) and the water content of the moving body (very low, low, moderate, high, very high, and extremely high):

1 **Falls** involve the detachment, fall, rolling, and bouncing of rock, ice, or soil fragments, singly or in clusters, usually from a steep slope. Soil fragments may fall from cohesive soil bodies, as in riverbanks. **Rockfalls** are more common, especially in landscapes with steep, towering rock slopes and cliffs (Figure 8.1a; Plates 8.1 and 8.2). Talus slopes commonly form in such landscapes (p. 168).

2 **Topples** involve the forward rotation of rock or soil about an axis or point sitting below the centre of gravity of the displaced mass (usually at or near the base of the slope). They characteristically occur in strongly jointed, fractured, or steeply dipping slab-like or columnar rock masses, such as columnar basalts, slates, and rocks with well-developed jointing. Soil may also topple.

3 **Slides** are a widespread form of mass movement. They take place along clear-cut shear planes and are usually ten times longer than they are wide. Two chief subtypes are rotational slides and translational slides. **Rotational slides**, also called **slumps**, occur along concave shear planes, normally under conditions of low to moderate water content, and are commonest on thick, uniform materials such as clays (Figure 8.1b; Plate 8.3). They include rock slumps and soil slumps. The landslide at the Quirang on the Trotternish escarpment, the Isle of Syke, is the largest rotational slide in Britain, extending over 2 km in width (Figure 1.10). It occurred because Palaeogene lava flows weighed down on the weaker underlying Jurassic sedimentary rocks, which sheared along north–south aligned faults and huge blocks slid seawards along a rotational glide plane (Ballantyne 2007). **Translational slides** occur along planar shear planes and include rock planar slides (block slides) and

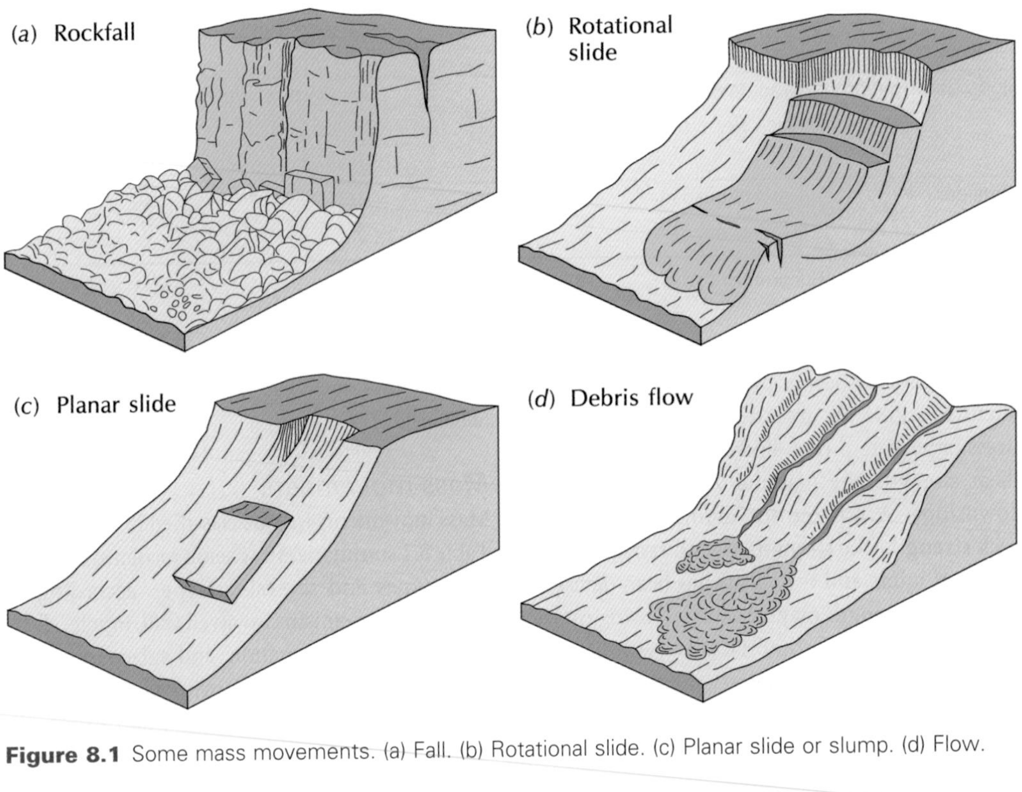

Figure 8.1 Some mass movements. (a) Fall. (b) Rotational slide. (c) Planar slide or slump. (d) Flow.

various types of soil slide (Figure 8.1c; Plate 8.4). They are usually shallow in comparison with the length of the mass movement, with a depth-to-length ratio of <0.1 (Guthrie 2013). Some of the largest and most damaging landslides on Earth are translational landslides.

4 **Spreads** occur when a cohesive rock or soil mass becomes liquefied owing to saturation or partial saturation. The trigger to liquefaction may be an earthquake or the removal of toe support, as for lake clays in the outside bend of a river. Once initiated, the **liquefaction spread** may extend many kilometres and retrograde inland as subsidence of the disturbed ground removes support for the ground still intact (Guthrie 2013). It can be extremely rapid (Plates 8.5 and 8.6). **Rock slope spread** or **block spread** involves the displacement and rotation of rigid blocks of stronger rock overlying softer material that suffers severe plastic deformation. This rock slope spread is very slow and very common in horizontally bedded, weak sedimentary strata.

5 **Flows** involve shear through the soil, rock, or snow and ice debris as it moves over a rigid bed (Plate 8.7). The rate of flow is slow at the base of the flowing body and increases towards the surface. Most movement occurs as turbulent motion. **Rock avalanches**, sometimes called **sturzstrom** (meaning rock slide stream), are the very rapid, massive, flow-like motion of fragmented rock from a large rock slide or rock fall, often down mountainsides. Glacial ice may be involved in such avalanches. Flows of **soil** go by various names, depending upon the material involved and its moisture status: dry flow, flowslide, debris flow, mud flow, debris flood, debris avalanche, earthflow (Plate 8.8), and peat flow. A **debris flow** is a fast-moving body of sediment particles with water or air

Plate 8.1 Rockfall at Kern Knotts on Great Gable, England. Some of the boulders at the base of the crag came down as a result of the Carlisle earthquake in December 1979. (*Photograph by Peter Wilson*)

Plate 8.2 Rockfall in cliffs near Burton Bradstock, Dorset, England. (*Photograph by Nick Scarle*)

Plate 8.3 Shallow rotational landslide, Rockies foothills, Wyoming, USA. (*Photograph by Tony Waltham Geophotos*)

Plate 8.4 A spectacular rockslide on the north side of Crag Fell at the northern end of Ennerdale Water in the English Lake District. Notice the headscarp extending across the hillside for over 1 km and attaining a maximum height of 40 m. In places, the slipped masses appear to be 'coherent' megablocks; elsewhere, fragmentation is marked by 6–8-m tall rock pinnacles. (*Photograph by Peter Wilson*)

or both that often has the consistency of wet cement. Debris flows occur as a series of surges lasting from a few seconds to several hours that move at 1 to 20 m/s. They may flow several kilometres beyond their source areas (Figure 8.1d). Some are powerful enough to destroy buildings and snap off trees that lie in their path. **Mud flows** triggered by water saturating the debris on the sides of volcanoes are called **lahars**. When Mount St Helens, USA, exploded on 18 May 1980 a huge **debris avalanche** mobilized a huge body of sediment into a remarkable lahar that ran 60 km from the volcano down the north and south forks of the Toutle River, damaging 300 km of road and 48 road bridges in the process.

6 **Slope deformation** occurs in several ways. **Mountain slope deformation** involves highly

Plate 8.5 Saint-Jude landslide, Quebec, Canada, a complex spread that occurred on May 10 2010. It resulted from high artesian water pressure under the river and riverbank erosion. A small rotational failure at the toe probably led to the progressive failure along horizontal surface and the rearrangement of the soil into horsts and grabens above a then remoulded clay layer. (*Photograph courtesy of the Ministère des Transports du Québec*)

stressed rock masses forming mountain slopes, with vertical relief of the order of 1 km or more, deforming to create such topographic surface features such as scarps, benches, cracks, trenches, and bulges. This deformation is induced by gravity. The nature of the deformation depends on many factors, the most important of which appear to be weathering and alteration of the rock mass caused by climatic factors and the circulation of fluids within the mountain, which both depend upon the physicochemical and mechanical properties of the rock. The basic process appears to be that rock weathering and alteration progressively reduce the effective strength of an initially homogeneous and stable mountain so that it eventually undergoes increasing inelastic, gravity-driven deformation,

Plate 8.6 Horst and graben formed in Saint-Jude landslide. (*Photograph courtesy of the Ministère des Transports du Québec*)

Plate 8.7 The Oso flowslide: on 22 March 2014, a hillside made unstable by saturation from heavy rainfall collapsed, sending debris across the North Fork of the Stillaguamish River, Washington, USA, engulfing a rural neighbourhood, and killing forty-three people. The line of fallen trees grew at the top of the slope before the slide occurred. (*Photograph by Mark Reid, USGS*)

Plate 8.8 Slump and earthflow.
(*Photograph by Marli Miller*)

Plate 8.9 Bent tree trunks from soil creep, east Nevada, USA. These are known as pistol-butt trees.
(*Photograph by Marli Miller*)

Plate 8.10 Terracettes on Melbury Hill, Dorset, England.
(*Photograph by Nick Scarle*)

including the sagging of crests and the appearance of large fractures that may create landslides (Chemenda *et al.* 2009). It is also induced by fluvial incision destabilizing slopes (Tsou *et al.* 2015). **Soil slope deformation** is the deep-seated, slow to extremely slow deformation of valley or hillslopes formed of soils (usually cohesive soils). It often occurs in permafrost slopes with high ice content. **Soil creep** is the very slow movement of the top soil layers, normally <1 m deep, on a slope resulting from alternating phases of expansion and contraction, in conjunction with other processes, that create **heave** in the soil. Heave results from settling movements in the soil effected by biogenic mechanisms (burrowing animals, tree throw, and so on), solution, freeze–thaw cycles, warming–cooling cycles, wetting–drying cycles, and, in some hillslopes, the shrinking and swelling of clays and the filling of desiccation cracks from upslope. Material moves downslope during the cycles because expansion lifts material at right-angles to the slope but contraction drops it nearly vertically under the influence of gravity. Heave is classed as soil creep (finer material) or talus creep (coarser material). Soil creep is common under humid and temperate climates (Plate 8.9). It occurs mainly in environments with seasonal changes in moisture and soil temperature. **Terracettes** frequently occur on steep grassy slopes where soil creep is active, although shallow landslides may be an important factor in their formation (Plate 8.10). **Talus creep** is the slow downslope movement of talus and results chiefly from rockfall impact, but thermal expansion and contraction may play a role. **Frost creep** occurs when the expansion and contraction is brought about by freezing and thawing (p. 318). **Solifluction** (or soil fluction) is the very slow but intensive shallow soil creep involving the active layer in Alpine or polar permafrost that creates characteristic solifluction lobes (p. 326). It is the downslope movement of water-saturated soil over frozen ground, which acts as a sliding plane, during summer months in periglacial environments. It results from the combined action of frost creep and gelifluction, which is the slow saturated flowage of thawed ice-rich sediments (see p. 318).

7 **Subsidence** occurs in two ways: cavity collapse and settlement. First, in **cavity collapse**, rock or soil plummets into underground cavities, as in karst terrain (p. 417), in lava tubes, or in mining areas. In **settlement**, the ground surface is lowered progressively by compaction, often because of groundwater withdrawal or earthquake vibrations.

Gravity tectonics

Mass movements may occur on geological scales. Large rock bodies slide or spread under the influence of gravity to produce such large-scale features as **thrusts** and **nappes**. Most of the huge nappes in the European Alps and other intercontinental orogens are probably the product of massive **gravity slides**. Tectonic denudation is a term that describes the unloading of mountains by gravity sliding and spreading. The slides are slow, being only about 100 m/yr under optimal conditions (that is, over such layers as salt that offer little frictional resistance).

Hillslope transport processes

Surface processes: dry ravel

Dry ravel is the rolling, bouncing, and sliding of individual particles down a slope (Gabet 2003). It is a dominant hillslope sediment-transport process in steep arid and semi-arid landscapes, and includes the mobilization of particles during fires when sediment wedges that have accumulated behind vegetation collapse, as well as mobilization by bioturbation and by small landslides.

Surface processes: rainsplash and rainflow

Rainsplash and **sheet wash** are common in arid environments and associated with the generation of Hortonian overland flow (p. 210). There is a continuum from rainsplash, through rainflow, to sheet wash. Falling raindrops dislodge sediment to form 'splash', which moves in all directions through the air resulting in a net downslope transport of material. Experimental studies using a sand trough and simulated rainfall showed that on a 5° slope about 60 per cent of the sediment moved by raindrop impact moves downslope and 40 per cent upslope; on a 25° slope 95 per cent of the sediment moved downslope (Mosley 1973). Smaller particles are more susceptible to rainsplash than larger ones. The amount of splash depends upon many factors, including rainfall properties (e.g. drop size and velocity, drop circumference, drop momentum, kinetic energy, and rainfall intensity) and such landscape characteristics as slope angle and vegetation cover (see Salles *et al.* 2000). **Rain power** is a mathematical expression that unites rainfall, hillslope, and vegetation characteristics, and that allows for the modulation by flow depth (Gabet and Dunne 2003). It is a good predictor of the detachment rate of fine-grained particles.

Rainflow is transport caused by the traction of overland flow combined with detachment by raindrop impact, which carries particles further than rainsplash alone. **Sheet wash** carries sediment in a thin layer of water running over the soil surface. This is not normally a uniformly thick layer of water moving downslope; rather, the sheet subdivides and follows many flowpaths dictated by the microtopography of the surface. Sheet wash results from overland flow. On smooth rock and soil surfaces, a continuous sheet of water carries sediment downslope. On slightly rougher terrain, a set of small rivulets link water-filled depressions and bear sediment. On grassed slopes, sediment-bearing threads of water pass around stems; and, in forests with a thick litter layer, overland flow occurs under decaying leaves and twigs. The efficacy of sheet wash in transporting material is evident in the accumulation of fine sediment upslope of hedges at the bottom of cultivated fields.

Vegetation cover has a huge impact on erosion by rainsplash, rainflow, and sheet wash. Soils with very little or no cover of plants, leaf litter, or crop residues are far more vulnerable to erosion. Plants, surface litter, and organic residues serve to guard the soil from raindrop impact and splash, to slow down the flow rate of surface runoff, and to allow excess surface water to infiltrate the soil.

Subsurface processes: leaching and through-wash

Leaching involves the removal of weathered products in solution through the rock and the soil. Solution is an efficacious process in hillslope denudation. It does not always lead to surface lowering, at least at first, because the volume of rock and soil may stay the same. Solution takes place in the body of the regolith and along subsurface lines of concentrated water flow, including throughflow in percolines and pipes.

In well-vegetated regions, the bulk of falling rain passes into the soil and moves to the water table or moves underneath the hillslope surface as throughflow. Throughflow carries sediment in solution and in suspension. This process is variously called **through-wash**, **internal erosion**, and **suffossion**, which means a digging under or undermining (Chapuis 1992). Suspended particles and colloids transported this way will be about ten times smaller than the grains they pass through, and through-wash is important only in washing silt and clay out of clean sands, and in washing clays through cracks and roots holes. For instance, in the Northaw Great Wood, Hertfordshire, England, field evidence suggests that silt and clay have moved downslope through Pebble Gravel, owing to through-wash (Huggett 1976). Where throughflow returns to the surface at seeps, positive pore pressures may develop that grow large enough to cause material to become detached and removed. Throughflow may occur along percolines. It may also form pipes in the soil, which form gullies if they should collapse, perhaps during a heavy rainstorm.

Bioturbation

Geomorphologists have until recently tended to dismiss the effects of animals and plants on hillslope processes, this despite the early attribution of soil creep to the action of soil animals and plant roots (Davis 1898). However, animals and plants make use of the soil for food and for shelter and, in doing so, affect it in multifarious ways. For instance, the uprooting of trees may break up bedrock and transport soil downslope. Since the mid-1980s, the importance of **bioturbation** – the churning and stirring of soil by organisms – to sediment transport and soil production on hillslopes has come to the fore. Andre Lehre (1987) found that biogenic creep is more important than inorganic creep. Another study concluded that bioturbated areas on Alpine slopes in the Rocky Mountains of Colorado, USA, have sediment movement rates increased by one or two orders of magnitude compared with areas not subject to significant bioturbation (Caine 1986). A review in 2003 concluded that bioturbation is undeniably a key geomorphic factor in many landscapes (Gabet *et al.* 2003), a fact strongly supported by William E. Dietrich and J. Taylor Perron (2006).

Climate and hillslope processes

Extensive field measurements since about 1960 show that hillslope processes appear to vary considerably with climate (Young 1974; Saunders and Young 1983; Young and Saunders 1986). Soil creep in temperate maritime climates shifts about 0.5–2.0 mm/year of material in the upper 20–25 cm of regolith; in temperate continental climates rates run in places a little higher at 2–15 mm/year, probably owing to more severe freezing of the ground in winter. Generalizations about the rates of soil creep in other climatic zones are unforthcoming owing to the paucity of data. In Mediterranean, semi-arid, and savannah climates, creep is probably far less important than surface wash as a denuder of the landscape and probably contributes significantly to slope retreat only where soils are wet, as in substantially curved concavities or in seepage zones. Such studies as have been made in tropical sites indicate a rate of around 4–5 mm/year. Solifluction, which includes frost creep caused by heaving and gelifluction, occurs 10–100 times more rapidly than soil creep and affects material down to about 50 cm, typical rates falling within the range 10–100 mm/year. Wet conditions and silty soils favour solifluction: clays are too cohesive, and sands drain too readily. Solifluction is highly seasonal, most of it occurring during the summer months. The rate of surface wash, which comprises rainsplash and surface flow, is determined very much by the degree of vegetation cover, and its relation to climate is not clear. The range is 0.002–0.2 mm/year. It is an especially important denudational agent in semi-arid and (probably) arid environments, and makes a significant contribution to denudation in tropical rainforests. Solution (leaching) probably removes as much material from drainage basins as all other processes combined. Rates are not so well documented as for other geomorphic processes, but typical values, expressed as surface-lowering rates, are as follows: in temperate climates on siliceous rocks, 2–100 mm/millennium, and on limestones 2–500 mm/millennium. In other climates, data are fragmentary, but often fall in the range 2–20 mm/millennium and show little clear relationship with temperature or rainfall. On slopes where landslides are active, the removal rates are very high irrespective of climate, running at between 500 and 5,000 mm/millennium.

Transport-limited and supply-limited processes

It is common to draw a distinction between hillslope processes limited by the transporting capacity of sediment and hillslope processes limited by the supply of transportable material (Kirkby 1971; cf. p. 167). In **transport-limited processes**, the rate of soil and rock transport limits the delivery of sediment to streams. In other words, the supply of sediment exceeds the capacity to remove it, and transport processes and their spatial variation dictate hillslope form. Soil creep, gelifluction, through-wash, rainflow, rainsplash, and rillwash are all hillslope processes limited by transporting capacity. On **supply-limited** (or **weathering-limited**) **hillslopes**, the rate of sediment production by weathering and erosional detachment (through overland flow and mass movement) limits the delivery of sediment to streams. In other words, weathering and erosional processes dictate hillslope form. Leaching of solutes, landsliding, debris avalanches, debris flows, and rockfall are all hillslope processes limited by sediment supply.

The distinction between transport-limited and supply-limited processes is often blurred. Nonetheless, it is an important distinction because it affects the long-term evolution of hillslopes. Hillslopes and landscapes dominated by transport-limited removal typically carry a thick soil layer supporting vegetation, and slope gradients tend to reduce with time. Hillslopes and landscapes dominated by supply-limited removal often bear thin soils with little vegetation cover, and characteristically steep slopes tend to retreat maintaining a sharp gradient. Mathematical models of hillslope evolution support these findings, suggesting that the wearing back or wearing down of the mid-slope depends upon the processes in operation. As a generalization, surface wash processes lead to a back-wearing of slopes, whereas creep processes lead to a down-wearing of slopes (e.g. Nash 1981). Nonetheless, the pattern of slope retreat and slope decline is crucially dependent on conditions at the slope base, and especially on the transport capacity of streams.

A study of young fault scarps formed in alluvium in north-central Nevada, USA, showed that hillslope processes change as the scarps age (Wallace 1977) (Figure 8.2). The original fault scarps stand at 50° to 70°. At this stage, mass

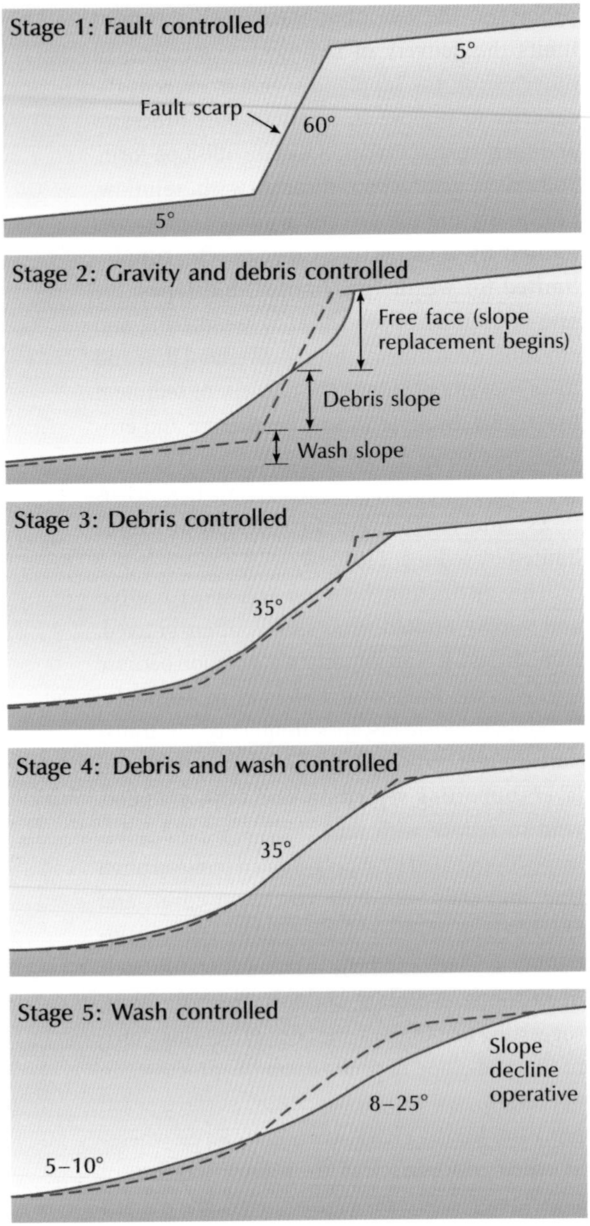

Figure 8.2 Proposed sequence of change on a fault scarp developed in alluvium, Nevada, USA. The changes are incremental, the dashed line shown at each stage representing the hillslope profile at the previous stage. *Source*: Adapted from Wallace (1977)

wasting is the dominant process, a free face develops at the scarp top, which retreats through debris fall, and material accumulates lower down. Later, the scarp slope adopts the angle of repose of the debris, which is about 35°. At this gentler gradient, wash erosion dominates hillslope development and further slope decline occurs.

Hillslope development

Slope processes fashion hillsides over hundreds of thousands to millions of years. It is therefore impossible to study hillslope evolution directly. Space–time substitution allows the reconstruction of long-term changes in hillslopes under special circumstances (p. 60). Mathematical models offer another means of probing long-term changes in hillslope form.

Michael J. Kirkby is a leading figure in the field of hillslope modelling. He used the **continuity equation** of debris moving on hillslopes and in rivers as a basis for hillslope models (Kirkby 1971). In one dimension, the equation of debris on a hillside is:

$$\frac{\delta h}{\delta t} = -\frac{dS}{dx}$$

where h is the height of the land surface and S is the sediment transport rate, which needs defining by a transport (process) equation for the process or processes being modelled. A general **sediment transport equation** is:

$$S = f(x)^m \left(\frac{dh}{dx}\right)^n$$

where $f(x)^m$ is a function representing hillslope processes in which sediment transport is proportional to distance from the watershed (roughly the distance of overland flow) and $(dh/dx)^n$ represents processes in which sediment transport is proportional to slope gradient. Empirical work suggests that $f(x)^m = xm$, where m varies according to the sediment-moving processes in operation, representative values being 0 for soil creep and

Box 8.1 HILLSLOPE MODELS

Michael J. Kirkby's (1985) attempts to model the effect of rock type on hillslope development, with rock type acting through the regolith and soil, nicely demonstrate the process of hillslope modelling. Figure 8.3 shows the components and linkages in the model, which are more precisely defined than in traditional models of hillslope development. Rock type influences rates of denudation by solution, the geotechnical properties of soil, and the rates of percolation through the rock mass and its network of voids to groundwater. Climate acts through its control of slope hydrology, which in turn determines the partitioning of overland and subsurface flow. With suitable process equations fitted, the model simulates the development of hillslopes and soils for a fixed base level. Figure 8.4 is the outcome of a simulation that started with a gently sloping plateau ending in a steep bluff and a band of hard rock dipping at 10° into the slope. The hard rock is less soluble, and has a lower rate of landslide retreat than the soft band, but has the same threshold gradient for landsliding. Threshold gradients, or angles close to them, develop rapidly on the soft strata. The hard rock is undercut, forming a free face within a few hundred years. After some 20,000 years, a summit convexity begins to replace the threshold slope above the hard band, the process of replacement being complete by 200,000 years when the hard band has little or no topographic expression. The lower slope after 200,000 years stands at an almost constant gradient of 12.4°, just below the landslide threshold. Soil development (not shown on the diagram) involves initial thickening on the plateau and thinning by landslides on the scarp. Soil distribution is uneven owing to the localized nature of landslides. Once the slope stabilizes, thick soils form everywhere except over the hard band.

From this simulation and another in which solution is the sole process, Kirkby makes a number of deductions that appear to correspond to features in actual landscapes. First, the geotechnical

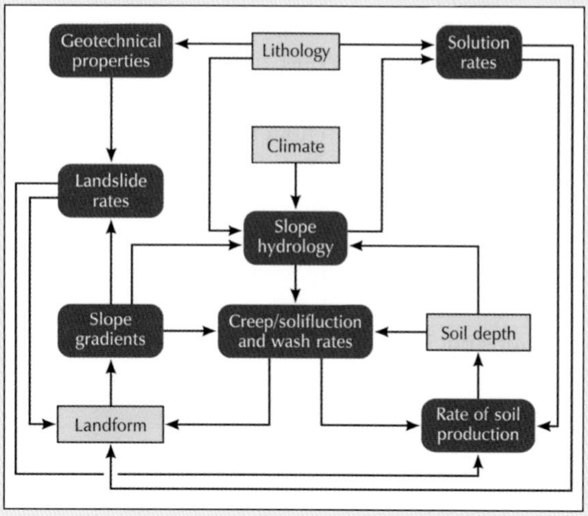

Figure 8.3 Components and linkages in Kirkby's model of hillslope evolution. *Source*: Adapted from Kirkby (1985)

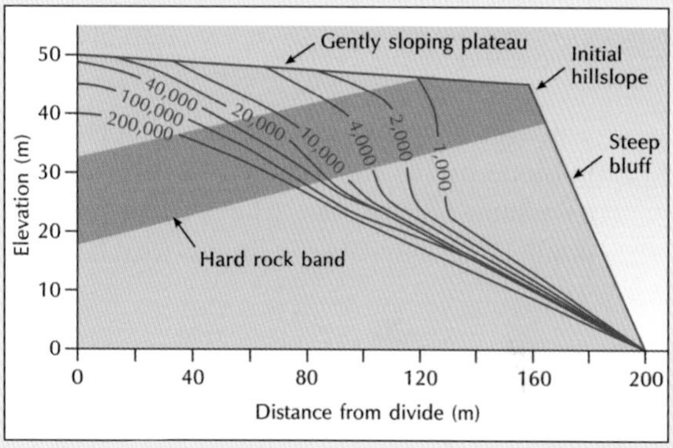

Figure 8.4 Simulation of hillslope change for an initial gently sloping plateau ending in a steep bluff with a band of hard rock dipping at 10° into the hillside. Time is in years. *Source*: Adapted from Kirkby (1985)

continued . . .

properties of rock, in particular the rate of decline towards the threshold gradient of landslides, are more important than solution in determining slope form. Only on slopes of low gradient and after long times (200,000 years and more) do solutional properties play a dominant role in influencing slope form. Second, gradient steepening and soil thinning over 'resistant' strata are strictly associated with the current location of an outcrop, though resistant beds, by maintaining locally steep gradients, tend to hold the less resistant beds close to the landslide threshold and so increase gradients everywhere. Third, gradients close to landslide threshold gradients commonly outlive landslide activity by many thousands of years and, because of this, may play a dominant role in determining regional relief in a tectonically stable area. Fourth, soils are generally thin under active landsliding and wash; thick soils tend to indicate the predominance of solution and creep or solifluction processes. Catenas in humid climates can be expected to develop thicker soils in downslope positions but in semi-arid areas, where wash keeps soils thin except on the lowest gradients, catenas can be expected to have deeper soils upslope and thinner soils downslope.

rainsplash and 1.3–1.7 for soil wash. The exponent n is typically 1.0 for soil creep, 1.0–2.0 for rainsplash, and 1.3–2.0 for soil wash (Kirkby 1971). For a hillslope catena, the solution of the equation takes the general form:

$$h = f(x, t)$$

This equation describes the development of a hillslope profile for specified slope processes, an assumed initial state (the original hillslope profile), and boundary conditions (what happens to material at the slope base, for example). Some of Kirkby's later models demonstrate the process, and some of the drawbacks, of long-term hillslope modelling (Box 8.1).

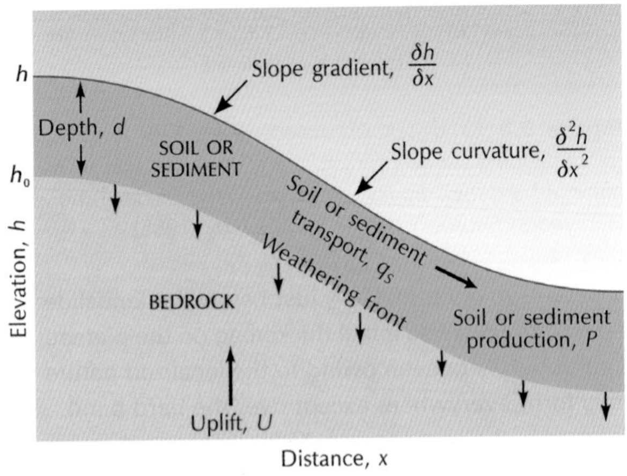

Figure 8.5 Components of numerical landscape models.

Hillslope models have become highly sophisticated. They still use the continuity equation for mass conservation, but now apply reasonably well established **geomorphic transport laws** (e.g. Dietrich and Perron 2006). The basis of such models, which include river systems as well as hillslopes, is the equation

$$\frac{dh}{dt} = U - P - \nabla \cdot q_s$$

which, in ordinary language, states that (Figure 8.5):

Rate of change in elevation (dz/dt)
= Uplift rate (U) – Soil production rate (P) – Sediment transport ($\nabla \cdot q_s$).

Figure 8.6 shows how a three-dimensional hillslope model of this kind explains the development of ridge-and-valley topography in soil-mantled terrain (Dietrich and Perron 2006).

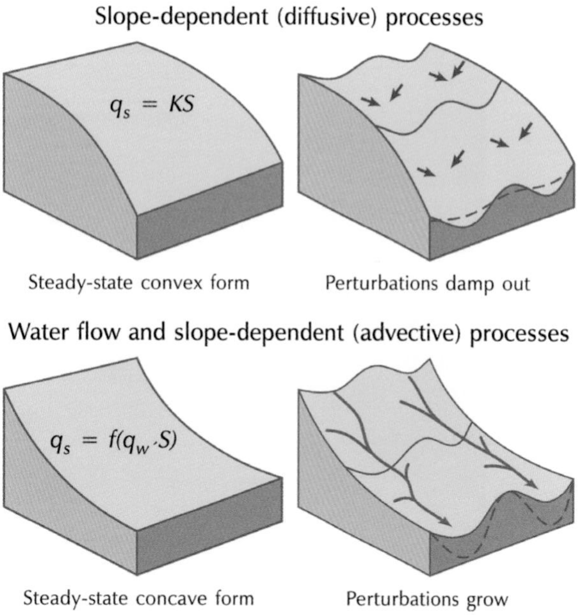

Figure 8.6 An explanation for the development of ridge-and-valley topography in soil-mantled terrain. Slope-dependent (diffusive) transport leads to convex hillslopes, and when the topography is laterally perturbed the transport direction (brown lines) causes the topographic highs to lower and topographic lows to fill in, resulting in smooth topography, as suggested by the dashed line. In contrast, advective transport, which depends on water flow and slope gradient, carries sediment downslope and produces concave hillslopes. Flow concentrations (blue flowpaths) resulting from lateral topographic perturbation lead to incision, as suggested by the dashed lines. The competition of these two processes leads to diffusion-dominated ridges and advection-dominated valleys. *Source*: Adapted from Dietrich and Perron (2006)

HILLSLOPE FORMS

Slope units

The term slope has two meanings. First, it refers to the angle of inclination of the ground surface, expressed in degrees or as a percentage. Second, it refers to the inclined surface itself. To avoid misunderstanding, the term **hillslope** usually applies to the inclined surface and the term **slope angle**, **slope gradient**, or simply **slope** to its inclination. All landforms consist of one or more slopes of variable inclination, orientation, length, and shape (Butzer 1976, 79). Most hillslope profiles consist of three slope units – an upper convex unit where gradient increases with length, a straight middle unit of constant gradient, and a concave lower unit where gradient decreases with length (Figure 8.7) (White 1966). The transition between these slope units may be smooth or abrupt (Figure 8.8). The middle unit is sometimes absent, giving a **concavo-convex slope profile**, as commonly found in English chalklands (Plate 8.11; see also p. 328).

The terms used to describe slope units vary. Anthony Young (1971) defined them as follows: a **slope unit** is either a segment or an element, whereas a **segment** is a portion of a slope profile on which the angle remains roughly the same, and an **element** is a portion of a slope profile on which the curvature remains roughly the same. Convex, straight, and concave hillslope units form a **geomorphic catena**, which is a sequence of linked slope units (cf. Speight 1974; Scheidegger 1986). Several schemes devised to describe hillslope profiles recognize these three basic units, although subunits are also distinguished (Figure 8.9). One scheme recognizes four slope units: the waxing slope, also called the convex slope or upper wash slope; the free face, also called the gravity or derivation slope; the constant slope, also called the talus or debris slope where scree is present; and the waning slope, also called the pediment, valley-floor basement, and lower wash slope (Wood 1942). A widely used system has five slope units – summit, shoulder, backslope, footslope, and toeslope (Figure 8.10) (Ruhe 1960). A similar

Plate 8.11 Concavo-convex slope on the chalk ridge, Isle of Purbeck, Dorset, England. The ruins of Corfe Castle lie in the middle ground. (*Photograph by Tony Waltham Geophotos*)

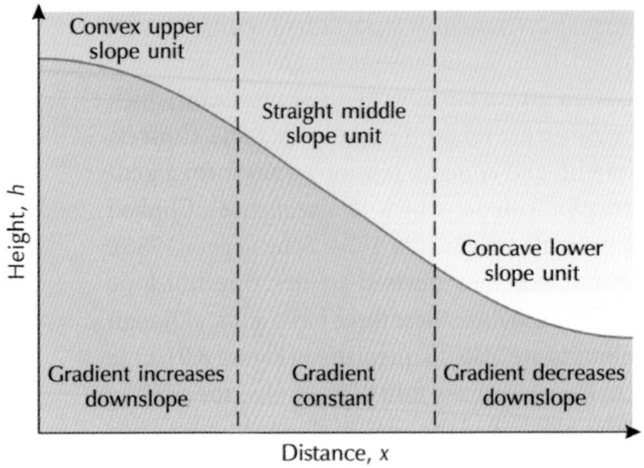

Figure 8.7 Three form elements of slopes.

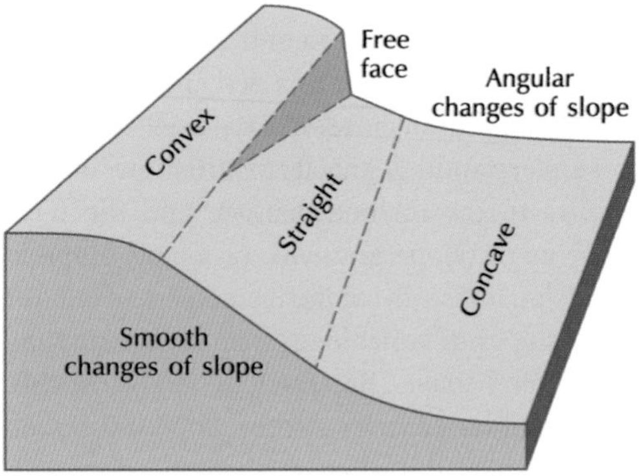

Figure 8.8 Abrupt and smooth transitions between slope elements.

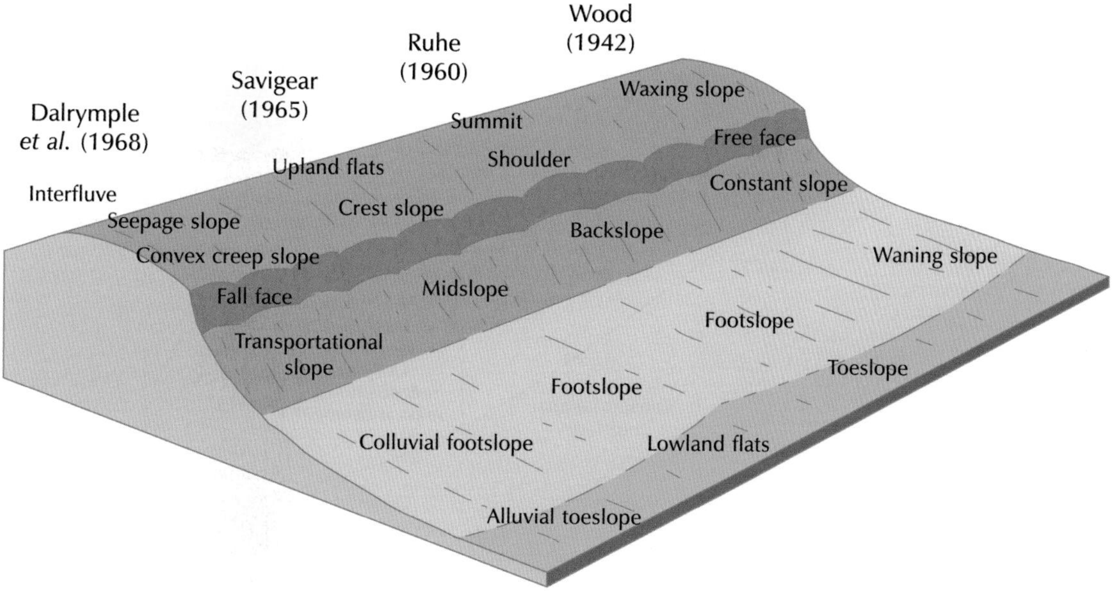

Figure 8.9 Systems for naming hillslope elements.

system uses different names – upland flats (gradient less than 2°), crest slope, midslope, footslope, and lowland flats (gradient less than 2°) (Savigear 1965). The nine-unit land-surface model embraces and embellishes all these schemes and distinguishes the following units – interfluve, seepage slope, convex creep slope, fall face, transportational slope, colluvial footslope, alluvial toeslope, channel wall, and channel bed (Figure 8.9; Dalrymple *et al.* 1968).

Different slope processes tend to dominate the various slope elements along a catena (Figure 8.11). On convex slope segments, commonly found on the upper parts of hillslope profiles, soil creep and rainsplash erosion dominate, at least when slopes are below the threshold for rapid mass wasting; subsurface movement of soil water is also important. Where convex segments are steeper than about 45°, fall, slide, and physical weathering are the chief processes. Straight (midslope) elements usually receive a large amount of material from upslope by mass wasting processes (including flow, slump, and slide), surface wash, and subsurface water movement. Concave slope elements are commonly sites of transport and deposition. They usually develop near the base of hillslope profiles in situations where waste material moving down the hillside through mass wasting and surface and subsurface water action comes to rest and rivers at the hillslope base do not remove it.

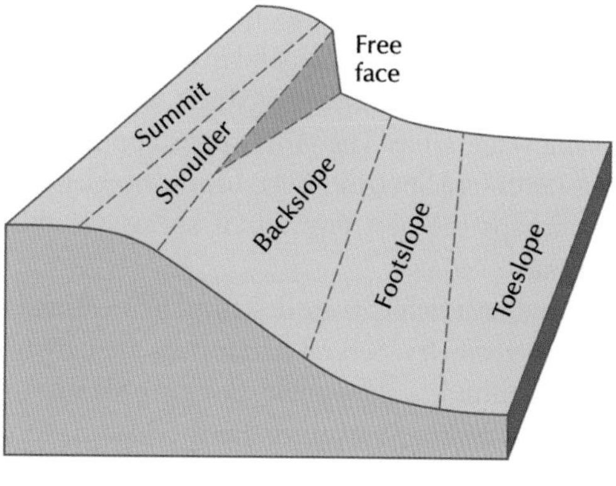

Figure 8.10 Ruhe's (1960) slope units.

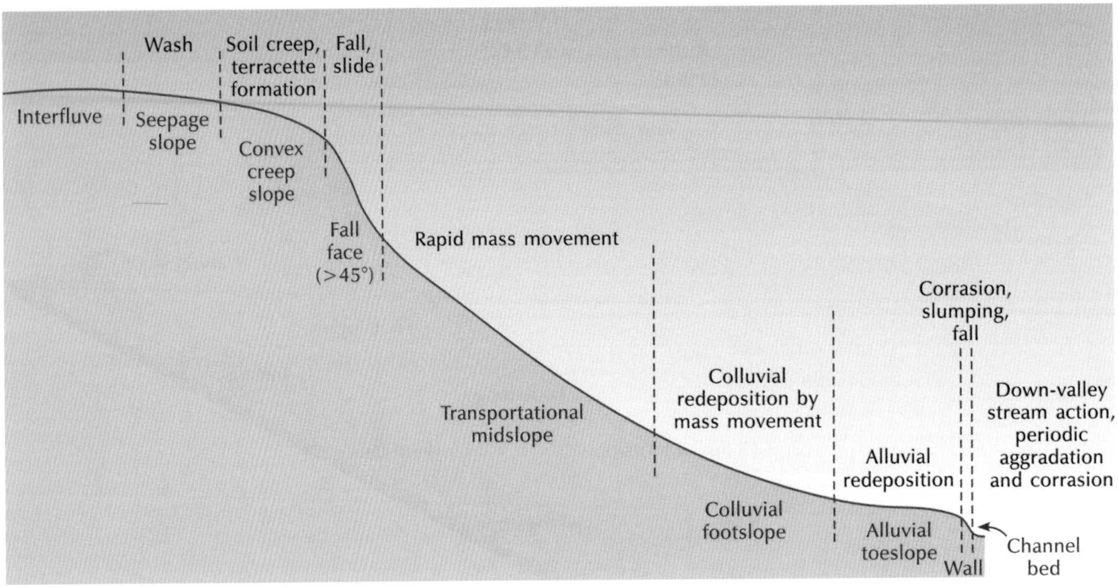

Figure 8.11 Hillslope processes typically associated with units in the nine-unit land-surface model.

Landform elements

From a geomorphological viewpoint, the ground surface is composed of landform elements. **Landform elements** are recognized as simply-curved geometric surfaces lacking inflections (complicated kinks) and are considered in relation to upslope, downslope, and lateral elements. Slope is essential in defining them. Landscape elements go by a plethora of names – facets, sites, land elements, terrain components, and facies. The 'site' (Linton 1951) was an elaboration of the 'facet' (Wooldridge 1932), and involved altitude, extent, slope, curvature, ruggedness, and relation to the water table. The other terms appeared in the 1960s (see Speight 1974). Landform element is perhaps the best term, as it seems suitably neutral.

Landform elements are described by local land-surface geometry. Several parameters are derivatives of altitude – slope angle, slope profile curvature, and contour curvature. Further parameters go beyond local geometry, placing the element in a wider landscape setting – distance from the element to the crest, catchment area per unit of contour length, dispersal area (the land area down-slope from a short increment of contour). Digital elevation models (DEMs) have largely superseded the classic work on landform elements and their descriptors. Topographic elements of a landscape can be computed directly from a DEM, and these are often classified into primary (or first-order) and secondary (or second-order) attributes (Moore *et al.* 1993). **Primary attributes** are calculated directly from the digital elevation data and the most commonly derived include slope and aspect (Table 8.2). **Secondary attributes** combine primary attributes and are 'indices that describe or characterise the spatial variability of specific processes occurring in the landscape' (Moore *et al.* 1993, 15); examples are irradiance and a wetness index (Table 8.2). Such methods allow modellers to represent the spatial variability of the processes, whereas in the past they could model them only as point processes. An enormous literature describes the use of DEMs to produce both primary and secondary attributes; an equally large

Table 8.2 Primary and secondary attributes that can be computed from DEMs

Attribute	Definition	Applications
Primary attributes		
Altitude	Height above mean sea level or local reference point	Climate variables (e.g. pressure, temperature), vegetation and soil patterns, material volumes, cut-and-fill and visibility calculations, potential energy determination
Slope	Rate of change of elevation – gradient	Steepness of topography, overland and subsurface flow, resistance to uphill transport, geomorphology, soil water content
Aspect	Compass direction of steepest downhill slope – azimuth of slope	Solar insolation and irradiance, evapotranspiration
Profile curvature	Rate of change of slope	Flow acceleration, erosion and deposition patterns and rate, soil and land evaluation indices, terrain unit classification
Plan curvature	Rate of change of aspect	Converging and diverging flow, soil water characteristics, terrain unit classification
Secondary attributes		
Wetness Index	ln $(A_s / \tan b)$ where A_s is specific catchment and b is slope	Index of moisture retention
Irradiance	Amount of solar energy received per unit area	Soil and vegetation studies, evapotranspiration

Source: Adapted from Huggett and Cheesman (2002, 20)

literature also considers how best to incorporate primary and secondary attributes into spatial models that simulate physical processes influenced and controlled by the nature of topography (e.g. Wilson and Gallant 2000).

Slope and aspect are two of the most important topographic attributes. **Slope** is a plane tangent to the terrain surface represented by the DEM at any given point. It has two components: (1) **gradient**, which is the maximum rate of change of altitude and expressed in degrees or per cent; and (2) **aspect**, the compass direction of the maximum rate of change (the orientation of the line of steepest descent expressed in degrees and converted to a compass bearing). Because slope allows gravity to induce the flow of water and other materials, it lies at the core of many geomorphological process models. For instance, slope and flowpath (i.e. slope steepness and length) are parameters in the dimensionless Universal Soil Loss Equation (USLE), which is designed to quantify sheet and rill erosion by water (p. 204).

The paper by Jozef Minár and Ian S. Evans (2008) provides an excellent discussion of approaches to land surface segmentation and the theoretical basis for terrain analysis and geomorphological mapping.

Landform classification

The toposphere contains a stupendous array of land forms. Unfortunately, landforms are notoriously difficult to classify quantitatively. Geomorphologists make a fundamental distinction between erosional landforms (sculptured by the action of wind, water, and ice) and depositional landforms (built by sediment accumulation). They also recognize basic differences between landforms in terrestrial, shallow marine, and deep marine environments, each of which fosters a distinct suite of geomorphic processes. However, many landform classifications use topographic form, and ignore geomorphic process. For example, one scheme for large-scale land-form classification uses three chief topographic characteristics (Hammond 1954). The first characteristic is the relative amount of gently sloping land (land with less than an 8 per cent slope). The second characteristic is the local relief (the difference between highest and lowest elevation in an area). The third characteristic is the 'generalized profile'. This defines the location of the gently sloping land – in valley bottoms or in uplands. In combination, these characteristics define the following landforms:

- Plains with a predominance of gently sloping land combined with low relief.
- Plains with some features of considerable relief. This group may be subdivided by the position of the gently sloping land into three types – plains with hills, mountains, and tablelands.
- Hills with gently sloping land and low-to-moderate relief.
- Mountains with little gently sloping land and high local relief.

There are many such schemes, all with their good and bad points. Modern research in this field combines terrain attributes to create some form of regional topographic classification (e.g. Giles 1998; Giles and Franklin 1998).

HUMANS AND HILLSLOPES

Hillslopes are the location of much human activity, and their study has practical applications. Knowledge of runoff and erosion on slopes is important for planning agricultural, recreational, and other activities. Land management often calls for slopes designed for long-term stability. Mine tailing piles, especially those containing toxic materials, and the reclamation of strip-mined areas also call for a stable slope design. This final section will consider the effects of humans upon hillslope soil erosion.

Soil erosion modelling

Soil erosion has become a global issue because of its environmental consequences, including pollution and sedimentation. Major pollution problems may occur from relatively moderate and frequent erosion events in both temperate and tropical climates. In almost every country of the world under almost all land-cover types the control and prevention of erosion are needed. Prevention of soil erosion means reducing the rate of soil loss to approximately the rate that would exist under natural conditions. It is crucially important and depends upon the implementation of suitable soil conservation strategies (Morgan 1995). **Soil conservation strategies** demand a thorough understanding of the processes of erosion and the ability to provide predictions of soil loss, which is where geomorphologists have a key role to play. Factors affecting the rate of soil erosion include rainfall, runoff, wind, soil, slope, land cover, and the presence or absence of conservation strategies.

Soil erosion is an area where process geomorphological modelling has had a degree of success. One of the first and most widely used empirical models was the **Universal Soil Loss Equation (USLE)** (Box 8.2). The USLE has been widely used, especially in the USA, for predicting sheet and rill erosion in national assessments of soil erosion. However, empirical models predict soil erosion on a single slope according to statistical relationships between important factors and are rather approximate. Models based on the physics of soil erosion were developed during the 1980s to provide better results. Two types of physically based model have evolved – lumped models and distributed models (see Huggett and Cheesman 2002, 156–9). **Lumped models** are non-spatial, predicting the overall or average response of a watershed. **Distributed models** are spatial, which means that they predict the spatial distribution of runoff and sediment movement over the land surface during individual storm events, as well as predicting total runoff and soil loss (Table 8.3). Many physically based soil-erosion models have benefited from GIS technology.

Box 8.2 THE UNIVERSAL SOIL LOSS EQUATION (USLE)

The USLE (Wischmeier and Smith 1978) predicts soil loss from information about (1) the potential erosivity of rainfall and (2) the erodibility of the soil surface. The equation is usually written as:

$$E = R \times K \times L \times S \times C \times P$$

where E is the mean annual rainfall loss, R is the rainfall erosivity factor, K is the soil erodibility factor, L is the slope length factor, S is the slope steepness factor, C is the crop management factor, and P is the erosion control practice factor. **The rainfall erosivity factor** is often expressed as a rainfall erosion index, EI_{30}, where E is rainstorm energy and I is rainfall intensity during a specified period, usually 30 minutes. **Soil erodibility**, K, is defined as the erosion rate (per unit of erosion index, EI_{30}) on a specific soil in a cultivated continuous fallow on a 9 per cent slope on a plot 22.6 m long. **Slope length**, L, and **slope steepness**, S, are commonly combined to produce a single index, LS, that represents the ratio of soil loss under a given slope steepness and slope length to the soil loss from a standard 9 per cent, 22.6-m-long slope. **Crop management**, C, is given as the ratio of soil loss from a field with a specific cropping-management strategy compared with the standard continuous cultivated fallow. **Erosion control**, P, is the ratio of soil loss with contouring strip cultivation or terracing to that of straight-row, up-and-down slope farming systems. The measurements of the standard plot – a slope length of 22.6 m (72½ feet), 9 per cent gradient, with a bare fallow land-use ploughed up and down the slope – seem very arbitrary and indeed are historical accidents. They are derived from the condition common at experimental field stations where measured soil losses provided the basic data for calibrating the equation. It was convenient to use a plot area of 1/100 acre and a plot width of 6 feet, which meant that the plot length must be 72½ feet.

To use the USLE, a range of erosion measurements must be made, which are usually taken on small bounded plots. The problem here is that the plot itself affects the erosion rate. On small plots, all material that starts to move is collected and measured. Moreover, the evacuation of water and sediment at the slope base may itself trigger erosion, with rills eating back through the plot, picking up and transporting new sources of sediment in the process. Another difficulty lies in the assumption that actual slopes are uniform and behave like small plots. Natural slopes usually have a complex topography that creates local erosion and deposition of sediment. For these reasons, erosion plots established to provide the empirical data needed to apply the USLE almost always overestimate the soil-loss rate from hillslopes by a factor twice to ten times the natural rate.

Hillslope erosion along trails

The trampling of humans (walking or riding) and other animals along trails may lead to **soil erosion**. Anyone who has walked along footpaths, especially those in hilly terrain, is bound to have firsthand experience of the problem. The problem has become acute over the last twenty or thirty years as the number of people using mountain trails, either on foot or in some form of off-road transport, has risen sharply. A study in Costa Rican forest confirmed that trails generate runoff more quickly, and erode sooner, than is the case in off-trail settings (Wallin and Harden 1996). This finding, which is typical of trail erosion

Table 8.3 Examples of physically based soil erosion models

Model	Use	References
Lumped or non-spatial models		
CREAMS (Chemicals, Runoff and Erosion from Agricultural Management Systems)	Field-scale model for assessing non-point-source pollution and the effects of different agricultural practices	Knisel (1980)
WEPP (Water Erosion Prediction Project)	Designed to replace USLE in routine assessments of soil erosion	Nearing *et al.* (1989)
EUROSEM (European Soil Erosion Model)	Predicts transport, erosion, and deposition of sediment throughout a storm event	Morgan (1994)
Distributed or spatial models		
ANSWERS (Areal Nonpoint Source Watershed Environment Response Simulation)	Models surface runoff and soil erosion within a catchment	Beasley *et al.* (1980)
LISEM (Limburg Soil Erosion Model)	Hydrological and soil erosion model, incorporating raster GIS (information stored on a spatial grid), that may be used for planning and conservation purposes	De Roo *et al.* (1996)

studies in all environments, underscores the need for careful management of ecotourism in trail-dependent activities. Strategies for combating trail erosion can work. Smedley Park lies in the Crum Creek watershed, Delaware County, near Media, Pennsylvania, USA. The trails in the park pass through several areas with fragile environments (Lewandowski and McLaughlin 1995). A strategy was devised using network analysis, which altered the efficiency of the trail system by more fully connecting sites with robust environments and reducing the potential for visitors to use environmentally fragile sites. Some of the severest erosion is associated with logging trails. In the Paragominas region of eastern Amazonia, tree damage in unplanned and planned logging operations was associated with each of five logging phases: tree felling, machine manoeuvring to attach felled boles to chokers, skidding boles to log landings, constructing log landings, and constructing logging roads (Johns *et al.* 1996).

The nature of **trail use** affects the degree of soil erosion. The comparative impact of hikers, horses, motorcycles, and off-road bicycles on water runoff and sediment yield was investigated on two trails – the Emerald Lake Trail and the New World Gulch Trail – in, and just outside, respectively, the Gallatin National Forest, Montana, USA (Wilson and Seney 1994). The results revealed the complex interactions that occur between topographic, soil, and geomorphic variables, and the difficulty of interpreting their impact on existing trails. In brief, horses and hikers (hooves and feet) made more sediment available than wheels (motorcycles and off-road bicycles), with horses producing the most sediment, and sediment production was greater on pre-wetted trails. In the northern Rocky Mountains, Montana, USA, trails across meadow vegetation bear signs of damage – bare soil and eroded areas – through human use (Weaver and Dale 1978). The meadows were principally Idaho fescue–Kentucky bluegrass (*Festuca idahoensis– Poa pratensis*) communities. Experiments were run on meadows underlain by deep sandy-loam soils at 2,070 m near Battle Ridge US Forest Ranger Service Station, in the Bridge Range. They involved getting hikers, horse riders, and a motorcyclist to pass up and down slopes of 15°. The hikers weighed 82–91 kg and wore hiking boots with cleated soles; the horses weighed 500–79 kg and had uncleated shoes; the motorcycle was a Honda 90 running in second gear at speeds below 20 km/hr. The experiments showed that horses and motorcycles do more damage (as measured by per-cent-bare area, trail width, and trail depth) on these trails than do hikers (Figure 8.12). Hikers, horses, and motorcycles all do more damage on sloping ground than on level ground. Hikers cause their greatest damage going downhill. Horses do more damage going uphill than downhill, but the difference is not that big. Motorcycles do much damage going downhill and uphill, but cut deep trails when going uphill.

SUMMARY

Hillslopes are the commonest landform. There are bare and soil-mantled varieties. Gravity and water (and sometimes wind) transport material over and through hillslopes. Weathered debris may move downslope under its own weight, a process called mass wasting. Gravity-driven mass wasting is determined largely by the relationships between stress and strain in Earth materials, and by the rheological behaviour of brittle solids, elastic solids, plastic solids, and liquids. Mass movements occur in seven ways: fall, topple, slide, spread, flow, slope deformation, and subsidence. Half-mountain-sized mass movements are the subject of gravity tectonics. Transport processes on hillslopes include surface processes (rainsplash, rainflow, sheet wash) and subsurface processes (leaching, through-wash, and mixing by organisms or bioturbation). Transport-limited processes, such as creep and rainsplash, are distinct from supply-limited processes, such as solute leaching and debris avalanching. Hillslopes with transport limitations tend to carry a thick soil mantle, and their slopes tend to decline with time. Hillslopes limited by the supply of material through weathering tend to be bare or have thin soils, and their slopes tend to retreat at a constant angle. Mathematical models based on the continuity equation for mass conservation and geomorphic transport laws provide a means of probing long-term hillslope development. A hillslope profile consists of slope units, which may be slope segments (with a roughly constant gradient) or slope elements (with a roughly constant curvature). A common sequence of slope elements, starting at the hilltop, is convex–straight–concave. These elements form a geomorphic catena. Different geomorphic processes dominate different slope elements along a catena.

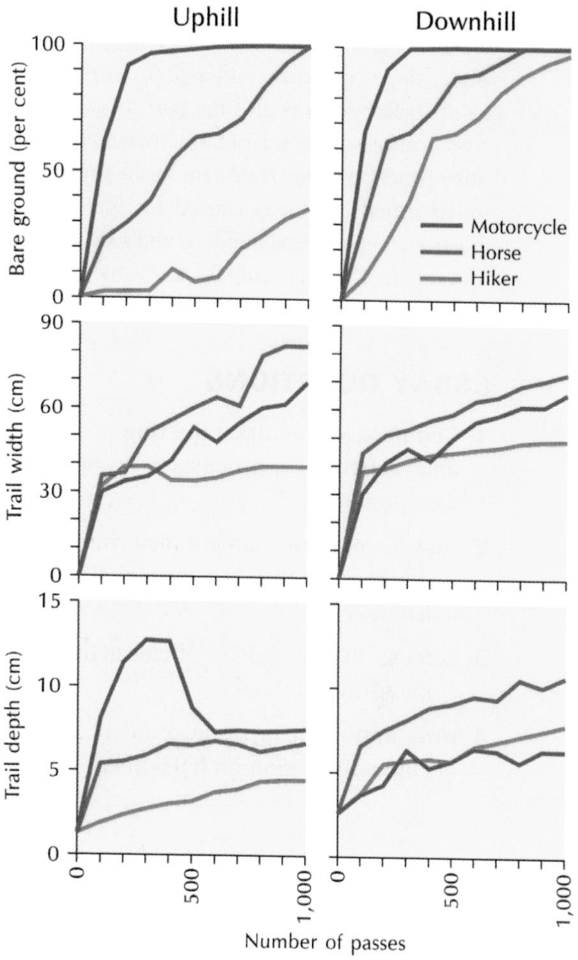

Figure 8.12 Experimental damage done by hikers, bikers, and horses moving uphill and downhill on trails in Bridge Range, Montana, USA, on a sloping 15° meadow site. *Source*: Adapted from Weaver and Dale (1978)

Landform elements are basic units of the two-dimensional land surface. Properties such as slope angle, slope curvature, and aspect define them. Land-surface form is also the basis of landform classification schemes. Human activities alter hillslope processes. This is evident in the erosion of soil-mantled hillslopes caused by agricultural practices, logging, road building, and so forth. The movement of people, animals, and vehicles along trails may also cause soil to erode.

ESSAY QUESTIONS

1 Compare and contrast the role of surface and subsurface processes in hillslope development.

2 How useful are mathematical models in understanding the long-term evolution of hillslopes?

3 Discuss the difficulties of classifying hillslope forms.

4 How important is slope gradient in predicting soil erosion on hillslopes?

FURTHER READING

Anderson, M. G. and Brooks, S. M. (eds) (1996) *Advances in Hillslope Processes*, 2 vols. Chichester: John Wiley & Sons.
A very good state-of-the-art (in the mid-1990s) and advanced text.

Carson, M. A. and Kirkby, M. J. (2009) *Hillslope Form and Process*. Cambridge: Cambridge University Press.
A digitally printed version of the 1972 classic.

Morgan, R. P. C. (2005) *Soil Erosion and Conservation*, 3rd edn. Oxford: Blackwell.
Probably the best introductory text on the topic.

Selby, M. J. (1993) *Hillslope Materials and Processes*, 2nd edn. With a contribution by A. P. W. Hodder. Oxford: Oxford University Press.
An excellent account of the geomorphology of hillslopes.

Thornes, J. B. (ed.) (1990) *Vegetation and Erosion: Processes and Environments*. Chichester: John Wiley & Sons.
A collection of essays that, as the title suggests, consider the effects of vegetation on soil erosion in different environments.

CHAPTER NINE

FLUVIAL LANDSCAPES

9

Running water wears away molehills and mountains, and builds fans, floodplains, and deltas. This chapter covers:

- Running water and fluvial processes
- Water-carved landforms
- Water-constructed landforms
- Fluvial landscapes and humans
- Past fluvial landscapes

RUNNING WATER IN ACTION: FLOODS

Plum Creek flows northwards over a sand bed between Colorado Springs and Denver in the USA, and eventually joins the South Platte River. On 16 June 1965, a series of intense convective cells in the region climaxed in an intense storm, with 360 mm of rain falling in four hours, and a flood (Osterkamp and Costa 1987). The flood had a recurrence interval of between 900 and 1,600 years and a peak discharge of 4,360 m^3/s, which was fifteen times higher than the 50-year flood. It destroyed the gauging station at Louviers and swept through Denver causing severe damage. The flow at Louviers is estimated to have gone from less than 5 m^3/s to 4,360 m^3/s in about 40 minutes. At peak flow, the water across the valley averaged from 2.4 to 2.9 m deep, and in places was 5.8 m deep. The deeper sections flowed at around 5.4 m/s. The flood had far-reaching effects on the geomorphology and vegetation of the valley floor. Rampant erosion and undercutting of banks led to bank failures and channel widening. The processes were aided by debris snagged on trees and other obstructions, which caused them to topple and encourage sites of rapid scouring. Along a 4.08-km study reach, the average channel width increased from 26 to 68 m. Just over half the woody vegetation was destroyed. Following a heavy spring runoff in 1973, the channel increased to 115 m in width and increased its degree of braiding.

FLUVIAL ENVIRONMENTS

Running water dominates fluvial environments, which are widespread except in frigid regions, where ice dominates, and in dry regions, where wind tends to be the main erosive agent. However, in arid and semi-arid areas, fluvial activity can be instrumental in fashioning landforms. Flash floods build alluvial fans and run out on to desert floors. In the past, rivers once flowed across many areas that today lack permanent watercourses.

Water runs over hillslopes as overland flow and rushes down gullies and river channels as streamflow. The primary determinant of overland flow and streamflow is runoff production. **Runoff** is a component of the land-surface water balance. In brief, runoff is the difference between precipitation and evaporation rates, assuming that soil water storage stays roughly constant. In broad terms, fluvial environments dominate where, over a year, precipitation exceeds evaporation and the temperature regime does not favour persistent ice formation. Those conditions cover a sizeable portion of the land surface. The lowest annual runoff rates, less than 5 cm, are found in deserts. Humid climatic regions and mountains generate the most runoff, upwards of 100 cm in places, and have the highest river discharges.

Runoff is not produced evenly throughout the year. Seasonal changes in precipitation and evaporation generate systematic patterns of runoff that are echoed in **streamflow**. Streamflow tends to be highest during wet seasons and lowest during dry seasons. The changes of streamflow through a year define a river regime. Each climatic type fosters a distinct river regime. In monsoon climates, for example, river discharge swings from high to low with the shift from the wet season to the dry season. Humid climates tend to sustain a year-round flow of water in **perennial streams**. Some climates do not sustain a year-round river discharge. **Intermittent streams** flow for at least one month a year when runoff is produced. **Ephemeral streams**, which are common in arid environments, flow after occasional storms but are dry the rest of the time.

FLUVIAL PROCESSES

Flowing water

Figure 9.1 is a cartoon of the chief hydrological processes that influence the geomorphology of hillslopes and streams. Notice that water flows over and through landscapes in unconcentrated and concentrated forms.

Splash, overland flow, and rill flow

Rainsplash results from raindrops striking rock and soil surfaces. An impacting raindrop compresses and spreads sideways. The spreading causes a shear on the rock or soil that may detach particles from the surface, usually particles less than 20 micrometres in diameter. If entrained by water from the original raindrop, the particles may rebound from the surface and travel in a parabolic curve, usually no more than a metre or so. Rainsplash releases particles for entrainment and subsequent transport by unconcentrated surface flow, which by itself may lack the power to dislodge and lift attached particles.

Unconcentrated surface flow (**overland flow**) occurs as inter-rill flow. **Inter-rill flow** is variously termed sheet flow, sheet wash, and slope wash. It involves a thin layer of moving water together with strands of deeper and faster-flowing water that diverge and converge around surface bulges causing erosion by soil detachment (largely the result of impacting raindrops) and sediment transfer. **Overland flow** is produced by two mechanisms:

1 **Hortonian overland flow** occurs when the rate at which rain is falling exceeds the rate at which it can percolate into the soil (the **infiltration rate**). Hortonian overland flow is more common on bare rock surfaces, and in deserts, where soils tend to be thin, bedrock outcrops common, vegetation scanty, and rainfall rates high. It can contribute large volumes of water to streamflow and cover large parts of an arid drainage basin, and is the basis of the 'partial area model' of streamflow generation.

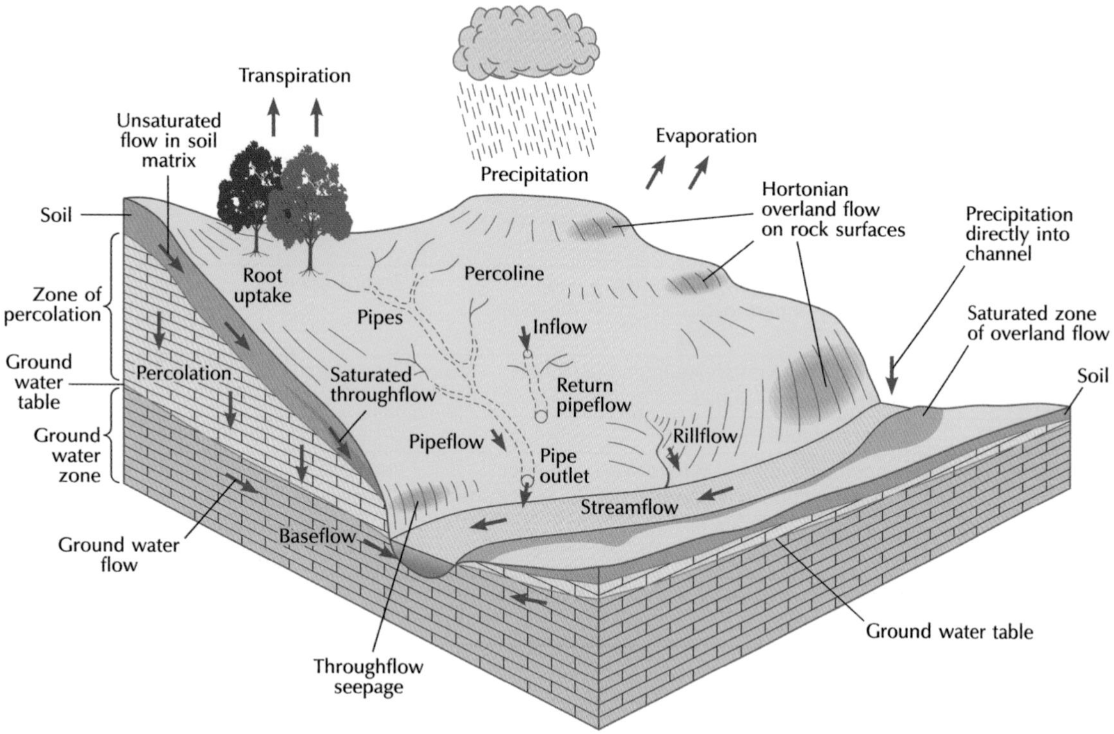

Figure 9.1 The chief hydrological processes that influence the geomorphology of hillslopes and streams. Water flows over and through landscapes in unconcentrated and concentrated forms.

2 **Saturation overland flow** or **seepage flow** occurs where the groundwater table sits at the ground surface. Some of the water feeding saturation overland flow is flow that has entered the hillside upslope and moved laterally through the soil as **throughflow**; this is called **return flow**. Rain falling directly on the hillslope may feed saturation overland flow.

Rill flow is deeper and speedier than inter-rill flow and is characteristically turbulent. It is a sporadic concentrated flow that grades into streamflow.

Subsurface flow

Flow within a rock or soil body may take place under unsaturated conditions, but faster **subsurface flow** is associated with localized soil saturation. Where the hydraulic conductivity of soil horizons decreases with depth, and especially when hardpans or clay-rich substrata are present in the soil, infiltrating water is deflected downslope as **throughflow**. Engineering hydrologists use the term **interflow** to refer to water arriving in the stream towards the end of a storm after having followed a deep subsurface route, typically through bedrock. **Baseflow** is water entering the stream from the water table or delayed interflow that keeps rivers in humid climates flowing during dry periods. Subsurface flow may take place as a slow movement through rock and soil pores, sometimes along distinct lines called **percolines**, or as a faster movement in cracks, soil pipes (**pipe flow**), and underground channels in caves.

Springs

Springs occur where the land surface and the water table cross. Whereas saturation overland flow is the seepage from a temporary saturation

Table 9.1 Springs

Type	Occurrence	Example
Waste cover	Dells and hollows where lower layers of soil or bedrock are impervious	Common on hillslopes in humid environments
Contact	Flat or gently dipping beds of differing perviousness or permeability at the contact of an aquifer and an aquiclude. Often occur as a spring line	Junction of Totternhoe Sands and underlying Chalk Marl, Cambridgeshire, England
Fault	Fault boundaries between pervious and impervious, or permeable and impermeable, rocks	Delphi, Greece
Artesian	Synclinal basin with an aquifer sandwiched between two aquicludes	Artois region of northern France
Karst	Karst landscapes	Orbe spring near Vallorbe, Switzerland
Vauclusian	U-shaped pipe in karst where water is under pressure and one end opens on to the land surface	Vaucluse, France; Blautopf near Blaubeuren, Germany
Thermal	Hot springs	Many in Yellowstone National Park, Wyoming, USA
Geyser	A thermal spring that spurts water into the air at regular intervals	Old Faithful, Yellowstone National Park

zone, springs arise where the water table is almost permanent. Once a spring starts to flow, it causes a dip in the water table that creates a pressure gradient in the aquifer. The pressure gradient then encourages water to move towards the spring. Several types of spring are recognized, including waste cover springs, contact springs, fault springs, artesian springs, karst springs, vauclusian springs, and geysers (Table 9.1).

Streamflow

Rivers are natural streams of water that flow from higher to lower elevations across the land surface. Their continued existence relies upon a supply of water from overland flow, throughflow, interflow, baseflow, and precipitation falling directly into the river. **Channelized rivers** are streams structurally engineered to control floods, improve drainage, maintain navigation, and so on. In some lowland catchments of Europe, more than 95 per cent of river channels have been altered by channelization.

Water flowing in an open channel (**open channel flow**) is subject to gravitational and frictional forces. Gravity impels the water downslope, while friction from within the water body (viscosity) and between the flowing water and the channel surface resists movement. **Viscosity** arises through cohesion and collisions between molecules (**molecular** or **dynamic viscosity**) and the interchange of water adjacent to zones of flow within eddies (**eddy viscosity**).

Water flow may be turbulent or laminar. In **laminar flow**, thin layers of water 'slide' over each other, with resistance to flow arising from molecular viscosity (Figure 9.2a). In **turbulent flow**, which is the predominant type of flow in stream channels, the chaotic flow-velocity fluctuations are superimposed on the main forward flow, and resistance is contributed by molecular viscosity and eddy viscosity. In most channels, a thin layer or laminar flow near the stream bed is surmounted by a much thicker zone of turbulent flow (Figure 9.2b). Mean flow velocity, molecular viscosity, fluid density, and the size of the flow section determine the type of flow. The size of the flow section may be measured as either the depth of flow or as the hydraulic radius. The **hydraulic**

radius, *R*, is the cross-sectional area of flow, *A*, divided by the wetted perimeter, *P*, which is the length of the boundary along which water is in contact with the channel (Figure 9.3):

$$R = \frac{A}{P}$$

In broad, shallow channels, the flow depth can approximate the hydraulic radius. The **Reynolds number**, *Re*, named after English scientist and engineer Osborne Reynolds, may be used to predict the type of flow (laminar or turbulent) in a stream (Box 9.1).

Box 9.1 REYNOLDS AND FROUDE NUMBERS

The **Reynolds number** is a dimensionless number that includes the effects of the flow characteristics, velocity, and depth, and the fluid density and viscosity. It may be calculated by multiplying the mean flow velocity, *v*, and hydraulic radius, *R*, and dividing by the kinematic viscosity, ν (nu), which represents the ratio between molecular viscosity, μ (mu), and the fluid density, ρ (rho) (and therefore inverted to give ρ/μ in the equation):

$$R_s = \frac{\rho v R}{\mu}$$

For stream channels at moderate temperatures, the maximum Reynolds number at which laminar flow is sustained is about 500. Above values of about 2,000, flow is turbulent, and between 500 and 2,000 laminar and turbulent flow are both present.

The **Froude number** is defined by the square root of the ratio of the inertia force to the gravity force, or the ratio of the flow velocity to the velocity of a small gravity wave (a wave propagated by, say, a tossed pebble) in still water. The Froude number is usually computed as:

$$F = \frac{v}{\sqrt{gd}}$$

where *v* is the flow velocity, *g* is the acceleration of gravity, *d* is the depth of flow, and $\sqrt{gd}$ is the velocity of the gravity waves. When $F < 1$ (but more than zero) the wave velocity is greater than the mean flow velocity and the flow is known as **subcritical** or **tranquil** or **streaming**. Under these conditions, ripples propagated by a pebble dropped into a stream create an egg-shaped wave that moves out in all directions from the point of impact. When $F = 1$ flow is critical, and when $F > 1$ it is **supercritical** or **rapid** or **shooting**. These different types of flow occur because changes in discharge can be accompanied by changes in depth and velocity of flow. In other words, a given discharge is transmittable along a stream channel either as a deep, slow-moving, subcritical flow or else as a shallow, rapid, supercritical flow. In natural channels, mean Froude numbers are not usually higher than 0.5 and supercritical flows are only temporary, since the large energy losses that occur with this type of flow promote bulk erosion and channel enlargement. This erosion results in a lowering of flow velocity and a consequential reduction in the Froude number of the flow through negative feedback. For a fixed velocity, streaming flow may occur in deeper sections of the channel and shooting flow in shallower sections.

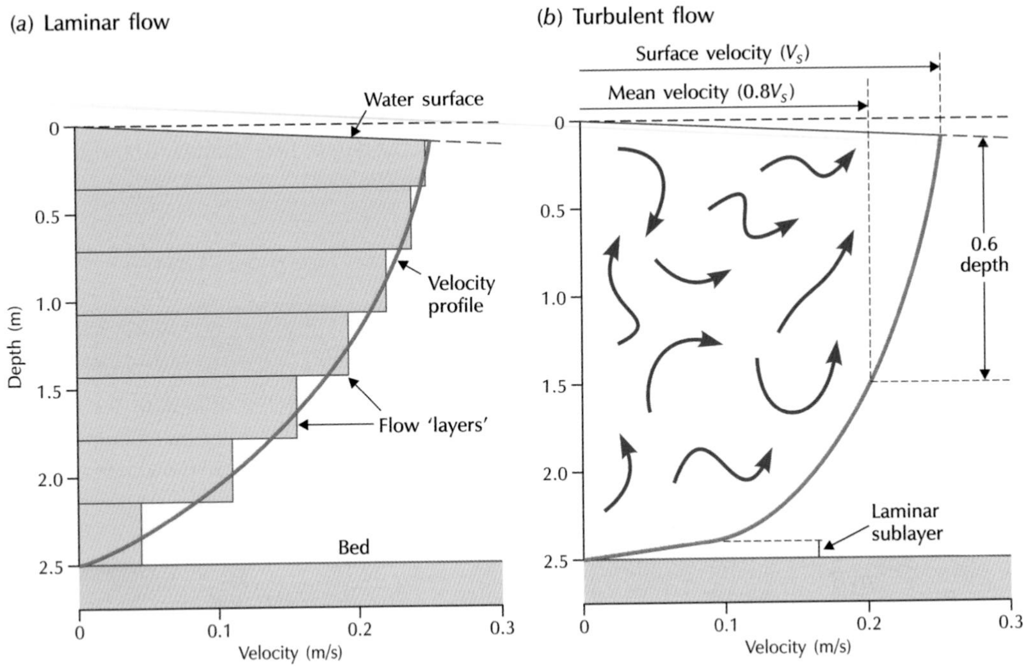

Figure 9.2 Velocity profiles of (a) laminar and (b) turbulent flow in a river.

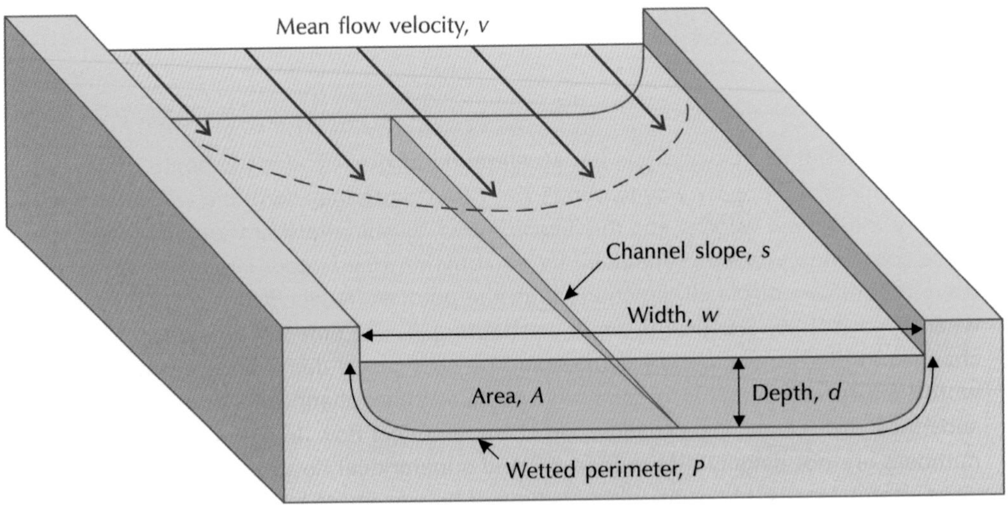

Figure 9.3 Variables used in describing streamflow.

In natural channels, irregularities on the channel bed induce variations in the depth of flow, so propagating ripples or waves that exert a weight or gravity force. The Froude number, *F*, of the flow, named after the English engineer and naval architect William Froude, can be used to distinguish different states of flow – subcritical flow and critical flow (Box 9.1). Plunging flow is a third kind of turbulent flow. It occurs at a waterfall, when water plunges in free fall over very steep, often vertical or overhanging rocks (Plate 9.1). The water falls as a coherent mass or as individual water strands or, if the falls are very high and the discharge low, as a mist resulting from the water dissolving into droplets.

Flow velocity controls the switch between subcritical and supercritical flow. A **hydraulic jump** is a sudden change from supercritical to subcritical flow. It produces a stationary wave and an increase in water depth (Figure 9.4a). A **hydraulic drop** marks a change from subcritical to supercritical flow and is accompanied by a reduction in water depth (Figure 9.4b). These abrupt changes in flow regimes may happen where there is a sudden change in channel bed form, a situation rife in mountain streams where there are usually large obstructions such as boulders.

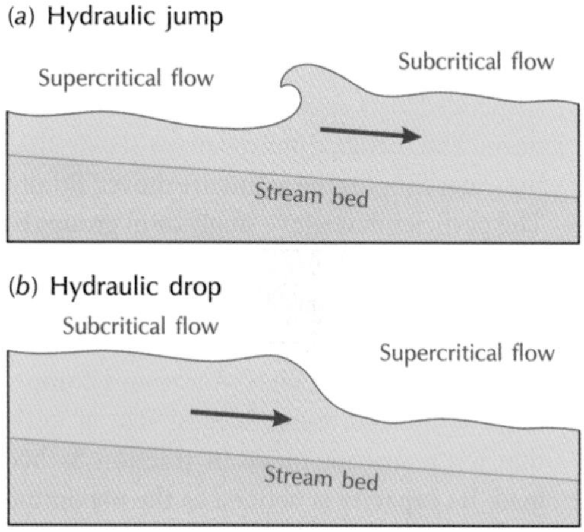

Figure 9.4 (a) Hydraulic jump. (b) Hydraulic drop.

Plate 9.1 Victoria Falls, a waterfall on the Zambezi River at Zambia–Zimbabwe border. The entire discharge of the river runs through the First Gorge (shown), which is about 110 m wide and 150 m long and cut into a basalt plateau. (*Photograph by Nick Scarle*)

Slope gradient, bed roughness, and cross-sectional form of the channel affect flow velocity in streams. It is very time-consuming to measure streamflow velocity directly, and empirical equations have been devised to estimate mean flow velocities from readily measured channel properties. The **Chézy equation**, named after the eighteenth-century French hydraulic engineer Antoine de Chézy, estimates velocity in terms of the hydraulic radius and channel gradient, and a coefficient expressing the gravitational and frictional forces acting upon the water. It defines mean flow velocity, $\bar{v}$, as:

$$\bar{v} = C\sqrt{Rs}$$

where *R* is the hydraulic radius, *s* is the channel gradient, and *C* is the Chézy coefficient representing gravitational and frictional forces. The **Manning equation**, which was devised by the American hydraulic engineer Robert Manning at the end of the nineteenth century, is a more commonly used formula for estimating flow velocity:

$$\bar{v} = \frac{R^{2/3}s^{1/2}}{n}$$

where R is the hydraulic radius, s the channel gradient, and n the **Manning roughness coefficient**, which is an index of bed roughness and is usually estimated from standard tables or by comparison with photographs of channels of known roughness. Manning's formula can be useful in estimating the discharge in flood conditions. The height of the water can be determined from debris stranded in trees and high on the bank. Only the channel cross-section and the slope need measuring.

Fluvial erosion and transport

Streams are powerful geomorphic agents capable of eroding, carrying, and depositing sediment. **Stream power** is the capacity of a stream to do work. It may be expressed as:

$$\Omega = \rho g Q s$$

where Ω (omega) is stream power per unit length of stream channel, ρ (rho) is water density, Q is stream discharge, and s is the channel slope. It defines the rate at which potential energy, which is the product of the weight of water, mg (mass, m, times gravitational acceleration, g), and its height above a given datum, h, is expended per unit length of channel. In other words, stream power is the rate at which a stream works to transport sediment, overcome frictional resistance, and generate heat. It increases with increasing discharge and increasing channel slope.

Stream load

All the material carried by a stream is its **load**. The **total load** consists of the dissolved load (solutes), the suspended load (grains small enough to be suspended in the water), and the bed load (grains too large to be suspended for very long under normal flow conditions). In detail, the three components of stream load are as follows:

1 The **dissolved load** or **solute load** comprises ions and molecules derived from chemical weathering plus some dissolved organic substances. Its composition depends upon several environmental factors, including climate, geology, topography, and vegetation. Rivers fed by water that has passed though swamps, bogs, and marshes are especially rich in dissolved organic substances. River waters draining large basins tend to have a similar chemical composition, with bicarbonate, sulphate, chloride, calcium, and sodium being the dominant ions (but see p. 94 for continental differences). Water in smaller streams is more likely to mirror the composition of the underlying rocks.

2 The **suspended load** consists of solid particles, mostly silts and clays, that are small enough and light enough to be supported by turbulence and vortices in the water. Sand is lifted by strong currents, and small gravel can be suspended for a short while during floods. The suspended load reduces the inner turbulence of the stream water, so diminishing frictional losses and making the stream more efficient. Most of the suspended load is carried near the stream bed, and the concentrations become lower in moving towards the water surface.

3 The **bed load** or **traction load** consists of gravel, cobbles, and boulders, which are rolled or dragged along the channel bed by traction. If the current is very strong, they may be bounced along in short jumps by saltation. Sand may be part of the bed load or part of the suspended load, depending on the flow conditions. The bed load moves more slowly than the water flows as the grains are moved fitfully. The particles may move singly or in groups by rolling and sliding. Once in motion, large grains move more easily and faster than small ones, and rounder particles move more readily than flat or angular ones. A stream's **competence** is defined as the biggest size of grain that a stream can move in traction as bed load. Its **capacity** is defined as the maximum amount of debris that it can carry in traction as bed load.

In addition to these three loads, the suspended load and the bed load are sometimes collectively called the **solid-debris load** or the **particulate load**. And the **wash load**, a term used by some hydrologists, refers to that part of the sediment load comprising grains finer than those on the channel bed. It consists of very small clay-sized particles that stay in more or less permanent suspension.

Stream erosion and transport

Streams may attack their channels and beds by corrosion, corrasion, and cavitation. **Corrosion** is the chemical weathering of bed and bank materials in contact with the stream water. **Corrasion** or **abrasion** is the wearing away of surfaces over which the water flows by the impact or grinding action of particles moving with the water body. **Evorsion** is a form of corrasion in which the sheer force of water smashes bedrock without the aid of particles. In alluvial channels, **hydraulicking** is the removal of loose material by the impact of water alone. **Cavitation** occurs only when flow velocities are high, as at the bottom of waterfalls, in rapids, and in some artificial conduits. It involves shockwaves released by imploding bubbles, which are produced by pressure changes in fast-flowing streams, smashing into the channel walls, hammer-like, and causing rapid erosion. The three main erosive processes are abetted by vortices that may develop in the stream and that may suck material from the streambed.

Streams may erode their channels downwards or sideways. **Vertical erosion** in an alluvial channel bed (a bed formed in fluvial sediments) takes place when there is a net removal of sands and gravels. In bedrock channels (channels cut into bedrock), vertical erosion is caused by the channel's bed load abrading the bed. **Lateral erosion** occurs when the channel banks are worn away, usually by being undercut, which leads to slumping and bank collapse.

The ability of flowing water to erode and transport rocks and sediment is a function of a stream's kinetic energy (the energy of motion). Kinetic energy, E_k, is half the product of mass and velocity, so for a stream it may be defined as

$$E_k = \frac{mv^2}{2}$$

where m is the mass of water and v is the flow velocity. If Chézy's equation (p. 215) is substituted for velocity, the equation reads

$$E_k = \frac{mCRs}{2}$$

This equation shows that kinetic energy in a stream is directly proportional to the product of the hydraulic radius, R (which is virtually the same as depth in large rivers), and the stream gradient, s. In short, the deeper and faster a stream, the greater its kinetic energy and the larger its potential to erode. The equation also conforms to the **DuBoys equation** defining the shear stress or tractive force, τ (tau), on a channel bed:

$$\tau = \gamma ds$$

where γ (gamma) is the specific weight of the water (g/cm^3), d is water depth (cm), and s is the stream gradient expressed as a tangent of the slope angle. A stream's ability to set a pebble in motion – its **competence** – is largely determined by the product of depth and slope (or the square of its velocity). It can move a pebble of mass m when the shear force it creates is equal to or exceeds the critical shear force necessary for the movement of the pebble, which is determined by the mass, shape, and position of the pebble in relation to the current. The pebbles in gravel bars often develop an imbricated structure (overlapping like tiles on a roof), which is particularly resistant to erosion. In an imbricated structure, the pebbles have their long axes lying across the flow direction and their second-longest axes aligned parallel to the flow direction and angled down upstream. Consequently, each pebble is protected by its neighbouring upstream pebble.

Only if a high discharge occurs are the pebbles set in motion again.

A series of experiments enabled Filip Hjulström (1935) to establish relationships between a stream's flow velocity and its ability to erode and transport grains of a particular size. The relationships, which are conveniently expressed in the oft-reproduced **Hjulström diagram** (Figure 9.5), cover a wide range of grain sizes and flow velocities. The upper curve is a band showing the critical velocities at which grains of a given size start to erode. The curve is a band rather than a single line because the critical velocity depends partly on the position of the grains and the way that they lie on the bed. Notice that medium sand (0.25–0.5 mm) is eroded at the lowest velocities. Clay and silt particles, even though they are smaller than sand particles, require a higher velocity for erosion to occur because they lie within the bottom zone of laminar flow and, in the case of clay particles, because of the cohesive forces holding them together. The lower curve in the Hjulström diagram shows the velocity at which particles already in motion cannot be transported further and fall to the channel bed. This is called the fall velocity. It depends not just on grain size but on density and shape, too, as well as on the viscosity and density of the water. Interestingly, because the viscosity and density of the water change with the amount of sediment the stream carries, the relationship between flow velocity and deposition is complicated. As the flow velocity reduces, so the coarser grains start to fall out, while the finer grains remain in motion. The result is differential settling and **sediment sorting**. Clay and silt particles stay in suspension at velocities of 1–2 cm/s, which explains why suspended load deposits are not dumped on streambeds. The region between the lower curve and the upper band defines the velocities at which particles of different sizes are transported. The wider is the gap between the upper and lower

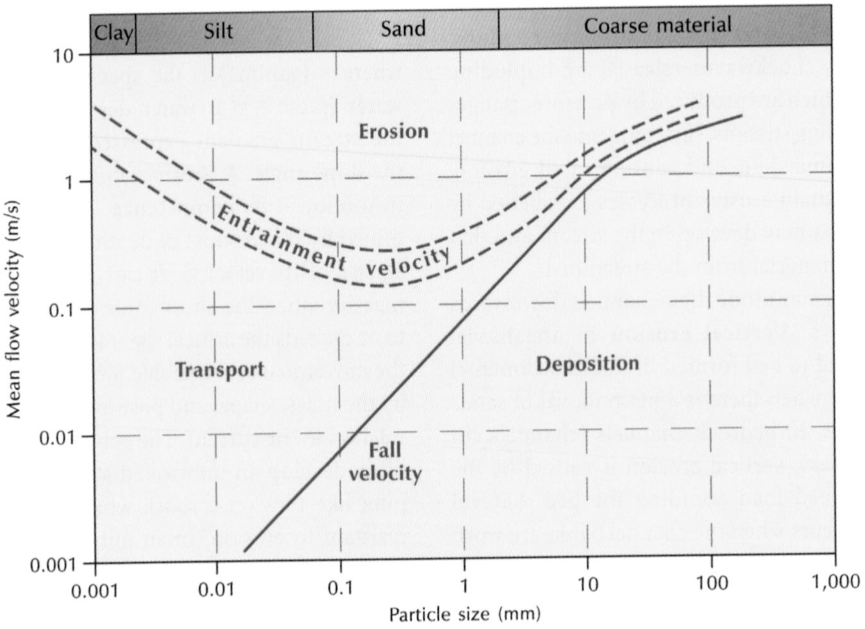

Figure 9.5 The Hjulström diagram showing the water velocity at which entrainment and deposition occur for particles of a given size in well-sorted sediments. *Source*: Adapted from Hjulström (1935)

lines, the more continuous is the transport. Notice that the gap for particles larger than 2 mm is small. In consequence, a piece of gravel eroded at just above the critical velocity will be deposited as soon as it arrives in a region of slightly lower velocity, which is likely to lie near the point of erosion. As a rule of thumb, the flow velocity at which erosion starts for grains larger than 0.5 mm is roughly proportional to the square root of the grain size. Or, to put it another way, the maximum grain size eroded is proportional to the square of the flow velocity.

It should be noted that the Hjulström diagram, based on laboratory conditions, is not easily applied to natural channels, where flow conditions may change rapidly, bed sediments are often of mixed calibre, and bank erosion is a source of sediment. Moreover, the Hjulström diagram applies only to erosion, transport, and deposition in alluvial channels. In bedrock channels, the bed load abrades the rock floor and causes vertical erosion. Where a stationary eddy forms, a small hollow is ground out that may eventually deepen to produce a **pothole** (Plate 9.2).

Plate 9.2 Pothole in bedrock bed of River Clough, near Sedbergh, Cumbria, England. The bedrock is Great Scar Limestone (Carboniferous). (*Photograph by Richard Huggett*)

Channel initiation

Stream channels can be created on a newly exposed surface or develop by the expansion of an existing channel network. Their formation depends upon water flowing over a slope becoming sufficiently concentrated for channel incision to occur. Once formed, a channel may grow to form a permanent feature. Robert E. Horton (1945) was the first to formalize the importance of topography to hillslope hydrology by proposing that a **critical hillslope length** was required to generate a channel (cf. p. 210). The critical length was identified as that required to generate a boundary shear stress of Hortonian overland flow sufficient to overcome the surface resistance and result in scour. In Horton's model, before overland flow is able to erode the soil, it has to reach a critical depth at which the eroding stress of the flow exceeds the shear resistance of the soil surface (Figure 9.6). Horton proposed that a '**belt of no erosion**' is present on the upper part of slopes because here the flow depth is not sufficient to cause erosion. However, subsequent work has demonstrated that some surface wash is possible even on slope crests, although here it does not lead to rill development because the rate of incision is slow and incipient rills are filled by rainsplash.

Further studies have demonstrated that a range of relationships between channel network properties and topography exist, although the physical processes driving these are not as well understood. In semi-arid and arid environments, the Hortonian overland-flow model provides a reasonable framework for explaining channel initiation, but it does not for humid regions. Thomas Dunne's (1980) research into humid channels showed that **spring sapping** from groundwater and throughflow may create channels. In humid regions, channel initiation is more related to the location of surface and subsurface flow convergence, usually in slope concavities and adjacent to existing drainage lines, than to a critical distance of overland flow. Rills can develop as a result of a sudden outburst of subsurface flow at the surface close to the base of a slope. So, channel development in humid

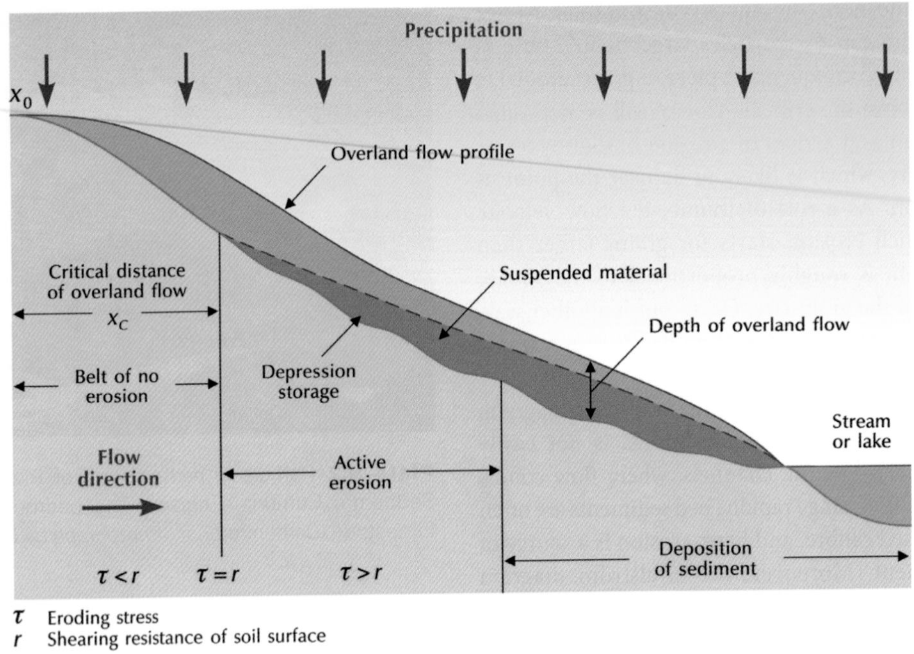

Figure 9.6 Horton's model of overland flow production. *Source*: Adapted from Horton (1945)

regions is very likely to occur where subsurface **pipes** are present. Pipe networks can help initiate channel development, either through roof collapse or by the concentration of runoff and erosion downslope of pipe outlets. Piping can also be important in semi-arid regions. Channel initiation may also take place where slope wash and similar mass movements dominate soil creep and creep-like processes (e.g. Smith and Bretherton 1972; Tarboton *et al.* 1992; Montgomery and Dietrich 1988, 1989). Recent work in the Higashi-gouchi catchment in the Akaishi Mountains of central Honshu, Japan, showed that surface and subsurface flows created most channel heads in the deeply incised subcatchments, although many landslides have also occurred around the channel heads (Imaizumi *et al.* 2009).

Fluvial deposition

Rivers may deposit material anywhere along their course, but they mainly deposit material in valley bottoms where gradients are low, at places where gradients change suddenly, or where channelled flow diverges, with a reduction in depth and velocity. The Hjulström diagram (p. 218) defines the approximate conditions under which solid-load particles are deposited upon the stream bed. Four types of fluvial deposit are recognized: **channel deposits**, **channel margin deposits**, **overbank floodplain deposits**, and **valley margin deposits** (Table 9.2). When studying stream deposition, it is useful to take the broad perspective of erosion and deposition within drainage basins. Stream erosion and deposition take place during flood events. As discharge increases during a flood, so erosion rates rise and the stream bed is scoured. As the flood abates, sediment is redeposited over days or weeks. Nothing much then happens until the next flood. Such **scour-and-fill cycles** shift sediment along the streambed. Scour-and-fill and channel deposits are found in most streams. Some streams actively accumulate sediment along much of their courses,

Table 9.2 Classification of valley sediments

Type of deposit	Description
Channel deposits	
Transitory channel deposits	Resting bed-load. Part may be preserved in more durable channel fills or lateral accretions
Lag deposits	Sequestrations of larger or heavier particles. Persist longer than transitory channel deposits
Channel fills	Sediment accumulated in abandoned or aggrading channel segments. Range from coarse bed-load to fine-grained oxbow lake deposits
Channel margin deposits	
Lateral accretion deposits	Point bars and marginal bars preserved by channel shifting and added to the overbank floodplain
Overbank floodplain deposits	
Vertical accretion deposits	Fine-grained sediment deposited from the load suspended in overbank flood-water. Includes natural levees and backswamp deposits
Splays	Local accumulations of bed-load materials spread from channel onto bordering floodplains
Valley margin deposits	
Colluvium	Deposits derived mainly from unconcentrated slope wash and soil creep on valley sides bordering floodplains
Mass movement deposits	Debris from earthflow, debris avalanches, and landslides, commonly intermixed with marginal colluvium. Mudflows normally follow channels but may spill over the channel bank

Source: Adapted from Benedict *et al.* (1971)

and many streams deposit material in broad expanses in the lower reaches but not in their upper reaches. **Alluviation** is large-scale deposition affecting much of a stream system. It results from fill preponderating scour for long periods. As a rule, scour and erosion dominate upstream channels, and fill and deposition dominate downstream channels. This pattern arises from steeper stream gradients, smaller hydraulic radii, and rougher channels upstream promoting erosion; and shallower gradients, larger hydraulic radii, and smoother channels downstream promoting deposition. In addition, flat, low-lying land bordering a stream that forms a suitable platform for deposition is more common at downstream sites.

Alluviation may be studied by calculating **sediment budgets** for alluvial or valley storage in a drainage basin. The change in storage during a time interval is the difference between the sediment gains and the sediment losses. Where gains exceed losses, storage increases with a resulting **aggradation** of channels or floodplains or both. Where losses exceed gains, channels and floodplains are eroded (**degraded**). It is feasible that gains counterbalance losses to produce a steady state. This condition is surprisingly rare, however. Usually, valley storage and fluxes conform to one of four common patterns under natural conditions (Trimble 1995): a quasi-steady-state typical of humid regions, vertical accretion of channels and aggradation of floodplains, valley trenching (arroyo cutting), episodic gains and losses in mountain and arid streams (Figure 9.7).

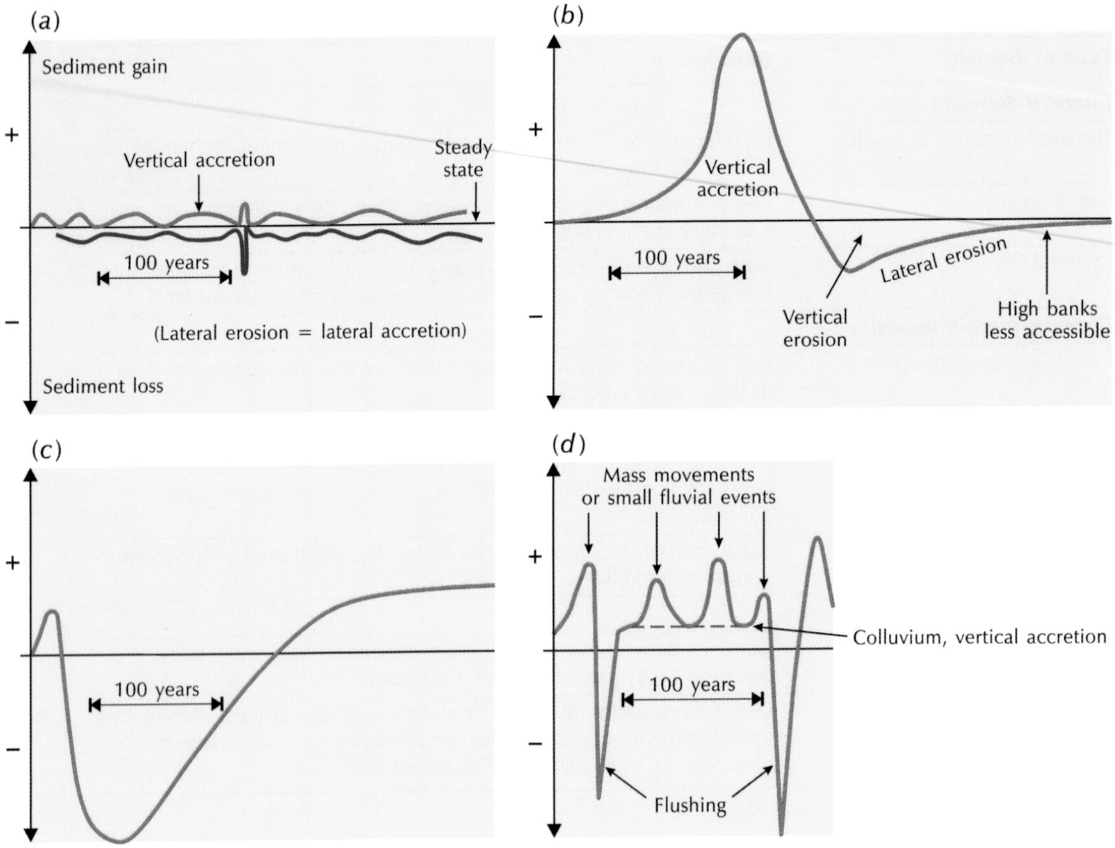

Figure 9.7 Four common patterns of valley sediment storage and flux under natural conditions. (a) Quasi-steady-state typical of humid regions. (b) Great sediment influx with later amelioration producing vertical accretion of channels and aggradation of floodplains. (c) Valley trenching (arroyo cutting). (d) High-energy instability seen as episodic gains and losses in mountain and arid streams. *Source*: Adapted from Trimble (1995)

FLUVIAL EROSIONAL LANDFORMS

The action of flowing water cuts rills, gullies, and river channels into the land surface.

Rills and gullies

Rills are tiny hillside channels a few centimetres wide and deep that are cut by ephemeral rivulets. They grade into gullies. An arbitrary upper limit for rills is less than a third of a metre wide and two-thirds of a metre deep. Any fluvial hillside channel larger than that is a gully. **Gullies** are intermediate between rills and arroyos, which are larger incised stream beds. They tend to be deep and long and narrow, and continuous or discontinuous. They are not as long as valleys but are too deep to be crossed by wheeled vehicles or to be 'ironed out' by ploughing. They often start at a head-scarp or waterfall. Gullies bear many local names, including dongas, *vocarocas*, ramps, and *lavakas*. Much current gullying appears to result from human modification of the land surface leading to disequilibrium in the hillslope system. **Arroyos**, which are also called **wadis**, **washes**,

dry washes, and **coulees**, are ephemeral stream channels in arid and semi-arid regions. They often have steep or vertical walls and flat, sandy floors. Flash floods course down normally dry arroyos during seasonal or irregular rainstorms, causing considerable erosion, transport, and deposition.

Bedrock channels

River channels may cut into rock and sediment. It is common to distinguish alluvial and bedrock channels, but many river channels form in a combination of alluvium and bedrock (Wohl 2015). Bedrock may alternate with thick alluvial fills, or bedrock may lie below a thin veneer of alluvium. Rivers with alternate bedrock and alluvial channels downstream include the Congo, Africa, the Colorado, USA, and the Dee, Scotland. Semi-controlled or channelized channels are a third category.

Bedrock channels are eroded into rock. They are resistant to erosion and tend to persist for long periods. They may move laterally in rock that is less resistant to erosion. The rate of river incision into bedrock is critical for studies of long-term landscape evolution and of the linkages between climate, erosion, and tectonics as it dictates the style and tempo of long-term landscape change in mountainous regions (Whipple 2004). Most rivers cut into bedrock in their upper reaches, where gradients are steep and their loads coarser. However, some rivers, such as many in Africa, flow in alluvium in their upper reaches and cut into bedrock in the lower reaches (cf. p. 123). Bedrock channels are not well researched, with most attention being given to such small-scale erosional features as scour marks and potholes in the channel bed. The long profiles of bedrock channels are usually more irregular than the long profiles of alluvial channels. The irregularities may result from the occurrence of more resistant beds, from a downstream steepening of gradient below a **knickpoint** caused by a fall of baselevel, from faulting, or from landslides and other mass movements dumping a pile of debris in the channel. Rapids and waterfalls often mark their position.

Given that many kinds of bedrock are resistant to erosion, it might seem improbable that bedrock channels would meander. However, incised meanders do form in horizontally bedded strata. They form when a meandering river on alluvium eats down into the underlying bedrock. **Intrenched meanders**, such as those in the San Juan River, Utah, USA, are symmetrical forms and evolve where downcutting is fast enough to curtail lateral meander migration, a situation that would arise when a large fall of baselevel induced a knickpoint to migrate upstream (Plate 9.3). **Ingrown meanders** are asymmetrical and result from meanders moving sideways at the same time as they slowly incise owing to regional warping. A **natural arch** or **bridge** forms where two laterally migrating meanders cut through a bedrock spur (p. 437).

Springs sometimes cut into bedrock. Many springs issue from alcoves, channels, or ravines that have been excavated by the spring water. The 'box canyons' that open into the canyon of the Snake River in southern Idaho, USA, were cut into basalt by the springs that now rise at the canyon heads.

Plate 9.3 Incised meander, a 350-m deep canyon of the San Juan River at Goosenecks, southern Utah, USA. (*Photograph by Tony Waltham Geophotos*)

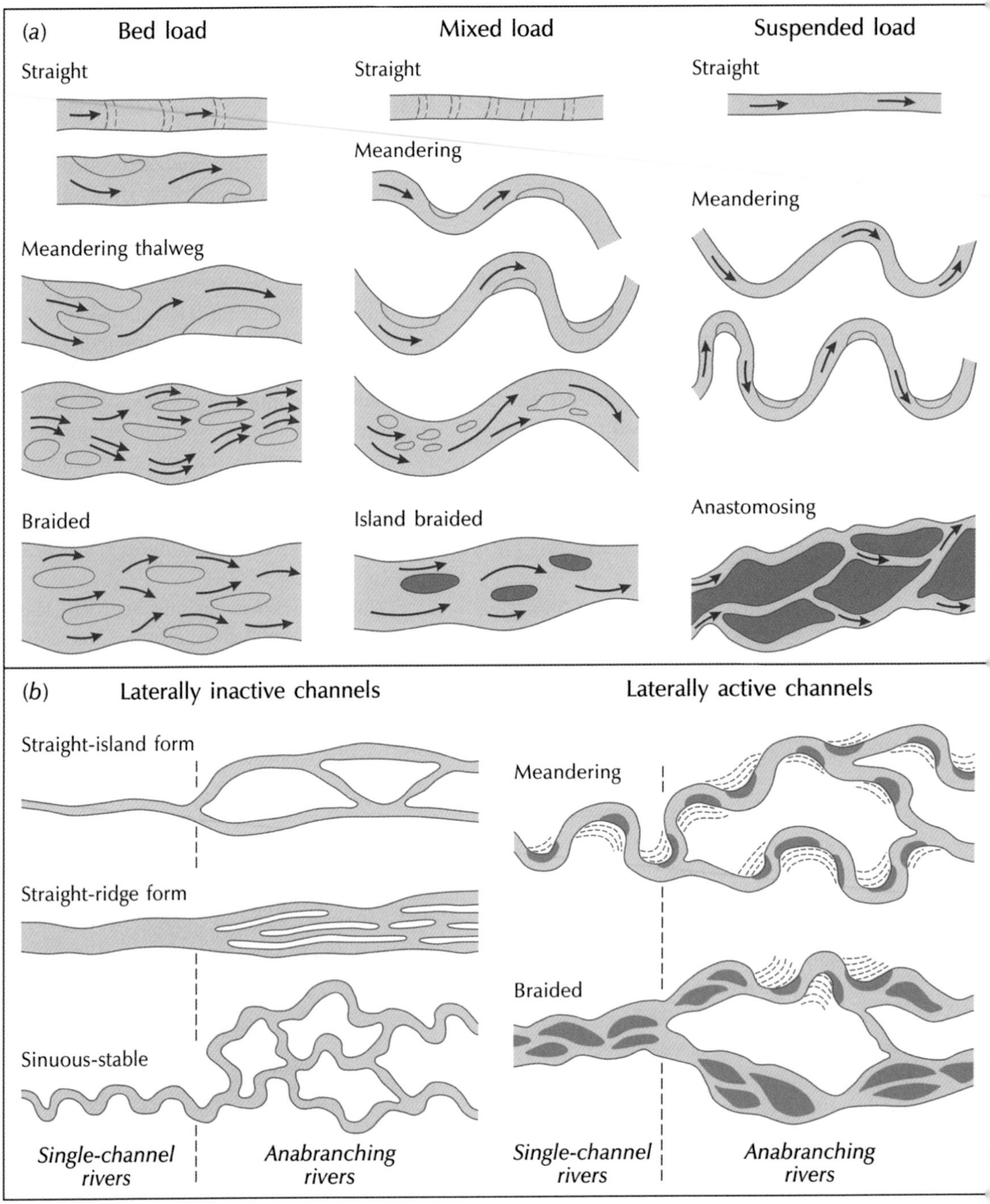

Figure 9.8 Classifications of channel patterns. (a) Channel form classified according to channel pattern (straight meandering, braided, and anastomosing) and sediment load (suspended load, suspended-load and bed-load mix bed load). (b) A classification of river patterns that includes single-channel and anabranching forms. *Sources*: (a Adapted from Schumm (1981, 1985b) and Knighton and Nanson (1993); (b) Adapted from Nanson and Knightor (1996)

Alluvial channels

Alluvial channels form in sediment that has been, and is being, transported by flowing water. They are very diverse owing to the variability in the predominant grain size of the alluvium, which ranges from clay to boulders. They may change form substantially as discharge, sediment supply, and other factors change because alluvium is normally unable to resist erosion to any great extent. In plan view, alluvial channels display four basic forms that represent a graded series – straight, meandering, braided, and anastomosing (Figure 9.8a) – with wandering channels sometimes recognized as an intermediate grade between meandering channels and braided channels, and with anabranching channels as another category (Figure 9.8b). However, there is no agreed classification and no agreed terminology for channel patterns. Paul Carling and his colleagues (2014) have suggested a new channel classification that tries to summarize what is widely agreed upon in defining channels according to planform (Figure 9.9).

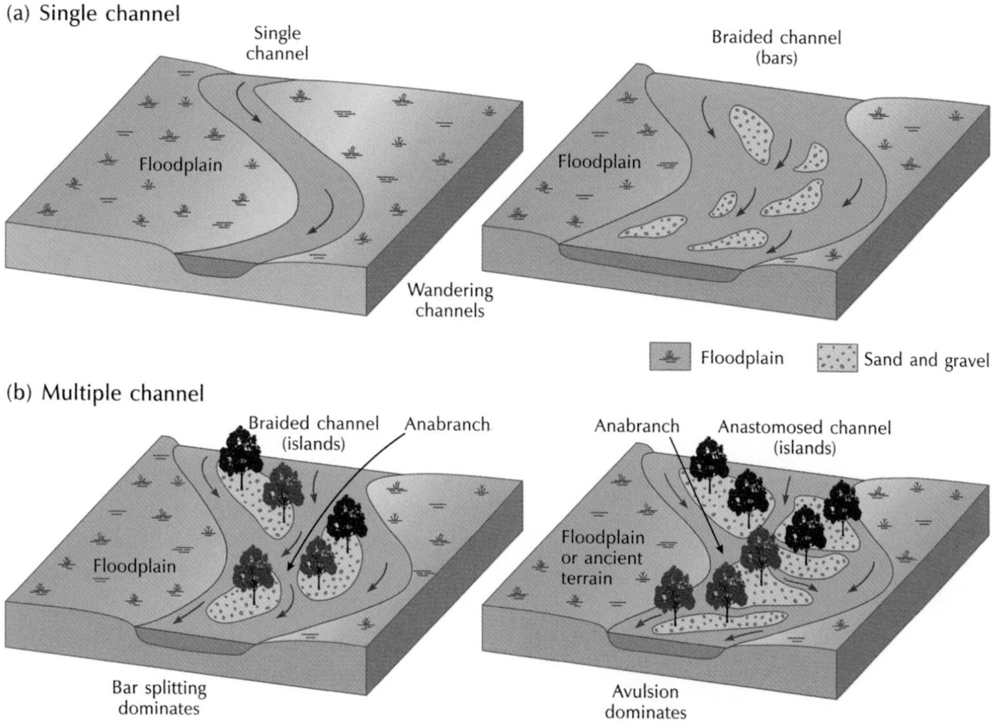

Figure 9.9 Another classification of channel planform. Single channels are considered to be developed within alluvial floodplains, although bedrock-confined and bedrock-constrained channels exist and are not shown on the diagram. 'Wandering' channels are shown as transitional between single channels and fully braided channels rather than a separate channel planform category. Braided channels are regarded as single channels, characterized by mobile, unvegetated bars that are inundated during near-bankfull flows when a single channel forms. Multiple channels can be separated on the basis of process: in the case of the braided channel (islands), accretionary alluvial islands cause channel splitting into anabranches; in anastomosed systems, avulsion across floodplain or other terrain leads to the development of anabranches and multiple-channel networks. In both cases the islands are not inundated during near-bankfull flows. *Source*: After Carling *et al.* (2014)

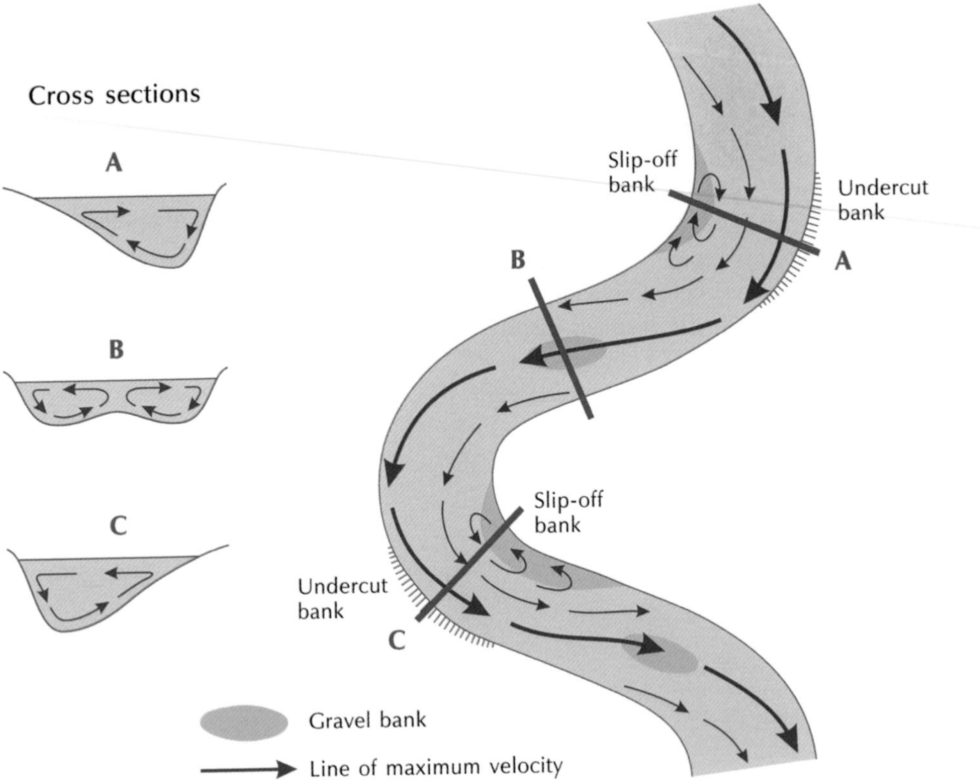

Figure 9.10 Water flow in a meandering channel.

Plate 9.4 Meanders on the River Bollin, Cheshire, England. (*Photograph by David Knighton*)

Straight channels

These are uncommon in the natural world. They are usually restricted to stretches of V-shaped valleys that are themselves straight owing to structural control exerted by faults or joints. **Straight channels** in flat valley-floors are almost invariably artificial. Even in a straight channel, the thalweg (the trace of the deepest points along the channel) usually winds from side to side, and the long profile usually displays a series of deeper and shallower sections (pools and riffles, p. 245) much like a meandering stream or a braided stream.

Meandering channels

Meandering channels wander snake-like across a floodplain (Plates 9.4 and Plate 9.5). The dividing line between straight and meandering is arbitrarily defined by a sinuosity of 1.5, calculated by

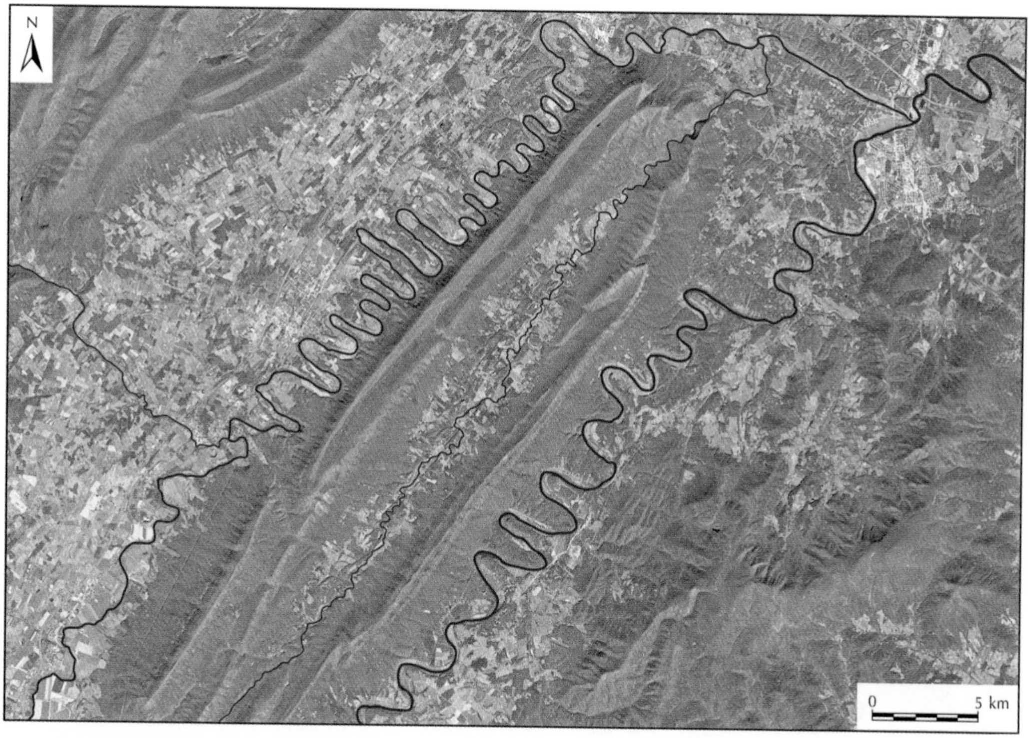

Plate 9.5 Meanders on the Shenandoah River, Virginia, USA. The chief rivers have been traced for clarity. (*Photograph courtesy of NASA*)

Plate 9.6 Meandering river and abandoned channels, Owens River, California, USA. (*Photograph by Marli Miller*)

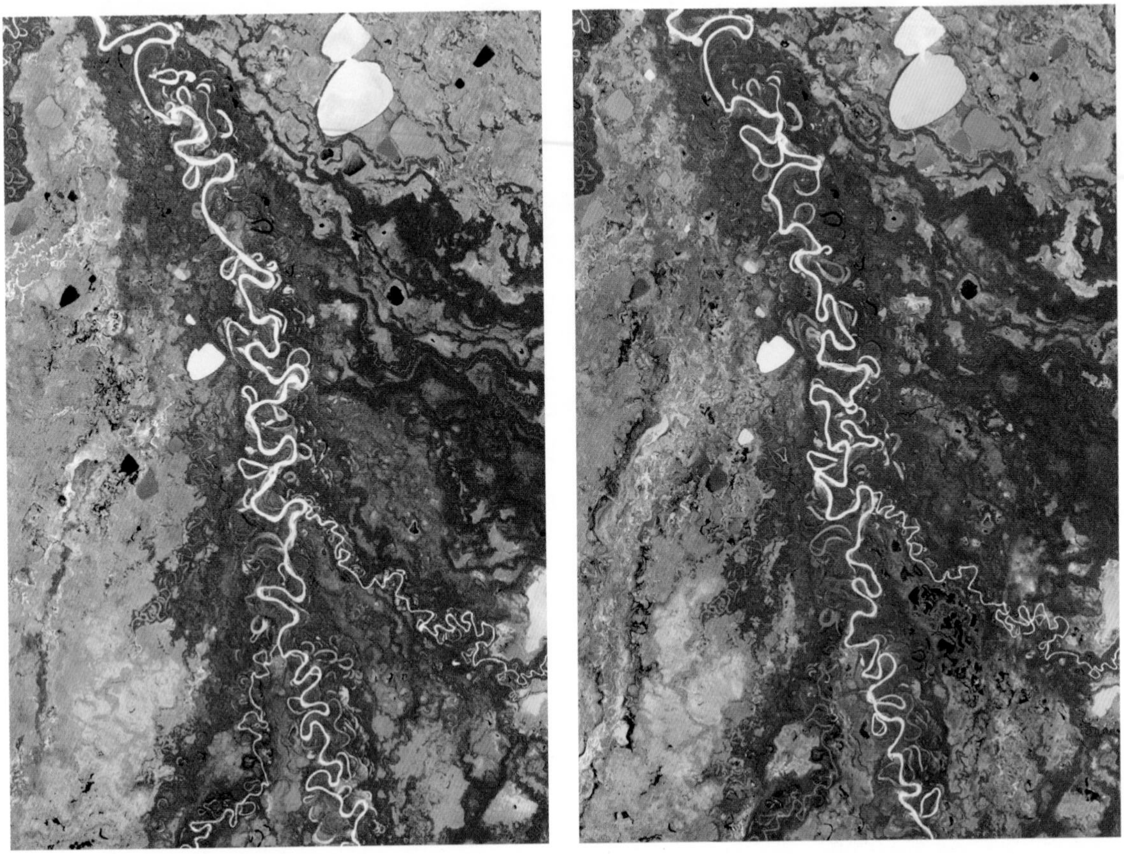

Plate 9.7 Wandering meanders in Amazonia: changes on the Rio Mamoré over nearly thirty years. The 11 June 1985 image was taken by the Thematic Mapper (TM) on the Landsat 5 satellite, the July 13 2014 by the Operational Land Imager (OLI) on Landsat 8. The river flows towards the north (left in these images) and receives a large amount of sediment at the confluence with the Rio Grande. The extra sediment enhances the growth of point bars, which shows as the lighter-coloured, vegetation-free areas along the inside bends of the riverbank. The point bars cause erosion and altered river flow that lead to a 1.7-fold increase in the rate of river migration downstream and a doubling of meander cutoff rates (Constantine *et al.* 2014). (*Photographs by NASA*)

dividing the channel length by the valley length. Water flows through meanders in a characteristic pattern (Figure 9.10). The flow pattern encourages erosion and undercutting of banks on the outside of bends and deposition, and the formation of point bars, on the inside of bends. The position of meanders changes, leading to the alteration of the course through cut-offs and channel diversion (avulsions). **Avulsions** are the sudden change in the course of a river leading to a section of abandoned channel, a section of new channel, and a segment of higher land (part of the floodplain) between them (Plates 9.6 and 9.7).

Meanders may cut down or incise. Plate 9.3 shows the famous incised meanders of the San Juan River, southern Utah, USA. **Cut-off incised meanders** may also form.

Meanders may be defined by several morphological parameters (Figure 9.11). Natural meanders are seldom perfectly symmetrical and regular owing to variations in the channel bed. Nonetheless, for most meandering rivers, the

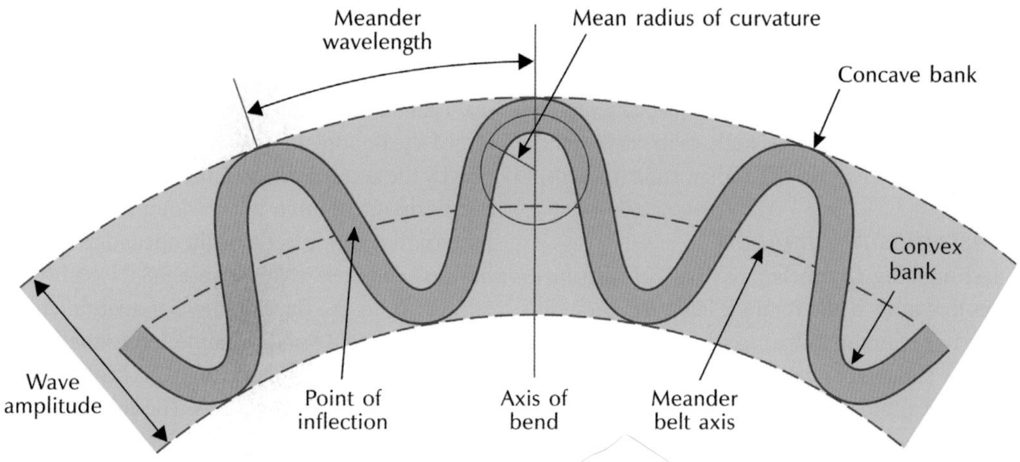

Figure 9.11 Parameters for describing meanders.

relationships between the morphometric parameters give a consistent picture: meander wavelength is about ten times channel width and about five times the radius of curvature.

Meandering is favoured where banks resist erosion, so forming deep and narrow channels. However, why rivers meander is not entirely clear. Ideas centre on: (1) the distribution and dissipation of energy within a river; (2) helical flow; and (3) the interplay of bank erosion, sediment load, and deposition. A consensus has emerged that meandering is caused by the intrinsic instabilities of turbulent water against a movable channel bank.

Braided channels

Braided channels (Plate 9.8) are essentially depositional forms that occur where the flow divides into a series of braids separated by islands or bars of accumulated sediment (see Best and Bristow 1993). The islands support vegetation and last a long time, while the bars are more impermanent. Once bars form in braided rivers, they are rapidly colonized by plants, so stabilizing the bar sediments and forming islands. However, counteracting the stabilization process is a highly variable stream discharge, which encourages alternate phases of degradation and aggradation in the channel and militates against vegetation establishment. Some braided rivers have twenty or more channels at one location.

Braided channels tend to form where (1) stream energy is high; (2) the channel gradient is steep; (3) sediment supply from hillslopes, tributaries, or glaciers is high and a big portion of coarse material is transported as bed load; and (4) bank material is erodible, allowing the channel

Plate 9.8 Braiding in Resurrection River, Alaska, USA. (*Photograph by Marli Miller*)

to shift sideways with relative ease. They are common in glaciated mountains, where channel slopes are steep and the channel bed is very gravelly. They form in sand-bed and silt-bed streams where the sediment load is high, as in parts of the Brahmaputra River on the Indian subcontinent.

Anastomosing channels

Anastomosing channels have a set of distributaries that branch and rejoin (Plate 9.9). They are suggestive of braided channels, but braided channels are single-channel forms in which flow is diverted around obstacles in the channel, while anastomosing channels are a set of interconnected channels separated by bedrock or by stable alluvium. The formation of anastomosing channels is favoured by an aggradational regime involving a high suspended-sediment load in sites where lateral expansion is constrained. Anastomosing channels are rare: River Feshie, Scotland, is the only example in the UK.

Anabranching channels

Anabranching rivers consist of multiple channels separated by vegetated and semi-permanent alluvial islands or alluvial ridges. The islands are cut out of the floodplain or are constructed in channels by the accretion of sediments. Anabranching is a fairly uncommon but a widespread channel pattern that may affect straight, meandering, and braided channels alike (Figure 9.8). Conditions conducive to the development of anabranching include frequent floods, channel banks that resist erosion, and mechanisms that block or restrict channels and trigger avulsions. The anabranching rivers of the Australian interior seem to be the outcome of low-angle slopes and irregular flow regimes. Those on the alluvial plains of south-western New South Wales form a complicated network along 100 km and more of the Edward and Murray Rivers; for instance, Beveridge Island is about 10 km long and lies between two roughly equal branches of the Murray River. Those on the

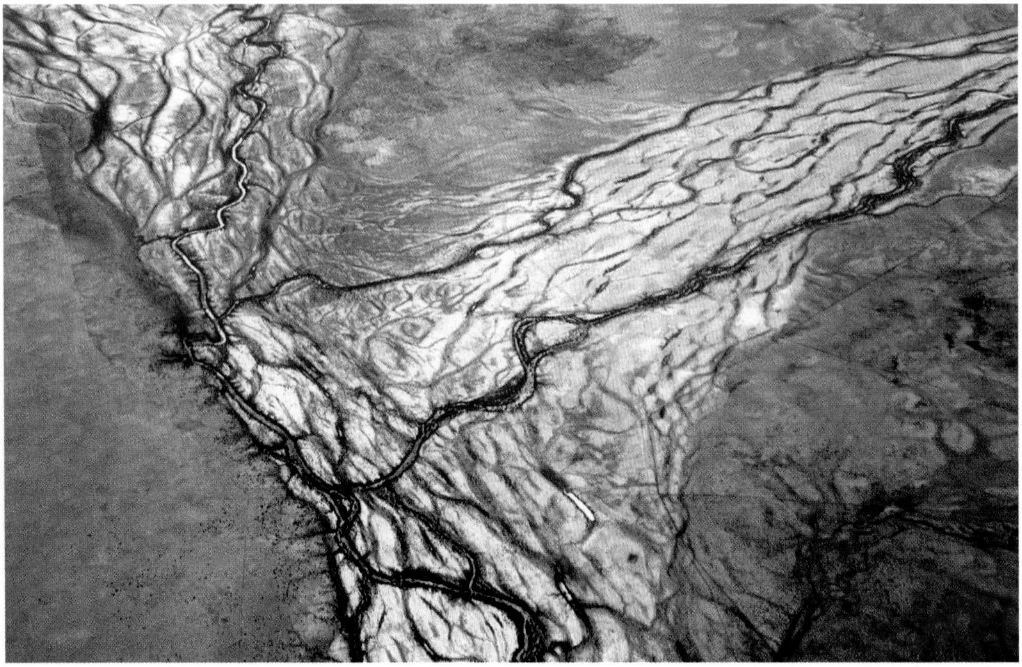

Plate 9.9 The junction of two anastomosing rivers, Queensland, Australia. (*Photograph by David Knighton*)

Northern Plains near Alice Springs appear to be a stable river pattern designed to preserve a throughput of relatively coarse sediment in low-gradient channels that characteristically have abundant vegetation in them and declining downstream discharges (Tooth and Nanson 1999).

Channels in mountains

Mountain drainage basins have their own characteristic set of channel forms. The basic channel processes are the same as in other streams, but mountain streams tend to be confined, hillslope processes and riparian vegetation may play a large role in their development, and they often contain much woody debris. There are seven channel-reach types: colluvial, bedrock, and five alluvial channel types – cascade, step–pool, plane bed, pool–riffle, and dune ripple (Figure 9.12). The form of the alluvial channels reflects specific roughness configurations adjusted to the relative magnitudes of sediment supply and transport capacity: steep alluvial channels (cascade and step–pool) have high transport capacities and a low supply of sediment and so are resilient to changes in discharge and in sediment supply; low-gradient alluvial channels (pool–riffle and dune ripple) have lower transport capacities and a greater supply of sediment, and so show significant and prolonged response to changes in sediment supply and discharge (Montgomery and Buffington 1997).

Hydraulic geometry

The controlling influence of discharge upon channel form, resistance to flow, and flow velocity is explored in the concept of **hydraulic geometry**. The key to this concept is the discharge equation:

$$Q = wdv$$

where Q is stream discharge (m^3/s), w is the stream width (m), d is the mean depth of the stream in a cross-section (m), and v is the mean flow velocity in the cross-section (m/s). Hydraulic geometry considers the relationships between the average channel form and discharge. It does so at-a-station (discharge changes at a specific point along a river) and downstream (discharge changes along a river). Discharge is the independent variable and channel form (width, depth, and velocity) are the dependent variables. At-a-station dependent variables are power functions of discharge (Leopold and Maddock 1953):

$$w = aQ^b$$
$$d = cQ^f$$
$$v = kQ^m$$

The exponents indicate the increase in hydraulic variable (width, depth, and velocity) per unit increase in discharge. Manning's roughness factor and slope can be added to the list of dependent variables (Singh 2003). Now, discharge is the product of width and depth (cross-sectional area) and velocity, so:

$$Q = wdv = (aQ^b)(cQ^f)(kQ^m)$$

which may be written

$$Q = wdv = ackQ^{b+f+m}$$

Therefore,

$$ack = 1 \text{ and } b + f + m = 1$$

The values of the exponents vary with location, climate, and discharge conditions. There seems to be a tendency for the river to establish steady state between the dominant discharge and the sediment load. Proceeding downstream on the same river, width, depth, and velocity all increase regularly with increasing discharge. The downstream relationships between the dependent hydraulic variables and the independent discharge are expressible as a similar set of equations to the at-a-station relationships:

$$w = hQ^r$$
$$d = pQ^s$$
$$v = nQ^t$$

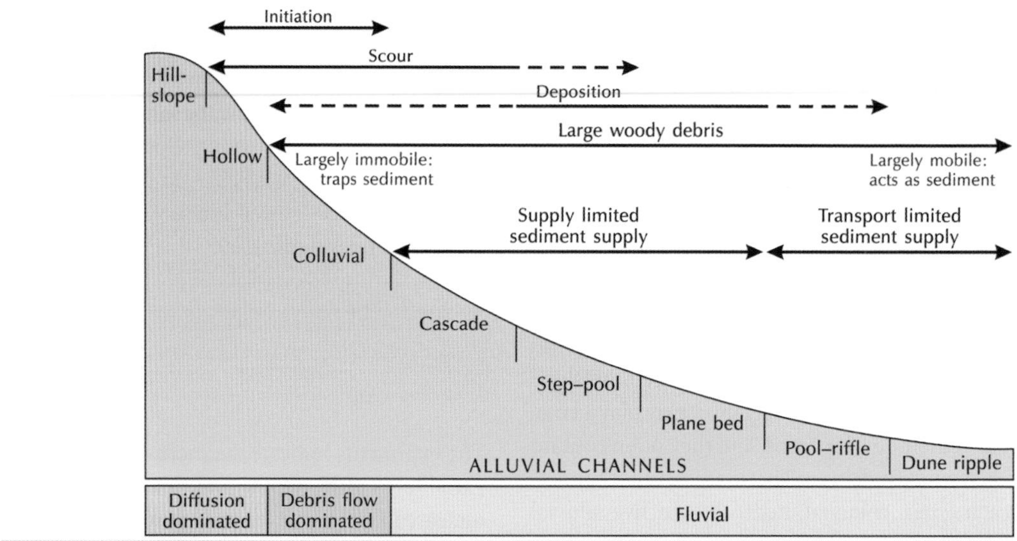

	Bedrock	Colluvial	Cascade	Step–pool	Plane bed	Pool–riffle	Dune ripple
Typical bed material	Rock	Variable	Boulder	Cobble-boulder	Gravel-cobble	Gravel	Sand
Bedform pattern	Irregular	Variable	Random	Vertically oscillatory	Featureless	Laterally oscillatory	Multilayered
Dominant roughness elements	Boundaries (bed and banks)	Grains	Grains, banks	Bedforms (steps, pools), grains, banks	Grains, banks	Bedforms (bars, pools), grains, banks, sinuosity	Sinuosity, bedforms (bars, ripples, dunes), grains, banks
Dominant sediment sources	Fluvial, hillslope, debris flows	Hillslope, debris flows	Fluvial, hillslope, debris flows	Fluvial, hillslope, debris flows	Fluvial, bank failure, debris flows	Fluvial, bank failure	Fluvial, bank failure
Sediment storage elements	Pockets	Bed	Lee and stoss sides of flow obstructions	Lee and stoss sides of flow obstructions	Overbank	Overbank, bedforms	Overbank, bedforms
Typical confinement	Confined	Confined	Confined	Confined	Variable	Unconfined	Unconfined
Typical pool spacing (channel widths)	Variable	Unknown	<1	1 to 4	None	5 to 7	5 to 7

(i) Colluvial (ii) Cascade (iii) Step–pool (iv) Plane bed (v) Pool–riffle

Figure 9.12 Channel forms in mountain streams. *Sources:* Adapted from Montgomery and Buffington (1997); (*Photographs by Dave Montgomery*)

As a rule of thumb, the mean velocity and width–depth ratio (*w*/*d*) both increase downstream along alluvial channels as discharge increases. If discharge stays the same, then the product *wdv* does not change. Any change in width or depth or velocity causes compensating changes in the other two components. If stream width reduces, then water depth increases. The increased depth, through the relationships expressed in the Manning equation (p. 215), leads to an increased velocity. In turn, the increased velocity may then cause bank erosion, so widening the stream again and returning the system to a balance. The compensating changes are conservative in that they operate to achieve a roughly continuous and uniform rate of energy loss – a channel's geometry is designed to keep total energy expenditure to a minimum. Nonetheless, the interactions of width, depth, and velocity are indeterminate in the sense that it is difficult to predict an increase of velocity in a particular stream channel. They are also complicated by the fact that width, depth, velocity, and other channel variables respond at different rates to changing discharge. Bedforms and the width–depth ratio are usually the most responsive, while the channel slope is the least responsive. Another difficulty lies in knowing which stream discharge a channel adjusts to. Early work by M. Gordon Wolman and John P. Miller (1960) suggested that the bankfull discharge, which has a five-year recurrence interval, is the dominant discharge, but recent research shows that as hydrological variability or channel boundary resistance (or both) becomes greater, then channel form tends to adjust to the less frequent floods. Such incertitude over the relationship between channel form and discharge makes reconstructions of past hydrological conditions from relict channels problematic. Despite problems associated with them (see Singh (2003) for an excellent discussion of these), the hydraulic geometry relationships have proved of immense practical value in predicting channel changes, in the design of stable canals and intakes, river flow control works, irrigation schemes, and river improvement works, and in many other ways.

Changes in hydrological regimes may lead to a complete alteration of alluvial channel form, or what Stanley A. Schumm called a '**river metamorphosis**'. Such a thoroughgoing reorganization of channels may take decades or centuries. Human interference within a catchment often triggers it, but it may also occur owing to internal thresholds within the fluvial system and happen independently of changes in discharge and sediment supply. A good example of this comes from the western USA, where channels incised when aggradation caused the alluvial valley floor to exceed a threshold slope (Schumm and Parker 1977). As the channels cut headwards, so the increased sediment supply caused aggradation and braiding in downstream reaches. When incision ceased, less sediment was produced at the stream head and incision began in the lower reaches. Two or three such aggradation–incision cycles occurred before equilibrium was accomplished.

River long profiles, baselevel, and grade

The **longitudinal profile** or **long profile** of a river is the gradient of its water-surface line from source to mouth. Streams with discharge increasing downstream have concave long profiles. This is because the drag force of flowing water depends on the product of channel gradient and water depth. Depth increases with increasing discharge and so, in moving downstream, a progressively lower gradient is sufficient to transport the bed load. Many river long profiles are not smoothly concave but contain flatter and steeper sections. The steeper sections, which start at **knickpoints**, may result from outcrops of hard rock, the action of local tectonic movements, sudden changes in discharge, or critical stages in valley development such as active headward erosion. The long profile of the River Rhine in Germany is shown

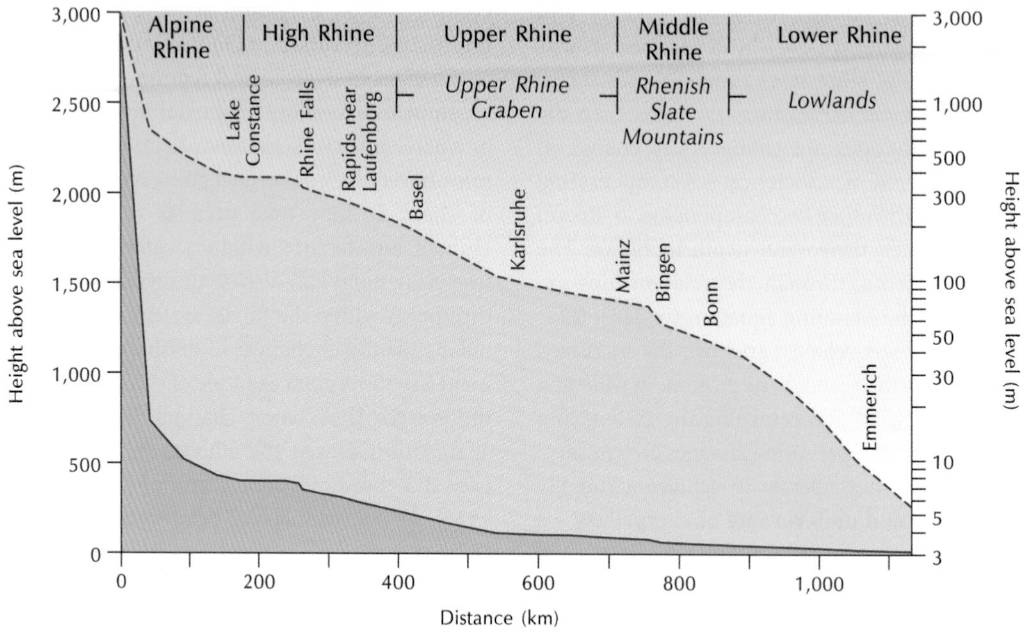

Figure 9.13 Long-profile of the River Rhine, shown on an arithmetic height scale (dashed line) and logarithmic height scale (solid line). *Source*: After Ahnert (1998, 174)

in Figure 9.13. Notice that the river is 1,236 km long and falls about 3 km from source to mouth, so the vertical distance from source to mouth is just 0.24 per cent of the length. Knickpoints can be seen at the Rhine Falls near Schaffhausen and just below Bingen. Most long profiles are difficult to interpret solely in terms of fluvial processes, especially in the case of big rivers, which are normally old rivers with lengthy histories, unique tectonic and other events which may have influenced their development. Even young rivers cutting into bedrock in the Swiss Alps and the Southern Alps of New Zealand have knickpoints, which seem to result from large rock-slope failures (Korup 2006).

Baselevel is the lowest elevation to which downcutting by a stream is possible. The ultimate baselevel for any stream is the water body into which it flows – sea, lake, or, in the case of some enclosed basins, playa, or salt lake (p. 250). Main channels also prevent further downcutting by tributaries and so provide a baselevel. Local baselevels arise from bands of resistant rock, dams of woody debris, beaver ponds, and human-made dams, weirs, and so on. The complex long profile of the River Rhine has three segments, each with a local baselevel. The first is Lake Constance, the second lies below Basel, where the Upper Rhine Plain lies within the Rhine Graben, and the third lies below Bonn, where the Lower Rhine embayment serves as a regional baselevel above the mouth of the river at the North Sea (Figure 9.13).

Grade, as defined by J. Hoover Mackin (1948), is a state of a river system in which controlling variables and baselevel are constant:

> A graded stream is one in which, over a period of years, slope is delicately adjusted to provide, with available discharge and with prevailing channel characteristics, just the velocity required for the transportation of the load provided by the drainage basin. The graded stream is a system in equilibrium; its diag-

> nostic characteristic is that any change in any of the controlling factors will cause a displacement of the equilibrium in a direction that will tend to absorb the effect of the change.
>
> (Mackin 1948, 471)

If the baselevel changes, then streams adjust their grade by changing their channel slope (through aggradation or degradation), or by changing their channel pattern, width, or roughness. However, as the controlling variables usually change more frequently than the time taken for the channel properties to respond, a graded stream displays a quasi-equilibrium rather than a true steady state.

Drainage basins and river channel networks

A river system can be considered as a network in which **nodes** (stream tips and stream junctions) are joined by **links** (streams). Stream segments or links are the basic units of stream networks. **Stream order** is used to denote the hierarchical relationship between stream segments and allows drainage basins to be classified according to size. Stream order is a basic property of stream networks because it relates to the relative discharge of a channel segment. Several stream-ordering systems exist, the most commonly used being those devised by Arthur N. Strahler and by Ronald L. Shreve (Figure 9.14). In **Strahler's ordering system**, a stream segment with no tributaries that flows from the stream source is denoted as a first-order segment. A second-order segment is created by joining two first-order segments, a third-order segment by joining two second-order segments, and so on. There is no increase in order when a segment of one order is joined by another of a lower order. Strahler's system takes no account of distance and all fourth-order basins are considered as similar. **Shreve's ordering system**, on the other hand, defines the magnitude of a channel segment as the total number of tributaries that feed it. Stream magnitude is closely related to the proportion of the total basin area contributing runoff, and so it provides a good estimate of relative stream discharge for small river systems.

Strahler's stream order has been applied to many river systems and it has been proved statistically to be related to a number of drainage-basin morphometry elements. For instance, the mean

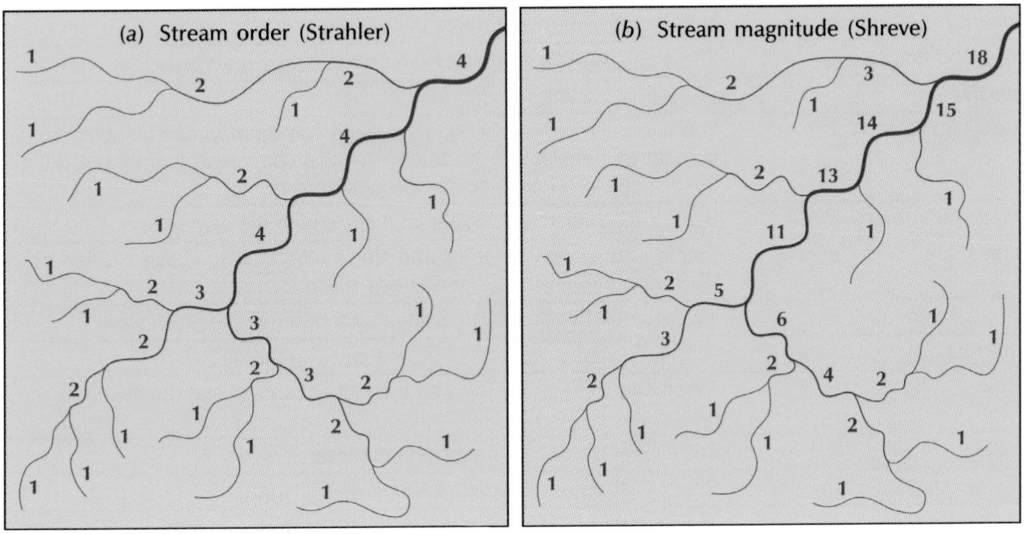

Figure 9.14 Stream ordering. (a) Strahler's system. (b) Shreve's system.

stream gradients of each order approximate an inverse geometric series, in which the first term is the mean gradient of first-order streams. A commonly used topological property is the **bifurcation ratio**, that is, the ratio between the number of stream segments of one order and the number of the next-highest order. A mean bifurcation ratio is usually used because the ratio values for different successive basins will vary slightly. With relatively homogeneous lithology, the bifurcation ratio is normally not more than five or less than three. However, a value of ten or more is possible in very elongated basins where there are narrow, alternating outcrops of soft and resistant strata.

The main geometrical properties of stream networks and drainage basins are listed in Table 9.3. The most important of these is probably drainage density, which is the average length of channel per unit area of drainage basin. **Drainage density** is a measure of how frequently streams occur on the land surface. It reflects a balance between erosive forces and the resistance of the ground surface, and is therefore related closely to climate, lithology, and vegetation. Drainage densities can range from less than 5 km/km^2 when slopes are gentle, rainfall low, and bedrock permeable (e.g. sandstones), to much larger values of more than 500 km/km^2 in upland areas where rocks are impermeable, slopes are steep, and rainfall totals are high (e.g. on unvegetated clay 'badlands' – Plate 9.10). Climate is important in basins of very high drainage densities in some semi-arid environments that seem to result from the prevalence of surface runoff and the relative ease with which new channels are created. Vegetation density is influential in determining drainage density, since it binds the surface layer preventing overland flow from concentrating along definite lines and from eroding small rills, which may develop into stream channels. Vegetation slows the rate of overland flow and effectively stores some of the water for short time periods. Drainage density also relates to the length of overland flow, which is approximately equal to

Table 9.3 Selected morphometric properties of stream networks and drainage basins

Property	Symbol	Definition
Network properties		
Drainage density	D	Mean length of stream channels per unit area
Stream frequency	F	Number of stream segments per unit area
Length of overland flow	L_g	The mean upslope distance from channels to watershed
Areal properties		
Texture ratio	T	The number of crenulations in the basin contour having the maximum number of crenulations divided by the basin perimeter length. Usually bears a strong relationship to drainage density
Circulatory ratio perimeter	C	Basin area divided by the area of a circle with the same basin
Elongation ratio	E	Diameter of circle with the same area as the drainage basin divided by the maximum length of the drainage basin
Lemniscate ratio	k	The square of basin length divided by four times the basin area
Relief properties		
Basin relief	H	Elevational difference between the highest and lowest points in the basin
Relative relief	R_{hp}	Basin relief divided by the basin perimeter
Relief ratio	R_h	Basin relief divided by the maximum basin length
Ruggedness number	N	The product of basin relief and drainage density

Source: Adapted from Huggett and Cheesman (2002, 98)

the reciprocal of twice the drainage density. And, importantly, it determines the distance from streams to valley divides, which strongly affects the general appearance of any landscape.

Early studies of stream networks indicated that purely random processes could generate fluvial systems with topological properties similar to natural systems (Shreve 1975; Smart 1978). Such random-model thinking has been extremely influential in channel network studies. However, later research has identified numerous regularities in stream network topology. These systematic variations appear to be a result of various factors, including the need for lower-order basins to fit together, the sinuosity of valleys and the migration of valley bends downstream, and the length and steepness of valley sides. These elements are more pronounced in large basins, but they are present in small catchments.

Plate 9.10 High drainage density in the Zabriskie Point badlands, Death Valley, California, USA. *(Photograph by Kate Holden)*

Folds, rivers, and drainage patterns

Geomorphologists once described individual streams according to their relationship with the initial surface upon which they developed. A **consequent stream** flowed down, and was a consequence of, the slope of the presumed original land surface. Streams that developed subsequently along lines of weakness, such as soft strata or faults running along the strike of the rocks, were **subsequent streams**. Subsequent streams carved out new valleys and created new slopes drained by **secondary consequent** or **resequent streams**, which flowed in the same direction as the consequent stream, and **obsequent streams**, which flowed in the opposite direction. This nomenclature is defunct, since it draws upon a presumed time-sequence in the origin of different streams. In reality, the entire land area drains from the start, and it is patently not the case that some parts remain undrained until main drainage channels have evolved. Modern stream nomenclature rests upon structural control of drainage development (Figure 6.16). In regions where a sequence of strata of differing resistance is tilted, streams commonly develop along the strike. **Strike streams** gouge out strike valleys, which are separated by strike ridges. Tributaries to the strike streams enter almost at right angles. Those that run down the dip slope are **dip streams** and those that run counter to the dip slope are **anti-dip streams**. The length of dip and anti-dip streams depends upon the angle of dip. Where dip is gentle, dip streams are longer than anti-dip streams. Where the dip is very steep, as in hogbacks, the dip streams and anti-dip streams will be roughly the same length, but often the drainage density is higher on the anti-dip slope and the contours are more crenulated because the anti-dip streams take advantage of joints in the hard stratum while dip streams simply run over the surface.

Most stream networks are adapted to regional slope and geological structures, picking out the main fractures in the underlying rocks. The high degree of conformity between stream networks and geological structure is evident in the nine chief drainage patterns (Morisawa 1985). A tenth category, **irregular** or **complex drainage**, which displays no unambiguous pattern, could be added – as could an eleventh, **deranged drainage**, which forms on newly exposed land, such as that exposed beneath a retreating ice sheet, where there is almost no structural or bedrock control and drainage is characterized by irregular stream

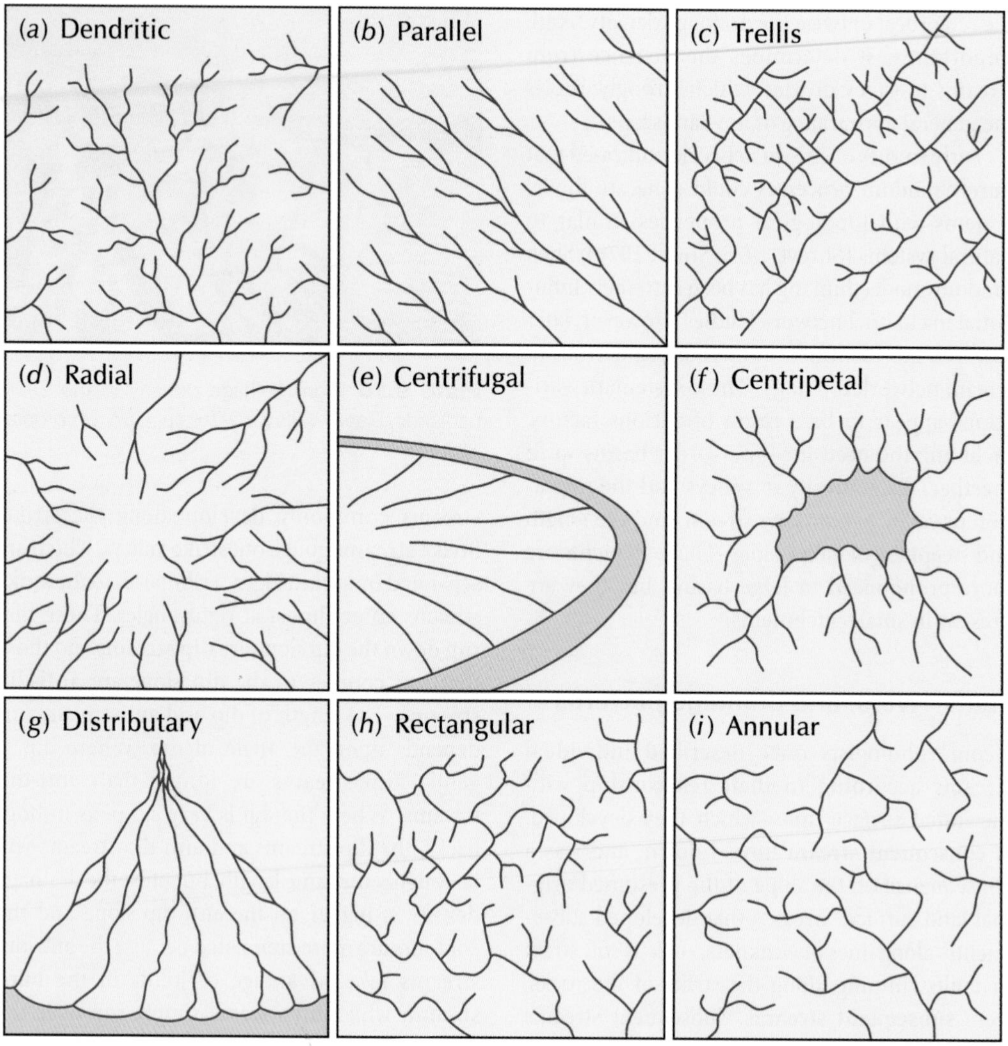

Figure 9.15 Drainage patterns controlled by structure or slope. *Source*: Mainly after Twidale and Campbell (2005, 191) and adapted from Twidale (2004, 173)

courses with short tributaries, lakes, and swamps. Figure 9.15 shows the major types of drainage pattern and their relationship to structural controls:

1 **Dendritic drainage** has a spreading, tree-like pattern with an irregular branching of tributaries in many directions and at almost any angle. It occurs mostly on horizontal and uniformly resistant strata and unconsolidated sediments and on homogeneous igneous rocks where there are no structural controls. Pinnate drainage, which is associated with very steep slopes, is a special dendritic pattern wherein the tributaries are more or less parallel and join the main stream at acute angles.
2 **Parallel drainage** displays regularly spaced and more or less parallel main streams with tributaries joining at acute angles. Parallel dip streams dominate the pattern. It develops where strata are uniformly resistant and the

regional slope is marked, or where there is strong structural control exerted by a series of closely spaced faults, monoclines, or isoclines.

3 **Trellis drainage** has a dominant drainage direction with a secondary direction parallel to it, so that primary tributaries join main streams at right angles and secondary tributaries run parallel to the main streams. It is associated with alternating bands of hard and soft dipping or folded beds or recently deposited and aligned glacial debris. Fold mountains tend to have trellis drainage patterns. An example is the Appalachian Mountains, north-east USA, where alternating weak and strong strata have been truncated by stream erosion.

4 **Radial drainage** has streams flowing outwards in all directions from a central elevated tract. It is found on topographic domes, such as volcanic cones and other sorts of isolated conical hills. On a large scale, radial drainage networks form on rifted continental margins over mantle plumes, which create lithospheric domes (Cox 1989; Kent 1991). A postulated Deccan plume beneath India caused the growth of a topographic dome, the eastern half of which is now gone (Figure 9.16a). Most of the rivers rise close to the west coast and drain eastwards into the Bay of Bengal, except those in the north, which drain north-eastwards into the Ganges, and a few that flow westwards or south-westwards (possibly along failed rift arms). Mantle plumes beneath southern Brazil and southern Africa would account for many features of the drainage patterns in those regions (Figure 9.16b–c).

5 **Centrifugal drainage** is similar to radial and occurs where, for example, gutters develop on the insides of meander loops on the tidal mudflats of coastal north-west Queensland, Australia.

6 **Centripetal drainage** has all streams flowing towards the lowest central point in a basin floor. It occurs in calderas, craters, dolines, and tectonic basins. A large area of internal drainage lies on the central Tibetan Plateau.

7 **Distributary drainage** typifies rivers debouching from narrow mountain gorges and running over plains or valleys, particularly during occasional floods when they overtop their banks. Many deltas display a similar pattern of drainage (p. 393).

8 **Rectangular drainage** displays a perpendicular network of streams with tributaries and main streams joining at right angles. It is less regular than trellis drainage, and is controlled by joints and faults. Rectangular drainage is common along the Norwegian coast and in portions of the Adirondack Mountains, USA. Angulate drainage is a variant of rectangular drainage and occurs where joints or faults join each other at acute or obtuse angles rather than at right angles.

9 **Annular drainage** has main streams arranged in a circular pattern with subsidiary streams lying at right angles to them. It evolves in a breached or dissected dome or basin in which erosion exposes concentrically arranged hard and soft bands of rock. An example is found in the Woolhope Dome in Herefordshire, England.

Investigations by Adrian E. Scheidegger revealed a strong **tectonic control** on drainage lines in some landscapes. In eastern Nepal, joint orientations, which strike consistently east to west, in large measure determine the orientation of rivers (Scheidegger 1999). In south-western Ontario, Canada, the Proterozoic basement (Canadian Shield), which lies under Pleistocene glacial sediments, carries a network of buried bedrock channels. The orientation of these channels shows a statistically significant relationship with the orientation of regional bedrock joints that formed in response to the mid-continental stress field. Postglacial river valleys in the area are also orientated in a similar direction to the bedrock joints. Both the bedrock channels and modern river channels bear the hallmarks of tectonically predesigned landforms (Eyles and

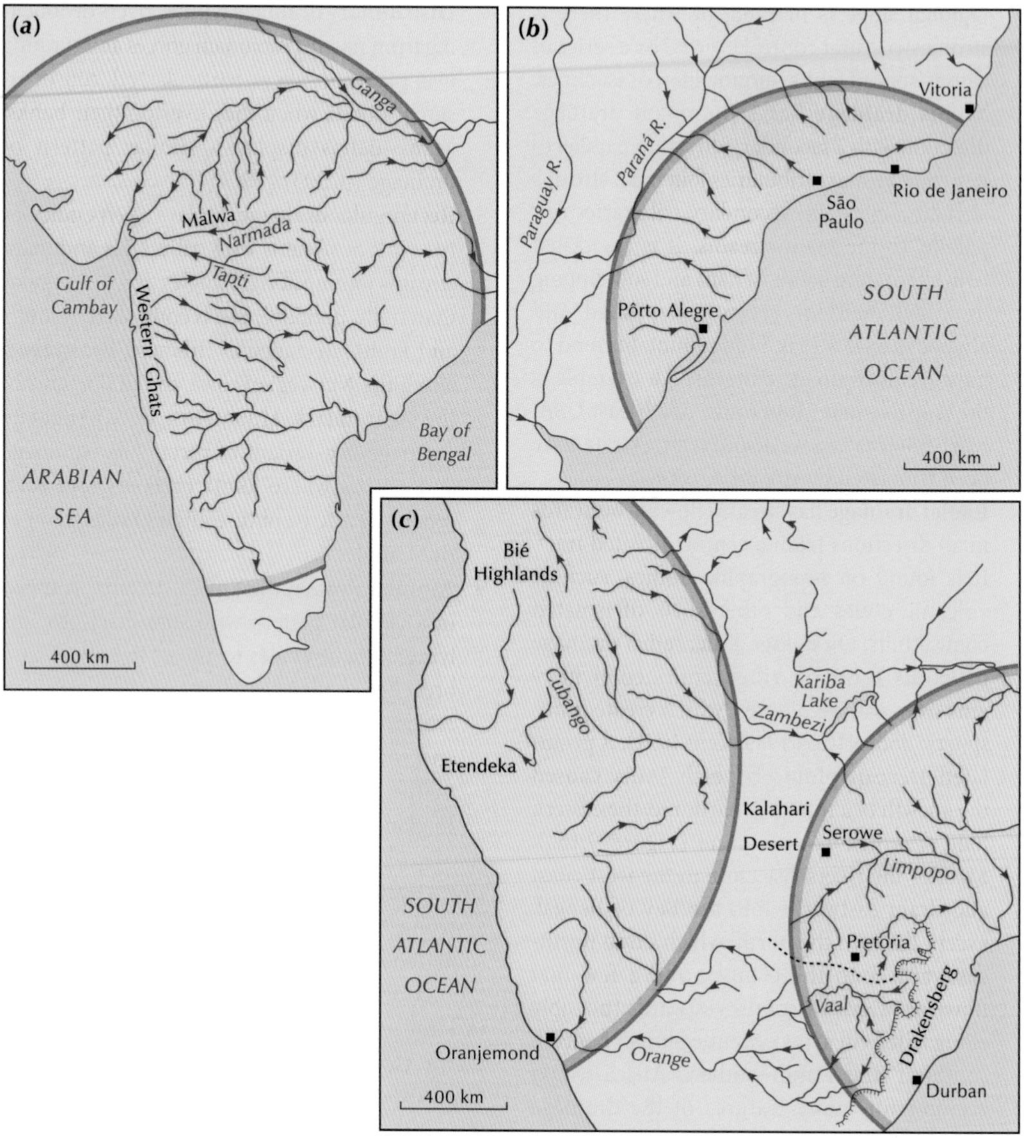

Figure 9.16 Drainage patterns influenced by mantle plumes. (a) The drainage pattern of peninsular India with the postulated Deccan plume superimposed. Most of the peninsula preserves dome-flank drainage. The Gulf of Cambay, Narmada, and Tapti systems exhibit rift-related drainage. (b) The drainage pattern of southern Brazil with superimposed plume. Dome-flank drainage is dominant except near Porto Alegre. (c) The drainage pattern in south-eastern and south-western Africa with the Paraná plume (left) and Karoo plume (right) superimposed. Rivers over the Paraná plume show an irregular dome-flank pattern drainage eastwards into the Kalahari. Notice that the Orange River gorge is formed where antecedent drainage has cut through younger uplift. Rivers over the Karoo plume display preserved dome-flank drainage west of the Drakensberg escarpment. The dotted line separates dome-flank drainage in the south from rift-related drainage in the north. *Source*: Adapted from Cox (1989)

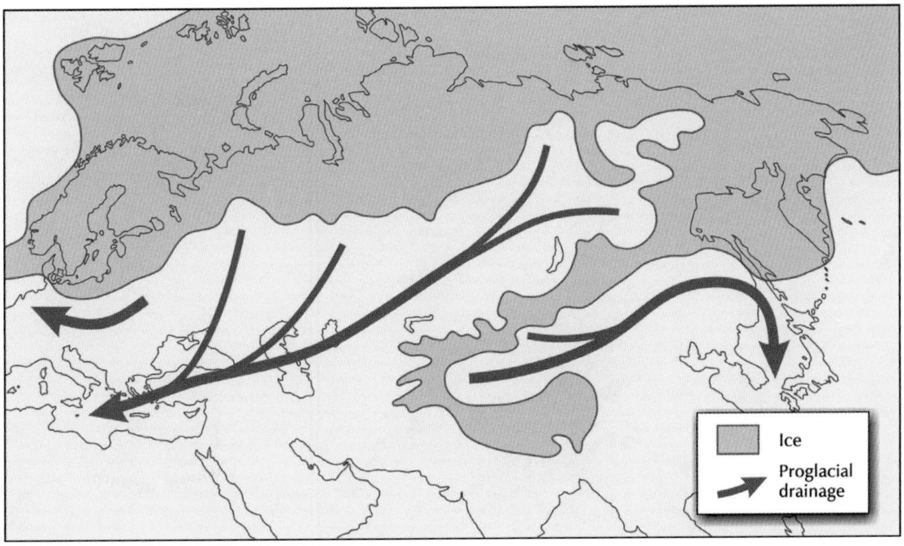

Figure 9.17 Proglacial drainage systems in northern Eurasia during the last glaciation.
Source: Adapted from Grosswald (1998)

Scheidegger 1995; Eyles *et al.* 1997; Hantke and Scheidegger 1999).

Structural and tectonic features, such as joints, faults, and lineaments (p. 154), may produce essentially **straight rivers**, that is, rivers with limited meander development (Twidale 2004). Joints and faults may produce short linear sections of rivers, typically a few tens of metres long. Longer straight rivers commonly follow regional lineament patterns, an example coming from central and northern Australia, where long sections of several alluvial rivers, including the Finke River, Georgina River, Thompson River, Darling River, and Lachlan River, track lineaments in the underlying bedrock. The Darling River, flowing over Quaternary alluvium, follows a lineament in Palaeozoic and Mesozoic bedrock between St George in south-east Queensland and near Menindee in western New South Wales, a distance of about 750 km.

Anomalous drainage patterns

Anomalous drainage bucks structural controls, flowing across geological and topographic units. A common anomalous pattern is where a major stream flows across a mountain range when just a short distance away is an easier route. In the Appalachian Mountains, north-east USA, the structural controls are aligned south-west to north-east but main rivers, including the Susquehanna, run north-west to south-east. Such **transverse drainage** has prompted a variety of hypotheses: diversion, capture or piracy, antecedence, superimposition, stream persistence, and valley impression.

Diverted rivers

Glacial ice, uplifted fault blocks, gentle folding, and lava flows may all cause major **river diversions.** Glacial ice is the most common agent of river diversions. Where it flows across or against the regional slope of the land, the natural drainage is blocked and proglacial or ice-dammed marginal lakes grow. Continental diversion of drainage took place during the last glaciation across northern Eurasia (Figure 9.17; cf. p. 305).

The Murray River was forced to go around the Cadell Fault Block, which was uplifted in the Late Pleistocene near Echuca, Victoria, Australia

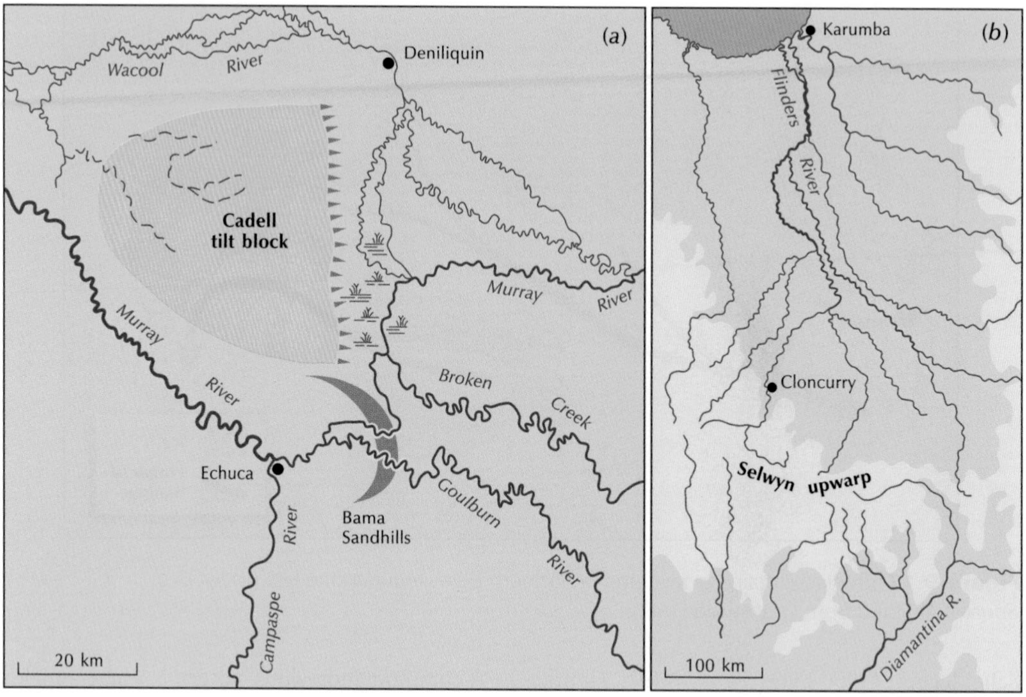

Figure 9.18 River diversions in Australia. (a) The diversion of the Murray River near Echuca, Victoria. (b) The diversion of the Diamantina River, north Queensland, owing to the Selwyn Upwarp. *Sources*: (a) Adapted from Bowler and Harford (1966) and (b) Adapted from Twidale and Campbell (2005, 110)

(Figure 9.18a). The Diamantina River, north-west Queensland, Australia, was diverted by Pleistocene uplift along the Selwyn Upwarp (Figure 9.18b). Faults may also divert drainage (see p. 155).

Captured rivers

Trellis drainage patterns, which are characteristic of folded mountain belts, result from the **capture** of strike streams by dip or anti-dip streams working headwards and breaching ridges or ranges. Capture is often shown by abrupt changes in stream course, or what are called **elbows of capture**.

Antecedent rivers

An **antecedent stream** develops on a land surface before uplift by folding or faulting occurs. When uplift does occur, the stream is able to cut down fast enough to hold its existing course and carves out a gorge in a raised block of land. The River Brahmaputra in the Himalaya is probably an antecedent river, but proving its antecedence is difficult. The problem of proof applies to most suspected cases of antecedent rivers.

Superimposed rivers

Superimposed drainage occurs when a drainage network established on one geological formation cuts down to, and is inherited by, a lower geological formation. The superimposed pattern may be discordant with the structure of the formation upon which it is impressed. A prime example comes from the English Lake District (Figure 9.19). The present radial drainage pattern is a response to the doming of Carboniferous, and possibly Cretaceous, limestones. The streams cut through the base of the Carboniferous limestone

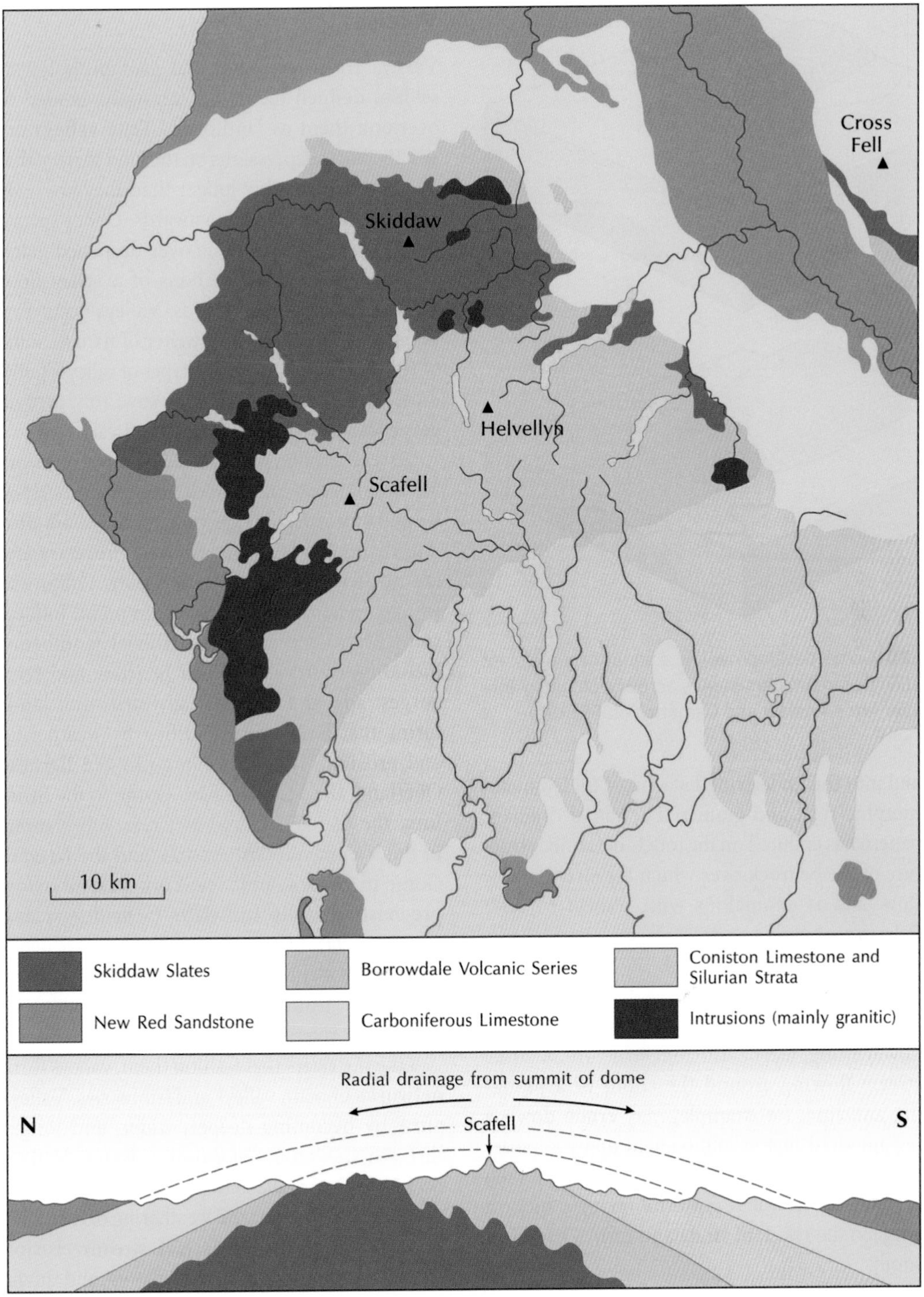

Figure 9.19 Superimposed drainage in the English Lake District. *Source*: Adapted from Holmes (1965, 564)

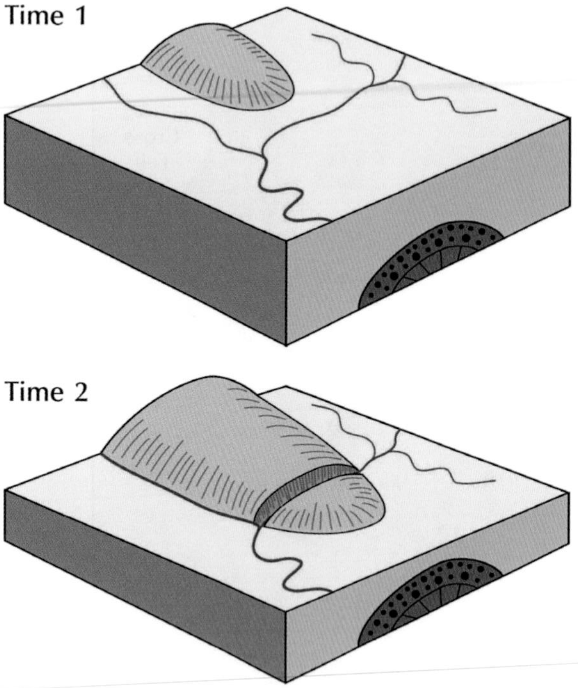

Figure 9.20 Gorge development in a snout of a resistant rock formation by stream persistence across a plunging anticline. *Source*: After Twidale and Campbell (2005, 195)

and into the underlying Palaeozoic folded metamorphic rock and granite. The radial drainage pattern has endured on the much-deformed structure of the bedrock over which the streams now flow, and is anomalous with respect to their Palaeozoic base.

Persistent rivers

Streams adjusted to a particular structure may, on downcutting, meet a different structure. A strike stream flowing around the snout of a plunging anticline, for example, may erode down a few hundred metres and be held up by a harder formation (Figure 9.20). The stream may then be diverted or, if it is powerful enough, incise a gorge in the resistant strata and form a breached snout.

Valleys

Valleys are so common that geomorphologists seldom defined them and, strangely, tended to overlook them as landforms. **True valleys** are simply linear depressions on the land surface that are almost invariably longer than they are wide with floors that slope downwards. Under special circumstances, as in some over-deepened glaciated valleys (p. 289), sections of a valley floor may be flat or slope upwards. Valleys occur in a range of sizes and go by a welter of names, some of which refer to the specific types of valley – gully, draw, defile, ravine, gulch, hollow, run, arroyo, gorge, canyon, dell, glen, dale, and vale.

As a rule, valleys are created by fluvial erosion, but often in conjunction with tectonic processes. Some landforms that are called 'valleys' are produced almost entirely by tectonic processes and are not true valleys – Death Valley, California, which is a half-graben, is a case in point. Indeed, some seemingly archetypal fluvial landforms, including river valleys, river benches, and river gorges, appear to be basically structural landforms that have been modified by weathering and erosion. The Aare Gorge in the Bernese Oberland, the Moutier–Klus Gorge in the Swiss Jura, the Samaria Gorge in Crete, hill-klamms in the Vienna Woods, Austria, and the Niagara Gorge in Ontario and New York state all follow pre-existing faults and clefts (Scheidegger and Hantke 1994). Erosive processes may have deepened and widened them, but they are essentially endogenic features and not the product of antecedent rivers.

Like the rivers that fashion them, valleys form **networks** of main valleys and tributaries. Valleys grow by becoming deeper, wider, and longer through the action of running water. Valleys deepen by hydraulic action, corrasion, abrasion, potholing, corrosion, and weathering of the valley floor. They widen by lateral stream erosion and by weathering, mass movements, and fluvial processes on the valley sides. They lengthen by headward erosion, by valley meandering, by extending over newly exposed land at their bottom ends, and by forming deltas.

Some valley systems are exceptionally old – the Kimberly area of Australia had been land throughout the Phanerozoic and was little affected by the ice ages (Ollier 1991, 99). The drainage system in the area is at least 500 million years old. Permian, Mesozoic, Mid- to Late Cretaceous, and Early Tertiary drainage has also been identified on the Australian continent.

FLUVIAL DEPOSITIONAL LANDFORMS

Alluvial bedforms

Riverbeds develop a variety of landforms generated by turbulence associated with irregular cross-channel or vertical velocity distributions that erode and deposit alluvium. The forms are **riffle–pool sequences** (Box 9.2) and **ripple–**

Box 9.2 POOLS AND RIFFLES

River channels, even initially straight ones, tend to develop deeper and shallower sections. These are called **pools** and **riffles** respectively (Plate 9.11). Experiments in flumes, with water fed in at a constant rate, produce pool-and-riffle sequences, in which the spacing from one pool to the next is about five times the channel width (Figure 9.21). Continued development sees meanders forming with alternate pools migrating to opposite sides. The meander wavelength is roughly two inter-pool spacings of ten channel widths, as is common in natural rivers.

Plate 9.11 Riffles and pools in a meandering section of the Poynton Brook, Poynton Coppice, Cheshire, England. (*Photograph by Richard Huggett*)

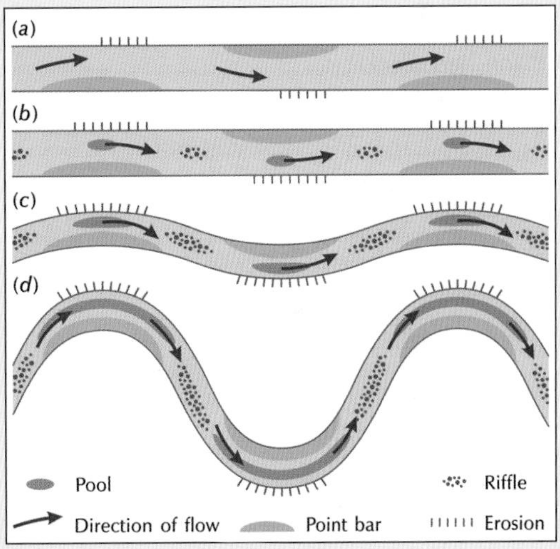

Figure 9.21 Pool-and-riffle sequences in river channels. (a) Alternating zones of channel erosion and accretion in response to faster and slower flow. (b) Pool spacing influencing the evolution of a straight channel into a meandering channel. (c) Additional pools form as the meandering channel lengthens. (d) Development of meandering channel with pools and riffles. *Source*: Adapted from Dury (1969)

antidune sequences (Figure 9.22). In steep headwater streams, steps often alternate with pools to create **step–pool sequences** in which form maximizes resistance to streamflow; maximum flow resistance appears to obtain when the steps are regularly spaced and the mean step steepness slightly exceeds the channel slope (Abrahams *et al.* 1995). It is possible that step–pools are analogous to meanders in the vertical dimension that form because a mountain stream, being unable to adjust energy expenditure in the plane dimension, instead adjusts it in the vertical to

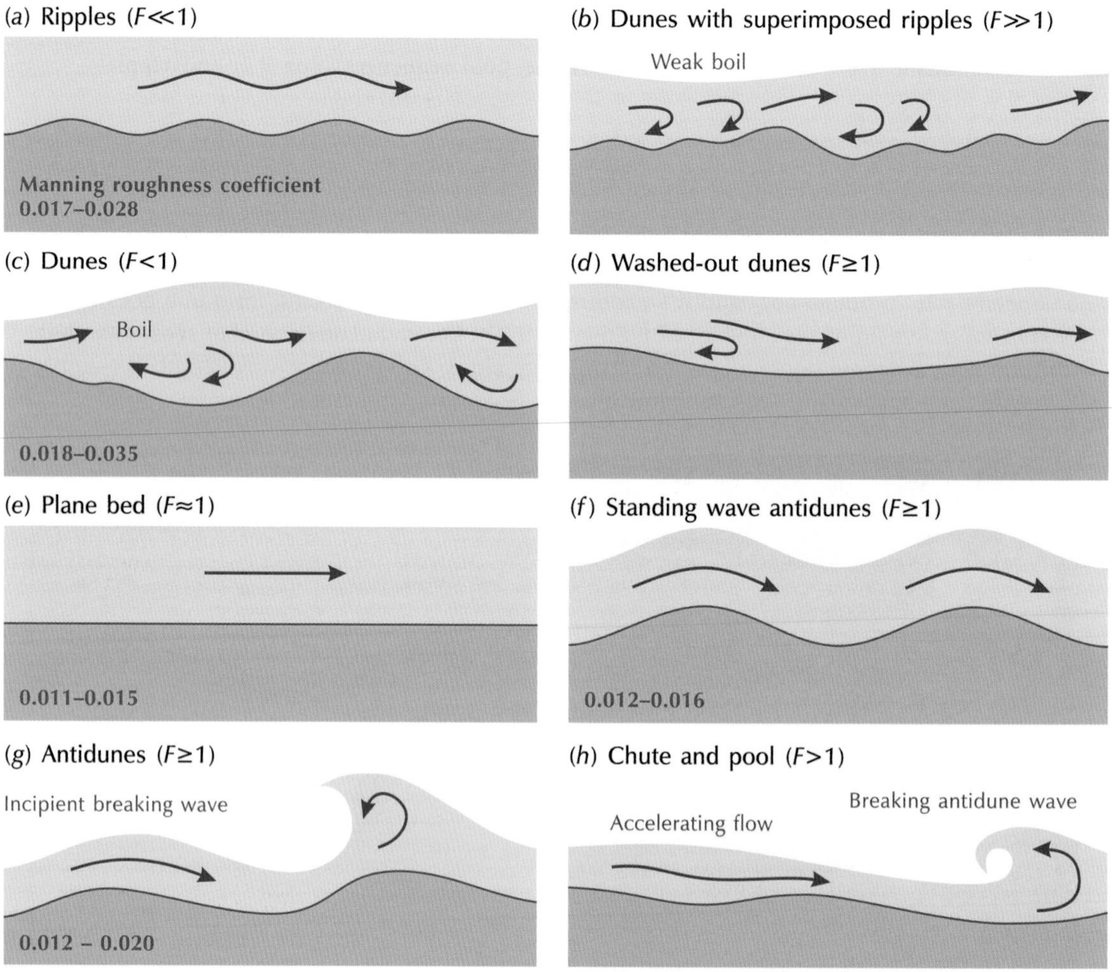

Figure 9.22 Bedforms in a sandy alluvial channel change as the Froude number, *F*, changes. At low flow velocities, ripples form that change into dunes as velocity increases. A further increase of velocity planes off bed undulations, and eventually a plane bed forms. The plane bed reduces resistance to flow, and sediment rates increase. The channel then stands poised at the threshold of subcritical and supercritical flow. A further increase of velocity initiates supercritical flow, and standing antidunes form. Flow resistance is low at this stage because the antidunes are in phase with the standing waves. The antidunes move upstream because they lose sediment from their downstream sides faster than they gain it through deposition. At the highest velocities, fast-flowing and shallow chutes alternate with deeper pools. *Source*: Adapted from Simons and Richardson (1963) and Simons (1969)

produce rhythmic gravel bedforms along the channel that may merge into riffle–pool sequences downstream (Chin 2002).

Floodplains

Most rivers, save those in mountains, are flanked by an area of moderately flat land called a **floodplain**, which is formed from debris deposited when the river is in flood. Small floods that occur frequently cover a part of the floodplain, while rare major floods submerge the entire area. The width of floodplains is roughly proportional to river discharge. The active floodplain of the lower Mississippi River is some 15 km across. Adjacent floodplains in regions of subdued topography may coalesce to form **alluvial plains**.

Floodplains come in several types. A well-known distinction is between convex floodplains and flat floodplains. **Convex floodplains** are the low-gradient floodplains of most large rivers, including those of the Rivers Mississippi, Amazon, and Nile. These floodplains are broad and have slightly convex cross-sections, the land sloping away from the riverbank to the valley sides (Figure 9.23a). The convexity is primarily a product of sedimentation. Bed load and suspended sediment are laid down in the low-water channel and along its immediate edges, while only suspended materials are laid down in the flood basins and backswamps. Bed load accumulates more rapidly than suspended load, and deposition is more frequent in and near to the channel than it is in overbank sites. In consequence, the channel banks and levees grow faster than the flood basins and may stand 1–15 m higher.

Flat floodplains are usually small and flat or gently concave in cross-section (Figure 9.23b). On these flat floodplains, natural levees are small or absent and the alluvial flats rise gently to the

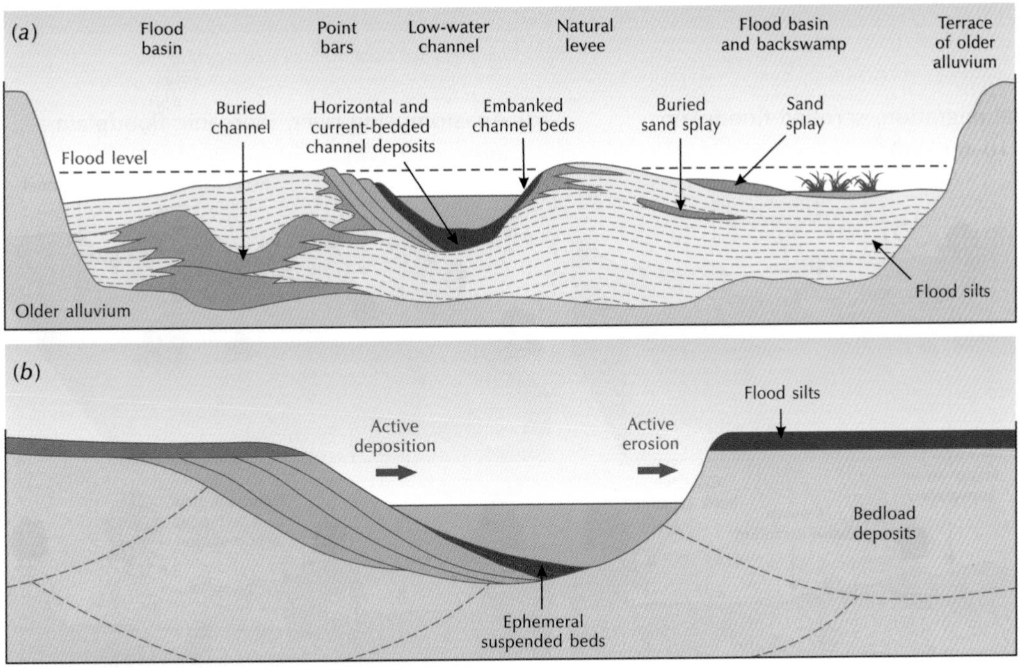

Figure 9.23 Sections through floodplains. (a) A convex floodplain. Point-bar deposits occur on inside meander bends and rarely opposite developing levees. The vertical exaggeration is considerable. (b) A flat floodplain. *Source*: Adapted from Butzer (1976, 155, 159)

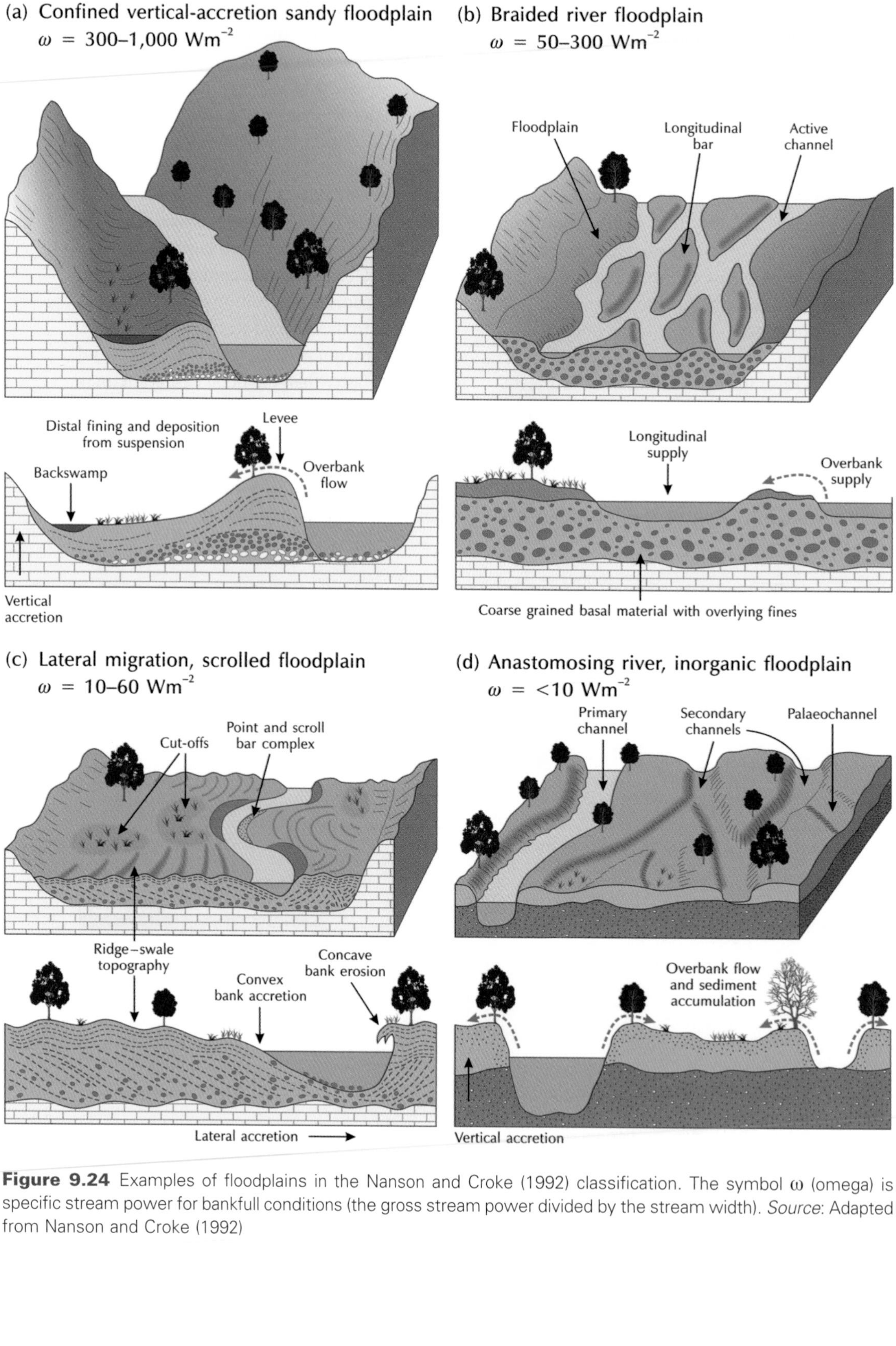

Figure 9.24 Examples of floodplains in the Nanson and Croke (1992) classification. The symbol ω (omega) is specific stream power for bankfull conditions (the gross stream power divided by the stream width). *Source*: Adapted from Nanson and Croke (1992)

valley sides. The concave form is encouraged by a small floodplain area that is liable to continual reworking by the stream. Most medium-sized rivers, and many major rivers, have flat floodplains formed chiefly by lateral accretion (sedimentation on the inside of meander bends). Flat floodplains may also form by alluviation in braided streams.

Gerald Nanson and Jacky Croke (1992) devised a genetic classification of floodplains based on floodplain association with the mainstream channel. They argued that, although floodplains form through a complex interaction of fluvial processes, their nature and evolution hinges upon the product of stream power and sediment character. These two factors determine a stream's ability to entrain and carry sediment and the floodplain alluvium's ability to resist channel erosion. On this basis, Nanson and Croke (1992) identified three chief types of floodplains (and thirteen derivative orders and suborders):

1 High-energy floodplains of non-cohesive deposits (with vertical accretion).
2 Medium-energy floodplains of non-cohesive deposits (with lateral accretion).
3 Low-energy floodplains of cohesive deposits (braiding, meandering, and/or anastomosing with vertical accretion).

Figure 9.24 shows some examples of these.

Lateral accretion, which creates ridge and swale topography, and overbank vertical accretion (both across broad floodplains and in partially constrained valleys) are the major floodplain-forming processes (Fryirs and Brierley 2013). Braided channel accretion, oblique accretion on channel margins, counterpoint accretion in the flow separation zone, and abandoned channel accretion are secondary mechanisms. Floodplain-transforming processes that modify the floodplain-building processes include lateral migration (bend extension, translation, or rotation), cut-offs (neck or chute cut-offs), avulsion, stripping, scouring of flood channels, and channel expansion.

The Nanson–Croke classification works well for many floodplains, but it is less effective for large ones. This is because large floodplains reflect the activities of several channels and subsystems, some of which are partially disconnected with mainstream activity; because they are complex, with floodplain-forming and floodplain-transforming processes displaying considerable variation over them; and because they are built of forms that have developed at different times and under different conditions, including the Quaternary and even earlier (Lewin and Ashworth 2014).

Alluvial fans

An **alluvial fan** is a cone-shaped body that forms where a stream flowing out of mountains debouches on to a plain (Plate 9.12). The alluvial deposits radiate from the **fan apex**, which is the point at which the stream emerges from the mountains. Radiating channels cut into the fan. These are at their deepest near the apex and shallow with increasing distance from the apex, eventually converging with the **fan surface**. The zone of deposition on the fan runs back from the break of slope between the fan surface and the flat land in front of the **fan toe**. It was once thought that deposition was induced by a break of slope in the stream profile at the fan apex, but it has been shown that only rarely is there a break of slope at that point. The steepness of the fan slope depends on the size of the stream and the coarseness of the load, with the steepest alluvial fans being associated with small streams and coarse loads. Fans are common in arid and semi-arid areas but occur in all climatic zones. They range greatly in size. Some in Queensland, Australia, are plain to see on topographic maps or satellite images, but cannot be recognized on the ground because they have radii of about 100 km and are so flat.

Plate 9.12 Dark alluvial fan abutting white playa deposits; road around toe of fan, Death Valley, California, USA. (*Photograph by Marli Miller*)

Plate 9.13 Bajada in Death Valley, California, USA. Note the light-coloured active channels and the dark interfluves where clasts are coated with rock varnish. (*Photograph by Marli Miller*)

Plate 9.14 Playa in Panamint Valley, California, USA. A bajada can be seen rising towards the mountains in the background. (*Photograph by Tony Waltham Geophotos*)

Alluvial fans are dynamic landforms. External environmental forcing by climate change, tectonic movements, and baselevel change, and internal feedbacks between process and form, control their evolution (Nicholas *et al.* 2009). Internal feedbacks include switches between sheet flow and channelized flow, driven by aggradation and degradation, which may bring about changes in sediment transport capacity. Numerical modelling demonstrates that internal feedbacks between fan size, aggradation rate, flow width, and sediment transport capacity can drive spectacular and long-term (millennial scale) fan entrenchment in the absence of external forcing, superimposed on which short-term (decadal to centennial scale) fluctuations in water and sediment supply lead to the formation of a complex sequence of unpaired terraces (Nicholas and Quine 2007).

Active alluvial fans tend to occur in arid and semi-arid environments. They are common in closed basins of continental interiors, which are called **bolsons** in North America. The bolsons are surrounded by mountains out of which floodwaters laden with sediment debouch into the basin. The coarser sediment is deposited to form alluvial fans, which may coalesce to form complex sloping plains known as **bajadas** (Plate 9.13). The remaining material – mainly fine sand, silt, and clay – washes out over the **playa** and settles as the water evaporates. The floor of the playa accumulates sediment at the rate of a few centimetres to a metre in a millennium. As water fills the lowest part of the playa, deposited sediment tends to level the terrain. Indeed, playas are the flattest and the smoothest landforms on the Earth (Plate 9.14). A prime example is the Bonneville salt flats in Utah, USA, which is ideal for high-speed car racing, although some playas contain large desiccation cracks so caution is advised. Playas are known as *salinas* in Australia and South America and *sabkhas* or *sebkhas* in Africa. Playas typically occupy about 2–6 per cent of the depositional area in a bolson. Many bolsons contained perennial lakes during the Pleistocene.

River terraces

A **terrace** is a roughly flat area that is limited by sloping surfaces on the upslope and downslope sides. River terraces are the remains of old valley floors that are left sitting on valley sides after river downcutting. Resistant beds in horizontally lying strata may produce flat areas on valley sides – **structural benches** – so the recognition of terraces requires that structural controls have been ruled out. River terraces slope downstream but not necessarily at the same grade as the active floodplain. **Paired terraces** form where the vertical downcutting by the river is faster than the lateral migration of the river channel (Figure 9.25a). **Unpaired terraces** form where the channel shifts laterally faster than it cuts down, so terraces are formed by being cut in turn on each side of the valley (Figure 9.25b).

The floor of a river valley is a precondition for river terrace formation. Two main types of river terrace exist that correspond to two types of valley floor: bedrock terraces and alluvial terraces.

Bedrock terraces

Bedrock or **strath** terraces start in valleys where a river cuts down through bedrock to produce a V-shaped valley, the floor of which then widens by lateral erosion (Figure 9.26). A thin layer of gravel often covers the flat, laterally eroded surface. Renewed downcutting into this valley floor then leaves remnants of the former valley floor on the slopes of the deepened valley as rock-floored terraces. Rock-floored terraces are pointers to prolonged downcutting, often resulting from tectonic uplift. The rock floors are cut by lateral erosion during intermissions in uplift.

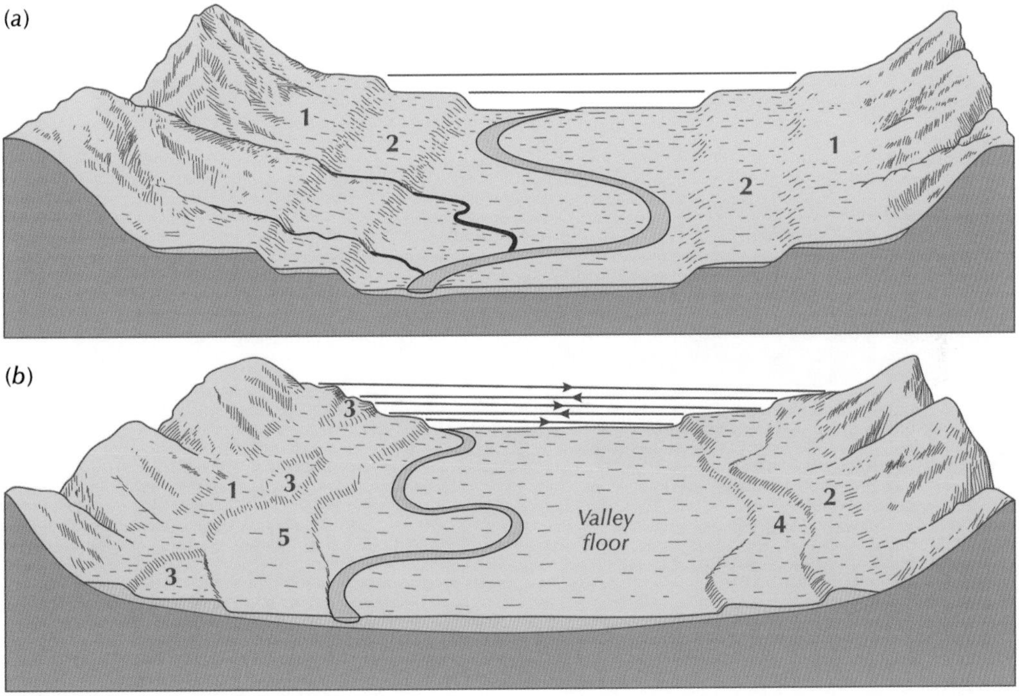

Figure 9.25 Paired and unpaired terraces. (a) Paired, polycyclic terraces. (b) Unpaired, noncyclic terraces. The terraces are numbered 1, 2, 3, and so on. *Sources*: Adapted from Sparks (1960, 221–3) and Thornbury (1954, 158)

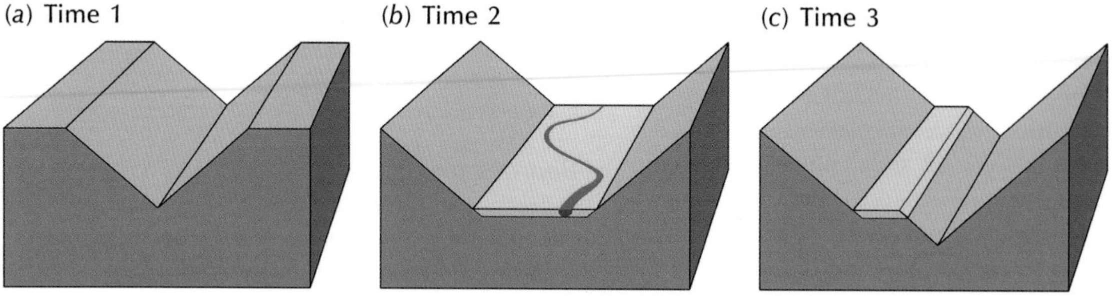

Figure 9.26 Strath (bedrock) terrace formation. (a) Original V-shaped valley cut in bedrock. (b) Lateral erosion cuts a rock-floored terrace. (c) Renewed incision cuts through the floor of the terrace.

Alluvial terraces

Alluvial or **accumulation** terraces are relicts of alluvial valley floors (Plate 9.15). Once a valley is formed by vertical erosion, it may fill with alluvium to create a floodplain. Recommenced vertical erosion then cuts through the alluvium, sometimes leaving accumulation terraces stranded on the valley sides. The suites of alluvial terraces in particular valleys have often had complicated histories, with several phases of accumulation and downcutting that are interrupted by phases of lateral erosion. They often form a **staircase**, with each tread (a terrace) being separated by risers. A schematic diagram of the terraces of the upper Loire River, central France, is shown in Figure 9.27.

Plate 9.15 Sequence of river terraces, Kadjerte River, Kyrgyzstan. (*Photograph by Marli Miller*)

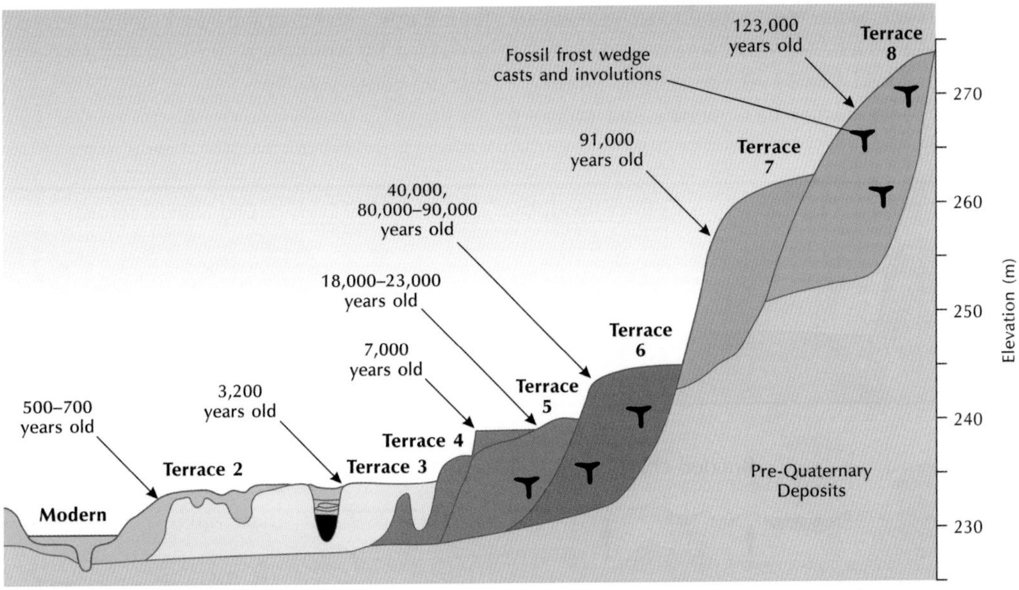

Figure 9.27 Terraces on the upper Loire River, France (diagrammatic). *Source*: Adapted from Colls *et al.* (2001)

Terrace formation and survival

Four groups of processes promote river terrace formation: (1) crustal movement, especially tectonic and isostatic movements; (2) eustatic sea-level changes; (3) climatic changes; and (4) stream capture. In many cases, these factors work in combination. River terraces formed by stream capture are a special case. If the upper reach of a lower-lying stream captures a stream with a high baselevel, the captured stream suddenly has a new and lower baselevel and cuts down into its former valley floor. This is a one-off process and creates just one terrace level. Crustal movements may trigger bouts of downcutting. Eustatic falls of sea level may lead to headward erosion from the coast inland if the sea-floor is less steep than the river. Static sea levels favour lateral erosion and valley widening. Rising sea levels cause a different set of processes. The sea level rose and fell by over 100 m during the Pleistocene glacial–interglacial cycles, stimulating the formation of suites of terraces in many coastal European river valleys, for instance.

Climatic changes affect stream discharge and the grain size and volume of the transported load (Figure 9.28). The classic terrace sequences on the Rivers Iller and Lech, in the Swabian–Bavarian Alpine foreland, are climatically controlled terraces produced as the climate swung from glacial to interglacial states and back again. The rivers deposited large tracts of gravel during glacial stages, and then cut into them during interglacial stages. Semi-arid regions are very susceptible to climatic changes because moderate changes in annual precipitation may produce material changes in vegetation cover and thus a big change in the sediment supply to streams. In the south-west USA, arroyos (ephemeral stream channels) show phases of aggradation and entrenchment over the last few hundred years, with the most recent phase of entrenchment and terrace formation lasting from the 1860s to about 1915.

Terraces tend to survive in parts of a valley that escape erosion. The slip-off slopes of meanders are such a place. The stream is directed away from the slip-slope while it cuts down and is not undercut by the stream. Spurs at the confluence of tributary valleys also tend to avoid being eroded. Some of the medieval castles of the middle Rhine, Germany – the castles of Gutenfels and Maus, for example – stand on small rock-floored terraces protected by confluence spurs on the upstream side of tributary valleys.

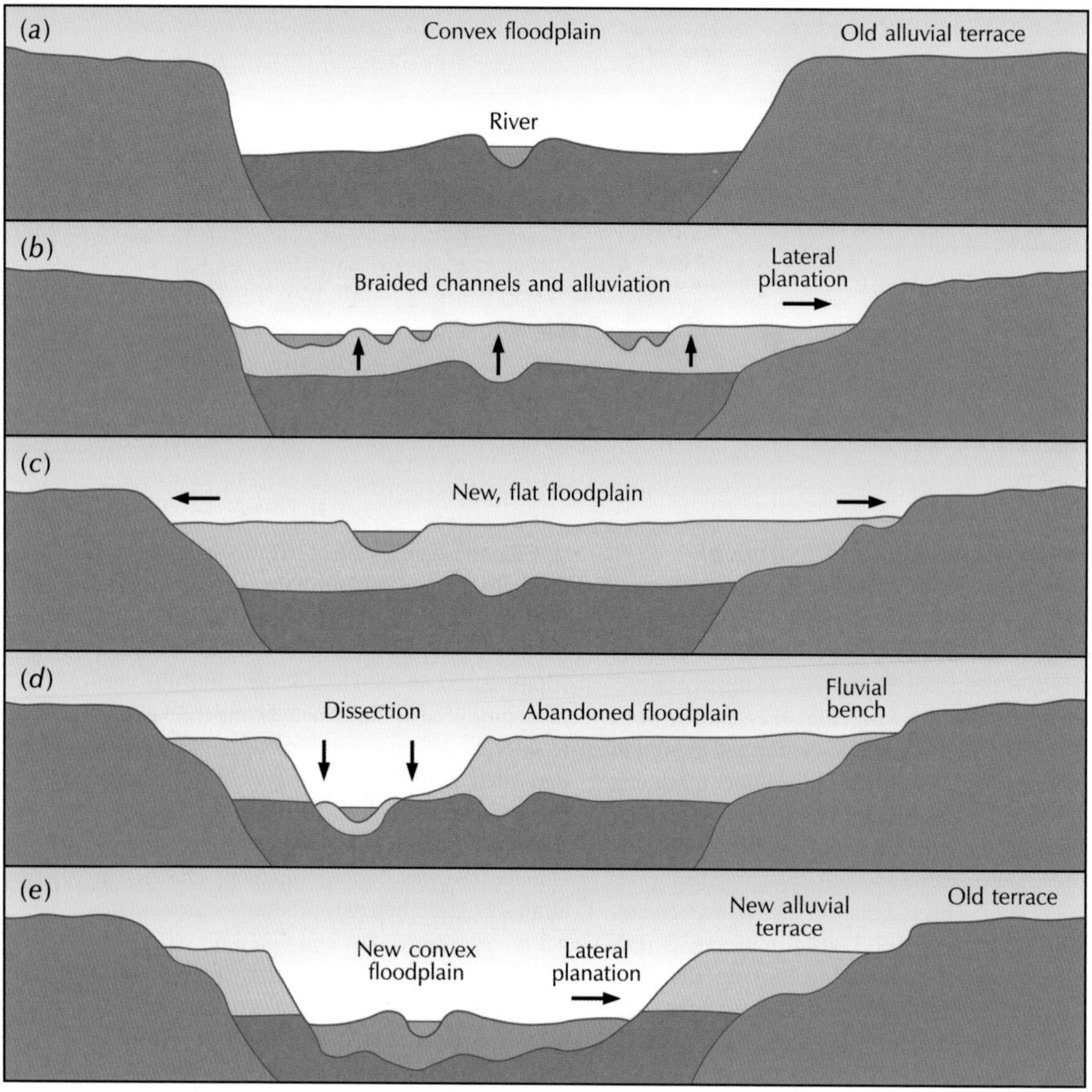

Figure 9.28 Alluvial terrace formation. (a) An initial convex floodplain. (b) Burial of the initial floodplain by coarser sediments through rapid alluviation of braided channels. (c) A stable, flat floodplain forms by alluviation and some lateral planation. (d) Another environmental change leads to dissection of alluvium and the abandonment of the flat floodplain. (e) A new convex floodplain is established by the alluviation of fine sediments and lateral planation. *Source*: Adapted from Butzer (1976, 170)

Lacustrine deltas

Lacustrine or **lake deltas** are accumulations of alluvium laid down where rivers flow into lakes. In moving from a river to a lake, water movement slows and with it the water's capacity and competence to carry sediment. Providing sediment is deposited faster than it is eroded, a lacustrine delta will form.

HUMAN IMPACTS ON THE FLUVIAL SYSTEM

Human agricultural, mining, and urban activities have caused changes in rivers. This section will consider three topics: the increased flux of fluvial sediments; the effect of dams on streamflow, sediment transfer, and channels; and river modification and management.

River sediment increase

In North America, agricultural land-use typically accelerates erosion tenfold to a hundredfold through fluvial and aeolian processes. Much of this high sediment yield is stored somewhere in the river system, mainly in channels, behind dams, and as alluvium and colluvium. Many other reports in the literature support this conclusion. With the maturation of farmlands worldwide, and with the development of better soil conservation practices, it is probable that the human-induced erosion is less than it was several decades ago (e.g. Trimble 1999). Overall, however, there has been a significant anthropogenic increase in the mobilization of sediments through fluvial processes. Global estimates of the quantities vary considerably: one study gave a range of 24–64 billion tonnes per year of bulk sediments, depending on the scenario used (Stallard 1998); another study calculated that as much as 200 billion tonnes of sediment move every year (Smith *et al.* 2001).

River channels and dams

Dams impose changes in streamflow and the transfer of sediment. A study of the impacts of 633 of the world's largest reservoirs (with a maximum storage capacity of 0.5 km^3 or more), and the potential impacts of the remaining >44,000 smaller reservoirs reveals the strong influence of dams on streamflow and sediment flux (Vörösmarty *et al.* 2003). It uses the residence time change (the time that otherwise free-flowing river water stays in a reservoir), in conjunction with a sediment retention function, as a guide to the amount of incoming sediment that is trapped. Across the globe, the discharge-weighted mean residence time change for individual impoundments is 0.21 years for large reservoirs and 0.011 years for small reservoirs. The large reservoirs intercept more than 40 per cent of global river discharge, and approximately 70 per cent of this discharge maintains a theoretical sediment-trapping efficiency in excess of 50 per cent. Half of all discharge entering large reservoirs shows a local sediment trapping efficiency of 80 per cent or more. Between 1950 and 1968, global sediment trapping in large reservoirs tripled from 5 per cent to 15 per cent; it doubled to 30 per cent between 1968 and 1985, but then stabilized. Several large basins such as the Colorado and Nile show almost complete trapping due to large reservoir construction and flow diversion. From the standpoint of sediment retention rates, the most heavily regulated drainage basins lie in Europe. Large reservoirs also strongly affect sediment retention rates in North America, Africa, and Australia–Oceania. Worldwide, artificial impoundments potentially trap more than 50 per cent of basin-scale sediment flux in regulated basins, with discharge-weighted sediment trapping due to large reservoirs of 30 per cent, and an additional contribution of 23 per cent from small reservoirs. Taking regulated and unregulated basins together, the interception of global sediment flux by all 45,000 registered reservoirs is at least 4–5 billion tonnes per year, or 25–30 per cent

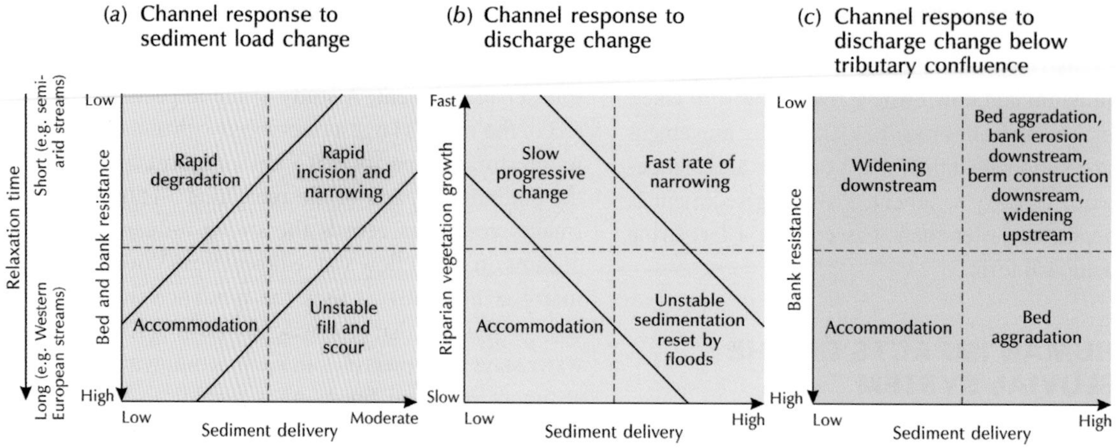

Figure 9.29 Domains of channel change in response to changing sediment load and discharge in different regions. Responses are to (a) A dominant reduction in sediment loads, (b) A dominant reduction in floods. (c) The special case of channel change below a tributary confluence in a regulated river dominated by flood reduction. *Source*: Adapted from Petts and Gurnell (2005)

of the total. There is an additional but unknown impact due to the still smaller 800,000 or so unregistered impoundments. The study shows that river impoundment is a significant component in the global fluxes of water and sediment.

Changes in streamflow and sediment transfer caused by dams lead to downstream changes in channel form. The degradation of rivers downstream of dams is a concern around the world. It has proved difficult to generalize about responses of channels downstream of dams. Figure 9.29 displays expected responses over a timescale of about fifty years to a reduction in sediment load (Figure 9.29a) and a reduction in flood magnitude (Figure 9.29b). Figure 9.29c shows the special case in which a tributary confluence is involved. In all cases, a change in a single process may produce any one of four channel responses.

River modification and management

Fluvial environments present humans with many challenges. Many European rivers are complex managed entities. In the Swiss Jura, changes in some rivers to improve navigation destabilized the channels and a second set of engineering works was needed to correct the impacts of the first (Douglas 1971). Within the Rhine Valley, the river channel is canalized and flows so swiftly that it scours its bed. To obviate undue scouring, a large and continuous programme of gravel replenishment is in operation. The Piave river, in the eastern Alps of Italy, has experienced remarkable channel changes following decreased flows and decreased sediment supply (Surian 1999). The width of the channel has shrunk to about 35 per cent of its original size, and in several reaches the pattern has altered from braided to wandering. In England, the channelization of the River Mersey through the south of Manchester has led to severe bank erosion downstream of the channelized section, and electricity pylons have had to be relocated (Douglas and Lawson 2001).

By the 1980s, increasing demand for environmental sensitivity in **river management**, and the realization that hard engineering solutions were not fulfilling their design life expectancy, or were transferring erosion problems elsewhere in river systems, produced a spur for changes in management practices. Mounting evidence and theory demanded a geomorphological approach to river management (e.g. Dunne and Leopold 1978;

Brookes 1985). Thus, to control bank erosion in the UK, two major changes in the practices and perceptions of river managers took place. First, they started thinking about bank erosion in the context of the sediment dynamics of whole river systems, and began to examine upstream and downstream results of bank protection work. Second, they started prescribing softer, more natural materials to protect banks, including both traditional vegetation, such as willow, osier, and ash, and new geotextiles to stimulate or assist the regrowth of natural plant cover (Walker 1999). River management today involves scientists from many disciplines - geomorphology, hydrology, and ecology - as well as conservationists and various user groups, such as anglers (e.g. Douglas 2000). Thus, in Greater Manchester, England, the upper Mersey basin has a structure plan that incorporates flood control, habitat restoration, and the recreational use of floodplains; while, in the same area, the Mersey Basin Campaign strives to improve water quality and river valley amenities, including industrial land regeneration throughout the region (Struthers 1997).

FLUVIAL LANDSCAPES IN THE PAST

The fluvial system responds to environmental change. It is especially responsive to tectonic changes, climatic changes, and changes in vegetation cover and land use. Some of the effects of tectonic processes on drainage and drainage patterns were considered earlier in the chapter. Climatic changes are evidenced in **misfit streams** (streams presently too small to have created the valleys they occupy), **entrenched meanders**, and **relict fluvial features** in deserts. Deserts with hyper-arid climates today contain landforms created by fluvial processes - alluvial spreads, pediments, and valleys carved out by streams. Wind erosion does not readily obliterate these features, and they linger on as vestiges of former moist episodes. The geomorphic effects of changing land use are evident in the evolution of some Holocene river systems. The Romans transformed fluvial landscapes in Europe and North Africa by building dams, aqueducts, and terraces (p. 59). A water diversion on the Min River in Sichuan, China, has been operating ceaselessly for over 2,000 years. In the northeastern USA, forest clearance and subsequent urban and industrial activities greatly altered rivers early in the nineteenth century. To expand upon these points, this section will look at the effects of glacial–interglacial cycles during the Pleistocene on fluvial landscapes, at the impact of Holocene climatic and vegetation changes in the USA, and at the complex Holocene history of river systems in Mediterranean valleys and in Germany.

Pleistocene changes

A study of Early and Middle Pleistocene fluvial and coastal palaeoenvironments in eastern England showed that changes in river energetics accorded with the relative importance of geomorphic processes operating in river catchments determined by orbital forcing (Rose *et al.* 2001) (cf. p. 280). The size distributions and lithologies of deposits indicate a shift from low-energy systems comprising mainly suspended-load sediments and locally important bedload sediments to higher-energy bedload and bedload assemblages containing much far-travelled material with a glacial input (Figure 9.30). This shift correlates with a switch from low-amplitude climatic change dominated by the 21,000-year precession cycle to moderate-amplitude climatic change dominated by the 41,000-year tilt cycle. The low-amplitude, high-frequency climate lasted through the Pliocene to about 2.6 million years ago, and the moderate-amplitude, moderate-frequency climate from 2.6 million years ago to about 900,000 years ago. It seems that the shift from low to moderate climatic variations, and especially the trend towards a colder climate, would have favoured the operation of cold climate processes, such as gelifluction and glaciation. Peak river discharges produced by seasonal meltwater under this climatic regime were able to carry coarse-grained sediment along river channels and through river catchments as bedload. The longer duration of the climatic variations would have given enough time for gelifluction and other slope processes to take material from hillside slopes to valley bottoms, and for glaciers to develop to a large size and subglacial material to reach the glacier margin. Such conditions would enable material in the upper reaches of river networks to arrive at the lower reaches. It seems likely that the nineteen orbitally forced cold episodes in the 800,000-yearlong period dominated by moderate-amplitude, moderate-frequency climatic variations allowed bedload to move from the upper Thames catchment in Wales and an inferred Ancaster river in the Pennines to the western coast of the North Sea in East Anglia. Similarly, in cold episodes during the next 1.3 million years, bedload moved through the river systems. The arrival of the Anglian glaciation some 480,000 years ago, with ice up to 1,000 m thick that reached as far as London and Bristol, was associated with large-magnitude, long-duration 100,000-year eccentricity cycle driven climatic changes. It radically altered the catchments and the topography.

Holocene changes

Alluviation in the USA

Early discussion of alluvial episodes in the USA engaged the minds of big names in geomorphology (Box 9.3). A modern review of the response of river systems to Holocene climates in the USA argued that fluvial episodes in regions of varying vegetation cover occurred roughly at the same times, and that the responsiveness of the rivers to climatic change increased as vegetation cover decreased (Knox 1984). Alluvial episodes occurred between roughly 8,000 and 6,000, 4,500 and 3,000, and 2,000 and 800 years ago. Before 8,000 years ago, changing vegetation and rapid climatic warming caused widespread alluviation. The magnitude of this alluvial episode generally rose to the west in parallel with increased drying and increased vegetation change. Between 8,000 and 7,500 years ago, erosion broke in upon alluviation. Although of minor proportions in the East and humid Mid-West, this erosion was severe in the South-West. For the next 2,000 years, warm and dry conditions in the southern South-West and parts of the East and South-East (caused by the persistent zonal circulation of the early Holocene epoch) led to a slowing of alluviation in all places except the South-West, where major erosion of valley fills occurred. Although the South-East was warm and wet at the time, it did not suffer erosion because forests were established. From 6,000 to 4,500 years ago, all the Holocene valley fills were eroded, except those in the South-West, where alluviation continued. The extensive erosive phase resulted from a climatic cooling that improved the vegetation cover, reduced sediment loads, and promoted trenching; and from the circulation of the atmosphere becoming increasingly meridional during summer, so bringing higher rainfall and larger floods. The

Figure 9.30 Early Pleistocene and early Middle Pleistocene fluvial and offshore palaeogeography of midland and eastern England and the adjacent North Sea basin. The heavy lines indicate main drainage trajectories based on 'long-established' valley systems and Early and early Middle Pleistocene sediments. Over this period, eastern England acted as a depositional centre for drainage from the Thames, Bytham, and Ancaster catchments. The distribution of lithostratigraphic units is given, along with the location of outcrop of distinctive indicator lithologies, that were transported by the river systems to the coastal zone. (a) The drainage system and extent of coastal deposits at the time of deposition of the Red and Norwich Crag Formations. (b) The drainage system and extent of coastal deposits at the time of the deposition of the Dobb's Plantation and How Hill Members of the Wroxham Crag Formation. (c) The drainage system and extent of coastal deposits at the time of deposition of the Mundesley Member of the Wroxham Crag Formation. *Source*: After Rose *et al.* (2001)

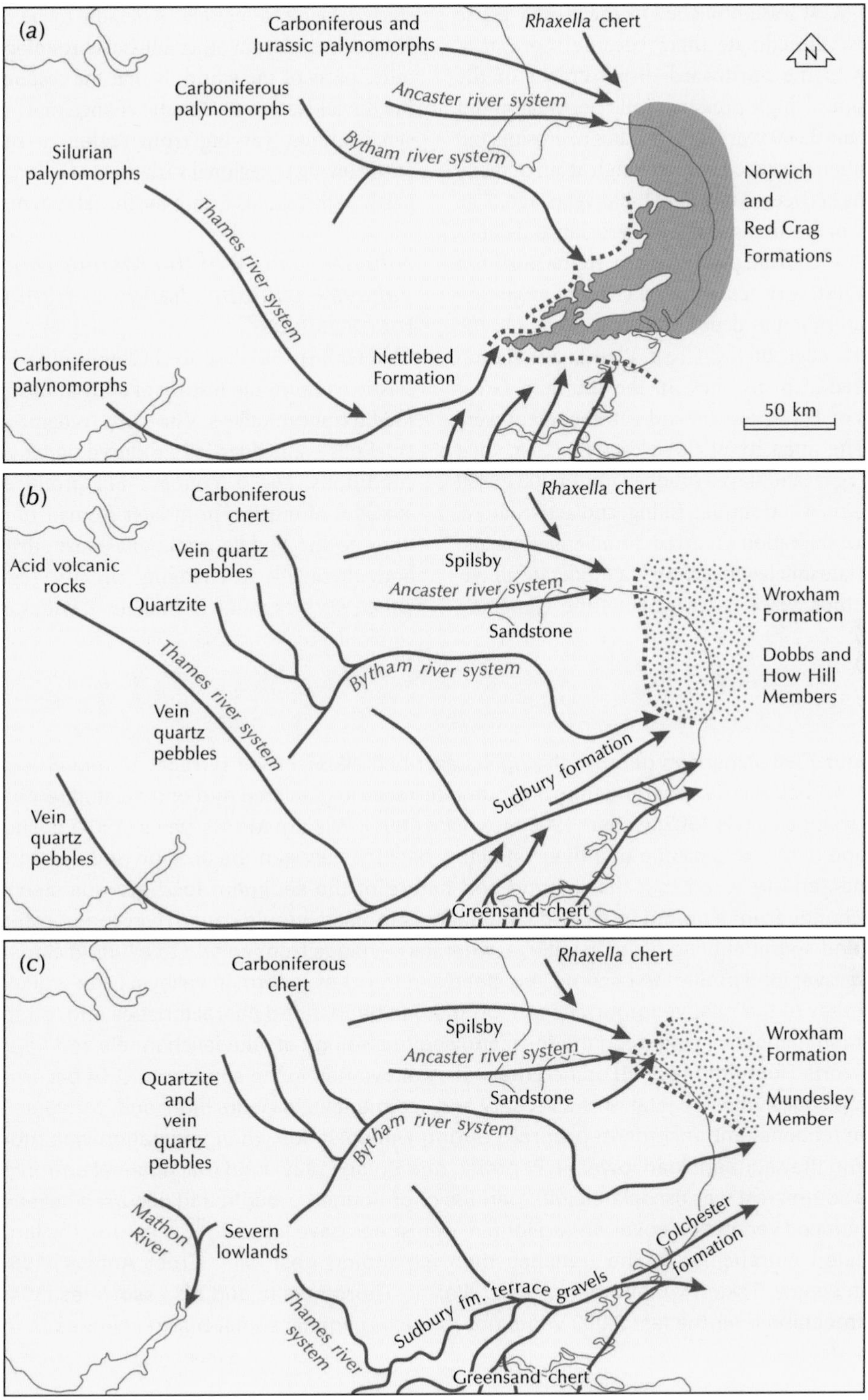
(a)
Carboniferous and Jurassic palynomorphs
Rhaxella chert
N
Carboniferous palynomorphs
Ancaster river system
Bytham river system
Silurian palynomorphs
Thames river system
Norwich and Red Crag Formations
Carboniferous palynomorphs
Nettlebed Formation
50 km
(b)
Carboniferous chert
Rhaxella chert
Vein quartz pebbles
Acid volcanic rocks
Spilsby
Ancaster river system
Quartzite
Wroxham Formation
Sandstone
Dobbs and How Hill Members
Bytham river system
Thames river system
Vein quartz pebbles
Sudbury formation
Vein quartz pebbles
Greensand chert
(c)
Carboniferous chert
Rhaxella chert
Spilsby
Wroxham Formation
Ancaster river system
Quartzite and vein quartz pebbles
Sandstone
Mundesley Member
Bytham river system
Mathon River
Severn lowlands
Colchester formation
Sudbury fm. terrace gravels
Thames river system
Greensand chert

South-West was untouched by the erosive phase because the climate there became more arid, owing to the northward displacement of the subtropical high-pressure cell. Between about 4,500 and 3,000 years ago, the rates of erosion and deposition slackened but were high again in many regions between 3,000 and 1,800 years ago. The nature of the intensification of erosion and deposition varied from place to place. In the northern Mid-West, very active lateral channel migration with erosion and deposition took place. On the western edge of the Great Plains, alluviation occurred at many sites. In the southern Great Plains of Texas, erosion and entrenchment were rife. The intensity of fluvial activity then died down again and stayed subdued until 1,200 to 800 years ago, when cutting, filling, and active lateral channel migration occurred. From 800 years ago to the late nineteenth century, a moderate alluviation took place, after which time trenching started in most regions. A lesson to be learnt from this, and from other alluvial chronologies in other parts of the world, is that the response of the fluvial system to climatic change may not be synchronous, varying from region to region, partly owing to regional variations of climate and partly to thresholds within the fluvial system itself.

Alluvial history of the Mediterranean valleys – climatic change or human malpractices?

Chapter 3 (p. 59) described Claudio Vita-Finzi's classic work on the history of alluvial fills in the Mediterranean valleys. Vita-Finzi recognized two chief fills – an Older Fill produced under glacial conditions, and a Younger Fill produced by episodes of erosion from later Roman Imperial times to the Middle Ages. Vita-Finzi attributed both these fills to changing climatic regimes. Other workers point to human activities as the

Box 9.3 OLD IDEAS ON NORTH AMERICAN ALLUVIATION

Prior to about 1890, American geomorphologists ascribed **alluvial river terraces** to movements of the Earth's crust. A little later, and they attributed terraces in glaciated and unglaciated regions to climatic change (Davis 1902; Gilbert 1900; Johnson 1901). William Morris Davis (1902) posited that the slope of the long profile of a river reflects a balance between the erosion and transport of sediments, and believed that the volume and nature of the sediment load are adjusted to climate. A change from a humid to an arid climate, he surmised, would cause river long profiles to steepen and aggradation to occur in valleys; whereas a change from an arid to a humid climate would cause river long profiles to become less steep and trenches to form in valleys. Later workers were divided as to the relative importance of, on the one hand, flood characteristics and, on the other, sediment supply in explaining the form and sedimentology of alluvial channels and floodplains. Ellsworth Huntington (1914) opined that valley alluviation in the southwest USA occurred during dry episodes when vegetation was scanty and sediment yields were high; and, conversely, degradation (channel entrenchment) occurred during wet episodes when vegetation was more abundant and the sediment load lower. In contrast, Kirk Bryan (1928) held that channel entrenchment in the south-west was associated with periods of prolonged drought and occurred because the much-reduced vegetation cover during long dry episodes gave large floods. In turn, the large floods initiated entrenchment, the trenches then expanding upstream. Ernst Antevs (1951) endorsed this view. Taking yet another tack, C. Warren Thornthwaite and his associates (1942) attributed trenching over the last 2,000 years not to major climatic shifts, but to changes in the intensity of storms.

primary cause of the Younger Fill (see Macklin and Woodward 2009). Explanations for the Medieval Fill in the area around Olympia, Greece – the site of the ancient Olympian Games – illustrate the arguments for climatic versus human causes.

Olympia sits to the north of the Alphéios valley where the Kládheos stream enters (Figure 9.31). The sacred site of Altis lies just eastward of the Kládheos, close to the foot of Kronos hill. Excavations at the site revealed stone buildings, including the Temple of Zeus, a Hippodrome, and a Byzantine fortress. The archaeological remains lie beneath 5–6 m of silt, which appears to have begun accumulating after AD 600. In antiquity, the Kládheos stream seems to have occupied a lower level than it does today, a basal conglomerate, possibly of early Pleistocene date, indicating its bed. A pipe built during the reign of the Emperor Hadrian in AD 130 to drain the athletes' baths, the kitchens, and the sanitation facilities could not have functioned without sewage backing up unless the average levels of the Alphéios and Kládheos were about 2 m lower in antiquity than today (Büdel 1982, 343). During the deposition of the Medieval (Younger) Fill, the Kládheos flowed at a higher level than today, its floodplain burying the Olympian ruins and the Byzantine fortress. Some time after the Medieval Fill ceased forming, possibly in the fourteenth or fifteenth century, the Kládheos cut down to near its original level, breaching a Roman confining wall now mainly on its west side. At the same time, the Alphéios shifted northwards, eating into the remains of the Hippodrome and forming a cliff in the tail of the Kládheos sediments that defines the edge of a Medieval terrace (Figure 9.31). These changes seem to have stopped by the mid-eighteenth century.

Julius Büdel (1982, 345) believed that human activities caused the changes in fluvial activity. He argued that the fairly uniform conditions of the Alphéios bed from about 1000 BC to AD 500 reflected a long period of political and agricultural stability. The phase of medieval alluviation, he contended, stemmed from the destruction of a well-ordered peasant agriculture, the lapse of a sacred truce that allowed people from a wide area to flock to the Games every four years, and an exodus of the populace to safer areas. However, this argument seems the wrong way round: while the population of the area was rising and the land was used more intensively, the landscape was stable, but once the population declined erosion set in (Grove and Rackham 2001, 292). A scrutiny of the wider region of Olympia places the question of erosion in a different perspective (Table 9.4). First, the Ládon, a tributary of the Alphéios, connects through underground passages to Phenéos, a large karst basin. The underground channels block and unblock owing to earthquakes and the washing in and out of trees from the surrounding forests. When blocked, a lake forms in the Phenéos. If the lake should reach 100 m before decanting, the catastrophic discharge would uproot trees in the Ládon and Alphéios valleys and carry them downstream. Gravel-pits near Alphioússa, 5 km downstream from Olympia, contain tree trunks with roots attached, some lying about 2 m below the surface and some on the surface, the latter being radiocarbon dated to the last 300 years. Second, a site at Górtys, which lies on an upstream tributary of the Alphéios, shows three phases of slope erosion and alluviation: prehistoric, early Byzantine, and several centuries later. This additional evidence shows the complexities of invoking a single cause for alluviation in all catchments. On the Alphéios, at least two catastrophic events dislodged huge quantities of gravel, uprooted trees, and carried them downstream. These events little affected the Kládheos, although the Alphéios gravels could have encouraged the trapping of finer sediment in the Kládheos. Near Górtys, on the Loúsios river, a tributary of the Alphéios, two phases of historical deposition occurred, each followed by downcutting. Given the tectonic instability of this region, it is perhaps not surprising that different areas suffer massive erosion at different times (Grove and Rackham 2001, 295).

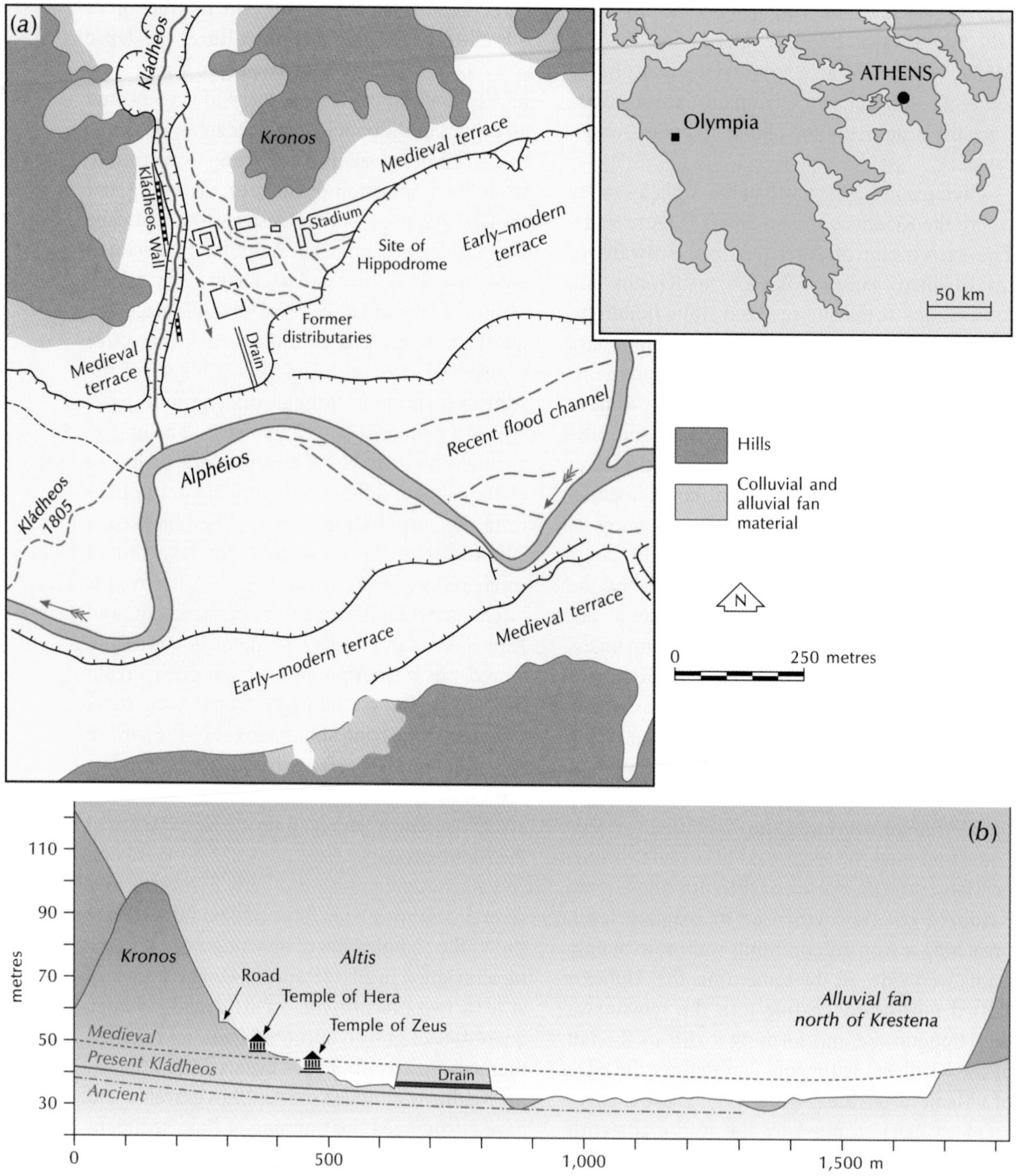

Figure 9.31 Olympia. (a) The archaeological site (Altis) with the Kládheos stream running alongside it to enter the Alphéios from the north. (b) North–south cross-section of Kronos Hill and the Alphéios terraces and valley floor. *Source*: Adapted from Büdel (1963)

Table 9.4 Timetable of events in the Alphéios basin, Peloponnese, Greece

Time	Site		
	Olympia	**Phenéos karst basin**	**Górtys**
1766–1990 AD	Archaeological digs: 1829, 1975–81, 1936–41 and later	High lake levels in late eighteenth century, 1821–34, 1838–1907	Archaeological dig: 1954
1500–1766 AD	Kládheos cuts down; Alphéios moves north		Loúsios cuts down several metres
1200–1500 AD	Kládheos cuts down		Partial burial of Chapel of St Andrew
1000–1200 AD			Chapel of St Andrew built on terrace step
800–1000 AD	Kládheos deposits 3–6 m of sediment (Medieval fill) on Altis site		Loúsios cuts down a few metres
600–800 AD			Thermal baths buried
393–600 AD	Cult and Games abandoned; Zeus temple overthrown; Christian basilica and other buildings built		
776 BC–393 AD	Great sanctuary; Olympian Games held; Temple of Zeus and other temples and treasuries built	At least five cycles of lake filling and emptying	Thermal baths on 'Holocene' terrace
1000–776 BC	Beginnings of Altis sanctuaries		
2000–1000 BC	Bronze Age settlement; Alphioússa gravel terrace starts to form; pre-1700 BC structures buried		
Earlier prehistoric	Settlement starts		Accumulation of 'Holocene' terrace of Loúsios river

Source: Adapted from Grove and Rackham (2001, 296)

Karl Butzer (2005) favours an interpretation of the fluvial history of Olympia based on extreme precipitation events associated with intervals of high climatic variability triggering or exacerbating a landscape already destabilized by human activity. He contends that such events lead to the erosion of susceptible slopes by sheetfloods or gullying. The eroded material then forms poorly sorted colluvium on concave footslopes, any excess sediment being carried into streams during heavy bouts of rain. A stable phase ensues if an abatement of human activity allows vegetation to recover. However, should human pressures resume, secondary episodes of landscape disequilibrium would see renewed erosion, the entrenchment of channels, and braided streams.

Human impacts on Erhai Lake catchment, China

By studying palaeoenvironmental and documentary records from the catchment of Erhai Lake in Yunnan, China, John Dearing (2008)

found complex associations between land use, soil erosion, and monsoon intensity over the past 3,000 years. Analysis of the records for land use and soil erosion suggested the landscape system possesses alternative steady states separated by thresholds (Figure 9.32). The 'initial' state (shown in green on the diagram) is a steady non-degraded state in which changes of land use (disturbed land) do not cause significant soil erosion. Then around 1,400 calendar years BP, the landscape system crossed an internal threshold relating to climate or land use (there was considerable agricultural expansion at the time) and entered a 600-year transition state with rapid gully erosion. It eventually moved into a steady degraded state (shown in purple on the diagram) in which soil erosion rate responds linearly to land use intensity that has lasted till the present day. In this degraded state, soil erosion decreases with decreasing land-use intensity decreases, but does not retrace the 'transition state' back to the 'initial' non-degraded state. This is because the soil erosion alters soil fertility, water fluxes, and so forth on a permanent basis. In consequence, the landscape system moves to a future state with continued erosion that eventually strips soil cover and exposes bedrock, or else, with reforestation, to a future state similar to the initial steady non-degraded steady state.

Human impacts in the Lippe valley, Germany

The Holocene history of the River Lippe shows how human activities can materially alter a fluvial system (Herget 1998). The Lippe starts as a karst spring at the town of Bad Lippspringe and flows westwards to the lower Rhine at Wessel. The Lippe Valley contains a floodplain and two

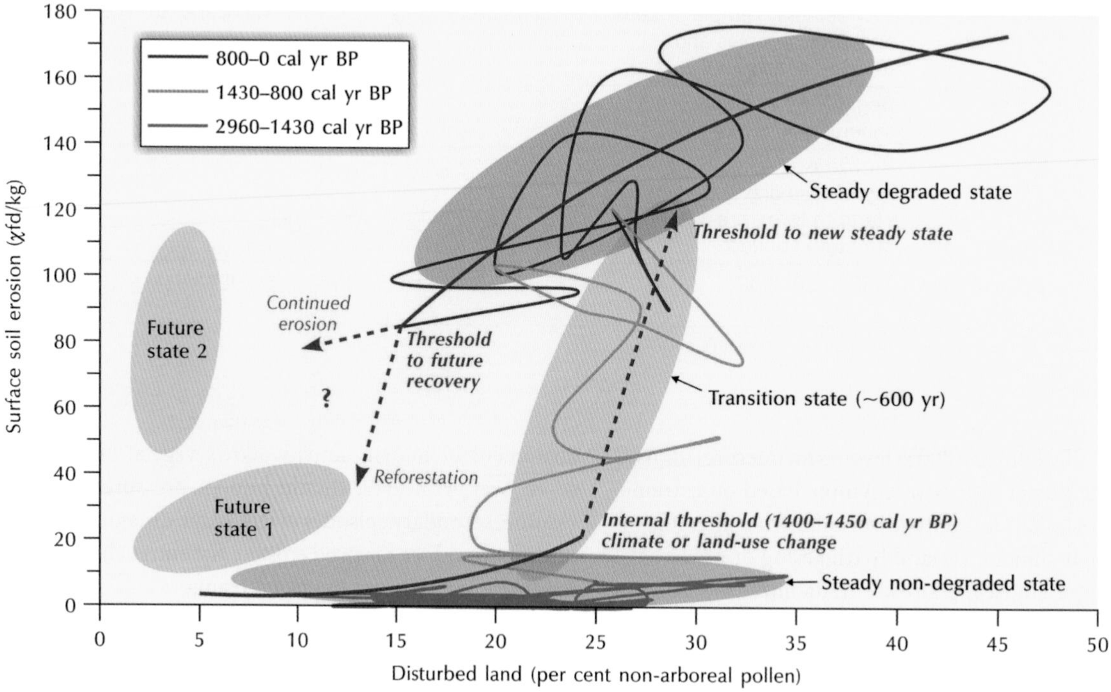

Figure 9.32 Landscape stability in the Erhai lake catchment, Yunnan, China, showing alternative steady states. χfd (chi-fd) stands for frequency dependent magnetic susceptibility, which is a measure of surface soil erosion. *Source*: Adapted from Dearing (2008) and Dotterweich (2013)

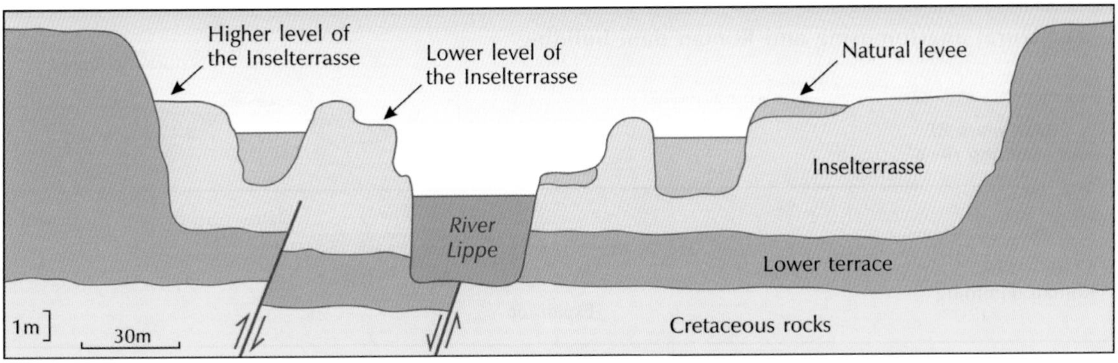

Figure 9.33 Schematic cross-section of the lower Lippe Valley in north-west Germany, showing terraces. *Source*: Adapted from Herget (1998)

Holocene terraces, the younger being called the Aue or Auenterrasse and the older the Inselterrasse. Both these sit within an older terrace – the 115,000–110,000-year-old Weichselian Lower Terrace (Figure 9.33). The Inselterrasse ('island terrace') is a local feature of the lower Lippe Valley west of Lünen. It began to accumulate about 8,000 years ago and stopped accumulating about AD 980, and survived as separate terrace islands left by abandoned channels. The Aue (or 'towpath') runs from the headwaters, where it is quite wide, to the lower valley, where it forms a narrow strip paralleling the river channel. It is younger than the Inselterrasse. The characteristics of the Holocene valley bottom are not typical of valley bottoms elsewhere in central Europe in at least four ways. First, the Inselterrasse is confined to the lower Lippe Valley; second, it is in places split into two levels that are not always easy to distinguish; third, the Aue is just a narrow strip in the lower reaches; and fourth, it lies above the average flood level, while the Inselterrasse is periodically flooded and in historical times was frequently flooded.

Human activities in the valley may explain these features, but two interpretations are possible (Figure 9.34):

1 *Natural river anastomosing and Roman dam building.* Under natural conditions, the River Lippe anastomosed with discharge running through several channels. The valley bottom was then a single broad level. Evidence for this interpretation comes from the lower valley, where some of the abandoned channels are too narrow and shallow to have conveyed the mean discharge, and several channels could easily have formed in the highly erodible sandy sediments. Later, during their campaign against the German tribes, the Romans used the river to transport supplies. Although there is no archaeological evidence for this, they may have dammed some channels, so concentrating discharge into a single channel that would then broaden and deepen and start behaving as a meandering river.
2 *Natural meandering river with modification starting in medieval times.* Under natural conditions, the Lippe actively meandered across the floodplain, eroding into meanders and eroding avulsion channels during floods, and the Aue consisted of several small channels that carried discharge during floods. Starting about 1,000 years ago, several meanders were artificially cut to shorten the navigation route and new towpaths built in sections of the avulsion channels. Shipbuilding started in Dorsten in the twelfth century, and it is known that a towpath was built next to the river at variable heights to move the ships. The artificial cutting

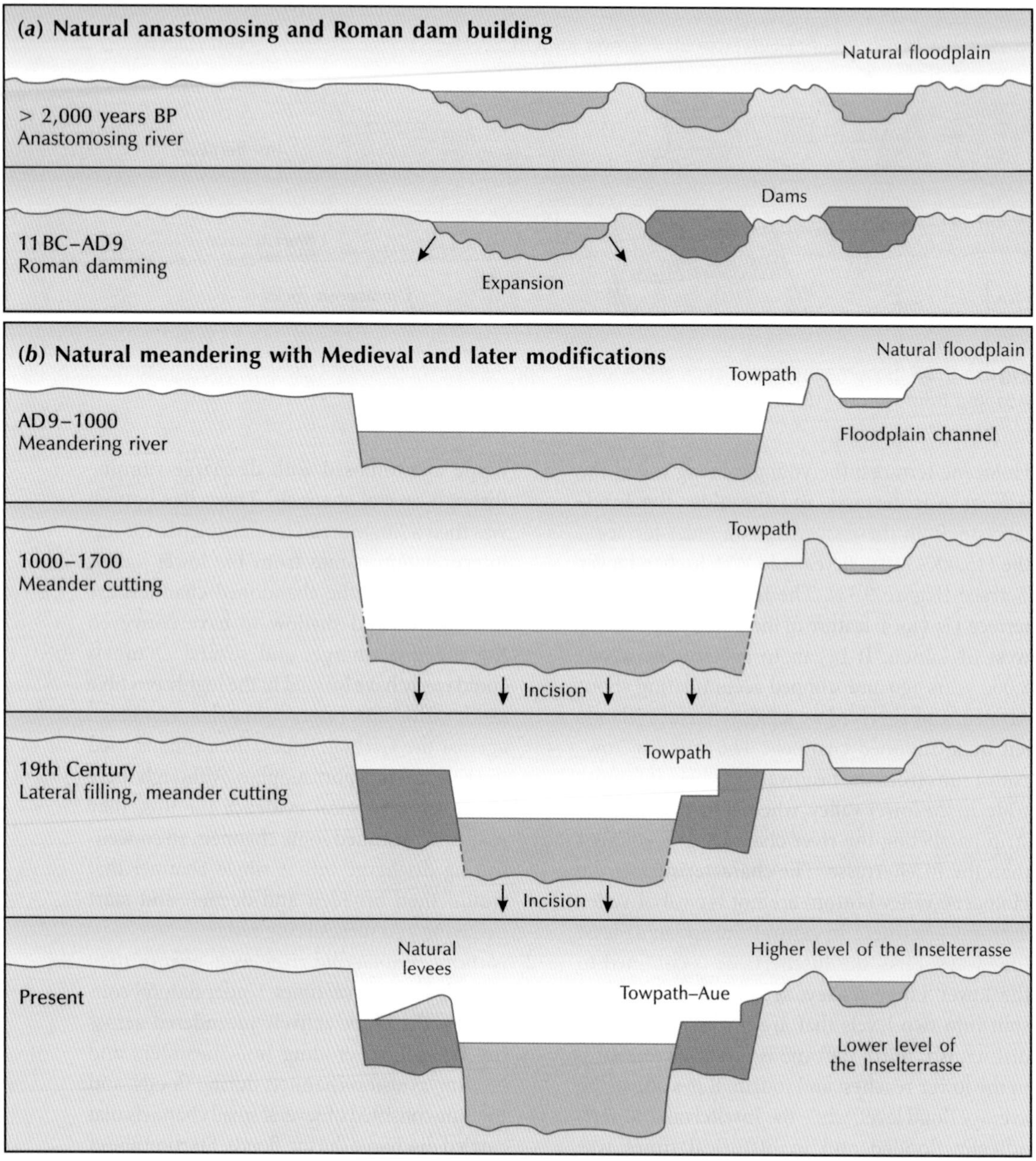

Figure 9.34 Evolution of Holocene terraces in the Lippe Valley. (a) Interpretation 1. (b) Interpretation 2. *Source*: Adapted from Herget (1998)

> steepened the channel gradient and encouraged meander incision. In the nineteenth century, a higher water level was needed for navigation on the river, and sediments from sections with steep embankments and natural levees were used to narrow the channel. The result was another bout of channel incision and the building of a new towpath. Recently, the towpath has widened owing to flood erosion, and the river is building a new terrace between the higher level of the Inselterrasse and the Weichselian Lower Terrace.

This example shows how difficult it can be to reconstruct the history of river valleys, and how humans have affected rivers for at least 2,000 years.

River changes in Swinhope Burn, 1815–1991

Swinhope Burn is a tributary of the upper River Wear, in the northern Pennines, England. It is a gravel-bed stream with a catchment area of 10.5 km^2 (Warburton and Danks 1998; see also Warburton *et al.* 2002). Figure 9.35 shows the historical evidence for changes in the river pattern from 1815 to 1991. In 1815 the river meandered with a sinuosity similar to that of the present meanders (Plate 9.16). By 1844 this meandering pattern had broken down to be replaced by a relatively straight channel with a bar braid at the head, which is still preserved in the floodplain. By 1856, the stream was meandering again, which pattern persists to the present day. The change from meandering to braiding appears to be associated with lead mining. A small vein of galena cuts across the catchment, and there is a record of 326 tonnes of galena coming out of Swinhope mine between 1823 and 1846. It is interesting that, although the mining operations were modest, they appear to have had a major impact on the stream channel.

SUMMARY

Flowing water is a considerable geomorphic agent in most environments, and a dominant one in fluvial environments. Water runs over the land surface, through the soil and rock (sometimes emerging as springs), and along rills and rivers. Streams are particularly effective landform-makers. They conduct material along their beds, keep finer particles in suspension, and carry a burden of dissolved substances. They wear away their channels and beds by corrosion, corrasion, and cavitation, and they erode downwards and sideways. They lay down sediments as channel deposits, channel margin deposits, overbank floodplain deposits, and valley margin deposits. Episodes of continued deposition and valley filling (alluviation) often alternate with periods of erosion and valley cutting. Flowing water carves many erosional landforms, including rills and gullies, bedrock channels, and alluvial channels. River profiles, drawn from source to mouth, are normally concave, although they often possess knickpoints marked by steeper gradients. Rivers form networks that may be described by several geometrical and topological properties. River systems commonly display distinct drainage patterns that often reflect the structure of underlying folded sedimentary beds. Valleys are an overlooked erosional landform. Flowing water deposits sediment to build many depositional landforms. The smallest of these are features on channel beds (riffles and dunes, for example). Larger forms are floodplains, alluvial fans, river terraces, and lake deltas. Human agricultural, mining, and urban activities cause changes in rivers. Overall, they increase the flux of fluvial sediments. Dams affect streamflow, sediment transfer, and channel form downstream. Human actions modify many rivers, which need managing. Fluvial geomorphology lies at the heart of modern river management. Flowing water is sensitive to environmental change, and especially to changes of climate, vegetation cover, and land-use. Many river valleys record a history of

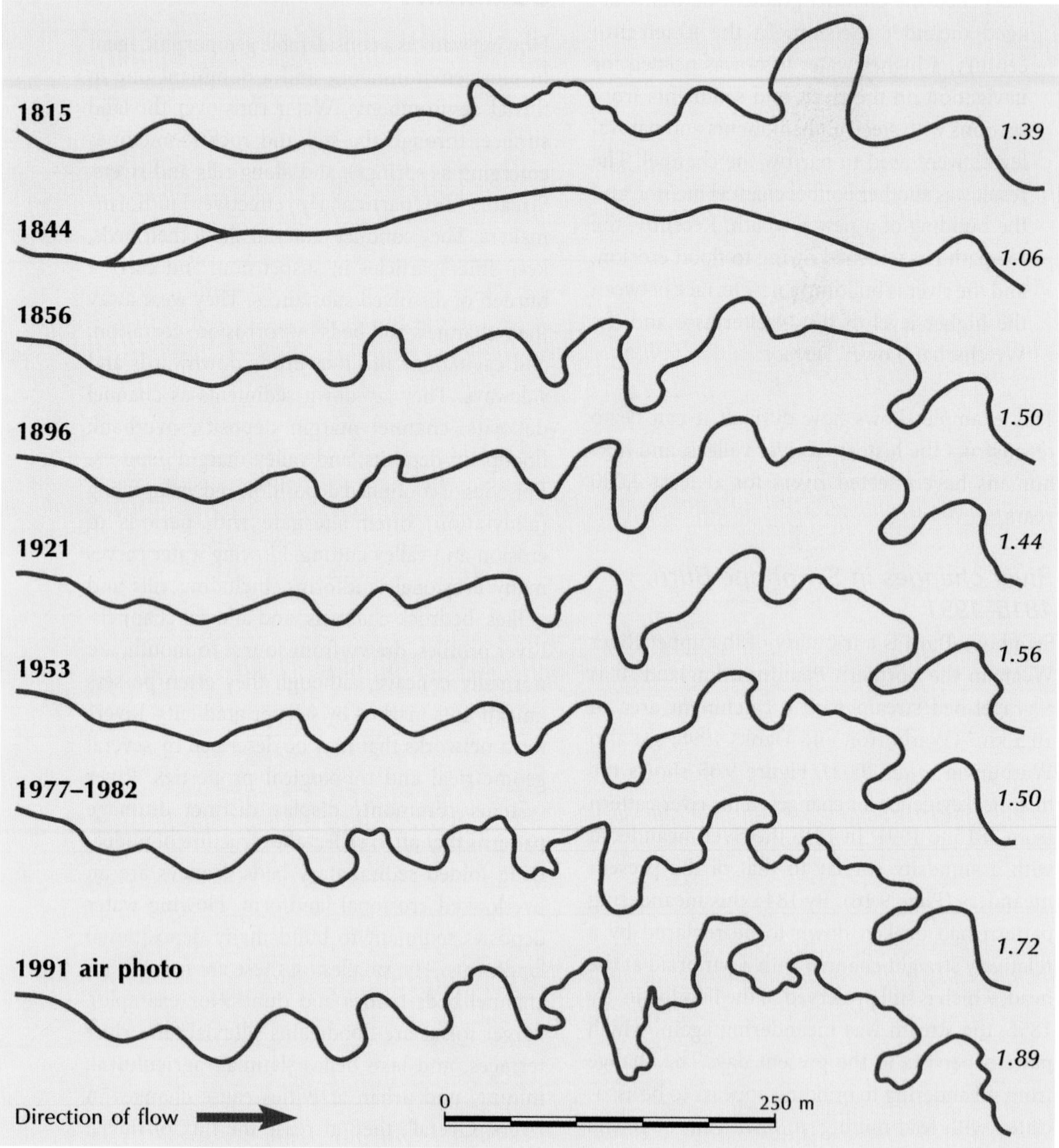

Figure 9.35 Channel change in Swinhope Burn, Upper Weardale, Yorkshire. The diagram shows the channel centre-line determined from maps, plans, and an air photograph. Sinuosity values shown in italics. *Source*: After Warburton and Danks (1998) and Warburton *et al.* (2002)

Plate 9.16 River meanders in Swinhope Burn, northern Pennines, England.
(*Photograph by Jeff Warburton*)

changing conditions during the Quaternary, induced by changing climates and changing land-use, that have produced adjustments in the fluvial system. Fluvial system response to environmental change is usually complex. Large changes occur in the wake of shifts from glacial to interglacial climates. Changes in historical times, as deciphered from sequences of alluvial deposits, suggest that the response of the fluvial system to climatic change may vary from place to place, partly owing to regional variations of climate and partly to thresholds within the fluvial system itself. In places where human occupancy has affected geomorphic processes, as in the Mediterranean valleys, it is difficult to disentangle climatic effects from anthropogenic effects.

ESSAY QUESTIONS

1. How would you convince a sceptical friend that rivers carved the valleys through which they flow?
2. Why do river channel patterns vary?
3. To what extent have humans modified fluvial landscapes?
4. Discuss the problems of interpreting historical changes in fluvial systems.

FURTHER READING

Acreman, M. (2000) *The Hydrology of the UK: A Study of Change*. London: Routledge.
Not strictly geomorphology, but highly relevant to the subject.

Bridge, J. S. (2003) *Rivers and Floodplains: Forms, Processes, and Sedimentary Record*. Oxford: Blackwell Science.
A useful text for more advanced readers.

Brookes, A. J. and Shields, F. D. (1996) *River Channel Restoration: Guiding Principles for Sustainable Projects*. Chichester: John Wiley & Sons.
If you are interested in applied fluvial geomorphology, try this.

Charlton, R. (2008) *Fundamentals of Fluvial Geomorphology*. Abingdon: Routledge.
An ideal book for undergraduate geomorphologists.

Jones, J. A. A. (1997) *Global Hydrology: Process, Resources and Environmental Management*. Harlow, Essex: Longman.
Gives a hydrological context for fluvial processes.

Knighton, A. D. (1998) *Fluvial Forms and Processes: A New Perspective*, 2nd edn. London: Arnold.
A top-rate book on fluvial geomorphology.

Kondolf, M. and Pigay, H. (2002) *Methods in Fluvial Geomorphology*. New York: John Wiley & Sons.
Discusses an integrated approach to river restoration.

Leopold, L. B., Wolman, M. G., and Miller, J. P. (1964) *Fluvial Processes in Geomorphology*. San Francisco, Calif., London: W. H. Freeman. (Published by Dover Publications, New York, 1992.)
The book that process geomorphologists used to rave about. Worth dipping into but not always easy reading.

Lewin, J., Macklin, M. G., and Woodward, J. C. (eds) (1996) *Mediterranean Quaternary River Environments*. Rotterdam: Balkema.
A useful set of case studies, although not easy for the novice.

Robert, A. (2003) *River Processes: An Introduction to Fluvial Dynamics*. London: Arnold.
A very good treatment of physical processes in alluvial channels.

Thorne, C. R., Hey, R. D., and Newson, M. D. (1997) *Applied Fluvial Geomorphology for River Engineering and Management*. Chichester: John Wiley & Sons.
Another book that considers applied aspects of the subject.

CHAPTER TEN

GLACIAL AND GLACIOFLUVIAL LANDSCAPES

10

Sheets, caps, and rivers of ice flow over frozen landscapes; seasonal meltwater courses over landscapes at the edges of ice bodies. This chapter covers:

- Ice and where it is found
- Processes associated with ice
- Glaciated valleys and other landforms created by ice erosion
- Drumlins and other landforms created by ice deposition
- Eskers and other landforms created by meltwater
- Ice-conditioned landforms
- Humans and icy landscapes

MELTWATER IN ACTION: GLACIAL SUPERFLOODS

The Altai Mountains in southern Russia consist of huge intermontane basins and high mountain ranges, some over 4,000 m. During the Pleistocene, the basins were filled by lakes wherever glaciers grew large enough to act as dams. Research in this remote area has revealed a fascinating geomorphic history (Rudoy 1998). The glacier-dammed lakes regularly burst out to generate glacial superfloods that have left behind exotic relief forms and deposits – giant current ripple-marks, swells and terraces, spillways, outburst and oversplash gorges, dry waterfalls, and so on. These features are 'diluvial' in origin, meaning they were produced by a large flood. They are allied to the Channeled Scabland features of Washington State, USA, which were produced by catastrophic outbursts from glacial Lake Missoula. The outburst superfloods discharged at a rate in excess of 1 million cubic metres per second, flowed at dozens of metres a second, and some stood more than 100 metres deep. The super-powerful diluvial waters changed the land surface in minutes, hours, and days. Diluvial accumulation, diluvial erosion, and diluvial evorsion were widespread. Diluvial accumulation built up ramparts and terraces (some of which were made of deposits 240 m thick), diluvial berms (large-scale counterparts of boulder-block ramparts and spits – 'cobblestone pavements' – on big modern rivers), and giant ripple-marks with wavelengths up to 200 m and heights up to 15 m (Plate 10.1). Some giant ripple-marks in the foothills of the Altai, between Platovo and Podgornoye, which lie 300 km from the site of the flood outbursts, point to a mean flood velocity of 16 m/s, a flood depth of 60 m, and a discharge of no less than 600,000 m^3/s. Diluvial super-erosion led to the formation of deep outburst gorges, open-valley spillways, and diluvial valleys and oversplash gorges where water could not be contained within the valley and plunged over the local watershed. Diluvial eversion, which occurred beneath mighty waterfalls, forced out hollows in bedrock that today are dry or occupied by lakes.

Plate 10.1 Giant current ripples in the Kuray Basin, Altai Mountains, southern Siberia. (*Photograph by Alexei N. Rudoy*)

GLACIAL ENVIRONMENTS

The totality of Earth's frozen waters constitutes the **cryosphere**. The cryosphere consists of ice and snow, which is present in the atmosphere, in lakes and rivers, in oceans, on the land, and under the Earth's surface (Figure 10.1). It constitutes less than 2 per cent of the total water in the hydrosphere, but glaciers and permanent snow account for just over two-thirds of all fresh water (Table 10.1). At present, glaciers cover about 10 per cent of the Earth's land surface, and pack or sea ice coats about 7 per cent of the ocean surface (during winter conditions, when such ice is at its maximum extent). Most of the glacier ice is confined to polar latitudes, with 99 per cent being found in Antarctica, Greenland, and the islands

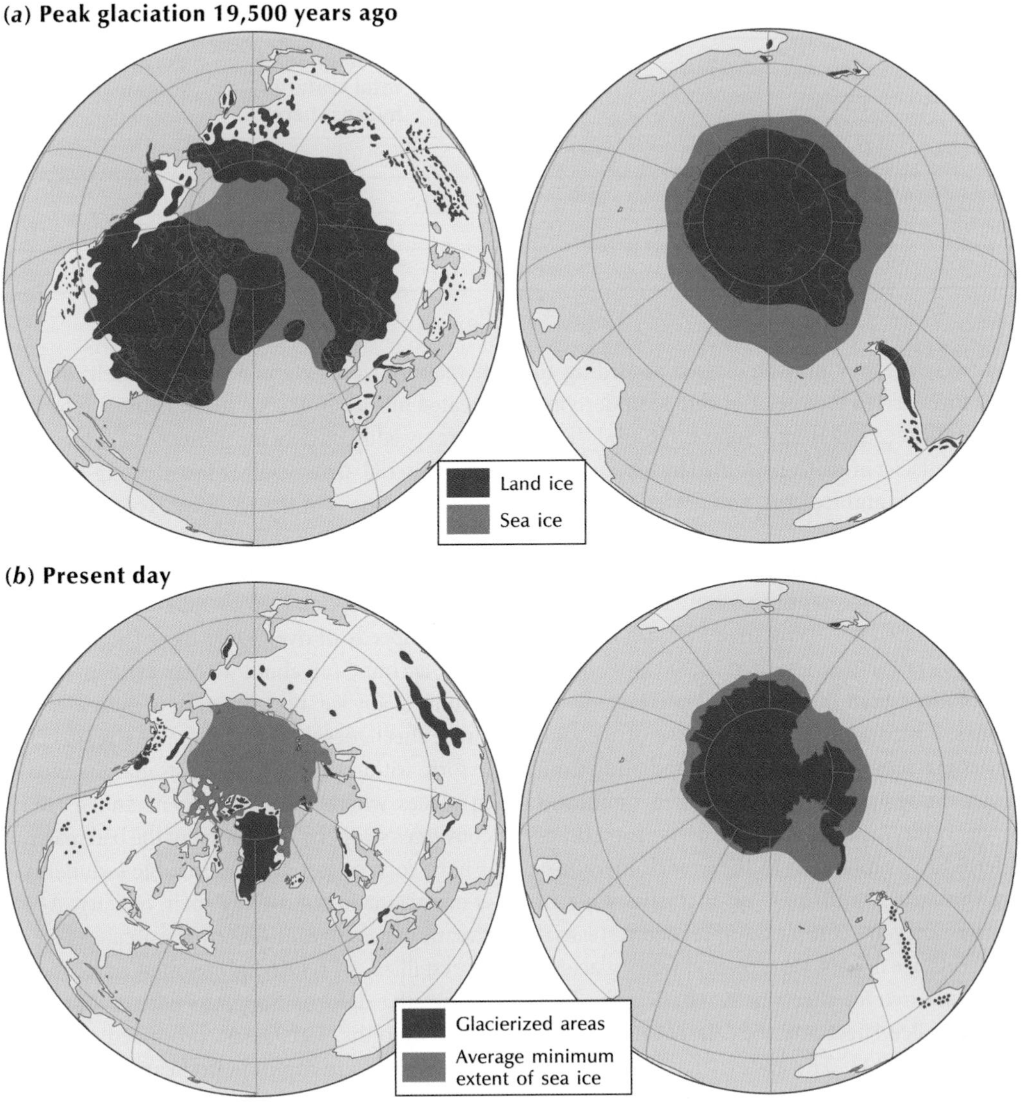

Figure 10.1 Distribution of ice.

of the Arctic archipelago. At the height of the last glaciation, currently estimated to have occurred between 26,500 and 19,000–20,000 years ago (Clark *et al.* 2009), ice covered some 32 per cent of the Earth's land surface. Continuous and discontinuous zones of permanently frozen ground underlie another 22 per cent of the Earth's land surface, but volumetrically they account for less than 1 per cent of all fresh water (Table 10.1). These **permafrost zones** contain ground ice and will be dealt with in the next chapter.

Glaciers

Glaciers are large masses of ice formed of compressed snow that move slowly under their

Table 10.1 Water in the cryosphere

Water	Water volume (km^3)	Percentage of total water in hydrosphere	Percentage of freshwater
Total water in hydrosphere	1,386,000,000	100	–
Total freshwater	35,029,000		100
Glacier ice and permanent snow	24,064,000	1.74	68.70
Ground ice and permafrost	300,000	0.022	0.86

Source: Adapted from Laycock (1987) and Shiklomanov (1993)

own weight. They may be classed according to their form and to their relationship to underlying topography (Sugden and John 1976, 56). Two types of glacier are unconstrained by topography: (1) ice sheets and ice caps, and (2) ice shelves.

Ice sheets, ice caps, and ice shelves

Ice sheets and **ice caps** are essentially the same, the only difference being their size: ice caps are normally taken to be less than 50,000 km^2 and ice sheets more than 50,000 km^2. They include **ice domes**, which are domelike masses of ice, and **outlet glaciers**, which are glaciers radiating from an ice dome and commonly lying in significant topographic depressions. Ice sheets, sometimes referred to as **inlandsis** in the French literature, are the largest and most all-inclusive scale of glacier. They are complexes of related terrestrial ice sheets, ice domes, ice caps, and valley glaciers. There are two ice sheets in Antarctica: the East Antarctic Ice Sheet and the West Antarctic Ice Sheet (Box 10.1). The eastern ice sheet covers some 10,350,000 km^2 and includes three domes – the Argus Dome, the Titan Dome (close to the South Pole), and the Circe Dome. The ice is some 4,776 m thick under the Argus Dome. Many parts of this ice sheet attain altitudes in excess of 3,000 m. The Transantarctic Chain separates the western ice sheet from the eastern ice sheet. It covers some 1,970,000 km^2, and the Ross Sea, the Weddell Sea, and the Antarctic Peninsula bound it.

Ice at the base of an ice sheet is generally warmer than ice at the cold surface, and in places, it may be warm enough to melt. Meltwater so created lubricates the ice sheet, helping it to flow more speedily, as does the presence of deformable bed material. The result is fast-flowing currents – **ice streams** – in the ice sheet. Ice streams are characteristically hundreds of kilometres long, tens of kilometres wide (with a maximum of around 50 km), and up to 2,000 m thick; some flow at speeds of over 1,000 m/yr (Figure 10.3). They account for about 10 per cent of the ice volume in any ice sheet, but most of the ice leaving an ice sheet goes through them. Ice streams tend to form within an ice sheet near its margin, usually in places where water is present and ice flow converges strongly. The nature of the bed material – hard rock or soft and deformable sediments – is important in controlling their velocity. At ice stream edges, stream deformation causes ice to recrystallize, so rendering it softer and concentrating the deformation into narrow bands or shear margins. Crevasses, produced by rapid deformation, are common in shear margins. The fastest-moving ice streams have the heaviest crevassing. Terrestrial and marine ice streams exist. Terrestrial ice streams lie on a bed that slopes uphill inland. Marine ice streams ground farther below sea level on a bed that slopes downhill into marine subglacial basins. In Antarctica (Box 10.1), ice streams are the most dynamic part of the ice sheet, and drain most of the ice. Ice streams may play two major roles in the global climate system. First, by being the chief determinant of ablation rates, they partly regulate the response of their parent ice sheets to climate

Box 10.1 ANTARCTICA

Antarctica (Figure 10.2) is the fifth-largest continent, but the highest (with an average elevation exceeding 2,000 m, over twice that of Asia), the coldest, and the windiest. With an area of about 14,000,000 km^2, it is bigger than Australia (9,000,000 km^2) and the subcontinent of Europe (10,200,000 km^2). Ice and snow cover 13,720,000 km^2 of the continent, and just 280,000 km^2, or about 2 per cent, is ice-free. With a very low snowfall, most of Antarctica is strictly a desert, with the ice sheet containing almost 70 per cent of global fresh water and 90 per cent of global ice reserves. Huge icebergs break off each year from the floating ice shelves and half of the surrounding ocean freezes over in winter, more than doubling the size of the continent.

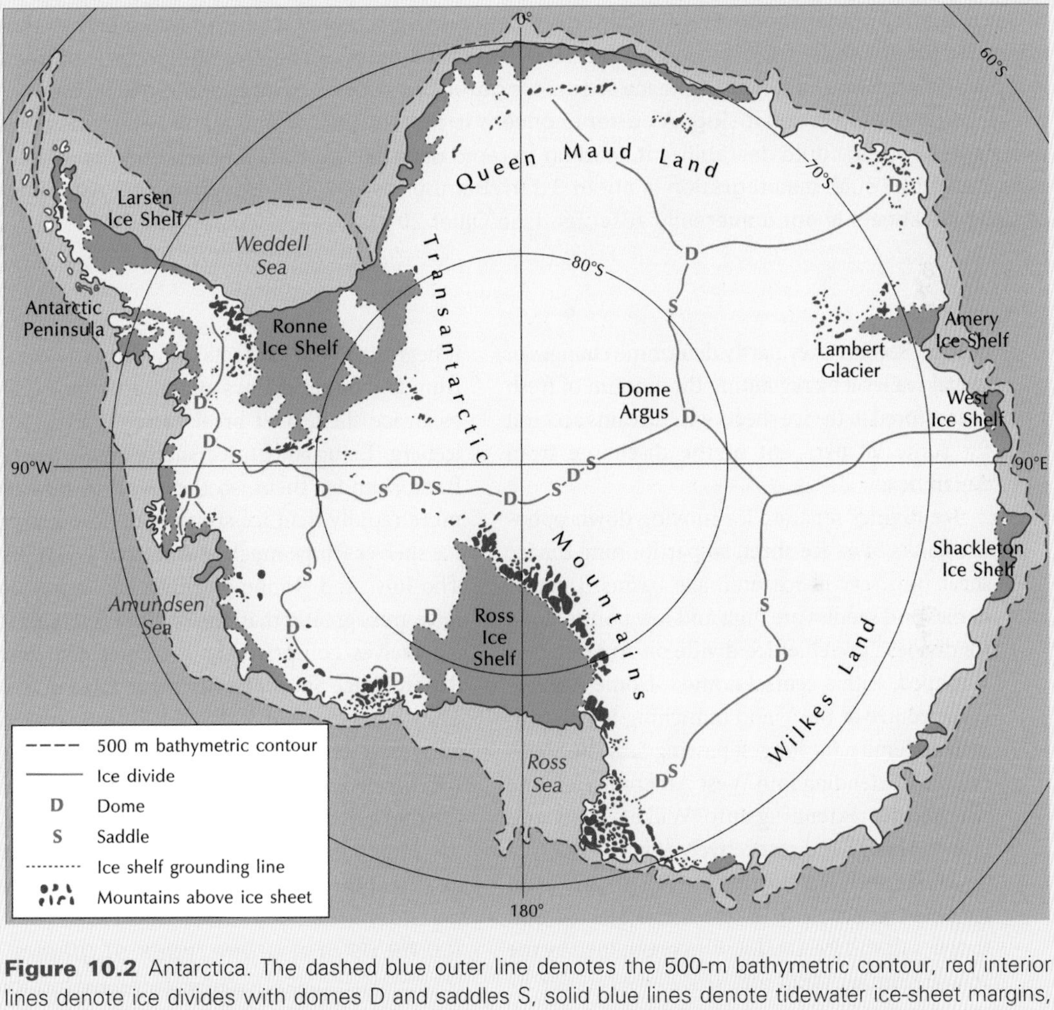

Figure 10.2 Antarctica. The dashed blue outer line denotes the 500-m bathymetric contour, red interior lines denote ice divides with domes D and saddles S, solid blue lines denote tidewater ice-sheet margins, dotted blue lines denote ice shelf grounding lines, and brown areas denote mountains above the ice sheet. *Source*: Adapted from Hughes *et al.* (1985)

continued . . .

The Antarctic ice sheet is in places more than 4,500 m thick. The ice lies in deep subglacial basins and over high subglacial plateaux. The Transantarctic Mountains separate the two main ice sheets of East Antarctica and West Antarctica. These ice sheets have different characteristics. The East Antarctic Ice Sheet is similar to the ice sheet covering Greenland in that they both cover landmasses and are frozen to the bedrock. Its ice ranges in thickness from approximately 2,000 to 4,000 metres. Terrestrial ice streams drain the edge of the East Antarctic Ice Sheet, which contains several ice shelves, including the Amery Ice Shelf. Only the Lambert Glacier runs deeply into the heart of the ice sheet. The West Antarctic Ice Sheet lies on a generally rugged bedrock floor, much of which lies below sea level. If the ice were to melt, this floor would rise through isostatic compensation. Apart from several islands, this area would remain below sea level.

In part owing to the ruggedness of the bedrock floor, the surface of the West Antarctic Ice Sheet has an irregular topography. Floating ice shelves in protected embayments, such as the Ronne Ice Shelf, fringe it. Marine ice controls the seaward drainage. The West Antarctic Ice Sheet is the world's only remaining marine ice sheet. It is grounded over deep interior subglacial basins, which helps to stop it collapsing. A current concern is that, in theory, the ice sheet is unstable and a small retreat could destabilize it, leading to rapid disintegration. The estimated rise of sea level caused by such disintegration is about 3.3 m (Bamber *et al.* 2009). However, the instability of large ice sheets is not universally accepted (see Ollier 2010).

change. Second, they partly determine changes of global sea level by regulating the amount of fresh water stored in the ice sheets – ice streams account for some 90 per cent of the discharge from Antarctica.

Ice divides separate ice moving down opposite flanks of an ice sheet, so partitioning the ice sheet into several ice drainage basins. Interior domes and saddles are high and low points along ice divides. The chief ice divide on Antarctica is Y-shaped, with a central dome – Dome Argus – at the centre of the Y and branching ice dives at each extremity, the longest passing near the South Pole and extending into West Antarctica and the two shorter extending into Wilkes Land and Queen Maud Land respectively (Figure 10.2).

An **ice shelf** is a floating ice cap or part of an ice sheet attached to a terrestrial glacier that supplies it with ice. It is loosely constrained by the coastal configuration and it deforms under its own weight. Ice is less dense than water and, because near the coast ice sheets generally rest on a bed below sea level, there comes a point where it begins to float. It floats in hydrostatic equilibrium and either it stays attached to the ice sheet as an ice shelf, or it breaks away (calves) as an iceberg. Being afloat, ice shelves experience no friction under them, so they tend to flow even more rapidly than ice streams, up to 3 km/year. Ice shelves fringe much of Antarctica (Box 10.1). The Ross and Ronne–Filchner ice shelves each have areas greater than the British Isles. Antarctic ice shelves comprise about 11 per cent of the Antarctic Ice Sheet and discharge most of its ice. They average about 500 m thick, compared with an average of 2,000 m for grounded Antarctic ice. All current ice shelves in Antarctica are probably floating leftovers of collapsed marine portions of the larger grounded Antarctic Ice Sheet that existed at the height of the last glaciation.

Ice fields and other types of glacier

Several types of glacier are constrained by topography including ice fields, niche glaciers, cirque glaciers, valley glaciers, and other small glaciers. **Ice fields** are roughly level areas of ice in which

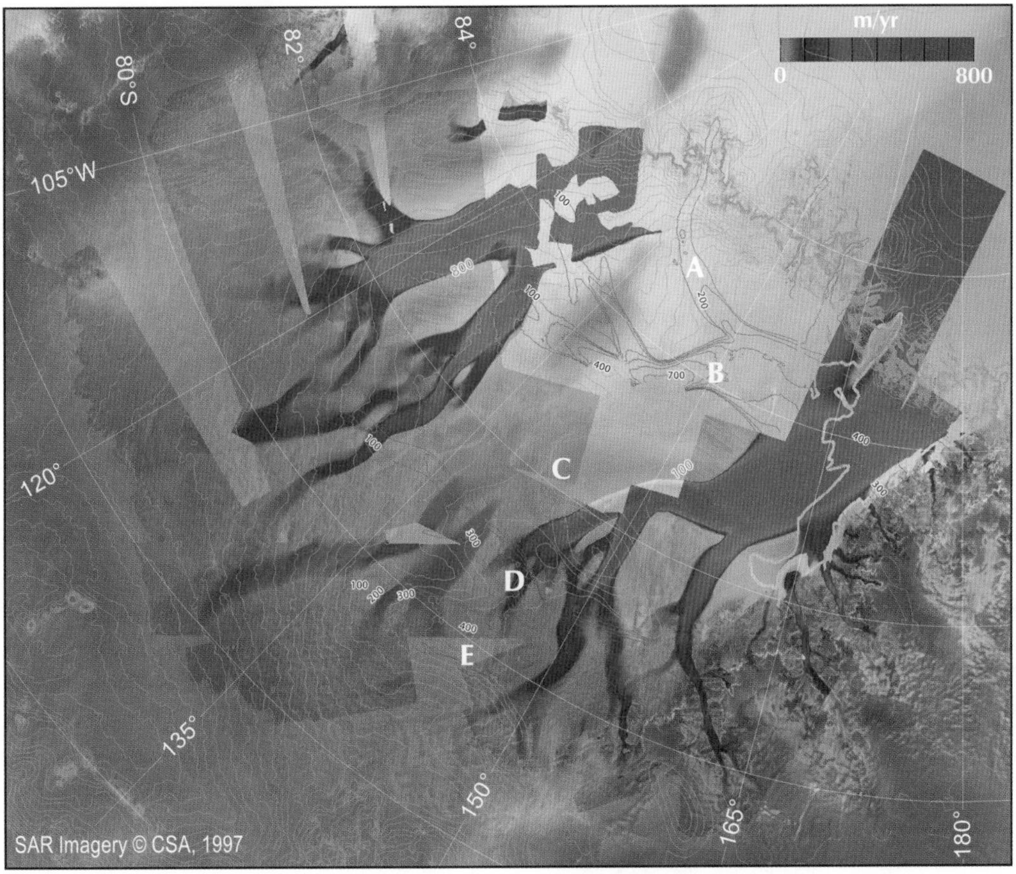

Figure 10.3 Satellite image showing the arterial character of West Antarctic ice streams. The labels identify the Mercer (A), Whillans (B), Kamb (C), Bindschadler (D), and MayAyeal (E) ice streams. Satellite imagery from the Canadian Space Agency. *Source*: Slightly adapted by Garry Clarke (2005) after Joughlin *et al.* (2002), copyright by the American Geophysical Union (2002)

underlying topography controls flow. Figure 10.4 shows the North Patagonian Ice Field and the glacial landforms associated with it. **Mountain glaciers** form in high mountainous regions, often flowing out of ice fields spanning several mountain peaks or a mountain range. **Hanging glaciers**, or **ice aprons**, cling to steep mountainsides. They are common in the Alps, where they often trigger avalanches, owing to their association with steep slopes. **Niche glaciers** are very small, occupying gullies and hollows on north-facing slopes (in the northern hemisphere) and looking like large snowfields. They may develop into a cirque glacier under favourable conditions. **Cirque** or **corrie glaciers** are small ice masses occupying armchair-shaped bedrock hollows in mountains (Plate 10.2). **Valley glaciers** sit in rock valleys and rock cliffs overlook them (Plate 10.3). They commonly begin as a cirque glacier or an ice sheet. Tributary valley glaciers may join large valley glaciers to create a valley-glacier network. **Piedmont glaciers** form where valley glaciers leave mountains and spread on to a flat land as large lobes of spreading ice, an example being the

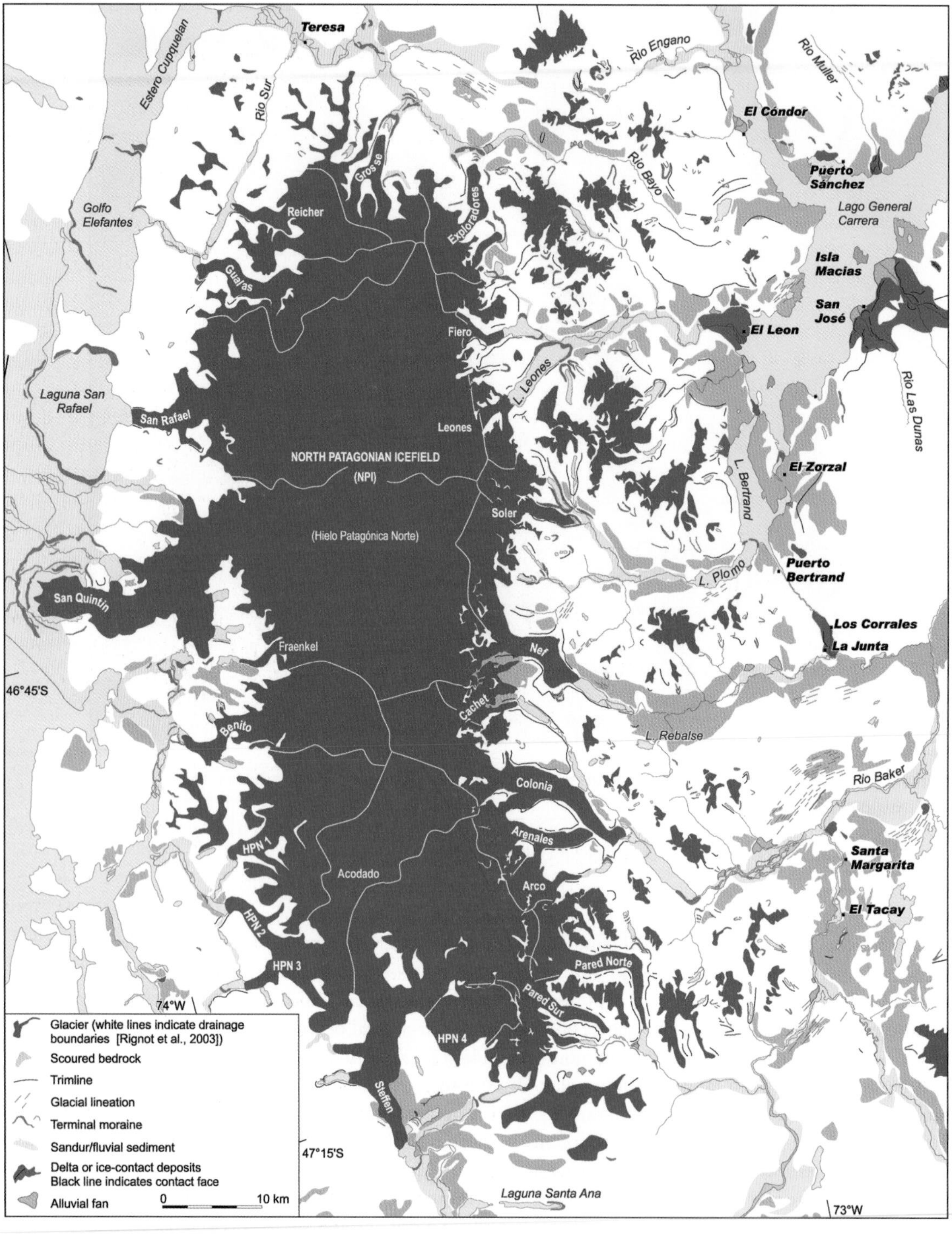

Figure 10.4 Glacial geomorphological map of the North Patagonian Ice Field constructed from a visual interpretation of remotely sensed images from satellites. *Source*: After Glasser *et al.* (2005)

Malaspina Glacier, Alaska. **Tidewater glaciers** are valley glaciers that flow into the sea, where they produce many small icebergs that may pose a danger to shipping.

The map of the North Patagonian Ice Field (Figure 10.4) shows how remotely sensed images provide a means of charting glacial landscapes as a whole. Remote sensing and associated techniques do not rule out the need for field investigation, but they do enable the detailed mapping of glaciers and glacial features, as well as ice and snow properties, over large areas that may include much inaccessible terrain. It is worth adding here that GIS is proving a huge boon for glacial geomorphologists, who use it to integrate data from several sources, to manage information across a variety of scales, and to identify previously unrecognized spatial and temporal relationships (Napieralski *et al.* 2007). And GIS-based analyses connected with numerical modelling have boosted understanding of glacial landscape evolution, have enabled new quantitative and systematic investigations of spatial and temporal patterns of glacial landforms and processes, and have promoted the development of insights and concepts unlikely to have emerged using only traditional methods (Napieralski *et al.* 2007).

Plate 10.2 Cirque glaciers feeding glacier with clean ice and snow above dirty summer ice, separated by balance line, Glacière de la Lex Blanche, Mont Blanc, Italy. (*Photograph by Tony Waltham Geophotos*)

Plate 10.3 Merging valley glaciers with medial moraines, Meade Glacier, Juneau Icefield, Alaska, USA. (*Photograph by Tony Waltham Geophotos*)

Quaternary glaciations

It is important to realize that the current distribution of ice is much smaller than its distribution during glacial stages over the last million years or so. Oxygen isotope data from deep-sea cores (and loess sequences) has revealed a sequence of alternating frigid conditions and warm interludes known as glacial and interglacial stages (Figure 10.5), which were driven by cycles in the Earth's orbital parameters, often referred to as Milankovitch or Croll–Milankovitch cycles (Box 10.2). The coldest conditions occurred at high latitudes, but the entire Earth seems to have cooled down, with snowlines lower than at present even in the tropics (Figure 10.6). **Palaeoglaciology** deals with the reconstruction of these Quaternary, and older, ice sheets, mainly by analysing the nature and distribution of glacial landforms (see Glasser and Bennett 2004).

Quaternary **glacial–interglacial cycles** have caused distinctive changes in middle- and high-latitude landscapes. At the extremes, cold and dry climates alternated with warm and moist climates. These changes would have affected

Box 10.2 MILANKOVITCH CYCLES

The Earth turns about its rotatory or spin axis while revolving around the Sun on the ecliptic (the plane of its orbit). However, the gravitational jostling of the planets, their satellites, and the Sun leads to orbital variations occurring with periods in the range 10,000 to 500,000 years that perturb Earth's climate. Four orbital variations are important in Milutin Milankovitch's theory, although Milankovitch was unaware of the fourth of these:

1 Earth's orbit is a nearly circular ellipse that has the barycentre (centre of mass) of the Solar System at one focus. The eccentricity of the orbit measures the divergence of the orbital ellipse from a circle. Variations in Earth's orbital eccentricity display periods of about 100,000 years (short eccentricity cycle) and 400,000 years (long eccentricity cycle).
2 Earth's axis of rotation tilts. At present, the angle of tilt from the equatorial plane (technically called the equinoctial plane) is about 23.5°. The Earth's axial tilt causes the march of the seasons: if the spin axis stood bolt upright, there would be no seasons. The tilt of the spin axis slowly oscillates between 22° and 24° 30′. The oscillations have a major periodicity of 41,000 years.
3 The orientation of the Earth's rotatory axis gradually alters relative to reference frame of the stars. In other words, the celestial poles (the points where the Earth's spin axis, when extended, pierce the celestial sphere) change. The North Pole slowly rotates or precesses in the opposite direction to the Earth's rotation. In doing so, it traces out a circle that, when joined to the Earth's centre of mass, describes a precessional cone. The South Pole moves in the same manner. This slow movement of the rotation axis is axial precession. It takes 25,800 years for the spin axis to precess once round the precessional cone relative to a fixed perihelion. The average periodicity of precession is 21,700 years, with major periods of 19,000 and 23,000 years.
4 The inclination of the orbital plane compared to the invariable plane of the Solar System varies by about 2° over a 100,000-year cycle, which may accentuate the climatic effects of the short eccentricity cycle.

These orbital forcings do not change the total amount of solar energy received by the Earth during the course of a year, but they do modulate the seasonal and latitudinal distribution of solar energy. In doing so, they wield a considerable influence over climate (Table 10.2). Orbital variations in the 10,000–500,000-year frequency band appear to have driven climatic change during the Pleistocene and Holocene. Orbital forcing has led to climatic change in middle and high latitudes, where ice sheets have waxed and waned, and to climatic change in low latitudes, where water

Table 10.2 Orbital forcing cycles

Cycle	Approximate period (years)
Tilt	41,000
Precession	19,000 and 23,000
Short eccentricity and orbital plane inclination	100,000
Long eccentricity	400,000

budgets and heat budgets have marched in step with high-latitude climatic cycles. Quaternary loess deposits, sea-level changes, and oxygen-isotope ratios of marine cores record the 100,000-year cycle of eccentricity. The precessional cycle (with 23,000- and 19,000-year components) and the 41,000-year tilt cycle ride on the 100,000-year cycle. They, too, generate climatic changes that register in marine and terrestrial sediments. Oxygen isotope ratios (δO^{18}) in ocean cores normally contain signatures of all the Earth's orbital cycles, though the tilt cycle, as it affects seasonality, has a stronger signature in sediments deposited at high latitudes.

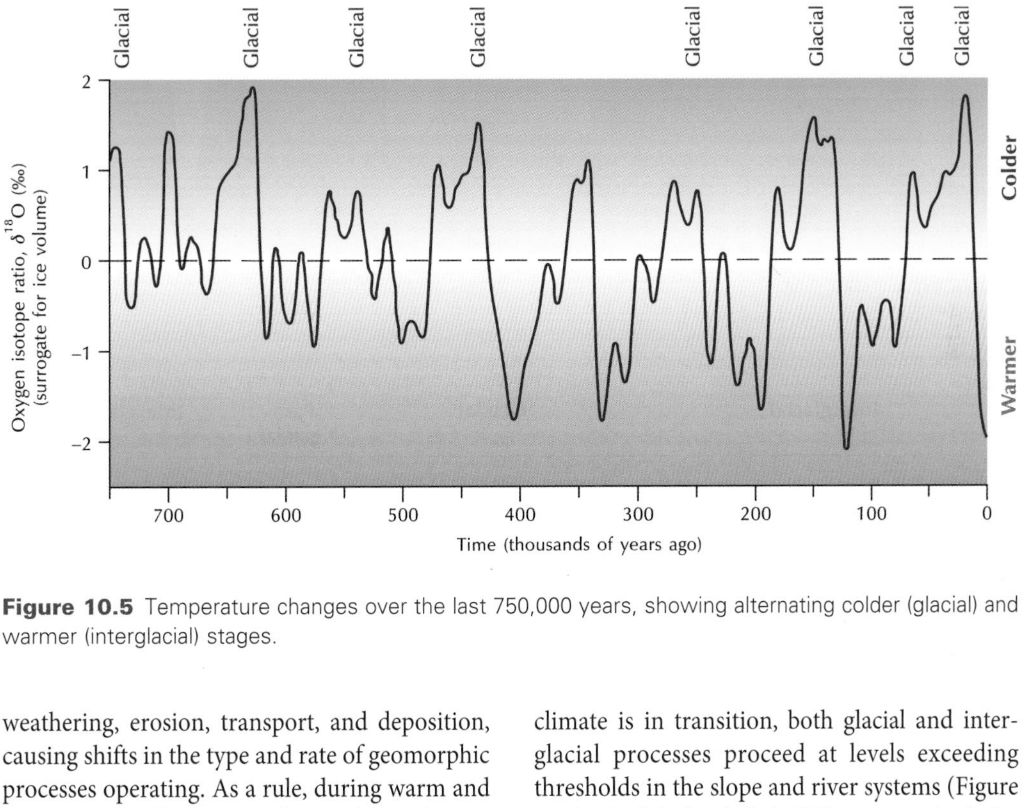

Figure 10.5 Temperature changes over the last 750,000 years, showing alternating colder (glacial) and warmer (interglacial) stages.

weathering, erosion, transport, and deposition, causing shifts in the type and rate of geomorphic processes operating. As a rule, during warm and wet interglacials, strong chemical weathering processes (such as leaching and piping) would have led to deep soil and regolith formation. During cold and dry glacials, permafrost, ice sheets, and cold deserts developed. The landforms and soils produced by glacial and by interglacial process regimes are generally distinctive, and are normally separated in time by erosional forms created in the relatively brief transition period from one climatic regime to another. When the climate is in transition, both glacial and interglacial processes proceed at levels exceeding thresholds in the slope and river systems (Figure 10.7). Leslek Starkel (1987) summarized the changes in a temperate soil landscape during a glacial–interglacial cycle. During a cold stage, erosion is dominant on the upper part of valley-side slopes, while in the lower reaches of valleys abundant sediment supply leads to overloading of the river, to deposition, and to braiding. During a warm stage, erosion thresholds are not normally exceeded, most of the slopes are stable, and soil formation proceeds, at least once the paraglacial

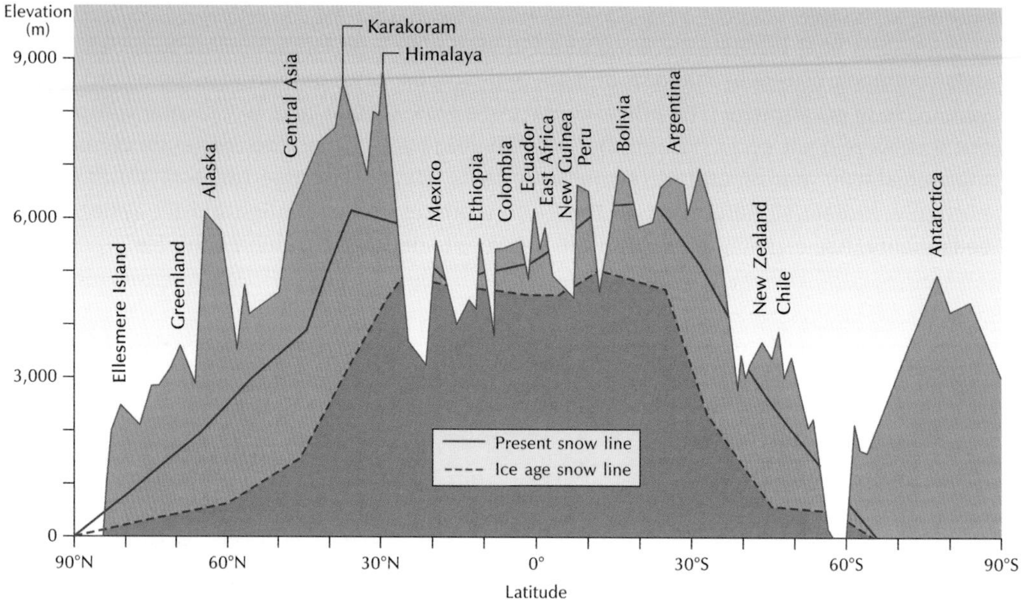

Figure 10.6 Latitudinal cross-section of the highest summits and highest and lowest snowlines. *Source*: Adapted from Barry and Ives (1974)

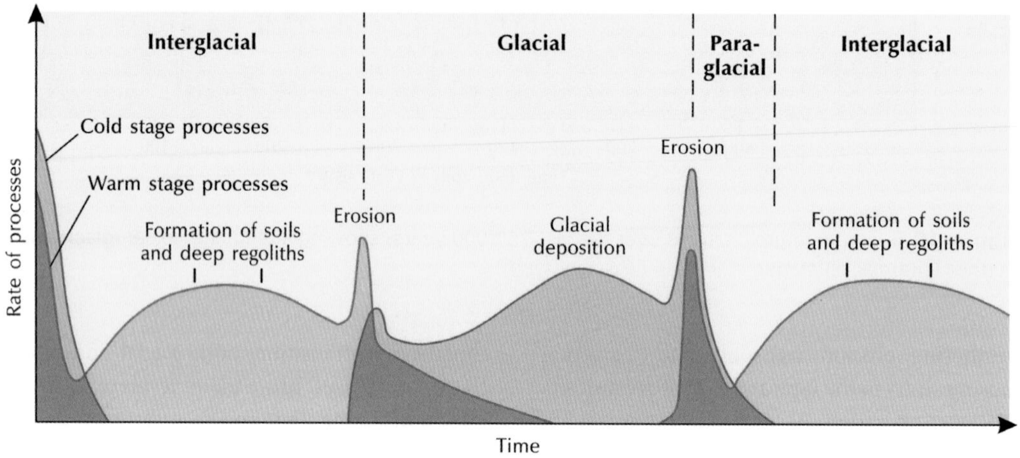

Figure 10.7 Suggested changes in geomorphic systems during a glacial–interglacial cycle. *Source*: Adapted from Starkel (1987)

period ends (p. 308). Meandering channels tend to aggrade, and erosion is appreciable only in the lowest parts of undercut valley-side slopes and in headwater areas. All these changes create distinct sequences of sediments in different parts of the fluvial system. Equivalent changes occurred in arid and semi-arid environments. For instance, gullying eroded talus deposits formed during prolonged mildly arid to semi-arid pluvial climatic modes, leaving talus flatiron relicts during arid to extremely arid interpluvial climatic modes (Gerson and Grossman 1987). In north-western

Texas and eastern New Mexico, a vast sheet of Quaternary loess (Blackwater Draw Formation) covers more than 100,000 km^2 and lies up to 27 m thick; it records more than 1.4 million years of aeolian sedimentation (Holliday 1988, 1989). Six buried soils in the formation reveal that stable landscapes obtained under subhumid to semiarid conditions, similar to those of the past several tens of thousands of years, whereas regional wind deflation and aeolian deposition prevailed during periods of prolonged drought.

GLACIAL PROCESSES

Ice, snow, and frost are solid forms of water. Each is a powerful geomorphic agent. It is convenient to discuss frost and snow processes in the periglacial landscape chapter and focus here on processes associated with flowing ice in glaciers.

Glacier mass balance

A glacier will form whenever 'a body of snow accumulates, compacts, and turns to ice' (Bennett and Glasser 2009, 41). The process of glacier formation may occur in any climate where snow falls at a faster rate than it melts. The more rapid the accumulation of snow and its conversion to ice, the quicker a glacier will form. Once formed, the survival of a glacier depends upon the balance between the rate of accumulation and the rate of ablation (ice loss). This mass balance of a glacier depends strongly on climate and determines the net gain or loss of ice on all kinds of glacier.

A **glacier mass balance** is an account of the inputs and outputs of water occurring in a glacier over a specified time, often a year or more. A **glacier balance year** is the time between two consecutive summer surfaces, where a summer surface is the date when the glacier mass is lowest. Mass balance terms vary with time and may be defined seasonally. The winter season begins when the rate of ice gain (**accumulation**) exceeds the rate of ice loss (**ablation**), and the summer season begins when the ablation rate exceeds the accumulation rate. By these definitions, the glacier balance year begins and ends in late summer or autumn for most temperate and subpolar regions. Snowfall accounts for most ice accumulation, but contributions may come from rainfall freezing on the ice surface, hail, the condensation and freezing of saturated air, the refreezing of meltwater and slush, and avalanching of snow from valley sides above the glacier. In temperate regions, ablation results mainly from melting, but it is also accomplished by evaporation, sublimation, wind and stream erosion, and calving into lakes and the sea. In Antarctica, calving is nearly the sole mechanism of ice loss.

The changes in the form of a glacier during an equilibrium balance year are shown in Figure 10.8. The upper part of the glacier is a snow-covered accumulation zone and the lower part is an ablation zone. **Firn** or **névé** is the term for snow that survives a summer melt season and begins its conversion to glacier ice. The **firn line** is the dividing line between the accumulation and ablation zones. For a glacier that is in equilibrium, the net gains of water in the accumulation zone match the net losses of water in the ablation zone and the glacier retains its overall shape and volume from year to year. If there is either a net gain or

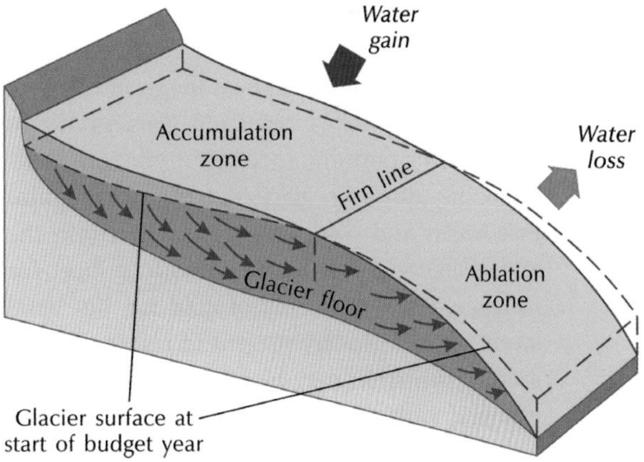

Figure 10.8 Glacier mass balance: schematic changes in the geometry of a glacier during an equilibrium budget year. *Source*: Adapted from Marcus (1969)

a net loss of water from the entire glacier, then attendant changes in glacier shape and volume and in the position of the firn line will result.

Mass balances may also be drawn up for continental ice sheets and ice caps. In an ice sheet, the accumulation zone lies in the central, elevated portion and a skirting ablation zone surrounds it at lower elevation. In Antarctica, the situation is more complicated because some ice streams suffer net ablation in the arid interior and net accumulation nearer to the wetter coasts.

It is important to distinguish between active ice and stagnant ice. **Active ice** moves downslope and is replenished by snow accumulation in its source region. **Stagnant ice** is unmoving, no longer replenished from its former source region, and decays where it stands.

Cold-based and warm-based glaciers

Glaciers are often classed as warm (or temperate) and cold (or polar), according to the temperature of the ice. A key idea in understanding the difference between warm and cold glaciers is the **pressure melting point**. The melting point of ice within a glacier varies with depth owing to pressure changes - the greater the depth of overlying ice the higher the pressure. The melting point at the base of a 2000-m-thick ice sheet is −1.6°C, rather than 0°C. **Warm glaciers** have ice at pressure melting point except near the surface, where cooling occurs in winter. **Cold glaciers** have a considerable portion of ice below pressure melting point. However, glaciologists now recognize that warm and cold ice may occur within the same glacier or ice sheet. The Antarctic sheet, for instance, consists mainly of cold ice, but basal layers of warm ice are present in places. A more useful distinction may be between **warm-based glaciers**, with a basal layer at pressure melting point, and **cold-based glaciers**, with a basal layer below pressure melting point. The presence of thick ice, slow ice movement, no summer melting, and severe winter freezing favour the formation of cold-based glaciers; whereas the presence of thin ice, fast ice movement, and much summer melting promote the growth of warm-based glaciers.

The basal thermal regime of a glacier is hugely important to geomorphology because it controls the pattern of erosion and deposition within the ice. Glaciers of cold ice are frozen to their beds, no meltwater is present at the interface between ice and bed, and no basal sliding occurs. Glaciers of warm ice have a constant supply of lubricating meltwater at their beds that encourages basal sliding. Warm ice glaciers therefore have the potential to flow much faster than cold ice glaciers and to erode their beds.

Ice flow

Ice in a glacier flows because gravity causes it to deform. The slope of a glacier from its origin to its end sets up the gravitational potential. Three mechanisms cause ice to flow, all of which are a response to shear stress: internal deformation (creep and large-scale folding and faulting), basal sliding, and subglacial bed deformation.

Internal deformation

Creep occurs because individual planes of hydrogen atoms slide on their basal surfaces. In addition, crystals move relative to one another owing to recrystallization, crystal growth, and the migration of crystal boundaries. Flow rates are speeded by thicker ice, higher water contents, and higher temperatures. For this reason, flow rates tend to be swiftest in warm ice. Warm ice is at the pressure melting point and contrasts with cold ice, which is below the pressure melting point. For a given stress, ice at 0°C deforms a hundred times faster than ice at −20°C. These thermal differences have led to a distinction between warm and cold glaciers, even though cold and warm ice may occur in the same glacier. Details of glacier flow are given in Box 10.3.

Where creep cannot accommodate the applied stresses in the ice, **faults** and **folds** may develop.

Box 10.3 GLACIER FLOW

Glaciers flow because gravity produces compressive stresses within the ice. The compressive stress depends on the weight of the overlying ice and has two components: the hydrostatic pressure and the shear stress. **Hydrostatic pressure** depends on the weight of the overlying ice and is spread equally in all directions. **Shear stress** depends upon the weight of the ice and the slope of the ice surface. At any point at the base of the ice, the shear stress, τ_0, is defined as

$$\tau_0 = \rho_i gh \sin \beta$$

where ρ_i is ice density, g is the acceleration of gravity, h is ice thickness, and β is the ice-surface slope. The product of ice density and the gravitational acceleration is roughly constant at 9 kN/m^3, so that the shear stress at the ice base depends on ice thickness and ice-surface slope. The shear stress at the base of glaciers lies between 50 and 150 kN/m^2.

Under stress, ice crystals deform by basal glide, which process occurs in layers running parallel to the crystals' basal planes. In glaciers, higher stresses are required to produce basal glide because the ice crystals are not usually orientated for basal glide in the direction of the applied stress. Ice responds to applied stress as a pseudoplastic body (see Figure 4.5). Deformation of ice crystals begins as soon as a shear stress is applied, but the response is at first elastic and the ice returns to its original form if the stress is removed. With increasing stress, however, the ice deforms plastically and attains a nearly steady value beyond the elastic limit or yield strength. In this condition, the ice continues to deform without an increase in stress and is able to creep or flow under its own weight. **Glen's power flow law** gives the relationship between shear strain and applied stress in ice:

$$\dot{\varepsilon} = A_i \tau n$$

where $\dot{\varepsilon}$ (epsilon dot) is the strain rate, A_i is an ice hardness 'constant', τ (tau) is the shear stress, and n is a constant that depends upon the confining pressure and the amount of rock debris in the ice – it ranges from about 1.3 to 4.5 and is often around 3. A_i is controlled by temperature, by crystal orientation, and by the impurity content of the ice. Its effect is that cold ice flows more slowly than warm ice, because a 20°C change in temperature generates a hundredfold increase in strain rate for a given shear stress. With an exponent $n = 3$, a small increase in ice thickness will have a large effect on the strain rate as it will cube the shear stress. With no basal sliding, Glen's flow law dictates that the surface velocity of a glacier varies with the fourth power of ice thickness and with the third power of the ice-surface gradient.

Crevasses are tensional fractures that occur on the surface. They are normally around 30 m deep in warm ice, but may be much deeper in cold ice. Shear fractures, which result from ice moving along slip planes, are common in thin ice near the glacier snout. Fractures tend not to occur under very thick ice where creep is operative.

Basal sliding

Ice may **slip** or **slide** over the glacier bed. Sliding cannot take place in a cold-ice glacier, because the glacier bottom is frozen to its bed. In a warm-ice glacier, sliding is common and is aided by lubricating meltwater, which if under pressure will also help to bear the weight of the overlying ice. Enhanced basal creep, whereby increased stress on

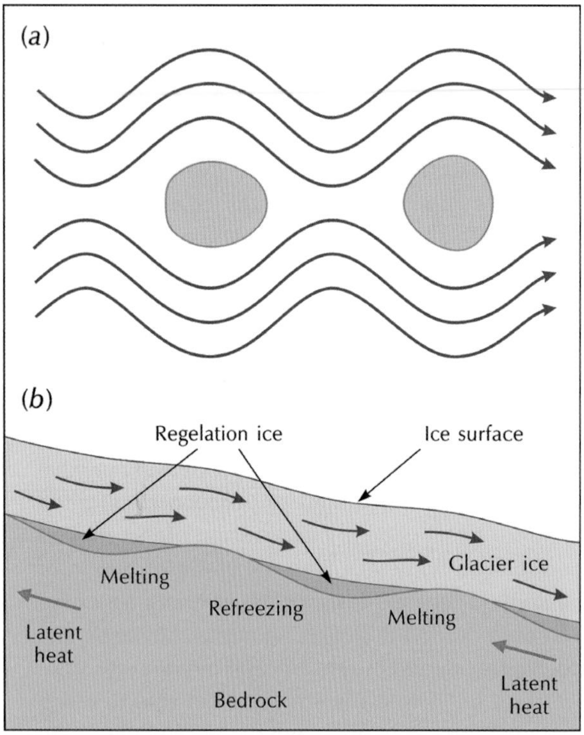

Figure 10.9 Basal sliding in ice. (a) High stresses upstream of obstacles in the glacier bed cause the ice to deform and flow around them. (b) Obstacles are also bypassed by pressure melting on the upstream side of obstacles and meltwater refreezing (relegation) on the downstream side. *Sources*: (a) Adapted from Weertman (1957); (b) Adapted from Kamb (1964)

the stoss- (up-ice) side of obstacles raises the strain rate and allows ice to flow around the obstacle, assists the slippage of ice over irregular beds in warm-based and cold-based glaciers. It is an extension of the normal ice-creep process. Also, under warm-based glaciers, water may melt as pressures rise on striking an obstacle and refreeze (a process called **regelation**) as pressures fall in the lee of the obstacle (Figure 10.9). Such pressure melting appears to work best for obstacles smaller than about 1 m.

Subglacial bed deformation

In some situations, glaciers may also move forward by **deforming their beds**: soft and wet sediments lying on plains may yield to the force exerted by the overlying ice. So, it would be wrong to suppose that the beds of all glaciers are passive and rigid layers over which ice moves. Where the bed consists of soft material (till), rather than solid bedrock, the ice and bed form a coupled system in which the bed materials deform in response to applied stress from the ice and so contribute to glacier motion. Thus the ice itself creeps and may slide over the till, ploughing the upper layers of till as it does so. The moving ice causes shear stress within the body of till, which itself may move along small fault lines near its base.

Glacial erosion

Three chief processes achieve glacial erosion: quarrying or plucking (the crushing and fracturing of structurally uniform rock and of jointed rock), abrasion, and meltwater erosion (p. 300). The bottom of the glacier entrains the material eroded by abrasion and fracturing.

1 **Quarrying** or **plucking**. This involves two separate processes: the fracturing of bedrock beneath a glacier, and the entrainment of the fractured or crushed bedrock. Thin and fast-flowing ice favours quarrying because it encourages extensive separation of the ice from its bed to create subglacial cavities and because it focuses stresses at sites, such as bedrock ledges, where ice touches the bed. In uniform rocks, the force of large clasts in moving ice may crush and fracture structurally homogeneous bedrock at the glacier bed. The process creates **crescent-shaped features**, **sheared boulders**, and **chattermarks** (p. 294). Bedrock may also fracture by pressure release once the ice has melted. With the weight of ice gone, the bedrock is in a stressed state and joints may develop, which often leads to exfoliation of large sheets of rock on steep valley sides. Rocks particularly prone to glacial fracture are those that possessed joint systems before the advent

of ice, and those prone to erosion are stratified, foliated, and faulted. The joints may not have been weathered before the arrival of the ice; but, with an ice cover present, freeze–thaw action at the glacier bed may loosen blocks and subglacial meltwater may erode the joint lines. The loosening and erosion facilitate the quarrying of large blocks of rock by the sliding ice to form rafts. Block removal is common on the down-glacier sides of roches moutonnées (p. 293).

2 **Glacial abrasion**. This is the scoring of bedrock by subglacial sediment or individual rock fragments (clasts) sliding over bedrock. The clasts scratch, groove, and polish the bedrock to produce **striations** (fine grooves) and other features (Plates 10.4 and 10.5), as well as grinding the bedrock to mill fine-grained materials (less than 100 micrometres diameter). Smoothed bedrock surfaces, commonly carrying striations, testify to the efficacy of glacial abrasion. **Rock flour** (silt-sized and clay-sized particles), which finds its way into glacial meltwater streams, is a product of glacial abrasion. The effectiveness of glacial abrasion depends upon at least eight factors (cf. Hambrey 1994, 81). (1) The presence and concentration of basal-ice debris. (2) The velocity at which the glacier slides. (3) The rate at which fresh debris is carried towards the glacier base to keep a keen abrading surface. (4) The ice thickness, which defines the normal stress at the contact between entrained glacial debris and substrate at the glacier bed. All

Plate 10.4 Striations on Tertiary gabbro with erratics, Loch Coruisk, Isle of Skye, Scotland. (*Photograph by Mike Hambrey*)

Plate 10.5 Glacially polished rock with striations from Laurentian ice sheet, shore of Lake Superior, Canada. (*Photograph by Tony Waltham Geophotos*)

other factors being constant, the abrasion rate increases as the basal pressure rises. Eventually, the friction between an entrained debris particle and the glacier bed rises to a point where the ice starts to flow over the glacier-bed debris and the abrasion rate falls. And, when the pressure reaches a high enough level, debris movement, and hence abrasion, stops. (5) In warm-based glaciers, the basal water pressure, which partly counteracts the normal stress and buoys up the glacier. (6) The difference in hardness between the abrading clasts and the bedrock. (7) The size and shape of the clasts. (8) The efficiency with which eroded debris is removed, particularly by meltwater.

Quarrying and abrasion can occur under cold-based glaciers, but they have a major impact on glacial erosion only under temperate glaciers where released meltwater lubricates the glacier base and promotes sliding.

Glacial debris entrainment and transport

Two processes incorporate detached bedrock into a glacier. Small rock fragments adhere to the ice when refreezing (regelation) takes place, which is common on the downstream side of bedrock obstacles. Large blocks are entrained as the ice deforms around them and engulfs them. Warm-based glaciers also entrain sediments derived from earlier ice advances, such as till, alluvium, and talus, by freezing on to the glacier sole.

Moving ice is a potent erosive agent only if sediment continues to be entrained and transported (Figure 10.10). **Subglacial debris** is carried along the glacier base. It is produced by basal melting in 'warm' ice and subsequent refreezing (regelation), which binds it to the basal ice. Creep may also add to the subglacial debris store, as may the squeezing of material into subglacial cavities in warm-based glaciers and the occurrence of thrust as ice moves over large obstacles. **Supraglacial debris** falls on to the ice surface from rock walls and other ice-free areas. It is far more common on valley and cirque glaciers than over large ice sheets. It may stay on the ice surface within the ablation zone, but it tends to become buried in the accumulation zone. Once buried, the debris is called **englacial debris**, which may re-emerge at the ice surface in the ablation zone or become trapped with subglacial debris, or it may travel to the glacier snout. Where compression near the glacier base leads to slip lines in the ice, which is common in the ablation zone, subglacial debris may be carried into an englacial position.

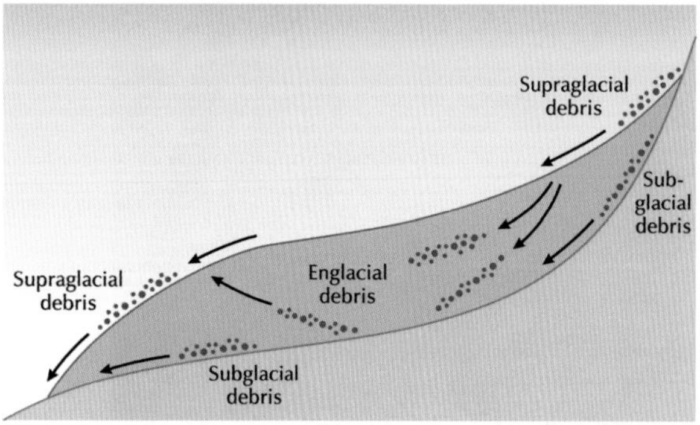

Figure 10.10 Transport by ice: supraglacial, englacial, and subglacial paths. *Source*: Adapted from Summerfield (1991, 271)

Glacial deposition

A host of processes bring about the deposition of glacial sediments. The mechanisms involved may be classified according to location relative to a glacier – subglacial, supraglacial, and marginal. **Subglacial deposition** is effected by at least three mechanisms: (1) undermelt, which is the deposition of sediments from melting basal ice; (2) basal lodgement, which is the plastering of fine sediments on to a glacier bed; and (3) basal flowage, which is in part an erosional process and involves the pushing of unconsolidated water-soaked sediments into basal ice concavities and the streamlining of till by overriding ice. **Supra-**

glacial deposition is caused by two processes: melt-out and flowage. Melt-out, which is the deposition of sediments by the melting of the ice surface, is most active in the snout of warm glaciers, where ablation may reduce the ice surface by 20 m in one summer. Flowage is the movement of debris down the ice surface. It is especially common near the glacier snout and ranges from a slow creep to rapid liquid flow. **Marginal deposition** arises from several processes. Saturated till may be squeezed from under the ice, and some supraglacial and englacial debris may be dumped by melt-out.

Proglacial sediments form in front of an ice sheet or glacier. The sediments are borne by meltwater and deposited in braided river channels and proglacial lakes. The breaching of glacial lakes may lay down glacial sediments over vast areas (p. 271).

EROSIONAL GLACIAL LANDFORMS

Glaciers and ice sheets are very effective agents of erosion. Large areas of lowland, including the Laurentian Shield of North America, bear the scars of past ice movements. More spectacular are the effects of glacial erosion in mountainous terrain, where ice carries material wrested from bedrock to lower-lying regions (Figure 10.4).

Glacial erosion moulds a panoply of landforms. One way of grouping these landforms is by the dominant formative process: abrasion, abrasion and rock fracture combined, rock crushing, and erosion by glacier ice and frost shattering (Table 10.3). Notice that abraded landforms are 'streamlined', landforms resulting from the combined effects of abrasion and rock fracture are partly streamlined, while the landforms resulting from rock fracture are not streamlined. The remaining group of landforms is residual, representing the ruins of an elevated mass of bedrock after abrasion, fracturing by ice, frost-shattering, and mass movements have operated.

Abrasional landforms

Glacial abrasion produces a range of streamlined landforms that range in size from millimetres to thousands of kilometres (Table 10.3). In sliding over obstacles, ice tends to abrade the up-ice side or stoss-side and smooth it. The down-ice side or leeside is subject to bedrock fracture, the loosening and displacement of rock fragments, and the entrainment of these fragments into the sliding glacier base. In consequence, the downstream surfaces tend to be rough and are described as plucked and quarried.

Scoured regions

The largest abrasive feature is a low-amplitude but irregular relief produced by the **areal scouring** of large regions such as broad portions of the Laurentian Shield, North America. Scoured bedrock regions usually comprise a collection of streamlined bedrock features, rock basins, and stoss and lee forms (see below and Figure 10.4). In Scotland, parts of the north-west Highlands were scoured in this way to give '**knock and lochan' topography**; the 'knocks' are rocky knolls and the 'lochans' are lakes that lie in depressions.

Glacial troughs – glaciated valleys and fjords

Glacial troughs are dramatic landforms (Plates 10.6 and 10.7). Either valley glaciers erode them or they develop beneath ice sheets and ice caps where ice streaming occurs. Most glacial troughs have, to varying degrees, a U-shaped cross-section, and a very irregular long-profile with short and steep sections alternating with long and flat sections. The long, flat sections often contain rock basins filled by lakes. In glacial troughs where a line of basins holds lakes, the lakes are called **paternoster lakes** after their likeness to beads on a string (a rosary). The irregular long-profile appears to result from uneven over-deepening by the ice, probably in response to variations in the resistance of bedrock rather than to any peculiarities of glacier flow. Paraglacial stress release of valley-side slopes, associated with the

Plate 10.6 Glacial trough with a hanging valley to the right, Yosemite, California, USA. (*Photograph by Mike Hambrey*)

departure of ice during interglacial stages, helps to fashion the shape of glacial troughs (p. 308).

There are two kinds of glacial trough: **glaciated valleys** and **fjords**. A glaciated-valley floor lies above sea level, while a fjord floor lies below sea level and is a glaciated valley drowned by the sea. In most respects, glaciated valleys and fjords are similar landforms. Indeed, a glaciated valley may pass into a fjord. Many fjords, and especially those in Norway, are deeper in their inner reaches because ice action was greatest there. In their outer reaches, where the fjord opens into the sea, there is often a shallow sill or lip. The Sognefjord, Norway, is 200 km long and has a maximum depth of 1,308 m. At its entrance, it is just 3 km wide and is 160m deep, and its excavation

Table 10.3 Landforms created by glacial erosion

Landform	Description
Abrasion by glacier ice – streamlined relief forms (mm to 1000s km)	
Areal scouring	Regional expanses of lowland bedrock, up to 1000s km in extent, scoured by ice. Sometimes contain sets of parallel grooves and bedrock flutes
Glaciated valley	Glacial trough, the floor of which is above sea level. Often U-shaped
Fjord	Glacial trough, the floor of which is below sea level. Often U-shaped
Hanging valley	Tributary valley whose floor sits above the floor of the trunk valley
Breached watershed	Col abraded by a valley glaciers spilling out of its confining trough
Dome	Dome-shaped structure found in uniform bedrock where ice has abraded an obstacle to leave a smoothed rock hillock that has been subject to exfoliation after the ice has left
Whaleback or rock drumlin	Glacially streamlined erosional feature 100–1000 m long, intermediate in size between a roche moutonnée and a flyggberg
Striation	Scratch on bedrock or clast made by ice (or other geomorphic agents such as landslides, tectonic disturbance, and animals)
Polished surface	Bedrock surface made shiny by a host of tiny scratches scored by fine-grained clasts
Groove	A furrow cut into bedrock by fragments of rock (clasts) held in advancing ice
Plastically moulded forms (p-forms)	Smooth and complex forms on rock surfaces. They include cavetto forms (channels on steep rock faces) and grooves (on open flat surfaces). *Sichelwannen* and Nye channels (curved and winding channels) are also p-forms, but probably produced mainly by meltwater erosion (Table 10.5)
Abrasion and rock fracturing by glacier ice – partly streamlined relief forms (1 m to 10 km)	
Trough head	Steep, rocky face at the head of many glaciated valleys and fjords
Rock or valley step	Bedrock steps in the floor of glacial troughs, possibly where the bedrock is harder and often where the valley narrows
Riegel	Low rock ridge, step, or barrier lying across a glaciated-valley floor

Table 10.3 *continued*

Landform	Description
Cirque	Steep-walled, semi-circular recess or basin in a mountain
Col	Low pass connecting two cirques facing in opposite directions
Roche moutonnée	Bedrock feature, generally less than 100 m long, the long-axis of which lies parallel to the direction of ice movement. The up-ice (stoss) side is abraded, polished, and gently sloping, and the down-ice (lee) side is rugged and steep
Flyggberg	Large (>1000 m long) streamlined bedrock feature, formed through erosion by flowing ice. The up-ice (stoss) side is polished and gently sloping, whereas the down-ice (lee) side is rough, irregular, and steep. A flyggberg is a large-scale roche moutonnée or whaleback. The name is Swedish.
Crag-and-tail or lee-side cone	An asymmetrical landform comprising a rugged crag with a smooth tail in its lee
Rock crushing – non-streamlined relief forms (1 cm to 10s cm)	
Lunate fracture	Crescent-shaped fractures with the concavity facing the direction of ice flow
Crescentic gouge	Crescent-shaped features with the concavity facing away from the direction of ice flow
Crescentic fracture	Small, crescent-shaped fractures with the concavity facing away from the direction of ice flow
Chattermarks	Crescent-shaped friction cracks on bedrock, produced by the juddering motion of moving ice
Erosion by glacier ice, frost shattering, mass movement – residual relief forms (100 m to 100 km)	
Arête	Narrow, sharp-edged ridge separating two cirques
Horn	Peak formed by the intersecting walls of three or more cirques. An example is the Matterhorn in the European Alps
Nunatak	Unglaciated 'island' of bedrock, formerly or currently surrounded by ice

Source: Adapted from Hambrey (1994, 84)

required the removal of about 2,000 km^3 of rock (Andersen and Borns 1994). Skelton Inlet, Antarctica, is 1,933 m deep.

Breached watersheds and hanging valleys are of the same order of size as glacial troughs, but perhaps generally a little smaller. **Breached watersheds** occur where ice from one glacier spills over to an adjacent one, eroding the intervening col in the process. Indeed, the eroding may deepen the col to such an extent that the glacier itself is diverted. **Hanging valleys** are the vestiges of tributary glaciers that were less effective at eroding bedrock than the main trunk glacier, so that the tributary valley is cut off sharply where it meets the steep wall of the main valley (Plate 10.6), often with a waterfall coursing over the edge.

Plate 10.7 Glacial trough with valley glaciers at head, East Greenland. (*Photograph by Mike Hambrey*)

Domes and whalebacks

Various glacially abraded forms are less than about 100 m in size. **Domes** and **whalebacks** (**rock drumlins, tadpole rocks, streamlined hills**) form where flowing ice encounters an obstruction and, unable to obliterate it, leaves an upstanding, rounded hillock.

Striated, polished, and grooved bedrock

Striated, polished, and grooved surfaces are all fashioned by rock material carried by flowing ice. Large clasts (about 1 cm or bigger) erode by scratching and create striations and grooves. Finer material (less than a centimetre or so), and especially the silt fractions, erodes by polishing bedrock surfaces. **Striations** are finely cut, U-shaped grooves or scratches, up to a metre long or more, scored into bedrock by the base of a sliding glacier. They come in a multiplicity of forms, some of which, such as rattails, indicate the direction of ice flow. Large striations are called **grooves**, which attain depths and widths of a few metres and lengths of a few hundred metres (Plate 10.8). Glacial valleys may be thought of as enormous grooves. Grooves form through glacial abrasion or the generation of meltwater under pressure. Bedrock bearing a multitude of tiny scratches has a polished look. The finer is the abrading material, the higher is the polish. Striations are equivocal evidence of ice action, especially in the geological record, as such other processes as avalanches and debris flows are capable of scratching bedrock. **Rock basins** are depressions with diameters in the range several metres to hundreds of metres, carved into bedrock, commonly found in association with roches moutonnées. They form where rocks contain structural weaknesses exploitable by glacial erosion.

Plate 10.8 Two-metre deep striated groove carved by the Laurentide ice sheet, Whitefish Falls, Ontario, Canada. (*Photograph by Mike Hambrey*)

Plastically moulded forms

Some glaciated rock surfaces carry complex, smooth forms known as **plastically moulded forms**, or **p-forms** (Plates 10.9 and 10.10). The origin of these puzzling features is debatable. Possibilities are glacial abrasion, the motion of saturated till (till slurry) at the bottom or sides of a glacier, and meltwater erosion, especially meltwater under high pressure beneath a glacier. If a meltwater origin is certain, then the features are **s-forms**.

Abrasion-cum-rock-fracture landforms

In combination, glacial abrasion and rock fracture produce partly streamlined landforms that range in size from about 1 m to 10 km (Table 10.3).

Trough heads, valley steps, and riegels

Trough heads (or **trough ends**) and **valley steps** are similar to roches moutonnées (see below) but larger. Trough heads are steep and rocky faces that mark the limit of over-deepening of glacial troughs. Their 'plucked' appearance suggests that they may follow original breaks of slope related to hard rock outcrops. In sliding over the break of slope, the ice loses contact with the ground, creating a cavity in which freeze–thaw processes aid the loosening of blocks. The ice reconnects with the ground further down the valley. Where another hard rock outcrop associated with an original break of slope is met, a rock or valley step develops by a similar process. However, the formation of trough heads and rock steps is little researched and far from clear.

A **riegel** is a rock barrier that sits across a valley, often where a band of hard rock outcrops. It may impound a lake.

Cirques

Cirques are typically armchair-shaped hollows that form in mountainous terrain, though their form and size are varied (see Figure 10.11). The classical shape is a deep rock basin, with a steep headwall at its back and a residual lip or low bedrock rim at its front, and often containing a lake. A terminal moraine commonly buries the lip. Cirques possess several local names, including **corrie** in England and Scotland and **cwm** in Wales (see photograph of Cwm Idwal, Figure 1.10). They form through the conjoint action of warm-based ice and abundant meltwater. Corries are commonly deemed to be indisputable indicators of past glacial activity, and geomorphologists use them to reconstruct former regional snowlines (Box 10.4).

Stoss and lee forms

Roches moutonnées, flyggbergs, and crag-and-tail features are all asymmetrical, being streamlined on the stoss-side and 'craggy' on the leeside. They are the productions of glacial abrasion and quarrying. **Roches moutonnées** are common in glacially eroded terrain. They are named after the wavy wigs (moutonnées) that were popular in Europe at the close of the eighteenth century (Embleton and King 1975a, 152). Roches moutonnées are probably small hills that existed before the ice came and glacial action modified them. They vary from a few tens to a few hundreds of metres long, are best developed in jointed crystalline rocks, and cover large areas (Plate 10.11). In general, they provide a good pointer to the direction of past ice flow if used in conjunction with striations, grooves, and other features. **Flyggbergs** are large roches moutonnées, more than 1,000 m long. **Crag-and-tail features** are tadpole-shaped landforms of upstanding resistant rocks eroded on the rugged stoss-side (the crag) with softer rocks, sometimes bearing till, in the protected and smooth leeside. In East Lothian, Scotland, deep glacial erosion has produced several crags of resistant volcanic necks and plugs intruded into relatively soft Carboniferous sedimentary rocks; North Berwick Law is an excellent

Plate 10.9 Plastically moulded forms (p-forms) and striations on a roche moutonnée near calving front of Columbia Glacier, Prince William Sound, Alaska, USA. (*Photograph by Mike Hambrey*)

Plate 10.10 Subglacially formed p-forms and pothole, cut in Proterozoic schists, Loch Treig, Grampian Highlands, Scotland. (*Photograph by Mike Hambrey*)

Box 10.4 CIRQUES

Cirques usually start as depressions excavated by streams, or as any hollow in which snow collects and accumulates (**nivation hollow**). Snow tends to accumulate on the leeside of mountains, so cirques in the Northern Hemisphere tend to face north and east. In the steep terrain of alpine regions, it is usual for cirques to show poor development and to slope outwards. In less precipitous terrain, as in the English Lake District, they often have rock basins, possibly with a moraine at the **lip**, that frequently hold lakes (**tarns**). Despite their variable form and size, the ratio of length to height (from the lip of a mature cirque to the top of the **headwall**) is surprisingly constant, and lies within the range 2.8 : 1 to 3.2 : 1 (Manley 1959). The largest known cirque is Walcott Cirque, Victoria Land, Antarctica, which is 16 km wide and 3 km high. Some cirques have a composite character. Many British mountains have **cirques-within-cirques**. In Coire Bà, one of the largest cirques in Britain, which lies on the east face of Black Mount, Scotland, several small cirques cut into the headwall of the main cirque. **Cirque staircases occur**. In Snowdon, Wales, Cwm Llydaw is an over-deepened basin with a tarn and sheer headwall. Cwm Glaslyn, a smaller cirque, which also holds a tarn, breaches the headwall partway up. And above Cwm Glaslyn lies an incipient cirque just below the summit of Y Wyddfa. It is unclear if such staircases represent the influence of different snowlines or the exploitation of several stream-cut hollows or geological sites.

example. Small crag-and-tail features occur where resistant grains or mineral crystals protect rock from glacial abrasion. An example is found on slate in North Wales, where pyrite crystals have small tails of rock that indicate the orientation and direction of ice flow (Gray 1982), and on carbonate rocks in Arctic Canada, where limestone ridges less than 5 cm high and 25 cm long form in the lee of more resistant chert nodules (England 1986).

Plate 10.11 Roche moutonnée, shaped in granite by glacial flow, ice moving right to left, Yosemite National Park, California, USA. (*Photograph by Marli Miller*)

Rock-crushed landforms

Small-scale, crescent-shaped features, ranging in size from a few centimetres up to a couple of metres, occur on striated and polished rock surfaces. These features are the outcome of rock crushing by debris lodged at the bottom of a glacier. They come in a variety of forms and include lunate fractures, crescentic gouges, crescentic fractures, and chattermarks. **Lunate features** are fractures shaped like crescents with the concavity facing the direction of ice flow. **Crescentic gouges** are crescent-shaped gouges, but unlike lunate features they face away from the direction of ice flow. **Crescentic fractures** are similar to crescentic gouges but are fractures rather than gouges. **Chattermarks** are also crescent-shaped. They are friction marks on bedrock formed as moving ice judders and are comparable to the rib-like markings sometimes left on wood and metal by cutting tools (Plate 10.12).

Residual landforms

Arêtes, cols, and horns

In glaciated mountains, abrasion, fracturing by ice, frost-shattering, and mass movements erode the mountain mass and in doing so sculpt a set of related landforms: arêtes, cols, and horns (Figure 10.11). These landforms tend to survive as relict features long after the ice has melted. **Arêtes** form where two adjacent cirques eat away at the intervening ridge until it becomes a knife-edge, serrated ridge. Frost shattering helps to give the ridge its serrated appearance, upstanding pinnacles on which are called **gendarmes** ('policemen'). The ridges, or arêtes, are sometimes breached in places by **cols**. If three or more cirques eat into a mountain mass from different sides, a **pyramidal peak** or **horn** may eventually form. The classic example is the Matterhorn on the Swiss–Italian border.

Nunataks

Nunataks are rock outcrops, ranging from less than a kilometre to hundreds of kilometres in size, surrounded by ice. They include parts of mountains where ice has not formed, or entire mountain ranges, including the Transantarctic Mountains on Antarctica (see Figure 10.2), that have escaped ice formation everywhere but their flanks.

DEPOSITIONAL GLACIAL LANDFORMS

Debris carried by ice is eventually dumped to produce an array of landforms (Table 10.4). It is expedient to group these landforms primarily according to their position in relation to the ice (supraglacial, subglacial, and marginal) and secondarily according to their orientation with respect to the direction of ice flow (parallel, transverse, and non-orientated).

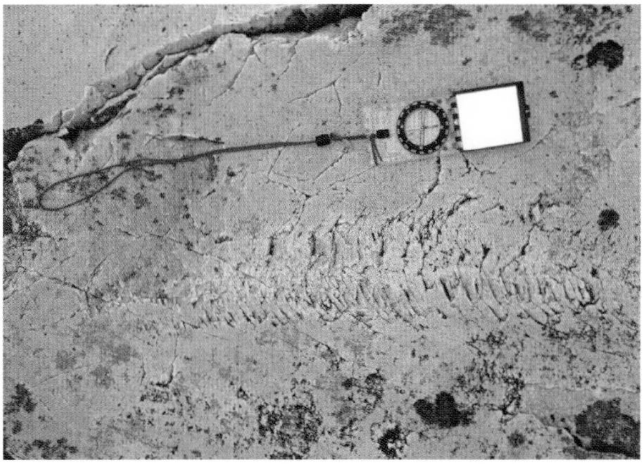

Plate 10.12 Chattermarks on Cambrian quartzite, An Teallach, north-west Highlands, Scotland. (*Photograph by Mike Hambrey*)

Supraglacial landforms

Debris on a glacier surface lasts only as long as the glacier, but it produces eye-catching features in current glacial environments. **Lateral moraines** and **medial moraines** lie parallel to the glacier. **Shear** or **thrust moraines**, produced by longitudinal compression forcing debris to the surface, and rockfalls, which spread debris across a glacier, lie transversely on the glacier surface. **Dirt cones**, **erratics** (Plate 10.4), and **crevasse fills** have no particular orientation with respect to the ice movement.

Many features of supraglacial origin survive in the landscape once the ice has gone. The chief such forms are lateral moraines and moraine dumps, both of which lie parallel to the ice flow, and hummocky moraines and erratics, which have no particular orientation. Lateral moraines are impressive landforms. They form from frost-shattered debris falling from cliffs above the glacier and from debris trapped between the glacier and the valley sides (Figure 10.11c). Once the ice has gone, lateral moraines collapse.

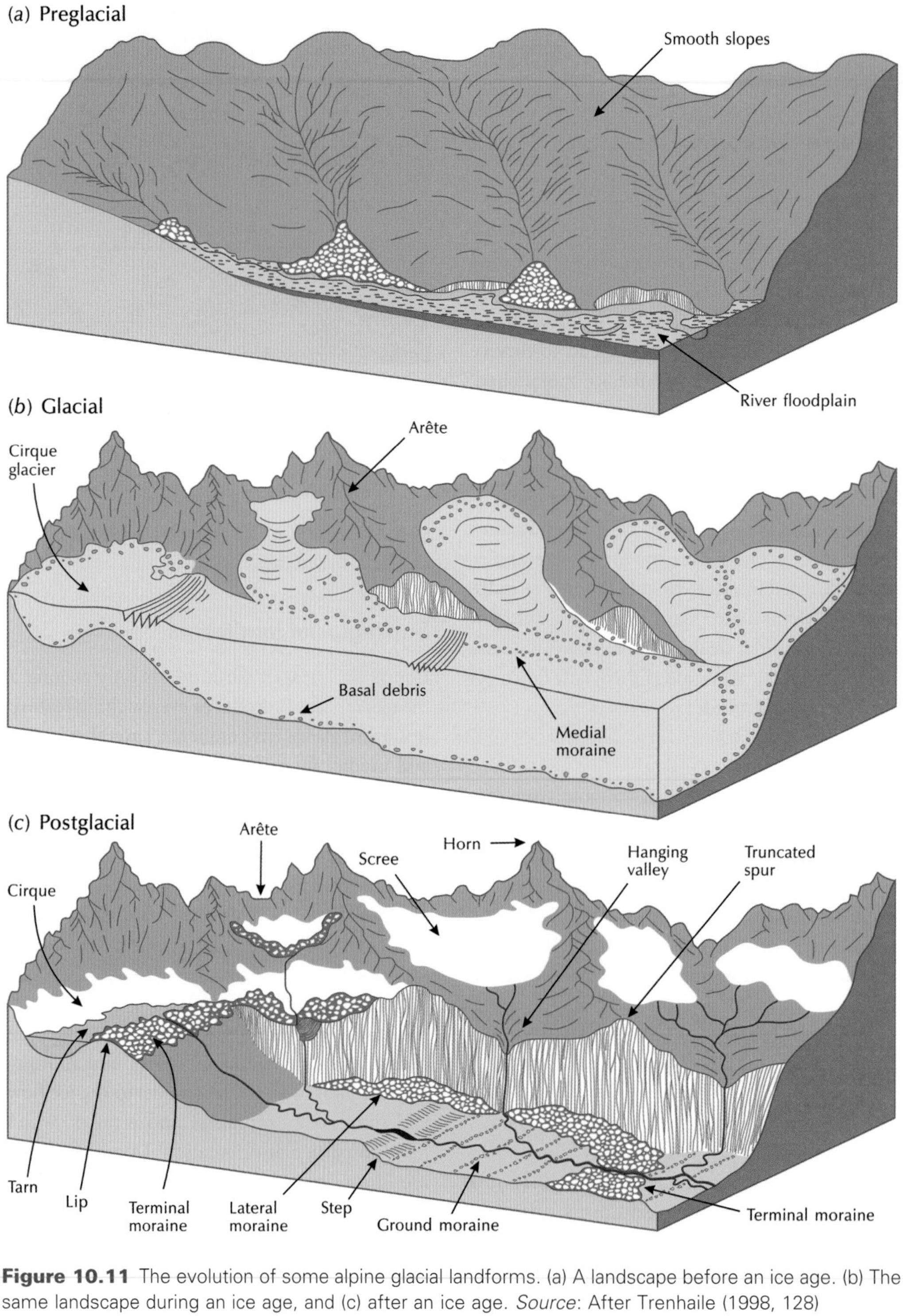

Figure 10.11 The evolution of some alpine glacial landforms. (a) A landscape before an ice age. (b) The same landscape during an ice age, and (c) after an ice age. *Source*: After Trenhaile (1998, 128)

Table 10.4 Landforms created by glacial deposition

Orientation with ice flow	Landform	Description
Supraglacial (still accumulating)		
Parallel	Lateral moraine	A moraine, often with an ice core, formed along the side of a valley glacier
	Medial moraine	A moraine formed by the coalescence of two lateral moraines at a spur between two valley glaciers
Transverse	Shear or thrust moraine	Ridges of debris from the base of a glacier brought to the surface by longitudinal compression
	Rockfall	Rockslides from the valley-side slopes deposit lobes of angular debris across a glacier
Non-orientated	Dirt cone	Cones of debris derived from pools in supraglacial streams
	Erratic	A large, isolated angular block of rock carried by a glacier and deposited far from its source
	Crevasse fill	Debris washed into an originally clean crevasse by surface meltwater streams
Supraglacial during deposition		
Parallel	Lateral moraine	A moraine, often with an ice core, formed along the side of a valley glacier (in part subglacial)
	Moraine dump	A blanket of debris near the glacier snout where several medial moraines merge
Non-orientated	Hummocky (or dead ice/ disintegration) moraine	A seemingly random assemblage of hummocks, knobs, and ridges (composed of till and ill-sorted clastic sediments) that contains kettles, depressions, and basins
	Erratic	A large rock fragment (clast) transported by ice action and of different composition from the local rocks
Subglacial during deposition		
Parallel	Drumlin	An elongated hill with an oval, egg-shaped, or cigar-shaped outline
	Drumlinoid ridge (drumlinized ground moraine)	Elongated, cigar-shaped ridges, and spindle forms. Formed under ice in conditions unsuited to individual drumlin formation
	Fluted moraine (flute)	Large furrows, up to about 2 m in wavelength, resembling a ploughed field. Found on fresh lodgement till (till laid in ground moraine under the ice) surfaces and, occasionally, glaciofluvial sand and gravel
	Crag-and-tail ridge	A tail of glacial sediments in the lee of a rock obstruction
Transverse	De Geer (washboard) moraine	A series of small, roughly parallel ridges of till lying across the direction of ice advance. Often associated with lakes or former lakes
	Rogen (ribbed, cross-valley) moraine	A crescentic landform composed chiefly of till, orientated with its long axis normal to ice flow and its horns pointing in the down-ice direction
Non-orientated	Ground moraine	A blanket of mixed glacial sediments (primarily tills and other diamictons), characteristically of low relief
	Till plain	Almost flat, slightly rolling, gently sloping plains comprising a thick blanket of till

continued . . .

Table 10.4 *continued*

Orientation with ice flow	Landform	Description
	Gentle hill	A mound of till resting on an isolated block of bedrock
	Hummocky ground moraine	(See hummocky moraine above)
	Cover moraine	A thin and patchy layer of till that reveals the bedrock topography in part (a blanket) or in full (a veneer)
Ice marginal during deposition		
Transverse	End moraines	Any moraine formed at a glacier snout or margin
	Terminal moraine	An arcuate end moraine forming around the lobe of a glacier at its peak extent
	Recessional moraine	An end moraine marking a time of temporary halt to glacial retreat and not currently abutting a glacier
	Push moraine	An end moraine formed by sediment being bulldozed by a glacier snout. Some push moraines show annual cycles of formation and comprise a set of small, closely spaced ridges
Non-orientated	Hummocky moraine	(See hummocky moraine above)
	Rockfall, slump, debris flow	Discrete landforms produced by each type of mass movement

Source: Mainly adapted from Hambrey (1994)

But even in Britain, where glaciers disappeared 10,000 years ago, traces of lateral moraines are still visible as small steps on mountainsides (Plate 10.13). Moraine dumps rarely survive glacial recession.

Hummocky moraines, also called **dead-ice moraines** or **disintegration moraines**, are seemingly random assemblages of hummocks, knobs, and ridges of till and other poorly sorted clastic sediments, dotted with kettles, depressions, and basins frequently containing lacustrine sediment. Most researchers regard the majority of hummocky moraines as the product of supraglacial deposition, although some landforms suggest subglacial origins. Far-travelled **erratics** are useful in tracing ice movements.

Subglacial landforms

A wealth of landforms form beneath a glacier. It is convenient to class them according to their orientation with respect to the direction of ice movement (parallel, transverse, and non-orientated). Forms lying parallel to ice flow are drumlins, drumlinized ridges, flutes, and crag-and-tail ridges. **Drumlins** are elongated hills, some 2–50 m high and 10–20,000 m long, with an oval, an egg-shaped, or a cigar-shaped outline. They are composed of sediment, sometimes with a rock core (Plate 10.14), and usually occur as **drumlin fields**, giving rise to the so-called 'basket of eggs' topography because of their likeness to birds' eggs. They are perhaps the most characteristic features of landscapes created by glacial deposition. The origin of drumlins is debatable, and at least four hypotheses exist (Menzies 1989). First, they may be material previously deposited beneath a glacier that subglacial meltwater moulds. Second, they may be the result of textural differences in subglacial debris. Third, they may result from active basal meltwater carving cavities beneath an ice mass and afterwards filling in space with a range of stratified sediments. Catastrophic meltwater floods underneath Pleistocene ice sheets may have fashioned some large drumlin fields, the form of which is redolent of bedforms created by turbulent airflow and turbulent water flow (Shaw *et al.* 1989; Shaw 1994). A fourth and currently popular hypothesis is the subglacial deformation of till (for a review, see Bennett and Glasser 2009, 268–83). It may be that several different sets of processes can create drumlins, and if this should be so, it would provide an excellent example of equifinality.

De Geer and Rogen moraines lie transversely to the direction of ice flow. **De Geer moraines** or **washboard moraines** are a series of small and roughly parallel ridges of till that are ordinarily associated with lakes or former lakes. **Rogen moraines**, also called **ribbed moraines** and **cross-valley moraines**, are crescent-shaped landforms composed largely of till that are formed by subglacial thrusting. They grade into drumlins.

Various types of **ground moraine** display no particular orientation with respect to ice flow. A ground moraine is a blanket of mixed glacial sediments – mainly tills and other diamictons – formed beneath a glacier. Typically, ground moraines have low relief. Four kinds of ground moraine are recognized: till plain, gentle hill, hummocky ground moraine, and cover moraine. Till plains (or till sheets) are the thickest type and cover moraine the thinnest. The most representative, and by far the most common, form of deposit in lowland areas is a till sheet or till plain, usually gently undulating and sometimes with drumlins. A review of **subglacial tills** argues that they form through a range of processes – deformation, flow, sliding, lodgement, and ploughing – that act to mobilize and carry sediment and lay it down in a great variety of forms, ranging from glaciotectonically folded and faulted stratified material to texturally uniform diamicton (Evans *et al.* 2006). Moreover, owing to the fact that glacier beds are mosaics of deformation and sliding and warm- and cold-based conditions, most subglacial tills are likely to be hybrids created by a range of processes active in the subglacial traction zone. Nonetheless, glacial geologists can identify three distinct till types (Evans *et al.* 2006):

Plate 10.13 Line of angular boulders marking remnants of a lateral moraine in Coire Riabhach, Isle of Skye, Scotland. Cuillin ridge – an arête – may be seen in the background. The peak in the right background is Sgurr nan Gillean – a horn. (*Photograph by Mike Hambrey*)

Plate 10.14 Rock-cored drumlin, New Zealand. (*Photograph by Neil Glasser*)

1 **Glaciotectonite** – rock or sediment deformed by subglacial shearing or deformation (or both) and retaining some structural characteristics of the parent material.
2 **Subglacial traction till** – sediment released directly from the ice by pressure melting or

liberated from the substrate (or both) and then disaggregated and completely or largely homogenized by shearing sediment that is laid down by a glacier sole while sliding over or deforming its bed (or both).

3 **Melt-out till** – sediment released by the melting of stagnant or slowly moving debris-rich glacier ice, and directly deposited without later transport or deformation.

Plate 10.15 A pair of lateral moraines from a valley glacier in the Cordillera Blanca, Peru. Former ice flow is towards the viewer. (*Photograph by Neil Glasser*)

Plate 10.16 Nepal Himalaya, Langtang Himal, Kyimoshung Tsang glacier, showing ice retreat over the last 400 years, leaving bare stony ground inside terminal moraine left in Little Ice Age. (*Photograph by Tony Waltham Geophotos*)

Ice-margin landforms

Landforms produced at the ice margin include different types of end moraine, all of which form around a glacier snout. A lateral moraine lies at the sides of a glacier (Plate 10.15). A **terminal moraine** is an arcuate end moraine that forms around the lobe of a glacier at its farthest limit (Plate 10.16; see also Figure 10.4). A **recessional moraine** marks a time of temporary halt to glacial retreat and is not currently touching a glacier. A **push moraine** is formed by sediment being bulldozed by a glacier snout, especially a cold glacier. Some push moraines show annual cycles of formation and comprise a set of small, closely spaced ridges.

Other ice-marginal landforms, which have no preferred orientation with respect to ice flow, are **hummocky moraine** and various forms resulting from mass movements (rockfalls, slumps, and debris flows). A hummocky moraine formed near the ice margin is similar to a hummocky moraine produced elsewhere, but it includes irregular heaps of debris that fall from an ice mass in the ice-marginal zone and debris from dead ice that becomes detached from the main ice mass.

GLACIOFLUVIAL LANDFORMS

Meltwater shifts huge quantities of sediment. Indeed, more sediment may leave a glacial system in meltwater than in ice. Sediment-charged meltwater under a glacier is a potent erosive agent, especially towards the glacier snout. After leaving a glacier, meltwater may erode sediments, as well as laying down debris to create ice-marginal and proglacial depositional landforms (Table 10.5).

Subglacial landforms

Channels

Some glacial landscapes contain a range of channels cut into bedrock and soft sediments. The largest of these are **tunnel valleys**, such as those in East Anglia, England, which are eroded into chalk and associated bedrock. They can be 2–4 km wide, over 100 m deep, and 30–100 km long, and sediments – usually some combination of silt, clay, gravel, and peat – often fill them to varying depths. As to their formation, three mechanisms may explain these tunnels (Ó'Cofaigh 1996): (1) the creep of deformable subglacial sediment into a subglacial conduit, and the subsequent removal of this material by meltwater; (2) subglacial meltwater erosion during deglaciation; and (3) erosion by the catastrophic release of subglacial meltwater. Where the meltwater is under pressure, the water may be forced uphill to give a reversed gradient, as in the **Rinnen** of Denmark. Subglacial gorges, which are often several metres wide compared with tens of metres deep, are carved out of solid bedrock.

Table 10.5 Glaciofluvial landforms

Formative process	Landform	Description
Subglacial		
Erosion by subglacial water	Tunnel valley (*Rinnen*)	A large, subglacial meltwater channel eroded into soft sediment or bedrock
	Subglacial gorge	Deep channel eroded in bedrock
	Nye (bedrock) channel	Meltwater channel cut into bedrock under high pressure
	Channel in loose sediment	Meltwater channel eroded in unconsolidated or other types of glacial deposit
	Glacial meltwater chute	Channel running down a steep rock slope marginal to a glacier
	Glacial meltwater pothole	Circular cavity bored into bedrock by meltwater
	Sichelwannen ('sickle-shape troughs')	Crescentic depressions and scallop-like features on bedrock surfaces caused largely by meltwater, with cavitation being a key process
Deposition in sub-glacial channels, etc	Esker	Lengthy, winding ridge or series of mounds, composed mainly of stratified or semi-stratified sand and gravel etc.
	Nye channel fill	Debris plugging a Nye channel
	Moulin kame	Mound of debris accumulated at the bottom of a moulin
Ice marginal (ice contact)		
Ice-marginal stream erosion	Meltwater (or hillside) channel	Meltwater channel tending to run along the side of a cold glacier
	Overflow channel	Meltwater channel cut by marginal stream overtopping low cols at or below the ice-surface level
Ice-contact deposition from meltwater or in lakes or both	Kame	Flat-topped deposit of stratified debris
	Kame field	Large area covered with many individual kames
	Kame plateau	Broad area of ice-contact sediments deposited next to a glacier but not yet dissected
	Kame terrace	Kame deposited by a stream flowing between the flank of a glacier and the valley wall, left stranded on the hillside after the ice goes
	Kame delta (delta moraine)	Flat-topped, fan-shaped mound formed by meltwater coming from a glacier snout or flank and discharging into a lake or the sea

continued . . .

Table 10.5 *continued*

Formative process	Landform	Description
	Crevasse fill	Stratified debris carried into crevasses by supraglacial meltwater
Proglacial		
Meltwater erosion	Scabland topography, coulee, spillway	Meltwater features in front of a glacier snout. Water collected in ice-marginal or proglacial lakes may overflow through spillways
Meltwater deposition	Outwash plain or sandur (plural sandar)	Plain formed of material derived wholly or partially from glacial debris transported or reworked by meltwater and other streams. Most sandar are composed wholly of outwash, but some contain inwash as well
	Valley train	Collection of coarse river-sediment and braided rivers occupying the full width of a valley with mountains rising steep at either side
	Braided outwash fan	Debris fan formed where rivers, constrained by valleys, disembogue onto lowlands beyond a mountain range
	Kettle (kettle hole, pond)	Bowl-shaped depression in glacial sediment left when a detached or buried block of ice melts. Often contains a pond
	Pitted plain	Outwash plain pitted with numerous kettle holes

Source: Adapted from Hambrey (1994)

Eskers

Eskers are the chief landform created by subglacial meltwater and form by the infilling of subglacial or englacial channels or by sedimentation in supraglacial channels (Figure 10.12; Plate 10.17). Minor forms include sediment-filled Nye channels and moulin kames, which are somewhat fleeting piles of debris at the bottom of a **moulin** (a pothole in a glacier that may extend from the surface to the glacier bed). Esker is an Irish word and is now applied to long and winding ridges formed mostly of sand and gravel and laid down in a meltwater tunnel underneath a glacier. Some eskers form at ice margins, and are not to be confused with kames and kame terraces (see below), which are ice-contact deposits at the ice margin. In the past, confusion has beset the use of these terms, but the terminology was clarified in the 1970s (see Price 1973 and Embleton and King 1975a). Eskers can run uphill; sometimes they split, sometimes they are beaded. They may run for a few hundred kilometres and be 700 m wide and 50 m high, although they are typically an order of magnitude smaller.

Ice-margin landforms

Meltwater and overflow channels

Erosion by meltwater coursing alongside ice margins produces meltwater channels and overflow channels. **Meltwater channels** tend to run along the side of glaciers, particularly cold glaciers. They may be in contact with the ice or they may lie between an ice-cored lateral moraine and the valley side. After the ice has retreated, they can often be traced across a hillside.

Overflow channels are cut by streams at the ice margin overtopping low cols lying at or below the same level as the ice. Lakes may form before the overflow occurs. Until the mechanisms of subglacial drainage were understood, channels found in formerly glaciated temperate regions were ascribed to meltwater overflow, but many of

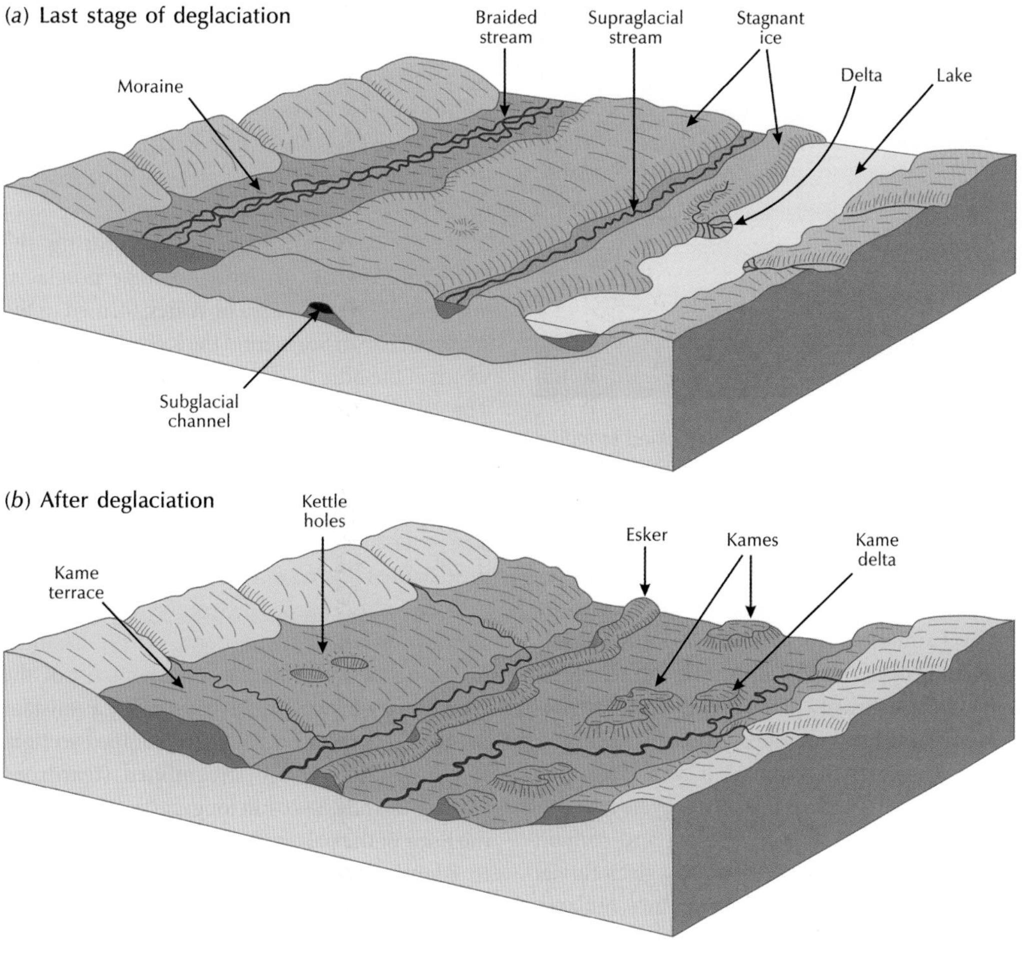

Figure 10.12 Subglacial and ice-margin landforms. (a) A landscape at the final stage of deglaciation. (b) A landscape after deglaciation. Please note that, although the diagram may imply that eskers, kames, kame terraces, and so forth form under conditions of stagnant ice, these features commonly form in association with active glaciers. *Source*: Adapted from Flint (1971, 209)

these channels are now known to have been wrought by subglacial erosion.

Kames

The main depositional landforms associated with ice margins are **kames** of various kinds (Figure 10.12). Crevasse-fillings, which comprise stratified debris that entered crevasses through supraglacial streams, are minor landforms. Kames commonly occur with eskers. They are flat-topped and appear as isolated hummocks, as broader plateau areas, or, usually in proglacial settings, as broken terraces. Individual kames range from a few hundred metres to over a kilometre long, and a few tens of metres to over a hundred metres wide. They have no preferred orientation with respect to the direction of ice flow. If many individual kames cover a large area, the term 'kame field' is at times applied.

Kame terraces develop parallel to the ice-flow direction from streams flowing along the sides of a stable or slowly receding ice margin. They

Plate 10.17 Esker made up of slightly deformed stratified sands and gravels near the ice margin of Comfortlessbreen, Svalbard, Norway. (*Photograph by Mike Hambrey*)

consist of similar material to kames and they slope down-valley in accordance with the former ice level and often slope up the adjacent hillside.

Kame deltas or **delta moraines** are related to kames but are usually much bigger. They are flat-topped, fan-shaped mounds formed by meltwater coming from a glacier snout or flank and running into a proglacial lake or the sea. They lie at right-angles to the direction of ice flow and contain debris from the ice itself, as well as glaciofluvial debris. The three Salpausselkä moraines, Finland, are probably the biggest delta-moraine complexes in the world. They are associated with a lake impounded by the Fennoscandian ice sheet, which covered the southern Baltic Sea region.

Proglacial landforms

Scablands and spillways

Meltwater streams issuing from a glacier are usually charged with sediment and fast-flowing. They deposit the sediment in front of a glacier, and streams become clogged, leading to braiding. Lakes are common in this proglacial environment, and tend to fill and overflow through **spillways** during the summer. The impounding sediments are often soft and, once breached, are cut through quickly, lowering the lake level. Although uncommon today, large proglacial lakes were plentiful near the southern limits of the Pleistocene ice sheets and many abandoned spillways are known (Figure 10.13). Where huge glacial lakes broke through their containing dams, the rush of water produced scablands (p. 271).

Recent research suggests that during the Elsterian–Anglian glaciation, a huge proglacial lake, about twice the area of Wales, formed in the southern North Sea behind the Kent–Calais ridge, which at the time connected England with France. The lake breached the ridge, probably around 350,000–450,000 years ago, forming the Straits of Dover and sending a huge megaflood water down the English Channel (Gupta *et al.* 2007). The overflow water appears to have created the Channel River, seen today as a system of submerged palaeochannels. Geomorphological evidence of a catastrophic origin for the Channel River includes the 'box shape' character of the main palaeochannel, which has roughly a constant depth and width over a 300 km mapped section. It also includes large streamlined (teardrop-shaped) islands, 1–20 km long, carved in chalk on the floor of the palaeochannels, which are similar to islands found in the Channeled Scablands (Collier *et al.* 2015).

Jökulhlaups are outbursts of meltwater stored beneath a glacier or ice sheet as a subglacial lake. The best-known jökulhlaups occurred in the last century, with major ones in 1918 (Katla) and 1996 (Skeiðarásandur). Skeiðarásandur jökulhlaup resulted from the rapid melting of some 3.8 km^3 of ice after a volcanic eruption on 30 September 1996 underneath the Vatnajökull ice cap, Iceland (Gudmundsson *et al.* 1997). The ensuing flood involved a discharge of about 20,000 m^3/s, running at its peak at around 6 m/s and capable of transporting ice blocks at least 25 m large (van Loon 2004). It destroyed part of the main road along the southern coast of Iceland, including a bridge over the Skeiðarásandur (Plate 10.18). Catastrophic though the Skeiðarásandur jökulhlaup was, it was tame in comparison with

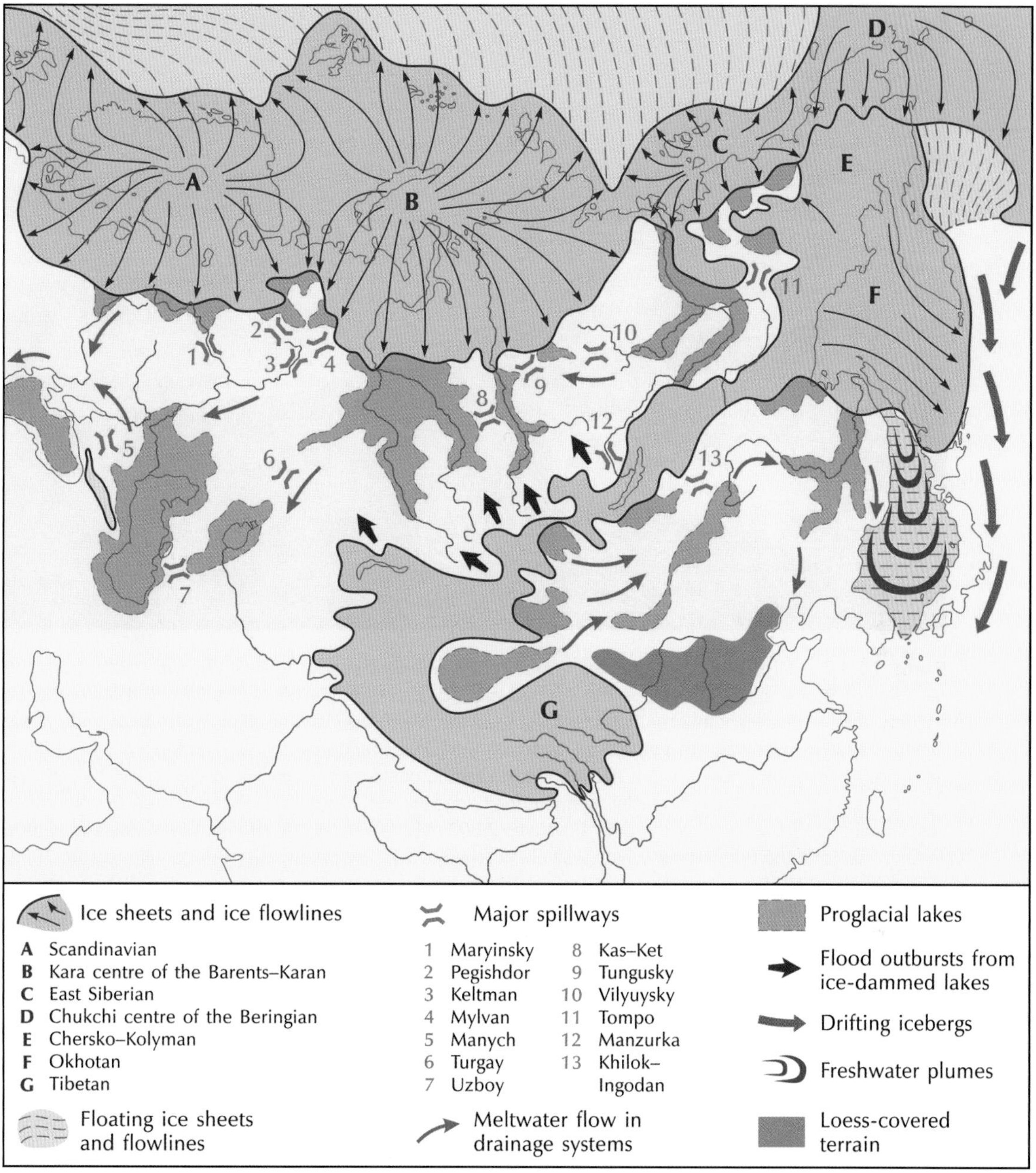

Figure 10.13 Glacial spillways in northern Eurasia. For a more recent and better-dated reconstruction of late-Quaternary ice-sheet history in northern Eurasia, see Svendsen *et al.* (2004). *Source*: Adapted from Grosswald (1998)

Plate 10.18 A large jökulhlaup spreading out across Skeiðarársandur (glacial outwash plain) from the terminus of Skeiðarárjökull (visible in the left background), an outlet glacier of Vatnajökull. It started as an accumulation of glacial meltwater in the Grímsvötn caldera from the Gjálp eruption in late September/early October 1996 in northwestern Vatnajökull, although the jökulhlaup did not occur until several weeks later, on 5 November 1996. Notice the breaches in the road across Skeiðarárjökull, and especially the removal of the bridge over the Gígjukvísl glacier river, but the bridge in the foreground over the glacier river Núpsvötn is still intact, probably because the peak flow was 2,300 m^3/s there, 33,000 m^3/s at Gígjukvísl, and 22,000 m^3/s at Skeiðará, the bridge over which was also carried away. The photo was taken on November 1996. (*Photograph by US Geological Survey*)

the 1918 Katla jökulhlaup, which involved a flood of about 300,000 m^3/s of water that carried 25,000 tons of ice and an equal amount of sediment every second (Tómasson 1996). Similar jökulhlaups are recorded since the Last Glacial Maximum (Box 10.5).

Outwash plains, valley trains, and braided outwash fans

Much of the vast quantity of sediment normally carried by meltwaters is laid down in the proglacial environment. Where glaciers end on land, systems of braided rivers, called **outwash plains** or **sandar** (singular **sandur**) develop (Plate 10.19; see also Figure 10.4). In south-eastern Iceland, outwash plains may be as wide as they are long and full of active braids. When jökulhlaups occur, the entire plain may be flooded. In mountainous terrain, braided river systems may extend across the full width of the valley floor, with mountains rising steeply from either edge. Such elongated and flat systems are called **valley trains**. Good examples come from the Southern Alps, New Zealand. **Braided outwash fans** occur where river systems hemmed in by valleys discharge on to lowlands beyond a mountain range. Many examples are found north of the European Alps.

Box 10.5 MULTIPLE JÖKULHLAUPS IN ICELAND

The Jökulsá á Fjöllum is one of Iceland's largest rivers, draining much of the 8,100 km^2 Vatnajökull ice cap which sits in the south of the island, and flowing 206 km north across central Iceland to the Arctic Ocean (Figure 10.14). Since the Last Glacial Maximum, the Jökulsá á Fjöllum has

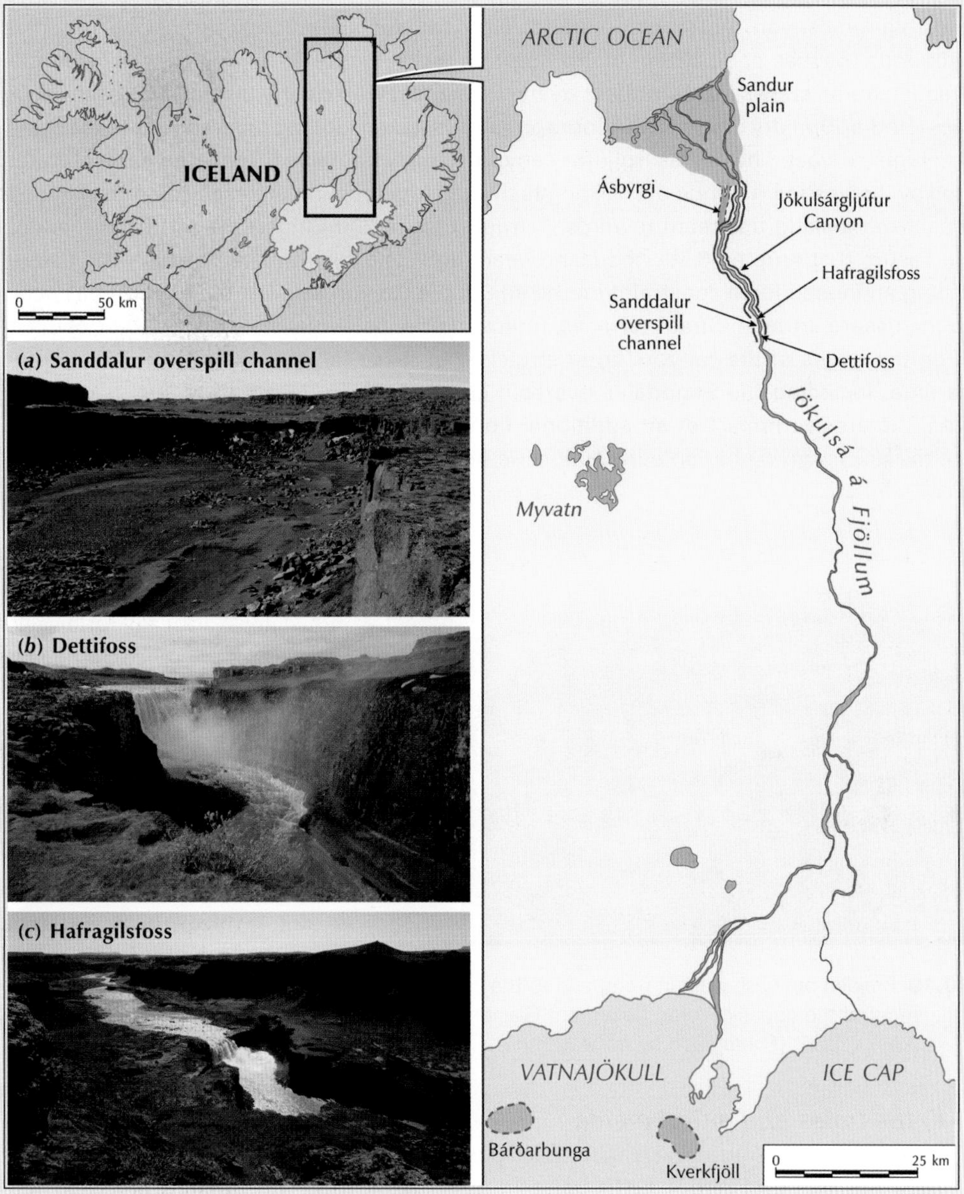

Figure 10.14 Evidence of jökulhlaups in Iceland. (*Photographs by Edwin Baynes*)

continued . . .

experienced multiple jökulhlaups of varying magnitude, with a peak discharge for the largest flood estimated at 900,000 m^3/s (Carrivick *et al.* 2013). Jökulhlaups occur along the Jökulsá á Fjöllum as a result of either subglacial volcanism beneath Vatnajökull from one or more of the Kverkfjöll, Grimsvötn, or Bárðarbunga volcanic centres), or from the release of floodwater from an ice-dammed lake to the south of Kverkfjöll.

Evidence for the impact of these jökulhlaups exists in the upper reach of the 28-km long Jökulsárgljúfur canyon, north-east Iceland, with canyon apex containing signs of Holocene palaeofloods (Baynes *et al.* 2015a, b). Photograph (a) on Figure 10.14, which was taken looking upstream (south), shows the Sanddalur overspill channel with clearly displayed fluvially sculpted surfaces and a 20-m dry cataract. Photograph (b) was, taken looking upstream towards Dettifoss, a 54-m high cataract. The Jökulsárgljúfur canyon actually contains three large vertical waterfalls formed by the upstream retreat of knickpoints through the toppling of basalt columns. Photograph (c) was taken looking upstream towards Hafragilsfoss (a 20-m high cataract) from the location of the fissure that erupted 8,500 thousand years ago. The canyon cuts through the fissure, so providing an independent constraint on the age of the canyon as all of the erosion has occurred since the fissure erupted. Strath terraces, indicating the palaeolocation of the river bed, can be seen on the edges of the canyon. Long stretches of these terraces have been exposed at the same time, including the Sanddalur overspill channel. Downstream from the Jökulsárgljúfur canyon, the erosive impact of an additional flood in the early Holocene is evident at Ásbyrgi, where the discharge during an early Holocene flood event was an estimated 39,000 m^3/s.

Plate 10.19 Braided outwash plain in front, and to the side, of the snout of the debris-covered Casement Glacier, Glacier Bay, Alaska, USA. (*Photograph by Mike Hambrey*)

Kettle holes and pitted plains

Many braided-river plains carry water-filled pits. These pits are called **kettles**, **kettle holes**, or **ice pits**. They form as a block of 'dead' ice decays and is buried. The ice block may be an ice remnant left stranded when the glacier retreated or a lump of ice washed down a stream during a flood. The water-filled kettles are called **kettle lakes** (Plate 10.20). An outwash plain pocked with many kettle holes is called a **pitted plain**.

PARAGLACIAL LANDFORMS

Paraglacial processes occur after a glacier retreats, exposing a landscape susceptible of rapid change. They do not involve glacial ice; rather they modify landforms conditioned by glaciation and deglaciation to fashion **paraglacial landforms** (see Ballantyne 2002; Slaymaker 2011). Once ice disappears, several changes occur in former glacial landscapes. First, rock slopes steepened by valley glaciers become unstable and vulnerable to slope failure and rockfall, although the timing of the rock-slope failure is crucial to pinpointing the causes. A rapid response after deglaciation would suggest the failure occurs because the ice no longer acts as a buttress; a delayed response would suggest a different cause. A study of five rock-slope failures in quartzite on the Isle of Jura, Scotland, dated using beryllium-10, showed a delay of 720–2,240 years after deglaciation, which tends to refute the debutressing hypothesis (Ballantyne *et al.* 2014). In fact, timing of the Jura rock-slope failures coincides with the period of maximum glacio-isostatic crustal uplift, suggesting that they were triggered by uplift-driven seismic events acting on rock masses that could have been fractured earlier when pressure was released as the ice melted.

Second, slopes bearing a mantle of drift but no vegetation become subject to rapid reworking by debris flows, snow avalanches, and slope wash. Third, glacier forelands become exposed to wind erosion and frost action. Fourth, rivers pick up and redistribute large amounts of unconsolidated sediment of glacial origin, later depositing it in a range of terrestrial, lacustrine, and marine environments. Fifth, wind entrains finer sediments, particularly silts, and may bear them thousands of kilometres and deposit them as loess deposits (p. 352). This accelerated paraglacial geomorphic activity follows deglaciation and lasts up to 10,000 years, until the landscape adjusts to non-glacial conditions.

June M. Ryder (1971a, b) coined the term 'paraglacial' to describe alluvial fans in British Columbia, Canada, formed through the reworking of glacial sediment by rivers and debris flows after the Late Pleistocene deglaciation. Michael Church and Ryder (1972, 3059) then formalized the idea by defining 'paraglacial' as 'nonglacial processes that are directly conditioned by glaciation', which includes proglacial processes and processes occurring 'around and within the margins of a former glacier that are the direct result of the former presence of ice'. Moreover, they recognized a '**paraglacial period**' – the time during which paraglacial processes operate. Later, they extended the notion to include all periods of glacier retreat, and not just the Late Pleistocene deglaciation (Church and Ryder 1989).

Plate 10.20 Small kettle-hole lake in end-moraine complex of Saskatchewan Glacier (seen in background) in the Canadian Rockies. (*Photograph by Mike Hambrey*)

Colin K. Ballantyne (2002) recognized six paraglacial 'land systems' – rock slopes, drift-mantled slopes, glacier forelands, and alluvial, lacustrine, and coastal systems – each containing a variety of paraglacial landforms and sediment facies (Figure 10.15). Taken together, he regarded these landforms and sediments – talus accumulations, debris cones, alluvial fans, valley fills, deltas, coastal barrier structures, and so forth – as storage components within an interrupted sediment cascade. The cascade has four primary sources of material – rockwalls, drift-mantled slopes, valley-floor glaciogenic deposits, and coastal glaciogenic deposits. And it has four terminal sediment sinks – alluvial valley-fill deposits, lacustrine deposits, coastal and nearshore deposits, and shelf and offshore deposits.

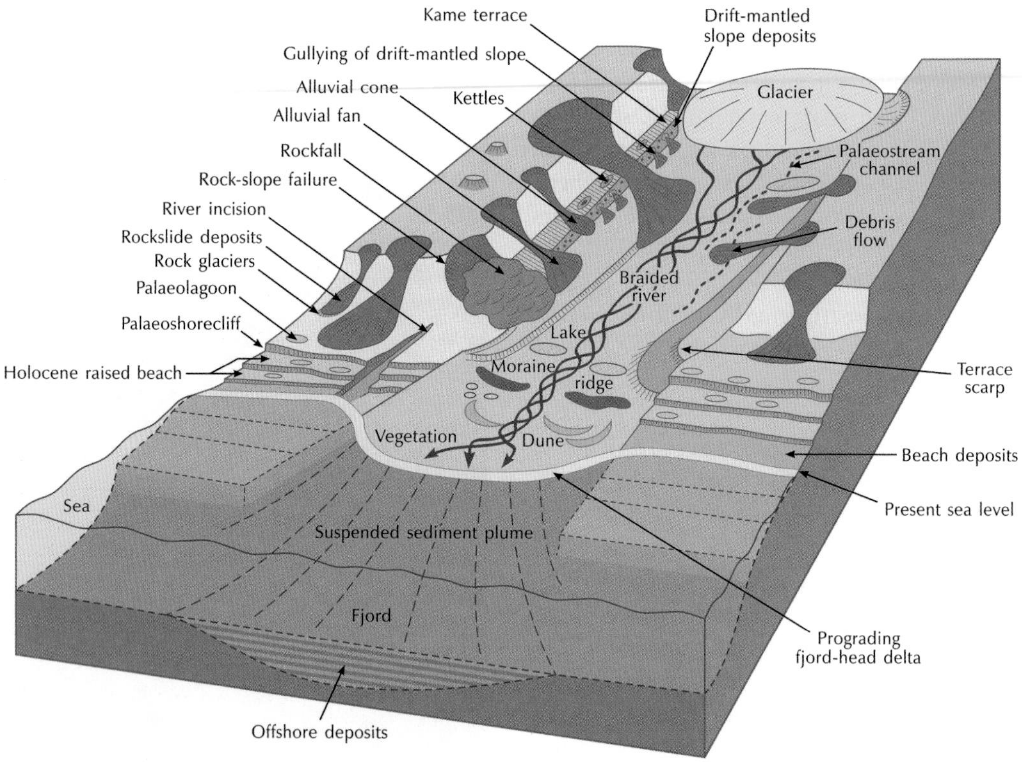

Figure 10.15 Paraglacial landform systems. *Source*: After Mercier (2008)

HUMANS AND GLACIAL ENVIRONMENTS

Glacial landscapes are productions of frigid climates. During the Quaternary, the covering of ice in polar regions and on mountain tops waxed and waned in synchrony with swings of climate through glacial–interglacial cycles. Humans can live in glacial and periglacial environments but only at low densities. Direct human impacts on current glacial landscapes are small, even in areas where tourism is popular. Indirect human impacts, which work through the medium of climatic change, are substantial: global warming appears to be melting the world's ice and snow. Over the last 100 years, mean global temperatures have risen by about 0.6°C, about half the rise occurring in the last 25 years. The rise is higher in high latitudes. For example, mean winter temperatures at sites in Alaska and northern Eurasia have risen by 6°C over the last 30 years (Serreze *et al.* 2000), which is why glacial environments are so vulnerable to the current warming trend.

Relict glacial landscapes, left after the last deglaciation some 10,000 years ago, are home to millions of people in Eurasia and North America. The relict landforms are ploughed up to produce crops, dug into for sand and gravel, and covered by concrete and tarmac. Such use of relict landscape raises issues of landscape conservation. The other side of the coin is that knowledge of Quaternary sediments and their properties can aid human use of relict glacial landscapes (Box 10.6).

Another aspect of human impact on glacial landscapes is the issue of **global warming**. Warmer temperatures alter glacier mass balances, with more melting occurring. The melting causes

Box 10.6 WASTE DISPOSAL SITES IN NORFOLK, ENGLAND

An understanding of the Quaternary sediments aids the designing of waste disposal sites in south Norfolk, England (Gray 1993). Geologically, south Norfolk is a till plain that is dissected in places by shallow river valleys. It contains very few disused gravel pits and quarries that could be used as landfill sites for municipal waste. In May 1991, Norfolk County Council applied for planning permission to create an aboveground or 'landraise' waste disposal site of 1.5 million cubic metres at a disused US Second World War airfield at Hardwick. The proposal was to dig a 2–4-m-deep pit in the Lowestoft Till and overfill it to make a low hill standing up to 10 m above the plain. The problem of leachate leakage from the site, which might contaminate groundwater and rivers, was to be addressed by relying on the low permeability of the till and reworking the till around the edges of the site to remove potentially leaky sand-lenses in its upper layers. In August 1993, after a public inquiry into the Hardwick site, planning permission was refused, partly because knowledge of the site's geology and land drainage was inadequate and alternative sites were available. Research into the site prompted by the proposal suggested that leachate containment was a real problem and that Norfolk County Council was mistaken in believing that the till would prevent leachates from leaking. It also identified other sites in south Norfolk that would be suitable landfill sites, including the extensive sand and gravel deposits along the margins of the River Yare and its tributaries. Landraising in a till plain is also unwelcome on geomorphological grounds, unless perhaps the resulting hill should be screened by woodland. A lesson from this case study is that knowledge of Quaternary geology is central to the planning and design of landfill in areas of glacial sediments.

the glaciers to shrink and to thin, their snouts withdrawing. More glaciers have retreated than have advanced since around 1850, the end of the Little Ice Age (Zemp *et al.* 2008). Over recent decades, the melting trend has increased, which many researchers attribute to human-induced climate warming. Should the predictions of 1.4° to 5.8°C temperature rises during the present century prove accurate, then melting will proceed apace. Already, the Alps have lost about half of their glacial terrain since the 1850s. Not all glaciers are in retreat and the mass balance patterns are varied. For example, some glaciers in maritime climates – Patagonia, Iceland, southeast Alaska, as well as coastal parts of Norway and New Zealand – show high mass turnovers, low equilibrium lines, and firn and ice at melting temperatures. On the other hand, some glaciers in dry-continental climates – northern Alaska, Arctic Canada, sub-Arctic Russia, parts of the Andes near the Atacama Desert, and in many central Asian mountain chains – show low mass turnover, equilibrium lines at high altitudes, and firn and ice well below melting temperatures (Zemp *et al.* 2009).

SUMMARY

Ice covers about 10 per cent of the land surface, although 20,000 years ago it covered 32 per cent. Most of the ice is in polar regions. Glaciers come in a variety of forms and sizes: ice sheets, ice caps, ice shelves, ice shields, cirque glaciers, valley glaciers, and other small glaciers. Glaciers have an accumulation zone, where ice is produced, and an ablation zone, where ice is destroyed. Ice abrades and fractures rock, picks up and carries large and small rock fragments, and deposits entrained material. Glaciers carry rock debris at the glacier base (subglacial debris), in the ice (englacial debris), and on the glacier surface (supraglacial debris). They also deposit sediment under, on, and by the side of the moving ice. Meltwater issuing from glacier snouts lays down proglacial sediments. Erosion by ice creates a wealth of landforms by abrasion, by fracture, by crushing, and by eroding a mountain mass. Examples include glacially scoured regions, glacial troughs, striated bedrock, trough heads, cirques, flyggbergs, crescentic gouges, horns, and nunataks. Debris laid down by ice produces an equal variety of landforms. Supraglacial deposits form lateral moraines, medial moraines, dirt cones, erratics, and many more features. Subglacial forms include drumlins and crags-and-tails. Terminal moraines, push moraines, hummocky moraines, and other forms occur at ice margins. Meltwater, which issues from glaciers in copious amounts during the spring, cuts valleys and deposits eskers beneath the ice, produces meltwater channels and kames at the edge of the ice, and fashions a variety of landforms ahead of the ice, including spectacular scablands and spillways, outwash plains, and, on a much smaller scale, kettle holes. A variety of paraglacial landforms develop immediately glaciers melt. Humans interact with glacial landscapes. Their current industrial and domestic activities may, through global warming, shrink the cryosphere and destroy Quaternary landforms. Conversely, knowledge of Quaternary sediments is indispensable in the judicious use of glacially derived resources (such as sands and gravels) and in the siting of such features as landfill sites.

ESSAY QUESTIONS

1 Examine the geomorphological evidence for Milankovitch climate cycles.

2 How does ice flow?

3 How does ice fashion landforms?

4 Appraise the evidence for catastrophic glaciofluvial events.

FURTHER READING

Ballantyne, C. K. (2002) Paraglacial geomorphology. *Quaternary Science Reviews* 21, 1935–2017.
A superb and well-illustrated review of paraglacial geomorphology.

Benn, D. I. and Evans, D. J. A. (2010) *Glaciers and Glaciation*. London and New York: Routledge.
An excellent text.

Bennett, M. R. and Glasser, N. F. (2009) *Glacial Geology: Ice Sheets and Landforms*, 2nd edn. Chichester: Wiley-Blackwell.
Another excellent text.

Hambrey, M. J. (1994) *Glacial Environments*. London: UCL Press.
Beautifully illustrated and readable treatise on glacial landforms and processes.

Hubbard, B. and Glasser, N. F. (2005) *Field Techniques in Glaciology and Glacial Geomorphology*. Chichester: John Wiley & Sons.
An excellent book for those interested in measuring glacial landforms and processes.

Knight, J. and Harrison, S. (eds) (2009) *Periglacial and Paraglacial Processes and Environments* (Geological Society special publications 320). London: Geological Society.
Includes up-to-date chapters on paraglacial landforms and processes.

Martini, I. P., Brookfield, M. E., and Sadura, S. (2001) *Principles of Glacial Geomorphology and Geology*. Upper Saddle River, N.J.: Prentice Hall.
Up-to-date, accessible, and non-mathematical treatment that provides a good bridge between fundamental studies and advanced reading.

Sugden, D. E. and John, B. S. (1976) *Glaciers and Landscape: A Geomorphological Approach*. London: Edward Arnold.
Well worth a look, even after forty years.

CHAPTER ELEVEN

PERIGLACIAL LANDSCAPES

11

Frozen ground without an icy cover bears an assortment of odd landforms. This chapter covers:

- Ice in frosty landscapes
- Frost, snow, water, and wind action
- Pingos, palsas, and other periglacial landforms
- Humans in periglacial environments
- Past periglaciation

A WINDOW ON THE PERIGLACIAL WORLD

In 1928, the airship *Graf Zeppelin* flew over the Arctic to reveal:

> the truly bizarre landscape of the polar world. In some areas there were flat plains stretching from horizon to horizon that were dotted with innumerable and inexplicable lakes. In other regions, linear gashes up to a mile or more long intersected to form giant polygonal networks. This bird's-eye view confirmed what were then only incidental surface impressions that unglaciated polar environments were very unusual.
>
> (Butzer 1976, 336)

PERIGLACIAL ENVIRONMENTS

The Polish geomorphologist Walery von Lozinzki first used the term 'periglacial' in 1909 to describe frost weathering conditions in the Carpathian Mountains of Central Europe. In 1910, the idea of a 'periglacial zone' was established at the Eleventh Geological Congress in Stockholm to describe climatic and geomorphic conditions in areas peripheral to Pleistocene ice sheets and glaciers. This periglacial zone covered tundra regions, extending as far south as the latitudinal tree-line. In modern usage, periglacial refers to a wider range of cold but non-glacial conditions, regardless of their proximity to a glacier. It includes regions at high latitudes and below the altitudinal and latitudinal tree-lines: polar deserts and semi-deserts, the High Arctic and ice-free areas of Antarctica, tundra zones, boreal forest zones, and high alpine periglacial zones, which extend in mid-latitudes and even low latitudes. The largest alpine periglacial zone is the Qinghai–Xizang (Tibet) Plateau of China. Periglacial environments characteristically experience intense frosts during winter months and snow-free ground during summer months. Four distinct climates produce such conditions – polar lowlands, subpolar lowlands, mid-latitude lowlands, and highlands (Washburn 1979, 7–8).

1 **Polar lowland climates** have a mean temperature of the coldest month less than 3°C. They are associated with zones occupied by ice caps, bare rock surfaces, and tundra vegetation.
2 **Subpolar lowland climates** also have a mean temperature of the coldest month less than 3°C, but the temperature of the warmest month exceeds 10°C. In the Northern Hemisphere, the 10°C isotherm for the warmest month sits roughly at the latitudinal tree-line, and subpolar lowland climates are associated with the northern boreal forests.
3 **Mid-latitude lowland climates** have a mean temperature of the coldest month less than 3°C, but the mean temperature is more than 10°C for at least four months of the year.
4 **Highland climates** are cold owing to high elevation. They vary considerably over short distances owing to aspect. Daily temperature changes tend to be great.

Permafrost

Continuous and discontinuous zones of permanently frozen ground, known as **permafrost**, currently underlie some 25 per cent of the Earth's land surface. Permafrost is soil or rock that remains frozen for two or more consecutive years. It is not the same as frozen ground, as depressed freezing points allow some materials to stay unfrozen below 0°C and considerable amounts of liquid water may exist in frozen ground. Permafrost underlies large areas of the Northern Hemisphere Arctic and subarctic. It ranges from thin layers that have stayed frozen between two successive winters to frozen ground hundreds of metres thick and thousands of years old. It develops where the depth of winter freezing is greater than the depth of summer thawing, so creating a zone of permanently frozen ground. **Continuous** and **discontinuous permafrost zones** are recognized (Figure 11.1). Some authors have subdivided the zone of discontinuous permafrost into two, three, or four subzones. In North America, a tripartite sequence of widespread permafrost, sporadic permafrost, and isolated patches of permafrost is typical; in Russia, massive island permafrost, islands permafrost, and sporadic permafrost zones are a common sequence (Heginbottom 2002). A **suprapermafrost layer**, which is the ground that lies above the **permafrost table**, tops all types of permafrost. It consists of an active layer and an unfrozen layer or talik. The **active layer** is the layer of seasonal freezing and thawing of the ground above permafrost (Figure 11.2). The depth of the active layer varies from about 10 cm to 3 m. In the continuous permafrost zone, the active layer usually sits directly upon the permafrost table. In the discontinuous permafrost zone, the active layer may not reach the permafrost table and the permafrost itself consists of patches of ice. Lying within, below, or sometimes above the permafrost are **taliks**, which are unfrozen areas of irregular shapes. In the discontinuous permafrost, chimney-like taliks may puncture the frozen ground. Closed taliks are completely engulfed by frozen ground, while open taliks are connected with the active layer. Open taliks normally occur near lakes and other bodies of standing water, which provide a source of heat. Closed taliks result from lake drainage, past climates, and other reasons.

As well as occurring in Arctic and Antarctic regions (**polar** or **latitudinal permafrost**), permafrost also occurs in the alpine zone (**mountain**

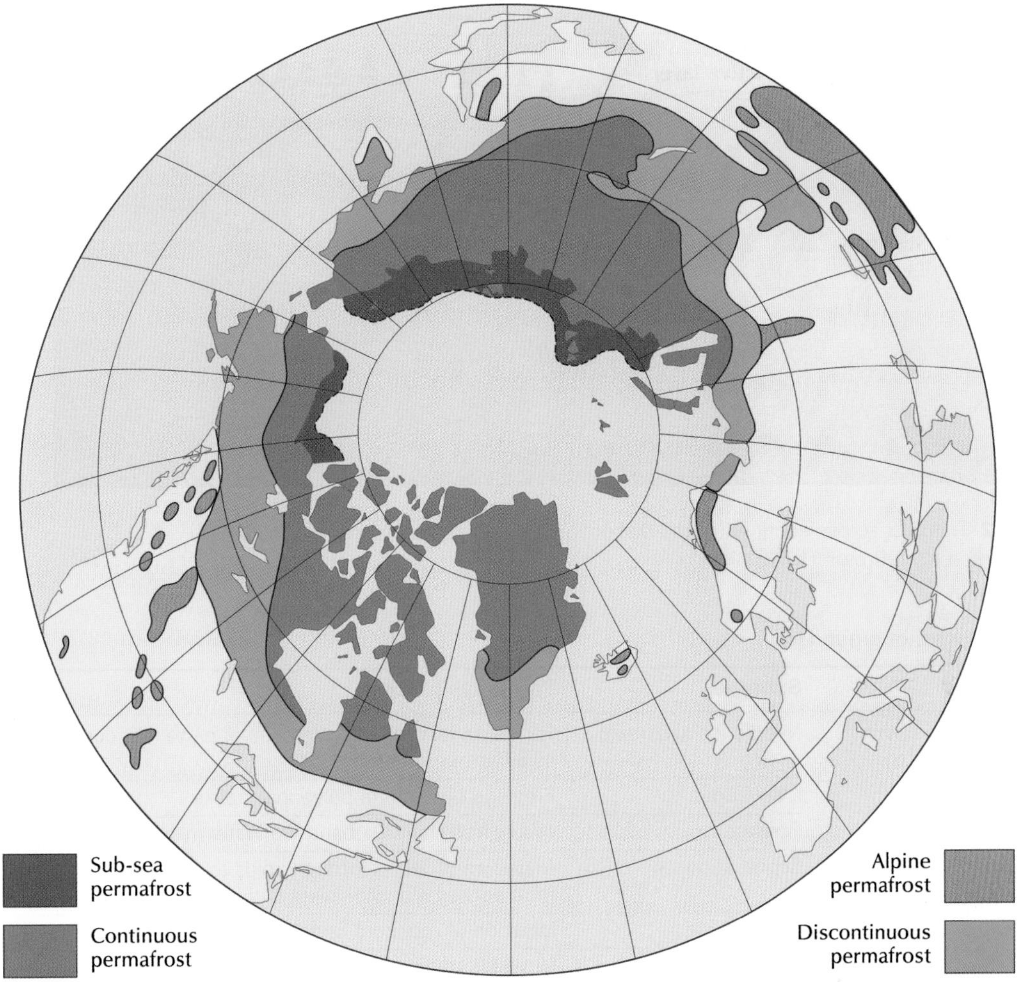

Figure 11.1 Distribution of permafrost in the Northern Hemisphere. Isolated areas of alpine permafrost, which are not shown, are found in high mountains of Mexico, Hawaii, Japan, and Europe. *Source*: Adapted from Péwé (1991)

permafrost), on some plateaux (**plateau permafrost**), and under some seas (**marine permafrost**) (Figure 11.1).

Ground ice

Ground ice is ice in frozen ground. It has a fundamental influence upon periglacial geomorphology, affecting landform initiation and evolution (Thorn 1992). It comes in a variety of forms (Table 11.1): **soil ice** (needle ice, segregated ice, and ice filling pore spaces); **vein ice** (single veins and ice wedges); **intrusive ice** (pingo ice and sheet ice); **extrusive ice**, which is formed subaerially, as on floodplains; **ice from sublimation**, which is formed in cavities by crystallization from water vapour; and **buried ice** (buried icebergs and buried glacier ice) (Embleton and King 1975b, 34). Some ground ice lasts for a day, forming under present climatic conditions, some of it for

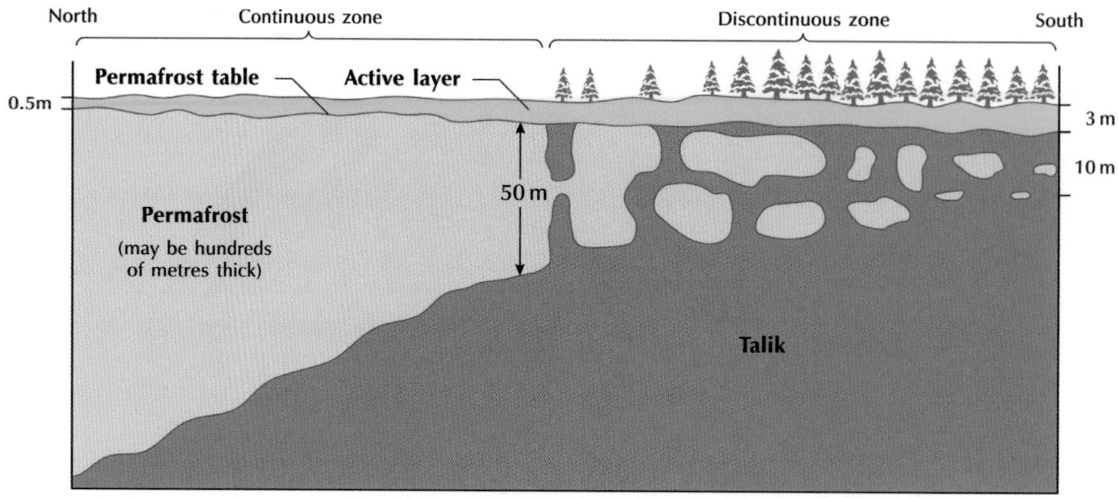

Figure 11.2 Transect across continuous and discontinuous permafrost zones in Canada.
Source: Adapted from Brown (1970, 8)

Table 11.1 Types of ground ice

Type	Subtype	Formative process
Epigenetic (formed within pre-existing sediments)	Needle ice (pipkrake)	Forms under stones or patches of earth that cool rapidly as air temperatures fall
	Ice wedges	Freezing of water in polygonal cracks
	Pore ice	*In situ* freezing of subsurface water in voids
	Segregation ice	Migration of water through voids to a freezing surface to form segregation ice layers and lenses
	Intrusive ice	Injection of moisture under pressure into sediments
	Aggradational ice	Upwards migration of the permafrost table, combining many segregated ice lenses, owing to a change in the environment
Syngenetic ice (formed in accumulating sediments)	Buried ice	Burial of snowbanks, stagnant glacial ice, or drift ice by deltaic, alluvial, or other sediments

thousands of years, forming under past climates and persisting as a relict feature. Almost all the water in permafrost occurs as ground ice, which can account for up to 90 per cent of the ground volume, although some areas of permafrost contain little ground ice and are 'dry'. Ice-rich permafrost occurs in the continuous and the discontinuous permafrost zones.

PERIGLACIAL PROCESSES

Most geomorphic processes occurring in periglacial zones occur in other climatic zones as well. Fluvial activity in particular is often the dominant process in periglacial landscapes. Some processes, and notably those related to the freezing and thawing of water, are highly active under periglacial conditions and may produce distinctive landforms.

Frost and snow processes

The freezing of water in rock, soil, and sediment gives rise to several processes – frost shattering, heaving and thrusting, and cracking – that are intense in the periglacial zone. Water in the ground may freeze *in situ* within voids, or it may migrate through the voids (towards areas where temperatures are sub-zero) to form discrete masses of segregated ice. **Segregated ice** is common in sediments dominated by intermediate grain sizes, such as silt. Coarse sediments, such as gravel, are too permeable and very fine-grained sediments, such as clay, too are impermeable and have too high a suction potential (the force with which water is held in the soil body) for segregation to occur. Frost action is crucially determined by the occurrence of freeze–thaw cycles at the ground surface. Freeze–thaw cycles are mainly determined by air temperature fluctuations, but they are modulated by the thermal properties of the ground-surface materials, vegetation cover, and snow cover.

Frost weathering and shattering

Frost weathering was covered in an earlier section (p. 161). Many periglacial landscapes are carpeted by angular rock debris, the origin of which is traditionally attributed to **frost shattering**. However, frost shattering requires freeze–thaw cycles and a supply of water. Field investigations, which admittedly are not yet large in number, indicate that such conditions may not be as common as one might imagine. Other processes, such as hydration shattering (caused by pressure of adsorbed water between grains in rocks on silicate mineral surfaces) and salt weathering (p. 162) in arid and coastal sites, may play a role in rock disintegration. It is also possible that, especially in lower-latitude glacial environments, the pervasive angular rock debris is a relict of Pleistocene climates, which were more favourable to frost shattering.

Frost heaving and thrusting

Ice formation causes **frost heaving**, which is a vertical movement of material, and **frost thrusting**, which is a horizontal movement of material. Heaving and thrusting normally occur together, though heaving is probably predominant because the pressure created by volume expansion of ice acts parallel to the direction of the maximum temperature gradient, which normally lies at right-angles to the ground surface. Surface stones may be lifted when needle ice forms. Needle ice or pipkrake forms from ice crystals that extend upwards to a maximum of about 30 mm (cf. Table 11.1). Frost heaving in the active layer seems to result from three processes: ice-lens growth as downward freezing progresses; ice-lens growth near the bottom of the active layer caused by upward freezing from the permafrost layer; and the progressive freezing of pore water as the active layer cools below freezing point. Frost heaving displaces sediments and appears to occasion the differential vertical movement of sedimentary particles of different sizes. In particular, the upward passage of stones in periglacial environments is a widely observed phenomenon. The mechanisms by which this process arises are debatable. Two groups of hypotheses have emerged: the frost-pull hypotheses and the frost-push hypotheses. In essence, **frost-pull** involves all soil materials rising with ground expansion on freezing, followed by the collapse of fine material on thawing while larger stones are still supported on ice. When the ice eventually melts, the fine materials support the stones. **Frost-push** results from ice forming beneath clasts (individual fragments of rock), owing to their higher thermal conductivity (which means that they cool down more quickly than the surrounding soil matrix), and then pushing them towards and eventually through the ground surface; the soil matrix collapses into the spaces beneath the clasts during the spring ice melt. The frost-push mechanism works under laboratory conditions but applies to stones near the surface. The frost-pull mechanism is in all likelihood the more important under natural circumstances.

Mass displacement

Frost action may cause local vertical and horizontal movements of material within soils. Such **mass displacement** may arise from cryostatic pressures within pockets of unfrozen soil caught between the permafrost table and the freezing front. However, differential heating resulting from annual freezing and thawing would lead to a similar effect. It is possible that, towards the foot of slopes, positive pore-water pressures would bring about mass displacement to form periglacial **involutions** in the active layer. Periglacial involutions consist of interpenetrating layers of sediment that originally lay flat.

Frost cracking

At sub-zero temperatures, the ground may crack by thermal contraction, a process called **frost cracking**. The polygonal fracture patterns so prevalent in periglacial environments largely result from this mechanism, though similar systems of cracks are made by drying out (**desiccation cracking**) and by differential heaving (**dilation cracking**).

Plate 11.1 A nivation hollow in Old Man Range, South Island, New Zealand. (*Photograph by Stefan Grab*)

Solifluction

Most kinds of mass movement occur in periglacial environments, but **solifluction** ('soil flow') is of paramount significance (p. 192). The term solifluction originally referred to a slow flowage of saturated regolith near the ground surface under the influence of gravity, as first observed in the Falkland Islands. Today, solifluction is widely seen as a process of cold climates involving frost creep and gelifluction. **Frost creep** moves some material downslope during alternate freeze–thaw cycles. Save in relatively dry environments, the bulk of material moves through **gelifluction**, which is the slow flowage of saturated regolith. It is especially important where regolith commonly becomes saturated owing to restricted drainage associated with a permafrost layer or seasonally frozen water table, and to moisture delivered by the thawing of snow and ice. The saturation creates high pressures in the soil pores and a drop in mechanical stability (liquefaction), so that the soil starts to flow downhill, even on slopes as shallow as 0.5°.

Nivation

This process is associated with late-lying or perennial **snow patches**. It is a local denudation brought about by the combined effects of frost action (freeze–thaw weathering, particularly the annual freeze), chemical weathering, gelifluction, frost creep, and meltwater flow (see Thorn and Hall 2002). It is most vigorous in subarctic and alpine environments, where it leads to the forming of nivation hollows as snow patches eat into hillsides. Snow patches often start in a small existing depression. Once initiated under a snow patch, a **nivation hollow** (Plate 11.1) increases its size and tends to collect more snow each year, so providing an example of positive feedback (p. 34).

Weathering, water, and wind processes in periglacial environments

Weathering

Geomorphologists have traditionally assumed that chemical weathering is subdued under periglacial climates, owing to the low temperatures, the storage of much water as ice for much of the year, and the low levels of biological activity. However, studies on comparative rates of chemical and mechanical weathering in periglacial environments are few. One study from northern Sweden indicated that material released by chemical weathering and removed in solution by streams accounted for about half of the denudational loss of all material (Rapp 1986). Later studies suggest that, where water is available, chemical weathering can be a major component of the weathering regime in cold environments (e.g. Hall *et al.* 2002). Geomorphic processes characteristic of periglacial conditions include frost action, mass movement, nivation, fluvial activity, and aeolian activity.

Fluvial action

Geomorphologists once deemed fluvial activity a relatively inconsequential process in periglacial environments due to the long period of freezing, during which running water is unavailable, and to the low annual precipitation. However, periglacial landscapes look similar to fluvial landscapes elsewhere and the role of fluvial activity in their creation has been re-evaluated. To be sure, river regimes are highly seasonal with high discharges sustained by the spring thaw. This high spring discharge makes fluvial action in periglacial climates a more potent force than the low precipitation levels might suggest, and even small streams are capable of conveying coarse debris and high sediment loads. In Arctic Canada, the River Mecham is fed by an annual precipitation of 135 mm, half of which falls as snow. Some 80–90 per cent of its annual flow occurs in a 10-day period, during which peak velocities reach up to 4 m/s and the whole river bed may be in motion.

Aeolian action

Dry periglacial environments are prone to wind erosion, as witnessed by currently arid parts of the periglacial environments and by areas marginal to the Northern Hemisphere ice sheets during the Pleistocene epoch. Strong winds, freeze-dried sediments, low precipitation, low temperatures, and scant vegetation cover promote much aeolian activity. Erosional forms include faceted and grooved bedrock surfaces, deflation hollows (p. 341) in unconsolidated sediments, and ventifacts (p. 344). Wind is also responsible for loess accumulation (p. 352).

PERIGLACIAL LANDFORMS

Many periglacial landforms originate from the presence of ice in the soil. The chief such landforms are ice and sand wedges, frost mounds of sundry kinds, thermokarst and oriented lakes, patterned ground, periglacial slopes, and cryoplanation terraces and cryopediments. They are conveniently discussed under the headings ground-ice landforms, ground-ice degradation landforms, and landforms resulting from seasonal freezing and thawing.

Ground-ice landforms

Ice and sand wedges

Ice wedges are V-shaped masses of ground ice that penetrate the active layer and run down into the permafrost (Figure 11.3). In North America, they are typically 2–3 m wide, 3–4 m deep, and formed in pre-existing sediments. Some in the Siberian lowlands are more than 5 m wide, 40–50 m long, and formed in aggrading alluvial deposits. In North America, active ice wedges are associated with continuous permafrost; relict wedges occur in the discontinuous permafrost zone. Ice wedges form during winter, when water in the ground freezes. Once the temperature falls

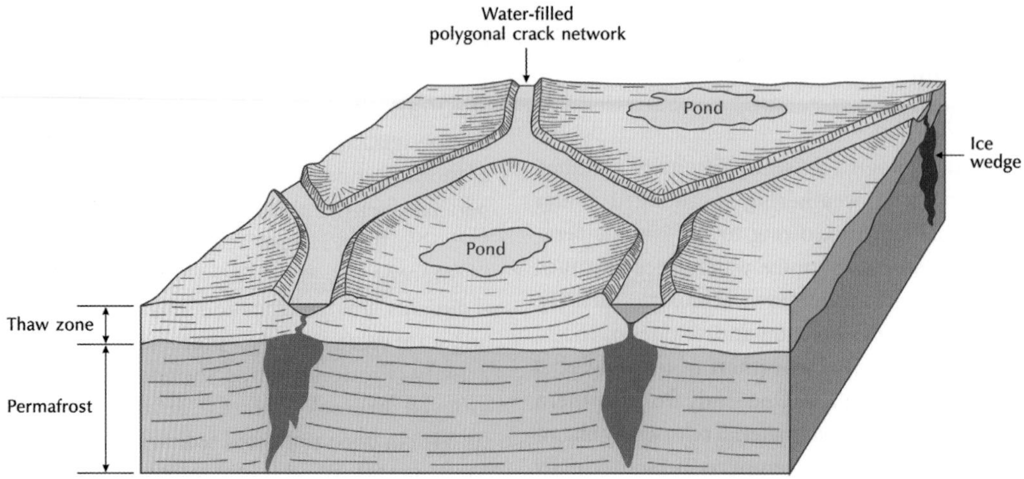

Figure 11.3 Ice-wedges, ice-wedge polygons, and raised rims. *Source*: Adapted from Butzer (1976, 342)

to −17°C or lower, the ice acts as a solid and contacts to create surface cracks that later fill with snowmelt that freezes. The ice wedges may grow each year. **Sand wedges** form by the filling in of winter contraction cracks. **Ice wedge pseudomorphs** form where thawing and erosion of an ice wedge produces an empty trough, which fills with loess or sand.

Perennial frost mounds

The expansion of water during freezing, plus hydrostatic or hydraulic water pressures (or both), creates a host of multifarious landforms collectively called '**frost mounds**' (see French 1996, 101–8). The chief long-lived mounds are pingos, palsas, and peat plateaux, while short-lived mounds include earth hummocks (p. 324), and seasonal forms include frost blisters, and icing mounds and icing blisters.

Pingos (also called hydrolaccoliths or cryolaccoliths) are large, perennial, conical, ice-cored mounds that are common in some low-lying permafrost areas dominated by fine-grained sediments, with the ice forming from injected water (Box 11.1). Their name is the Inuit word for a hill. Relict or inactive pingos occur in central Alaska, the Alaskan coastal plain, and the floor of the Beaufort Sea, in the Canadian Arctic. Active pingos occur in central Alaska and coastal Greenland, and the north of Siberia, particularly in deltas, estuaries, and alluvial areas.

A **palsa** is a low peat hill, commonly conical or dome-shaped, standing some 1–10 m high and having a diameter of 10–50 m. Palsas (or palsen) have a core of frozen peat or silt (or both), small ice crystals, and a multitude of segregated thin ice lenses and partings. They often form islands within bogs. Those lacking a peaty cover are mineral permafrost mounds (**lithalsas** or **mineral palsas**). **Peat plateaux** are larger landforms formed by the coalescence of palsas.

Many tundra landscapes contain small mounds, with or without ice cores or ice lenses. The variety of these features suggests that they may have more than one origin. The North American literature describes them as low, circular mounds, rarely standing more than 2 m high and normally in the range 15–50 m in diameter. They stand out as relatively dry sites and owls use them as perches. The Russian literature dubs them **bugors** (the Russian word for knolls) and bugor-like forms, and describes them as gently rising oval mounds or hydrolaccoliths that occur in scattered groups within

Box 11.1 PINGOS

Pingos are approximately circular to elliptical in plan (Plate 11.2). They stand 3 to 70 m high and are 30 to 7,500 m in diameter. The summit commonly bears dilation cracks, caused by the continuing growth of the ice core. Where these cracks open far enough, they may expose the ice core, causing it to thaw. This process creates a collapsed pingo, consisting of a nearly circular depression with a raised rim. Young pingos may grow vertically around 20 cm a year, but older pingos grow far less rapidly, taking thousands of years to evolve. The growth of the ice at the heart of a pingo appears to result from pressure exerted by water being forced upwards. Water may be forced upwards in at least two ways, depending on the absence (hydrostatic or closed-system pingos) or presence (hydraulic or open-system pingos) of a continuing source of unfrozen water after the formation of the initial core. First, in **hydrostatic** or **closed-system pingos**, a lake may be in-filled by sediment and vegetation, so reducing the insulation of the underlying, unfrozen ground (Figure 11.4a). Freezing of the lake surface will then cause permafrost to encroach from the lake margins, so trapping a body of water that is under hydrostatic pressure. The pressure causes the water to rise and spread sideways, eventually encountering ground temperatures cold enough to freeze it, at which point it expands and causes the overlying sediments and vegetation to dome. The same process would occur when a river is diverted or a lake drained. This mechanism for the origin at cryostatic pressure is supported by pingos in the Mackenzie Delta region, North West Territories, in Arctic Canada, where 98 per cent of 1,380 pingos recorded lie in, or near to, lake basins. A second plausible mechanism for forcing water upwards arises in

Plate 11.2 Pingo beside Tuktoyaktuk, an Inuit village on the Mackenzie Delta on the Arctic coast of Northwest Territories, Canada. The houses all stand on piles bored into the permafrost. (*Photograph by Tony Waltham Geophotos)*

continued . . .

hydraulic or **open-system pingos** (Figure 11.4b). Groundwater flowing downslope through taliks under hydrostatic pressure towards the site of a pingo may find a crack in the permafrost and freeze as it forces its way towards the surface. However, unconfined groundwater is unlikely to generate enough hydrostatic force to raise a pingo, and the open-system mechanisms may occur under temporary closed-system conditions as open taliks are frozen in winter.

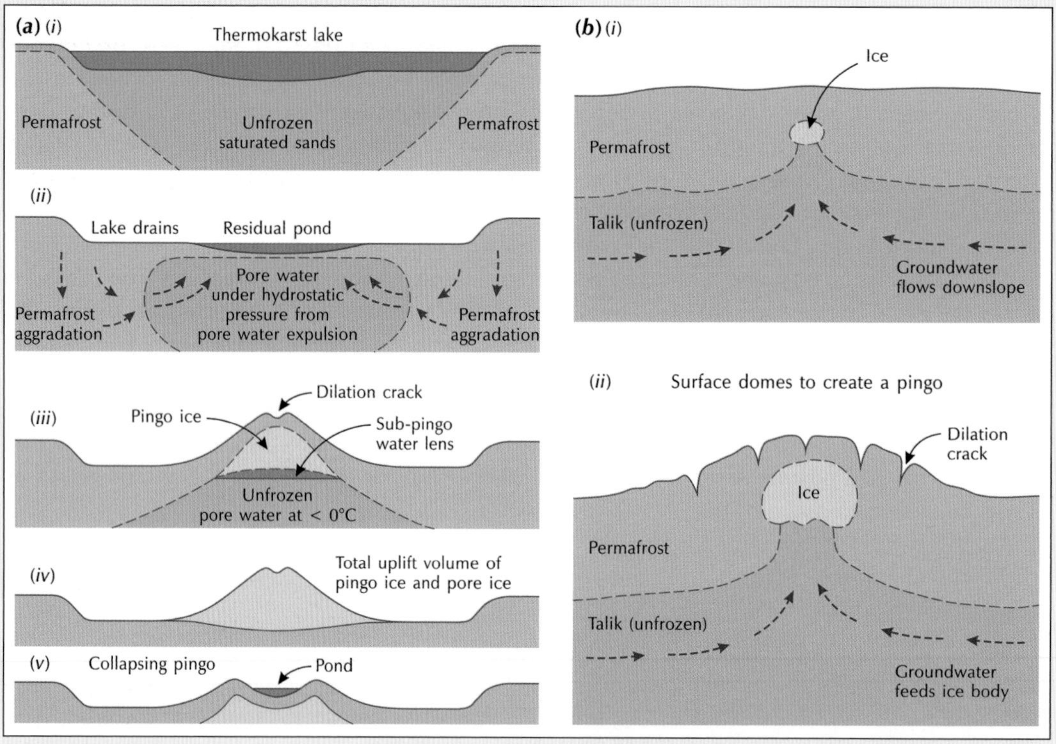

Figure 11.4 Pingo formation. (a) Hydrostatic or closed-system pingo produced after the infilling of a lake. (b) Hydraulic or open-system pingo. *Source*: (a) Adapted from Mackay (1998, 8)

the active layer. They are 5–10 m high, 50–80 m wide, and 100–5,000 m long and resemble pingos and palsas. The origin of all these small tundra mounds is unclear, as they bear no apparent relationship to topography. Localized ice segregation owing to subtle thermal differences in soil and vegetation may be the key. Even smaller hydrolaccoliths, which are never more than 1 m high or about 4 m in diameter, occur in parts of the North American Arctic, including Southampton Island, in Northwest Territories, Canada, and Alaska, USA. These features seem to result from the segregation of ice.

String bogs, also called **patterned fens**, occur in muskeg. They are alternations of thin, string-like strips or ridges of peat, mainly *Sphagnum* moss, which may contain ice for at least part of the year and may include true palsas, and vegetation with shallow, linear depressions and ponds. The ridges stand some 1.5 m high, are 1–3 m wide, and are tens of metres long. The linear features often lie at right-angles to the regional slope. It is not certain how string bogs form. Possible formative processes include geliflunction, frost thrusting of ridges from adjacent ponds, differential frost heaving, ice-lens growth, and differential thawing of permafrost, and may involve hydrological and botanical factors.

Seasonal frost mounds

Smaller mounds than palsas contain ice cores or ice lenses. Seasonal **frost blisters**, common in Arctic and subarctic regions, may grow a few metres high and a few to around 70 m long during winter freeze-back, when spring water under high pressure freezes and uplifts soil and organic sediments. They are similar to palsas but form in a different way, grow at a faster rate, and tend to occur in groups as opposed to singly. **Icings** or **ice mounds** are sheet-like masses of ice formed during winter by the freezing of successive flows of water seeping from the ground, flowing from springs, or emerging through fractures in river ice. They may grow up to 13 m thick. They store water above ground until it is released in spring and summer, when they boost runoff enormously. Icings in stream valleys block spring runoff, promoting lateral erosion by the re-routed flow. By so widening the main channel, they encourage braiding. **Icing blisters** are ice mounds created by groundwater injected at high pressure between icing layers.

Ground-ice degradation landforms

Thermokarst is irregular terrain characterized by topographic depressions with hummocks between them. It results mainly from the thawing of ground ice, material collapsing into the spaces formerly occupied by ice. Thermokarst features may also be fashioned by flowing water released as the ice thaws. The thawed water is relatively warm and causes thermal and mechanical erosion of ice masses exposed along cliffs or in stream banks. The term thermokarst reflects the resulting landform's likeness to a karst landscape in limestone regions. Thermokarst features may result from climatic warming, but they are often part of the natural variability in the periglacial environment. Any modification of surface conditions can give rise to them, including vegetation disturbance, cliff retreat, and river-course changes.

Plate 11.3 Thermokarst thaw lakes, Mackenzie Delta, Northwest Territories, Canada. (*Photograph by Tony Waltham Geophotos*)

Thaw lakes are prevalent in thermokarst landscapes (Plate 11.3). Many thaw lakes are elliptical in plan, with their long axes pointing in the same direction, at right-angles to the prevailing wind during periods of open water. The alignment may relate to zones of maximum current, littoral drift, and erosion, but its causes are far from fully studied. **Oriented thaw lakes** are common in permafrost regions, but oriented lakes occur in other environments, too.

Landforms resulting from seasonal freezing and thawing

Patterned ground

In the periglacial zone, the ground surface commonly bears a variety of cells, mounds, and ridges that create a regular geometric pattern. Such ground patterning occurs in other environments, but it is especially common in periglacial regions, where the patterns tend to be more prominent. The main forms are circles, polygons,

Plate 11.4 Stone circles, Kongsfjord, Spitsbergen. (*Photograph by Wilfred H. Theakstone*)

nets, and stripes (Washburn 1979, 122–56). All these may occur in sorted or non-sorted forms. In sorted forms, coarser material is separated from finer material, whereas in non-sorted forms there is no segregation of particles by size and the patterns are disclosed by microtopography or vegetation or both. The various forms usually connect, with a transition from polygons, circles, and nets on flattish surfaces grading into steps and then stripes as slopes become steeper and mass movements become important.

1 **Circles** occur individually or in sets. They are usually 0.5 to 3 m in diameter. **Sorted circles** have fine material at the centre and a rim of stones, the stones being large in larger circles (Plate 11.4). The debris island is a particular type of sorted stone circle in which a core of fine material is girded by blocks and boulders on steep, debris-covered slopes. **Non-sorted circles** are dome-shaped, lack stony borders, and are fringed by vegetation. Circles are not restricted to areas of permafrost, and unsorted sorts are recorded from non-periglacial environments.
2 **Polygons** occur in sets. **Non-sorted polygons** range in size from about a metre across to large tundra or ice-wedge polygons that may be a hundred metres or more across. **Sorted polygons** are at most 10 m across and the borders of the polygons are formed of stones with finer material between them (Plate 11.5a). They are usually associated with flat land, while non-sorted polygons may occur on relatively steep slopes. Furrows or cracks edge non-sorted polygons (Figure 11.3). The best-developed polygons occur in regions with frosty climates, but polygons are known from hot deserts. **Ice-wedge polygons** are exclusively found in permafrost zones, the ice-wedges often occurring at the edges of large, non-sorted polygons. Two kinds of ice-wedge polygons are recognized. The first is a saucer-shaped polygon with a low centre, which may hold standing water in summer, and marginal ridges on either side of the ice-wedge trough. The second has a high centre hemmed by ice-wedge troughs. Both types form through repeated cracking of permafrost, and freezing of meltwater in cracks (p. 318).
3 **Nets** are a transitional form between circles and polygons. They are typically small with a diameter of less than a couple of metres. **Earth hummocks** (also called **thúfur** and **pounus**) consist of a domed core of mineral soil crowned by vegetation and are a common type of unsorted net. They are about 0.5 m high and 1–2 m in diameter and form mainly in fine-grained material in cold environments

Plate 11.5 (a) Stone polygons in active layer over permafrost in old lake bed, Tangle Lakes, Denali Highway, Alaska, USA. (b) Stone stripes, Disko Island, western Greenland. (*Photographs by Tony Waltham Geophotos*)

Plate 11.6 Earth hummocks, Drakensberg, Lesotho. (*Photograph by Stefan Grab*)

Plate 11.7 Non-sorted striped ground (elongated earth hummocks), Rock and Pillar Range, South Island, New Zealand. (*Photograph by Stefan Grab*)

where ample moisture and seasonal frost penetration permanently displace surface materials. Earth hummocks occur mainly in polar and subpolar regions, but examples are known from alpine environments. They are present and periodically active in the alpine Mohlesi Valley of Lesotho, southern Africa (Grab 1994, 2005) (Plate 11.6).

4 **Stripes**, which are not confined to periglacial environments, tend to develop on steeper slopes than steps (p. 327). **Sorted stripes** are composed of alternating stripes of coarse and fine material downslope (Plate 11.5b). Sorted stripes at High Pike in the northern English Lake District occur at 658 m on a scree with an aspect of 275° and a slope angle of 17–18° (Warburton and Caine 1999). These stripes are formed at a relatively low altitude, possibly because the scree has a large proportion of fine material susceptible to frost action and is free of vegetation. The sorted stripes are still active. **Non-sorted stripes** are marked by lines of vegetation lying in slight troughs with bare soil on the intervening slight ridges (Plate 11.7).

The origin of patterned ground is not fully clear. Three sets of processes seem important –

sorting processes, slope processes, and patterning processes (Figure 11.5). The main patterning processes are cracking, either by thermal contraction (frost cracking), drying (desiccation cracking), or heaving (dilation cracking), of which only frost cracking is confined to periglacial environments. Patterning may also result from frost heaving and mass displacement. Frost heaving is also an important source of sorting, helping to segregate the large stones by shifting them upwards and outwards leaving a fine-grained centre. As many forms of patterned ground are so regular, some geomorphologists have suggested that convective cells form in the active layer. The cells would develop because water is at its densest at 4°C. Water at the thawing front is therefore less dense than the overlying, slightly warmer water and rises. Relatively warm descending limbs of the convective cells would cause undulations in the interface between frozen and unfrozen soil that might be echoed in the ground surface topography. How the echoing takes place is uncertain, but frost heaving is one of several possible mechanisms. Stripe forms would, by this argument, result from a downslope distortion of the convective cells. Another possibility is that convective cells develop in the soil itself, and evidence for a cell-like soil circulation has been found. But the processes involved in patterned ground formation are complex, and all the more so because similar kinds of patterned ground appear to be created by different processes (an example of equifinality – see p. 60), and the same processes can produce different kinds of patterned ground. For instance, patterned ground occurs in deserts.

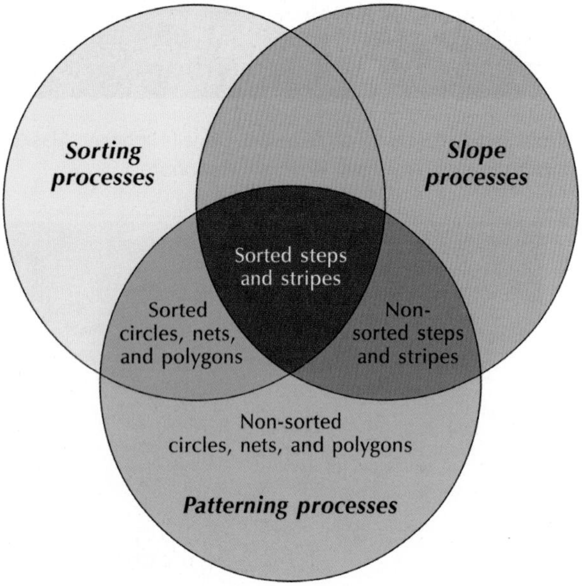

Figure 11.5 Relationships between patterned ground and sorting processes, slope processes, and patterning processes. *Source*: Adapted from Washburn (1979, 160)

Plate 11.8 Solifluction lobes, Drakensberg, South Africa. (*Photograph by Stefan Grab*)

Solifluction landforms

Solifluction (frost-creep and gelifluction) is an important periglacial process and forms sheets, lobes, terraces, and ploughing boulders. Such landforms are more common in Low Arctic, subarctic and alpine environments than in High Arctic polar deserts, which are too dry to promote much solifluction. Tongue-like lobes are common in the tundra and forest tundra, where some vegetation patches occur (Plate 11.8). **Solifluction lobes** tend to form below snow patches. Typically, they are tongued-shaped features, 10 to 100 m long, 5 to 50 m wide, with steep frontal margins or risers, which may stand 1.5 m high. Frost-sorting processes often bring about a concentration of clasts around a lobe's outer margins, which are called stone-banked lobes; lobes lacking marginal clasts are turf-banked lobes. Areas of widespread solifluction lobes are **solifluction sheets**, which can produce smooth terrain with low slope gradients (1° to 3°) where vegetation is scanty. **Terraces** are common on lower slopes of valleys (Plate 11.9). **Steps** are terrace-like landforms that occur on relatively steep slopes. They develop from circles, polygons, and nets, and run either parallel to hillside contours or become elongated downslope to create lobate forms. In unsorted steps, the rise of the step is well vegetated and the tread is bare. In sorted steps, the step is edged with larger stones. The lobate varieties are called stone garlands. No step forms are limited to permafrost environments. **Ploughing boulders** or **ploughing blocks** move down slopes through the surrounding soil, leaving a vegetated furrow in their wake and building a lobe in their van (Plate 11.10).

Rock glaciers are lobes or tongues of frozen, angular rock and fine debris mixed with interstitial ice and ice lenses (Plate 11.11). They occur in high mountains of polar, subpolar, mid-latitude, and low-latitude regions. Active forms tend to be found in continental and semi-arid climates, where ice glaciers do not fill all suitable sites. They range from several hundred metres to more than a kilometre long and up to 50 m thick. They flow slowly, at a 1 m or so a year. They are the commonest permafrost landforms in many alpine environments. Recent work has shown that all active rock glaciers contain a deforming ice core, usually 50–90 per cent of rock glacier volume. Their formation is debatable, but basic ingredients seem to be a cold climate, a copious supply of rock debris, and a slope. Three possibilities are the burial of a glacier by debris to leave an ice core (glacigenic ice core origin), the sinking of meltwater and rain into debris to form interstitial ice

Plate 11.9 Solifluction terrace at Okstindan, northern Norway. Notice that the vegetation in the foreground, which lies immediately in front of the solifluction lobe, is different from the vegetation on the lobe itself. (*Photograph by Wilfred H. Theakstone*)

Plate 11.10 Ploughing boulder with furrow, levee, and frontal lobe, Rock and Pillar Range, New Zealand. (*Photograph by Stefan Grab*)

Plate 11.11 Active rock glacier, Swiss Alps. (*Photograph by Stefan Grab*)

(glacigenic permafrost origin), and the accumulation of debris in an environment where average annual temperature is zero degrees or less and the ratio between the debris input and precipitation creates a suitable mix (Barsch 1996; Clark *et al.* 1998). It is possible that all three processes produce rock glaciers, which would then provide a fine example of equifinality (p. 60).

Periglacial hillslopes

Periglacial slopes are much like slopes formed in other climatic regimes, but some differences arise owing to frost action, a lack of vegetation, and the presence of frozen ground. Slope profiles in periglacial regions seem to come in five forms (French 1996, 170–80). Type 1, which is the best-known slope form from periglacial regions, consists of a steep cliff above a concave debris (talus) slope, and gentler slope below the talus (Figure 11.6a). Type 2 are rectilinear debris-mantled slopes, sometimes called **Richter slopes**, in which debris supply and debris removal are roughly balanced (Figure 11.6b). They occur in arid and ice-free valleys in parts of Antarctica and in the unglaciated northern Yukon, Canada. Type 3 comprises frost-shattered and gelifluction debris with moderately smooth, **concavo-convex profiles** (Figure 11.6c). Residual hillside tors may project through the debris on the upper valley sides. Such profiles are often identified as relict periglacial forms dating from the Pleistocene, but they are not widely reported from present-day periglacial regions. Type 4 profiles are formed of gently sloping **cryoplanation terraces** (also called 'goletz' terraces, altiplanation terraces, nivation terraces, and equiplanation terraces) in the middle and upper portions of some slopes that are cut into bedrock on hill summits or upper hillslopes (Figure 11.6d). Cryoplanation terraces range from 10 m to 2 km across and up to 10 km in length. The risers between the terraces may be 70 m high and slope at angles of 30° or more where covered with debris or perpendicularly where cut into bedrock. Cryoplanation terraces occur chiefly in unglaciated northern Yukon and Alaska, and in Siberia. They are attributed to nivation and scarp recession through gelifluction (e.g. Nelson 1998), but substantive field research into their formation is very limited (see Thorn and Hall 2002). Type 5 profiles are rectilinear **cryopediments**, which are very gently concave erosional surfaces that usually cut into the base of valley-side or mountain slopes, and are common in very dry periglacial regions (Figure 11.6e). Unless they cut across geological structures, they are difficult to distinguish from structural benches (p. 145). Lithological and structural controls are important in their development, which occurs in much the same way as cryoplanation terraces except that slope wash, rather than gelifluction, is more active in aiding scarp recession. The processes involved in their formation appear to be bedrock weathering by frost action combined with gravity-controlled cliff retreat and slope replacement from below. In profile types 3 and 4, residual hilltop or summit tors surrounded by gentler slopes are common on the interfluves. Many authorities argue that periglacial slopes evolve to become smoother and flatter, as erosion is concentrated on the higher section and deposition on the lower section.

HUMANS AND PERIGLACIAL ENVIRONMENTS

Attempts to develop periglacial regions face unique and difficult problems associated with building on an icy substrate (Box 11.2). Undeterred, humans have exploited tundra landscapes for 150 years or more, with severe disturbances occurring after the Second World War with the exploration for petroleum and other resource development (e.g. Bliss 1990). **Permafrost degradation** occurs where the thermal balance of the permafrost is broken, either by climatic changes or by changing conditions at the ground surface. The main effect is the deepening of the active layer, which causes subsidence and thermokarst development in ice-rich permafrost.

In the Low Arctic, mineral exploration has led to the melting of permafrost. Under natural conditions, peat, which is a good insulator, tends to prevent permafrost from melting. Where the peat layer is disturbed or removed, as by the use of tracked vehicles along summer roads, permafrost melt is encouraged. Ground-ice melting and subsequent subsidence produce **thermokarst**, which resembles karst landscapes (cf. p. 415). In the Tanana Flats, Alaska, USA, ice-rich permafrost that supports birch forest is thawing rapidly, the forests changing to minerotrophic floating mat fens (Osterkamp *et al.* 2000). A hundred years ago or more at this site, some 83 per cent of 260,000 ha

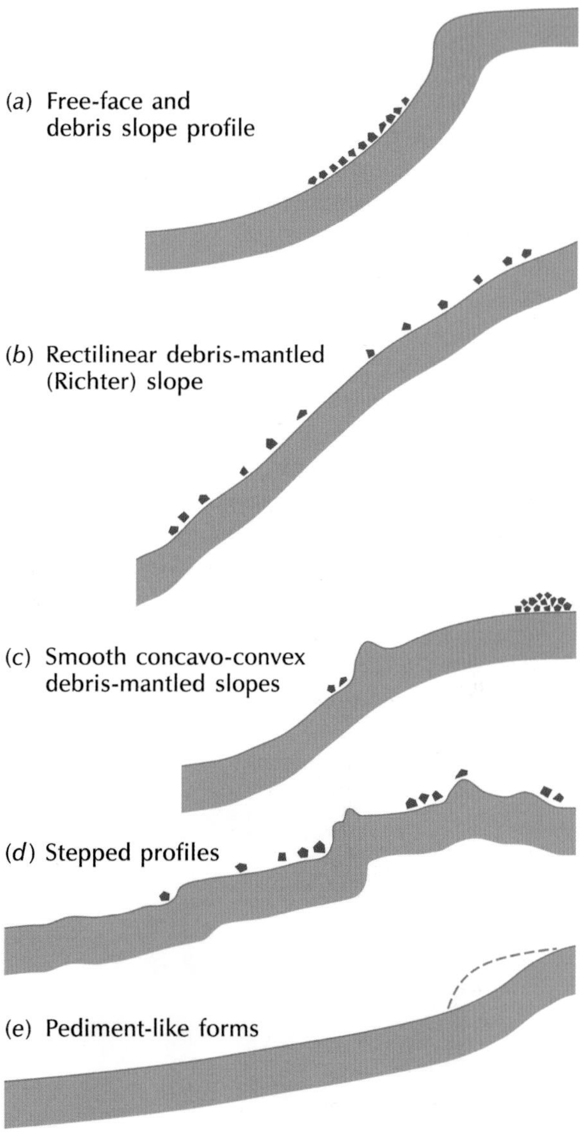

Figure 11.6 Types of periglacial slopes. (a) Cliff above a debris slope. (b) Rectilinear, debris-mantled or Richter slope. (c) Smooth concavo-convex profile with frost-shattered and solifluction debris. (d) Stepped profiles: cryoplanation or altiplanation terraces. (e) Pediment-like forms, or cryopediments. *Source*: Adapted from French (2007, 217)

Box 11.2 PROBLEMS OF DEVELOPMENT ON PERMAFROST

Buildings, roads, and railways erected on the ground surface in permafrost areas face two problems (e.g. French 1996, 285–91). First, the freezing of the ground causes frost heaving, which disturbs buildings, foundations, and road surfaces. Second, the structures themselves may cause the underlying ice to thaw, bringing about heaving and subsidence, and they may sink into the ground (Plate 11.12). To overcome this difficulty, the use of a pad or some kind of fill (usually gravel) may be placed upon the surface. If the pad or fill is of the appropriate thickness, the thermal regime of the underlying permafrost is unchanged. Structures that convey significant amounts of heat to the permafrost, such as heated buildings and warm oil pipelines, require the taking of additional measures. A common practice is to mount buildings on piles, so allowing an air space below between the building and the ground surface in which cold air may circulate (Plate 11.2). Even so, in ground subject to seasonal freezing, the pile foundations may move, pushing the piles upwards. In consequence, bridges, buildings, military installations, and pipelines may be damaged or destroyed if the piles are not placed judiciously. Other measures include inserting open-ended culverts into pads and the laying of insulating matting beneath them. In addition, where the cost is justified, refrigeration units may be set around pads or through pilings. Pipes providing municipal services, such as water supply and sewage disposal, cannot be laid underground in permafrost regions. One solution, which was used at Inuvik, in the Canadian Northwest Territories, is to use utilidors. Utilidors are continuously insulated aluminium boxes that run above ground on supports, linking buildings to a central system.

Plate 11.12 Subsidence due to thawing of permafrost, Dawson, Klondike, Alaska, USA. (*Photograph by Tony Waltham Geophotos*)

The Trans-Alaska Pipeline System (TAPS), which was finished in 1977, is a striking achievement of construction under permafrost conditions. The pipeline is 1,285 km long and carries crude oil from Prudhoe Bay on the North Slope to an ice-free port at Valdez on the Pacific Coast. It was originally planned to bury the pipe in the ground for most of the route, but as the oil is carried at 70–80°C this would have melted the permafrost and the resulting soil flow would have damaged the pipe. In the event, about half of the pipe was mounted on elevated beams held up by 120,000 vertical support members (VSMs) that were frozen firmly into the permafrost using special heat-radiating thermal devices to prevent their moving. This system allows the heat from the pipe to be dissipated into the air, so minimizing its impact on the permafrost.

Few roads and railways have been built in permafrost regions. Most roads are unpaved. Summer thawing, with concomitant loss of load-bearing strength in fine-grained sediments, and winter frost-heaving call for the constant grading of roads to maintain a surface smooth enough for driving. Paved roads tend to become rough very quickly, most of them requiring resurfacing every three to five years. Railways are difficult to build and expensive to keep up in permafrost regions. The Trans-Siberian Railway, and some Canadian railways in the north of the country (e.g. the Hudson Bay railway), cross areas where the ground ice is thick. At these sites, year-round, costly maintenance programmes are needed to combat the effects of summer thawing and winter frost-heaving and keep the track level. The Hudson Bay railway has been operating for over sixty years. For all that time, it has faced problems of thaw settlement along the railway embankment and the destruction of bridge decks by frost heave. Heat pipes help to minimize thaw subsidence but they are very expensive.

was underlain by permafrost. About 42 per cent of this permafrost has been affected by thermokarst development within the last 100 to 200 years. The thaw depths are typically 1–2 m, with some values as high as 6 m. On the Yamal Peninsula of northwest Siberia, land-use and climatic changes since the 1960s, when supergiant natural gas fields were discovered, have led to changes in the tundra landscape (Forbes 1999). Extensive exploration meant that large areas were given over to the construction of roads and buildings. Disturbance associated with this development has affected thousands of hectares of land. The increasing amount of land given over to roads and buildings, together with the associated disturbed land, has driven a fairly constant or increasing reindeer population on to progressively smaller patches of pasture. In consequence, the patches have suffered excessive grazing and trampling of lichens, bryophytes, and shrubs. In many areas, sandy soils have been deflated (see p. 338). The human- and reindeer-induced disturbance may easily initiate thermokarst formation and aeolian erosion, which would lead to significant further losses of pasture.

Thermokarst is less likely to develop in the High Arctic, owing to the lower permafrost temperatures and the generally lower ice content. Nonetheless, gully erosion can be a serious problem in places lacking a peat cover. For instance, snow piled up when clear areas for airstrips and camps are ploughed melts in the spring. The meltwater runs along minor ruts caused by vehicles. In a few years, erosion may turn these minor ruts into sizeable gullies. A trickle of water may become a potent erosive force that transforms the tundra landscape into a slurry of mud and eroding peat. Restoration work is difficult because gravel is in short supply and a loss of soil volume occurs during the summer melt. In any case, gravel roads, although they will prevent permafrost melt and subsidence if they are thick enough, have deleterious side-effects. For instance, culverts designed to take water under the roads may fill with gravel or with ice in the winter. In three sites within the Prudhoe Bay Oil Field, studied from 1968 to 1983, blocked drainage-ways have led to 9 per cent of the mapped area being flooded and 1 per cent of the area being thermokarst (Walker *et al.* 1987). Had not the collecting systems, the camps, and the pipeline corridors been built in an environmentally acceptable manner, the flooding and conversion to thermokarst might have been far greater. Water running parallel to the roads and increased flow from the culverts may lead to combined thermal and hydraulic erosion and the production of thermokarst.

Future enhanced **global warming**, with its associated changes in temperature and precipitation regimes, will have a huge impact on the climatically determined environments where periglacial processes occur, and above all in upland and glaciated catchments (see Knight and Harrison 2009). It seems likely that sediment production and supply will decrease over time as the land area under 'periglacial friendly' climates shrinks. Should human activities extend and make warmer the current interglacial, then sediment fluxes from the headwaters of mid-latitude glaciated basins will decrease radically, leading to sediment starvation and, eventually, to cannibalization of river lowlands and coastal fringes (Knight and Harrison 2009). In high-latitude areas, permafrost melt and reduced sea-ice protection is already boosting coastal erosion and sediment supply (Lawrence *et al.* 2008). And, to be sure, global warming is already causing a decrease in the continuity and interconnectedness of permafrost and associated periglacial processes (Lunardini 1996; Lemke *et al.* 2007). Much of the discontinuous permafrost in Alaska is now extremely warm, usually within 1–2°C of thawing. Ice at this temperature is highly susceptible to thermal degradation, and any additional warming during the current century will result in the formation of new thermokarst (Osterkamp *et al.* 2000). In the Yamal Peninsula, a slight warming of climate, even without the human impacts on the landscape, would produce massive thermokarst erosion (Forbes 1999).

RELICT PERIGLACIAL FEATURES

Areas fringing the Northern Hemisphere ice sheets and other areas that were appreciably colder during the Quaternary are rich in relict features of periglaciation. The blockfields (p. 168) of the Appalachian Mountains, eastern USA, are considered fossil periglacial landforms, and in Norway, some Tertiary blockfields have been identified that seem to have formed under a Mediterranean-type climate. Studies in Europe have yielded a large number of relict periglacial features (Box 11.3). Periglacial landforms also survive from previous cold periods. Siltstones with fossil root traces and surface mats of fossil plants occur in the mid-Carboniferous Seaham Formation near Lochinvar, New South Wales, Australia (Retallack 1999). They represent ancient soils of tundra and bear signs of freeze–thaw banding and earth hummocks.

SUMMARY

Periglacial landscapes experience intense frosts during winter and snow-free ground during the summer. They are underlain by either continuous or patchy permafrost (permanently frozen ground), which at present lies beneath about 22 per cent of the land surface. Several geomorphic processes operate in periglacial environments. Frost action is a key process. It causes weathering, heaving and thrusting, mass displacement, and cracking. Solifluction (frost creep and gelifluction) dominates mass movements. Nivation combines several processes to form hollows under snow patches. Fluvial and aeolian action may also be very effective land-formers in periglacial environments. Periglacial landforms, some of them bizarre, include ground-ice land-

Box 11.3 RELICT PERIGLACIAL FEATURES IN ENGLAND

England possesses many landforms formed under periglacial conditions and surviving as relicts. A few examples will illustrate the point.

'Head' is used to describe deposits of variable composition that were mainly produced by a gelifluction or solifluction moving material from higher to lower ground. Head deposits are widespread in eastern England and are a relict periglacial feature (Catt 1987). They occur on lower scarp and valley slopes and overlie a variety of bedrock types. Thick **Coombe Deposits** lie on the floors of dry chalkland valleys. They consist of frost-shattered bedrock that has been carried down slopes greater than 2° by rolling, frost creep, or mass sliding over melting ice lenses or a permafrost table. The more extensive thin spreads of stony fine loams – clay vale head deposits – that cover the floors of clay vales occur on very gentle slopes (often less than 1°) or almost level ground but contain stones from hard rock escarpments several kilometres away. They appear to be cold climate mudflows initiated on steep slopes (7–10°) that fluvial activity has reworked a little.

Non-sorted **frost-wedge polygons** and stripes are found over large areas of the chalk outcrop in eastern England, including many areas covered by Coombe Deposits. They are readily apparent in soil and crop marks in aerial photographs. Near Evesham, in southern England, polygonal patterns with meshes 8 m across have been noted. Remnants of pingos occur in the south of Ireland, beyond the limits of the last glaciation (Coxon and O'Callaghan 1987). The pingo remnants are large (10–100 m in diameter) and occur as individuals, as small groups, and as large clusters. The tors, rock platforms, and debris slopes on the Stiperstones in Shropshire appear to have formed concurrently under periglacial conditions (Clark 1994). The landscape is thus inherited. The crest-line cryoplanation platforms are probably the clearest of the remnant and they display manifest relationships with the tors and debris slopes.

forms (ice wedges and a range of frost mounds – pingos, palsa, peat plateaux, string bogs, frost blisters, icing mounds and icing blisters), ground-ice degradation landforms (thermokarst and oriented lakes), and landforms resulting from seasonal freezing and thawing (patterned ground and periglacial slopes). Patterned ground is a geometrical arrangement of circles, polygons, nets, steps, and stripes. Periglacial slopes include cryoplanation terraces. Human activities in periglacial environments and global warming are leading to permafrost degradation and the formation of thermokarst. Many current periglacial features are vestiges of frigid conditions during the Quaternary ice ages.

ESSAY QUESTIONS

1 How distinctive are periglacial landforms?

2 How does patterned ground form?

3 Examine the problems of living in periglacial environments.

4 How will global warming affect periglacial landscapes?

FURTHER READING

Ballantyne, C. K. and Harris, C. (1994) *The Periglaciation of Great Britain.* Cambridge: Cambridge University Press.
A very good book that includes an introduction to the idea of periglaciation.

French, H. M. (2007) *The Periglacial Environment*, 3rd edn. Chichester: John Wiley & Sons.
The best recent account of periglacial landforms and processes.

French, H. M. (ed.) (2004) *Periglacial Geomorphology* (Geomorphology: Critical Concepts in Geography, vol. V). London: Routledge.
A valuable collection of essays on various aspects of periglaciation.

Knight, J. and Harrison, S. (eds) (2009) *Periglacial and Paraglacial Processes and Environments* (Geological Society, London, Special Publications, Vol. 320). London: Geological Society.
A fascinating selection of up-to-date essays.

Washburn, A. L. (1979) *Geocryology: A Survey of Periglacial Processes and Environments.* London: Edward Arnold.
Another good account of periglacial landscapes, but dated.

Williams, P. J. and Smith, M. W. (1989) *The Frozen Earth: Fundamentals of Geocryology.* Cambridge: Cambridge University Press.
Well worth a look.

CHAPTER TWELVE

AEOLIAN LANDSCAPES

12

Wind is a forceful instrument of erosion and deposition where conditions are dry and the ground surface bare. This chapter covers:

- Places where wind is an important geomorphic agent
- Wind processes
- Landforms fashioned by wind erosion
- Landforms fashioned by wind deposition
- Humans and wind processes
- Windy landscapes in the past

WIND IN ACTION

As an agent of transport, and therefore of erosion and deposition, the work of the wind is familiar wherever loose surface materials are unprotected by a covering of vegetation. The raising of clouds of dust from ploughed fields after a spell of dry weather and the drift of wind-swept sand along a dry beach are known to everyone. In humid regions, except along the seashore, wind erosion is limited by the prevalent cover of grass and trees and by the binding action of moisture in the soil. But the trials of exploration, warfare and prospecting in the desert have made it hardly necessary to stress the fact that in arid regions the effects of the wind are unrestrained. The 'scorching sand-laden breath of the desert' wages its own war on nerves. Dust-storms darken the sky, transform the air into a suffocating blast and carry enormous quantities of material over great distances. Vessels passing through the Red Sea often receive a baptism of fine sand from the desert winds of Arabia; and dunes have accumulated in the Canary Islands from sand blown across the sea from the Sahara.

(Holmes 1965, 748–9)

AEOLIAN ENVIRONMENTS

Wind is a geomorphic agent in all terrestrial environments. It is a potent agent only in dry areas with fine-grained soils and sediments and little or no vegetation. The extensive sand seas and grooved bedrock in the world's arid regions attest to the potency of aeolian processes. More local wind action is seen along sandy coasts and over bare fields, and in alluvial plains containing migrating channels, especially in areas marginal to glaciers and ice sheets. In all other environments, wind activity is limited by a protective cover of vegetation and moist soil, which helps to bind soil particles together and prevent their being winnowed out and carried by the wind, and only in spaces between bushes and on such fast-drying surfaces as beaches can the wind free large quantities of sand.

Deserts are regions with very low annual rainfall (less than 300 mm), meagre vegetation, extensive areas of bare and rocky mountains and plateaux, and alluvial plains, that cover about a third of the Earth's land surface (Figure 12.1). Many deserts are hot or tropical, but some polar regions, including Antarctica, are deserts because they are dry. Aridity forms the basis of classifications of deserts. Most classifications use some combination of the number of rainy days, the total annual rainfall, temperature, humidity, and other factors. In 1953, Peveril Meigs divided desert regions on Earth into three categories according to the amount of precipitation they receive:

1 Extremely arid lands have at least 12 consecutive months without rainfall.
2 Arid lands have less than 250 mm of annual rainfall.
3 Semi-arid lands have a mean annual precipitation of between 250 and 500 mm.

Arid and extremely arid land are deserts; semi-arid grasslands mostly prairies or steppes. The

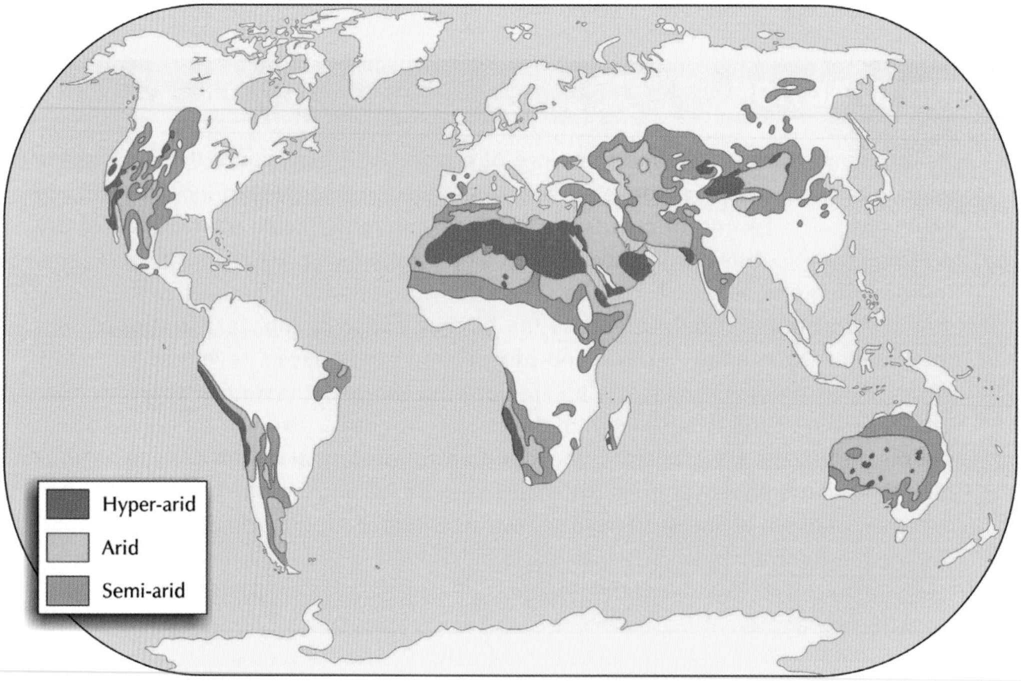

Figure 12.1 The world's deserts. *Source*: Adapted from Thomas (1989)

Table 12.1 Degrees of aridity defined by an aridity index

Aridity type	Aridity index	World land area (per cent)
Hyper-arid	<0.05	7.5
Arid	0.05–0.20	12.1
Semi-arid	0.20–0.50	17.7
Dry subhumid	0.50–0.65	9.9

United Nations Environment Programme (UNEP) uses a different index of aridity, defined as

$$AI = \frac{PE}{P}$$

where *PE* is the potential evapotranspiration and *P* is the average annual precipitation (Middleton and Thomas 1997). Four degrees of aridity derive from this index (Table 12.1).

Although wind action is an important process in shaping desert landforms, desert landform assemblages vary in different tectonic settings. Table 12.2, which shows the proportion of landforms in the tectonically active south-west USA and in the tectonically stable Sahara, brings out these regional differences.

Table 12.2 Landforms assemblages in deserts of the south-west USA and the Sahara

Landform	South-west USA (per cent)	Sahara (per cent)
Desert mountains	38.1	43.0
Playas	1.1	1.0
Desert flats	20.5	10.0
Bedrock fields (including hamadas)	0.7	10.0
Regions bordering through-flowing rivers	1.2	1.0
Dry washes (ephemeral stream beds)	3.6	1.0
Alluvial fans and bajadas	31.4	1.0
Sand dunes	0.6	28.0
Badlands	2.6	2.0
Volcanic cones and fields	0.2	3.0

Source: Adapted from Cooke *et al.* (1993, 20)

AEOLIAN PROCESSES

Air is a dusty gas. It moves in three ways: (1) as **streamlines**, which are parallel layers of moving air; (2) as **turbulent flow**, which is irregular movements of air involving up-and-down and side-to-side currents; and as (3) **vortices**, which are helical or spiral flows, commonly around a vertical central axis. Streamlined objects, such as aircraft wings, split streamlines without creating much turbulence. Blunt objects, such as rock outcrops and buildings, split streamlines and stir up turbulent flow, the zones of turbulence depending on the shape of the object.

Air moving in the lower 1,000 m of the atmosphere (the boundary layer) is affected by the frictional drag associated with the ground surface. The drag hampers motion near the ground and greatly lessens the mean wind speed. In consequence, the wind-speed profile looks much like the velocity profile of water in an open channel and increases at a declining rate with height, as established in wind-tunnel experiments by the English engineer and professional soldier Brigadier Ralph Alger Bagnold (1941). The **wind-velocity profile** (Figure 12.2) may be written as:

$$u_z = \frac{u^*}{\kappa} \ln \frac{z}{z_0}$$

where u_z is the wind speed at height z, z is height above the ground, κ (kappa) is the Kármán constant (which is usually taken as ≈0.4), z_0 is roughness length (which depends on grain size), and u^* is the shear or friction, defined as:

$$u^* = \sqrt{(\tau_0/\rho_a)}$$

where τ_0 (tau-zero) is the shear force per unit area and ρ_a (rho-a) is the air density.

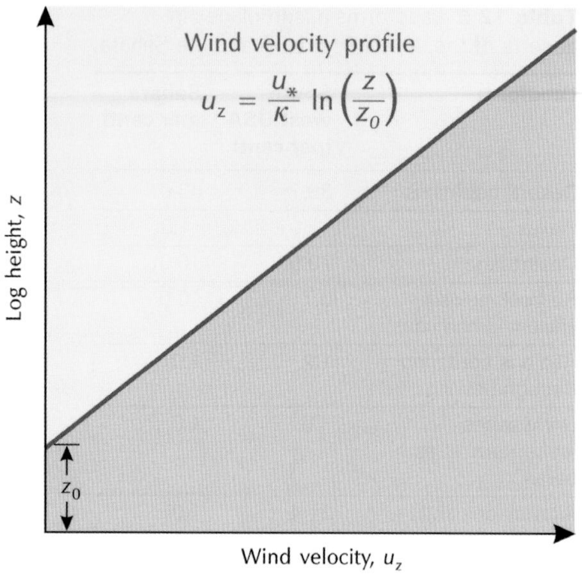

Figure 12.2 Wind velocity profile. Note that the relationship between wind velocity and height is a straight line because the height axis is logarithmic; in the diagram showing the velocity profile in water (Figure 9.2), the height axis is arithmetic and the relationship is a curve.

In moving, air behaves much like water. As air is about a thousand times less dense than water, it cannot transport such large particles. Nonetheless, the wind is an agent of erosion and transport. The ability of wind to erode, entrain, and convey rock and soil particles depends upon the nature of the wind, the nature of the ground surface, and the nature of the soil or rock. Crucial wind factors are the wind velocity and the degree of turbulence, with air density and viscosity playing lesser roles. Ground-surface factors include vegetation cover, roughness, obstacles, and topographic form. Soil factors include moisture content, structure, and density.

Wind erosion

Wind erosion engages two processes – deflation and abrasion. **Deflation** is the removal of loose particles by the wind. Smaller sedimentary particles are more susceptible to wind erosion than larger particles. Particles of about 100 micrometres diameter are the most vulnerable to wind erosion. Above that size, increasingly higher velocities are needed to entrain increasingly large grains and to keep them airborne. Below that diameter, and especially for clay particles, greater wind velocities are needed to surmount the cohesional forces binding individual grains together. Deflation of sand-sized particles is localized, and it takes a long time to move sand great distances. Silt and clay, on the other hand, are far more readily lifted by turbulence and carried in suspension in the atmosphere, the finest material being transported great distances. The world's hot deserts are a leading source of atmospheric dust. Even temperate areas may produce dust. In south-eastern Australia, a windblown dust, locally called **parna**, covers wide areas. **Soil erosion** by wind is well documented and well known (p. 355).

Wind without grains is an impotent geomorphic agent; wind armed with grains may be a powerful erosive agent. **Abrasion** is the cannonading of rock and other surfaces by particles carried in the wind – a sort of natural 'sand blasting'. Rocks and boulders exposed at the ground surface may be abraded by sand and silt particles. Abrasion rates appear to be highest where strong winds carry hard sand grains from soft and friable rocks upwind. Sand particles are carried within a metre or two of the ground surface, and abrasion is not important above that height.

Wind transport

Before the wind can transport particles, it must lift them from the ground surface. Particles are raised by 'lift', which is produced by the **Bernoulli effect** and the local acceleration of wind, and bombardment by particles already in the air. The Bernoulli effect arises from the fact that wind speed increases swiftly away from the ground surface, so that a surface particle sits in a pressure gradient, the top of the particle experiencing a lower pressure than the bottom of the particle.

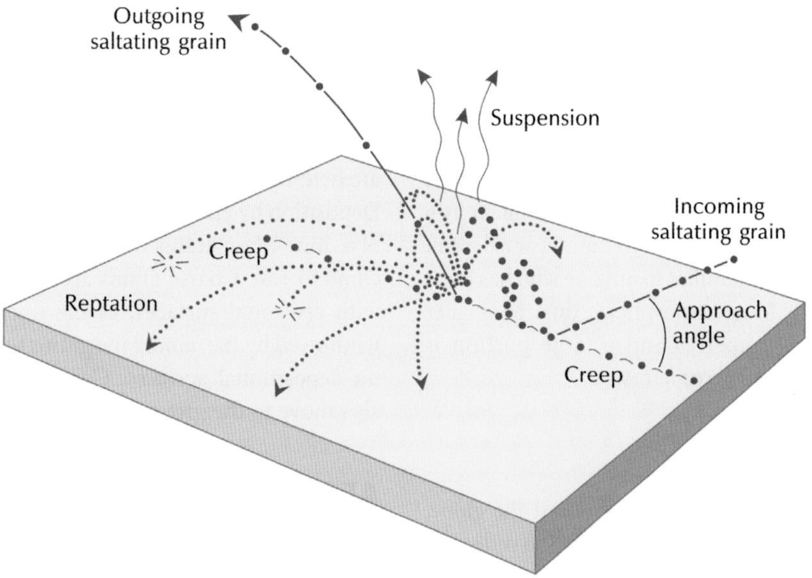

Figure 12.3 Modes of grain transport by wind. *Source*: Adapted from Livingstone and Warren (1996, 13)

The Bernoulli effect is boosted where airflow accelerates around protruding objects. However, the most effective mechanism for getting particles airborne is bombardment by particles already in flight. So the movement of particles is slow when a wind starts, as only lift is operative, but it picks up by leaps and bounds once saltation and associated bombardment come into play.

Wind transport encompasses four processes – saltation, reptation, suspension, and creep (Figure 12.3):

1 **Saltation**. Sand grains bound, land, and rebound, imparting renewed impetus to other sand grains. Such motion is confined to short distances and heights of about 2 m.
2 **Reptation**. On hitting the surface, saltating grains release a small splash-like shower of particles that make small hops from the point of impact. This process is reptation.
3 **Suspension**. Particles of silt and clay lifted into the atmosphere become suspended and may be carried great distances. Sand particles may be lifted into the lower layers of the atmosphere, as in sandstorms, but will fall out near the point of takeoff. Dust particles may be carried around the globe. Dust storms may carry 100 million tonnes of material for thousands of kilometres. A dramatic dust storm, which carried an estimated 2 million tonnes of dust, engulfed Melbourne, Australia, on 8 February 1983 (Raupach *et al.* 1994).
4 **Creep** and related near-surface activity. Coarse sand and small pebbles inch forward by rolling and sliding with the momentum gained from the impact of jumping sand particles and down the tiny crater-slopes produced by an impacting particle.

It should be stressed that saltation is the key process. Once saltation cuts in, it powers all the other processes, especially creep and reptation. Even the entrainment of fine particles destined to become suspended is mainly induced by jumping grains.

The dividing line between saltation and suspension appears to lie at about particles of 100 micrometres diameter. Particles smaller than

100 micrometres have fall velocities lower than the upward velocity of the turbulent wind and so stay in the air until the wind abates, which may be thousands of kilometres from the point of entrainment. Indeed, dust particles can be carried around the world (in less than 80 days!) (p. 81). **Dust** is a somewhat loose term but can be taken as a suspension of solid particles in the air (or a deposit of such particles, familiar to anyone who has done housework). Most atmospheric dust is smaller than 100 micrometres and a large portion is smaller than 20 micrometres.

Wind deposition

Wind moves much sediment around the globe, although by no means so much as the sediment moved by rivers. Some of this sediment, representing 10 per cent of that carried by rivers, is delivered to the oceans. The rest falls on land. In Israel, the average fall is 0.25 kg/m^2/yr but falls of as much as 8.3 kg/m^2/yr are recorded after storms.

Wind deposition may take place in three ways (Bagnold 1941): (1) sedimentation, (2) accretion, and (3) encroachment. **Sedimentation** occurs when grains fall out of the air or stop creeping forward. For sand grains, this happens if the air is moving with insufficient force to carry the grains forwards by saltation or to move other grains by creep. For silt and clay, this happens if particles are brought to the ground by air currents or if the air is still enough for them to settle out (**dry deposition**), or if they are brought down by rain (**wet deposition**). Wet deposition appears to be significant where dust plumes pass over humid regions and out over the oceans. It is the main process bringing down Saharan dust in the Mediterranean region (Löye-Pilot and Martin 1996). Wet deposition may give rise to blood rains and red rains. Measured deposition rates on land range from 3.5 t/km^2/yr to 200 t/km^2/yr (Goudie 1995; Middleton 1997). **Accretion** occurs when grains being moved by saltation hit the surface with such force that some grains carry on moving forward as surface creep, but the majority come to rest where they strike. Accretion deposits are thus moulded by the combined action of saltation and surface creep. **Encroachment** takes place when deposition occurs on a rough surface. Under these conditions, grains moving as surface creep are held up, while saltating grains may move on. Deposition by encroachment occurs on the front of a dune when grains roll down the surface and come to rest. Coarse grains are often associated with erosional surfaces, as the fine grains are winnowed by the wind. Fine grains tend to occur on depositional surfaces. Coarse particles may also move to the ground surface from below.

AEOLIAN EROSIONAL FORMS

Landforms resulting from wind erosion are seldom preserved except in arid areas. In alluvial plains and beaches, subsequent action by rivers and by waves erases traces of aeolian erosion. In arid areas, other denudational agents are often weak or absent and fail to destroy erosional landforms. The chief erosional forms in drylands caused by wind erosion are lag deposits, desert pavements, ventifacts, yardangs, and basins (see Livingstone and Warren 1996; Breed *et al.* 1997; Goudie 1999).

Lag deposits and stone pavements

Deflation winnows silt and fine sand, lowering the level of the ground surface and leaving a concentrated layer of rock and coarse sand that acts as a protective blanket. Such thin veneers of gravel, or coarser material, that overlie predominantly finer materials are called **lag deposits** (Plate 12.1). Lag deposits cover a significant proportion of the world's deserts, but they also occur in other environments with little vegetation, including mountains and periglacial zones. The coarse material has several local names – gibber in Australia, desert armour in North America, and *hammada*, *serir*, and *reg* in the Arab world.

Lag deposits may result from the deflation of poorly sorted deposits, such as alluvium, that contain a mix of gravel, sand, and silt. The wind removes the finer surface particles, leaving a blanket of material too coarse to undergo deflation. The blanket shields the underlying finer materials from the wind. However, other processes can lead to the concentration of coarse particles on bare surfaces – surface wash, heating and cooling cycles, freezing and thawing cycles, wetting and drying cycles, and the solution and recrystallization of salts.

Where the stone cover is continuous (and the particles generally flat), surfaces covered by lag deposits are called **stone pavements**, but they go by a variety of local names – desert pavements in the USA, gibber plains in Australia, *gobi* in Central Asia, and *hammada*, *reg*, or *serir* in the Arab world. *Hammada* is rocky desert, in which the lag consists of coarse, mechanically weathered regolith. *Serir* is pebbly desert with a lag of rounded gravel and coarse sand produced by deflation of alluvial deposits.

Plate 12.1 Lag deposits lying on a stone pavement, Dhakla, Western Desert, Egypt. (*Photograph by Tony Waltham Geophotos*)

Deflation hollows and pans

Deflation can scour out large or small depressions called **deflation hollows** or **blowouts**. Blowouts are the commonest landforms produced by wind erosion. They are most common in weak, unconsolidated sediments. In size, they range from less than a metre deep and a few metres across, through enclosed basins a few metres deep and hundreds of metres across (pans), to very large features more than 100 m deep and over 100 km across. They are no deeper than the water table, which may be several hundred metres below the ground surface.

Pans are closed depressions that are common in many dryland areas and that seem to be at least partly formed by deflation (Figure 12.4; Plate 12.2). In size, they range from a few metres wide and only centimetres deep, to kilometres across and tens of metres deep. The largest known pan, which was discovered in eastern Australia, is 45 km wide. Pans are prominently developed in southern Africa, on the High Plains of the USA,

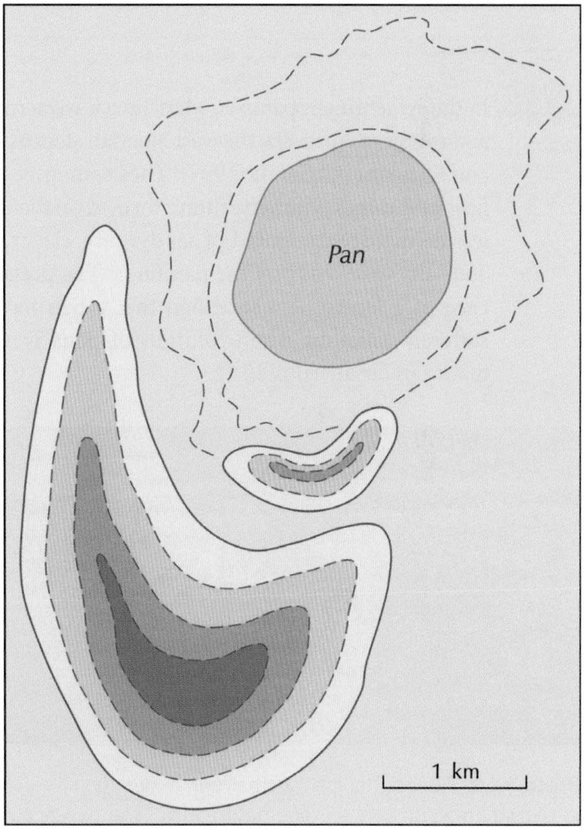

Figure 12.4 A pan in southern Africa. *Source*: Adapted from Grove (1969)

Box 12.1 THE ORIGIN OF PANS

A uniquely aeolian origin for pans is disputable. Recent research indicates that a range of processes may lead to pan formation. Deflation may top the list, but excavation by animals and karst-type solution may play a role in some cases. Pan formation appears to run along the following lines (Goudie and Wells 1995). First, certain environmental conditions are prerequisites to pan formation. Low effective precipitation and sparse vegetation cover are the main necessary conditions, but salt accumulation helps as it curbs vegetation growth. Second, the local ground surface and sedimentary cover must be susceptible of erosion. Vulnerable materials include sands and sandstones, clays and shales, and marls. These materials are susceptible only where more than a thin layer of a resistant deposit such as calcrete does not cap them. Once an initial depression is created, several processes may assist its growth. Deflation is the chief process but it may be enhanced by animals' overgrazing and trampling the ground and by salt weathering (p. 162), which may attack bedrock. A depression will not continue to grow unless it is protected from fluvial processes by being isolated from an effective and integrated fluvial system. Such protection may be afforded by low slope angles, episodic desiccation and dune encroachment, dolerite intrusions, and tectonic disturbance.

in the Argentinian pampas, Manchuria, western and southern Australia, the west Siberian steppes, and Kazakhstan (Goudie 1999). They sometimes have **clay dunes** or **lunette dunes** formed on their leeside that are composed of sandy, silty, clayey, and salty material from the pan floor. The presence of a lunette is a sure sign that a pan has suffered deflation. The evolution of pans is a matter of debate (Box 12.1).

Plate 12.2 Floor of Rooipan, a small pan or deflation hollow in the south-west Kalahari, southern Africa. The pan accumulates limited rainfall (less than 150 mm per annum in this area) in the wet season, but receives additional moisture by groundwater seepage. (*Photograph by Dave Thomas*)

Deflation appears to have played a starring role in scooping out great **erosional basins**, such as the large oasis depressions in the Libyan Desert. However, such large basins are almost certain to have had a complex evolution involving processes additional to deflation, including tectonic subsidence. The deepest of such basins is the Qattara Depression in northern Egypt, which is cut into Pliocene sediments. At its lowest point, the Qattara Depression lies 134 m below sea level.

Yardangs

Yardangs are normally defined as spectacular streamlined, sharp and sinuous ridges that extend parallel to the wind, and are separated by parallel depressions. They are sometimes said to resemble upturned ships' hulls. Yet the form of yardangs varies. Two size classes are distinguished – mega-yardangs and yardangs. **Mega-yardangs**, which are over 100 m long and up to 1,000 m wide, are reported only from the central Sahara and Egypt, some good examples occurring in the Boukou area near the Tibesti Mountains of Chad.

In the Qaidam Basin, Central Asia, eight forms of yardang occur: mesas, sawtooth crests, cones, pyramids, very long ridges, hogbacks, whalebacks, and low streamlined whalebacks (Halimov and Fezer 1989). Yardangs have been reported from Central Asia (the Taklimakan Desert, China), the Near East (the Lut Desert, south-eastern Iran; the Khash Desert, Afghanistan; the Sinai Peninsula; and Saudi Arabia), several localities across the Saharan region, North America (the Mojave Desert, California), and South America (the Talara and Paracas–Ica regions, Peru). The yardangs in the Lut Basin, Iran, are among the largest on the planet. They stand up to 80 m tall and are carved out of the Lut Formation, which consists of fine-grained, horizontally bedded, silty clays and limey gypsum-bearing sands. Yardangs also occur on Mars (Plate 12.3).

Yardangs are fashioned from sediments by abrasion and deflation, although gully formation, mass movements, and salt weathering may also be involved. Yardang evolution appears to follow a series of steps (Halimov and Fezer 1989; Goudie 1999). First, suitable sediments (e.g. lake beds and swamp deposits) form under humid conditions. These sediments then dry out and are initially eaten into by the wind or by fluvial gullying. The resulting landscape consists of high ridges and mesas separated by narrow corridors that cut down towards the base of the sediments. Abrasion then widens the corridors and causes the ridge noses to retreat. At this stage, slopes become very steep and mass failures occur, particularly along desiccation and contraction cracks. The ridges are slowly converted into cones, pyramids, sawtooth forms, hogbacks, and whalebacks. Once the relief is reduced to less than 2 m, the whole surface is abraded to create a simple aerodynamic form – a low streamlined whaleback – which is eventually reduced to a plain surface.

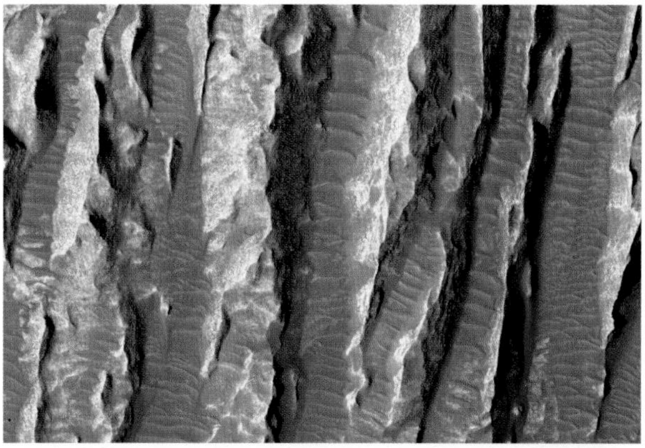

Plate 12.3 Yardangs in the Arsinoes Chaos region of Mars, seen as brightly curved terrain. Transverse sand ridges lie between the yardangs. (*Photograph by NASA/JPL-Caltech/University of Arizona*)

Plate 12.4 *Zeugen*, Farafra, Western Desert, Egypt. The limestone pillars are undercut by sand-blasting. (*Photograph by Tony Waltham Geophotos*)

Zeugen

Zeugen (singular *Zeuge*, from the German meaning 'witness' – the German geologist Johannes Walther (1900) wrote of *Zeugenberge* or 'witness mountains' formed by deflation) are residual blocks of rock, ranging between about 10 cm to 30 m high, produced by the wind eating away strata, and especially soft strata close to the ground (Plate 12.4). In some cases, harder strata overlying softer aid the differential weathering near the ground. *Zeugen* are similar in appearance and formation to **perched** or **mushroom rocks**. Exceptionally, where sand-laden wind is funnelled

Plate 12.5 Ventifacts, eroded by wind into vesicular basalt, Death Valley, California, USA. (*Photograph by Marli Miller*)

by topography, even hard rocks may be fluted, grooved, pitted, and polished by sand blasting. An example comes from Windy Point, near Palm Springs, in the Mojave Desert, California, where strong winds have sand-blasted gneiss (see Twidale and Campbell 2005, 227).

Ventifacts

Cobbles and pebbles on stony desert surfaces often bear facets called **ventifacts**. The number of edges or keels they carry is sometimes connoted by the German terms *Einkanter* (one-sided), *Zweikanter* (two-sided), and *Dreikanter* (three-sided). The pyramid-shaped *Dreikanter* are particularly common (Plate 12.5). The abrasion of more than one side of a pebble or cobble does not necessarily mean more than one prevailing wind direction. Experimental studies have shown that ventifacts may form even when the wind has no preferred direction. And, even where the wind does tend to come from one direction, a ventifact may be realigned by dislodgement.

The mechanisms by which ventifacts form are debatable, despite over a century of investigation (see Livingstone and Warren 1996, pp. 30–2), but abrasion by dust and silt, rather than by blasting by sand, is probably the chief cause. Interestingly, the best-developed ventifacts come from polar and periglacial regions, where, owing partly to the higher density of the air and partly to the higher wind speeds, larger particles are carried by the wind than in other environments.

AEOLIAN DEPOSITIONAL FORMS

Sand accumulations come in a range of sizes and forms. Deposition may occur as **sheets of sand** (dune fields and sand seas) or **loess** or as characteristic **dunes**. It is a popular misconception that the world's deserts are vast seas of sand. Sandy desert (or *erg*) covers just 25 per cent of the Sahara, and little more than a quarter of the world's deserts. Smaller sand accumulations and dune-fields are found in almost all the world's arid and semi-arid regions.

Sand accumulations, in sand seas and in smaller features, usually evolve bedforms. They are called bedforms because they are produced on the 'bed' of the atmosphere by fluid movement – airflow. They often develop regular and repeating patterns in response to the shearing force of the wind interacting with the sediment on the ground surface. The wind moulds the sediment into various landforms. In turn, the landforms modify the airflow. A kind of equilibrium may become established between the airflow and the evolving landforms, but it is readily disrupted by changes in sand supply, wind direction, wind speed, and, where present, vegetation.

Dune formation

Traditionally, geomorphologists studied dune form and the texture of dune sediments. Since around 1980, emphasis has shifted to investigations of sediment transport and deposition and of their connection to dune inception, growth, and maintenance. Research has involved field work and wind-tunnel experiments, as well as mathematical models that simulate dune formation and development (see Nickling and McKenna

Neuman 1999). Nonetheless, it is still not fully clear how wind, blowing freely over a desert plain, fashions dunes out of sand. The interactions between the plain and the flow of sand in which regular turbulent patterns are set up are probably the key. Plainly, it is essential that wind velocity is reduced to allow grains to fall out of the conveying wind. Airflow rates are much reduced in the lee of obstacles and in hollows. In addition, subtle influences of surface roughness, caused by grain size differences, can induce aerodynamic effects that encourage deposition. Deposition may produce a sand patch. Once a sand patch is established, it may grow into a dune by trapping saltating grains, which are unable to rebound on impact as easily as they are on the surrounding stony surface. This mechanism works only if the sand body is broader than the flight lengths of saltating grains. A critical lower width of 1–5 m seems to represent the limiting size for dunes. On the leeside of the dune, airflow separates and decelerates. This change enhances sand accumulation and reduces sand erosion, so the dune increases in size. The grains tend to be trapped on the slip face, a process aided by wind compression and consequent acceleration over the windward slope. The accelerated airflow erodes the windward slope and deposits the sand on the lee slope. As the sand patch grows it becomes a dune. Eventually, a balance is reached between the angle of the windward slope, the dune height, the level of airflow acceleration, and so the amount of erosion and deposition on the windward and lee slopes. The dune may move downwind (Figure 12.5).

Figure 12.6 is a speculative model of the conditions conducive to the formation of different dune types, which are discussed below (Livingstone and Warren 1996, 80). The two axes represent the two main factors controlling dune type. The first represents an unspecified measure of the amount of sand available for dune formation, while the second axis represents the variability of wind direction.

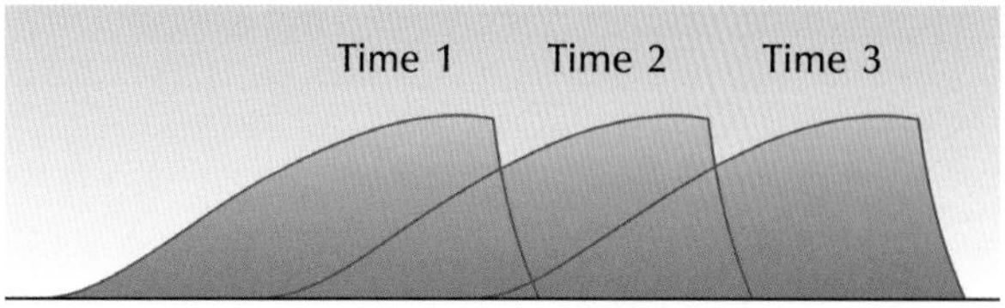

Figure 12.5 The downwind progress of a transverse dune. *Source*: Adapted from Livingstone and Warren (1996, 73)

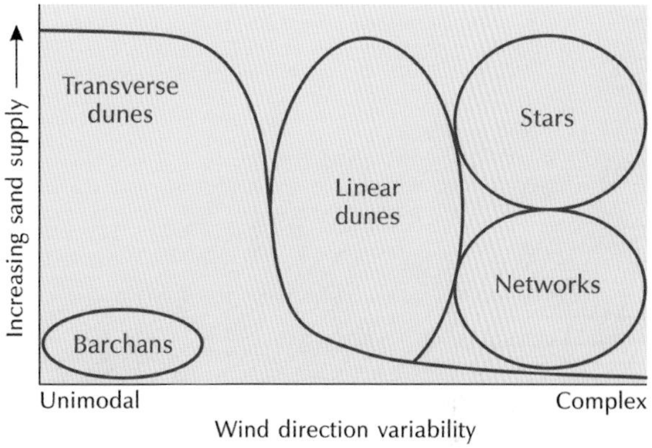

Figure 12.6 Dune types in relation to the variability of wind direction and sand supply. *Source*: Adapted from Livingstone and Warren (1996, 80)

Dune types

Some researchers believe that aeolian bedforms form a three-tiered hierarchy. Nicholas Lancaster (1995) identified three superimposed bedforms, the first two of which occur in all sand seas: (1) wind ripples; (2) individual simple dunes or superimposed dunes on compound and complex dunes; and (3) compound and complex dunes or draa.

Ripples

Wind ripples are the smallest aeolian bedform. They are regular, wave-like undulations lying at right-angles to the prevailing wind direction. The size of ripples increases with increasing particle size, but they typically range from about 10 to 300 mm high and are typically spaced a few

Plate 12.6 Rippled linear dune flank in the northern Namib Sand Sea, Namibia (*Photograph by Dave Thomas*)

Plate 12.7 Mega-ripples formed on a hard sebkha surface in the United Arab Emirates. (*Photograph by Dave Thomas*)

centimetres to tens of metres apart (Plates 12.6 and 12.7). Wind ripples develop in minutes to hours and quickly change if wind direction or wind speed alters.

Seemingly simple aeolian bedforms, ripples have withstood attempts to explain them. Several hypotheses have been forthcoming, but most are flawed (see Livingstone and Warren 1996, 27). According to what is perhaps the most plausible model (Anderson 1987; Anderson and Bunas 1993), ripple initiation requires an irregularity in the bed that perturbs the population of reptating grains. By simulating the process, repeated ripples occurred after about 5,000 saltation impacts with a realistic wavelength of about six mean reptation wavelengths. In a later version of the model (Anderson and Bunas 1993), two grain sizes were included. Again, ripples developed and these bore coarser particles at their crests, as is ordinarily the case in actual ripples.

Free dunes

Dunes are collections of loose sand built piecemeal by the wind (Figure 12.7). They usually range from a few metres across and a few centimetres high to 2 km across and 400 m high. Typical sizes are 5–30 m high and spacing at 50–500 m intervals. The largest dunes are called **draa** or **mega-dunes** and may stand 400 m high and sit more than 500 m apart, with some displaying a spacing of up to 4 km.

Dunes may occur singly or in dune fields. They may be active or else fixed by vegetation. And they may be free dunes or dunes anchored in the lee of an obstacle (impeded dunes). The form of free dunes is determined largely by wind characteristics, while the form of anchored dunes is strongly influenced by vegetation, topography, or highly local sediment sources. Classifications of dune forms are many and varied, with local names often being used to describe the same forms. A recent classification is based upon dune formation and identifies two primary forms – free and anchored – with secondary forms being established according to morphology or orientation,

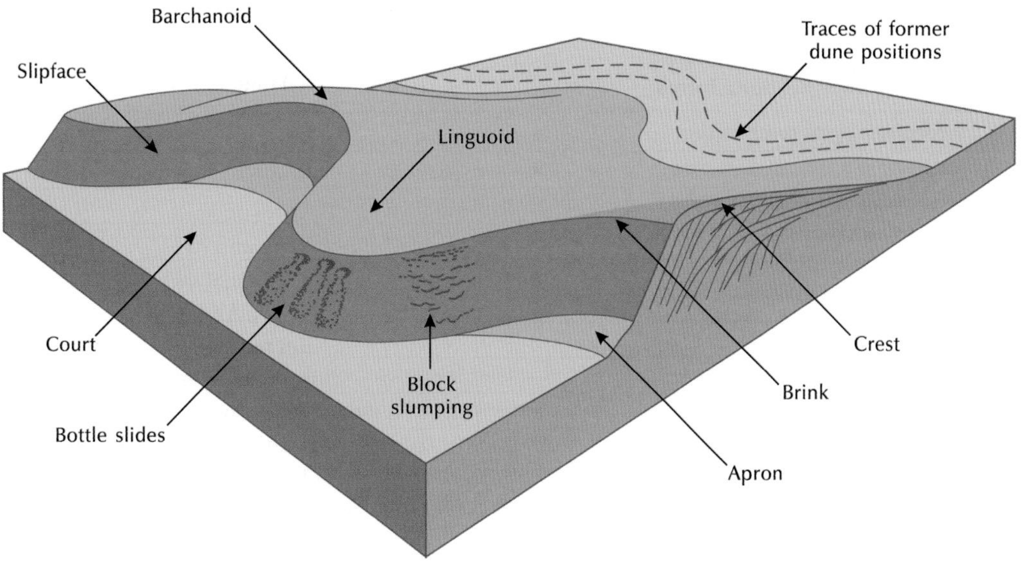

Figure 12.7 The main features of a dune. *Source*: Adapted from Livingstone and Warren (1996, 65)

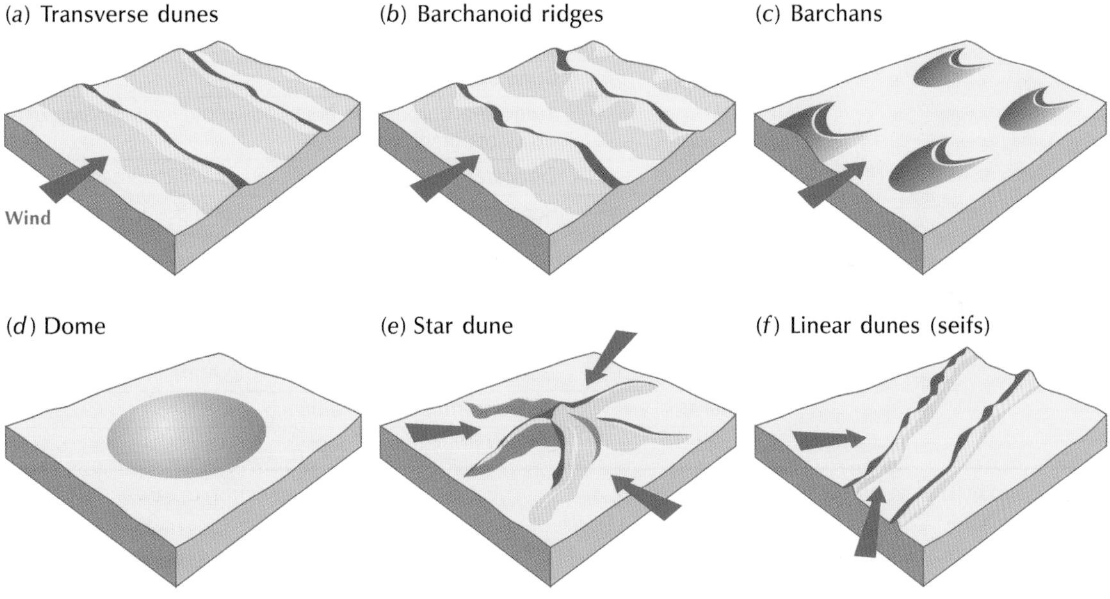

Figure 12.8 Types of free dunes. *Source*: Adapted from McKee (1979)

Table 12.3 A classification of dunes

Primary dune forms	Criteria for subdivision	Secondary dune forms	Description
Free	***Morphology or orientation***		
	Transverse	Transverse	Asymmetric ridge
		Barchan	Crescentic form
		Dome	Circular or elliptical mound
		Reversing	Asymmetric ridge with slip faces on either side of the crest
	Linear	Seif	Sharp-crested ridge
		Sand ridge	Rounded, symmetric ridge, straight or sinuous
	Star	Star	Central peak with three or more arms
		Network	Confused collection of individual dunes whose slip faces have no preferred orientation
	(Sheets)	Zibar	Coarse-grained bedform of low relief and possessing no slip face
		Streaks or stringers or sand sheets	Large bodies of sand with no discernible dune forms
Anchored	***Vegetation and topography***		
	Topography	Echo	Elongated ridge lying roughly parallel to, and separated from, the windward side of a topographic obstacle
		Climbing dune or sand ramp	Irregular accumulation going up the windward side of a topographic obstacle
		Cliff-top	Dune sitting atop a scarp
		Falling	Irregular accumulation going down the leeward side of a large topographic obstacle
		Lee	Elongated downwind from a topographic obstacle
		Fore	Roughly arcuate with arms extending downwind around either side of a topographic obstacle
		Lunette	Crescent-shaped opening upwind
	Vegetation	Vegetated sand mounds	Roughly elliptical to irregular in plan, streamlined downwind
		Parabolic	U-shaped or V-shaped in plan with arms opening upwind
		Coastal	Dunes formed behind a beach
		Blowout	Circular rim around a depression

Source: Based on Livingstone and Warren (1996, 74–101)

in the case of free dunes, and vegetation and topography, in the case of anchored dunes (Livingstone and Warren 1996, 75) (Table 12.3).

Free dunes may be classed according to orientation (transverse) or form (linear, star, and sheet) (Figure 12.8). All types of **transverse dune** cover about 40 per cent of active and stabilized sand seas. The transverse variety (Table 12.3) is produced by unidirectional winds and forms asymmetric ridges that look like a series of barchan dunes whose horns are joined, with their slip faces all facing roughly in the same direction. **Barchans** are isolated forms that are some 0.5–100 m high and 30–300 m wide (Plate

12.8). They rest on firm desert surfaces, such as stone pavements, and move in the direction of the horns, sometimes as much as 40 m/yr. They form under conditions of limited sand supply and unidirectional winds. Other transverse dune types are domes and reversing dunes. **Domes** lack slip faces but have an orientation and pattern of sand transport allied to transverse dunes. **Reversing dunes**, which have slip faces on opposite sides of the crest that form in response to wind coming from two opposing directions, are included in the transverse class because net sand transport runs at right-angles to the crest.

Linear dunes have slip faces on either side of a crest line, but only one of them is active at any time, and sand transport runs parallel to the crest. They may be divided into sharp-crested *seifs*, also called *siefs* and *sayfs* (Plate 12.9), and more rounded sand ridges. Both are accumulating forms that either trap downwind sand from two directions or lie parallel to the dominant wind. Linear dunes occur in all the world's major sandy deserts. They stand from less than a couple of metres high to around a couple of hundred metres high and may extend for tens of kilometres. They often run parallel but many meander with varied spacing and may join at 'Y' or 'tuning fork' junctions.

Dune networks and **star dunes** possess a confused set of slip faces that point in several directions. Dune networks, which are very widespread, usually occur in a continuous sand cover. They are composed of dunes no more than a few metres high and spaced 100 m or so apart. Star dunes bear several arms that radiate from a central peak (Plate 12.10). They may be up to 400 m high and spaced between about 150 and at least 5,000 m. Found in many of the world's major sand seas, star dunes cover a large area only in the Great Eastern Sand Sea of Algeria.

Sheets of sand come in two varieties – zibars and streaks. **Zibars** are coarse-grained bedforms of low relief with no slip faces. Their surfaces consist exclusively of wind ripples and local shadow and shrub-coppice dunes. They are

Plate 12.8 Barchan sand dunes moving to right, Luderitz diamond fields, Namibia. (*Photograph by Tony Waltham Geophotos*)

Plate 12.9 Linear dunes in the Mesquite Flat Dunes, which include crescent dunes and star dunes, Death Valley, California, USA. (*Photograph by Marli Miller*)

Plate 12.10 Star dunes in the Ibex Dunes, Death Valley, California, USA. (*Photograph by Marli Miller*)

common on sand sheets and upwind of sand seas. **Streaks**, also called **sand sheets** or **stringers**, are large bodies of sand that bear no obvious dune forms. They occupy larger areas of sand seas than accumulations with dunes.

Anchored dunes

Several types of dune are controlled by vegetation, topography, or local sediment sources. These **anchored** or **impeded dunes** come in a variety of forms (Table 12.3; Figure 12.9). Topographic features cause several distinct types of anchored dune. **Lee dunes** and **foredunes** are connected to the pattern of airflow around obstacles. Wind-tunnel experiments have shown that the growth of **climbing dunes** (Plate 12.11) and **echo dunes** depends upon the slope of the obstacle. When the upwind slope of an obstacle is less than around 30°, sand blows over it. When it is above 30°, then sand is trapped and a climbing dune or sand ramp forms. If it exceeds 50°, then an echo dune forms at an upwind distance of some thrice the height of the obstacle. **Cliff-top dunes** may form in the zone of slightly lower wind velocity just beyond the crest of an obstacle. **Falling dunes** form in the lee of an obstacle, where the air is calmer. If the obstacle is narrow, then sand moving around the edges may form lee dunes that extend downwind. **Lunettes** are crescent-shaped dunes that open upwind and are associated with pans (p. 342).

Plants may act as foci for dune formation, and three types of dune are associated with vegetation. The commonest type of plant-anchored dune is **vegetated sand mounds**, also known as *nabkha*, *nebkha*, shrub dunes, coppice dunes, hummock dunes, and phreatophyte mounds (Plate 12.12). These form around a bush or clump of grass, which acts as an obstacle for sand entrapment. **Parabolic dunes**, or '**hairpin**' **dunes**, are

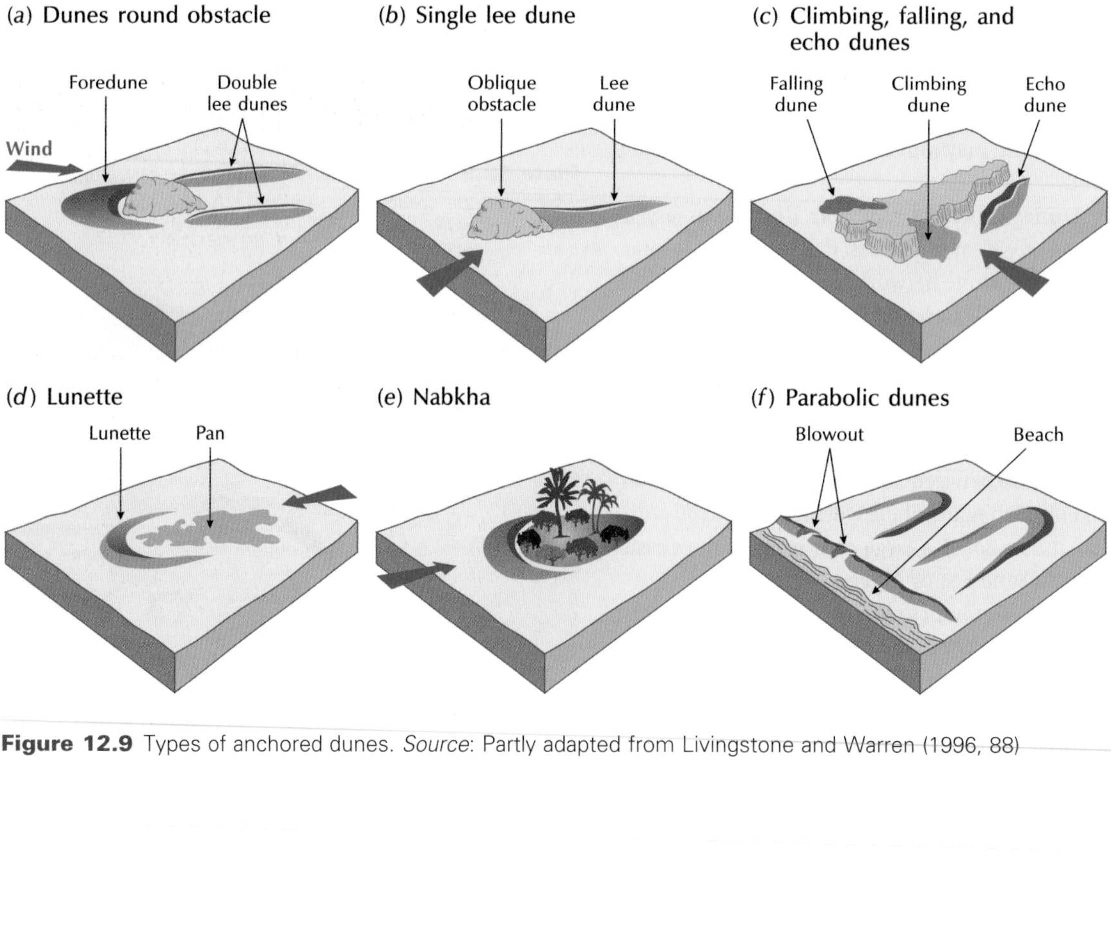

Figure 12.9 Types of anchored dunes. *Source*: Partly adapted from Livingstone and Warren (1996, 88)

U-shaped or V-shaped in plan with their arms opening upwind. They are common in vegetated desert margins. In the Thar Desert, India, they may attain heights of many tens of metres. They are also found in cold climates, as in Canada and the central USA, and at coastal sites. As to their formation, it is generally thought that parabolic dunes grow from blowouts. **Blowouts** are depressions created by the deflation of loose sand partly bound by plant roots. They are bare hollows within vegetated dunes and are very common in coastal dunes and in stabilized (vegetated) dunes around desert margins.

Plate 12.11 Small climbing dunes in the Mohave Desert, south-western USA. (*Photograph by Dave Thomas*)

Plate 12.12 *Nebkha* dunes formed from gypsum-rich sands in central Tunisia. Note that the palm trees in the background are growing on an artesian spring mound. (*Photograph by Dave Thomas*)

Dunefields and sand seas

Dunefields are accumulations of sand, occupying areas of less than 30,000 km^2 with at least ten individual dunes spaced at distances exceeding the dune wavelength (Cooke *et al.* 1993, 403). They contain relatively small and simple dunes. They may occur anywhere that loose sand is blown by the wind, even at high latitudes, and there are thousands of them. In North America, dunefields occur in the south-western region, and in intermontane basins such as Kelso and Death Valley, California.

Sand seas differ from dunefields in covering areas exceeding 30,000 km^2 and in bearing more complex and bigger dunes. In both sand seas and dunefields, ridges or mounds of sand may be repeated in rows, giving the surface a wavy appearance. About 60 per cent of sand seas are dune-covered, while others may be dune-free and comprise low sand sheets, often with some vegetation cover (Plates 12.13 and 12.14). Sand seas have several local names: ergs in the northern Sahara, *edeyen* in Libya, *qoz* in the Sahara, *koum* or *kum* and *peski* in Central Asia, and *nafud* or *nefud* in Arabia. They are regional accumulations of windblown sand with complex ancestry that are typically dominated by very large dunes (at least 500 m long or wide or both) of compound or complex form with transverse or pyramidal shapes (Figure 12.10). They also include accumulations of playa and lake deposits between the dunes and areas of fluvial, lake, and marine sediments. Sand seas are confined to areas where annual rainfall is less than 150 mm within two latitudinal belts, one 20°–40° N and the other 20°–40° S. The largest sand sea is the Rub' al Khali (the 'Empty Quarter') in Saudi Arabia, which is part of a 770,000-km^2 area of continuous dunes. About fifty comparable, if somewhat less extensive, sand seas occur in North and southern Africa, Central and Western Asia, and central Australia. In South America, the Andes constrain the size of sand seas, but they occur in coastal Peru and north-west

Plate 12.13 Northernmost edge of the Namib Sand Sea from 'Visitors Dune'. (*Photograph by Abi Stone*)

Plate 12.14 Northern part of the Namib Sand Sea looking across the Tsondab Flats. The hill in the middle is eroded Tsondab Sandstone and a proto-*zeugen*. (*Photograph by Abi Stone*)

Argentina and contain very large dunes. In North America, the only active sand sea is in the Gran Desierto of northern Sonora, northern Mexico, which extends northwards into the Yuma Desert of Arizona and the Algodones Dunes of south-eastern California. The Nebraska Sand Hills are a sand sea that has been fixed by vegetation. A single sand sea may store vast quantities of sand. The Erg Oriental in north-east Algeria, with an area of 192,000 km^2 and average thickness of 26 m, houses 4,992 km^3 of sand. The Namib Sand Sea is more modest, storing 680 km^3 of sand (Lancaster 1999). Sand seas that have accumulated in subsiding basins may be at least 1,000 m thick, but others, such as the ergs of linear dunes in the Simpson and Great Sandy Deserts of Australia, are as thick as the individual dunes that lie on the alluvial plains.

Dunefields and sand seas occur largely in regions lying downwind of plentiful sources of dry, loose sand, such as dry river beds and deltas, floodplains, glacial outwash plains, dry lakes, and beaches. Almost all major ergs are located downwind from abandoned river courses in dry areas lacking vegetation that are prone to persistent wind erosion. Most of the Sahara sand supply, for instance, probably comes from alluvial, fluvial, and lacustrine systems fed by sediments originating from the Central African uplands, which are built of Neogene beds. The sediments come directly from deflation of alluvial sediments and, in the cases of the Namib, Gran Desierto, Sinai, Atacama, and Arabian sand seas, indirectly from coastal sediments. Conventional wisdom holds that sand from these voluminous sources moves downwind and piles up as very large dunes in places where its transport is curtailed by topographic barriers that disrupt airflow or by airflow being forced to converge. By this process, whole ergs and dunefields may migrate downwind for hundreds of kilometres from their sand sources.

Loess

Loess is a terrestrial sediment composed largely of windblown silt particles made of quartz. It covers some 5–10 per cent of the Earth's land surface, much of it forming a blanket over pre-existing topography that may be up to 400 m thick (Figure 12.11; Plate 12.15). On the Chinese loess plateau, thicknesses of 100 m are common, with 330 m recorded near Lanzhou. In North America, thicknesses range from traces (< 1 m) to a maximum of 40–50 m in western Nebraska and western Iowa. Loess is easily eroded by running water and possesses underground pipe

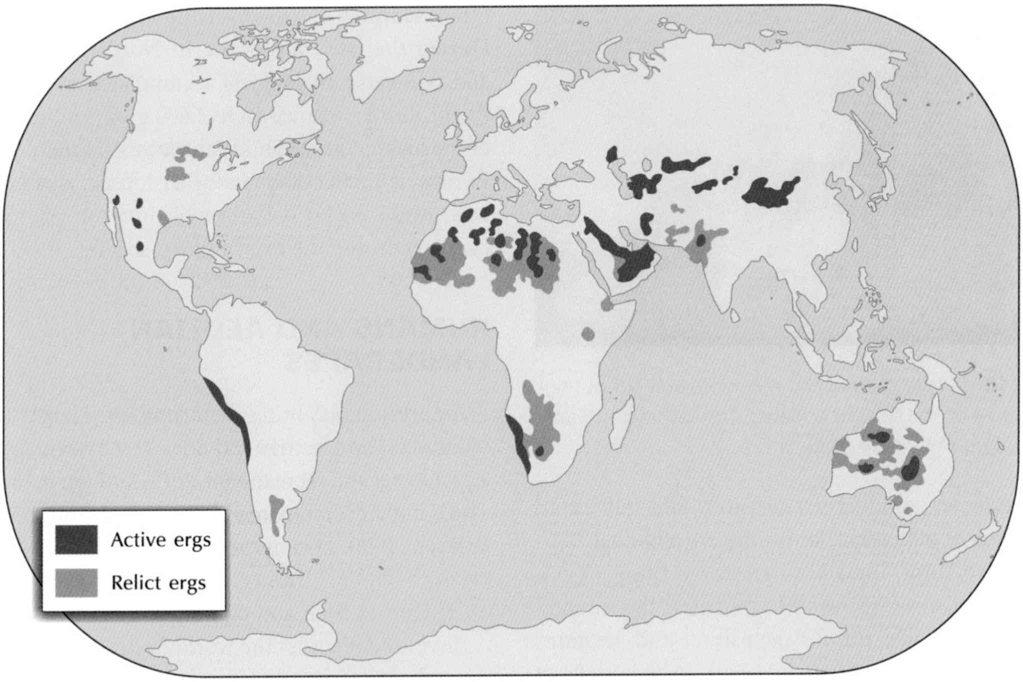

Figure 12.10 World distribution of active and relict ergs. *Sources*: Adapted from Sarnthein (1978) and Wells (1989)

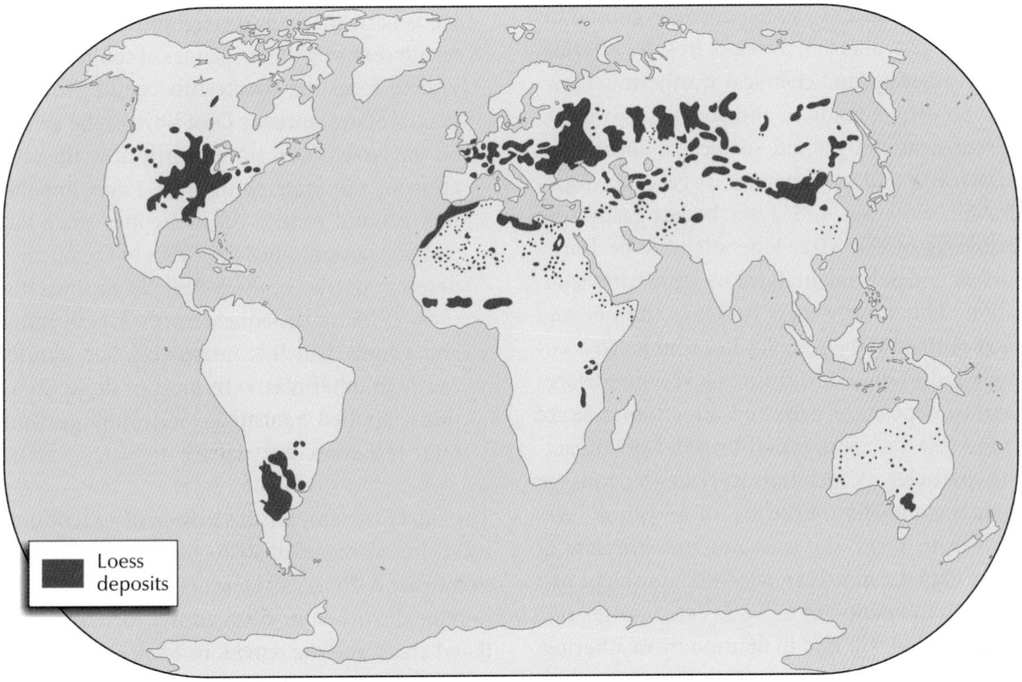

Figure 12.11 World distribution of loess. *Source*: Adapted from Livingstone and Warren (1996, 58)

Plate 12.15 Section through an approximately 15-m-thick loess exposure on the Columbia Plateau in Washington State, USA. (*Photograph by Kate Holden*)

systems, pseudo-karst features, and gullies. In areas of high relief, landslides are a hazard.

To form, loess requires three things: (1) a source of silt; (2) wind to transport the silt; and (3) a suitable site for deposition and accumulation (Pye and Sherwin 1999). In the 1960s, it was thought that glacial grinding of rocks provided the quartz-dominated silt needed for loess formation. It is now known that several other processes produce silt-sized particles – comminution by rivers, abrasion by wind, frost weathering, salt weathering, and chemical weathering. However produced, medium and coarse silt is transported near the ground surface in short-term suspension and by saltation. Vegetation, topographic obstacles, and water bodies easily trap materials of this size. Fine silt may be borne further and be brought down by wet or dry deposition. This is why loess becomes thinner and finer-grained away from the dust source. To accumulate, dust must be deposited on rough surfaces because deposits on a dry and smooth surface are vulnerable to resuspension by wind or impacting particles. Vegetation surfaces encourage loess accumulation. Even so, for a 'typical' loess deposit to form, the dust must accumulate at more than 0.5 mm/year, which is equivalent to a mass accumulation of 625 g/m^2/yr. A lower rate of deposition will lead to dilution by weathering, by mixing by burrowing animals, by mixing with other sediments, and by colluvial reworking. During the late Pleistocene in North America and Western Europe, loess accumulated at more than 2 mm/yr, equivalent to 2,600 g/m^2/yr. Much of the loess in humid mid-latitudes, especially in Europe, is a relict of the Late Pleistocene, when it was produced by deflation of outwash plains (sandar) during the retreat of ice-sheets.

HUMANS AND AEOLIAN LANDSCAPES

Wind erosion may bring about long-term impacts on humans and human activities. It may damage agricultural and recreational lands, and, on occasions, impair human health. As Livingstone and Warren (1996, 144) put it:

> There has been and continues to be massive investment across the world in the control of aeolian geomorphological processes. It has happened in Saharan and Arabian oases for thousands of years; on the Dutch coast since the fourteenth century; on the Danish sandlands particularly in the eighteenth and nineteenth centuries; in the Landes of southwestern France from the nineteenth century; in the United States since the Dust Bowl of the 1930s; on the Israeli coast since shortly after the creation of the State in the late 1940s; on the Russian and central Asian steppes since the Stalinist period; since the 1950s in the oil-rich desert countries of the Middle East; since the early 1970s in the Sahel, North Africa, India and China; and less intensively but significantly in other places. In most of these situations, applied aeolian geomorphology won huge resources and prestige.

The chief problems are the erosion of agricultural soils, the raising of dust storms, and the activation of sand dunes, all of which may result from human disturbance, overgrazing, drought, deflated areas, and the emissions of alkali-rich dust (see Livingstone and Warren 1996, 144–71).

Cases of wind erosion

The **Dust Bowl** of the 1930s is the classic example of wind erosion (Box 12.2). Even greater soil-erosion events occurred in the Eurasian steppes in the 1950s and 1960s. On a smaller scale, loss of soil by wind erosion in Britain, locally called blowing, is a worse problem than erosion by water. The light sandy soils of East Anglia, Lincolnshire, and east Yorkshire, and the light peats of the Fens are the most susceptible. **Blows** can remove up to 2 cm of topsoil containing seeds, damage crops by sandblasting them, and block ditches and roads. Blowing is recorded as long ago as the thirteenth century, but the problem worsened during the 1960s, probably owing to a change in agricultural practices. Inorganic fertilizers replaced farmyard manure, heavy machinery was brought in to cultivate and harvest some crops, and hedgerows were grubbed to make fields better-suited to mechanized farming. Intensively cultivated areas with light soils in Europe are generally prone to wind erosion and the subject of the European Union research project on **Wind Erosion and European Light Soils (WEELS)** (e.g. Riksen and De Graaff 2001). This international project began in 1998 and looked at sites in England, Sweden, Germany, and the Netherlands where serious wind-erosion problems occur. The damage recorded depended very much on landscape factors and land-use. Most on-site damage, mainly in the form of crop losses and the cost of reseeding, occurred in sugar beet, oilseed rape, potato, and maize fields. In the cases of sugar beet and oilseed rape, the costs may be as much as €500 per hectare every five years, although farmers are fully aware of the risk of wind erosion and take preventive measures. In Sweden, measures taken to reduce wind erosivity include smaller fields, autumn sowing, rows planted on wind direction, mixed cropping, and shelterbelts. And measures taken to reduce soil erodibility include minimum tillage, manuring, applying rubber emulsion, watering the soil, and pressing furrows.

Modelling wind erosion

Researchers have devised empirical models, similar in form to the Universal Soil Loss Equation (p. 205), to predict the potential amount of wind erosion under given conditions and to serve as guide to the management practices needed to control the erosion. The **Wind Erosion Equation (WEQ)**, originally developed by William S. Chepil, takes the form:

$$E = (I, C, K, L, V)$$

where E is the soil loss by wind, I is the erodibility of the soil (vulnerability to wind erosion), C is a factor representing local wind conditions, K is the soil surface roughness, L is the width of the field in the direction of the prevailing wind, and V is a measure of the vegetation cover. Although this equation is similar to the USLE, its components cannot be multiplied together to find the result. Instead, graphical, tabular, or computer solutions are required. Originally designed to predict wind erosion in the Great Plains, the WEQ has been applied to other regions in the USA, especially by the Natural Resources Conservation Service (NRCS). However, the WEQ suffered from several drawbacks. It was calibrated for conditions in eastern Kansas, where the climate is rather dry; it was only slowly adapted to tackle year-round changes in crops and soils; it was unable to cope with the complex interplay between crops, weather, soil, and erosion; and it over-generalized wind characteristics.

Advances in computing facilities and databases have prompted the development of a more refined **Wind Erosion Prediction System (WEPS)**, which is designed to replace WEQ. This computer-based model simulates the spatial and temporal variability of field conditions and soil erosion and deposition within fields of varying shapes and edge types and complex topographies. It does so by using the basic processes of wind erosion and the processes that influence the erodibility of the soil. Another **Revised Wind Erosion Equation (RWEQ)** has been used in conjunction

Box 12.2 THE DUST BOWL

The natural vegetation of the Southern Great Plains of Colorado, Kansas, New Mexico, Oklahoma, and Texas is prairie grassland that is adapted to low rainfall and occasional severe droughts. During the 'Dirty Thirties', North American settlers arrived from the east. Being accustomed to more rainfall, they ploughed up the prairie and planted wheat. Wet years saw good harvests; dry years, which were common during the 1930s, brought crop failures and dust storms. In 1934 and 1935, conditions were atrocious. Livestock died from eating excessive amounts of sand, human sickness increased because of the dust-laden air. Machinery was ruined, cars were damaged, and some roads became impassable (Plate 12.16). A report of the time evokes the starkness of the conditions:

> The conditions around innumerable farmsteads are pathetic. A common farm scene is one with high drifts filling yards, banked high against buildings, and partly or wholly covering farm machinery, wood piles, tanks, troughs, shrubs, and young trees. In the fields near by may be seen the stretches of hard, bare, unproductive subsoil and sand drifts piled along

Plate 12.16 Buried machinery in a barn lot in Dallas, South Dakota, USA, May 1936. (*Photograph by United States Department of Agriculture*)

> fence rows, across farm roads, and around Russian-thistles and other plants. The effects of the black blizzards [massive dust storms that blotted out the Sun and turned day into night] are generally similar to those of snow blizzards. The scenes are dismal to the passerby; to the resident they are demoralizing
>
> (Joel 1937, 2)

The results were the abandonment of farms and an exodus of families, remedied only when the prairies affected were put back under grass. The effects of the dust storms were not always localized:

> On 9 May [1934], brown earth from Montana and Wyoming swirled up from the ground, was captured by extremely high-level winds, and was blown eastward toward the Dakotas. More dirt was sucked into the airstream, until 350 million tons were riding toward urban America. By late afternoon the storm had reached Dubuque and Madison, and by evening 12 million tons of dust were falling like snow over Chicago – 4 pounds for each person in the city. Midday at Buffalo on 10 May was darkened by dust, and the advancing gloom stretched south from there over several states, moving as fast as 100 miles an hour. The dawn of 11 May found the dust settling over Boston, New York, Washington, and Atlanta, and then the storm moved out to sea. Savannah's skies were hazy all day 12 May; it was the last city to report dust conditions. But there were still ships in the Atlantic, some of them 300 miles off the coast, that found dust on their decks during the next day or two.
>
> (Worster 1979, 13–14)

with GIS databases to scale up the field-scale model to a regional model (Zobeck *et al.* 2000). An **integrated wind-erosion modelling system**, built in Australia, combines a physically based wind-erosion scheme, a high-resolution atmospheric model, a dust-transport model, and a GIS database (Lu and Shao 2001). The system predicts the pattern and intensity of wind erosion, and especially dust emissions from the soil surface and dust concentrations in the atmosphere. It can also be used to predict individual duststorm events.

Desertification

In 1949, Auguste Aubréville, a French forester, noticed that the Sahara Desert was expanding into surrounding savannahs and coined the term **desertification** to describe the process. The term became widely known in the 1970s when a ruinous drought in the Sahel region of Africa led to the United Nations Conference on Desertification (UNCOD) in 1977, which showed that the process was probably occurring in all the world's drylands. The topic has since generated a huge literature, a legion of definitions, a collection of world maps, and much controversy. In essence, the process of desertification degrades land in arid, semi-arid, and dry subhumid areas, reducing the land's capacity to accept, store, and recycle water, energy, and nutrients. The primary causes of desertification are climatic variations, ecological change, and socio-economic factors, although the details of cause and effect are complex. At root, desertification occurs because dryland ecosystems are vulnerable to certain climatic changes and overexploitation and unsuitable land use – drought, poverty, political instability, deforestation, overgrazing by livestock, overcultivation, and bad irrigation practices can all weaken the land's fertility and allow degradation to take hold. Soil compaction and crusting, quarrying, and desert warfare may also be causative factors in some cases. Whatever its causes, desertification directly affects over 250 million people, and puts at risk some 1 billion people in over a hundred countries, which is why it has generated so much research, to which physical geographers have made valuable contributions.

Wind erosion can be an important factor in desertification. In the Sahel region of Africa centred on southern Mali, wind erosion of degraded soils leads to high burdens of atmospheric dust that travels thousands of kilometres over Africa and the tropical Atlantic, altering radiation and water balances. Cyril Moulin and Isabelle Chiapello (2006) established a direct correlation between dust optical thickness (a measure of dust content in the air) and the severity of wind erosion over the last two decades.

AEOLIAN LANDSCAPES IN THE PAST

'The Earth's most imposing aeolian landforms are inherited rather than products of contemporary processes' (Livingstone and Warren 1996, 125). Why should this be? The answer seems to lie in the changing windiness of the planet and in the changing distribution of arid desert environments.

A drier and windier world

The Earth is calm at present. During periods of the Pleistocene, and notably around the last glacial maximum, some 20,000 years ago, it was much windier and, in places, drier. Many aeolian features are inherited from those windy times in the Pleistocene when episodes of aeolian accumulation occurred in the world's drylands. Some sand seas expanded considerably and accumulated vast quantities of sand. Areas of expansion included the Sahel in northern Africa, the Kalahari in southern Africa, the Great Plains in the central USA, and large parts of Hungary and central Poland. Grass and trees now fix many of these inherited sand accumulations. Inherited Pleistocene landforms include the largest desert dunes, mega-yardangs as seen in the Tibesti region of the Sahara, and loess deposits, some 400 m thick, that cover about 10 per cent of the land area. High winds of the Pleistocene were also the main contributors to the large thickness of dust on ocean floors.

How do geomorphologists distinguish ancient dune systems from their modern counterparts? Several lines of biological, geomorphic, and sedimentological evidence are used to interpret the palaeoenvironments of aeolian deposits (e.g. Tchakerian 1999) (Table 12.4). Dune surface vegetation is a piece of biological evidence. Geomorphic evidence includes dune form, dune mobility, dune size, and dune dating. Sedimentological evidence includes granulometric analysis, sedimentary structures, grain roundness, palaeosols and carbonate horizons, silt and clay particles, dune reddening, scanning electron microscopy of quartz grain microfeatures, and aeolian dust.

By using methods of palaeoenvironmental reconstruction and dating, reliable pictures of Pleistocene changes in the world's drylands are emerging. The Kalahari sand sea was once much larger, covering 2.5 million km^2. This Mega-Kalahari sand sea now consists mainly of linear dunes bearing vegetation interspersed with dry lakes (Thomas and Shaw 1991). Luminescence dating shows that the three chief linear dunefields present in the Mega-Kalahari – the northern, southern, and eastern – were active at different times during the late Quaternary (Stokes *et al.* 1997). In the south-western portion of the sand sea, two dune-building (arid) episodes occurred, one between 27,000 and 23,000 years ago and the other between 17 and 10 million years ago. In the north-eastern portion, four dune-building episodes occurred at the following times: 115,000–95,000 years ago, 46,000–41,000 years ago, 26,000–22,000 years ago, and 16,000–9,000

Table 12.4 Evidence used in reconstructing dune palaeoenvironments

Evidence	Explanation
Biological evidence	
Dune vegetation	Presence of dune vegetation indicates reduced aeolian activity and dune stabilization
Geomorphological evidence	
Dune form	Degraded or wholly vegetated dunes in areas not presently subject to aeolian activity (with mean annual rainfall less than 250 mm) indicate relict dunes
Dune mobility	A 'dune mobility index' (Lancaster 1988) indicates whether dunes are active or inactive[a]
Dune size	Mega-dunes may form only during sustained high winds, as blew in the tropical deserts around the peak of the last ice age around 20,000 years ago
Dune dating	Relative or absolute dating techniques may be used to fix the age of a dune, luminescence dating being a promising approach in environments where organic remains are very limited
Sedimentological evidence	
Granulometric analysis	Standard granulometric measures – mean grain size, sorting (standard deviation), skewness, and kurtosis – (measuring the 'peakedness' of a distribution) may sometimes be used to distinguish ancient from modern dunes
Sedimentary structures	Primary structures may be altered or destroyed by processes after deposition, but may help in identifying past aeolian beds
Grain roundness	Active aeolian sand grains tend to be sub-rounded to sub-angular; ancient sand grains tend to be more rounded, but roundness also varies with dune type
Palaeosols and carbonate horizons	When found in aeolian accumulations, these suggest periods of geomorphic stability and act as useful dating markers
Silt and clay particles	Ancient dunes tend to contain a higher proportion of silt and clay particles than active dunes
Dune reddening	Ancient dune sediments tend to be redder than modern dune sediments, though factors determining the redness of sediments are complex and ambiguous
Quartz surface microfeatures	Scanning electron microscope analysis of sand grains may help to identify aeolian sediments and to distinguish between different depositional environments
Aeolian dust	May be found in alluvial fans, soils, and marine sediments

Notes:

a The dune mobility index, M, is defined as the length of time the wind blows above the threshold velocity for sand transport (5 m/s), W, multiplied by the precipitation–potential evapotranspiration ratio, P/PE: $M = W/(P/PE)$. Lancaster (1988) suggests four classes of dune activity: (1) inactive dunes ($M < 50$); (2) dune crests only active ($50 < M < 100$); (3) dune crests active, lower windward and slip faces and interdune depressions vegetated ($100 < M < 200$); (4) fully active dunes ($M > 200$)

Source: Adapted from discussion in Tchakerian (1999)

years ago. The arid, dune-building phases lasted some 5,000 to 20,000 years, while the intervening humid periods lasted longer – between 20,000 and 40,000 years. Figure 12.12 shows the compounded nature of large, complex, linear dunes in the Akchar Erg, Mauritania (Kocurek *et al.* 1991). The dune core consists of Pleistocene sand laid down 20,000 to 13,000 years ago. When rainfall increased, from 11,000 to 4,500 years ago, vegetation stabilized the dunes, soil formation altered the dune sediments, and lakes formed between the dunes. Renewed dune formation after 4,000 years ago cannibalized existing aeolian sediments on the upwind edge of the sand sea. The active

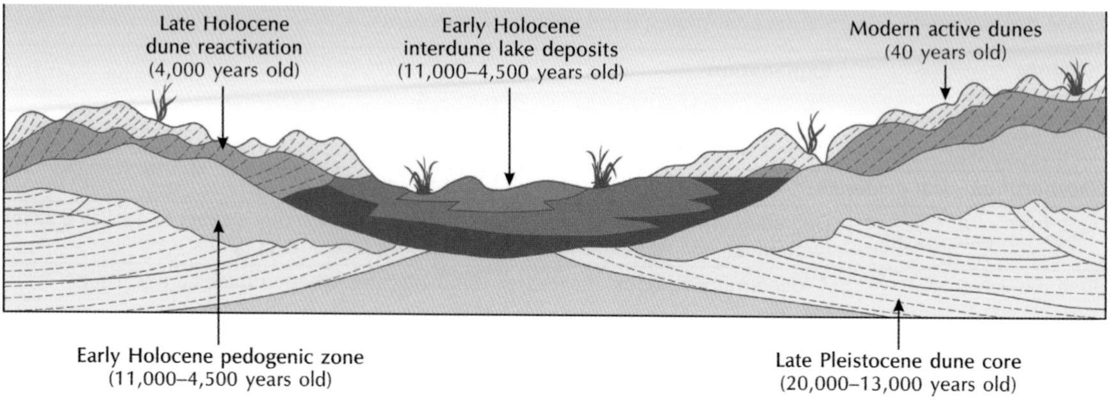

Figure 12.12 Amalgamated deposits of linear dunes in the Akchar Sand Sea, Mauritania. *Source*: Adapted from Kocurek *et al.* (1991)

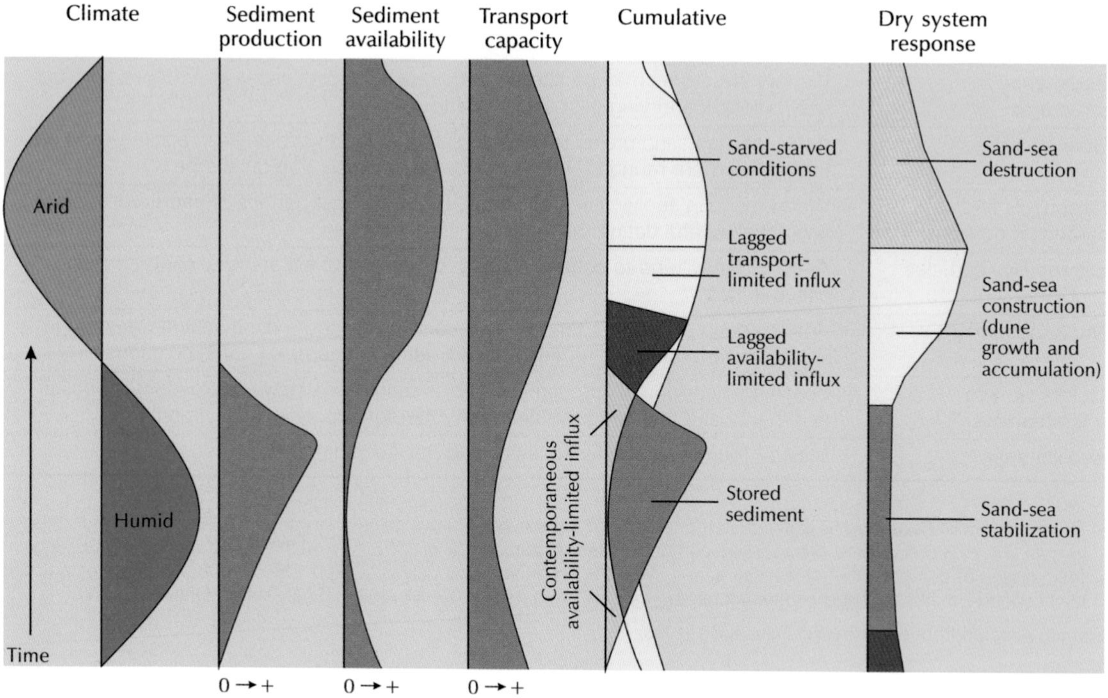

Figure 12.13 Process–response model for Saharan sand seas based on sediment production, sediment availability (supply), and transport capacity. The system is driven by a climatic cycle from humid to arid, shown on the left. An explanation is given in the text. *Source*: Adapted from Kocurek (1998)

Box 12.3 DUNE PATTERNS IN SAND SEAS: THE HISTORICAL DIMENSION

The dune forms in a sand sea are primarily a response to wind conditions and sand supply. Nonetheless, the pattern of dunes in many sand seas is much more intricate and requires more complex explanations (Lancaster 1999). Recent research points to the significance of sea-level and climatic changes in affecting sediment supply, sediment availability, and wind energy. The upshot of such changes is the production of different generations of dunes. So the varied size, spacing, and nature of dunes in sand seas catalogue changes in sand supply, sand availability, and sand mobility that have produced many superimposed generations of dune forms, each of a distinct type, size, alignment, and composition. In addition, the large dunes that characterize sand seas – compound dunes and complex dunes, megadunes, and draa – commonly seem to be admixtures of several phases of dune building, stabilization, and reworking. The indications are that, rather than being solely the production of contemporary processes, the form of sand seas is partly inherited, and to unlock the historical processes involved requires investigations of past conditions affecting sand accumulation. Vast sand accumulations take much time to grow. Ergs with very large dunes, as in the Arabian Peninsula, North Africa, and central Asia, may have taken a million or more years to form (Wilson 1971). Certainly, cycles of climatic change during the Quaternary period, involving swings from glacial to interglacial conditions, have played a key role in influencing sediment supply, availability, and mobility. Furthermore, different sand seas may react differently to sea-level and climatic changes. A crucial factor appears to be the size of the sand source. Where the sand supply is small, as in the Simpson Desert and the Akchar sand sea of Mauritania, the chief control on aeolian accumulation is sediment availability, and sand seas suffer multiple episodes of dune reworking. Where sand supply is plentiful, as in the Gran Desierto, Namib, and Wahiba sand seas, the accumulation of sand is effectively unlimited and multiple dune generations are likely to develop. A third possibility, which applies to the Australian Desert, is where sand accumulation is limited by the transporting capacity of the wind.

crescentic dunes that cap the older, linear dunes date from the last forty years.

It is possible that major dune production episodes relate to Croll–Milankovitch climatic cycles, which induce swings from glacial to interglacial climates. Gary Kocurek (1998, 1999) has presented a model relating the two (Figure 12.13). The key feature of the model is the interplay of sediment production, sediment availability, and transport capacity through a humid–arid cycle. During the humid period, geomorphic processes produce sediment, but this becomes available only during the arid period. The wind is capable of transporting sediment throughout the cycle, but its transport capacity is higher during the humid phase. The combined effects of these changes are complex. The humid phase sees sediment production and storage, with some sediment influx limited by availability. As the humid phase gives way to the arid phase, sediment influx increases as availability increases. It goes on increasing to the peak of the arid phases as transporting capacity rises to a maximal level. As the arid phase starts to decline, the lack of sediment production leads to sand-starved conditions. The dune-fields respond to these changes as follows.

During the humid phase, the dunes stabilize. As the arid phase kicks in, dune-building occurs using sediments released by increased availability and then increased transport capacity. Once the sediment supply dries up, the dunes are destroyed. This plausible model requires detailed field-testing.

The pattern of dunes within sand seas appears to involve several factors with a historical dimension (Box 12.3).

Ancient aridity

The distribution of desert climates has shifted during geological time. Most modern areas of aridity began during late Tertiary times, and especially in the Mid- to Late Miocene, as the climates of subtropical regions took on a modern aspect.

In the more distant past, the geological record of aeolian sandstones and evaporite deposits furnishes evidence for extensive deserts at several times in the Earth's history. The oldest aeolian deposits discovered so far come from Precambrian rocks in the Northwest Territories of Canada and from India. In Britain, the oldest aeolian deposits are of Devonian age and were formed when Britain lay south of the Equator in an arid and semi-arid palaeoclimatic belt. Remnants of large star dunes have been identified in Devonian sandstones of Scotland and fossil sand seas in Ireland. The best-known aeolian sandstones in Britain and northern Europe occur in Permo-Triassic rocks deposited when Britain had moved north of the Equator and into another arid climatic zone. The Rotliegendes (Early Permian) sandstones of the North Sea basin trap oil and gas. Quarry sections at Durham, England, reveal large linear mega-dunes with smaller features superimposed. Some of the Triassic sandstones of Cheshire and Lancashire are also aeolian deposits.

SUMMARY

Wind erodes dry, bare, fine-grained soils and sediments. It is most effective in deserts, sandy coasts, and alluvial plains next to glaciers. Wind erodes by deflating sediments and sandblasting rocks. Particles caught by the wind bounce (saltation), hop (reptation), 'float' (suspension), or roll and slide (creep). Wind deposits particles by dropping them or ceasing to propel them along the ground. Several landforms are products of wind erosion. Examples are lag deposits and stone pavements, deflation hollows and pans, yardangs and *Zeugen*, and ventifacts. Sand accumulations range in size from ripples, through dunes, to dunefields and sand seas. Dunes may be grouped into free and anchored types. Free dunes include transverse dunes, seifs, star dunes, and zibars. Anchored dunes form with the help of topography or vegetation. They include echo dunes, falling dunes, parabolic dunes, and coastal dunes. Dunefields and sand seas are collections of individual dunes. The largest sand sea – the Rub' al Khali of Saudi Arabia – occupies 770,000 km^2. Loess is an accumulation of windblown silt particles and covers about 5–10 per cent of the land surface. Wind erosion can often be a self-inflicted hazard to humans, damaging agricultural and recreational land and harming human health. Several models predict wind erosion at field and regional scales, the latest examples combining physical processes with GIS databases and atmospheric models. Many aeolian landforms are inherited from the height of the last ice age when the planet was drier and windier. The geological record registers colder times when aridity prevailed.

ESSAY QUESTIONS

1 How does wind shape landforms?

2 How do sand dunes form?

3 Discuss the problems and remedies of soil erosion by wind.

4 Compare and contrast sediment transport by wind and by water.

FURTHER READING

Cooke, R. U., Warren, A., and Goudie, A. S. (1993) *Desert Geomorphology*. London: UCL Press.
A comprehensive and clear account of form and process in arid and semi-arid environments.

Goudie, A. S., Livingstone, I., and Stokes, S. (eds) (1999) *Aeolian Environments, Sediments and Landforms*. Chichester: John Wiley & Sons.
Perhaps a little heavy for the neophyte, but full of excellent papers.

Lancaster, N. (1995) *Geomorphology of Desert Dunes*. London: Routledge.
If you are interested in sand dunes, then look no further.

Livingstone, I. and Warren, A. (1996) *Aeolian Geomorphology: An Introduction*. Harlow, Essex: Longman.
The best introduction to the subject. A must for the serious student.

Thomas, D. S. G. (ed.) (2011) *Arid Zone Geomorphology: Process, Form and Change in Drylands*, 3rd edn. Chichester: John Wiley & Sons.
An excellent collection of essays that is full of interesting ideas and examples.

13

CHAPTER THIRTEEN

COASTAL LANDSCAPES

The relentless buffeting of coasts by waves and their perpetual washing by currents fashion a thin line of unique landforms. This chapter covers:

- Waves, tides, and currents
- Coastal processes
- Cliffs, caves, and other erosional coastal landforms
- Beaches, barriers, and other depositional coastal landforms
- Humans and coasts
- Past coastal landscapes

CLIFF RETREAT: THE BEACHY HEAD ROCKFALL

Beachy Head and the Seven Sisters cliffs to the west are made of a hard chalk and stand up to 160 m above sea level along the southern coast of England. On 10 and 11 January 1999, the upper part of Beachy Head collapsed. The fallen portion was about 70 m long and 17 m deep with a mass of 50,000–100,000 tonnes, which buried the cliff base and is seen as 5–10 m of accumulated chalk rubble with blocks up to 4 m in diameter. The failure occurred along one of the vertical joints or minor fault planes that are common in the chalk. To the east of the main slip, a fracture some 2 m deep and 1 m wide and extending up to 10 m from the cliff edge has appeared. It runs down most of the cliff face. The toe of the chalk rubble erodes at high tide, when it produces a sediment plume that runs from just east of Beachy Head Cave to the unmanned Beachy Head Lighthouse. The cause of the rockfall is not certain, but 1998 was a wetter year than normal, and during the fortnight before the fall heavy rain fell on most days. The wet conditions may have increased the pore pressures in the chalk and triggered the rockfall. Events such as the Beachy Head rockfall have been occurring for thousands of years along southern and eastern English coastlines and have led to cliff retreat.

COASTAL ENVIRONMENTS

Classifying coasts

Coasts are difficult to classify, chiefly because a range of disciplines studies them, each discipline with its own interest in coasts, and because they cross several geographical scales (for example, a beach between two headlands, a single coastline, a full continental coastline) and several timescales (days and years, centuries, millennia, and millions of years). Rhodes W. Fairbridge (2004) suggested that a description of a given coast demands a minimum of three terms covering (Table 13.1): **coastal material** (hard or soft, soluble or non-soluble otherwise); **coastal agencies** (erosive, constructive, physical, chemical, biological, and their geographical setting – latitude, exposure, fetch); and **historical factors** (geotectonic, glacio-isostatic, eustatic, steric, human timescales). This scheme embraces the main coastal features and processes of interest to geomorphologists, many of which are discussed below.

Waves

Waves are undulations formed by wind blowing over a water surface. Turbulence in airflow generating pressure variations on the water causes them. Once formed, waves help to disturb the airflow and are partly self-sustaining. Energy is transferred from the wind to the water within the wave-generation area. The amount of energy transfer depends upon the wind speed, the wind duration (how long the wind blows), and the fetch (the extent of water over which the wind blows). **Sea waves** are formed by the wind within the generation area. They often have short crests and steep cross-sections, and are irregular. In mid-ocean, prolonged strong winds associated with severe storms and blowing over hundreds of kilometres produce waves more than 20 m high that travel up to 80 km/hr. On passing out of the generation area, sea waves become **swell waves** (or simply **swell**) and they are more regular with longer periods and longer crests. They may travel thousands of kilometres across oceans.

Waves formed in water deep enough for **free orbital motion** to occur are called **waves of oscillation**. The motion is called 'free orbital'

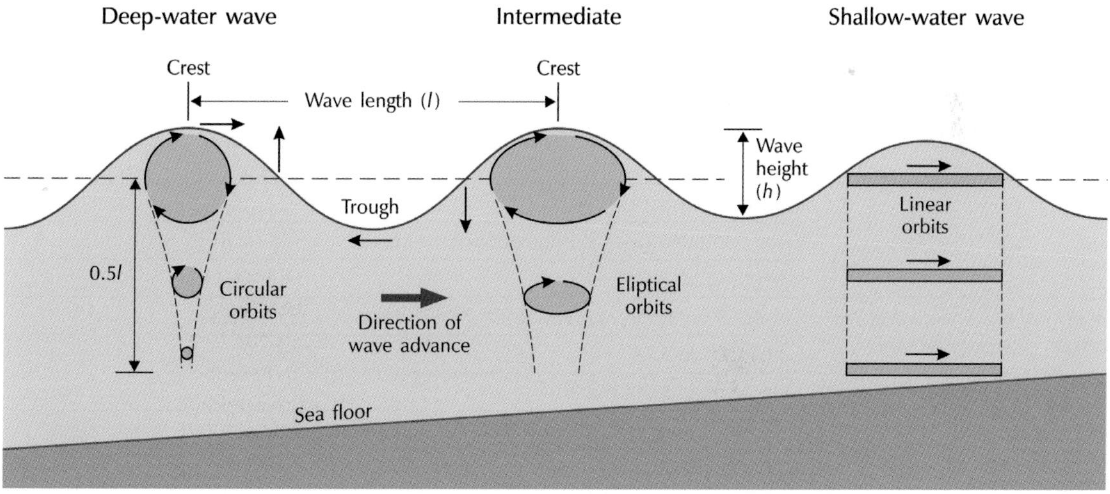

Figure 13.1 Terms associated with waves, including the orbital motion of waves in deep, intermediate, and shallow water. *Source*: Adapted from Komar (1998, 166)

Table 13.1 Fairbridge's classification of coasts

Basic elements	Explanation
Coastal material	
Soft, weakly consolidated, easily erodible (relatively soluble in seawater and rainwater; create mudflats, beaches, bluffs, and low cliffs)	Relatively insoluble: detrital products such as mud, silt, sand, gravel, boulders (loose)
	Relatively soluble: reef limestones; bioclastic carbonate debris (foraminifera, calcareous algae, mollusca, coral); beachrock and aeolianite
	Pre-weathered hard rocks: 'grusification' or reduction in hot-wet tropics to grus or crumble, leaving unweathered corestones within easily eroded saprolite
	Hard concretions (such as cherts or flints) released by differential wave erosion to create cobble or 'shingle' beaches
	Volcanic materials (interlayered lavas, pumice, ash, or lapilli (p. 135)), reduced by wave action to boulders, black sands, and so on
Hard rock and cliffed coasts	Longevity of hard-rock coasts
	Anomalous hard-rock boulders due to diachronous sea-ice transport
	Landsliding, with rotational slip
	Landsliding on volcanic cones, with control of atoll form
	Fault-controlled cliffs
	Fossil or 'dead' cliffs (*falaises mortes* in French – cliffs no longer in contact with water)
Physical setting	
Latitude	Solar radiation, seasonality, and weathering potential
Climate	Prevailing winds, storms, sea ice
Fetch	Open water for wave approach
Offshore bathymetry	Wave regime and longshore currents
Tides	Diurnal, fortnightly, seasonal, 18.6-year lunar nodal
Tsunami potential	Volcanoes, submarine slides
Homogeneity	Beach extent, headland frequency
Erosive agencies	
Physical agencies	Abrasion
	Hydraulic impact
	Wind and tide-driven ice floes and icebergs
	Ice-foot (glacial)
Chemical agencies (water, carbon dioxide, methane)	Subaerial weathering preparation (mainly feldspars and micas)
	Carbonate rocks
Biological agencies	Mangrove and salt marsh
	Limestone and uplifted coral reef undercuts populated by borers and scrapers
	Barnacles, footing solution
	Echinoids and boring molluscs
	Kelp and other algal holdfasts

Table 13.1 *continued*

Basic elements	Explanation
History	
Plate margin setting	Extensional, passive, trailing edge, or pull-apart plate margins
	Collision or intermediate-type (back-arc) plate margins
Isostatic readjustments	Geotectonic, vertical motions associated with plate rupture
	Glacio-isostatic, crustal response to glacial loading and unloading (including marginal bulge effect)
	Hydro-isostatic crustal response to water loading
Geoidal readjustments	Earth's spin rate
	Mass loading
	Atmospheric pressure
	Winds and currents
Steric changes	Volumetric response to thermal and salinity changes
Eustatic changes	Tectono-eustatic
	Sedimento-eustatic
	Glacio-eustatic
Artificial or human-made coasts	

Source: Based on Fairbridge (2004)

because the chief movement of the water is roughly circular in the direction of flow, moving forwards on the crest, upwards on the front, backwards in the trough, and downwards on the back (Figure 13.1). Water moves slowly in the direction of wave propagation because water moves faster on the crest than in the troughs. Oscillatory waves form **wave trains**. **Solitary waves** or **waves of translation**, in contrast, involve water moving in the direction of propagation without any compensatory backward motion. They are single, independent units and not associated with wave trains. They lack the distinct crests and troughs of oscillatory waves and appear as weals separated by almost flat water surfaces and are effective transporters and eroders of sediments and rocks. The breaking of oscillatory waves often generates them.

Once waves approaching a coastline 'feel bottom', they slow down. The waves crowd together, and their fronts steepen. **Wave refraction** occurs because the inshore part of a wave crest moves more slowly than the offshore part, owing to the shallow water depth, and the offshore part swings forwards and the wave crests tend to run parallel to the depth contours. Figure 13.2 shows wave refraction near a submarine canyon and a headland.

Eventually, the waves lunge forward or break to form surf. In breaking, waves of oscillation convert to waves of translation and rush up the beach as **swash**. After having attained its farthermost forward position, the water runs down the seaward slope as **backwash**. Four types of breaking wave are recognized: spilling, plunging, collapsing, and surging (Figure 13.3). **Spilling breakers** give the appearance of foam cascading down from the peaking wave crest. **Plunging breakers** have waves curling over and a mass of water collapsing on to the sea surface (Plate 13.1). **Collapsing breakers** have wave crests peaking as if about to plunge, but the base of the wave then rushes up the shore as a thin layer of foaming water. **Surging breakers** retain a smooth wave

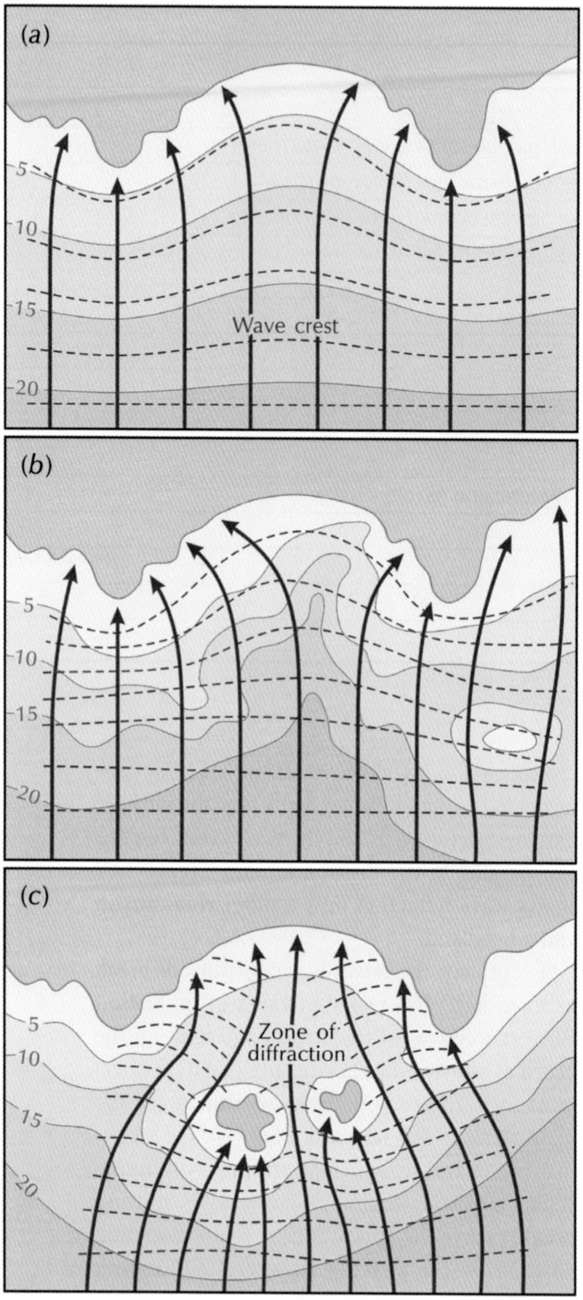

Figure 13.2 Wave refraction approaching headlands and bays above differing offshore topography. *Source*: Adapted from Bird (2000, 11)

form with no prominent crest as they slide up the shore, entraining little air in the act. The occurrence of these waves depends upon the deep-water wave height and the bottom slope. For a given deep-water wave height, waves will spill, plunge, collapse, and surge with increasing bottom slope. Spilling waves require a slope of less than about 11°, plunging waves up to 36°, collapsing waves up to 50°, and surging waves more than 50°.

Breaking waves are either **constructive** or **destructive**, depending on whether they cause a net shoreward or a net seaward movement of beach material. As a rule of thumb, surging, spilling, and collapsing breakers create a strong swash and gentle backwash and tend to be constructive, washing sediment on to a beach. Plunging waves have a relatively short swash and longer backwash, and tend to be destructive, removing material from a beach.

Nearshore currents

Currents created in the nearshore zone have a different origin from ocean currents, tidal currents, and wind-induced currents. **Nearshore currents** are produced by waves. They include longshore currents, rip currents, and offshore currents. **Longshore** or **littoral currents** are created when waves approach a coastline obliquely. They dominate the surf zone and travel parallel to the coast. **Rip currents**, or **rips**, are fed by longshore currents and develop at more or less regular intervals perpendicularly to the beach and flow through the breaker zone. They are strong currents and dangerous to swimmers. **Onshore currents** are slower and develop between rip currents. Even where waves approach a coastline head on, a nearshore circulation of longshore currents, rip currents, and onshore currents may evolve.

Tsunamis

Tsunamis are commonly produced by faulting of the sea floor, and much less commonly by

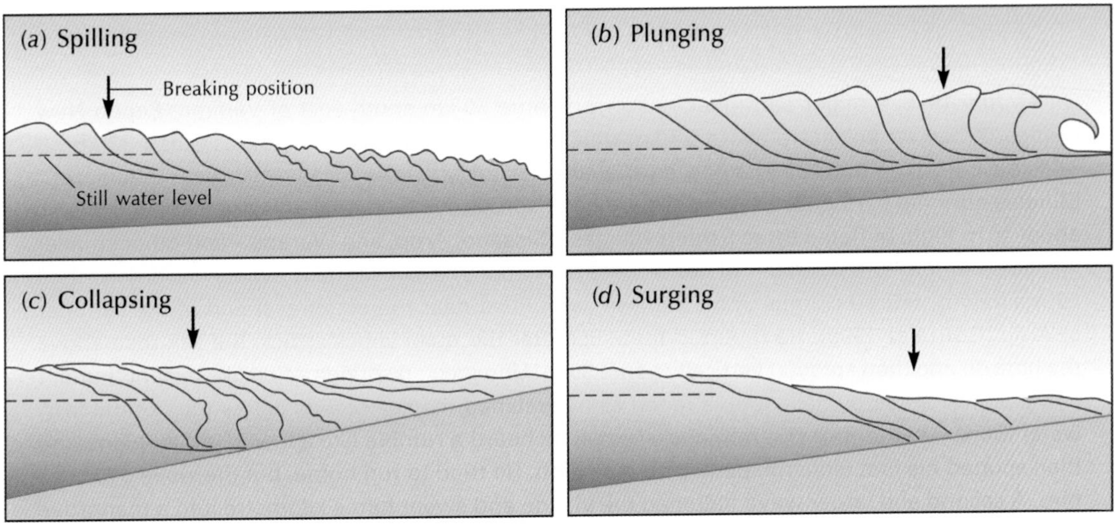

Figure 13.3 Kinds of breaking waves derived from high-speed moving pictures. *Source*: Adapted from Komar (1998, 210)

volcanic eruptions, landslides or slumping, or by impacting asteroids and comets. They are also referred to as tidal waves, although they bear no relation to tides and are named after the Japanese word meaning 'harbour wave'. The pushing up of water by sudden changes in the ocean floor generates a tsunami. From the site of generation, a tsunami propagates across the deep ocean at up to 700 km/hr. While in the deep ocean a tsunami is not perceptible as it is at most a few metres high with a wavelength about 600 times longer than its height. On approaching land, a tsunami slows down to around 100 km/hr and grows in height by a factor of about ten. It rushes ashore, either as a tide-like flood, or, if wave refraction and shoaling allow, a high wall of water.

Tsunamis occur on a regular basis. The historical average of reported tsunamis is fifty-seven tsunamis per decade, but in the period 1990–99 eighty-two were reported, ten of which were generated by earthquakes associated with plate collisions around the Pacific Rim and killed more than 4,000 people (Box 13.1).

Tides

Tides are the movement of water bodies set up by the gravitational interaction between celestial bodies, mainly the Earth, the Moon, and the Sun. They cause changes of water levels along coasts. In most places, there are **semi-diurnal tides** – two highs and two lows in a day. **Spring tides**, which are higher than normal high tides, occur every 14–75 days when the Moon and the Sun are in alignment. **Neap tides**, which are lower than normal low tides, alternate with spring tides and occur when the Sun and the Moon are positioned at an angle of 90° with respect to the Earth. The form of the wave created by tides depends upon several factors, including the size and shape of the sea or ocean basin, the shape of the shoreline, and the weather. Much of the coastline around the Pacific Ocean has mixed tides, with highs and lows of differing magnitude in each 24-hour period. Antarctic coasts have diurnal tides with just one high and one low every 24 hours.

Tidal ranges have a greater impact on coastal processes than tidal types. Three tidal ranges are distinguished – **microtidal** (less than 2 m),

Box 13.1 THE 1998 PAPUA NEW GUINEAN TSUNAMI

On 17 July 1998, a major earthquake occurred some 70 km south-east of Vanimo, Papua New Guinea. It had an epicentre about 20 km offshore and a depth of focus of less than 33 km. It registered a magnitude of 7.1. The earthquake stirred up three locally destructive tsunamis. Minutes after the earthquake rocked the area, the successive tsunamis, the largest of which was about 10 m high, buffeted three fishing villages – Sissano, Arop, and Warapu – and other smaller villages along a 30-km stretch of coast west of Atape. The subsequent events were described by a survivor, retired colonel John Sanawe, who lived near the south-east end of the sandbar at Arop (González 1999). He reported that, just after the main shock struck, the sea rose above the horizon and then sprayed vertically some 30 m. Unexpected sounds – first like distant thunder and then like a nearby helicopter – faded as he watched the sea recede below the normal low-water mark. After four or five minutes' silence, he heard a rumble like a low-flying jet plane and then spotted his first tsunami, perhaps 3–4 m high. He tried to run home, but the wave overtook him. A second and larger wave flattened the village and swept him a kilometre into a mangrove forest on the inland shore of the lagoon. Other villagers were not so lucky. Some were carried across the lagoon and became impaled on broken mangrove branches. Many were struck by debris. Thirty survivors eventually lost limbs to gangrene, and saltwater crocodiles and wild dogs preyed on the dead before help could arrive. The rush of water swept away two of the villages, one on the spit separating the sea from Sissano lagoon. A priest's house was swept 200 m inland. At Warapu and at Arop no house was left standing, and palm and coconut trees were torn out of the ground. In all, the tsunamis killed more than 2,200 people, including 240 children, and left more than 6,000 people homeless. About 18 minutes after the earthquake, the sea was calm again and the sand bar barren, with bare spots marking the former site of structures.

Plate 13.1 Plunging waves, Narrabeen Beach, Sydney, Australia. (*Photograph by Andrew Short*)

mesotidal (2 to 4 m), and **macrotidal** (more than 4 m) – corresponding to small, medium, and large tidal ranges (Figure 13.4). A large tidal range tends to produce a broad intertidal zone, so waves must cross a wide and shallow shore zone before breaking against the high-tide line. This saps some of the waves' energy and favours the formation of salt marshes and tidal flats. The greatest tidal ranges occur where the shape of the coast and the submarine topography effect an oscillation of water in phase with the tidal period. The tidal range is almost 16 m in the Bay of Fundy, north-eastern Canada. Some estuaries, such as the Severn Estuary in England, with high tidal ranges develop **tidal bores**, which are usually single waves up to several metres high that form as incoming tidal flow suffers drag on entering

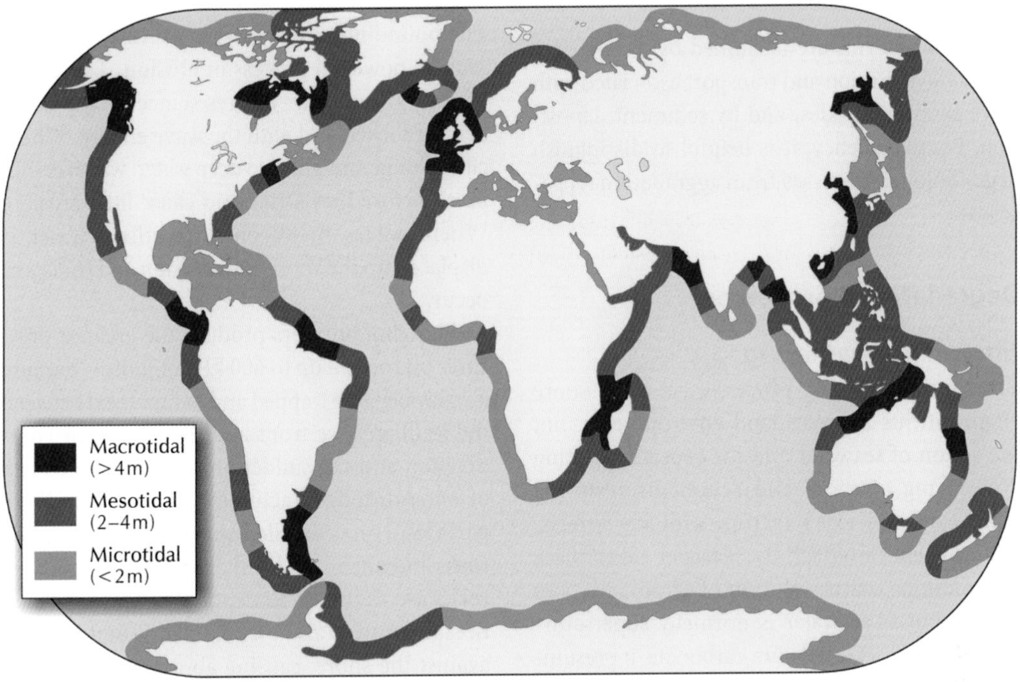

Figure 13.4 Global pattern of tidal range. The ranges indicated are for spring tides. *Source*: Adapted from Davies (1980, 51)

shallower water. Tidal bores run at up to 30 km/hr and are effective agents of erosion. Small tidal ranges encourage a more unremitting breaking of waves along the same piece of shoreline, which deters the formation of coastal wetlands.

Tides also produce tidal currents that run along the shoreline. They transport and erode sediment where they are strong, as in estuaries. Currents associated with rising or flood tides and falling or ebb tides often move in opposite directions.

Wave and tide dominance

Coasts are commonly classed as wave-dominated (with microtidal ranges) or tide-dominated (with mesotidal ranges). Each type tends to produce a distinct coastal morphology. However, while there is little doubt that the relative effects of waves and tides are extremely important in understanding coastal landforms, wave energy conditions are also significant (Davis and Hayes 1984). This is evident in the fact that some wave-dominated coasts have almost no tidal range, whilst some tide-dominated coasts have very small tidal ranges.

The interplay of waves and tides has a huge control over beach formation. Important factors involved are breaking wave height, wave period, spring tidal range, and sediment size. The three chief types are wave-dominated beaches, tide-modified beaches, and tide-dominated beaches (Anthony and Orford 2002; Short 2006; Short and Woodroffe 2009). In brief, wave-dominated beaches occur where waves accompany microtidal ranges. Tide-modified beaches occur in areas of higher tide range exposed to persistent waves. Tide-dominated beaches occur where very low waves accompany areas of a higher tide range. These beach types will be explored more fully later in the chapter.

COASTAL PROCESSES

Coastal landforms are fashioned by weathering, by sediment erosion and transport associated with wave action and tides, and by sediment deposition. For expediency, it is helpful to distinguish degradational processes from aggradational processes.

Degradational processes

Shoreline weathering

The same weathering processes act upon shore environments as upon land environments, but the action of seawater and the repeated wetting and drying of rocks and sediments resulting from tides are extra factors with big effects. Direct chemical attack by seawater takes place on limestone coasts: solution of carbonate rocks occurs, but as seawater is normally supersaturated with respect to calcium carbonate, it presumably takes place in rock pools, where the acidity of the seawater may change. Salt weathering is an important process in shoreline weathering, being most effective where the coastal rocks are able to absorb seawater and spray. As tides rise and fall, so the zone between the low-water mark and the highest limit reached by waves and spray at high tide is wetted and dried. Water-layer weathering is associated with these wetting and drying cycles. Biological erosion, or **bioerosion**, is the direct action of organisms on rock. It is probably more important in tropical regions, where wave energy is weak and coastal substrates are home to a multitude of marine organisms. Tactics employed by organisms in the erosive process are chemical, mechanical, or a mixture of the two. Algae, fungi, lichens, and animals without hard parts are limited to chemical attack through secretions. Algae, and especially cyanobacteria, are probably the most important bioerosional instruments on rock coasts. Many other animals secrete fluids that weaken the rock before abrading it with teeth and other hard parts. Grazing animals include gastropods, chitons, and echinoids (p. 167).

Wave erosion

The pounding of the coast by waves is an enormously powerful process of erosion. The effects of waves vary with the resistance of the rocks being attacked and with the wave energy. Where cliffs plunge straight into deep water, waves do not break before they strike and cause little erosion. Where waves break on a coastline, water is displaced up the shore, and erosion and transport occur.

Plunging breakers produce the greatest pressures on rocks – up to 600 kPa or more – because air may become trapped and compressed between the leading wave front and the shore. Air compression and the sudden impact of a large mass of water dislodge fractured rock and other loose particles, a process called **quarrying**. Well-jointed rocks and unconsolidated or loosely consolidated rocks are the most susceptible to wave erosion. Breaking waves also pick up debris and throw it against the shore, causing abrasion of shoreline materials. Some seashore organisms erode rocks by boring into them – some molluscs, boring sponges, and sea urchins do this (p. 106).

Aggradational processes

Sediment transport and deposition

Coastal sediments come from land inland of the shore or littoral zone, the offshore zone and beyond, and the coastal landforms themselves. In high-energy environments, cliff erosion may provide copious sediment, but in low-energy environments, which are common in the tropics, such erosion is minimal. For this reason, few tropical coasts form in bedrock and tropical cliffs recede slowly, although fossil beaches and dunes are eroded by waves. Sediment from the land arrives through mass movement, especially where cliffs are undercut. In periglacial environments, thermal erosion of ice-rich sediments (permafrost) combines with wave action to cause rapid coastal retreat and abundant supplies of sediment in some places. Nevertheless, the chief sediment source is fluvial erosion. Globally, rivers contribute a hundred times more sediment to coasts than marine erosion, with a proportionally greater contribution in the tropics and lower contribution at higher latitudes. Onshore transport of sediments may carry previously eroded beach material or fluvial sediments from the offshore zone to the littoral zone. Very severe storm waves, storm surges, and tsunamis may carry sediments from beyond the offshore zone. During the Holocene, sediment deposited on exposed continental shelves and then submerged by rising sea levels has been carried landwards. In some places, this supply of sediment appears to have dried up and some Holocene depositional landforms are eroding.

Tides and wave action tend to move sediments towards and away from shorelines. However, owing to the effects of longshore currents, the primary sediment movement is along the coast, parallel to the shoreline. This movement, called **longshore drift**, depends upon the wave energy and the angle that the waves approach the coast. Longshore drift is maximal when waves strike the coast at around 30°. It occurs below the breaker zone where waves are steep, or by beach drift where waves are shallow. **Beach drift** occurs as waves approaching a beach obliquely run up the shore in the direction of wave propagation, but their backwash moves down the steepest slope, normally perpendicular to the shoreline, under the influence of gravity (Figure 13.5). Consequently, particles being moved by swash and backwash follow a parabolic path that slowly moves them along the shore. Wherever beach drift is impeded, coastal landforms develop. Longshore currents and beach drifting may act in the same or opposite directions.

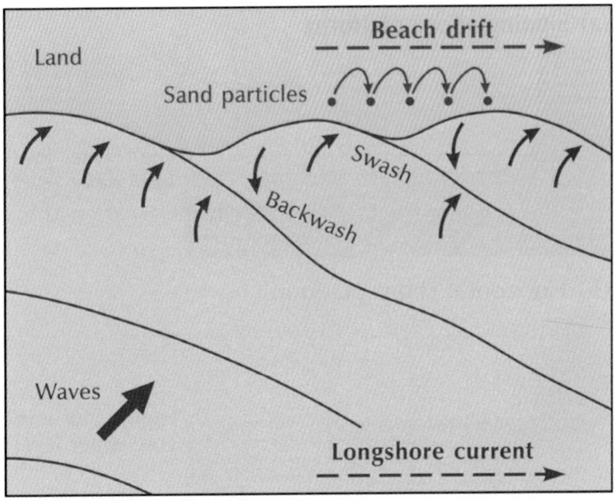

Figure 13.5 Beach drift. *Source*: Adapted from Butzer (1976, 226)

Biological activity

Some marine organisms build, and some help to build, particular coastal landforms. Corals and other carbonate-secreting organisms make coral reefs, which can be spectacularly large. The Great Barrier Reef extends along much of the northeast coast of Australia. Corals grow in the tropics, extratropical regions being too cold for them. Coral reefs cover about 2 million km^2 of tropical oceans and are the largest biologically built formation on Earth. Calcareous algae produce carbonate encrustations along many tropical shores.

Salt-tolerant plants colonize salt marshes. Mangroves are a big component of coastal tropical vegetation. With other salt-tolerant plants, they help to trap sediment in their root systems. Plants stabilize coastal dunes.

COASTAL EROSIONAL LANDFORMS

Erosional landforms dominate rocky coasts, but are also found in association with predominantly depositional landforms. Tidal creeks, for instance, occur within salt marshes. For the purposes of discussion, it seems sensible to deal with erosional features based in depositional environments under the 'depositional landform' rubric, and to isolate rocky coasts as the quintessential landforms of destructive wave action.

Shore platforms and plunging cliffs

Rocky coasts fall into three chief types – two varieties of **shore platform** (sloping shore platform

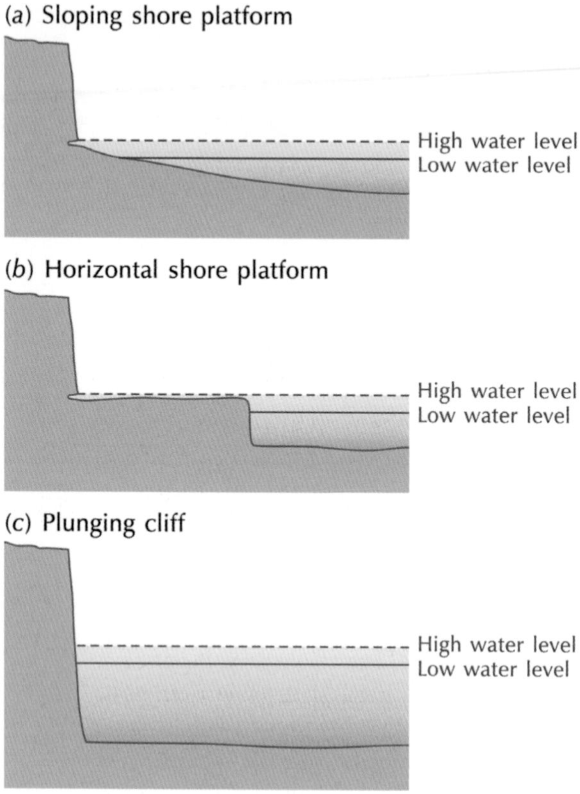

Figure 13.6 Three major forms on rocky coasts: shore platforms and plunging cliffs

Plate 13.2 Horizontal shore platform at low tide, Atia Point, Kaikoura Peninsula, South Island, New Zealand. Higher Pleistocene coastal terraces are also visible, the highest standing at 108 m. (*Photograph by Wayne Stephenson*)

and horizontal shore platform) and **plunging cliff** (Figure 13.6). Variants of these basic types reflect rock types and geological structures, weathering properties of rocks, tides, exposure to wave attack, and the inheritance of minor changes in relative sea level (Sunamura 1992, 139).

Horizontal platforms are flat or almost so (Plate 13.2). They go by a host of names: abrasion or denuded benches, coastal platforms, low-rock terraces or platforms, marine benches, rock platforms, shore benches, shore platforms, storm-wave platforms, storm terraces, wave-cut benches, and wave-cut platforms. Some of these terms indicate causal agents, e.g. 'wave-cut' and 'abrasion'. Because the processes involved in platform evolution are not fully known, the purely descriptive term 'shore platform' is preferable to any others from the wide choice available. **Sloping platforms** are eye-catching features of rocky coasts. As their name intimates, they slope gently between about 1° and 5°. They are variously styled abrasion platforms, beach platforms, benches, coastal platforms, shore platforms, submarine platforms, wave-cut benches, wave-cut platforms, wave-cut terraces, and wave ramps.

Shore platforms can form only if cliffs recede through cliff erosion, which involves weathering and the removal of material by mass movement. Two basic factors determine the degree of cliff erosion: the force of the assailing waves and the resisting force of the rocks. Rock resistance to wave attack depends on weathering and fatigue effects and upon biological factors. Tidal effects are also significant as they determine the height

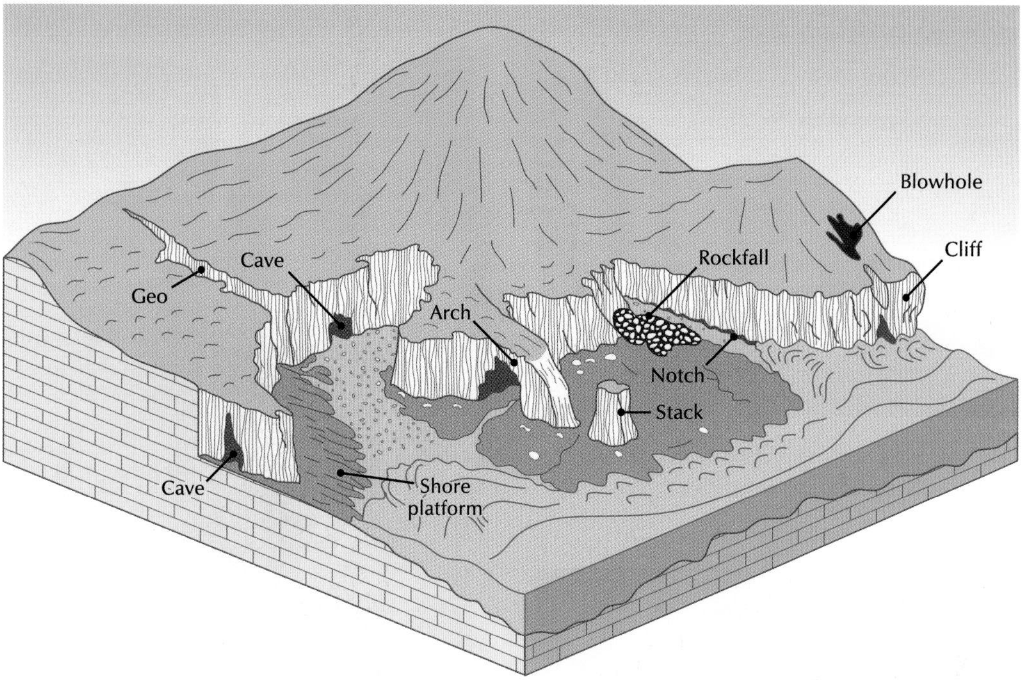

Figure 13.7 Erosional features of a rocky coast. *Source*: Adapted from Trenhaile (1998, 264)

of wave attack and the kind of waves doing the attacking, and as they may influence weathering and biological activities. The tide itself possesses no erosive force.

Plunging-cliff coasts lack any development of shore platforms. Most plunging cliffs are formed by the drowning of pre-existing, wave-formed cliffs resulting from a fall of land level or a rise of sea level.

Landforms of cliffs and platforms

Several coastal features of rocky coasts are associated with the shore platforms and plunging cliffs (Figure 13.7), including cliffs, notches, ramps and ramparts, and several small-scale weathering (including solution pools and tafoni, p. 172) and erosional features. Indeed, shore platforms, cliffs, stacks, arches, caves, and many other landforms routinely form conjointly.

Cliffs, notches, ramps, ramparts, and potholes

Cliffs are steep or vertical slopes that rise precipitously from the sea or from a basal platform (Plate 13.3). About 80 per cent of the world's oceanic coasts are edged with cliffs (Emery and Kuhn 1982). Cliff-base **notches** are sure signs of cliff erosion (Plate 13.4). Shallow notches are sometimes called **nips**. The rate at which notches grow depends upon the strength of the rocks in which the cliff is formed, the energy of the waves arriving at the cliff base, and the amount of abrasive material churned up at the cliff–beach junction.

Ramps occur at cliff bases and slope more steeply than the rest of the shore platform. They occur on sloping and horizontal shore platforms. Horizontal shore platforms may carry ridges or **ramparts**, perhaps a metre or so high, at their seaward margins.

Plate 13.3 Chalk cliffs with horizontal shore platform at Flamborough Head, Yorkshire, England. (*Photograph by Nick Scarle*)

Plate 13.4 Wave-cut, cliff-base notch in limestone, Ha Long Bay, Vietnam. (*Photograph by Tony Waltham Geophotos*)

Plate 13.5 Geo at Huntsman's Leap fault cleft, Castlemartin, South Dyfed, Wales. (*Photograph by Tony Waltham Geophotos*)

Marine **potholes** are roughly cylindrical or bowl-shaped depressions in shore platforms that the swirling action of sand, gravel, pebbles, and boulders associated with wave action grind out.

Caves, arches, stacks, and related landforms

Small bays, narrow inlets, sea caves, arches, stacks, and allied features usually result from enhanced erosion along lines of structural weakness in rocks. Bedding planes, joints, and fault planes are all vulnerable to attack by erosive agents. Although the lines of weakness are eroded out, the rock body still has sufficient strength to stand as high, almost perpendicular slopes, and as cave, tunnel, and arch roofs.

A **gorge** is a narrow, steep-sided, and often spectacular cleft, usually developed by erosion along vertical fault planes or joints in rock with a low dip. They may also form by the erosion of dykes, the collapse of lava tunnels in igneous rock, and the collapse of mining tunnels. In Scotland, and sometimes elsewhere, gorges are known as **geos** or **yawns** (Plate 13.5), and on the granitic peninsula of Land's End in Cornwall, south-west England, as **zawns**.

A **sea cave** is a hollow excavated by waves in a zone of weakness on a cliff (Plate 13.6). The cave

Plate 13.6 Sea caves, Flamborough Head, Yorkshire, England. (*Photograph by Nick Scarle*)

Plate 13.7 Durdle Door, a limestone sea arch on the Jurassic coast near Lulworth, Dorset, England (*Photograph by Nick Scarle*)

Plate 13.8 Sea stack by Bempton Cliffs, Yorkshire, England. (*Photograph by Nick Scarle*)

Plate 13.9 Sea stacks at Bedruthan Steps, Cornwall, England. (*Photograph by Tony Waltham Geophotos*)

depth is greater than the entrance width. Sea caves tend to form at points of geological weakness, such as bedding planes, joints, and faults. Fingal's Cave, Isle of Staffa, Scotland, which is formed in columnar basalt, is a prime example. It is 20 m high and 70 m long. A **blowhole** may form in the roof of a sea cave by the hydraulic and pneumatic action of waves, with fountains of spray emerging from the top. If blowholes become enlarged, they may collapse. An example of this is the Lion's Den on the Lizard Peninsula of Cornwall, England.

Where waves attack a promontory from both sides, a hollow may form at the promontory base,

Plate 13.10 A wave-dominated beach with well-developed transverse bars and adjacent deeper rip channels, Lighthouse Beach, New South Wales, Australia. (*Photograph by Andrew Short*)

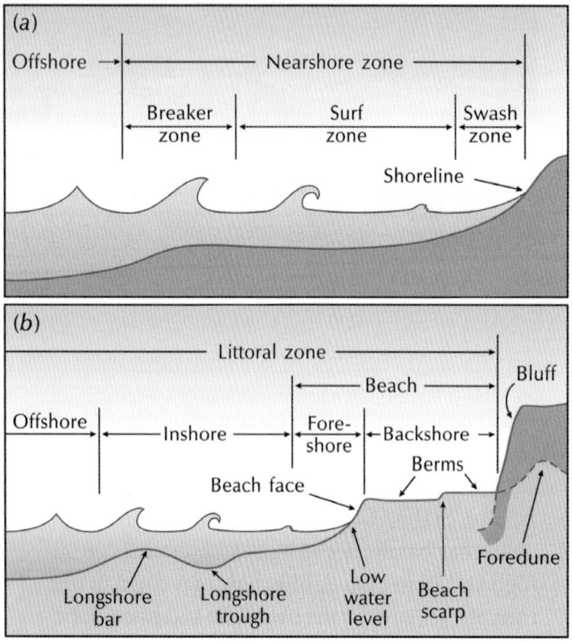

Figure 13.8 Terminology used to describe (a) wave and current processes in the nearshore, and (b) the beach profile. *Source*: Adapted from Komar (1998, 46)

often at a point of geological weakness, to form a sea arch (Plate 13.7). If an arch is significantly longer than the width of its entrance, the term '**sea tunnel**' is more appropriate. Merlin's Cave, at Tintagel, Cornwall, England, is a 100-m-long sea tunnel that has been excavated along a fault line. The toppling of a sea arch produces a **sea stack** (Plates 13.8 and 13.9). Old Harry Rocks are a group of stacks that were once part of the Foreland, which lies on the chalk promontory of Ballard Down in Dorset, England. On the west coast of the Orkney Islands, Scotland, the Old Man of Hoy is a 140-m stack separated from towering cliffs formed in Old Red Sandstone. Patrick Limber and Brad Murray (2015) provide an interesting analysis of sea stack formation using numerical modelling.

COASTAL DEPOSITIONAL LANDFORMS

Beaches

Beaches are the most significant accumulations of sediments along coasts. They form in the zone where wave processes affect coastal sediments. In composition, they consist of a range of organic and inorganic particles, mostly sands or shingle or pebbles. **Pebble beaches** are more common at middle and high latitudes, where pebbles are supplied by coarse glacial and periglacial debris.

Sand beaches are prevalent along tropical coasts, probably because rivers carry predominantly fine sediments and cliff erosion donates little to littoral deposits in the tropics (Plate 13.10). Under some conditions, and notably in the tropics, beach sediments may, through the precipitation of calcium carbonate, form **beachrock**.

Beach profile

Beaches have characteristic profiles, the details of which are determined by the size, shape, and composition of beach material, by the tidal range, and by the kind and properties of incoming waves (Figure 13.8).

Beach profiles all consist of a series of ridges and troughs, the two extreme forms being steep, storm profiles and shallow, swell profiles, with all grades in between. The most inland point of the beach, especially a storm beach, is the **berm**, which marks the landward limit of wave swash. Over the berm crest lies the **beach face**, the gradient of which is largely controlled by the size of beach sediment. Fine sand beaches slope at about 2°, and coarse pebble beaches slope at as much as 20°, the difference being accounted for by the high permeability of pebbly sediment, which discourages backwash. On shallow-gradient beaches, a submerged **longshore bar** often sits offshore, separated from the beach by a **trough**. Offshore bars, which are more common on swell beaches, seem to result from the action of breaking waves and migrate to and from the shoreline in response to changing wave characteristics. Similarly, the beach profile changes as wave properties run through an annual cycle. Beach-profile shape adjusts swiftly to changes in the wave 'climate', which commonly changes seasonally. Often, a beach will have two or more berms, the higher ones recording the action of very large storms in the past.

Basic beach types reflect varying degrees of wave and tide dominance. Andrew Short (2006; Short and Woodroffe 2009) established the following categories for Australian beaches, but they have general applicability (Figure 13.9):

1 **Wave-dominated beaches** range from dissipative to reflective (Figure 13.9a). **Dissipative beaches** tend to occur on high-energy coasts where waves regularly exceed 2.5 m and the dominant beach material is fine sand. These factors combine to maintain a low gradient surf zone up to 500 m wide with usually two and occasionally three shore-parallel bars, separated by subdued troughs. Waves start breaking several hundred metres offshore as spilling breakers on the outer bar, then reform in the outer trough to break again and again on the inner bar or bars; in so doing, they dissipate their energy across the wide surf zone. **Reflective beaches** lie at the lower energy end of the wave-dominated beach spectrum; they are typically relatively steep and narrow and built of coarser sand (0.4 mm). Intermediate types are **long-shore bar and trough**, **rhythmic bar and beach**, **transverse bar and rip**, and **low tide terrace**.
2 **Tide-modified beaches** occur mainly where higher tide ranges and lower waves result in the spring tide range being three to ten times greater than the average breaker wave height (Figure 13.9b). **Reflective plus low tide terrace beaches** tend to occur with short-period waves averaging 0.45 m in height and a tide range averaging up to ten times the wave height (that is, 4.5 m). They are characteristically a relatively steep cusped reflective high tide beach, generally composed of medium sand (0.45 mm). **Reflective plus low tide rips beaches** are the highest energy of the tide-affected types and form where waves average 0.7 m, sand is medium, and tides average 2.5 m. At high tide, the waves pass over the bar without breaking until the beach face, where they usually maintain a relatively steep beach with cusps. **Ultradissipative beaches** occur in higher energy (waves averaging 0.6 m high) tide-modified locations, where the beaches are composed of fine sand. They possess a very wide (200–400 m) intertidal zone, with a low to moderate gradient high-tide beach and a very low gradient to almost horizontal low-tide beach. The low beach gradient means that waves break across a relatively wide, shallow surf zone as a series of spilling breakers that continually dissipate the wave energy, hence the name 'ultradissipative'.
3 **Tide-dominated beaches** occur mainly where the spring tide range is ten to fifty times greater than the average breaker wave (Figure 13.9c). They consist of a low high-tide beach fronted increasingly by inter- to low-tide tidal flats, the latter grading with lower energy into true tidal flats. The three types are **reflective plus ridged**

sand flats, **beach plus sand flats**, and **beach plus tidal mudflats**.

Beach cusps and crescentic bars

Viewed from the air, beaches possess several distinctive curved plan-forms that show a series of regularly spaced secondary curved features (Figure 13.10). The primary and secondary features range in size from metres to more than 100 km. **Beach cusps** are crescent-shaped scallops lying parallel to the shore on the upper beach face and along the seaward margins of the berm with a spacing of less than about 25 m (Plate 13.11). Most researchers believe that they form

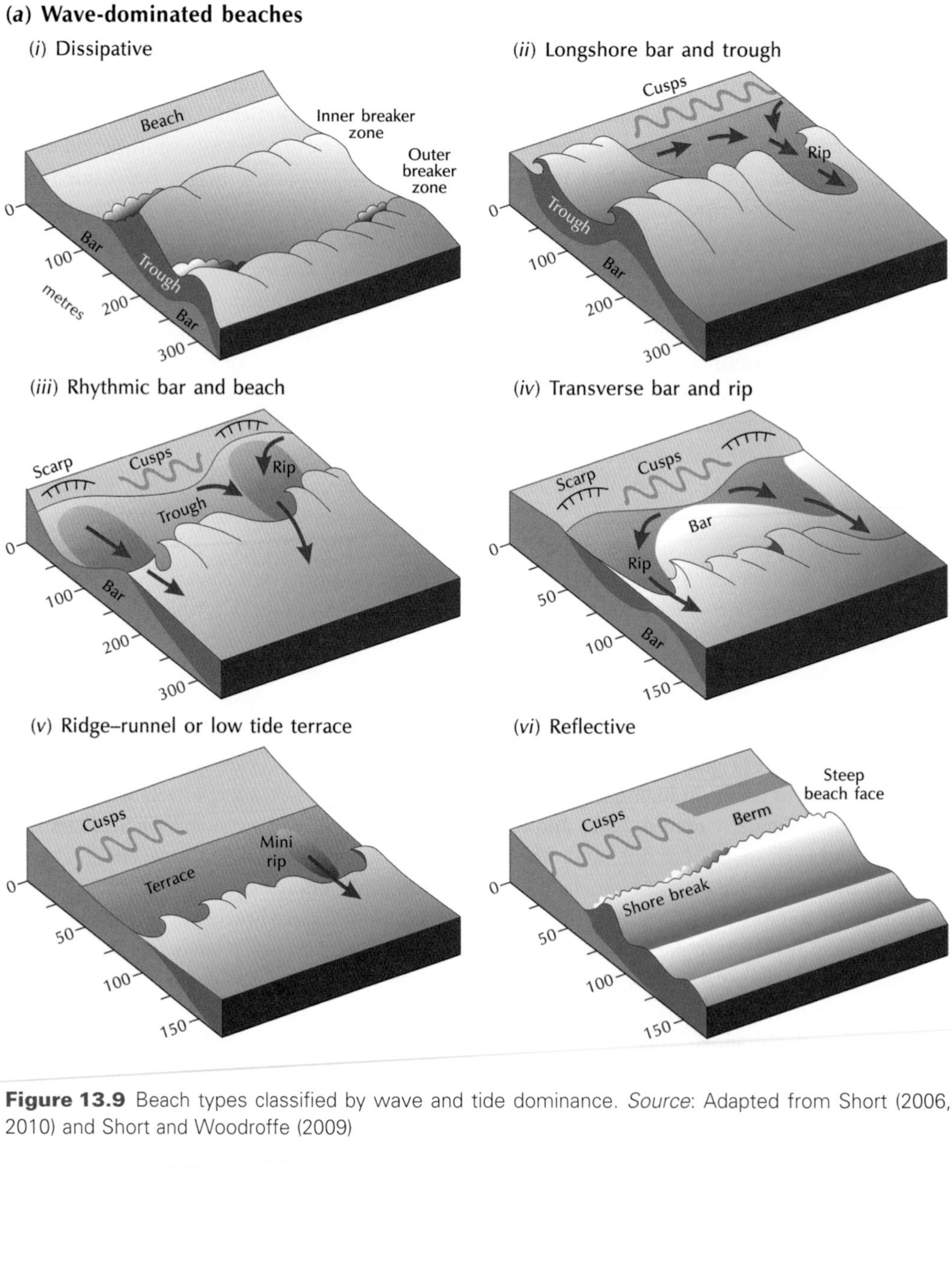

Figure 13.9 Beach types classified by wave and tide dominance. *Source*: Adapted from Short (2006, 2010) and Short and Woodroffe (2009)

when waves approach at right-angles to the shore, although a few think that oblique waves cause them. Their mode of formation is disputed, and they have been variously regarded as depositional features, erosional features, or features resulting from a combination of erosion and deposition.

Inner and **outer crescentic bars** are sometimes called **rhythmic topography**. They have wavelengths of 100–2,000 m, although the majority are somewhere between 200 and 500 m. Inner bars are short-lived and associated with rip currents and cell circulations. Their horns often extend across surf-zone shoals into very large

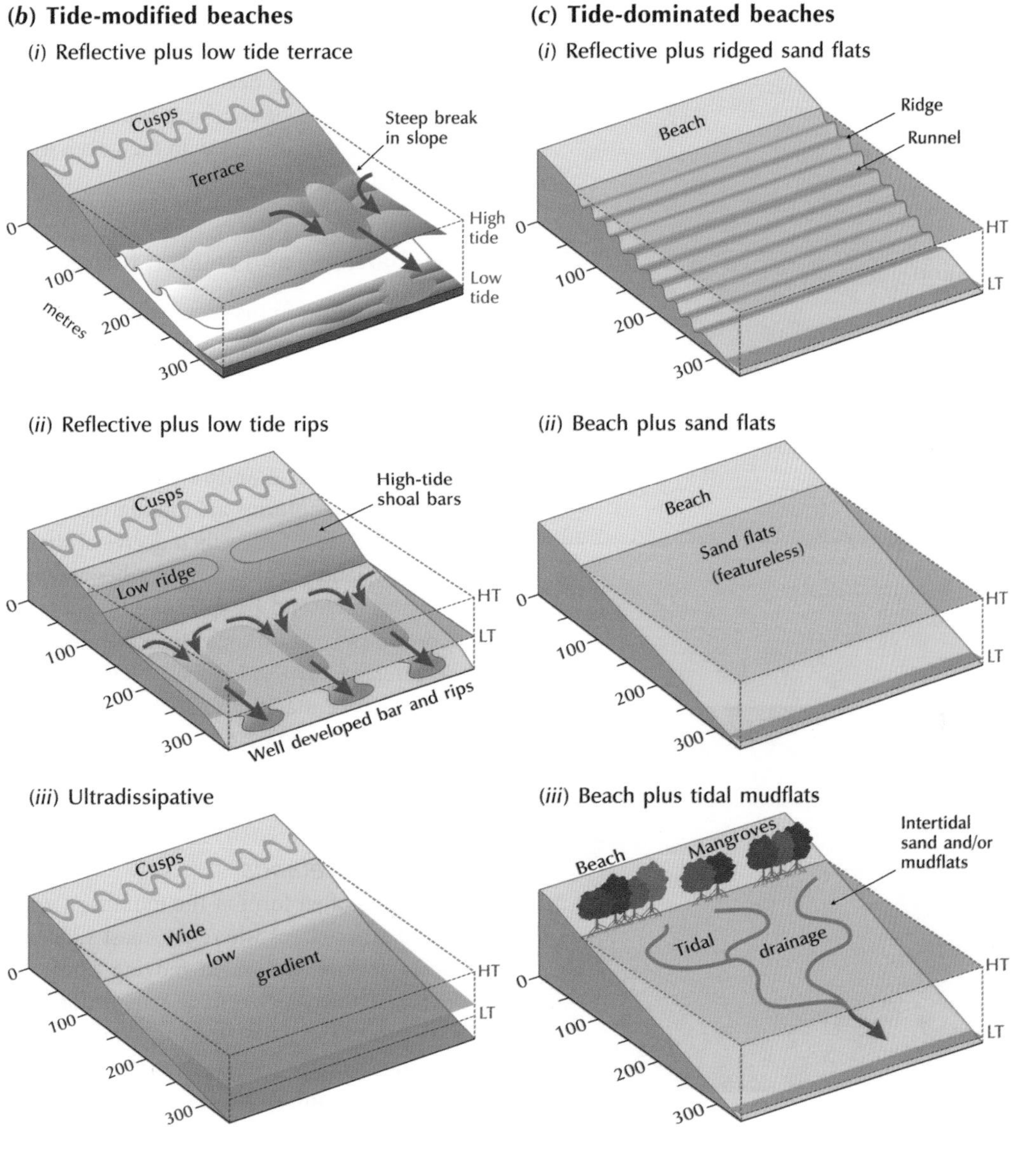

Figure 13.9 *continued*

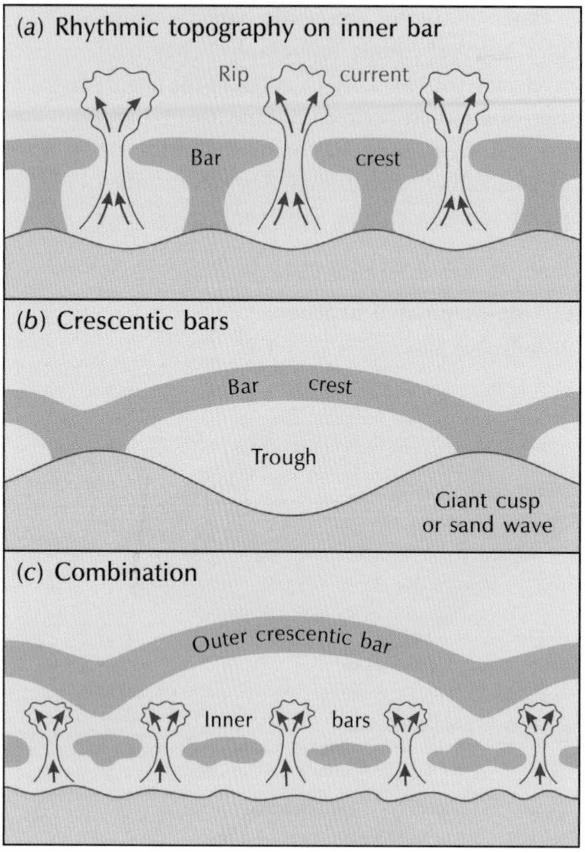

Figure 13.10 Cusps and crescentic bars. *Source*: Adapted from Komar (1998, 475)

Plate 13.11 Beach cusps on Seatown beach, Dorset, England. (*Photograph by Nick Scarle*)

shoreline cusps known as sand waves, which lie parallel to the shore and have wavelengths of about 200–300 m. Outer crescentic bars may be detached from the shore and are more stable than inner crescentic bars.

Many coasts display an orderly sequence of **capes** and **bays**. The bays usually contain **bayhead** or **pocket beaches** (Figure 13.11). In some places, including parts of the east coast of Australia, asymmetrically curved bays link each headland, with each beach section recessed behind its neighbour. These are called **headland bay beaches**, or **fish-hook beaches**, or **zetaform beaches**, owing to their likeness in plan-view to the Greek letter zeta, ζ (Figure 13.12).

Spits, barriers, and related forms

Accumulation landforms occur where the deposition of sediment is favoured (Figure 13.11). Suitable sites include places where obstructions interrupt longshore flow, where the coast abruptly changes direction, and in sheltered zones ('wave shadows') between islands and the mainland. Accumulation landforms are multifarious. They may be simply classified by their degree of attachment to the land (Table 13.2). Beaches attached to the land at one end are **spits** of different types and **forelands**. Spits are longer than they are wide, while forelands are wider than they are long. Beaches that are attached to the land at two ends are **looped barriers** and **cuspate barriers**, **tombolos**, and **barrier beaches**. Beaches detached from the land are **barrier islands**.

Spits and forelands

Barrier spits often form at the mouths of estuaries and other places where the coast suddenly changes direction. Sediment moving along the shore is laid down and tends to extend along the original line of the coast. Some spits project into the ocean and then curve round to run parallel to the coast. An example is Orfordness on the east coast of England, where the River Alde has been deflected some 18 km to the south. **Recurved**

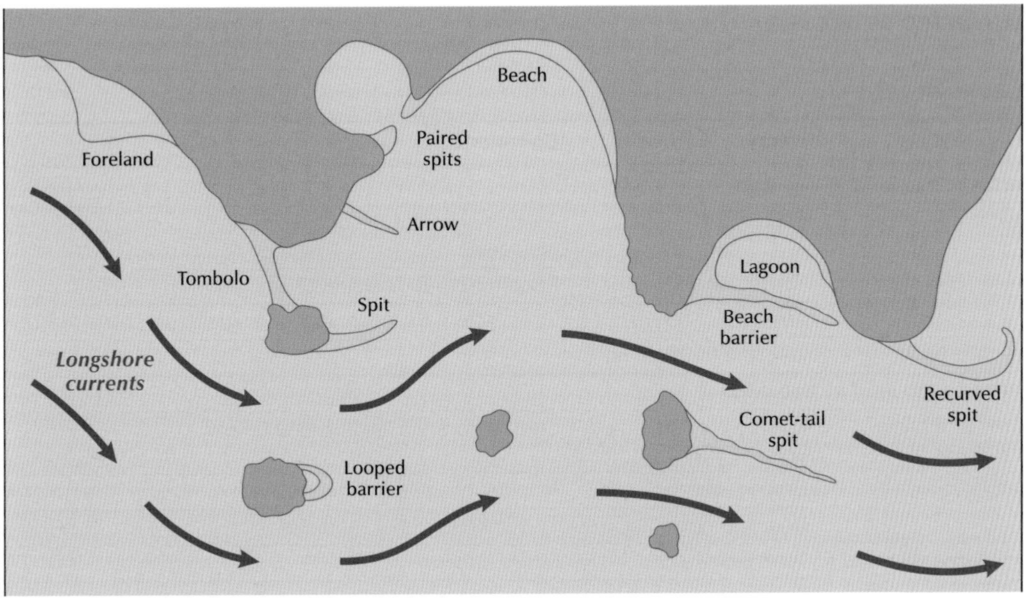

Figure 13.11 Depositional coastal landforms, shown diagrammatically.

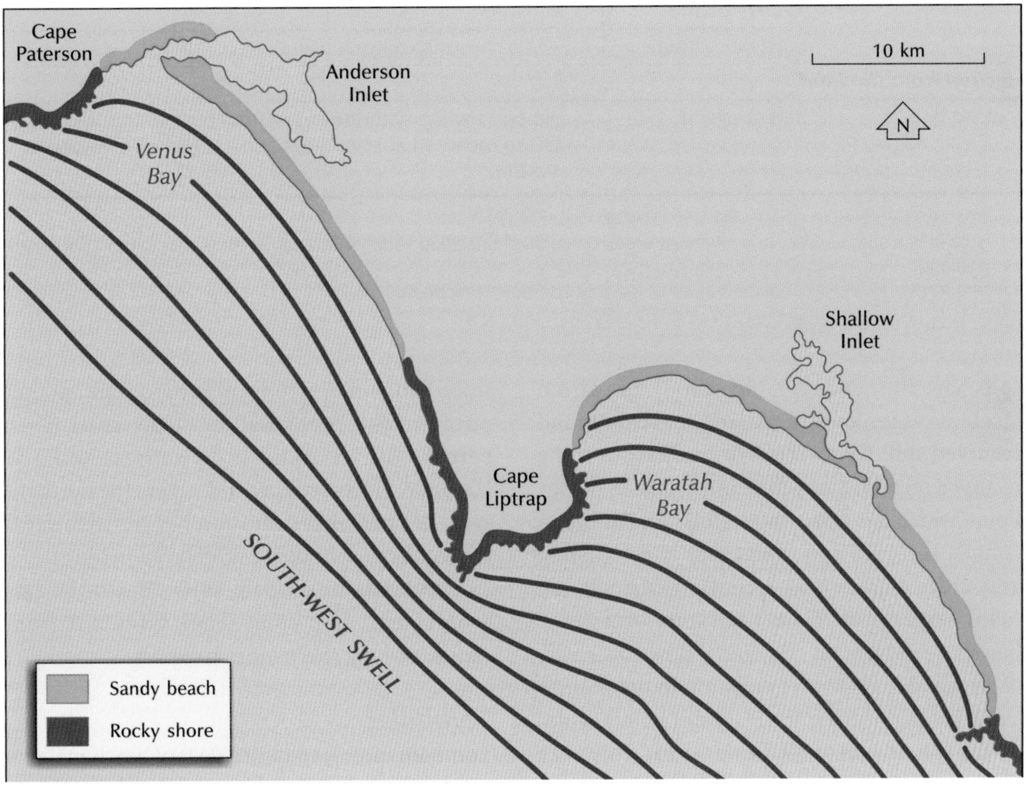

Figure 13.12 Zetaform beaches: Venus Bay and Waratah Bay, Victoria, Australia. *Source*: Adapted from Bird (2000, 119)

Table 13.2 Beach types

Form	Name	Comment
Beaches attached to land at one end		
Length greater than width	Barrier spit	A continuation of the original coast or running parallel to the coast[a]
	Comet-tail spit	Stretch from the leeside of an island
	Arrow	Stretch from the coast at high angles[b]
Length less than width	Foreland (cuspate spit)	–
Beaches attached to land at two ends		
Looped forms stretching out from the coast	Looped barriers Cuspate barriers	Stretch from the leeside of an island
	Looped spit	A spit curving back onto the land
	Double-fringing spit	Two joined spits or tombolos
Connecting islands with islands or islands with the mainland (tombolos)	Tombolo	Single form
	Y-tombolo	Single beach looped at one end
	Double tombolo	Two beaches
Closing off a bay or estuary (barrier beaches)	Baymouth barrier	At the mouth (front) of a bay
	Midbay barrier	Between the head and mouth of a bay
	Bayhead barrier	At the head (back) of a bay
Forms detached from the land		
A discrete, elongated segment	Barrier island	No connection with the land. Runs parallel to the coast. Often recurved at both ends and backed by a lagoon or swamp

Notes:

a A winged headland is a special case. It involves an eroding headland providing sediment to barrier spits that extend from each side of the headland.

b A flying spit is a former tombolo connected to an island that has now disappeared.

Source: Adapted from Trenhaile (1998, 244)

spits have their ends curving sharply away from incoming waves towards the land, and **compound recurved spits** have a series of landward-turning recurved sections along their inner side. Blakeney Point, which lies in north Norfolk, England, is a famous recurved spit. Spits that have twisting axes, created in response to shifting currents, are called '**serpentines**'. **Comet-tail spits** form where longshore movement of material down each side of an island leads to accumulation in the island's lee, as has happened at the Plage de Grands Sables on the eastern end of the Île de Groix, which lies off the coast of Brittany, France. **Arrows** are spit-like forms that grow seawards from a coast as they are nourished by longshore movement on both sides. Sometimes spits grow towards one another owing to the configuration of the coast. Such **paired spits** are found at the entrance to Poole Harbour, in Dorset, England, where the northern spit, the Sandbanks peninsula, has grown towards the southern spit, the South Haven peninsula.

Forelands or cuspate spits tend to be less protuberant than spits. They grow out from coasts, making them more irregular.

Tombolos

Tombolos are wave-built ridges of beach material connecting islands to the mainland or islands to islands. They come in single and double varieties. Chesil Beach in Dorset, England, is part of a **double tombolo** that attaches the Isle of Purbeck to the Dorset mainland (Plate 13.12). Tombolos tend to grow in the lee of islands, where a protection is afforded from strong wave action and where waves are refracted and convergent. **Y-shaped tombolos** develop where comet-tail spits merge with cuspate forms projecting from the mainland or where a cuspate barrier extends landwards or seawards. A **tombolino** or **tie-bar** is a tombolo that is partly or completely submerged by the sea at high tide.

Barriers and barrier beaches

Coastal barriers and barrier islands form on beach material deposited offshore, or across the mouths of inlets and embayments. They extend above the level of the highest tides, in part or in whole, and enclose lagoons or swamps. They differ from bars, which are submerged during at least part of the tidal cycle.

Coastal barriers are built of sand or gravel. Looped barriers and cuspate barriers result from growing spits touching an opposite shore, another spit, or an island. **Looped barriers** grow in the lee of an island when two comet-tail spits join. **Cuspate barriers** (cuspate forelands) resemble forelands except that the building of beach ridges parallel to their shores has enlarged them and they contain lagoons or swampy areas. An example is Dungeness in Kent, England, which is backed by marshland. If the lagoons or swamps drain and fill with sediment, cuspate barriers become forelands. Cuspate barriers form by a spit curving back to the land (a looped spit), or else by two spits or tombolos becoming joined to an island, which then vanishes (double-fringing spit).

Barrier beaches seal off or almost seal off the fronts, middles, or heads of bays and inlets. They are the product of single spits growing across bays or from pairs of converging spits built out by opposing longshore currents. They may also possibly form by sediment carried into bays by wave action independently of longshore movement.

Plate 13.12 Chesil Beach, Dorset, England. (*Photograph by Nick Scarle*)

Barrier islands

Barrier islands are elongated offshore ridges of sand paralleling the mainland coast and separated for almost their entire length by lagoons, swamps, or other shallow-water bodies, which are connected to the sea by channels or tidal inlets between islands. They are also called **barrier beaches**, **barrier bars**, and **offshore bars**. Sections of long barrier-island chains may be large spits or barrier beaches still attached to land at one end. As to their formation, some barriers are sections of long spits that have become detached, while some may simply be 'overgrown' bars (Figure 13.13). Others may have been formed by the rising sea levels over the last 10,000 years and perhaps have grown on former dunes, storm ridges, and berms, with lagoons forming as the land behind the old beaches was flooded. Barrier beaches may also have formed by the accumulation of sediment carried landwards by wave action as sea level rose.

Interestingly, tectonic plate margins strongly influence the occurrence of barrier coasts (barrier spits, barrier beaches, and barrier islands). Of all the world's barrier coasts, 49 per cent occur on passive margins, 24 per cent on collisional margins, and 27 per cent on marginal seacoasts.

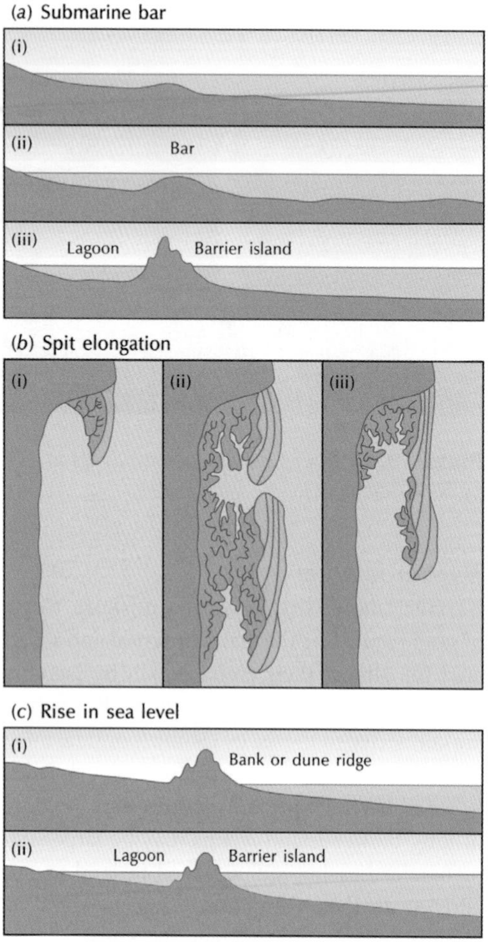

Figure 13.13 Ways in which barrier islands may form. (a) The growth of a submarine bar. (b) The elongation of a spit. (c) The submergence of a beach ridge or dunes by a rising sea-level. *Source*: Adapted from Hoyt (1967)

Beach ridges and cheniers

Sandy **beach ridges** mark the position of former shorelines, forming where sand or shingle have been stacked up by wave action along a prograding coast. They may be tens of metres wide, a few metres high, and several kilometres long. Beach ridge plains may consist of 200 individual ridges and intervening swales.

Cheniers are low and long ridges of sand, shelly sand, and gravel surrounded by low-lying mudflats or marshes. They were first described from southwestern Louisiana and south-eastern Texas, USA, where five major sets of ridges sit on a 200-km-long and 20–30-km-wide plain. These ridges bear rich vegetation and are settled by people. The word 'chenier' is from a Cajun expression originating from the French word for oak (*chêne*), which species dominates the crests of the higher ridges. Cheniers can be up to 1 km wide, 100 km long, and 6 m high. **Chenier plains** consist of two or more ridges with marshy or muddy sediments between. Most cheniers are found in tropical and subtropical regions, but they can occur in a wide range of climates (Figure 13.14). They cannot form in coasts with high wave energy as the fine-grained sediments needed for their growth are carried offshore (Figure 13.15).

Coastal sand dunes

Coastal dunes are heaps of windblown sediment deposited at the edge of large lakes and seas. With few exceptions, they are made from sediment blown off a beach to accumulate in areas sheltered from the action of waves and currents. Small, crescentic dune fields often form at the back of bays enclosed by rocky headlands, while larger prograding dune fields form on straight, sandy coasts that are exposed to prevailing and dominant onshore winds. They shield the land from extreme waves and tides and are stores of sediment that may replenish beaches during and after storms. Dunes may also occur on cliff tops. Coastal dunes are similar to desert dunes, but the **foredune** (the first dune formed behind the beach) is a prominent feature resulting from the interaction of nearshore processes, wind, sediments, and vegetation. The 'valleys' between the dune ridges are dune **slacks**.

Coastal dunes are mainly composed of medium-sized to fine quartz grains that are well

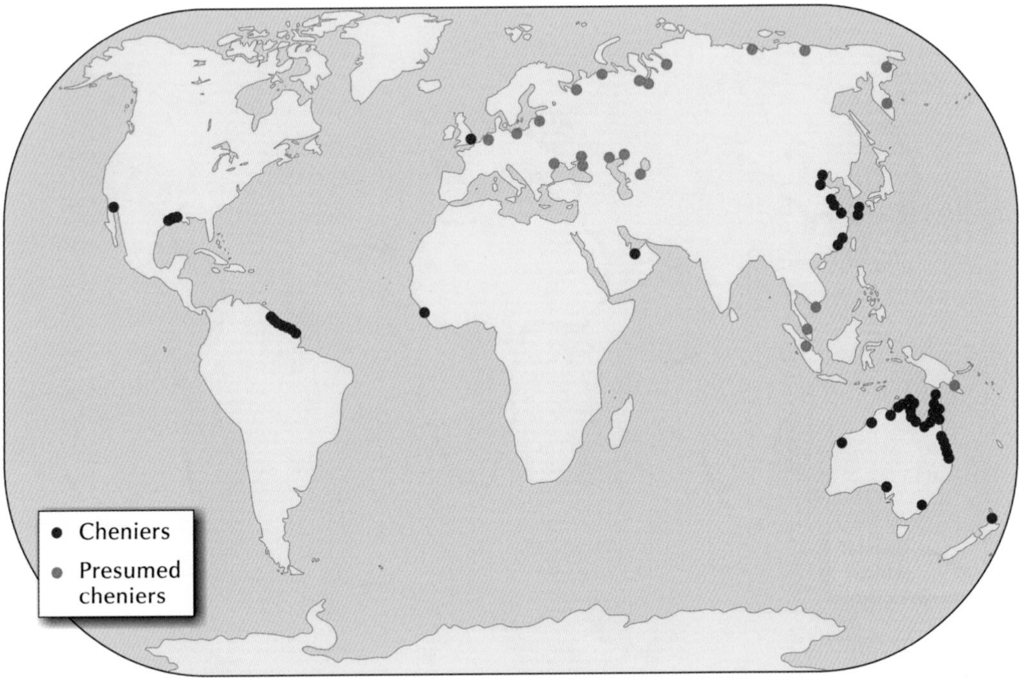

Figure 13.14 World distribution of cheniers. *Source*: Adapted from Augustinus (1989)

to very well sorted, but calcium carbonate is common in warm tropical and Mediterranean regions. They are found in a range of environments (Carter *et al.* 1990) (Figure 13.16). The largest dune systems occur in mid-latitudes, behind high-energy to intermediate wave-energy coasts and facing the prevailing and dominant westerly winds. Dunes also develop on east-facing swell and trade-wind coasts, but they are less common and smaller in polar and tropical regions. The occurrence and nature of coastal dunes are the outcome of a set of interacting factors (Box 13.2).

Blowouts are shallow, saucer-shaped depressions or deep and elongated troughs occurring in dunefields (cf. p. 351). They are begun by wave erosion, overwash, a lack of aeolian deposition, or deflation of vegetation or poorly vegetated areas. Once started, they are enlarged by wind scour and slumping, and avalanching on the sidewalls.

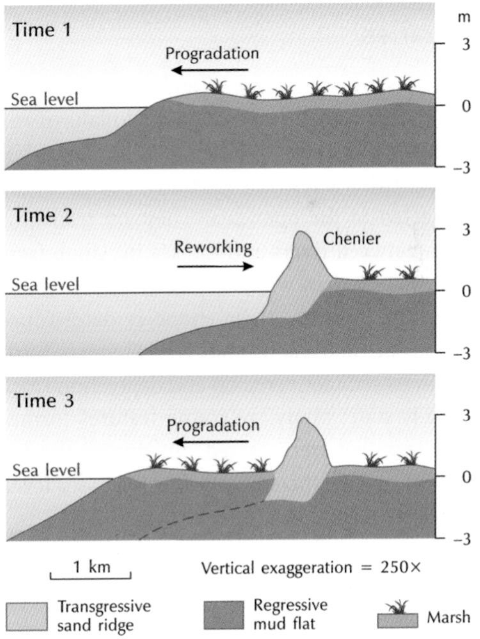

Figure 13.15 Formation of a chenier plain. *Source*: Adapted from Hoyt (1969)

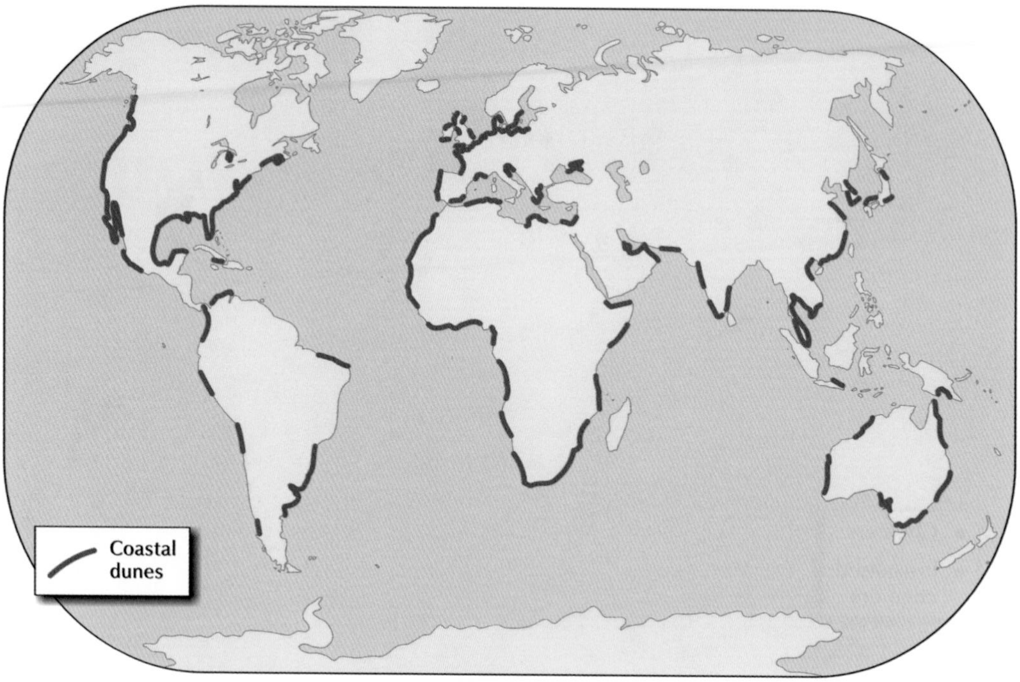

Figure 13.16 World distribution of coastal dunes. *Source*: Adapted from Carter *et al.* (1990)

Estuaries

Estuaries are tidal inlets, often long and narrow, that stretch across a coastal alluvial plain or run inwards along a river to the highest point reached by the tide. They are partially enclosed but connected to the open sea. They are transition zones between rivers and the sea in which fresh river water mixes with salty ocean water. Early in their evolution, their shape is determined by coastal topography, but this changes fairly rapidly as sediment erosion and deposition reach a steady state. Many are young features formed in valleys that were carved out during the last glacial stage and then drowned by rising sea levels during the Holocene epoch. Figure 13.18 is a physiographic classification due to Rhodes W. Fairbridge (1980).

Tidal flats, salt marshes, and mangals

Currents associated with tides carry copious amounts of sediment inside areas of shallow water. The ebb and flow of tidal currents fashions a range of coastal landforms.

Tidal flats

Tidal flats are banks of mud or sand that are exposed at low tide (Plate 13.13). They are not actually flat but slope very gently towards the sea from the high-tide level down to a little below the low-tide level. Three basic units may be identified in tidal flats: the **high-tide flat** (a gently sloping surface that is partly submerged at high tide); the **intertidal slope** (a steeper but still gently inclined zone lying between the high-tide flat and the lower tidal limit); and the **subtidal slope**, which is submerged even at low tide (Figure 13.19).

Box 13.2 COASTAL DUNES: NATURE AND OCCURRENCE

Coastal dunes are fashioned by the interplay of wind, waves, vegetation, and sediment supply (Pye 1990). Figure 13.17 depicts six basic cases. Rapidly prograding beach ridge plains with little dune development form where the beach sand budget is positive and wind energy is low (Figure 13.17a). Parallel foredune ridges occur under similar circumstances save that wind energy is higher and sand-trapping vegetation is present, leading to slower beach progradation (Figure 13.17b). Irregular 'hummock' dunes with incipient blowouts and parabolic dunes form on moderately prograding coasts where the beach-sand sediment budget is positive, wind energy is moderate, and there is an ineffectual or patchy vegetation cover (Figure 13.17c). Single foredune ridges, which grow upwards with no change of shore position, occur when sand supplied to the beach is delivered to the dunes and trapped by vegetation (Figure 13.17d). Slowly migrating parabolic dunes, blowouts, and salt-scalded vegetation occur behind beaches that slowly retreat landwards when the sand supplied to the beach is slightly less than that supplied to the dunes (Figure 13.17e). Transgressive sand sheets of low relief form when little or no sand is supplied to the beach and wind energy is high (Figure 13.17f). Under these conditions, the beach is rapidly lowered and dune vegetation destroyed, which increases exposure to storms and initiates coastal retreat.

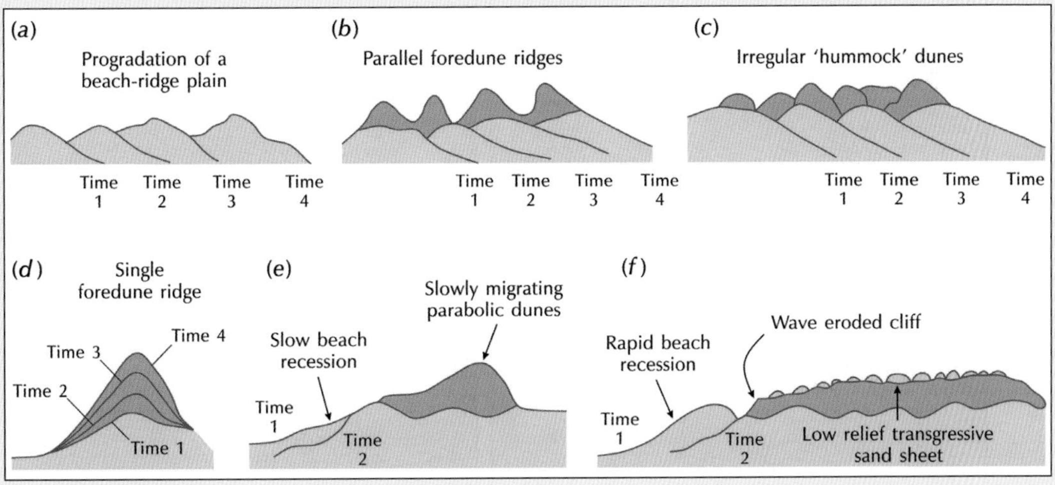

Figure 13.17 Factors affecting dune morphology. For explanation, see text. *Source*: Adapted from Pye (1990)

Tidal flats end at the edge of the sea or in major **tidal channels**, the floors of which lie below the lowest tide levels. As well as major tidal channels, **tidal creeks** flow across tidal flats. These are shallower than tidal channels and run down to low-tide level. On muddy tidal flats, tidal creeks often display a dendritic pattern with winding courses and point bars. On sandy tidal flats, tidal creeks have ill-defined banks, straight courses, and few tributaries.

Tidal flats are built up from clay-sized and fine silt-sized sediments carried to the coast by rivers. On meeting salt water, particles of clay and silt flocculate (form clot-like clusters) to become larger aggregates. They then settle out as mud in quiet coastal waters such as lagoons and

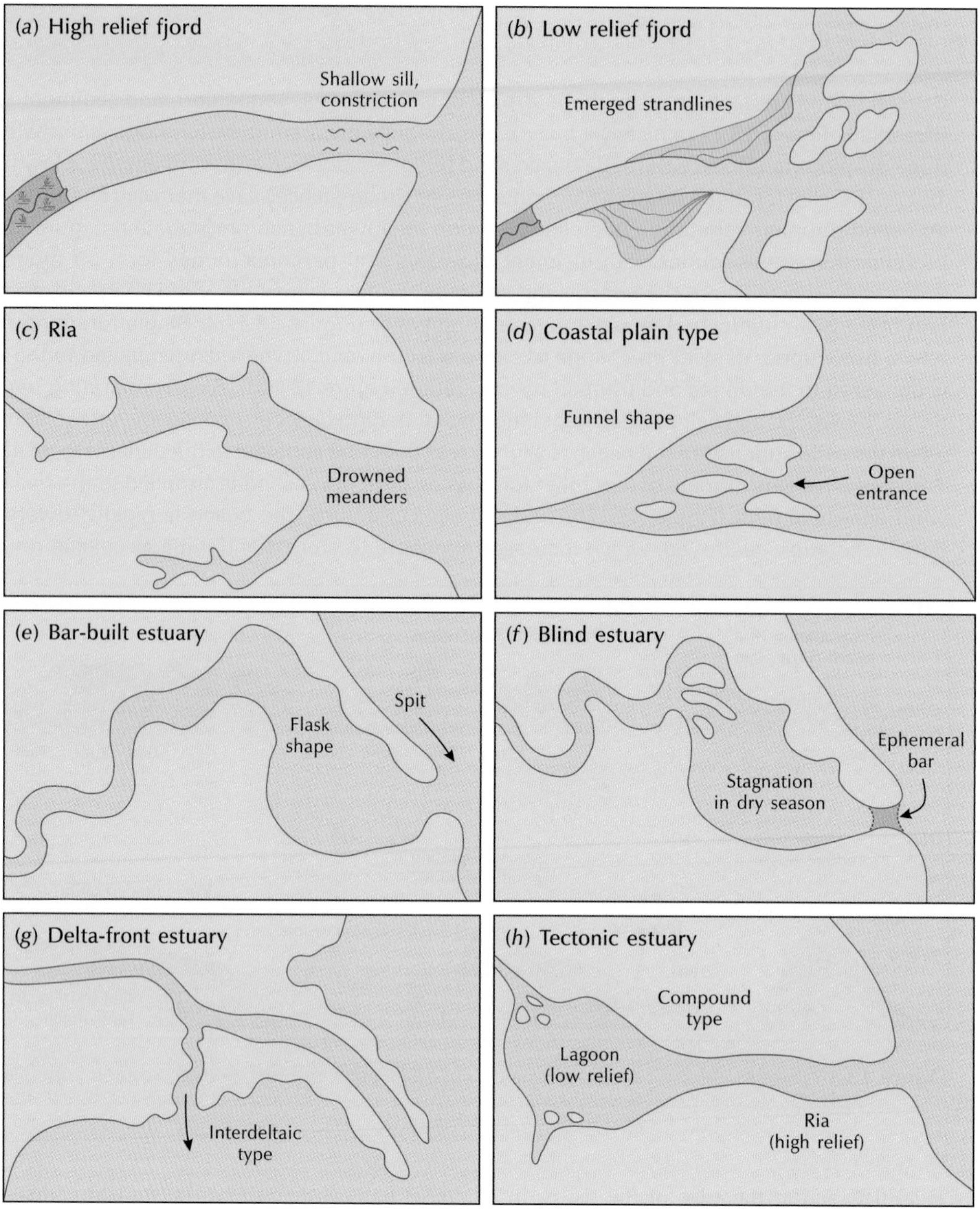

Figure 13.18 Types of estuary: a physiographic classification. Fjords are drowned glacial troughs (p. 289). Rias are erstwhile river valleys drowned by Holocene sea-level rise. They may include mudflats and have barrier spits at their mouths. Coastal plain estuaries are, as their name suggests, estuaries in coastal plains. Bar-built or barrier estuaries have barriers that enclose broad and shallow lagoons. Blind estuaries are closed by an ephemeral bar and stagnate in dry seasons. Delta-front estuaries are associated with river deltas. Tectonic estuaries are formed by folding, faulting, or other tectonic processes. Part of the San Francisco Bay, California, estuary comes under this heading. *Source*: Adapted from Fairbridge (1980)

sheltered estuaries. The mud is carried in by the incoming tide and deposited before the tide reverses. If the mud continues to build upwards, a part of the tidal flat will be exposed just above normal high-tide level. This area is then open to colonization by salt-tolerant plants, and salt marshes or mangroves may develop.

Salt marshes

Salt marshes are widespread in temperate regions, and are not uncommon in the tropics (Figure 13.20). They start to form when tidal flats are high enough to permit colonization by salt-tolerant terrestrial plants. Depending on their degree of exposure, salt marshes stretch from around the mean high-water, neap-tide level to a point between the mean and extreme high-water, spring-tide levels. Their seaward edge abuts bare intertidal flats, and their landward edge sits where salt-tolerant plants fail to compete with terrestrial plants. Salt marsh sediments are typically heavy or sandy clay, silty sand, or silty peat. Many salt marshes contain numerous shallow depressions, or pans, that are devoid of vegetation and fill with water at high spring tides.

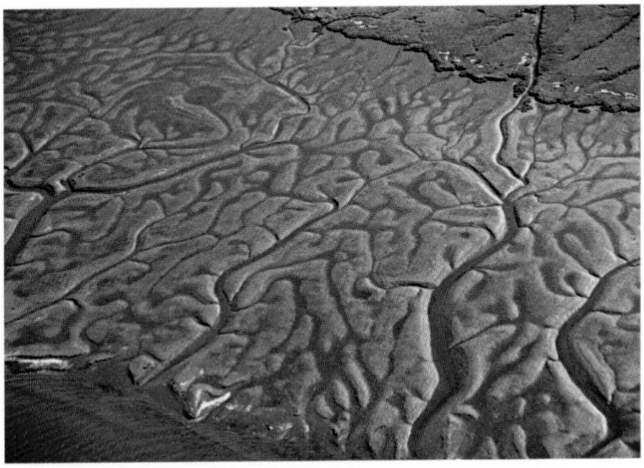

Plate 13.13 Tidal creeks and tidal channels on a tidal flat, Coos Bay, Oregon, USA. (*Photograph by Marli Miller*)

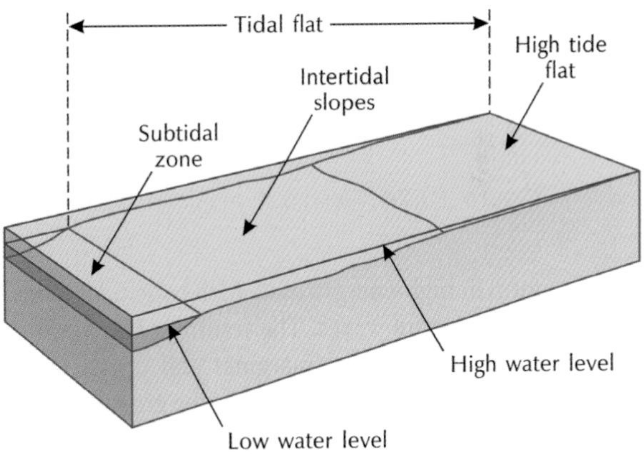

Figure 13.19 Tidal flats and their morphological units based on low- and high-tide positions. *Source*: Adapted from Davies (1980, 170)

Mangals

'**Mangrove**' is a general term for a variety of mainly tropical and subtropical salt-tolerant trees and shrubs inhabiting low inter-tidal areas. **Mangals** are communities of mangroves – shrubs and long-lived trees and with associated lianas, palms, and ferns – that colonize tidal flats in the tropics, and occur in river-dominated, tide-dominated, and wave-dominated coastal environments (Woodroffe 1990). They specifically favour tidal shorelines with low wave energy, and in particular brackish waters of estuaries and deltas (Figure 13.20). Some mangrove species are tolerant of more frequent flooding than salt marsh species, and so mangals extend from around the high spring-tide level to a little above mean sea level. They often contain **lagoons** and pools, but not the pans of salt marshes. Like salt marshes, mangals have creek systems, although their banks are often formed of tree roots.

Marine deltas

Marine deltas are formed by deposition where rivers run into the sea. So long as the deposition rate surpasses the erosion rate, a delta will grow. Deltas are found in a range of coastal environments. Some deltas form along low-energy coasts with low tidal ranges and weak waves. Others

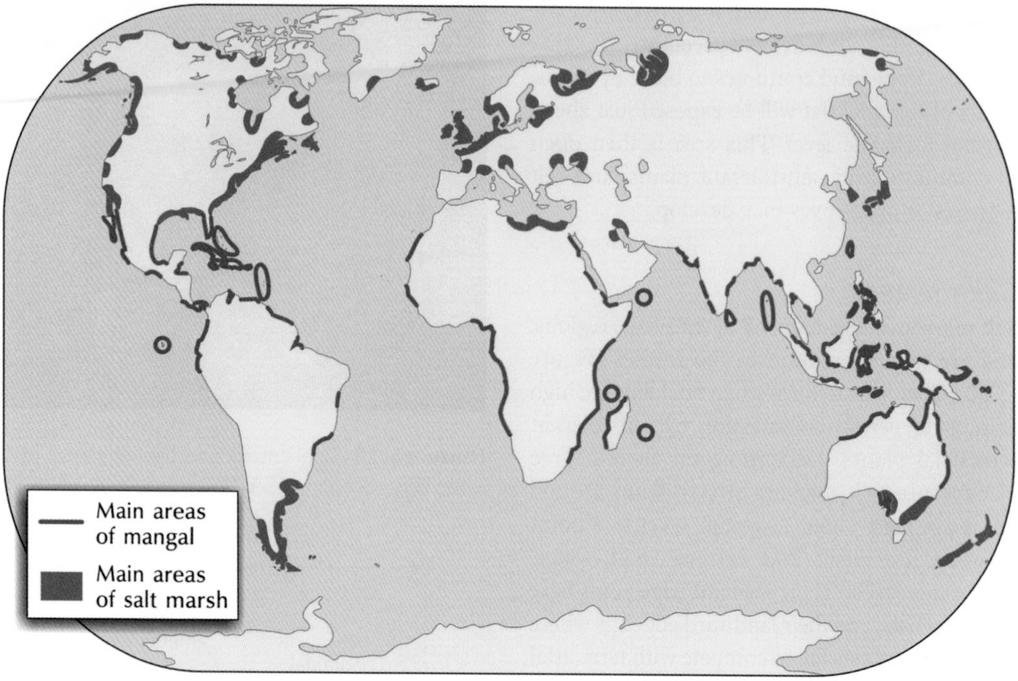

Figure 13.20 World distribution of salt marshes and mangals. *Source*: Adapted from Chapman (1977)

form in high-energy coasts with large tidal ranges and powerful waves. The trailing-edge coasts of continents (passive margins) and coasts facing marginal seas appear to favour the growth of large deltas.

Some deltas are triangular in plan, like the Greek letter delta, Δ, after which they were named almost 2,500 years ago by Herodotus. But deltas come in a multiplicity of forms, their precise shape depending upon the ability of waves to rework and redistribute the incoming rush of river-borne sediment. Six basic types are recognized (Box 13.3).

Coral reefs and atolls

A **coral reef** is a ridge or mound built of the skeletal remains of generations of coral animals, upon which grow living coral polyps. Reefs typically grow in shallow, clear waters of tropical oceans. The Great Barrier Reef, in the Coral Sea off the north-east Australian coast is, at over 2,600 km long, the world's largest living reef, and indeed the largest living organic feature. It comprises more than 3,000 individual reefs and hundreds of small coral islands, ranging in size from about 10 ha to 10,000 ha, formed along the edge of the continental shelf.

An **atoll** is a ring of coral reef and small sandy islands that encircles a shallow lagoon. Atolls are common in the tropical Pacific Ocean, where such groups of islands as the Marshall Islands and Kiribati are chains of atolls. They form when volcanic islands move away from the heating anomaly that creates them, as in the Hawaiian island hot-spot trace (p. 121), and they begin to subside beneath sea level. Reefs initially form as a fringe in the shallow waters around a volcanic island, as in Tahiti. With time, the island erodes and subsides. However, the reef continues growing upwards to create an offshore barrier reef separated from the main island by a lagoon,

Box 13.3 TYPES OF DELTA

Based on their overall morphology, six basic delta types are recognized that may be classified according to the importance of river, wave, and tidal processes (Wright 1985; Figure 13.21). The characteristics of the six types are as follows (Trenhaile 1997, 227–8):

- **Type 1 deltas** are dominated by river processes. They are elongated distributary mouth bar sands, aligned roughly at right-angles to the overall line of the coast. The protrusions are called bar-finger sands. In the modern birdfoot delta of the Mississippi River, seaward progradation of the principal distributaries has formed thick, elongated bodies of sand up to 24–32 km long and 6–8 km wide. Type 1 deltas form in areas with a low tidal range, very low wave energy, low offshore slopes, low littoral drift, and high, fine-grained suspended load. Examples are the deltas of the Mississippi in the USA, the Paraná in Brazil, the Dnieper in the Ukraine, and the Orinoco in Venezuela.
- **Type 2 deltas** are dominated by tides. They have broad, seaward-flaring, finger-like channel sand protuberances. Sandy tidal ridges produced by tidal deposition and reworked river sediments at distributary mouths front them. They occur in areas with a high tidal range and strong tidal currents, low wave energy, and low littoral drift. Examples are the deltas of the Ord in Western Australia, the Indus in Pakistan, the Colorado in the USA, and the Ganges–Brahmaputra in Bangladesh.
- **Type 3 deltas** are affected by waves and tidal currents. The tidal currents create sand-filled river channels and tidal creeks running approximately at right-angles to the coast, while the waves redistribute the riverine sand to produce beach–dune ridge complexes and barriers running parallel to the coast. Type 3 deltas are common in areas of intermediate wave energy, moderate to high tides, and low littoral drift. Examples are the Irrawaddy Delta in Burma, the Mekong Delta in Vietnam, and the Danube Delta in Romania.
- **Type 4 deltas** consist of finger-like bodies of sand deposited as distributary mouth bars, or they may coalesce to form a broad sheet of sand. They prograde into lagoons, bays, or estuaries sheltered by offshore or baymouth barriers. Their development is encouraged by intermediate wave energy, low offshore slopes, low sediment yields, and a low tidal range. Examples are the Brazos Delta in Texas and the Horton Delta in Canada.
- **Type 5 deltas** have extensive beach ridges and dune fields. These extensive sand sheets are shaped by wave redistribution of river-borne sands. They form where moderate to high wave energy is unremitting, where littoral drift is low, and where offshore slopes are moderate to steep. Examples are the deltas of the São Francisco in Brazil and the Godavari in India.
- **Type 6 deltas** form on coasts totally dominated by wave action. The waves straighten the coasts, and deltas consist of numerous sandy spit barriers running parallel to the coastline that alternate with fine-grained, abandoned channel fills. They are found in environments with strong waves, unidirectional longshore transport, and steep offshore slopes. Examples are the deltas of the Shoalhaven in New South Wales, Australia, and the Tumpat in Malaysia.

continued . . .

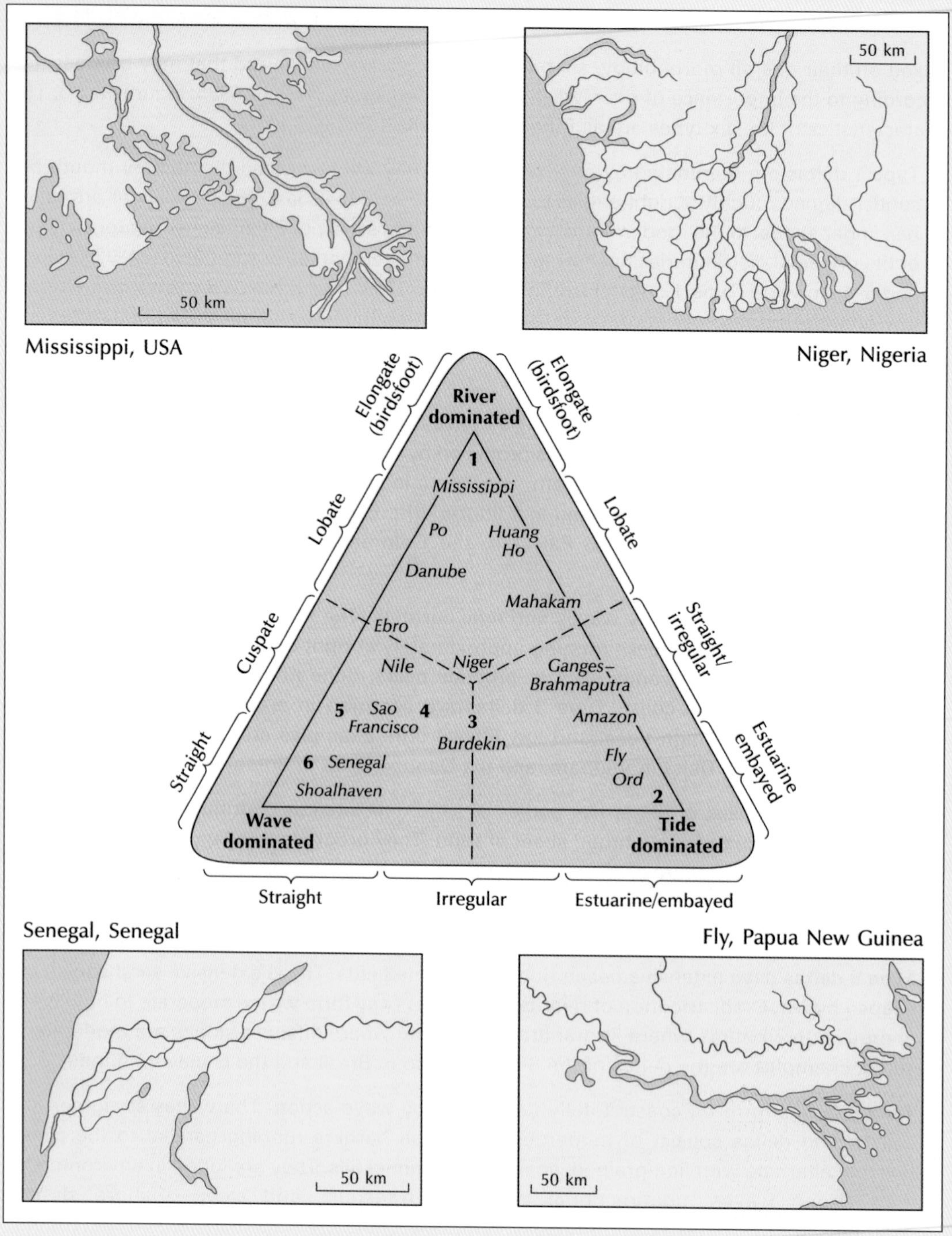

Figure 13.21 Types of deltas. *Source*: Partly adapted from Wright (1985)

as in the case of Bora Bora in the Society Islands. The lifting of a reef above sea level creates a **raised atoll**. These often have spectacular cave landscapes.

HUMANS AND COASTS

Humans affect erosion and deposition along coasts. They do so through increasing or decreasing the sediment load of rivers, by building protective structures, and indirectly by setting in train climatic processes that lead to sea-level rise. Three important issues focus around beach erosion and beach nourishment, the effect of rising sea levels over the next century, and the effect of changing wave climate.

Beach erosion and beach nourishment

To combat **beach erosion**, especially where it threatens to undermine and ruin roads and buildings, humans have often built **sea walls**. The idea is that a sea wall will stop waves attacking the eroding coast, commonly a retreating cliff, and undermining a slumping bluff or a truncated dune. Sea walls often start as banks of earth, but once these are damaged they are usually replaced by stone or concrete constructions. Other options are boulder ramparts (also called revetments or riprap) and artificial structures such as tetrapods, which are made of reinforced concrete. Solid sea walls, and even boulder barriers and other artificial structures, are effective and reflect breaking waves seawards, leading to a backwash that scours the beach of material. Such is the demand for counter-measures against coastal erosion that the world's coastline is littered with a battery of artificial structures. Some structures are successful, but the unsuccessful ones stand in ruins. Some have helped to maintain beaches, but others, by promoting eroding backwash, simply worsen beach erosion.

In an effort to prevent beach loss, the dumping of sand or gravel on the shore has become a common practice, mainly in the USA, Western Europe, and Australia. Such beach nourishment aims to create a beach formation, depleted by erosion, that 'will protect the coastline and persist in the face of wave action' (Bird 2000, 160). Many **beach nourishment** programmes were implemented at seaside resorts, where beaches are needed for recreational use (Box 13.4). Recently, the value of a beach in absorbing wave energy has been realized, and nourishment beaches are sometimes used to defend against further cliff erosion or damage to coastal roads and buildings. The key to a successful beach nourishment programme is a thorough comprehension of coastal geomorphology. Before developing and implementing a programme, it is necessary to find out the movement of sand and gravel in relation to the wave regimes and the effects of any artificial structures on the section of shore to be treated. It is also necessary to understand why the beach is eroding and where the sediment has gone – landward, seaward, or alongshore. The modelling of beach forms and processes can be helpful at this stage, but an experimental approach, based on accumulated experience, is often more productive. The sediment used to nourish a beach should be at least as coarse as the natural beach sediment and durable, but it may come from any source. More material than is necessary to restore the beach (so allowing for onshore, offshore, and alongshore losses) is usually dumped to form a beach terrace that is worked on by waves and currents to form a natural beach profile, often with sand bars just offshore. The restored beach may be held in place by building a retaining backwater or a series of groynes. In some cases, a beach can be nourished by dumping material where it is known that longshore or shoreward drift will carry it to the shore. Nourished beaches normally still erode and occasionally need replacing. More details of beach management are found in Bird (1996).

Many coastal geomorphologists are coming to accept that humans are now an integral part of the coastal environment and as such, coastal

Box 13.4 BEACH EROSION VERSUS BEACH NOURISHMENT – A DELAWARE CASE STUDY

It costs money to nourish beaches, and any beach nourishment programme has to consider the economics of letting beaches retreat compared with the economics of sustaining them. Take the Atlantic coastline of Delaware, eastern USA (Daniel 2001). Delaware's coastline combines high shoreline-property values with a growing coastal tourism industry. It is also a dynamic coastline, with storm damage and erosion of recreational beaches posing a serious threat to coastal communities. Local and state officials are tackling the problem. A comprehensive management plan, called Beaches 2000, considered beach nourishment and retreat. The goal of Beaches 2000 is to safeguard Delaware's beaches for the citizens of Delaware and out-of-state beach visitors. Since Beaches 2000 was published, Delaware's shorelines have been managed through nourishment activities, which have successfully maintained beach widths. Coastal tourism, recreational beach use, and real-estate values in the area continue to grow. The possibility of letting the coastline retreat was considered in the plan, but shelved as an option for the distant future. One study estimated the land and capital costs of letting Delaware's beaches retreat inland over the next fifty years (Parsons and Powell 2001). The conclusion was that, if erosion rates remain at historical levels for the next fifty years, the cost would be $291,000,000 but would be greater should erosion rates accelerate. In the light of this figure, beach nourishment makes economic sense, at least over the fifty-year time period.

environments are best understood by considered the human–coast system (Nordstrom 1994; Lazarus *et al.* 2016). It is vital to understand how human decisions bring about geomorphic responses and how those responses are acted upon by human decision-makers. Figure 13.22 portrays beach nourishment as a coastal exemplar of a coupled human–landscape system.

The effects of rising sea levels in the twenty-first century

A current worry is how coastlines will respond to **rising sea levels** during the present century. The rise will result from water additions to the oceans owing to the melting of glaciers and from the thermal expansion of seawater. Estimates of mean sea-level rise are about 44 cm by the year 2100 for the best-case scenario and 63 cm for the worst-case scenario (Church *et al.* 2013). The predicted rises will put coastal ecosystems, already under considerable pressure from development (60 per cent of the world's population living within 100 km of the coast), under further risk. Some areas are particularly vulnerable to sea-level rise; for example, in the South Pacific region where many island nations are threatened. These islands, which include Tonga, Fiji, Samoa, and Tuvalu, are low-lying and likely to see increased flooding and inundation, with other environmental impacts expected to include beach erosion, saltwater intrusion, and the distribution of the communications and other infrastructure.

Geomorphic effects of sea-level rise are varied. Inevitably, submerging coastlines, presently limited to areas where the land is subsiding, will become widespread and emerging coastlines will become a rarity. Broadly speaking, low-lying coastal areas will be extensively submerged and their high- and low-tide lines will advance landwards, covering the present intertidal zone. On steep, rocky coasts, high- and low-tide levels will simply rise, and the coastline stay in the same position. It seems likely that the sea will continue

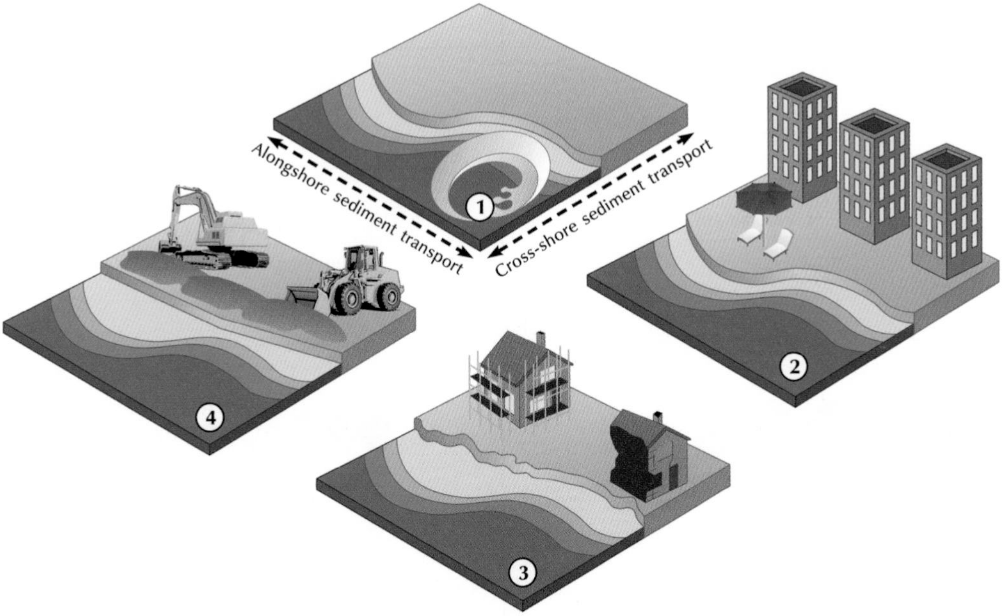

Figure 13.22 Beach nourishment as a coupled human–landscape system. (1) Natural littoral processes of alongshore and cross-shore sediment transport create spatial patterns of beach accretion and erosion. (2) Coastal development built to benefit economically from the natural capital of a wide beach. (3) The developed beach becomes exposed to damage from coastal hazards. (4) Risk exposure drives investment in hazard mitigation and shoreline protection. Where beach erosion is persistent, this cycle repeats through stages 1 to 4 on a multi-annual to decadal cycle. *Source*: Adapted from Werner and McNamara (2007) and Lazarus *et al.* (2016)

to rise, with little prospect of stabilization. If it does so, then coastal erosion will accelerate and become more prevalent as compensating sedimentation tails off. The rising seas will reach, reshape, and eventually submerge 'raised beaches' created during the Pleistocene interglacials. Forms similar to those found around the present world's coasts would not develop until sea level stabilized, which would presumably occur either when the measures adopted to counterbalance increasing greenhouse gases worked, or else when all the world's glaciers, ice sheets, and snowfields had melted, occasioning a global sea-level rise of more than 60 m (Bird 2000, 276).

Figure 13.23 summarizes specific effects of rising sea levels on different types of coast. Cliffs and rocky shores were largely produced by the tolerably stable sea levels that have dominated over the last 6,000 years. Rising sea levels will submerge shore platforms and rocky shores, allowing larger waves to reach cliffs and bluffs, so accelerating their erosion on all but the most resistant rocks (Figure 13.23a). Some eastern British cliffs are retreating about 100 cm a year, and this rate will increase by 35 cm a year for every 1 mm rise of sea level (Clayton 1989). Cliff notches will enlarge upwards as the rising sea eats into successively higher levels. Rising sea levels are also likely to increase the occurrence of coastal landslides and produce new and extensive slumps, especially where rocks dip towards the sea. The slump material will add to sediment supply for beaches, perhaps in part compensating for the rising sea level. The rise of sea level by 1 to 2 mm per year over the last few decades has caused beach erosion in many places around the world.

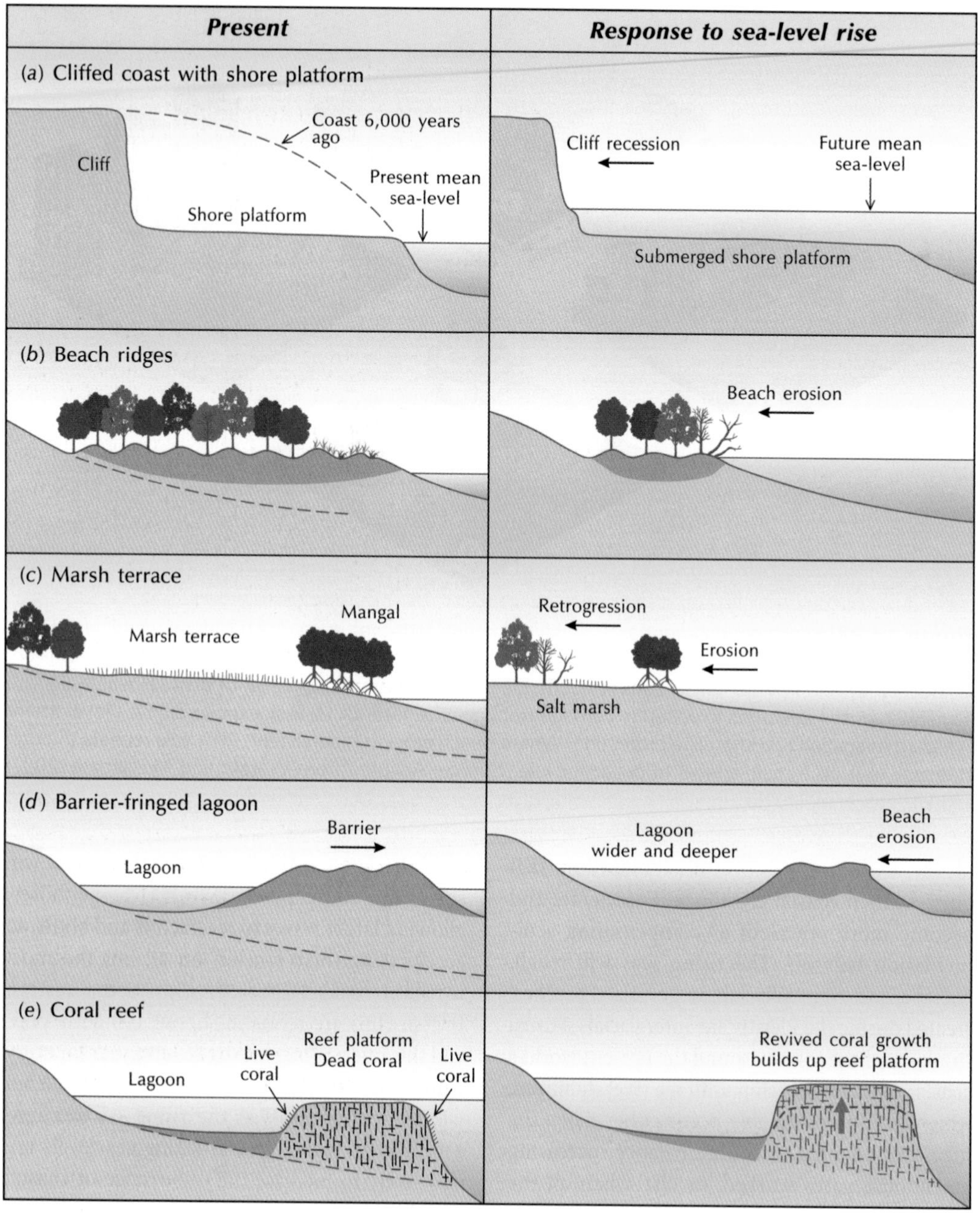

Figure 13.23 Coastal changes brought about by a rising sea level. (a) Cliffed coast with shore platform. (b) Beach ridges. (c) Marsh terrace. (d) Barrier-fringed lagoon. (e) Coral reef. *Source*: Adapted from Bird (2000, 278)

Accelerating sea-level rise will greatly exacerbate this problem. The seaward advance of prograding beaches will stop and erosion set in (Figure 13.23b). Where the beach is narrow, with high ground behind it, the beach may rapidly disappear unless nearby cliff erosion provides enough replenishment of sediment. Beaches fronting salt marshes and mangals will probably be eroded and over-washed. Beaches ahead of sea walls will be eroded or be removed by the scour resulting from the reflection of incident waves. Beaches will persist wherever the supply of sand or shingle is sustained, or where additional material is provided by cliff erosion or increased sediment load from rivers. Most present beaches will probably be lost as sea levels rise, but on coastal plains with coastal dunes new beaches may form by the shoreward drifting of sediment up to the new coastline, along the contour on which submergence stops.

Salt marshes, mangals, and intertidal areas will all be submerged beneath rising sea levels (Figure 13.23c). Small cliffs on the seaward margins of salt marshes and mangrove terraces will erode faster than at present. Continued submergence will see the seaward and landward margins move inland. In low-lying areas, this may produce new salt marshes or mangals, but steep-rising hinterlands will cause a narrowing and perhaps eventual disappearance of the salt marsh and mangal zone. The loss of salt marshes and mangals will not occur in areas where sediment continues to be supplied at a rate sufficient for a depositional terrace to persist. And modelling suggests the salt marshes of mesotidal estuaries, such as the Tagus estuary in Portugal, do not appear vulnerable to sea-level rise in all but the worst-case scenario with several industrialized nations not meeting the terms of the Kyoto Protocol (Simas *et al.* 2001). Inner salt marsh or mangal edges may expand inland, the net result being a widening of the aggrading salt-marsh or mangrove terrace. Intertidal areas – sandflats, mudflats, and rocky shores – will change as the sea level rises. The outer fringe of the present intertidal zone will become permanently submerged. As backing salt marshes and mangals are eroded and coastal lowland edges cut back, they will be replaced by mudflats or sandflats, and underlying rock areas will be exposed to form new rocky shores.

Estuaries will generally widen and deepen as sea level goes up, and may move inland. Coastal lagoons will become larger and deeper, and their shores and fringing swamp areas suffer erosion (Figure 13.23d). The enclosing barriers may be eroded and breached to form new lagoon inlets that, with continued erosion and submergence, may open up to form marine inlets and embayments. New lagoons may form where rising sea levels cause the flooding of low-lying areas behind dune fringes on coastal plains. They may also form where depressions are flooded as rising water tables promote the development of seasonal or permanent lakes and swamps. Wherever there is a supply of replenishing sediment, the deepening and enlargement of estuaries and lagoons may be countered.

Corals and algae living on the surface of intertidal reef platforms will be spurred into action by a rising sea level and grow upwards (Figure 13.23e). However, reef revival depends upon a range of ecological factors that influence the ability of coral species to recolonize submerging reef platforms. In addition, the response of corals to rising sea levels will depend upon the rate of sea-level rise. An accelerating rate could lead to the drowning and death of some corals, and to the eventual submergence of inert reef platforms. Studies suggest that reefs are likely to keep pace with a sea-level rise of less than 1 cm/year, to be growing upwards when sea-level rise falls within the range by 1–2 cm/year, and to be drowned when sea-level rise exceeds 2 cm/year (Spencer 1995).

Wave climate change

Global warming is driving a change in wave climate that is also affecting some coastal systems. Cape Hatteras and Cape Lookout, North Carolina,

USA, have become increasingly asymmetrical over the past thirty years, probably in response to increases in Atlantic Ocean summer wave heights, owing to hurricanes, and resulting changes in the distribution of wave-approach angles (Johnson *et al.* 2015). If hurricane-generated wave heights continue to increase, large-scale shifts in patterns of coastal erosion and accretion are likely to extend beyond sensitive cuspate landforms features such as Capes Hatteras and Lookout, with the larger-scale coastline shape coming into equilibrium with the changing wave conditions.

Human response to climate change and sea-level rise

Climate change and sea-level rise are forcing coastal and shoreline management schemes to take on board coastal landform change occurring over decades and centuries. Models dealing with changes on these timescales are meeting with some success (e.g. van Maanen *et al.* 2015).

COASTAL LANDSCAPES IN THE PAST

Sea-level change

Sea level seldom remains unchanged for long as changes in ocean volume, or changes of the distribution of mass within oceans, alter it (Box 13.5). Tectonics is the ultimate control of sea level. In the case of tectono-eustasy, the control is direct. In the case of glacio-eustasy, the control is indirect: tectonics (and other factors) alter climate and climate alters sea level.

Sea level fluctuates over all timescales. Medium-term and long-term changes are recorded in sedimentary rocks and revealed by the technique of seismic stratigraphy (e.g. Vail *et al.* 1977, 1991). This technique offers a precise means of subdividing, correlating, and mapping sedimentary rocks. It uses primary seismic reflections. The reflections come from geological discontinuities between stratigraphic units that result from relative changes of sea level. The discontinuities are lithological transitions caused by abrupt changes in sediment delivery, and they can be correlated worldwide. They display six superimposed orders of cyclical sea-level change during the Phanerozoic aeon (Table 13.4). Each cycle has a distinct signature. First-order cycles reflect major continental flooding. Second-order cycles register facies changes associated with major transgression–regression cycles. The lower-order cycles record stratigraphic deposition sequences, systems tracts (sets of linked contemporaneous depositional systems), and parasequences (the building blocks of systems tracts). The changes in sea level resulting from the first- and second-order cycles can be as much as 250 m (Figure 13.24). Such very long-term changes of sea level provide a useful yardstick against which to discuss Quaternary sea-level changes and predicted rises of sea level over the next century.

Whatever the cause of sea-level change, higher and lower sea levels, especially those that occurred during the Quaternary, leave traces in landscapes (e.g. Butzer 1975; Bloom and Yonekura 1990; Gallup *et al.* 1994; Ludwig *et al.* 1996). **Marine terraces** and **drowned landscapes** record highstands and lowstands of sea level during Quaternary glacial and interglacial climates. High levels during interglacial stages alternate with low levels during glacial stages, glacio-eustatic mechanisms largely driving the system. Classical work around the Mediterranean Sea recognized a suite of higher levels corresponding to glacial stages (Figure 13.25).

Highstands of sea level

Many shorelines bear evidence of higher sea levels. Various types of raised shoreline – stranded beach deposits, beds of marine shells, ancient coral reefs, and platforms backed by steep cliff-like slopes – all attest to higher stands of sea level. Classic examples come from fringing coasts of formerly glaciated areas, such as Scotland, Scandinavia, and North America. An example is the *Patella* raised beach on the Gower Peninsula, South Wales

Box 13.5 CAUSES OF SEA-LEVEL CHANGE

Volumetric and mass distribution changes in the oceans cause sea-level change (Table 13.3). Ocean volume changes are eustatic or steric. **Eustatic change** results from water additions or extractions from the oceans (glacio-eustatic change), and from changes in ocean-basin volume (tectono-eustatic change). **Steric change** results from temperature or density changes in seawater. Much of the predicted sea-level rise during the twenty-first century will result from the thermal expansion of seawater. Ocean thermal expansion is about 20 cm/°C/1,000 m (Mörner 1994).

Table 13.3 Causes of eustatic change

Seat of change	Type of change	Approximate magnitude of change (m)	Causative process
Ocean basin volume	Tectono-eustatic	50–250	Orogeny, mid-ocean ridge growth, plate tectonics, sea-floor subsidence, other Earth movements
Ocean water volume	Glacio-eustatic	100–200	Climatic change
	Hydro-eustatic	Minor	Changes in liquid water stored in sediment, lakes, and clouds; additions of juvenile water; loss of connate water
Ocean mass distribution and surface 'topography'	Geoidal eustatic	Up to 18	Tides
		A few metres	Obliquity of the ecliptic
		1 m per millisecond of rotation	Rotation rate
		Up to 5	Differential rotation
		2 (during Holocene)	Deformation of geoid relief
	Climo-eustatic	Up to 5 for major ocean currents	Short-term meteorological, hydrological, and oceanographic changes

Source: From Huggett (1997b, 151), partly adapted from Mörner (1987, 1994)

Glacio-eustatic change

Glacio-eustatic change is tightly bound to climatic change. Globally, inputs from precipitation and runoff normally balance losses from evaporation. (Gains from juvenile water probably balance losses in buried connate water.) However, when the climate system switches to an icehouse state, a substantial portion the world's water supply is locked up in ice sheets and glaciers. Sea level drops during glacial stages, and rises during interglacial stages. Additions and subtractions of water from the oceans, other than that converted to ice, may cause small changes in ocean volume. This minor process might be termed hydro-eustasy.

Tectono-eustatic change

Geological processes drive tectono-eustatic change. Even when the water cycle is in a steady state, so that additions from precipitation balance losses through evaporation, sea level may change owing to volumetric changes in the ocean basins. An increasing volume of ocean basin

continued . . .

would lead to a fall of sea level and a decreasing volume to a rise of sea level. Decreasing volumes of ocean basin are caused by sedimentation, the growth of mid-ocean ridges, and Earth expansion (if it should have occurred); increasing volumes are caused by a reduced rate or cessation of mid-ocean ridge production.

Other eustatic effects

Geoidal eustasy results from processes that alter the Earth's equipotential surface, or **geoid**. The ocean geoid is also called the geodetic sea level. The relief of the geoid is considerable: there is a 180-m sea-level difference between the rise at New Guinea and the depression centred on the Maldives, which lie a mere 50–60 degrees of longitude from one another. There is also a geoid beneath the continents. The configuration of the geoid depends on the interaction of the Earth's gravitational and rotational potentials. Changes in geoid relief are often rapid and lead to swift changes in sea level.

On a short timescale, local changes in weather, hydrology, and oceanography produce relatively tame fluctuations of sea level. These fluctuations might be called climo-eustasy. They may involve up to 5 m of sea-level change for major ocean currents, but less than half that for meteorological and hydrological changes.

Isostatic change

Isostasy refers the principle of buoyancy or flotation of continents and oceans. It assumes a state of gravitational equilibrium between the lithosphere and underlying asthenosphere such that the tectonic plates 'float' at an elevation dependent on their thickness and density. During ice ages, areas covered by ice tend to 'sink' under the weight of ice. Once the ice has melted, the rebounding of the lithosphere restores isostatic equilibrium. The uplift involved in this process, called **glacial isostatic adjustment**, produces an apparent lowering of sea level in areas affected. The adjustment may involve downward and lateral movement of the lithosphere, as well as rebound, and can lead to global sea-level changes by deforming the shape of the planet (Peltier 1998, 1999).

(Bowen 1973). A shingle deposit lies underneath tills and periglacial deposits associated with the last glacial advance. The shingle is well cemented and sits upon a rock platform standing 3–5 m above the present beach. It probably formed around 125,000 years ago during the last interglacial stage, when the sea was 5 m higher than now.

Ancient coral reefs sitting above modern sea level are indicative of higher sea levels in the past. In Eniwetok atoll, the Florida Keys, and the Bahamas, a suite of ancient coral reefs correspond to three interglacial highstands of sea level 120,000 years ago, 80,000 years ago, and today (Broecker 1965). Similarly, three coral-reef terraces on Barbados match interglacial episodes that occurred 125,000, 105,000, and 82,000 years ago (Broecker *et al.* 1968).

Lowstands of sea level

Submerged coastal features record lower sea levels during the Quaternary. Examples are the drowned mouths of rivers (rias), submerged coastal dunes, notches and benches cut into submarine slopes,

Table 13.4 Quasi-periodic sea-cycles revealed by seismic stratigraphy

Cycle order	Duration (years)	Eustatic expression	Possible cause
First	More than 50 million	Major continental flooding cycle	Change in ocean volume owing to sea-floor spreading rates and continental dispersal
Second	3 million to 50 million	Transgressive–regressive cycles	Change in ocean volume owing to sea-floor spreading rates
Third	500,000 to 3 million	Sequence cycles	Unclear
Fourth	80,000 to 500,000	Systems tracts	Orbital forcing in the Croll–Milankovitch frequency band
Fifth	30,000 to 80,000	Episodic parasequences	Orbital forcing in the Croll–Milankovitch frequency band
Sixth	10,000 to 30,000	Episodic parasequences	Orbital forcing in the Croll–Milankovitch frequency band

Source: Partly adapted from Vail *et al.* (1991)

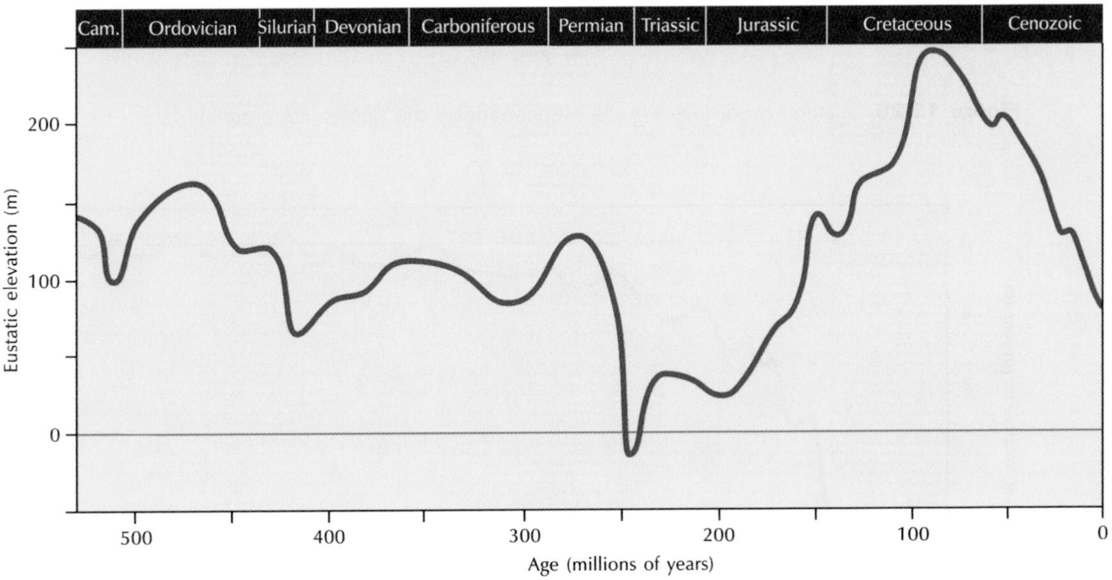

Figure 13.24 Proposed Phanerozoic eustatic curve. *Source*: The Palaeozoic section is adapted from an average estimated by Algeo and Seslavinsky (1995) and the Mesozoic and Cenozoic portion is adapted from Haq *et al.* (1987)

and the remains of forests or peat layers lying below modern sea level.

The lowering of the sea was substantial. During the Riss glaciation, a lowering of 137–59 m is estimated, while during the last glaciation (the Würm) a figure of 105–23 m is likely. A fall of 100 m or thereabouts during the last glaciation was enough to link several islands with nearby mainland: Britain to mainland Europe, Ireland to Britain, New Guinea to Australia, and Japan to China. It would have also led to the floors of the Red Sea and the Persian Gulf becoming dry land.

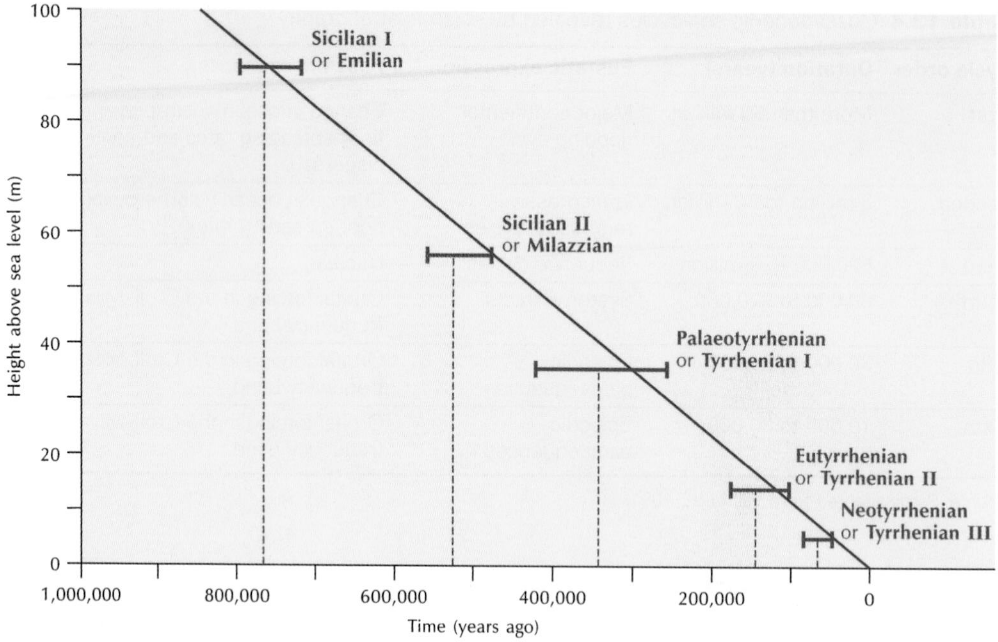

Figure 13.25 Quaternary sea-levels in the Mediterranean: the classic interpretation.

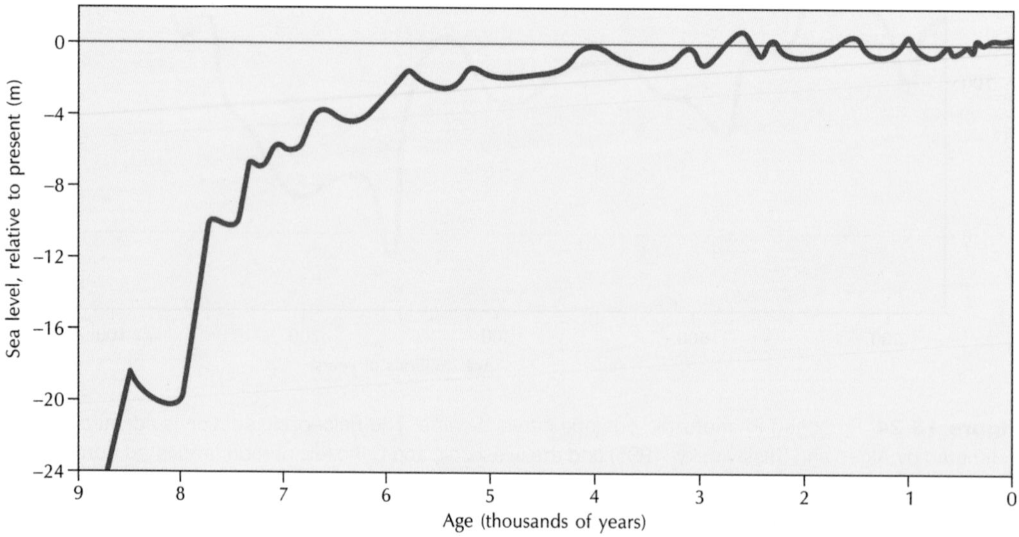

Figure 13.26 The Flandrian transgression in north-western Europe. Notice the rapid rise between about 9,000 and 7,000 years ago; this amounted to about 20 m, an average increase of 1 m per century. *Source*: Adapted from Mörner (1980)

Of particular interest to geomorphologists is the rise of sea level following the melting of the ice, which started around 12,000 years ago. This rise is known as the **Holocene** or **Flandrian transgression**. It was very rapid at first, up to about 7,000 years ago, and then tailed off (Figure 13.26). Steps on coastal shelves suggest that the rapid transgression involved stillstands, or even small regressions, superimposed on an overall rise. The spread of sea over land during this transgression would have been swift. In the Persian Gulf regions, an advance rate of 100–120 m a year is likely and even in Devon and Cornwall, England, the coastline would have retreated at about 8 m a year.

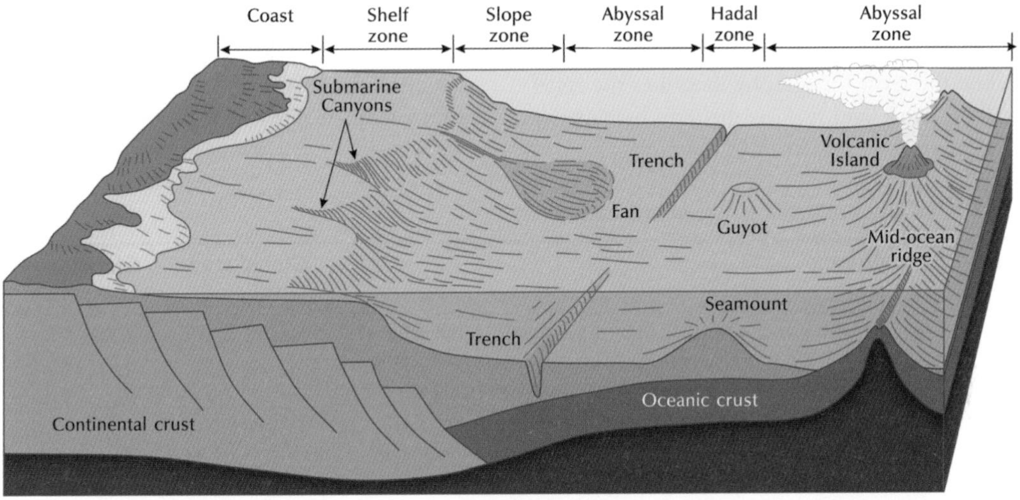

Shelf zone	Slope zone	Abyssal zone	Hadal zone
Low relief (<10 m)	Terraces	Abyssal plains (<300 m relief)	
Medium relief (10–50 m)		Abyssal hills (300–1,000 m relief)	
High relief (>50 m)		Abyssal mountains (>1,000 m relief)	
Shelf valleys		Continental rises	
Glacial troughs		Mid-ocean ridges	
Coral reefs		Rift valleys	
Basins (shelf perched)	Basins (slope perched)	Basins	Basins
Sills	Sills	Sills	Sills
	Escarpments	Escarpments	Escarpments
	Seamounts	Seamounts	Seamounts
	Guyots	Guyots	
	Canyons (shelf incising)	Canyons (shelf incising)	
	Canyons (blind)	Canyons (blind)	
	Ridges	Ridges	Ridges
	Troughs	Troughs	Troughs
		Trenches	Trenches
		Bridges	Bridges
	Fans	Fans	
	Plateaux	Plateaux	

Figure 13.27 Relief features of the seafloor.

SUBSEA LANDFORMS AND PROCESSES

Almost 71 per cent of the Earth's solid surface currently lies beneath the oceans. The geomorphology of the seafloor has practical applications in marine conservation, oil exploration, and hazard assessment (such as tsunami impacts), as well as being of academic interest.

Seafloor geomorphic features

While the term topography describes the relief of the Earth's land surface, the term bathymetry describes the relief of the solid Earth at the base of the oceans. Bathymetric maps use data from an echo sounder mounted beneath or over the side of a boat, 'pinging' a beam of sound downward at the seafloor, or from remote sensing LiDAR or LADAR systems, usually from the air. Bathymetric surveys reveal a huge variety of relief features on the seafloor, their formation the result of a complex mix of tectonic, volcanic, glacial, mass movement, and other processes (Figure 13.27).

In 2014, the first digital seafloor geomorphic features map (GSFM) of the global ocean was

Table 13.5 Global areas of seafloor geomorphic features

Geomorphic feature	Definition	Global area (km^2)	Global area (%)
Shelf area	A zone adjacent to a continent (or around an island) and extending from the low water line to a depth at which there is usually a marked increase of slope towards oceanic depths	32,242,540	8.91
Slope area	The deepening seafloor out from the shelf edge to the upper limit of the continental rise, or the point where there is a general decrease in steepness	19,606,260	5.42
Abyss area	Area of seafloor located at depths below the foot of the continental slope and above the depth of the hadal zone	306,595,900	84.70
Hadal area	Seafloor occurring at depths >6,000 m	3,437,930	0.95
Total		**361,883,510**	**100**
Basin	A depression, in the sea floor, more or less equi-dimensional in plan and of variable extent, defined by a closed bathymetric contour	158,529,660	43.8
Abyssal hills	Variation in relief over a 25-cell radius of 300–1,000 m	149,451,310	41.3
Abyssal plains	Variation in relief over a 25-cell radius of <300 m	100,863,730	27.9
Abyssal mountains	Variation in relief over a 25-cell radius of >1,000 m	57,678,740	15.9
Rise	Low gradient, evenly spaced, slope-parallel contours extending seawards from the foot of the continental slope generally confined to areas of sediment thickness >300 m	29,832,040	8.24
Escarpment	An elongated, characteristically linear, steep slope >5° and areal extent of >100 km^2, separating horizontal or gently sloping sectors of the sea floor in non-shelf areas	21,151,400	5.84
Plateau	Flat or nearly flat elevations of considerable areal extent, dropping off abruptly on one or more sides	18,486,610	5.11
Shelf – medium profile	Variation in relief over a five-cell radius of 10–50 m	14,447,690	3.99

Table 13.5 *continued*

Geomorphic feature	Definition	Global area (km^2)	Global area (%)
Shelf – low profile	Variation in relief over a five-cell radius of <10 m	9,799,870	2.71
Ridge	An isolated (or group of) elongated (length/width ratio >2), narrow elevation(s) of varying complexity having steep sides, >1,000 m in vertical relief	9,770,720	2.7
Fan	A relatively smooth, fan-like, depositional feature commonly found sloping away from the outer termination of a canyon or canyon system	8,303,160	2.29
Shelf – high profile	Variation in relief over a five-cell radius of >50 m	7,995,050	2.21
Seamount	A discrete (or group of) large isolated elevation(s), greater than 1000 m in relief above the sea floor, characteristically of conical form (length/width ratio <2)	7,859,200	2.17
Spreading ridge	The linked, major mid-oceanic mountain systems of global extent coinciding with the youngest ocean crust	6,699,460	1.85
Shelf valley	Valleys incised more than 10 m into the continental shelf	4,756,290	1.31
Canyon	Steep-walled, sinuous valleys with V-shaped cross sections, axes sloping outwards as continuously as river-cut land canyons and relief comparable to even the largest of land canyons. (a) total canyon area; (b) shelf incising, (c) blind, slope-confined	(a) 4,393,650 (b) 1,613,860 (c) 2,779,790	1.21 0.45 0.77
Glacial trough	Elongate troughs, typically trending across the continental shelf, attributed to glacial erosion during the Pleistocene ice ages	3,659,360	1.01
Trough	A long depression of the seafloor characteristically flat bottomed and steep sided, generally open at one end	2,841,420	0.785
Terrace	An isolated (or group of) relatively flat horizontal or gently inclined surface(s), sometimes long and narrow, which is (are) bounded by a steeper ascending slope on one side and by a steeper descending slope on the opposite side	2,303,490	0.637
Trench	A long narrow, characteristically very deep and asymmetrical depression of the sea floor, with relatively steep sides	1,967,350	0.544
Guyot	A seamount having a flat top >10 km^2 in areal extent and with a gradient of <2°	936,920	0.259
Rift valley	Valleys confined to the central axis of mid-ocean spreading ridges; they are elongate, local depressions flanked generally on both sides by ridges	710,060	0.196
Sill	A seafloor barrier of relatively shallow depth restricting water movement between basins	45,450	0.0126
Bridge	Features composed of blocks of material that partially infill trenches or troughs, forming a 'bridge' across them	8,270	0.00229
Coral reef	Coral reef	212,340	0.0587

Source: Adapted from Harris *et al.* (2014)

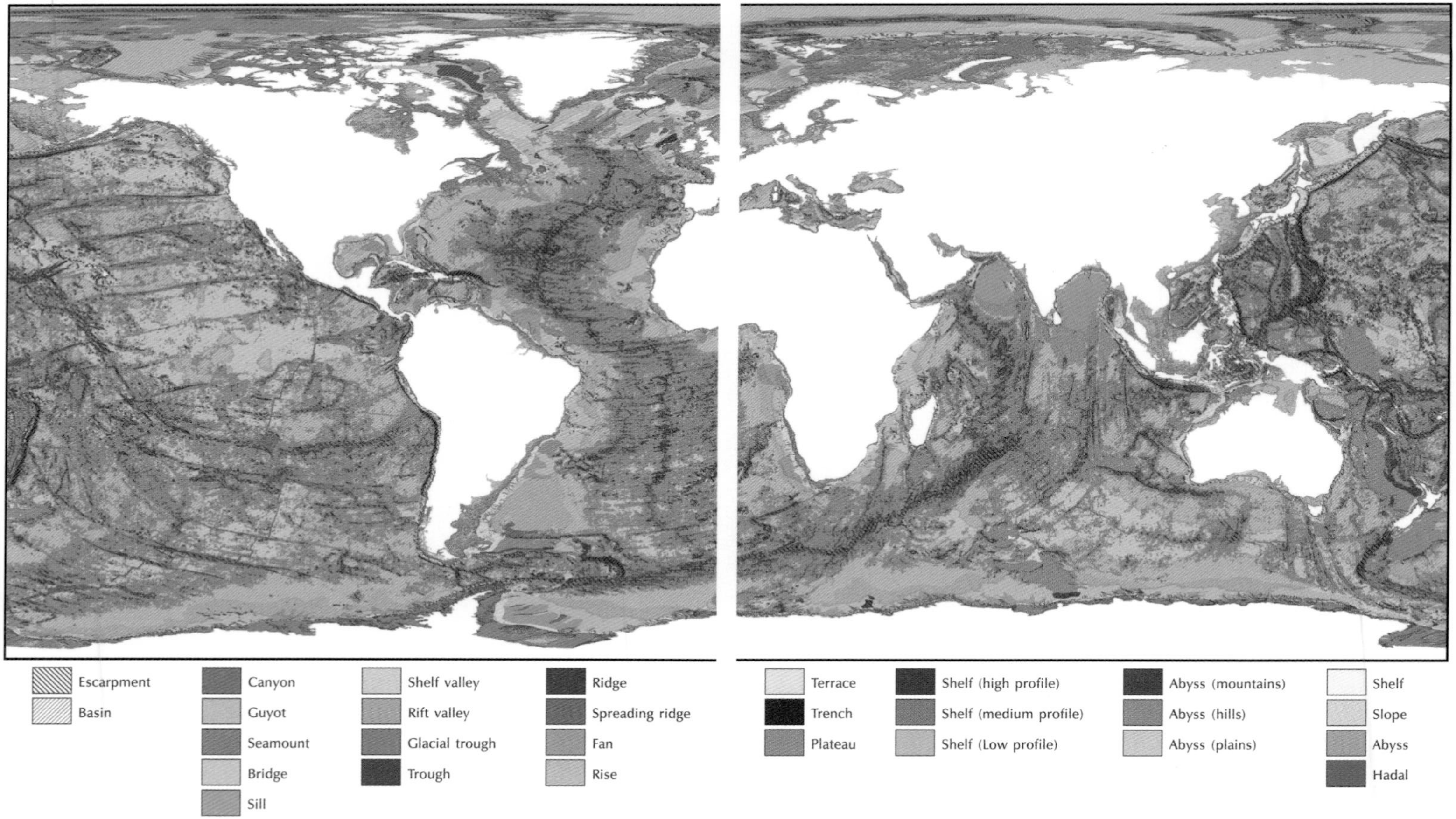

Figure 13.28 Geomorphic features map of the world's oceans. Dotted black lines mark boundaries between major ocean regions. Basins are not shown. *Source*: After Harris *et al.* (2014)

published (Harris *et al.* 2014). It used manual and ArcGIS methods based on the analysis and interpretation of a modified version of the SRTM30_PLUS global bathymetry grid (Becker *et al.* 2009) to map twenty-nine geomorphic features (Table 13.5; Figure 13.28).

Process implications of subsea landforms

The digital seafloor geomorphic features map (GSFM) promises advances in three key areas (Harris *et al.* 2014). First, in process studies, it should aid interpretations of features produced by a particular geological process. Second, in spatial analysis, it should allow a detailed study of the geomorphic composition of different ocean areas. Third, in seafloor mapping studies, it should improve methods for the interpretation and mapping of features.

By way of example, combining the GSFM with a dataset on seafloor spreading rates provided insights into geomorphic expressions of fast versus slow seafloor spreading rates, which appear to correlate with the occurrence of small versus large rift valley segment sizes, respectively. Second, the spatial analysis of submarine canyons, abyssal roughness, abyssal plains, and rises demonstrated that significant geomorphic differences occur between polar and non-polar margins that are attributable to continental glaciations. Third, recognition of seamounts as being a separate category of feature from ridges resulted in estimates of seamount number and area being much less than previous estimates that did not distinguish between ridges and seamounts.

SUMMARY

Waves and tsunamis buffet coasts, nearshore currents wash them, and tides wet them. Weathering and wave erosion destroy coastlines, while sediment deposition, reef-building corals, and the mangal and marsh builders create them. Rocky coasts are dominated by erosional landforms – shore platforms and plunging cliffs, caves, and arches and stacks, and many more. Some erosional landforms occur in predominantly depositional environments, as in tidal creeks cutting across salt marshes. Depositional landforms along coasts are many and varied. Beaches are the commonest features, with wave-dominated, tide-modified, and tide-dominated being the three chief types, but assorted species of spits and barriers are widespread. Other depositional landforms include beach ridges, cheniers, coastal sand dunes, estuaries, tidal flats, salt marshes, mangals, marine deltas, and coral reefs and atolls. Humans affect coastal erosion and deposition by increasing or decreasing the sediment load of rivers and by building protective structures. Many beaches in Western Europe, the USA, and Australia need feeding with sand to maintain them. The effects of a rising sea level over the next century following the warming trend are far-reaching and likely to impact severely on humans living at or near coasts. Sea-level changes are brought about by gains and losses of water to and from the oceans (glacioeustatic changes), from increases and decreases in oceanic basin volume (tectono-eustatic changes), and from fluctuations in ocean temperature or density (steric changes). High-stands and lowstands of sea level leave their marks on the land surface and beneath the waves. Stranded beach deposits, beds of marine shells, ancient coral reefs, and platforms backed by steep cliff-like slopes mark higher ocean levels. Submerged coastal features, including rias, notches, and benches cut into submarine slopes, and sunken forests mark lower levels. The rise of sea level associated with deglaciation may be very rapid, witness the Flandrian transgression. Over seven-tenths of the Earth's landforms lie beneath the ocean. The study of these features is in its infancy.

ESSAY QUESTIONS

1. How do currents and waves produce landforms?
2. Why do deltas display such a variety of forms?
3. Assess the likely consequences of a rising sea level during the present century for coastal landforms.
4. Compare and contrast terrestrial landforms and subsea landforms.

FURTHER READING

Bird, E. C. F. (2008) *Coastal Geomorphology: An Introduction*, 2nd edn. Chichester: John Wiley & Sons.
A highly readable, systematic coverage of form and process along coasts. Excellent.

Carter, R. W. G. (1992) *Coastal Environments*. London: Edward Arnold.
Covers applied coastal geomorphology, linking the physical and biological resources of coasts with their exploitation and use. Good for students interested in coastal management.

King, C. A. M. (1972) *Beaches and Coasts*, 2nd edn. London: Edward Arnold.
Dated, but is well worth digging out for the classic examples of coastal features.

Komar, P. D. (1998) *Beach Processes and Sedimentation*, 2nd edn. Upper Saddle River, N.J.: Prentice Hall.
A top-flight technical book on beach process, but some mathematical knowledge is needed.

Masselink, G. and Hughes, M. G. (2003) *An Introduction to Coastal Processes and Geomorphology*. London: Arnold.
An excellent introduction to the environments and processes occurring along the coastlines of the world.

Short, A. D. and Woodroffe, C. D. (2009) *The Coast of Australia*. Cambridge: Cambridge University Press.
An excellent account of Australian coasts, with wider applicability.

Sunamura, T. (1992) *Geomorphology of Rocky Coasts*. Chichester: John Wiley & Sons.
A good, if somewhat technical, account of form and process along rocky coasts.

Trenhaile, A. S. (1997) *Coastal Dynamics and Landforms*. Oxford: Clarendon Press.
An expansive treatment, brimful with detailed discussions and examples.

Woodroffe, C. D. (2002) *Coasts: Form, Process and Evolution*. Cambridge: Cambridge University Press.
A useful text for advanced undergraduates. It has a strong emphasis on coastal processes.

CHAPTER FOURTEEN

KARST LANDSCAPES

14

Acid attacking rocks that dissolve easily, and some rocks that do not dissolve so easily, create very distinctive and imposing landforms at the ground surface and underground. This chapter covers:

- The nature of soluble-rock terrain
- The dissolution of limestone
- Landforms formed on limestone
- Landforms formed within limestone
- Humans and karst
- Past karst

UNDERGROUND KARST: POOLE'S CAVERN, DERBYSHIRE

Poole's Cavern is a limestone cave lying under Grin Wood, almost 2 km from the centre of Buxton, a spa town in Derbyshire, England (Figure 14.1). The waters of the River Wye formed it. In about 1440, the highwayman and outlaw Poole reputedly used the cave as a lair and a base from which to waylay and rob travellers. He gave his name to the cave. Inside the cave entrance, which was cleared and levelled in 1854, is glacial sediment containing the bones of sheep, goats, deer, boars, oxen, and humans. Artefacts from the Neolithic, Bronze Age, Iron Age, and Roman periods are all present. Further into the cave is the 'Dome', a 12-m-high chamber that was probably hollowed out by meltwater coursing through the cavern at the end of the last ice age and forming a great whirlpool. Flowstone is seen on the chamber walls, stained blue-grey by manganese oxide or shale. A little further in lies the River Wye, which now flows only in winter as the river enters the cave from a reservoir overflow. The river sinks into the stream bed and reappears about 400 m away at Wye Head, although thousands of years ago it would have flowed out through the cave entrance. The river bed contains the 'Petrifying Well', a pool that will encrust such articles as bird's nests placed in it with calcite and 'turn them to stone'.

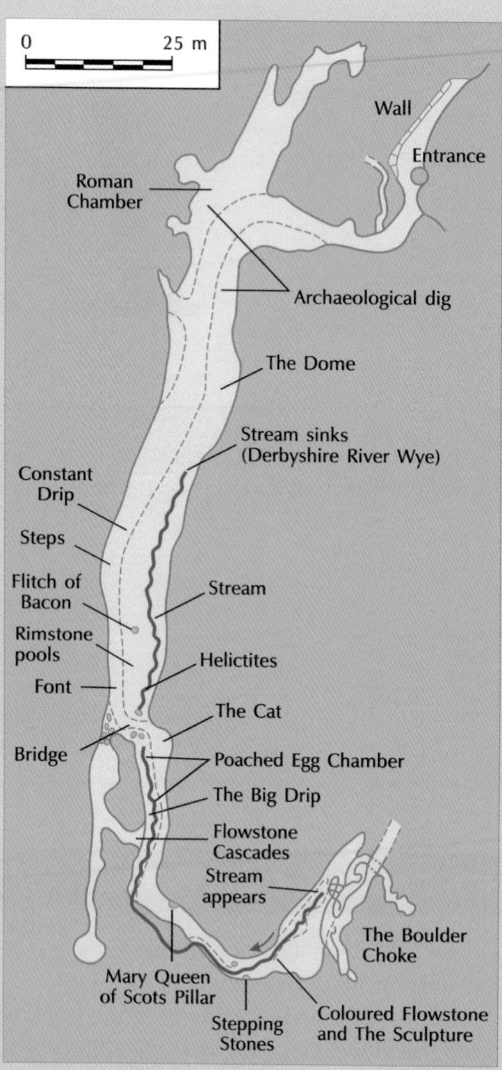

Figure 14.1 Plan of Poole's Cavern, Buxton, Derbyshire, England. *Source*: After Allsop (1992)

The 'Constant Drip' is a stalagmite that has grown over thousands of years, but, perhaps owing to an increased drip rate over recent years, it now has a hole drilled in it. Nearby is a new white flowstone formation that is made by water passing through old lime-tips on the hillside above. Further along hangs the largest stalactite in the cave – the 'Flitch of Bacon', so called owing to its resemblance to a half-side of that meat. It is almost 2 m long, but was longer before some vandalous visitor broke off the bottom section around 1840. Nearby, on the cave floor, are rimstone pools. The next chamber is the 'Poached Egg Chamber', which contains stalactites, straws, flowstones, columns, and curtains, all coloured in white, orange (from iron oxide), and blue-grey (from manganese oxide). These formations are created from lime waste from an old quarry tip above the cave. The iron has coated the tips of stalagmites to give them the appearance of poached eggs. At the far end of the Poached Egg Chamber are thousands of straws and stalactites, with a cascade of new flowstone on top of an old one known as the 'Frozen Waterfall'. Above this formation is the 'Big Drip', a 0.45-m-high stalagmite that is very active, splashing drips around its sides, so making itself thicker. At this point, bedding planes in the limestone show signs of cavern collapse. Turning to the left, the 'Mary Queen of Scots Pillar', a 2-m-high stalactite boss, presents itself. This feature is said to have been named by Mary Queen of Scots when she visited the cavern in 1582. In the last chamber, the River Wye can be seen emerging from the 15-m-high boulder choke that blocks the rest of the cavern system. A beautiful flowstone structure in this chamber was named the 'Sculpture' by a party of local schoolchildren in 1977, and above it is the 'Grand Cascade', another impressive flowstone formation stained with oxides of iron and manganese.

KARST ENVIRONMENTS

Karst is the German form of the Indo-European word *kar*, which means rock. The Italian term is **carso**, and the Slovenian *kras*. In Slovenia, *kras* or *krš* means 'bare stony ground' and is also a rugged region in the west of the country. In geomorphology, **karst** is terrain in which soluble rocks are altered above and below ground by the dissolving action of water and that bears distinctive characteristics of relief and drainage (Jennings 1971, 1). It usually refers to limestone terrain characteristically lacking surface drainage, possessing a patchy and thin soil cover, containing many enclosed depressions, and supporting a network of subterranean features, including caves and grottoes. However, all rocks are soluble to some extent in water, and karst is not confined to the most soluble rock types. Karst may form in evaporites such as gypsum and halite, in silicates such as sandstone and quartzite, and in some basalts and granites under favourable conditions (Table 14.1). Karst features may also form by other means – weathering, hydraulic action, tectonic movements, meltwater, and the evacuation of molten rock (lava). These features are called **pseudokarst** as solution is not the dominant process in their development (Table 14.1).

Extensive areas of karst evolve in carbonate rocks (limestones and dolomites), and sometimes in evaporites, which include halite (rock salt), anhydrite, and gypsum. Figure 14.2 shows the global distribution of exposed carbonate rocks.

Table 14.1 Karst and pseudokarst

Formed in	Formative processes	Examples
Karst		
Limestone, dolomite, and other carbonate rocks	Bicarbonate solution	Poole's Cavern, Buxton, England; Mammoth Cave, Kentucky, USA
Evaporites (gypsum, halite, anhydrite)	Dissolution	Mearat Malham, Mt Sedom, Israel
Silicate rocks (e.g. sandstone, quartzites, basalt, granite, laterite)	Silicate solution	Kukenan Tepui, Venezuela; Phu Hin Rong Kla National Park, Thailand; Mawenge Mwena, Zimbabwe
Pseudokarst		
Basalts	Evacuation of molten rock	Kazumura Cave, Hawaii
Ice	Evacuation of meltwater	Glacier caves, e.g. Paradise Ice Caves, Washington, USA
Soil, especially duplex profiles	Dissolution and granular disintegration	Soil pipes, e.g. Yulirenji Cave, Arnhemland, Australia
Most rocks, especially bedded and foliated ones	Hydraulic plucking, some exsudation (weathering by expansion on gypsum and halite crystallization)	Sea caves, e.g. Fingal's Cave, Isle of Staffa, Scotland
Most rocks	Tectonic movements	Fault fissures, e.g. Dan y Ogof, Wales; Onesquethaw Cave, New York, USA
Sandstones	Granular disintegration and wind transport	Rock shelters, e.g. Ubiri Rock, Kakadu, Australia
Many rocks, especially with granular lithologies	Granular disintegration aided by seepage moisture	Tafoni, rock shelters, and boulder caves, e.g. Greenhorn Caves, California, USA

Source: Partly after Gillieson (1996, 2)

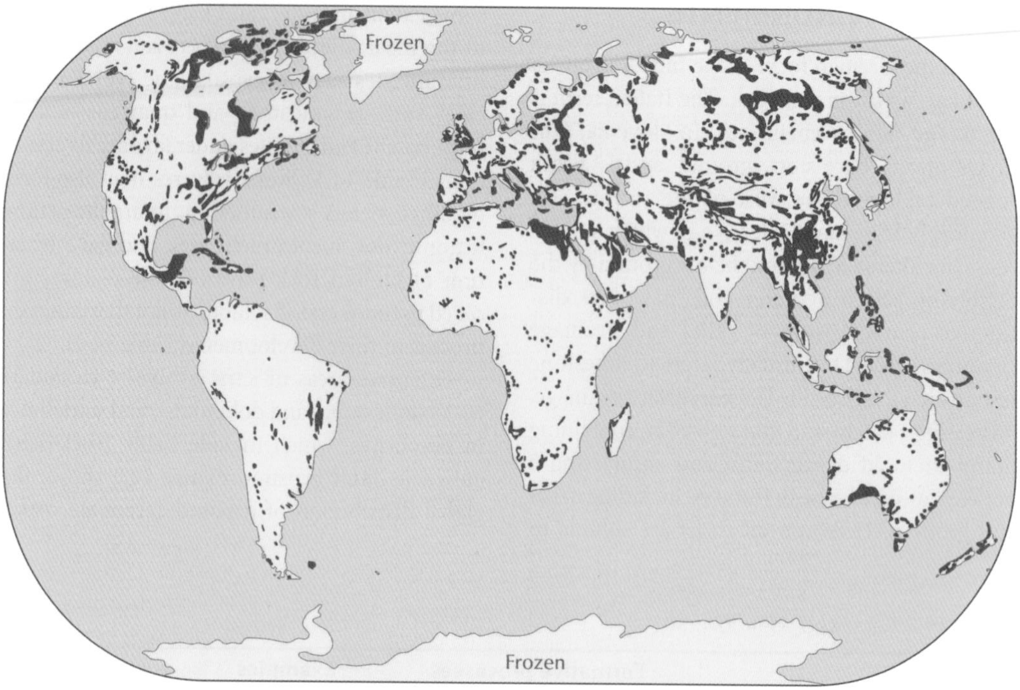

Figure 14.2 World distribution of carbonate rocks. *Source*: Adapted from Ford and Williams (1989, 4)

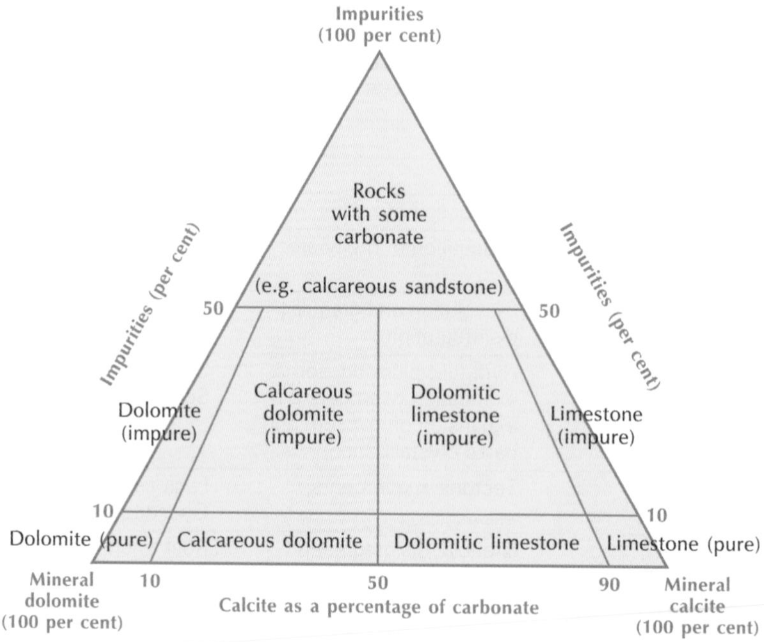

Figure 14.3 Classification of carbonate rocks. *Source*: Adapted from Leighton and Pendexter (1962)

Limestones and dolomites are a complex and diverse group of rocks (Figure 14.3). **Limestone** is a rock containing at least 50 per cent calcium carbonate ($CaCO_3$), which occurs largely as the mineral calcite and rarely as aragonite. Pure limestones contain at least 90 per cent calcite. **Dolomite** is a rock containing at least 50 per cent calcium–magnesium carbonate ($CaMg(CO_3)_2$), a mineral called dolomite. Pure dolomites (also called **dolostones**) contain at least 90 per cent dolomite. Carbonate rocks of intermediate composition between pure limestones and pure dolomites are given various names, including magnesian limestone, dolomitic limestone, and calcareous dolomite.

Karst features achieve their fullest evolution in beds of fairly pure limestone, with more than 80 per cent calcium carbonate, that are very thick, mechanically strong, and contain massive joints. These conditions are fulfilled in the classic karst area of countries bordering the eastern side of the Adriatic Sea. Chalk, although being a very pure limestone, is mechanically weak and does not favour the formation of underground drainage, which is a precondition for the evolution of medium-scale and large-scale surface-karst landforms.

KARST AND PSEUDOKARST PROCESSES

Few geomorphic processes are confined to karst landscapes, but in areas underlain by soluble rocks some processes operate in unique ways and produce characteristic features. Solution is often the dominant process in karst landscapes, but it may be subordinate to other geomorphic processes. Various terms are added to karst to signify the chief formative processes in particular areas. **True karst** denotes karst in which solutional processes dominate. The term **holokarst** is sometimes used to signify areas, such as parts of southern China and Indonesia, where karst processes create almost all landforms. **Fluviokarst** is karst in which solution and stream action operate together on at least equal terms, and is common in Western and Central Europe and in the mid-western United States, where the dissection of limestone blocks by rivers favours the formation of caves and true karst in interfluves. **Glaciokarst** is karst in which glacial and karst processes work in tandem, and is common in ice-scoured surfaces in Canada, and in the calcareous High Alps and Pyrenees of Europe. Finally, **thermokarst** is irregular terrain produced by the thawing of ground ice in periglacial environments and is not strictly karst or pseudokarst at all, but its topography is superficially similar to karst topography (see p. 329).

Karst drainage systems are a key to understanding many karst features (Figure 14.4). From a hydrological standpoint, karst is divided into the surface and near-surface zones, or epikarst, and the subsurface zones, or endokarst. **Epikarst** comprises the surface and soil (cutaneous zone), and the regolith and enlarged fissures (sub-cutaneous zone). **Endokarst** is similarly divided into two parts: the vadose zone of unsaturated water flow and the phreatic zone of saturated water flow. In the upper portion of the vadose zone, threads of water in the sub-cutaneous zone combine to form percolation streams, and this region is often called the percolation zone. Each zone has particular hydraulic, chemical, and hydrological properties, but the zones expand and contract with time and cannot be rigidly circumscribed.

The chief geomorphic processes characteristic of karst landscapes are solution and precipitation, subsidence, and collapse. Fluvial processes may be significant in the formation of some surface and subterranean landforms. Hydrothermal processes are locally important in caves. A distinction is often drawn between **tropical karst** and karst in other areas. The process of karstification is intense under tropical climates and produces such features as towers and cones (p. 430), which are not produced, at least not to the same degree, under temperate and cold climates. Discoveries in northwest Canada have shown that towers may

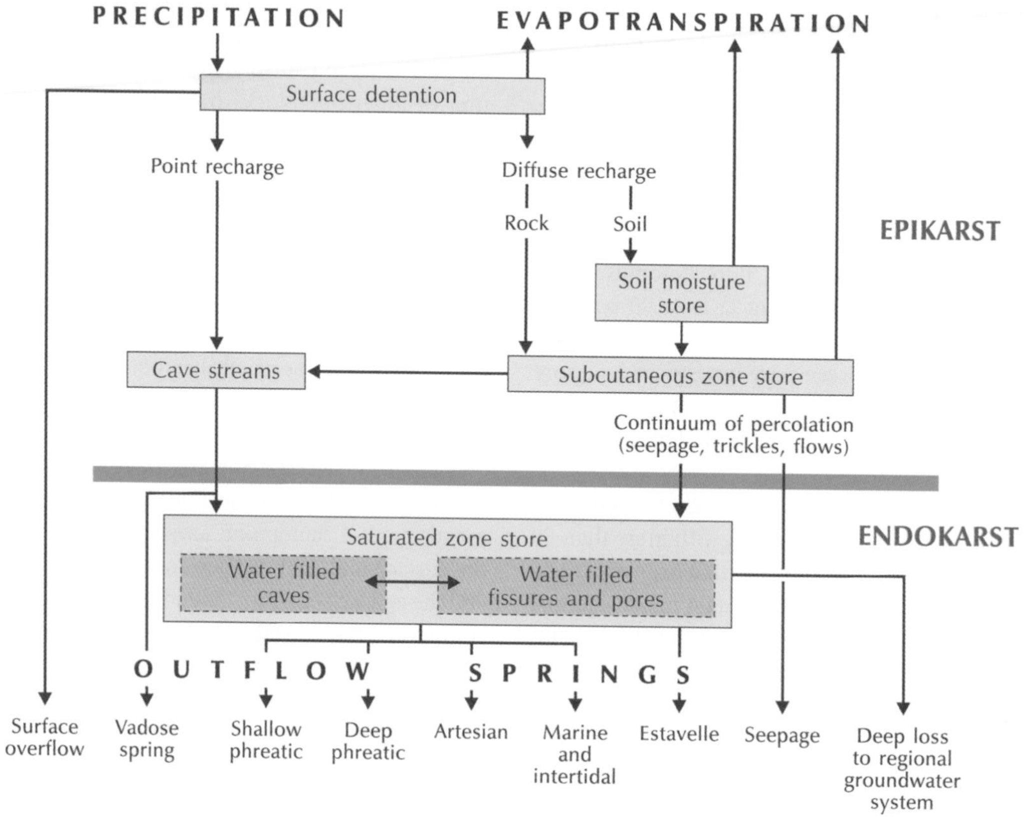

Figure 14.4 The karst drainage system: storages and flows. *Source*: Adapted from Ford and Williams (1989, 169)

form under cold climates (p. 430), but the widespread distribution of tropical karst testifies to the extremity of limestone solution under humid tropical climatic regimes.

SOLUTION AND PRECIPITATION

Limestone, dolomite, and evaporites

As limestone is the most widespread karst rock, its solution and deposition are important karst processes. With a saturation concentration of about 13 mg/l at 16°C and about 15 mg/l at 25°C, calcite has a modest solubility in pure water. However, it is far more soluble in waters charged with carbonic acid. It also appears to be more soluble in waters holding organic acids released by rotting vegetation, and is very soluble in waters containing sulphuric acid produced by the weathering of sulphide minerals such as pyrite and marcasite. Carbonic acid is the main solvent in karst landscapes, limestones readily succumbing to carbonation (p. 164). Dolomite rock behaves similarly to limestones in natural waters, although it appears to be slightly less soluble than limestone under normal conditions. Complexities are added with the presence of magnesium in dolomites. Evaporites, including gypsum, are much more soluble than limestone or dolomite but carbon dioxide is not involved in their solution. Gypsum becomes increasingly soluble up to a maximum of 37°C. It is deposited as warm water cools sufficiently and when evaporation leads to supersaturation.

Silicate rocks

Active sinkholes, dolines, and cave systems in quartzite must be produced by the excavation and underground transport of rock. As quartzite has a very low solubility, it is difficult to see how such processes could proceed. One possibility is that, rather than dissolving the entire rock, it is necessary only to dissolve the cementing material around individual quartz grains. Quartz grains have a solubility of less than 10 mg/l, while amorphous silica, which is the chief cement, has a solubility of 150 mg/l. With the cement dissolved, the quartzite would become mechanically incoherent, and loose grains could be removed by piping, so eroding underground passages. Alternatively, corrosion of the quartzite itself might produce the underground karst features. Corrosion of quartz is a slow process but, given sufficient time, this process could open underground passages. To be sure, some karst-like forms excavated in quartzites of the Cueva Kukenan, a Venezuelan cave system, consist of rounded columns some 2–3 m high. If these had been formed by cement removal, they should have a tapered cross-section aligned in the direction of flow. All the columns are circular, suggesting that corrosion has attacked the rock equally on all sides (see Doerr 1999). Also, thin sections of rocks from the cave system show that the individual grains are strongly interlocked by silicate overgrowths and, were any silica cement to be removed, they would still resist disintegration. Only after the crystalline grains themselves were partly dissolved could disintegration proceed.

Slow mass movements and collapse

It is expedient to distinguish between collapse, which is the sudden mass movement of the karst bedrock, and the slow mass movement of soil and weathered mantles (Jennings 1971, 32). The distinction would be artificial in most rocks, but in karst rocks solution ordinarily assures a clear division between the bedrock and the regolith.

Slow mass movements

Soil and regolith on calcareous rocks tend to be drier than they would be on impervious rocks. This fact means that lubricated mass movements (rotational slumps, debris slides, debris avalanches, and debris flows) are less active in karst landscapes. In addition, there is little insoluble material in karst rocks, and soils tend to be shallow, which reduces mass movement. Calcium carbonate deposition may also bond soil particles, further limiting the possibility of mass movement. Conversely, the widespread action of solution in karst landscapes removes support in all types of unconsolidated material, so encouraging creep, block slumps, debris slides, and especially soilfall and earthflow. As a rider, it should be noted that piping occurs in karst soil and regolith, and indeed may be stimulated by solutional processes beneath soils and regolith covers. Piping or tunnelling is caused by percolating waters transporting clay and silt internally to leave underground conduits that may promote mass movements.

Collapse

Rockfalls, block slides, and rock slides are very common in karst landscapes. This is because there are many bare rock slopes and cliffs, and because solution acts as effectively sideways as downwards, leading to the undercutting of stream banks.

Fluvial and hydrothermal processes

Solution is the chief player in cave formation, but corrosion by floodwaters and hydrothermal action can have significant roles. **Maze caves**, for instance, often form where horizontal, well-bedded limestones are invaded by floodwaters to produce a complicated series of criss-crossing passages. They may also form by hydrothermal action, either when waters rich in carbon dioxide or when waters loaded with corrosive sulphuric acid derived from pyrites invade well-jointed limestone.

SURFACE KARST FORMS

Early studies of karst landscapes centred on Vienna, with work carried out on the Dinaric karst, a mountain system running some 640 km along the eastern Adriatic Sea from the Isonzo River in northeastern Italy, through Slovenia, Croatia, Bosnia and Herzegovina, Montenegro and Serbia, to the Drin River, northern Albania. The Dinaric karst is still regarded as the 'type' area, and the Serbo-Croat names applied to karst features in this region have stuck, although most have English equivalents. However, the reader should be made aware that karst terms are very troublesome and the subject of much confusion. It may also be helpful to be mindful of a contrast often made between **bare karst**, in which bedrock is largely exposed to the atmosphere, and **covered karst**, in which bedrock is hardly exposed to the atmosphere at all. All degrees of cover, from total to none, are possible. Another basic distinction is drawn between **free karst**, which drains unimpeded to the sea, and **impounded karst**, which is surrounded by impervious rocks and has to drain through different hydrogeological systems to reach the sea.

Figure 14.5 illustrates diagrammatically some of the main karst landforms discussed in the following sections.

Karren

Karren is an umbrella term, which comes from Germany, to cover an elaborately diverse group of small-scale solutional features and sculpturing found on limestone and dolomite surfaces exposed at the ground surface or in caves. The French word *lapis* and the Spanish word *lapiaz* mean the same thing. Widespread, exposed tracts of karren on pavements and other extensive surfaces of calcareous rocks are termed *Karrenfeld* (**karren fields**). The terminology dealing with types of karren is bafflingly elaborate. The nomenclature devised by Joe Jennings (1971, 1985) brings some

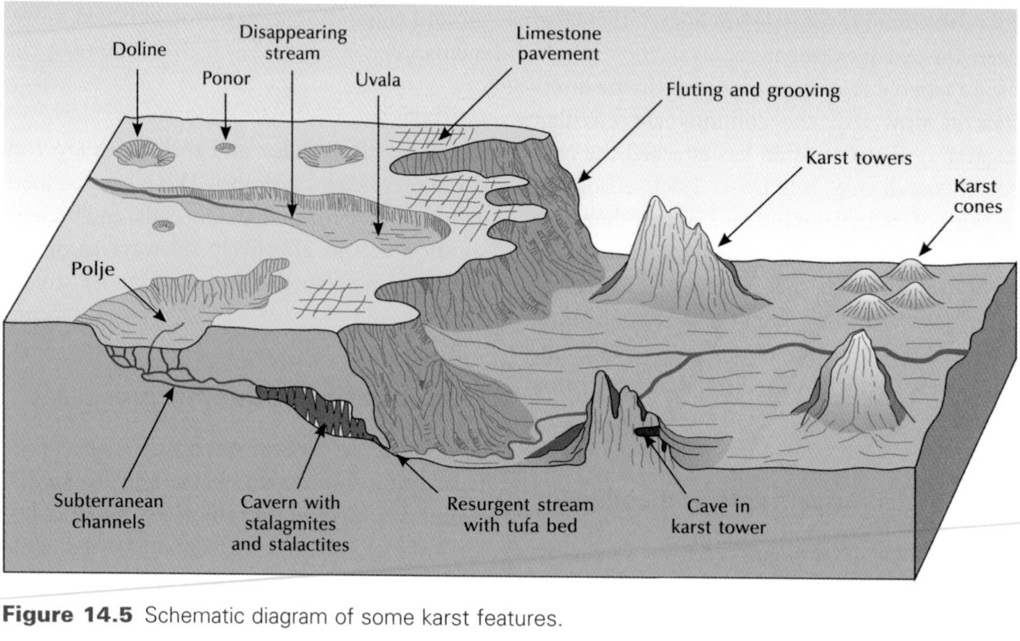

Figure 14.5 Schematic diagram of some karst features.

Table 14.2 Small limestone landforms produced by solution

Form	Comment
Bare limestone forms (surface wetting)	
Micropits and etched surfaces	Small pits produced by rain falling on gently sloping or flat bare rocks
Microrills	Rills no deeper or wider than about 1 mm and not longer than a few centimetres. Called *Rillenstein* when formed on stones and blocks
Solution ripples or fluted scallops	Shallow, ripple-like flutes formed on steep to vertical surfaces by flowing water normal to the direction of water flow. Prominent as a component of cockling patterns (a mixture of scallops, fluted scallops, or ripples) on steep and bare slopes
Solution flutes (*Rillenkarren*)	Longitudinal hollows that start at the slope crest and run down the maximum slope of fairly steep to vertical rock surfaces. They are of uniform fingertip width and depth, with sharp ribs between neighbouring flutes. May occur with rippling to give the rock a netted appearance
Solution bevels (*Ausgleichsflächen*)	Very smooth, flat or nearly so, forming tiny treads backed by steeper, fluted rises. A rare variant is the solution funnel step or heelprint (*Trittkarren* or *Trichterkarren*)
Solution runnels (*Rinnenkarren*)	Solution hollows, which result from Hortonian overland flow, running down the maximum slope of the rock, larger than solution flutes and increasing in depth and width down their length owing to increased water flow. Thick ribs between neighbouring runnels may be sharp and carry solution flutes
Decantation runnels	Forms related to solution runnels and include meandering runnels (*Mäanderkarren*) and wall solution runnels (*Wandkarren*). Produced by the dripping of acidulated water from an upslope point source. Channels reduce in size downslope
Decantation flutings	Packed channels, which often reduce in width downslope, produced by acidulated water released from a diffuse upslope source
Bare limestone forms (concentrated surface runoff)	
Microfissures	Small fissures, up to several centimetres long but no more than 1 cm deep, that follow small joints
Splitkarren	Solution fissures, centimetres to a few metres long and centimetres deep, that follow joints, stylolites, or veins. Taper with depth unless occupied by channel flow. May be transitional to pits, karren shafts, or grikes
Grikes (*Kluftkarren*)	Major solution fissures following joints or fault lines. The largest forms include bogaz, corridors, and streets
Clints (*Flackkarren*)	Tabular blocks between grikes
Solution spikes (*Spitzkarren*)	Sharp projections between grikes
Partly covered forms	
Solution pits	Round-bottomed or tapered forms. Occur under soil and on bare rock
Solution pans	Dish-shaped depressions formed on flat or nearly flat limestone, with sides that may overhang and carry solution flutes. The bottom of the pans may have a cover of organic remains, silt, clay, or rock debris
Undercut solution runnels (*Hohlkarren*)	Similar to runnels, become larger with depth resulting from damp conditions near the base associated with humus or soil accumulations
Solution notches (*Korrosionkehlen*)	Inward-curved recesses etched by soil abutting rock

continued . . .

Table 14.2 *continued*

Form	Comment
Covered forms	
Rounded solution runnels (*Rundkarren*)	Runnels formed under an 'acidulated' soil or sediment cover that smooths out the features
Cutters	An American term for soil-covered grikes that are widened at the top and taper with depth. Intervening clints are called subsoil pinnacles
Solution pipes, shafts, or wells	Cylindrical or conical holes developed along joint planes that connect to proto-caves or small caves. Shaft-like forms weathered below a deep and periodically saturated soil cover contain small caverns and are known as 'bone yard' forms. These are popularly used in ornamental rockeries
Polygenetic forms – assemblages of karren	
Karren fields (*Karrenfeld*)	Exposed tracts of karren that may cover up to several square kilometres
Limestone pavement	A type of karren field characterized by regular clints and grikes. They are called stepped pavements (*Schichttreppenkarst*) when benched
Pinnacle karst and stone forest	Topography with pinnacles, sometimes exposed by soil erosion, formed on karst rocks. Pinnacles may stand up to 45 m tall and 20 m wide at the base
Ruiniform karst	Karst with wide grikes and degrading clints exposed by soil erosion. Transitional to tors
Corridor karst (or labyrinth karst or giant grikeland)	Large-scale clint-and-grike terrains with grikes several metres or more wide and up to 1 km long
Coastal karren	A distinctive solutional topography on limestone or dolomite found around coasts and lakes

Source: After discussion in Jennings (1971, 1985) and Ford and Williams (1989, 376–7)

sense of order to a multilingual lexicon of confused and inconsistent usage. The basic forms are divided according to the degree of cover by soil and vegetation – bare ('free karren'), partly covered ('half-free karren'), and covered ('covered karren') (Bögli 1960). The bare forms are divided into those produced by surface wetting and those produced by concentrated surface runoff. Derek Ford and Paul Williams (1989, 376–7) offered a purely morphological classification of karren types because current understanding of karren-forming processes is too immature to build useful genetic classifications. However, their scheme, although using morphology as the basis for the major divisions, uses genetic factors for subdivisions. Jennings's classification underpins the following discussion, but a few types mentioned by Ford and Williams, and their 'polygenetic' class, are included (Table 14.2).

Bare forms

Bare forms produced by surface wetting comprise **pits**, **ripples**, **flutes**, **bevels**, and **runnels**, all of which are etched into bare limestone by rain hitting and flowing over the naked rock surface or dripping or seeping on to it. They are small landforms, the smallest, **micropits** and **microrills**, being at most 1 cm wide and deep, and the largest, **solution flutes** (***Rillenkarren***), averaging about 1.0–2.5 m wide and 15 m long. The smallest features are called **microkarren**. Solutional features of a few micrometres can be discerned under an electron microscope. Exposed karst rocks may develop relief of 1 mm or more within a few decades. The main bare forms resulting from surface wetting are **solution ripples**, **solution flutes** (***Rillenkarren***), **solution bevels** (***Ausgleichsflächen***), **solution runnels** (***Rinnenkarren***), and **decantation runnels** and

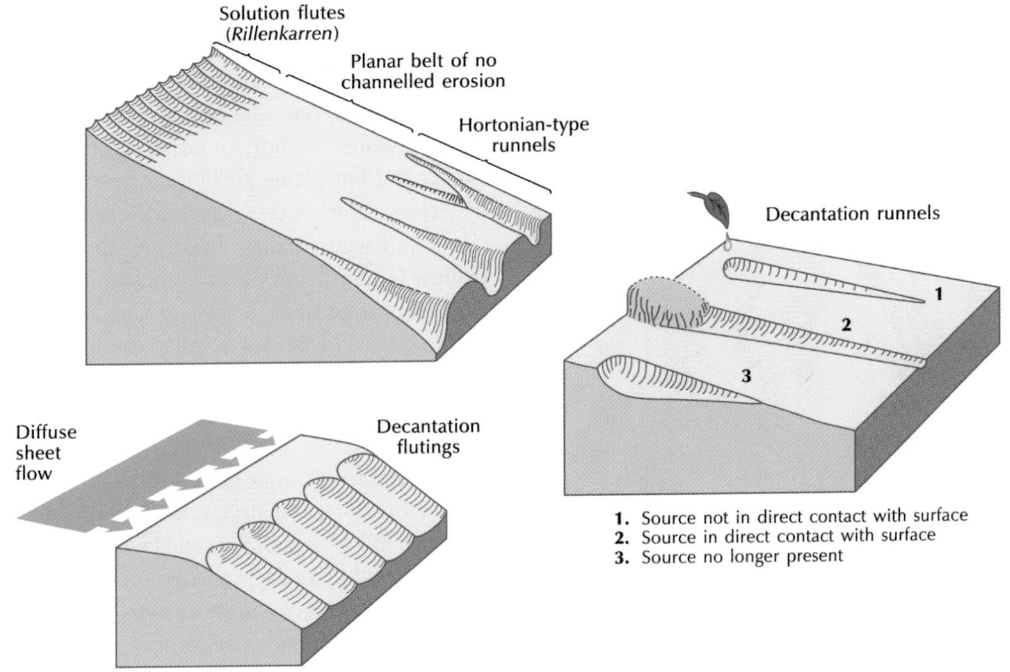

Figure 14.6 Solution flutes (*Rillenkarren*), decantation runnels, and decantation flutings. *Source*: After Ford and Williams (1989, 383)

flutings (Table 14.2; Figure 14.6; Plates 14.1 and 14.2).

Bare forms resulting from concentrated runoff are microfissures, splitkarren, grikes, clints, and solution spikes. **Microfissures** are solutional features following small joints. **Splitkarren** are larger solution channels that run along larger lines of weakness – joints, stylolites, and veins. **Grikes** (***Kluftkarren***), which are called solution slots in America, follow joints and cleavage planes, so may be straight, deep, and long, often occurring in networks (Plate 14.3). Grikes are the leading karren feature in most karren assemblages. Large openings may develop at joint intersections, some several metres deep and called karst wells, which are related to solution pipes and potholes. The intervening tabular blocks between grikes are called **clints** (***Flackkarren***) (Plate 14.3). Grikes in upright bedding planes are enlarged in the same ways as joints in flat bedding planes and

Plate 14.1 *Rillenkarren* formed on limestone in the Mortitx Valley, Serra de Tramunana, Mallorca, Spain. (*Photograph by Nick Scarle*)

Plate 14.2 Decantation runnels (*Rinnenkarren*) on marble near Pikhauga Ridge, Svartisen, Norway. (*Photograph by Derek C. Ford*)

Plate 14.3 Clints and grikes on a 'textbook' limestone pavement on the lip of Malham Cove, Yorkshire, England. Towards the cliff edge, the soils have always been thin; the grikes are simple linear features and the clints show little dissection. In the foreground, grike edges are rounded and clints dissected by subsoil *Rundkarren*. (*Photograph by Richard Huggett*)

are called **bedding grikes** (***Schichtfugenkarren***). However, residual blocks left between them commonly break into **pinnacles** or **solution spikes** (***Spitzkarren***) and beehives decorated by solution flutes. In horizontal strata, the near-surface bedding planes are likely to be opened up by seepage. This process may free the intervening clints and lead to their breaking up to form **shillow** (a term from northern England), which is roughly equivalent to the German *Trümmerkarren* and *Scherbenkarst*. All these forms are small. Grikes average about 5 cm across and up to several metres deep, clints may be up to several metres across, and solution spikes up to several metres long. Large-scale grikes, variously termed *bogaz*, corridors, and streets, are found in some areas and follow major joints and faults. ***Bogaz*** are up to 4 m wide, 5 m deep, and tens of metres long (Plate 14.4). **Karst corridors** and **streets** are even larger and take the form of gorges (p. 424).

Covered and partly covered forms

Partly covered forms develop in areas with a patchy soil, sediment, litter, or moss cover. **Solution pits** are round-bottomed or tapered forms, usually less than 1 m in diameter. Larger ones merge into **solution pans**. They occur under soil and on bare limestone. Along with shafts, they are the most widespread karren form. Many are transitional to shafts. Solution pans or solution basins are small depressions shaped like basins or dishes, usually with a thin cover of soil or algal or vegetal remains. They are no more than 3 m wide and 0.5 m deep, but many are much smaller. Some of the carbon dioxide released by the decaying organic matter dissolves in the water collected in the pans and boosts their dissolution. The Slav term for them is *kamenice* (singular *kamenica*), and the American term is tinajitas. **Undercut solution runnels** (***Hohlkarren***) are like runnels in form and size, except that they become wider with increasing depth, probably owing to accumulated organic matter or soil keeping the sides and base near the bottom damp. **Solution notches** (***Korrosionkehlen***) are about 1 m high and wide and 10 m long. They are formed where soil lies against projecting rock, giving rise to inward-curved recesses.

Covered forms develop under a blanket of soil or sediment, which acts like 'an acidulated sponge' (Jennings 1971, 48). Where it contacts the under lying limestone, the 'sponge' etches out its own array of landforms, the chief among which are rounded solution runnels and solution pipes. **Rounded solution runnels** (***Rundkarren***) are the same size as ordinary runnels but they are worn smooth by the active corrosion identified with acid soil waters. They are visible only when the soil or sediment blanket has been stripped off (Plate 14.3, foreground). **Cutters** are soil-covered clints that are widened at the top and taper with depth (Plate 14.5). **Solution pipes** (or **shafts** or **wells**) are up to 1 m across and 2–5 m deep, usually becoming narrower with depth, but many are smaller. They are cylindrical or conical holes, occurring on such soft limestones as chalk, as well as on the mechanically stronger and less permeable limestones. Solution pipes usually form along joint planes, but in the chalk of north-west Europe they can develop in an isolated fashion.

Plate 14.4 A *bogaz* on Mount Buila, southern Carpathian Mountains, Romania. (*Photograph by Laura Tîrlă*)

Polygenetic karst

Limestone pavements

Limestone pavements are karren fields developed in flat or gently dipping strata. They occur as extensive benches or plains of bare rock in horizontally bedded limestones and dolomites (Plate 14.3). Solution dissolves clefts in limestone and dolomite pavements that are between 0.5 and 25 m deep. The clefts, or grikes, separate surfaces (clints) that bear several solution features (karren). A survey in the early 1970s listed 573 pavements in the British Isles, most of them occurring on the Carboniferous limestone of the northern Pennines in the counties of North Yorkshire, Lancashire, and Cumbria (Ward and Evans 1976).

Debate surrounds the origin of pavements, some geomorphologists arguing that a cover of soil that is from time to time scoured by erosion encourages their formation. To be sure, the British pavements appear to have been produced by the weathering of the limestone while it was covered by glacial till. Later scouring by ice would remove any soil cover and accumulated debris. It may be no coincidence that limestone pavements are very common in Canada, where ice-scouring has occurred regularly (Lundberg and Ford 1994). Lesser pavements occur where waves, rivers in flood, or even sheet wash on pediments do the scouring instead of ice.

Plate 14.5 Cutters in a limestone fanglomerate in western Turkey. (*Photograph by Derek C. Ford*)

Plate 14.6 Pinnacle karst or shilin (shilin means 'stone forest' in Mandarin) exposed in a road-cut in Shilin National Park, Yunnan, China. The subsoil origin of the pinnacles is plainly seen. Their emergence is due to the general erosion of regional cover sediment. (*Photograph by Derek C. Ford*)

Pinnacle karst

Large *Spitzkarren* dominate **pinnacle karst**. In China, a famous example of pinnacle karst is the Yunnan Stone Forest (Plates 14.6 and 14.7). This is an area of grey limestone pillars covering about 350 km^3. The pillars stand 1–35 m tall with diameters of 1–20 m. **Arête-and-pinnacle karst**, which is found on Mount Kaijende in Papua New Guinea and Mount Api in Sarawak, consists of bare, net-like, saw-topped ridges with almost vertical sides that stand up to 120 m high. The spectacular ridges rise above forest-covered corridors and depressions. They seem to have formed by limestone solution without having previously been buried.

Ruiniform karst

This is an assemblage of exceptionally wide grikes and degrading clints that have been exposed by soil erosion (Plate 14.8). The clints stick out like 'miniature city blocks in a ruined townscape' (Ford and Williams 1989, 391). Ruiniform karst is found in the French Causses, where deforestation and soil erosion have occurred. On high crests, ruiniform karst is transitional to limestone tors.

Corridor karst

In places, grikes grow large to form a topography of aligned or criss-crossing corridors. The large grikes are called *bogaz*, corridors, *zanjones*, and streets. Grike-wall recession produces square-shaped or box valleys and large closed depressions called **platea**. Corridor karst landscapes are called **labyrinth karst**, **corridor karst**, or **giant grikeland**. It is large-scale clint-and-grike terrain but may have a complex history of development.

Grikelands form under tropical and temperate rainforest and in arid and semi-arid areas. Smaller-scale versions occur in the Nahanni limestone karst region of the Mackenzie Mountains, Canada (Brook and Ford 1978). Here, the labyrinth karst is stunning, with individual streets longer than 1 km and deeper than 50 m (Plate 14.9).

Plate 14.7 Pinnacle karst in Shilin National Park, Yunnan, China. *(Photograph by Tony Waltham Geophotos)*

Plate 14.8 A ruiniform assemblage (residual clint blocks) in flat-lying limestones near Padua, Italy. *(Photograph by Derek C. Ford)*

Plate 14.9a View of Nahanni labyrinth karst, showing intersecting networks of karst streets interspersed with karst platea, Mackenzie Mountains, Canada. *(Photograph by Paul Sanborn)*

Plate 14.9b Nahanni labyrinth karst at ground level. *(Photograph by Scott Smith, Agriculture and Agri-Food Canada)*

Coastal karren

Around coasts and lakes, limestone or dolomite outcrops often display a distinctive solutional topography, with features including intertidal and subtidal notches (also called nips; Plate 14.10) and a dense formation of pits, pans, micropits, and spikes (Plate 14.11). Boring and grazing organisms may help to form coastal karren, as may wave action, wetting and drying, salt weathering, and hydration.

Coral island karst

Carbonate sediments are the building material of the world's coral islands, all of which bear at least some karst features. For instance, Navassa Island, a 5-km^2 island in the Caribbean Sea between Haiti and Jamaica, may have started life as a small coral atoll. Some 5 million years ago, these coral reefs began to emerge, leading to the conversion of calcium carbonate sediments (aragonite) to calcium–magnesium carbonate rock (dolomite),

Plate 14.10 A limestone solution notch on a modern shore platform on the east coast of Okinawa, Japan. (*Photograph by Derek C. Ford*)

Plate 14.11 Limestone coastal karren pitting (sometimes called phytokarst or biokarst) on the west coast of Puerto Rico. (*Photograph by Derek C. Ford*)

the formation of a terrace around the island, and the onset of chemical weathering and the evolution of karst landforms, particularly caves and karst holes. Similarly, Yoron-Jima, a 21-km^2 carbonate island located in the central Ryukyu Island Arc of southern Japan, was raised above sea level in the Quaternary period. Subsequent karst processes have produced many closed depressions (Terry 2005).

Closed depressions

Dolines

The word **doline** derives from the Slovene word *dolina*, meaning a depression in the landscape. It is applied to the simpler forms of closed depressions in karst landscapes. **Sinkhole**, **swallet**, and **swallow hole** are English terms with rather loose connotations. Dolines resemble various shapes – dishes, bowls, cones, and cylinders. They range in size from less than a metre wide and deep to over hundreds of metres deep and several hundred metres or even a kilometre wide. The large forms tend to be complex and grade into other classes of closed depressions.

Several processes form dolines: surface solution, cave collapse, piping, subsidence, and stream removal of superficial covers. Although these processes frequently occur in combination and most dolines are polygenetic, they serve as a basis for a five-fold classification of dolines (Jennings 1985, 107; Ford and Williams 1989, 398) (Figure 14.7):

1 **Solution dolines** start where solution is concentrated around a favourable point such as joint intersections. The solution lowers the bedrock surface, so eating out a small depression (Figure 14.7a; Plates 14.12 and 14.13). The depression traps water, encouraging more solution and depression enlargement. Once begun, doline formation is thus self-perpetuating. However, insoluble residues and other debris may clog the doline floor, sometimes forming swampy areas or pools to form **pond dolines**. Dolines are one of the few karst landforms that develop in soft limestones such as chalk (e.g. Matthews *et al.* 2000).
2 **Collapse dolines** are produced suddenly when the roof of a cave formed by underground solution gives way and fractures or ruptures rock and soil (Figure 14.7b; Plate 14.14). Initially, they have steep walls, but, without further collapse, they become cone-shaped or bowl-shaped as the sides are worn down and the bottom is filled with debris. Eventually, they may be indistinguishable from other dolines except by excavation. The largest open

Plate 14.12 Small doline in steeply dipping limestone in the Rocky Mountain Front Ranges. The doline is formed on a cirque floor in the valley of Ptolemy Creek, Crownest Pass, Alberta, Canada. (*Photograph by Derek C. Ford*)

Plate 14.13 Doline in the Montello karst, Italy. A new vineyard sits in the background. (*Photograph by Ugo Sauro*)

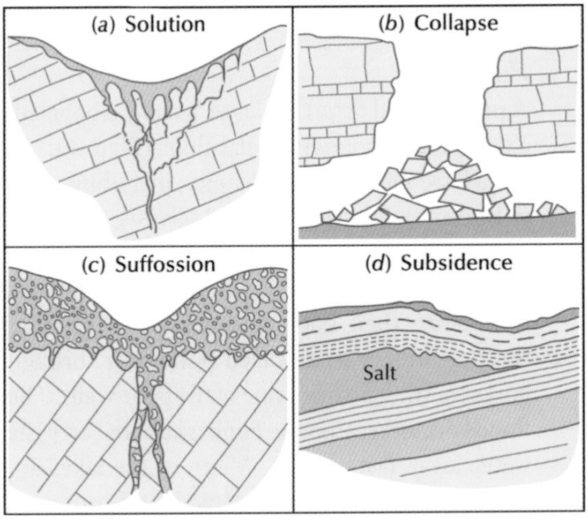

Figure 14.7 The main genetic classes of doline. (a) Solution doline. (b) Collapse doline. (c) Suffossion doline. (d) Subsidence doline. *Source*: Adapted from Ford and Williams (1989, 398)

Plate 14.14 Collapse dolines at the water table, Wood Buffalo National Park, Alberta, Canada. The dolines are created by collapses through dolostone into underlying gypsum. The diameters are 40–100 m. (*Photograph from Parks Canada archives*)

collapse doline is Crveno Jezero ('Red Lake') in Croatia, which is 421 m deep at its lowest rim and 518 m deep at its highest rim. If the collapse occurs into a water-filled cave, or if the water table has risen after the collapse occurred, the collapse doline may contain a lake, often deep, covering its floor. Such lakes are called **cenotes** on the Yucatán Peninsula, Mexico, and '***obruk***' lakes on the Turkish plateau. Some of the cenotes near the Mayan ruins of the northern Yucatán are very large. Dzitnup, at the Mayan ruins of Chichén Itzá, is a vertical-walled sinkhole some 60 m wide and 39 m deep, half-filled with water.

Subjacent karst-collapse dolines form even more dramatically than collapse dolines when beds of an overlying non-calcareous rock unit fall into a cave in the underlying limestone. An example is the Big Hole, near Braidwood, New South Wales, Australia. Here, a 115-m-deep hole in Devonian quartz sandstone is assumed to have collapsed into underlying Silurian limestone (Jennings 1967). As with collapse dolines, subjacent karst-collapse dolines start life as steep-walled and deep features but progressively come to resemble other dolines.

3 **Suffossion dolines** form in an analogous manner to subjacent karst-collapse dolines, with a blanket of superficial deposits or thick soil being washed or falling into widened joints and solution pipes in the limestone beneath (Figure 14.7c). In England, the 'shakeholes' of Craven, near Ingleborough, northern England, are conical suffossion dolines in glacial moraine laid upon the limestone during the ultimate Pleistocene glaciation (Sweeting 1950).

4 **Subsidence dolines** form gradually by the sagging or settling of the ground surface without any manifest breakage of soil or rock (Figure 14.7d). Natural dolines of subsidence origin are rare and are found where the dissolution of underground evaporite beds occurs, as in Cheshire, England, where salt extraction from Triassic rocks has produced depressions on the surface, locally known as **flashes**.

5 **Alluvial stream-sink dolines** form in alluvium where streams descend into underlying calcareous rocks. The stream-sink is the point at which a stream disappears underground. Several examples are found in the White Peak District of Derbyshire, England (Figure 14.8).

Karst windows

These are unroofed portions of underground caverns in which streams flow out of the cavern at one end, across the floor, and into a cavern at the other end. The openings may be mere peepholes or much larger.

Uvalas and egg-box topography

Uvalas, a word from Slovenia, are compound sinkholes or complex depressions composed of more than one hollow. They are larger than small dolines. Elongated forms follow strike lines or fault lines, while lobate forms occur on horizontal beds. Solution may play a big role in their formation, but, without further study, other processes cannot be discounted.

On thick limestone, where the water table is deep, solutional sinkholes may be punched downwards to form **egg-box topography**, known as ***fengcong*** in China, with sharp residual peaks along the doline rim and a local relief of hundreds of metres.

Polja

A **polje** (plural *polja*) is a large, usually elongated, closed depression with a flat floor (Plate 14.15). Polja have many regional names, including *plans* in Provence, France; *wangs* in Malaysia; and *hojos* in Cuba. Intermittent or perennial streams, which may be liable to flood and become lakes, may flow across their floors and drain underground through stream-sinks called **ponors** or through gorges cutting through one of the polje walls. The floods occur because the ponors cannot carry the water away fast enough. Many of the lakes are seasonal, but some are permanent features of polje floors, as in Cerknića Polje, Slovenia.

Polja come in three basic kinds: border polja, structural polja, and baselevel polja (Figure 14.9) (Ford and Williams 1989, 431–2). **Border polja** are fed by rivers from outside the karst region (allogenic rivers) that, owing to the position of the water table in the feed area and floodplain deposits over the limestone, tend to stay on the ground surface to cause lateral planation and alluviation. **Structural polja** are largely controlled by geology, often being associated with down-faulted inliers of impervious rocks in limestone terrain. They include the largest karst depressions in the world

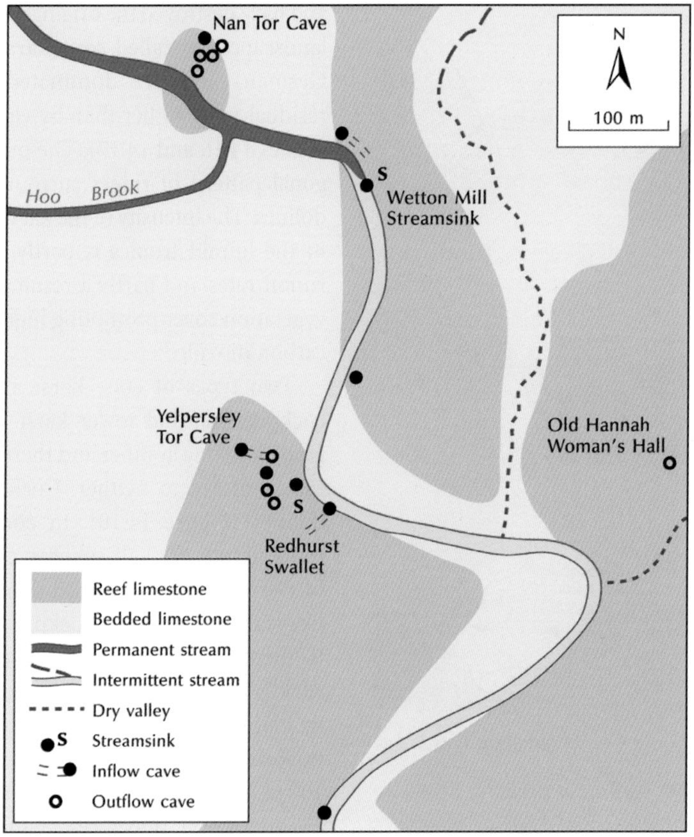

Figure 14.8 Stream-sinks on the River Manifold in the English Peak District. *Source*: Adapted from Warwick (1953)

and are the dominant type of polje in the Dinaric karst. **Baselevel polja** occur in limestone where a regional water table intersects the ground surface.

Cone karst

Tropical karst is one of the landform wonders of the world. Extensive areas of it occur in southern Mexico, Central America, the Caribbean, South-East Asia, southern China, South America, Madagascar, the Middle East, New Guinea, and northern Australia. Under humid tropical climates, karst landscapes take on a rather different aspect from 'classic' karst. In many places, owing to rapid and vigorous solution, dolines have grown large enough to interfere with each other

Plate 14.15 Dabarsko polje, Bosnia and Herzegovina, formed along a well-defined tectonic basin. (*Photograph by Amos Frumkin*)

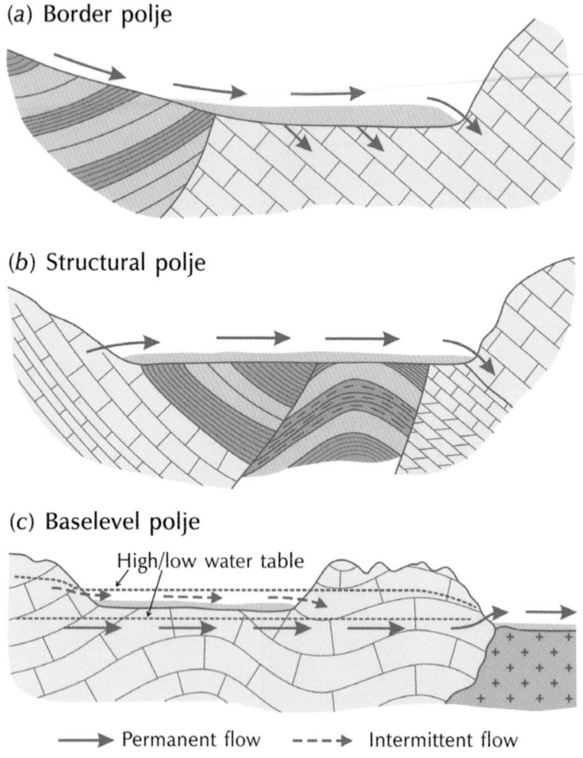

Figure 14.9 Types of polje. (a) Border polje. (b) Structural polje. (c) Baselevel polje. *Source*: After Ford and Williams (1989, 429)

Plate 14.16 Fengcong cone karst, hemispherical hills of limestone under soil cover, Gunung Sewu, Java, Indonesia. (*Photograph by Tony Waltham Geophotos*)

and have destroyed the original land surface. Such landscapes are called **cone karst** (***Kegelkarst*** in German) and are dominated by projecting residual relief rather than by closed depressions (Plates 14.16 and 14.17). The outcome is a polygonal pattern of ridges surrounding individual dolines. The intensity of the karstification process in the humid tropics is partly a result of high runoff rates and partly a result of thick soil and vegetation cover promoting high amounts of soil carbon dioxide.

Two types of cone karst are recognized – cockpit karst and tower karst – although they grade into one another and there are other forms that conform to neither. Cockpits are tropical dolines (Figure 14.10). In **cockpit karst**, the residual hills are half-spheres, called ***Kugelkarst*** in German, and the closed depressions, shaped like starfish, are called cockpits, the name given to them in Jamaica owing to their resembling cockfighting arenas. In **tower karst** (***Turmkarst*** in German), the residual hills are towers or **mogotes** (also called **haystack hills**), standing 100 m or more tall, with extremely steep to overhanging lower slopes (Plate 14.18). They sit in broad alluvial plains that contain flat-floored, swampy depressions. The residual hills may have extraordinarily sharp edges and form **pinnacle karst** (p. 424).

Studies in the Mackenzie Mountains, northwest Canada, have shattered the notion that cone karst, and especially tower karst, is a tropical landform (Brook and Ford 1978). Limestone in the Mackenzie Mountains is massive and very thick with widely spaced joints. Karst evolution in the area appears to have begun with the opening of deep dolines at 'weak' points along joints. Later, long and narrow gorges called karst streets formed, to be followed by a rectilinear network of deep gorges with other cross-cutting lines of erosion – **labyrinth karst**. In the final stage, the rock wall of the gorges suffered lateral planation, so fashioning towers.

Plate 14.17 Cone karst drowned by Holocene sea-level rise, islands of Wayang, Papua, Indonesia. (*Photograph by Lukas Plan*)

Plate 14.18 Fenglin tower karst, Yangshuo, Guilin, Guangxi, China. (*Photograph by Tony Waltham Geophotos*)

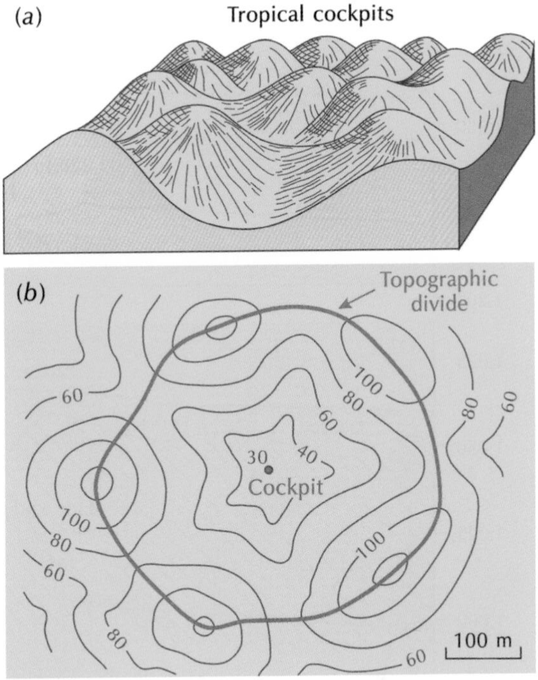

Figure 14.10 Tropical dolines (cockpits). (a) Block diagram. (b) Plan view. *Source*: Adapted from Williams (1969)

Fluvial karst

Although a lack of surface drainage is a characteristic feature of karst landscapes, several surface landforms owe their existence to fluvial action. Rivers do traverse and rise within karst areas, eroding various types of valley and building peculiar carbonate deposits.

Gorges

In karst terrain, rivers tend to erode **gorges** more frequently than they do in other rock types. In France, the Grands Causses of the Massif Centrale is divided into four separate plateaux by the 300–500-m-deep Lot, Tarn, Jonte, and Dourbie gorges. The gorges are commonplace in karst landscape because river incision acts more effectively than slope processes, which fail to flare back the valley-sides to a V-shaped cross-section.

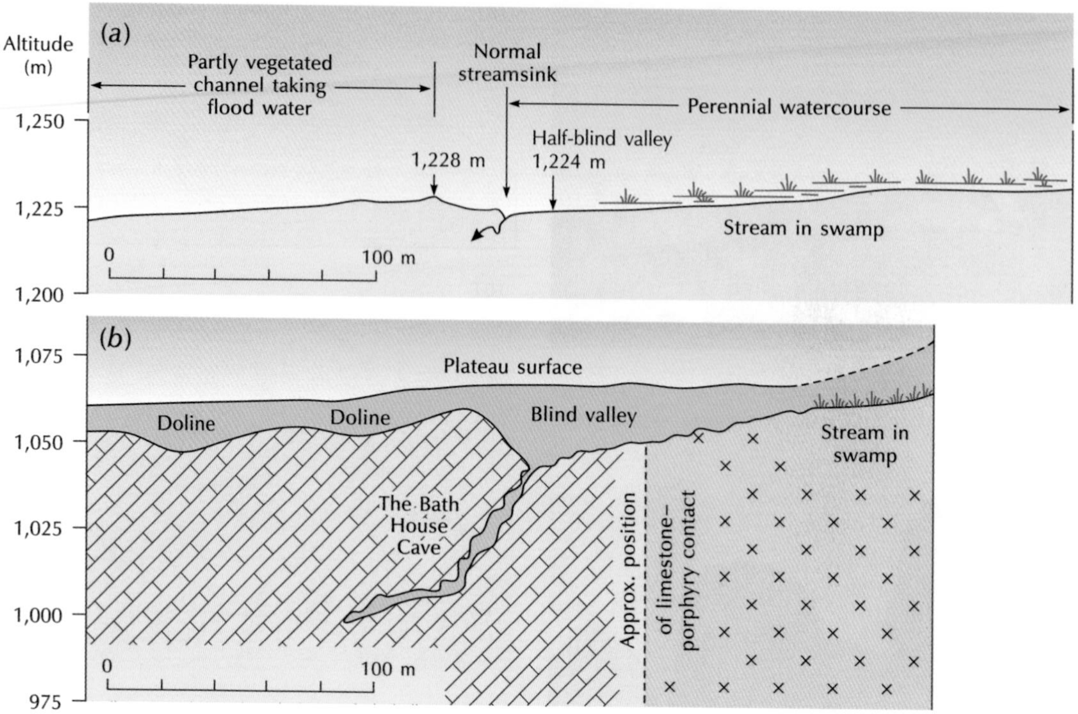

Figure 14.11 Blind and half-blind valleys in New South Wales, Australia. (a) A half-blind valley on Cooleman Plain. (b) A blind valley at Yarrangobilly. *Source*: Adapted from Jennings (1971, 110, 111)

Some gorges form by cavern collapse, but others are 'through valleys' eroded by rivers that manage to cross karst terrain without disappearing underground.

Blind and half-blind valleys

Rivers flowing through karst terrain may, in places, sink through the channel bed. The process lowers the bedrock and traps some of the sediment load. The sinking of the channel bed saps the power of the stream below the point of leakage. An upward step or threshold develops in the long profile of the stream, and the underground course becomes larger, diverting increasingly more flow. When large enough, the underground conduit takes all the flow at normal stages but cannot accommodate flood discharge, which ponds behind the step and eventually overspills it. The resulting landform is a **half-blind valley**.

A half-blind valley is found on the Cooleman Plain, New South Wales, Australia (Figure 14.11a). A small creek flowing off a granodiorite hill flows for 150 m over Silurian limestone before sinking through an earth hole. Beyond the hole is a 3-m-high grassy threshold separating the depression from a gravel stream bed that only rarely holds overflow. If a stream cuts down its bed far enough and enlarges its underground course so that even flood discharges sink through it, a **blind valley** is created that is closed abruptly at its lower end by a cliff or slope facing up the valley. Blind valleys carry perennial or intermittent streams, with sinks at their lower ends, or they may be dry valleys. Many blind valleys occur at Yarrangobilly, New South Wales, Australia. The stream here sinks into the Bath House Cave, underneath crags in a steep, 15-m-high counter-slope (Figure 14.11b).

Plate 14.19 Malham Cove, an 80-m-high limestone cliff with water from a sink on the limestone plateau emerging from a flooded tarn at the base, North Yorkshire, England. (*Photograph by Tony Waltham Geophotos*)

Steepheads

Steepheads or **pocket valleys** are steep-sided valleys in karst, generally short and ending abruptly upstream where a stream issues forth in a spring, or did so in the past. These cul-de-sac valleys are particularly common around plateau margins or mountain flanks. In Provence, France, the Fountain of Vaucluse emerges beneath a 200-m-high cliff at the head of a steephead. Similarly, if less spectacularly, the Punch Bowl at Burton Salmon, formed on Upper Magnesian Limestone, Yorkshire, England, is a steephead with a permanent spring issuing from the base of its headwall (Murphy 2000). Malham Cove, England, is also a steephead (Plate 14.19). Steepheads may form by headward recession, as spring sapping eats back into the rock mass, or by cave-roof collapse.

Dry valleys

Dry valleys are much like regular river valleys save that they lack surface stream channels on their floors. They occur on many types of rock but are noticeably common in karst landscapes. Eye-catching dry valleys occur where rivers flowing over impermeable rock sink on entering karst terrain, but their former courses are traceable above ground. In the Craven district, England, the Watlowes is a craggy dry valley in which the stream fed by Malham Tarn formerly flowed over the limestone to cascade over the 75-m cliff of Malham Cove, and indeed did so again in the very wet December of 2015 (Figure 14.12; Plate 14.19).

Extensive dry valley networks occur in some areas of karst. An impressive set is found in the White Peak, England. Here, a few major streams – the Rivers Manifold, Dove, and Wye – flow

Figure 14.12 Limestone features around Malham Cove, Craven, Yorkshire Dales, England. (a) Smelt Mill streamsink. (b) Malham Tarn. (c) Water Sinks below Malham Tarn. (d) Gordale Scar. (e) Janet's Foss. The material behind the waterfall is tufa. (f) Aire Head. (g) Malham Cove rising. (h) Malham Cove (compare with Plate 14.19). *Source*: Partly adapted from Jennings (1971, 91) (*Photographs by Richard Huggett*)

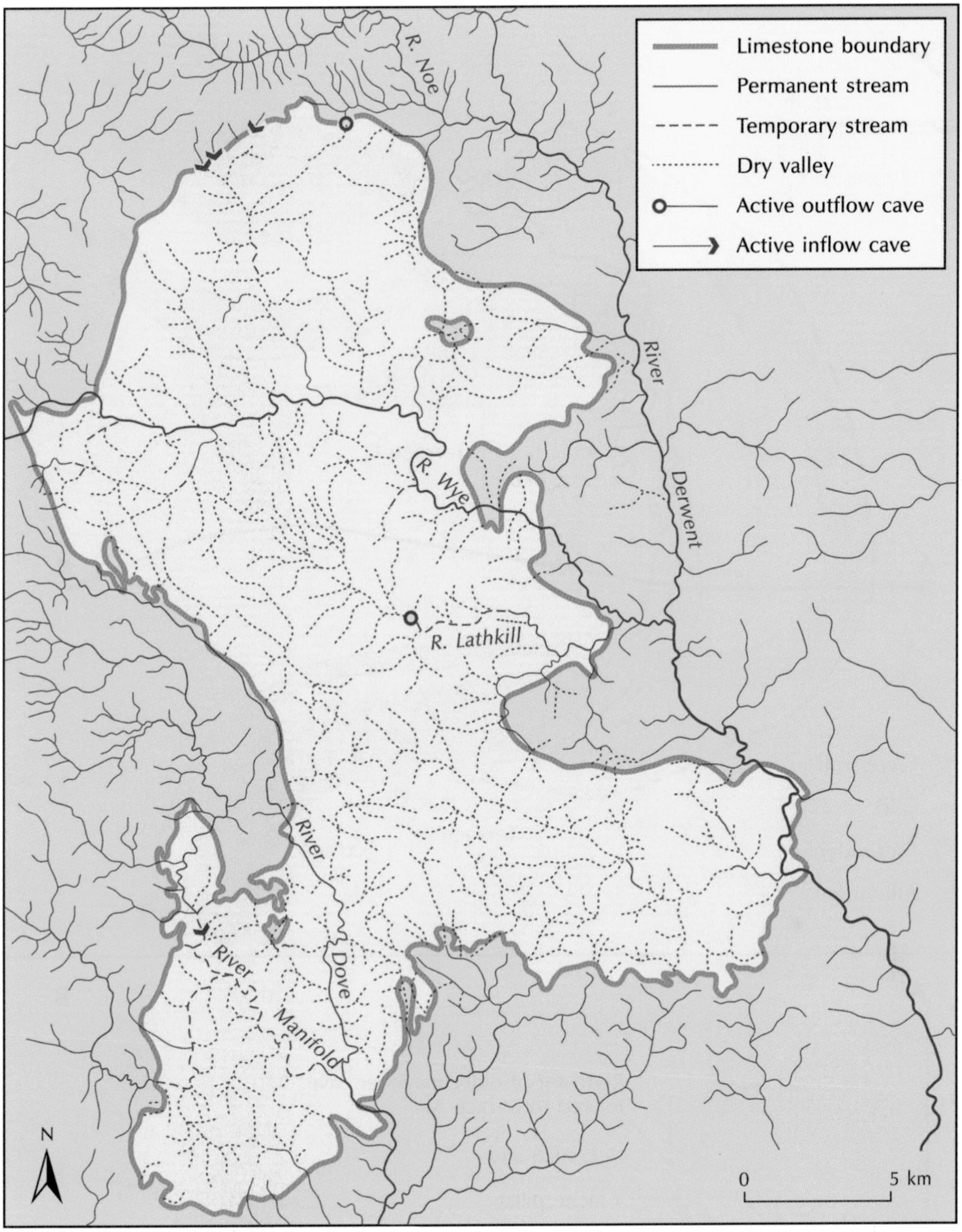

Figure 14.13 Dry valley systems in the White Peak, Peak District, England. *Source*: Adapted from Warwick (1964)

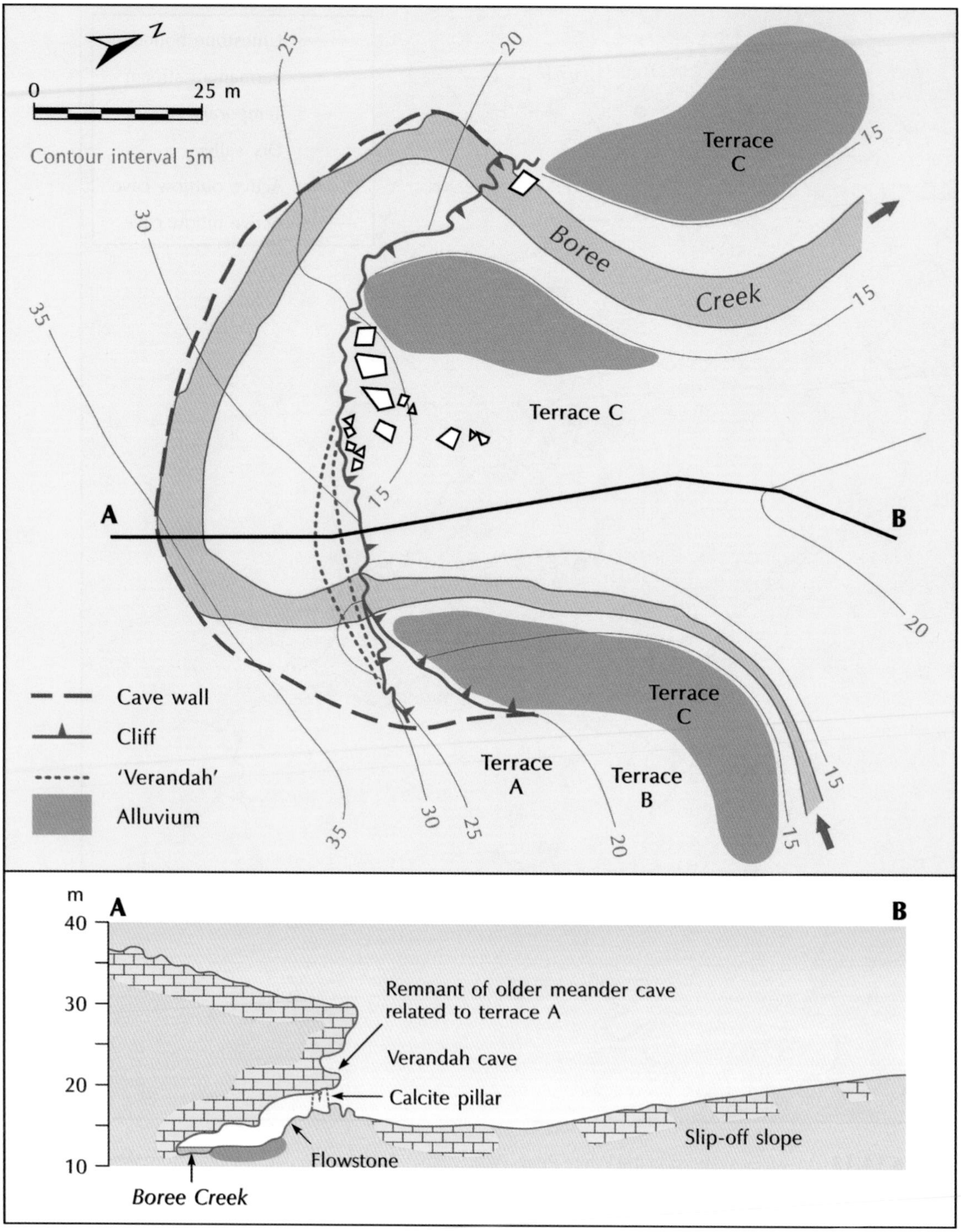

Figure 14.14 A meander cave on Boree Creek, Borenore, New South Wales, Australia.
Source: Adapted from Jennings (1971, 101)

across the region, but most other valleys are dry (Figure 14.13). Many of the dry valleys start as shallow, bowl-like basins that develop into rock-walled valleys and gorges. Other, smaller dry valleys hang above the major dry valleys and the permanent river valleys. The origin of such networks is puzzling but they appear to be the legacy of a former cover of impervious shales (Warwick 1964). Once the impervious cover was removed by erosion, the rivers cut into the limestone beneath until solution exploited planes of weakness and diverted the drainage underground. The 'hanging valleys', which are reported in many karst areas, resulted from the main valleys' continuing to incise after their tributaries ceased to have surface flow.

Meander caves

Meander caves are formed where the outer bend of a meander undercuts a valley-side. Now, stream debris does not hamper rivers from lateral erosion in karst landscapes as it does rivers on other rocks, because rivers carrying a large clastic load cannot move laterally by corrosion as easily as rivers bearing a small clastic load can by corrosion. For this reason, meander caves are better developed in karst terrain than elsewhere. A prime example is Verandah Cave, Borenore, New South Wales, Australia (Figure 14.14).

Natural bridges

Natural bridges are formed of rock and span ravines or valleys. They are productions of erosion and are commoner in karst terrain than elsewhere. Three mechanisms seem able to build natural bridges in karst areas. First, a river may cut through a very narrow band of limestone that crosses its path. Second, cave roofs may collapse leaving sections still standing. Third, rivers may capture each other by **piracy** (Figure 14.15). This happens where meander caves on one or both sides of a meander spur breach the wall of limestone between them (Figure 14.15).

Tufa and travertine deposits

Karst rivers may carry supersaturated concentrations of carbonates. When deposited, the carbonates may build landforms. Carbonate deposition occurs when (1) water is exposed to the atmosphere, and so to carbon dioxide, on emerging from underground; (2) when evaporation supersaturates the water; and (3) when plants secrete

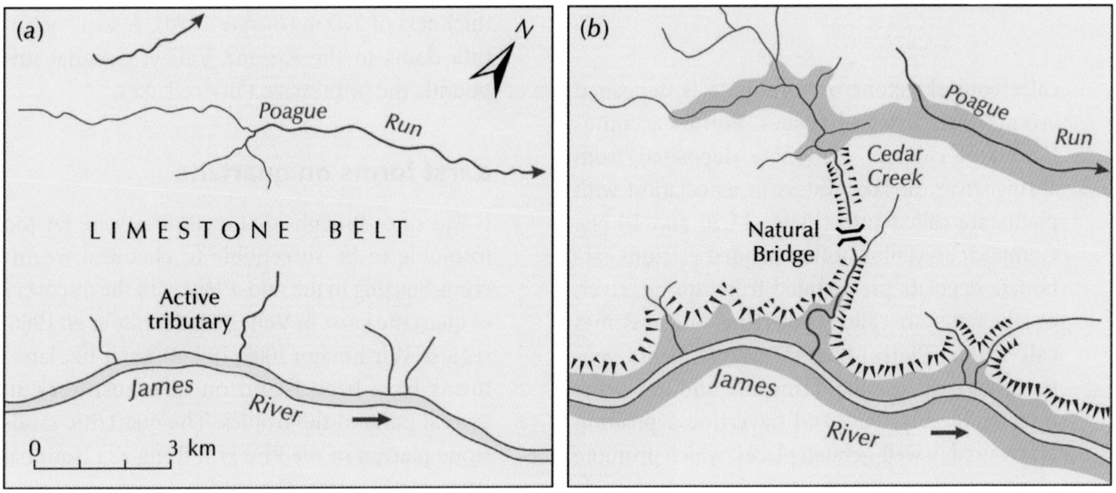

Figure 14.15 Natural Bridge, Cedar Creek, Virginia, USA. (a) Landscape before river piracy. (b) Landscape after river piracy. *Source*: Adapted from Woodward (1936)

Plate 14.20 Tufa deposited on the coarse limestone bedload of Gordale Beck, near Malham, North Yorkshire, England. (*Photograph by Richard Huggett*)

Plate 14.21 Tufa towers, Lake Mono, east-central California, USA. The towers reformed under water as calcium-bearing springs well up through the alkaline lake water that is rich in carbonates. A falling lake level has exposed the tufa towers, which cease to grow in the air. (*Photograph by Kate Holden*)

Plate 14.22 Hot spring travertine terraces at Pamukkale, Turkey. (*Photograph by Derek C. Ford*)

calcareous skeletons or carbonate is deposited around their external tissues. Porous accumulations of calcium carbonate deposited from spring, river, or lake waters in association with plants are called **tufa** (Plates 14.20 and 14.21). Compact, crystalline, often banded calcium carbonate deposits precipitated from spring, river, or lake water are called **travertine**, or sometimes **calc-sinter** (Plate 14.22). However, some geomorphologists use the terms tufa and travertine interchangeably. Tufa and travertine deposition is favoured in well-aerated places, which promote plant growth, evaporation, and carbon dioxide diffusion from the air. Any irregularity in a stream profile is a prime site. A barrier slowly builds up, on the front side of which frothing and bubbling encourage further deposition. The end result is that a dam and waterfall form across a karst river. The waterfall may move down the valley leaving a fill of travertine in its wake. Travertine may cover large areas. In Antalya, south-west Turkey, a travertine complex, constructed by the supersaturated calcareous waters of the Kirkgöz spring group, occupies 600 km^2 and has a maximum thickness of 270 m (Burger 1990). A sequence of tufa dams in the Korana Valley, Croatia, impounds the impressive Plitvice Lakes.

Karst forms on quartzite

It was once thought that quartzites were far too insoluble to be susceptible to chemical weathering. Starting in the mid-1960s with the discovery of quartzite karst in Venezuela (White *et al.* 1966; see also Wirthmann 2000, 104–9), karst-like landforms have been found on quartzose rock in several parts of the tropics. The quartzitic sandstone plateau of the Phu Hin Rong Kla National Park, north-central Thailand, bears features found in limestone terrain – rock pavements, karren fields, crevasses, and caves – as well as weathered polygonal crack patterns on exposed rock surfaces and **bollard-shaped rocks** (Doerr 2000). The crevasses, which resemble grikes, occur near the edge of the plateau and are 0.5–2 m wide, up to 30 m deep, and between 1 and 10 m apart. Smaller features are reminiscent of solution runnels and solution flutes. Caves up to 30 m long have been found in the National Park and were used for shelter during air raids while the area was a stronghold for the communists during the 1970s. Some of the caves are really crevasses that have been widened some metres below the surface, but others are underground passages that are not associated with enlarged vertical joints. In one of them, the passage is 0.5–1 m high and 16 m long. The bollard-shaped rock features are found near the plateau edge (Plate 14.23). They are 30–50 cm high with diameters of 20–100 cm. Their formation appears to start with the development of a **case-hardened surface** and its sudden cracking under tensile stresses to form a polygonal cracking pattern (cf. Williams and Robinson 1989; Robinson and Williams 1992). The cracks are then exploited by weathering. Further weathering deepens the cracks, rounding off the tops of the polygonal blocks, and eventually eradicates the polygonal blocks' edges and deepens and widens the cracks to form bollard-shaped rocks (Figure 14.16).

Karst-like landforms also exist on the surfaces of quartzite table mountains (Tepuis) in south-eastern Venezuela (Doerr 1999). At 2,700 m, the Kukenan Tepui is one of the highest table mountains in South America. The topography includes caves, crevasse-like fissures, sinkholes, isolated towers 3–10 m high, and shallow karren-like features. Evidence points to corrosion, rather than to erosive processes, as the formative agent of these landforms (see p. 217).

Plate 14.23 Bollard rocks formed in quartzitic sandstone, north-central Thailand. (*Photograph by Stefan Doerr*)

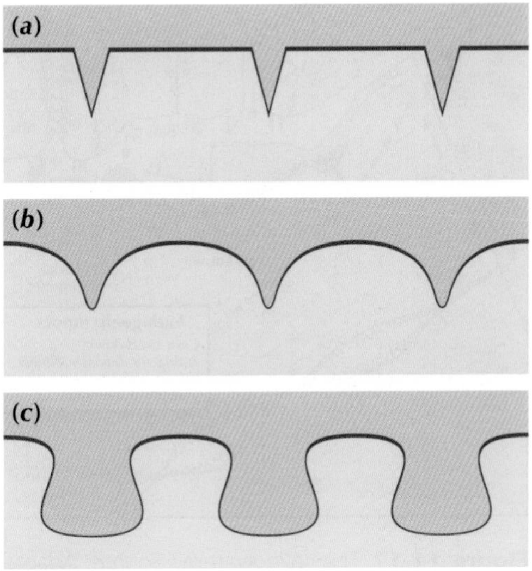

Figure 14.16 Proposed sequence of events leading to 'bollard' rock formation in quartzitic sandstone, north-central Thailand. (a) Polygonal cracks develop in a case-hardened surface that act as avenues of weathering. (b) Weathering deepens the cracks, forming a convex surface on each polygonal block. (c) Further weathering removes the edges of the polygonal blocks and deepens and widens the cracks. *Source*: Adapted from Robinson and Williams (1992) and Doerr (2000)

SUBTERRANEAN KARST FORMS

Waters from streams sinking into limestone flow through a karst drainage system – a network of fissures and conduits that carry water and erosion products to springs – where they reunite with the surface drainage system. In flowing through the karst drainage system, the water and its load abrade and corrode the rock, helping to produce cavern systems. These **subterranean landforms** contain a rich variety of erosional and depositional forms.

Erosional forms in caves

Caves are natural cavities in bedrock. They function as conduits for water flowing from a sink or a percolation point to a spring or to a seepage point (Figure 14.17). To form, caves need an initial cavity or cavities that channel the flow of rock-dissolving water. The origin of these cavities is debatable, with three main views taken:

1 The **kinetic view** sees tiny capillaries in the rock determining the nature of flow – laminar or turbulent. In capillaries large enough to permit turbulence, a helical flow accelerates solution of the capillary walls and positive feedback does the rest to form a principal cave conduit.
2 The **inheritance** or **inception horizon** view envisions a pre-existing small cavity or chain of vugs, which were formed by tectonic, diagenetic (mineralization), or artesian processes, being flooded and enlarged by karst groundwater, so forming a cave conduit.
3 The **hypergene view** imagines hydrothermal waters charged with carbon dioxide, hydrogen sulphide, or other acids producing heavily mineralized cavities, which are then overrun by cool karst waters to create larger and more integrated cavities or networks. All or any of these three processes may have operated in any cave during its history. In all cases, it is usually

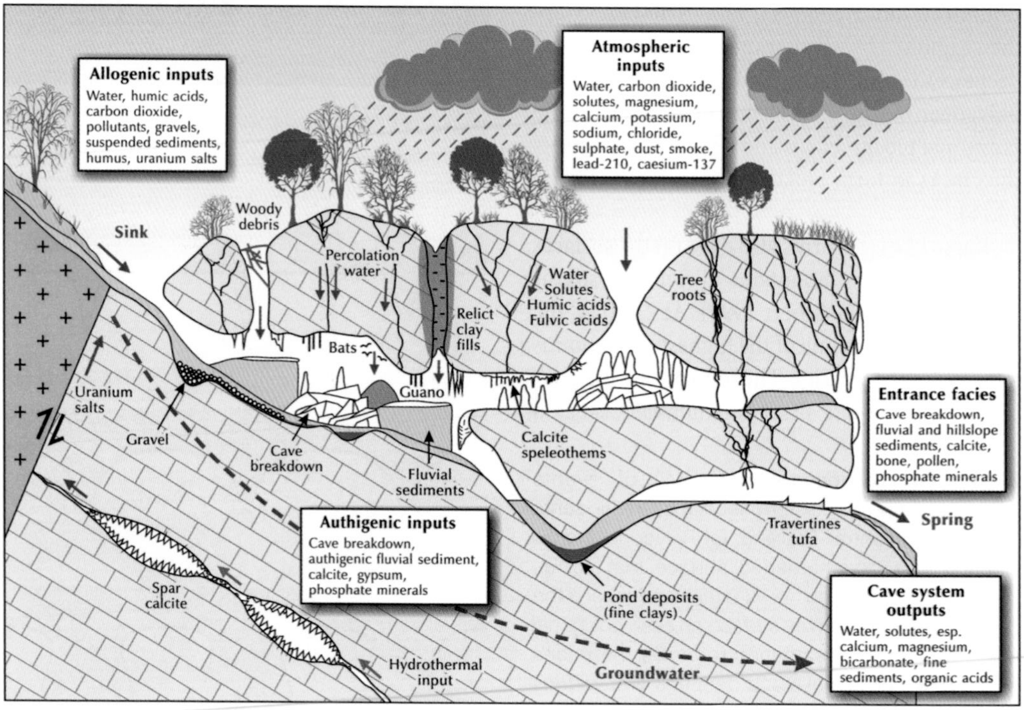

Figure 14.17 The cave system. *Source*: Adapted from Gillieson (1996, 7)

the case that, once an initial cave conduit forms, it dominates the network of passages and enlarges, becoming a primary tube that may adopt a variety of shapes (from a simple meandering tube to a highly angular or linear conduit) depending on rock structure.

Cave form

Cavern systems can be very extensive. Mammoth Cave, Kentucky, USA, comprises over 800 km of subterranean hollows and passages arranged on several levels, representing major limestone units with a vertical depth of 110 m. At 563,270 m, the cave system is the longest in the world. The form of caverns – their plan and cross-section – depends upon the purity of the limestone in which they are formed and the nature of the network of fissures dissecting the rock, as well as their hydrological setting.

The shape of caves is directed by lithology, by the pattern of joints, fractures, and faults, and by cave breakdown and evaporite weathering:

1 **Lithology**. Caves often sit at changes of lithology, with passages forming along or close to lithological junctions, for example the junctions between pure and impure limestones, between limestones and underlying shales, and between limestones and igneous rocks. Passages may have a propensity to form in a particular bed, which is then known as the inception horizon (Lowe 1992). For instance, in the Forest of Dean, England, caves start to form in interbedded sandstones and unconformities in the Carboniferous limestone.
2 **Joints**, **fractures**, and **faults**. Joint networks greatly facilitate the circulation of water in karst. Large joints begin as angular, irregular cavities that become rounded by solution. Cave formation is promoted when the joint spacing is 100–300 m, which allows flowing water to become concentrated. Some passages in most caves follow the joint network, and in extreme cases the passages follow the joint network fairly rigidly to produce a **maze cave**, such as Wind Cave, South Dakota, USA. Larger geological structures, and specifically faults, affect the complex pattern of caves in length and depth. Many of the world's deepest known shafts, such as 451-m-deep Epos in Greece, are located in fault zones. Individual cave chambers may be directed by faults, an example being Gaping Ghyll in Yorkshire, England. Lubang Nasib Bagsu (Good Luck Cave), Mulu, Sarawak is at 12 million cubic metres the world's largest known underground chamber and owes its existence to a combination of folding and faulting.
3 **Cave breakdown** and **evaporite weathering**. Limestone is a strong rock but brittle and fractures easily. Cave wall and ceiling collapse are important in shaping passages and chambers. Collapse is common near the cave entrance, where stress caused by unloading (p. 160) produces a denser joint network. Rock weathering by gypsum and halite crystallization (**exsudation**) may alter passage form. Water rich in soluble material seeping through the rocks evaporates upon reaching the cave wall. The expansion of crystals in the bedding planes or small fissures instigates sensational spalling.

Caves may also be classified in relation to the water table. The three main types are phreatic, vadose, and water table caves (Figure 14.18a, b, c). **Vadose caves** lie above the water table, in the unsaturated vadose zone, **water table** or **epiphreatic** or **shallow phreatic caves** lie at the water table, and **phreatic caves** below the water table, where the cavities and caverns are permanently filled with water. Subtypes are recognized according to the presence of cave loops (Figure 14.18d, e, f).

Speleogens

Cave forms created by weathering and by water and wind erosion are called **speleogens**. Examples are current markings, potholes and rock mills, rock pendants, and scallops.

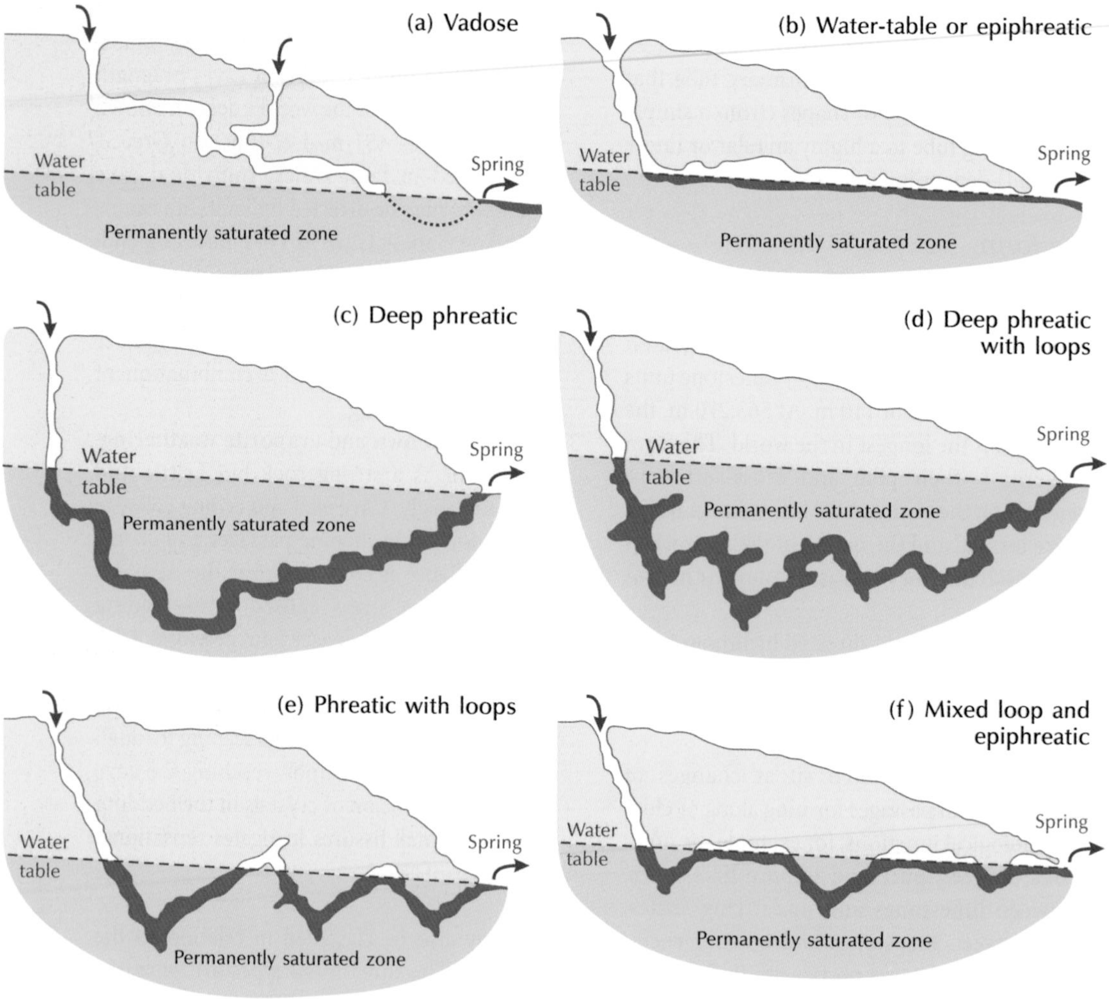

Figure 14.18 Types of caves. (a) Vadose). (b) Water-table or epiphreatic. (c) Deep phreatic. (e) Deep phreatic with loops. (f) Phreatic with loops. (f) Mixed loop and epiphreatic. *Source*: Adapted from Ford and Ewers (1978)

Potholes and **current markings** are gouged out by sediment-laden, flowing water in conjunction with some solutional erosion. The swirling motion of water is important in the formation of potholes. In the cave system behind God's Bridge rising in Chapel-le-Dale, North Yorkshire, England, grooves in bedrock, which look like rounded solution runnels, seem to be carved out by abrasion during times of high flow (Murphy and Cordingley 1999).

Rock pendants and **scallops** are products of solution. Rock pendants, which normally occur in groups, are smooth-surfaced protuberances in a cave roof. Scallops are asymmetrical, cuspate, oyster-shell-shaped hollows with a steep semicircular step on the upstream side and a gentle rise downstream ending in a point of the next downstream hollow (Plate 14.24). Scallop size varies inversely with the flow velocity of the water, and scallops may be used to assess flow conditions.

In the main passage of Joint Hole, Chapel-le-Dale, North Yorkshire, England, two contrasting-size populations of scallops were found (Murphy *et al.* 2000). Larger scallops occupy the walls and ceilings, and smaller scallops occupy the floor. The floor scallops suggest a higher velocity at the bottom of the conduit. Presumed solution features in the phreatic zone include spongework, bedding plane and joint anastomoses, wall and ceiling pockets, joint wall and ceiling cavities, ceiling half tubes, continuous rock spans, and mazes of passages (see Jennings 1971, 156–7).

Plate 14.24 Scallops in Joint Hole, Chapel-le-Dale, North Yorkshire, England, taken underwater. (*Photograph by Phil Murphy*)

Depositional forms in caves

Three types of deposit are laid down in caves: (1) cave formations or speleothems; (2) material weathered *in situ*; and (3) clastic sediments carried mechanically into the cave and deposited there (White 1976). Cave sediments are beyond the scope of this introductory text (see Gillieson 1996, pp. 143–66, for an excellent review), but the chemical precipitates known as **speleothems** will be discussed.

Most speleothems are made of carbonate deposits, with calcite and aragonite accounting for about 95 per cent of all cave minerals. The carbonates are deposited mainly by carbon dioxide loss (degassing) or by evaporation. Formations of carbonate may be arranged into three groups: dripstone and flowstone forms, eccentric or erratic forms, and sub-aqueous forms (White 1976).

Dripstone and flowstone

Dripstone is a deposit, usually composed of calcite, formed of drips from cave ceilings or walls. **Flowstone** is a deposit, again usually composed of calcite, formed from thin films or trickles of water over floors or walls. The forms fashioned by dripstone and flowstone are stalactites, stalagmites, draperies, and flowstone sheets. **Stalactites**, which develop downwards, grow from dripping walls and ceilings. The basic form is a **straw stalactite** formed by a single drop of water on the ceiling degassing and producing a ring of calcite about 5 mm in diameter that grows into a straw (Plates 14.25 and 14.26). The longest known straw stalactite is in Strong's Cave, Western Australia, and is 6.2 m. Leakage and blockage of a straw leads to the growth of a carrot-shaped stalactite. **Stalagmites** grow from the floor, their exact form (columnar or conical) depending upon drip rates, water hardness, and the cave atmosphere. A **column** forms where an upward-growing stalagmite joins a downward-growing stalactite. A study of six cave systems in Europe revealed that, for five sites with a good soil cover, stalagmite growth rate depends chiefly upon mean

Plate 14.25 Close-up of a straw stalactite, Ogof Ffynnon Ddu, Wales. (*Photograph by Clive Westlake*)

Plate 14.27 Curtains in Otter Hole, Chepstow, South Wales. (*Photograph by Clive Westlake*)

Plate 14.26 Straw stalactites, Pippikin Hole, Yorkshire Dales, England. (*Photograph by Tony Waltham Geophotos*)

annual temperature and the calcium content of the drip-water, but was unaffected by the drip rate (Genty *et al.* 2001). One site in the Grotte de Clamouse, which has little soil cover, failed to display a correlation between stalagmite growth rate and temperature, either because little carbon dioxide was produced in the thin soil or because calcite was precipitated before entering the system.

Water trickling down sloping walls or under a tapering stalactite produces **draperies** (curtains and shawls), which may be a single crystal thick (Plate 14.27). Varieties with coloured bands are called 'bacon'. **Flowstone sheets** are general sheets of flowstone laid down over walls and ceilings.

Plate 14.28 Helictites in Ogof Draenen, Pwll Ddu, South Wales. (*Photograph by Clive Westlake*)

Plate 14.29 Crystal pool, Ogof Ffynnon Ddu at Penwyllt, South Wales. Crystal deposits within the dam rim are visible. (*Photograph by Clive Westlake*)

Eccentric forms

Eccentric or **erratic** forms, which are speleothems of abnormal shape or attitude, include shields, helictites, botryoidal forms, anthodites, and moonmilk. **Shields** or **palettes** are made of two parallel plates with a small cavity between them through which water seeps. They grow up to 5 m in diameter and 4–10 cm thick. **Helictites** change their axis from the vertical during their growth, appearing to disobey gravity, to give a curving or angular, twig-like form (Plate 14.28). **Botryoidal forms** resemble bunches of grapes. They are a variety of **coralloid forms**, which are nodular and globular and look like coral. **Anthodites** are gypsum clusters that radiate from a central point. **Moonmilk** or **rockmilk** is a soft, white, plastic, moist form of calcite, and often shaped like a cauliflower.

Sub-aqueous forms

Sub-aqueous forms are rimstone pools, concretions, pool deposits, and crystal linings. **Rimstone pools** form behind **rimstone dams**, sometimes called **gours**, which build up in channels or on flowstones (Plate 14.29). In rimstone pools, a suite of deposits precipitates from supersaturated meteoric water flowing over the outflow rim and builds a rimstone dam. **Pool deposits** are any sediment or crystalline deposits in a cave pool.

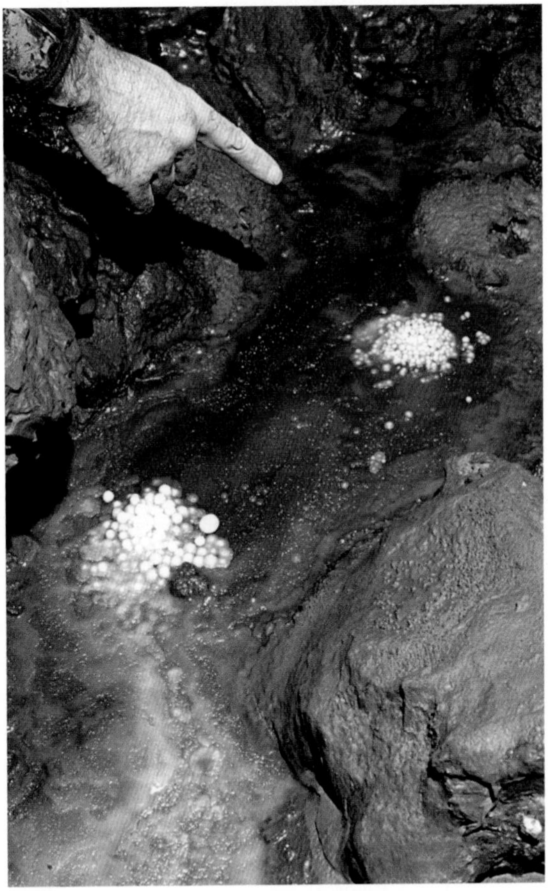

Plate 14.30 Cave pearls in Peak Cavern, Castleton, England. (*Photograph by Clive Westlake*)

Crystal linings are made of well-formed crystals and are found in cave pools with little or no overflow.

Pisoliths or **cave pearls** are small balls, ranging from about 0.2 mm to 15 mm in diameter, formed by regular accretions of calcite about a nucleus such as a sand grain (Plate 14.30). A few to thousands may grow in shallow pools that are agitated by drops of feedwater.

HUMAN IMPACTS ON KARST

Surface and subsurface karst are vulnerable to human activities. Visitors damage caves, and agricultural practices may lead to the erosion of soil cover from karst areas.

Soil erosion on karst

Karst areas worldwide tend to be susceptible to **soil erosion**. Their soils are usually shallow and stony, and, being freely drained, leached of nutrients.

When vegetation is removed from limestone soils or when they are heavily used, soil stripping down to bedrock is common. It can be seen on the Burren, Ireland, in the classic karst of the Dinaric Alps, in karst of China, in the cone karst of the Philippines, and elsewhere. In Greece, soil stripping over limestone began some 2,000 years ago. The limestone pavement above Malham Cove (Plate 14.3) may be a legacy of agricultural practices since Neolithic times, soils being thin largely because of overgrazing by sheep. Apart from resulting in the loss of an agricultural resource, soil stripping has repercussions in subterranean karst. The eroded material swiftly finds its way underground, where it blocks passages, diverts or impounds cave streams, and chokes cave life.

The prevention of soil erosion and the maintenance of critical soil properties depend crucially upon the presence of a stable vegetation cover. The Universal Soil Loss Equation or its more recent derivatives (p. 205) can predict soil erosion on karst terrain, but higher rates may be expected on karst as compared with most other soil types because features of the geomorphology conspire to promote even greater erosion than elsewhere. In most non-karst areas, soil erosion depends upon slope gradient and slope length, as well as the other factors in the USLE. It also depends partly on slope gradient and slope length in karst terrain but, in addition, the close connections between the surface drainage system and the underground conduit system produce a locally steeper hydraulic gradient that promotes erosive processes. Moreover, eroded material in karst areas has a greater potential to be lost down joints and fissures by sinkhole collapse, gullying, or soil stripping. An adequate vegetation cover and soil structure (which reduce erodibility) take on a greater significance in lessening this effect in karst areas than in most other places.

Humans and caves

Humans have long used caves for shelter, defence, sanctuaries, troglodytic settlements, a source of resources (water, food, guano, ore in mine-caves), and as spiritual sites. In the last few hundred years, caves have been used for the mining of cave formations and guano (especially during the American Civil War), for hydroelectric power generation from cave streams and springs (in China), for storage, and as sanatoria and tourist attractions. Evidence for the human occupancy of caves in China dates from over 700,000 years ago. Many caves are known to have housed humans at the start of the last glacial stage, and several have walls adorned with splendid paintings. Many caves in the Guilin tower karst, China, have walls at their entrances, suggesting that they were defended. Medieval fortified caves are found in Switzerland in the Grisons and Vallais. In Europe and the USA, some caves were used as sanatoria for tuberculosis patients on the erroneous premise that the moist air and constant temperature would aid recovery. Caves have also been widely used for cheese-making and rope manufacture, as in the entrance to Peak Cavern, Derbyshire, England.

Box 14.1 CAVE TOURISM

Some 20 million people visit caves every year. Mammoth Cave in Kentucky, USA, alone has 2 million visitors annually. Great Britain has some 20 show caves, with the most-visited receiving over 500,000 visitors every year. About 650 caves around the world have lighting systems, and many others are used for 'wild' cave tours where visitors carry their own lamps. Tourists damage caves and karst directly and indirectly through the infrastructure built for the tourists' convenience – car parking areas, entrance structures, paths, kiosks, toilets, and hotels. The infrastructure can lead to hydrological changes within the cave systems. Land surfaced with concrete or bitumen is far less permeable than natural karst, and the danger is that the feedwaters for stalactites may be dramatically reduced or stopped. Similarly, drains may alter water flow patterns and lead to changes in speleothem deposition. Drainage problems may be in part alleviated by using gravel-surfaced car parks and paths, or by including strips where infiltration may occur. Within caves, paths and stairs may alter the flow of water. Impermeable surfaces made of concrete or steel may divert natural water movement away from flowstones or stream channels, so leading to the drying out of cave formations or to increased sediment transport. These problems are in part overcome by the use of permeable steel, wooden, or aluminium walkways, frequent drains leading to sediment traps, and small barriers to water movement that approximate the natural flow of water in caves.

Cave tourists alter the cave atmosphere by exhaling carbon dioxide in respiration, by their body heat, and by the heat produced by cave lighting. A party of tourists may raise carbon dioxide levels in caves by 200 per cent or more. One person releases between 82 and 116 watts of heat, roughly equivalent to a single incandescent light bulb, which may raise air temperatures by up to 3°C. A party of tourists in Altamira Cave, Spain, increased air temperature by 2°C, trebled the carbon dioxide content from 0.4 per cent to 1.2 per cent, and reduced the relative humidity from 90 per cent to 75 per cent. All these changes led to widespread flaking of the cave walls, which affected the prehistoric wall paintings (Gillieson 1996, 242). A prolonged increase in carbon dioxide levels in caves can upset the equilibria of speleothems and result in solution, especially in poorly ventilated caves with low concentrations of the calcium ion in drip water (Baker and Genty 1998). Other reported effects of cave tourism include the colonization of green plants (mainly algae, mosses, and ferns) around continuous lighting, which is known as lampenflora, and a layer of dust on speleothems (lint from clothing, dead skin cells, fungal spores, insects, and inorganic material). The cleaning of cave formations removes the dust and lampenflora but also damages the speleothems. A partial solution is to provide plastic mesh walkways at cave entrances and for tourists to wear protective clothing. Recreational cavers may also adversely affect caves (Gillieson 1996, 246–7). They do so by carbide dumping and the marking of walls; the compaction of sediments with its concomitant effects on cave hydrology and fauna; the erosion of rock surfaces in ladder and rope grooves and direct lowering by foot traffic; the introduction of energy sources from mud on clothes and foot residues; the introduction of faeces and urine; the widening of entrances and passages by traffic or by digging; and the performing of cave vandalism and graffiti. The best way of limiting the impact of cave users is through education and the development of minimal-impact codes, which follow cave management plans drawn up by speleologists, to ensure responsible conduct (see Glasser and Barber 1995).

Kentucky bourbon from the Jack Daniels distillery relies partly on cave spring water.

Cave tourism started in the late eighteenth and early nineteenth centuries in Europe, when candle lanterns were used (e.g. Nicod 1998). Today, cave tourism is a growth industry: fibre-optic lights illuminate some caves, and electric trains transport tourists through the caverns. Tourism has an injurious impact on caves (Box 14.1). To combat the problems of cave tourism, **cave management** has evolved and is prosecuted by a body of government and private professionals. Several international groups are active in cave and karst management: the International Union of Speleology, the International Speleology Heritage Association, the International Geographical Union and the Commission for National Parks and Protected Areas, and the International Union for the Conservation of Nature and Natural Resources (IUCN).

Managing karst

Karst management is based on an understanding of karst geomorphology, hydrology, biology, and ecology. It has to consider surface and subsurface processes, since the two are intimately linked. The basic aims of karst management are to maintain the natural quality and quantity of water and air movement through the landscape, given the prevailing climatic and biotic conditions. The flux of carbon dioxide from the air, through the soils, to cave passages is a crucial karst process that must be addressed in management plans. In particular, the system that produces high levels of carbon dioxide in soil, which depends upon plant root respiration, microbial activity, and a thriving soil invertebrate fauna, needs to be kept running smoothly.

Many **pollutants** enter cave systems from domestic and municipal, agricultural, constructional and mining, and industrial sources. In Britain, 1,458 licensed landfill sites are located on limestone, many of which take industrial wastes. Material leached from these sites may travel to contaminate underground streams and springs for several kilometres. Sewage pollution is also common in British karst areas (Chapman 1993).

Limestone and marble are quarried around the world and used for cement manufacture, for high-grade building stones, for agricultural lime, and for abrasives. **Limestone mining** mars karst scenery, causes water pollution, and produces much dust. Quarrying has destroyed some British limestone caves and threatens to destroy others. In southern China, many small quarries in the Guilin tower karst extract limestone for cement manufactories and for industrial fluxes. In combination with vegetation removal and acid rain from coal burning, the quarrying has scarred many of the karst towers around Guilin city, which rise from the alluvial plain of the Li River. It is ironic that much of the cement is used to build hotels and shops for the tourists coming to see the limestone towers. In central Queensland, Australia, Mount Etna is a limestone mountain containing forty-six named caves, many of which are famous for their spectacular formations. The caves are home to some half a million insectivorous bats, including the rare ghost bat (*Macroderma gigas*). The mining of Mount Etna by the Central Queensland Cement Company has destroyed or affected many well-decorated caves. A public outcry led to part of the mountain being declared a reserve in 1988, although mining operations continue outside the protected area, where the landscape is badly scarred.

The IUCN World Commission on Protected Areas recognizes karst landscapes as critical targets for **protected area** status. The level of protection given in different countries is highly variable, despite the almost universal aesthetic, archaeological, biological, cultural, historical, and recreational significance of karst landscapes. Take the case of South-East Asia, one of the world's outstanding carbonate karst landscapes, with a total karst area of 458,000 km^2, or 10 per cent of the land area (Day and Urich 2000). Karstlands in this region are topographically diverse and include cockpit and cone karst, tower karst, and pinnacle karst, together with extensive dry valleys, cave systems, and springs. They include classic tropical karst landscapes: the Gunung Sewu of Java, the Chocolate Hills of Bohol, the pinnacles and caves of Gunong Mulu, and the karst towers of Vietnam and peninsular Malaysia. Human impacts on the South-East Asian karst landscapes are considerable: less than 10 per cent of the area maintains its natural vegetation. About 12 per cent of the regional karst landscape has been provided nominal protection by designation as a protected area, but levels of protection vary from country to country (Table 14.3). Protection is significant in Indonesia, Malaysia, the Philippines, and Thailand. Indonesia, for instance, has forty-four protected karst areas, which amount to 15 per cent of its total karst area. In Cambodia, Myanmar (Burma), and Papua New Guinea, karst conservation is minimal, but additional protected areas may be designated in these countries as well as in Vietnam and in Laos. Even so, South-East Asia's karstlands have an uncertain future. It should be stressed that the designation of karst as protected areas in South-East Asia is not based on the intrinsic or scientific value of the karst landscapes, but on unrelated contexts, such as biological diversity, timber resources, hydrological potential, or archaeological and recreational value. Nor, it must be said, does the conferral of a protected area status guarantee effective protection from such threats as forest clearance, agricultural inroads, or the plundering of archaeological materials.

The conservation of karst in the Caribbean is in a similar position to that in South-East Asia (Kueny and Day 1998). Some 130,000 km^2, more than half the land area of the Caribbean, is limestone karst. Much of it is found on the Greater Antilles, with other significant areas in the Bahamas, Anguilla, Antigua, the Cayman Islands, the Virgin Islands, Guadeloupe, Barbados, Trinidad and Tobago, and the Netherlands Antilles. Features include cockpits, towers, dry valleys, dolines, and caves. Humans have impacted on the karst landscapes and the necessity for protection at regional and international level is recognized. However, karst is in almost all cases protected by accident – karst areas happen to lie within parks, reserves, and sanctuaries set up to safeguard biodiversity, natural resources, or cultural and archaeological sites. Very few areas are given protected area status because of the inherent scientific interest of karst landscapes. At the regional level, 121 karst areas, covering 18,441 km^2 or 14.3 per cent of the total karst, are afforded protected area

Table 14.3 Protected karst areas in South-East Asia, 2000

Country	Karst area (km^2)	Protected karst area (km^2)	Protected karst area (%)	Number of protected areas
Cambodia	20,000	0	0	0
Indonesia	145,000	22,000	15	44
Laos	30,000	3,000	10	10
Malaysia	18,000	8,000	45	28
Myanmar (Burma)	80,000	650	1	2
Papua New Guinea	50,000	0	0	0
Philippines	35,000	10,000	29	14
Thailand	20,000	5,000	25	41
Vietnam	60,000	4,000	7	15
Total	458,000	52,650	12	154

Source: Adapted from Day and Urich (2000)

status. Higher levels of protection are found in Cuba, the Dominican Republic, and the Bahamas. Lower levels of protection occur in Jamaica, Puerto Rico, Trinidad, and the Netherlands Antilles, and minimal protection is established in the smaller islands. In Trinidad and Tobago, for instance, karst forms comprise karren, caves, springs, valley systems, and a range of sinkholes, including an area of polygonal cockpit karst (Day and Chenoweth 2004). Quarrying, and to a lesser degree logging and agriculture, have destroyed much of the karst in western Trinidad, whilst urban development and tourism have added to the damage in Trinidad and in the lowlands of western Tobago. Few of these karstlands lie within existing protected areas.

KARST IN THE PAST

Karst that formed in the geological past and survives to the present is surprisingly common. Such old karst is known as palaeokarst, although sometimes the term '**fossil karst**', which is rather ambiguous, is employed (see Bosák *et al.* 1989). Palaeokarst may be divided into buried karst and intrastratal karst.

Buried karst is karst formed at the ground surface and then covered by later sediments. **Intrastratal karst** is karst formed within bedding planes or unconformities of soluble rocks that are already buried by younger strata. An important distinction between buried karst and intrastratal karst is that buried karst is older than the covering rocks, while intrastratal karst is younger than the covering rocks. **Subjacent karst** is the most common form of intrastratal karst and develops in soluble rocks that lie below less soluble or insoluble strata. No intrastratal karst feature has ever belonged to a former karst landscape. A complication here is that, in many places, intrastratal karst is forming today. Palaeointrastratal karst is inactive or inert. The oldest known buried karst features are caves and cave deposits in the Transvaal, South Africa, which formed 2,200 million years ago (Martini 1981). In Quebec, Canada, Middle Ordovician dolines, rounded solution runnels, and solution pans have been discovered, exposed after survival beneath a blanket of later rocks (Desrochers and James 1988).

Another group of karst features was formed in the past when the climate and other environmental factors were different, but survive today, often in a degraded state, under conditions that are no longer conducive to their development. Such karst features are called **relict karst** and occur above and below ground. A cave system abandoned by the groundwater streams that carved it, owing to a lowering of the groundwater table, is an example of subterranean relict karst. Some caves are Tertiary in age, and some relict cave passages may even survive from the Mesozoic era (Osborne and Branagan 1988). Similar processes have operated over these timescales to produce deposits that can be investigated to reconstruct changing conditions. Other processes have operated to leave relict features that have no modern analogues (Gillieson 1996, 106). An example of a surface relict-karst feature is the stream-sink dolines found in the Northaw Great Wood, Hertfordshire, England, where the Cuffley Brook cuts down through London Clay and Reading Beds to reach Chalk. At the eastern end of the wood, where the Chalk lies just below alluvium, there are several stream-sink dolines standing higher than the present stream channel and were probably formed when the stream occupied a different and higher course. On a larger scale, the Qattara Depression, Egypt, may have started as a river valley, but karst processes mainly formed it during the Late Miocene period. It has subsequently been lowered by deflation (p. 338), but is partly a relict karst feature (Albritton *et al.* 1990).

To complicate matters even more, buried karst is sometimes re-exposed through the erosion of the covering strata to form **exhumed karst**. Near Madoc, Ontario, Canada, pure dolostones dating from the Grenville Orogeny, some 977 million years ago, today form a hilly terrain that is being exhumed from the Late Cambrian–Lower Ordovician cover rocks. Cone karst and cockpits, together with lesser dolines and grikes, have been identified in the exhumed surface (Springer 1983). If the environmental conditions on re-exposure are favourable, renewed karstification may proceed and create **rejuvenated karst**. The present upland surface of the Mendip Hills of Somerset, England, is the rejuvenated surface of a Triassic island, and some of the fissures on the Mendips may have been dolines or cenotes (Ford 1989). Similarly, the Yunnan Stone Forest (p. 424) started as a rugged tor-and-pediment topography that was buried by Tertiary sands and clays. Smooth and rounded pinnacles developed while the cover was present. Recent re-exposure is sharpening the pinnacles over an area of 35,000 ha.

SUMMARY

Karst is terrain with scant surface drainage, thin and patchy soils, closed depressions, and caves. Its distinctive features develop on fairly pure limestones, but also occur in evaporites and silicate rocks. It forms by the dissolution of limestone or other soluble rocks, in conjunction with creep, block slumps, debris slides, earthflows, soilfalls, rockfalls, block slides, and rock slides. Fluvial and hydrothermal processes may affect karst development. A multitude of landforms form on limestone: karren of many shapes and sizes, limestone pavements, pinnacles, karst ruins, corridors, and coastal karst features; also, a range of closed depressions: dolines, karst windows, uvalas, and polja. Cone karst is a tropical form of karst, two varieties of which are cockpit karst and tower karst. Labyrinth karst is an extratropical version of tower karst. Despite a scarcity of surface drainage in karst terrain, fluvial processes affect some karst landforms, including gorges, blind and half-blind valleys, steepheads, dry valleys, meander caves, natural bridges, and tufa and travertine deposits. Another multitude of landforms form within limestone in subterranean karst. Speleogens are erosional forms in caves. They include potholes and current-markings, rock pendants and scallops. Within caves, three types of deposit are found: cave formations or speleothems, material weathered *in situ*, and clastic sediments carried into caves and laid down there. Speleothems are multifarious, and may be grouped into dripstones (such as stalactites and stalagmites), eccentric forms (such as helictites and moonmilk), and subaqueous forms (such as rimstone pools and gours). Agricultural practices have led to the stripping of soil from some karst areas. The fascination of caves has produced a thriving cave tourist industry, but cave visitors may destroy the features they come to view. Karstlands, too, are threatened in many parts of the world and require protection. Karst surviving from the geological past – palaeokarst – is common.

ESSAY QUESTIONS

1 How distinctive are karst landscapes?

2 Discuss the role of climate in karst formation.

3 Why is palaeokarst so common?

4 Analyse the problems of karst management.

FURTHER READING

Ford, D. C. and Williams, P. W. (2007) *Karst Geomorphology and Hydrology*, revised edition. Chichester: John Wiley & Sons.
Simply the best book on karst.

Gillieson, D. (1996) *Caves: Processes, Development and Management*. Oxford: Blackwell.
A superb book on subterranean karst that includes chapters on management.

Jennings, J. N. (1985) *Karst Geomorphology*. Oxford and New York: Blackwell.
A classic by an author whose name is synonymous with karst geomorphology. A little dated but may still be read with profit.

Trudgill, S. (1985) *Limestone Geomorphology*. Harlow, Essex: Longman.
Includes a good discussion of karst processes.

15

CHAPTER FIFTEEN

LONG-TERM GEOMORPHOLOGY

Some landforms and landscapes are remarkably old, survivors from long-gone climatic regimes. Some landscapes evolve over geological timescales. This chapter covers:

- Old landforms and landscapes
- Evolving landscapes

LANDSCAPE INACTION

I have seen no inland rocks in Great Britain which seem to point so unequivocally to the action of the sea as the Brimham Rocks [Plate 15.1], about nine miles from Harrogate. They fringe an eminence, or upheaved island, partly spared and partly wrecked by the sea. A group of picturesque columns may be seen on the eastern shore of this ancient island, but the grand assemblage of ruins occurs on the north-western side . . .

First, a line of cliff . . . extending along the western and north-western part of the risen island of Brimham for more than half a mile. A detached part of this coastline, behind Mrs Weatherhead's farmhouse, shows a projecting arched rock with associated phenomena, which one familiar with sea-coast scenery could have no more hesitation in referring to wave-action than if he still beheld them whitened by the spray. Farther northwards the line of cliff in some places shows other characteristics of a modern sea-coast. Here an immense block of millstone grit has tumbled down through an undermining process – there a block seems ready to fall, but in that perilous position it would seem to have remained since the billows which failed to detach it retreated to a lower level. Along the base of the cliffs many blocks lie scattered far and near, and often occupy positions in reference to the cliffs and to each other which a power capable of transporting will alone explain. From the cliff-line passages ramify and graduate into the spaces separating the rocky pillars, which form the main attraction of this romantic spot. . . .

As we gaze on this wonderful group of insular wrecks, varying in form from the solemn to the grotesque, and presenting now the same general outlines with which they rose above the sea, we can scarcely resist contrasting the permanence of the 'everlasting hills' with the evanescence of man. Generation after generation of the inhabitants of the valleys within sight of the eminence on which we stand, have sunk beneath the sod, and their descendants can still behold in these rocky pillars emblems of eternity compared with their own fleeting career; but fragile, and transient, compared with the great cycle of geological events. Though the Brimham Rocks may continue invulnerable to the elements for thousands of years, their time will come, and that time will be when, through another submergence of the land, the sea shall regain ascendancy of these monuments of its ancient sway, completing the work of denudation it has left half-finished.

(Mackintosh 1869, 119–24)

Plate 15.1 Brimham Rocks, eroded remnants of Millstone Grit sandstone, Nidderdale, North Yorkshire, England. *(Photograph by Tony Waltham Geophotos)*

OLD LANDFORMS AND LANDSCAPES

Some geomorphologists, mainly the 'big names' in the field, have turned their attention to the long-term change of landscapes. Starting with William Morris Davis's 'geographical cycle' (p. 9), several theories to explain the prolonged decay of regional landscapes have been promulgated. Walther Penck offered a variation on Davis's scheme. According to the Davisian model, uplift and planation take place alternately. But, in many landscapes, uplift and denudation occur at the same time. The continuous and gradual interaction of tectonic processes and denudation leads to a different model of landscape evolution, in which the evolution of individual slopes is thought to determine the evolution of the entire landscape (Penck 1924, 1953). Three main slope forms evolve with different combinations of uplift and denudation rates. First, convex slope profiles, resulting from waxing development (*aufsteigende Entwicklung*), form when the uplift rate exceeds the denudation rate. Second, straight slopes, resulting from stationary (or steady-state) development (*gleichförmige Entwicklung*), form when uplift and denudation rates match one another. And, third, concave slopes, resulting from waning development (*absteigende Entwicklung*), form when the uplift rate is less than the denudation rate. Later work has shown that valley-side shape depends not on the simple interplay of erosion rates and uplift rates, but on slope materials and the nature of slope-eroding processes.

According to Penck's arguments, slopes may either recede at the original gradient or else flatten, according to circumstances. Many textbooks claim that Penck advocated 'parallel retreat of slopes', but this is a false belief (see Simons 1962). Penck (1953, 135–6) argued that a steep rock face would move upslope, maintaining its original gradient, but would soon be eliminated by a growing basal slope. If the cliff face was the scarp of a tableland, however, it would take a long time to disappear. He reasoned that a lower-angle slope, which starts growing from the bottom of the basal slope, replaces the basal slope. Continued slope replacement then leads to a flattening of slopes, with steeper sections formed during earlier stages of development sometimes surviving in summit areas (Penck 1953, 136–41). In short, Penck's complicated analysis predicted both **slope recession** and **slope decline**, a result that extends Davis's simple idea of **slope decline** (Figure 15.1). Field studies have confirmed that slope retreat is common in a wide range of situations. However, a slope that is actively eroded at its base (by a river or by the sea) may decline if the basal erosion should stop. Moreover, a tableland scarp retains its angle through parallel retreat until the erosion removes the protective cap rock, when slope decline sets in (Ollier and Tuddenham 1962).

Common to all these theories is the assumption that, however the land surface may appear at the outset, it will gradually be reduced to a low-lying plain that cuts across geological structures and rock types. These **planation surfaces** or **erosion surfaces** are variously styled peneplains, panplains, etchplains, and so forth (Table 15.1). Cliff Ollier (1991, 78) suggested that the term **palaeoplain** is preferable since it has no genetic undertones and simply means 'old plain'; the term **palaeosurface** is equally neutral. It is worth bearing in mind when discussing the classic theories of landscape evolution that palaeoplain formation takes hundreds of millions of years to accomplish, so that during the Proterozoic aeon enough time elapsed for but a few erosion surfaces to form. In southeastern Australia, the palaeoplain

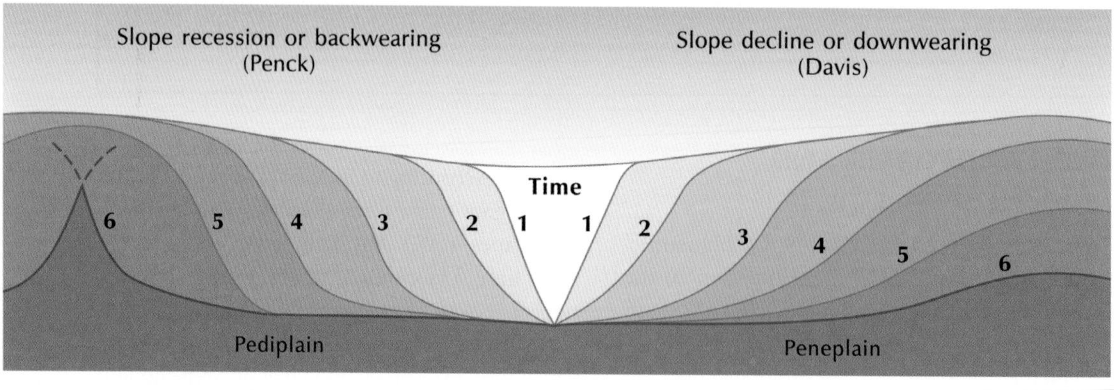

Figure 15.1 Slope recession, which produces a pediplain (p. 458) and slope decline, which produces a peneplain.
Source: Adapted from Gossman (1970)

Table 15.1 Types of palaeoplain

Landform or surface	Definition	References
Defined by genesis and appearance		
Erosion surface	A surface formed by removal of material through agents of erosion (glaciers, rivers, sea, wind), but not mass movements or weathering	Adams (1975)
Denudation surface	A surface created by denudational processes – weathering plus erosion (the weathering is crucial as it renders the bedrock removable)	Lidmar-Bergström (1988)
Peneplain	An almost featureless surface of low relief, in which occasional residual hills (monadnocks) may occur. Forms through the down-wearing of slopes to baselevel (sea level) after a bout of uplift	Davis (1899)
Pediplain	A flat area of low relief at the foot of an elevated feature, such as a hill or a mountain, occasionally broken up by residual hills (inselbergs). Forms as the end-product of a landscape fashioned by parallel slope retreat	Penck (1924)
Panplain	A broad and level surface fashioned by lateral corrasion of rivers leading to the coalescence of adjacent floodplains	Crickmay (1933, 1975)
Etch surface	A surface at the interface between weathered saprolite and unweathered bedrock	Büdel (1982), Thomas (1974, 1994)
Etchplain	A flattish surface created in tropical and subtropical environments when chemical weathering produces a thick regolith that erosion then strips	Thomas (1989a, 1989b)
Exhumed surface	A surface covered by, for example, Palaeozoic or Mesozoic cover rocks and later successively uncovered (exhumed)	Lidmar-Bergström (1988, 1995)
Defined by appearance		
Planation surface	A flat or nearly flat land surface fashioned by surface or near-surface processes acting on a rock mass	Adams (1975)
Defined by ancient age and preservation to the present		
Palaeosurface	'an identifiable topographic surface of either endogenic or exogenic origin . . . of demonstrable antiquity, which is, or was originally, of regional significance, and which as a consequence of its evolution, displays the effects of surface alteration resulting from a prolonged period of weathering, erosion, or non-deposition'	Widdowson (1997, 5)

Source: Adapted from Ebert (2009a)

first described by Edwin Sherbon Hills is still preserved along much of the Great Divide and is probably of Mesozoic age. In South America, where uplift has been faster, there are three or more erosion surfaces. Old erosion surfaces are commonly preserved in the geological record as unconformities.

Karin Ebert's (2009b) research in the Tjeuralako area of northern Sweden is a fine example of palaeosurface identification and interpretation. Using GIS analysis of DEM models, she recognized five palaeosurface generations (Figure 5.2). Each palaeosurface generation survives in a different way, the two highest as mountain peaks, the next two lower mainly on the plateau and representing a valley-in-valley pattern, and the fifth and lowest mainly as valley benches along the major valleys.

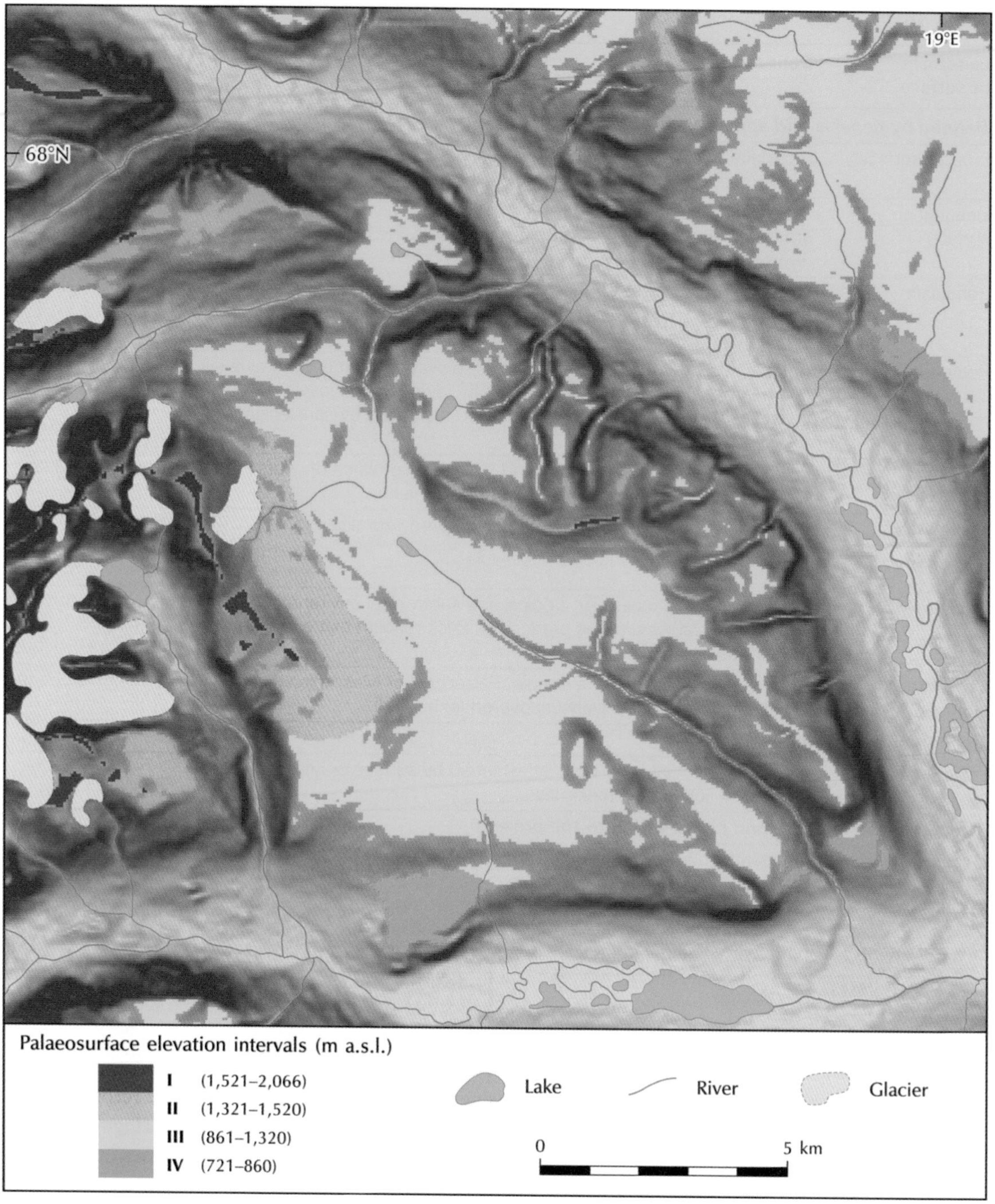

Figure 15.2 Surface and slope map of Tjeuralako Plateau, northern Sweden, and its surroundings with palaeosurface remnants and corresponding elevation classes shown in colour. *Source*: After Ebert (2009b)

In general, the palaeosurface remnants are best preserved at intermediate elevations, probably because their preservation under cold-based ice during glaciation was favoured in those localities. Figure 15.3 shows a tentative sequence of events that led to the formation of the five palaeosurfaces.

Palaeoplains and their development

Erosion surfaces

Denudation chronologists once eagerly sought erosion surfaces. However, the search for erosion surfaces became unfashionable, particularly in British geomorphological circles, during the second half of the twentieth century, with many geomorphologists questioning their existence. The current consensus is that they do exist, and a revival of interest in them is apparent. As Ollier (1981, 152) not so tactfully put it, 'Most people who are not blind or stupid can tell when they are in an area of relatively flat country: they can recognize a plain when they see one'. Of course, a plain may be depositional, constructed from successive layers of alluvial, lacustrine, marine, or other sediments. Erosional plains that cut across diverse bedrock types and geological structures are all planation surfaces of some kind. They occur in low-lying areas and at elevation. Elevated plains sometimes bear signs of an erosional origin followed by subsequent dissection. A good example is a **bevelled cuesta**. Here, the flat top or bevel on a cuesta is credible evidence that an upper erosion surface, sitting at about the level of the bevel, existed before differential erosion

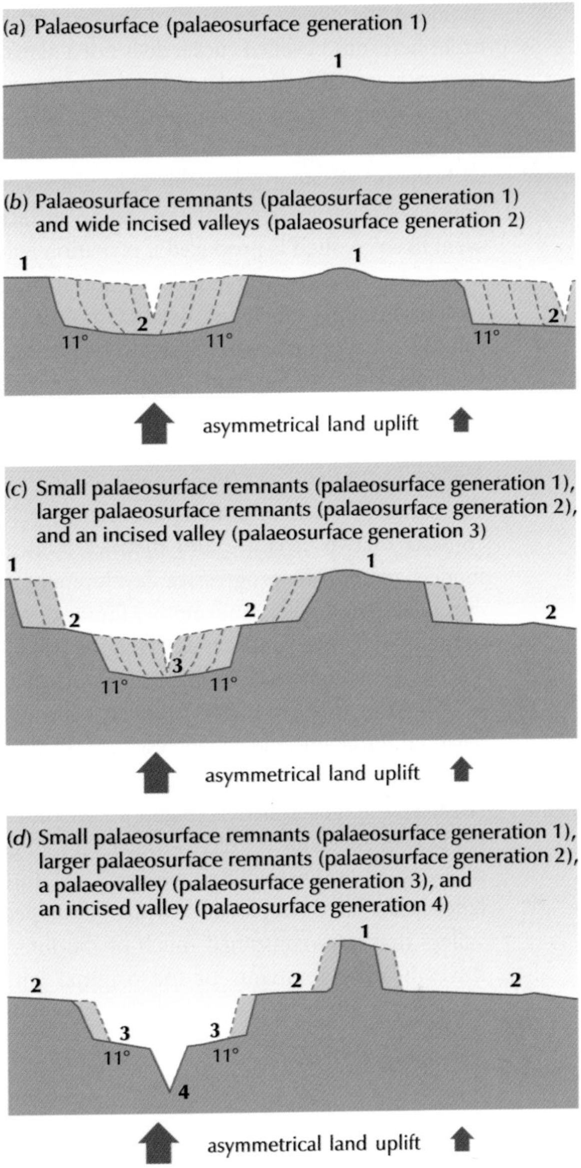

Figure 15.3 Karin Ebert's schematic development of surface generations in the Tjeuralako Plateau area of northern Sweden. Incision and slope backwearing adapted the landscape to a new baselevel after land uplift. Land uplift was asymmetrical in the northern Scandes. This caused an inclination of the landscape of up to 0.5° and had consequently only minor effects on palaeosurface inclination. Glacial erosion is not included. (a) A gently undulating palaeosurface, surface generation 1. (b) Watercourses incise wide valleys (surface generation 2) into the palaeosurface, caused by baselevel lowering after land uplift. Pediments develop through slope backwearing. The pediment angle is steepest at locations close to the escarpment. (c) A new valley is incised to the current base level (surface generation 3). Surface generations 1 and 2 are only left in remnants. Valley pediments close to the valley sides have the maximum pediment angle of 11°. (d) The valley centre is eroded through renewed valley incision after land uplift. A valley-in-valley pattern develops. Valley pediments of surface generation 3 are left as valley benches in the newly incised valley of surface generation 4. *Source*: After Ebert (2009b)

moulded the cuesta. A word of warning is in order here: one bevelled cuesta does not a planation surface make. An isolated bevel might have been a river terrace or some other small flat feature. Only when many bevelled cuestas occur, with the bevels all at about the same elevation, does the former existence of a planation surface seem likely. A **shelf** is produced if planation fails to remove the entire top of a cuesta and instead erodes a bench. A much discussed example is the early Pleistocene bench on the North Downs and Chiltern Hills of England. Plateaux are also elevated plains.

Peneplains

The Davisian system of landscape evolution (p. 9) consists of two separate and distinct cyclical models, one for the progressive development of erosional stream valleys and another for the development of whole landscapes (Higgins 1975). Valleys are thought to be V-shaped in youth, flat-bottomed in maturity, after lateral erosion has become dominant, and to possess very shallow features of extensive plains in old age, after lateral erosion has removed all hills (Figure 15.4). Young landscapes are characterized by much flat topography of the original uplifted peneplain. Mature landscapes have deeper and wider V-shaped valleys that have consumed much of the interfluves bearing remnants of the original land surface. Old landscapes are characterized by a **peneplain**, in which the interfluves are reduced to minor undulations (Figure 15.4).

Pediplains

Penck's model of slope retreat was adopted by Lester Charles King, who, in another model of landscape evolution, proposed that slope retreat produces pediments and that, where enough pediments form, a **pediplain** results (King 1953, 1967, 1983). King envisaged 'cycles of pedimentation'. Each cycle starts with a sudden burst of cymatogenic diastrophism and passes into a period of diastrophic quiescence, during which subaerial processes reduce the relief to a pediplain. However, cymatogeny and pediplanation are interconnected. As a continent is denuded, so the eroded sediment is deposited offshore. With some sediment removed, the continental margins rise. At the same time, the weight of sediment in offshore regions causes depression. The concurrent uplift and depression institutes the development of a major scarp near the coast that cuts back inland. As the scarp retreats, leaving a pediplain in its wake, it further unloads the continent and places an extra load of sediment offshore. Eventually, a fresh bout of uplift and depression will produce a new scarp. Thus, because of the cyclical relationship between continental unloading and the offshore loading, continental landscapes come

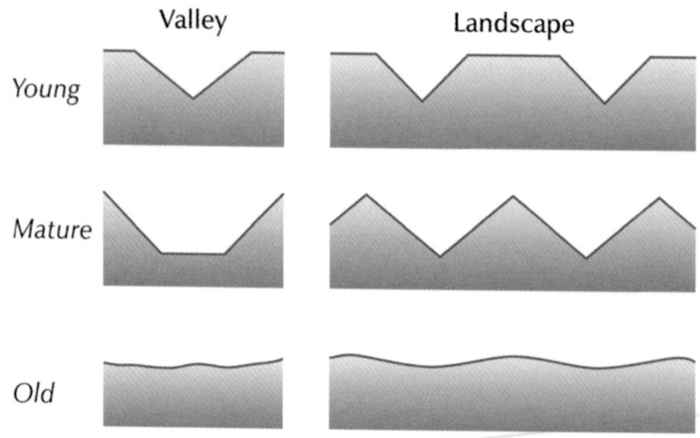

Figure 15.4 Traditional Davisian stage names for valley profiles and for landscape profiles. *Source*: Adapted from Ollier and Pain (1996, 204, 205)

to consist of a huge staircase of erosion surfaces (pediplains), the oldest steps of which occur well inland.

Panplains

Another variation on slope retreat concerns the notion of **unequal activity** espoused by Colin Hayter Crickmay (1933, 1975). Davis's, Penck's, and King's models of landscape evolution assume that slope processes act evenly on individual slopes. However, geomorphic agents act unequally. For this reason, a slope may recede only where a stream (or the sea) erodes its base. If this should be so, then slope denudation is largely achieved by the lateral corrasion of rivers (or marine erosion at a cliff foot). This will mean that some parts of the landscape will stay virtually untouched by slope recession. Some evidence supports this contention (p. 63). Crickmay opined that lateral planation by rivers creates panplains.

Etchplains and etched surfaces

Traditional models of landscape evolution assumed that mechanical erosion predominates. It was realized that chemical weathering reduces the mass of weathered material, but only on rocks especially vulnerable to solution (such as limestones) were chemical processes thought to have an overriding influence on landscape evolution. However, it now seems that forms of chemical weathering are important in the evolution of landscapes. **Groundwater sapping**, for instance, shapes the features of some drainage basins (e.g. Howard *et al.* 1988). And the solute load in three catchments in Australia comprised more than 80 per cent of the total load, except in one case where it comprised 54 per cent (Ollier and Pain 1996, 216). What makes these figures startling is that igneous rocks underlay the catchments. Information of this kind is making some geomorphologists suspect that chemical weathering plays a starring role in the evolution of nearly all landscapes.

In tropical and subtropical environments, chemical weathering produces a thick regolith that erosion then strips (Thomas 1989a, 1989b, 1994). This process is called **etchplanation**. It creates an **etched plain** or **etchplain**. The interface between the weathered saprolite and the unweathered bedrock is the **etch surface**, which is exposed after stripping takes place. The etchplain is largely a production of chemical weathering. In places where the regolith is deeper, weakly acid water lowers the weathering front, in the same way that an acid-soaked sponge would etch a metal surface. Some researchers contend that surface erosion lowers the land surface at the same rate that chemical etching lowers the weathering front (Figure 15.5). This is the theory of **double planation**. It envisages land surfaces of low relief being maintained during prolonged,

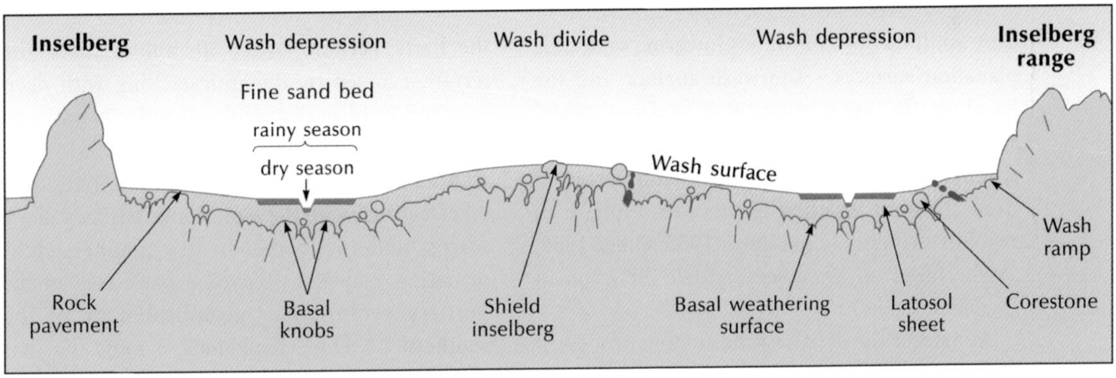

Figure 15.5 Double planation surfaces: the wash surface and the basal-weathering surface. *Source*: Adapted from Büdel (1982, 126)

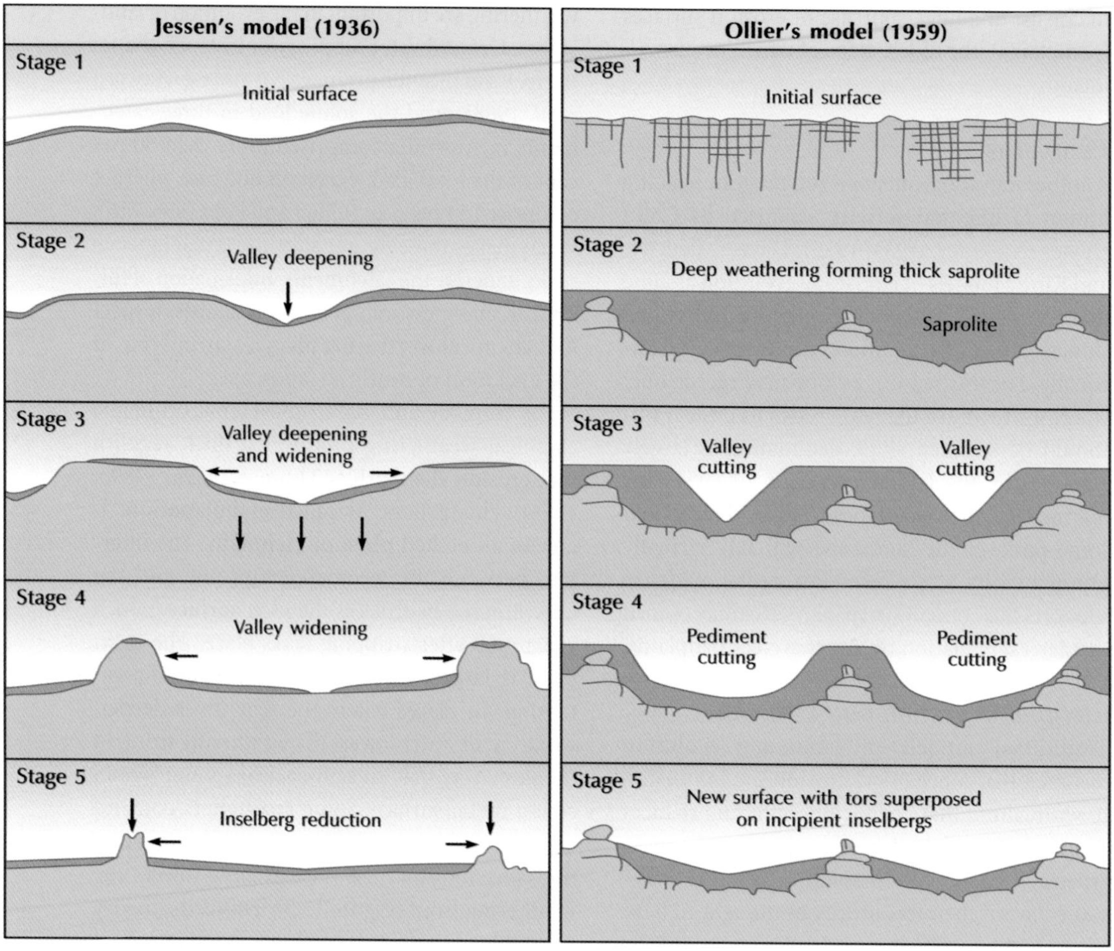

Figure 15.6 Theories of etchplanation. *Source*: Adapted from Jessen (1936) and Ollier (1995)

slow uplift by the continuous lowering of double planation surfaces – the **wash surface** and the **basal weathering surface** (**etch surface**) (Büdel 1957, 1982; Thomas 1965). A rival view, depicted schematically in Figure 15.6, is that a period of deep chemical weathering precedes a phase of regolith stripping (e.g. Linton 1955; Ollier 1959, 1960; Hill *et al.* 1995; see Twidale 2002 for an excellent review).

Whatever the details of the etching process, it is very effective in creating landforms, even in regions lying beyond the present tropics. The Scottish Highlands experienced a major uplift in the Early Tertiary. After 50 million years, the terrain evolved by dynamic etching with deep weathering of varied geology under a warm to temperate humid climate (Hall 1991). This etching led to a progressive differentiation of relief features, with the evolution of basins, valleys, scarps, and inselbergs. In like manner, etch-planation may have played a basic role in the Tertiary evolutionary geomorphology of the southern England Chalklands, a topic that has always generated much heat. There is a growing recognition that the fundamental erosional surface is a summit surface formed by etchplanation

Box 15.1 TERTIARY LANDSCAPE EVOLUTION IN SOUTH-EAST ENGLAND

The study of Tertiary landscape evolution in southern Britain nicely shows how emphasis in historical geomorphology has changed from land-surface morphology being the key to interpretation to a more careful examination of evidence for past geomorphic processes. As Jones (1981, 4–5) put it,

> [this] radical transformation has in large part resulted from a major shift in methodology, the heavily morphologically-biased approach of the first half of the twentieth century having given way to studies that have concentrated on the detailed examination of superficial deposits, including their faunal and floral content, and thereby provided a sounder basis for the dating of geomorphological events.

The key to Wooldridge and Linton's (1939, 1955) classic model of landscape evolution in Tertiary south-east England was three basic surfaces, each strongly developed on the chalkland flanks of the London Basin (Figure 15.7). First is an inclined, recently exhumed, marine-trimmed surface that fringes the present outcrop of Palaeogene sediments. Wooldridge and Linton called this the Sub-Eocene Surface (it is now more accurately termed the Sub-Palaeogene surface). Second is an undulating Summit Surface lying above about 210 m, mantled with thick residual deposits of 'Clay-with-Flints', and interpreted by Wooldridge and Linton as the remnants of a region-wide subaerial peneplain, as originally suggested by Davis in 1895, rather than a high-level marine plain lying not far above the present summits. Third is a prominent, gently inclined erosional platform, lying between about 150 and 200 m and cutting into the Summit Surface and seemingly truncating the Sub-Eocene Surface. As it bears sedimentary evidence of marine activity, Wooldridge and Linton interpreted it as a Pliocene marine plain. Wooldridge and Linton believed that the two higher surfaces – the Summit Surface and the marine platform – were not warped. They argued, therefore, that these surfaces must have formed after the tectonic episode that deformed the Sub-Eocene Surface, and that the summit plain had to be a peneplain fashioned during the Miocene and Pliocene epochs.

The Wooldridge and Linton model of Tertiary landscape evolution was the ruling theory until at least the early 1960s and perhaps as late as the early 1970s. Following Wooldridge's death in 1963, interest in the long-term landform evolution of Britain – or denudation chronology, as many geomorphologists facetiously dubbed it – waned. Critics accused denudation chronologists of letting their eyes deceive them: most purely morphological evidence is 'so ambiguous that theory feeds readily on preconception' (Chorley 1965, 151). However, alongside the denigration of and declining interest in denudation chronology, some geomorphologists reappraised the evidence for long-term landscape changes. This fresh work led in the early 1980s to the destruction of Wooldridge and Linton's 'grand design' and to the creation of a new framework that discarded the obsession with morphological evidence in favour of the careful examination of Quaternary deposits (Jones 1999, 5–6). The reappraisal was in part inspired by Philippe Pinchemel's (1954) alternative idea that the gross form of the chalk backslopes in southern England and northern France results from intersecting Palaeogene erosion surfaces that suffered exhumation and modification during the Neogene and the Quaternary times. Foremost among

continued . . .

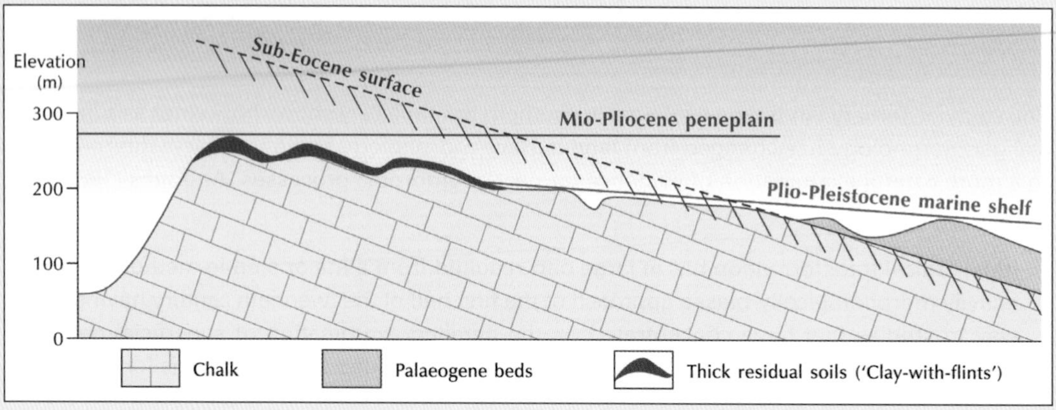

Figure 15.7 A sketch section of a chalk cuesta depicting the three basic surfaces of Wooldridge and Linton (1955). *Source*: Adapted from Jones (1981)

the architects of the new model of Tertiary landscape evolution in southern England were David Jones (1981) and Chris Green (1985). Jones (1999) confirmed this model in a region-wide synthesis, which, paraphrasing Jones's explanation, runs thus (Figures 15.8 and 15.9):

1 As a result of a combination of a eustatic fall in sea level and tectonic deformation, the deposition of a continuous and thick (up to 550 m) sheet of Upper Cretaceous chalk ceased in the Maastrichtian, and dry land had probably emerged by 65 million years ago.
2 Palaeocene denudation rapidly stripped up to 350 m of chalk, with the severest erosion on such uplift axes as the Weald and the Channel High, and in the west, where subaerial denudation under tropical climatic conditions quickly removed most of a sizeable chalk layer. A combination of eustatic fluctuations and tectonic movements led to a progressive encroachment of marine conditions from the east, starting with the Thanet Sands during the Palaeocene epoch around 57 million years ago, and ending in a nearly complete inundation by the London Clay sea in the Early Eocene epoch, some 53 million year ago (Pomerol 1989). The Palaeocene and Early Eocene sediments accumulated on a multifaceted or polycyclic, marine-trimmed Sub-Palaeogene Surface cut in chalk. The only exception is the extreme west, where the Upper Greensand had been exposed by the close of the Palaeocene epoch beneath a widespread etchplain, the lower parts of which were easily submerged by the transgressing London Clay sea.
3 Continuing pulses of tectonic deformation throughout the remainder of the Palaeogene period saw the further definition of the structural basins (London and Hampshire–Dieppe Basins) due to the progressive growth of the Weald–Artois Anticline and the Isle of Wight Monocline. Sub-aerial erosion on the axes of these upwarps led to the development of further facets of the Sub-Palaeogene Surface, while sedimentation was progressively limited to the basin areas and ultimately restricted to the Hampshire Basin in the Oligocene. Elsewhere, sub-aerial erosion under the hot climatic conditions of the Eocene epoch created an extensive etchplain with duricrusts over most of the present chalklands that sat at an elevation a few tens of metres

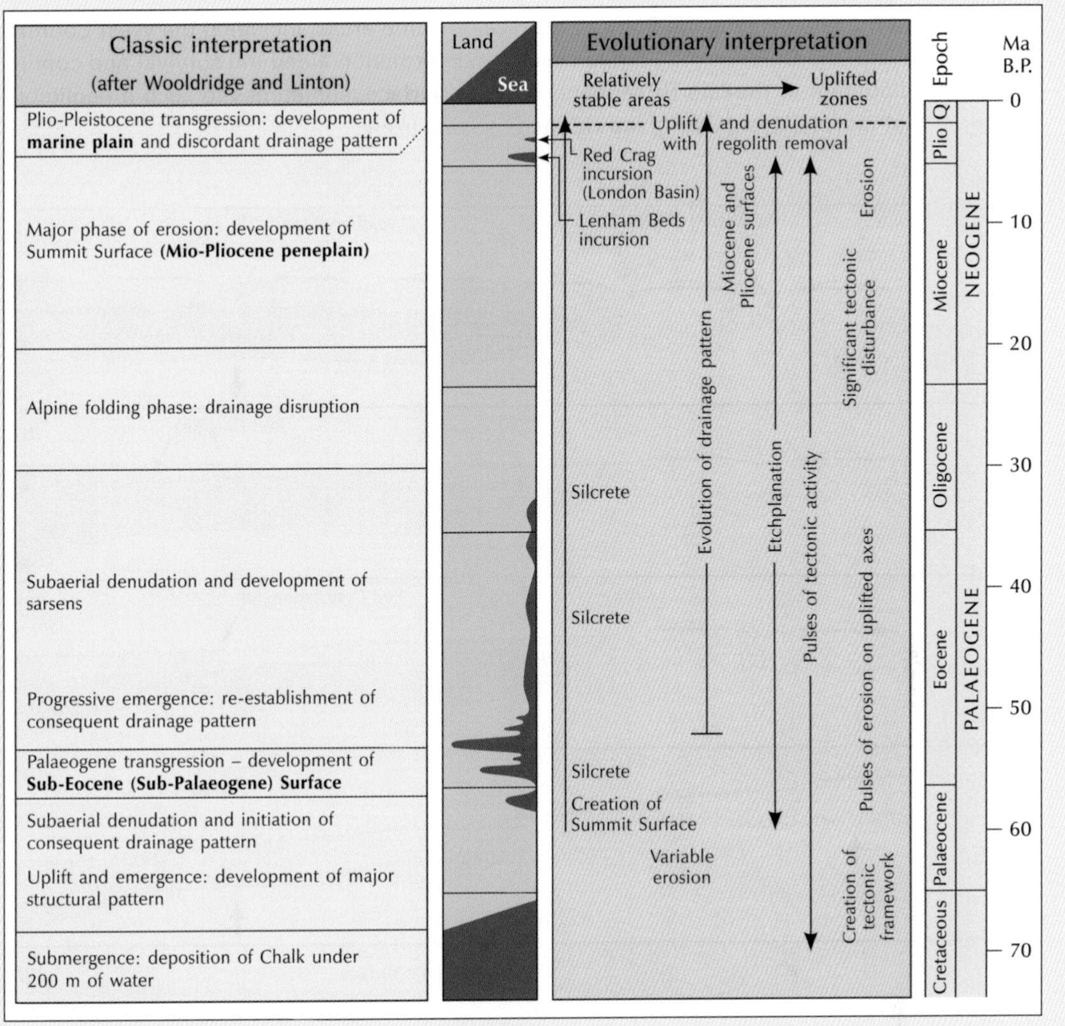

Figure 15.8 Classic and evolutionary interpretations of Tertiary landscape evolution in southern England. *Source*: Adapted from Jones (1999)

above the highest present summits. In the west, this surface had originated in the Palaeocene epoch and slowly evolved through a predominantly morphostatic episode, but to the east it had evolved through erosion of the previously deposited Palaeogene cover and chalk, especially in areas subject to tectonic movements.

4 Rather more pronounced tectonic deformation in the early Miocene epoch saw the further development of the structural pattern to almost its present form. The uplift of upwarped areas and major inversion axes generated both relief and erosion, so that denudation accelerated on the Channel High and Weald–Artois Anticline. Elsewhere, the lesser scale of tectonic movements resulted in the development of Neogene surfaces (Miocene and Pliocene) at the

continued . . .

expense of the late Palaeogene Summit Surface. In some areas, including the West Country tablelands and the Chilterns, the limited scale of deformation aided the survival and continuing development of the late Palaeogene Summit Surface, sometimes to such a negligible degree that morphostasis during the Neogene seems likely.

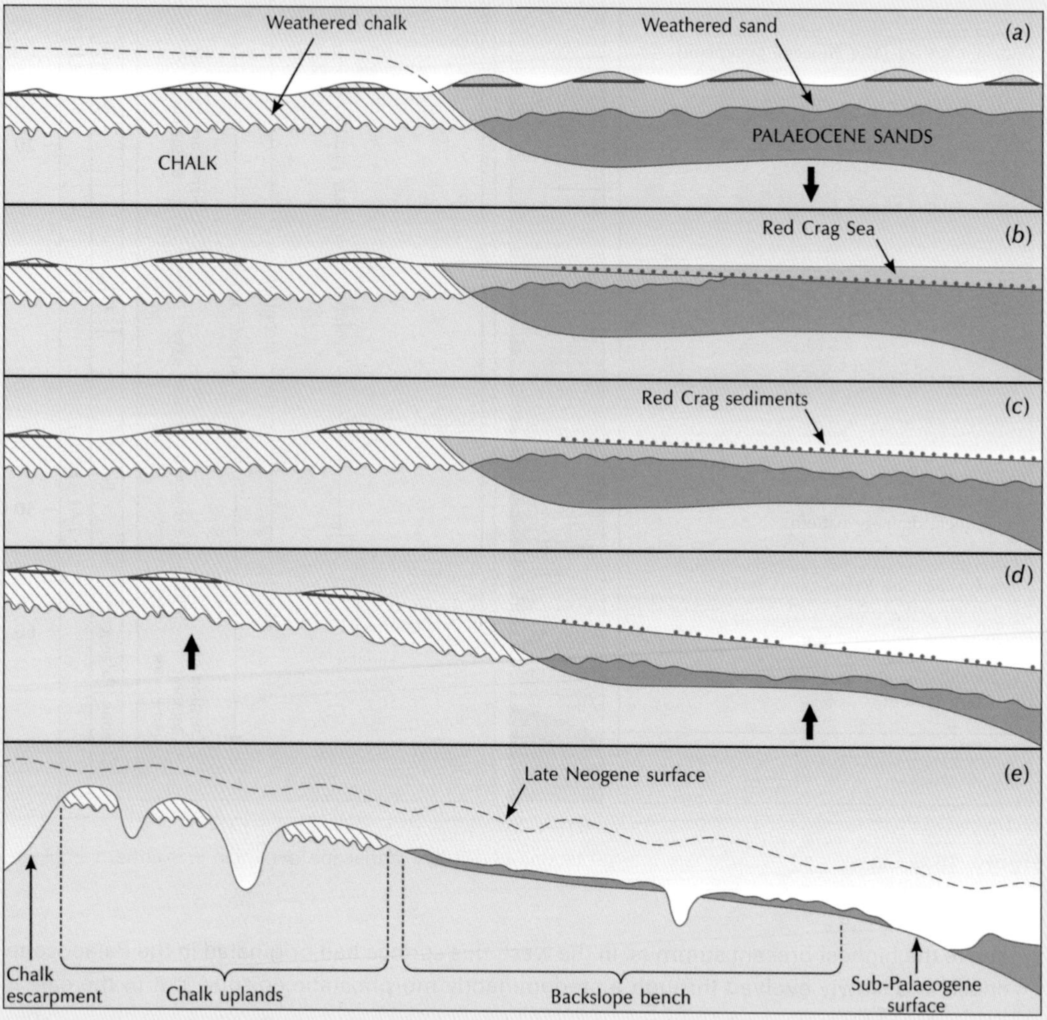

Figure 15.9 Cartoon depiction of a possible evolutionary model for the prominent backslope bench of the London Basin involving modified etchplanation. (a) Neogene duricrusted surface of low relief. (b) Red Crag incursion (see Figure 9.30) following downwarping to the east. (c) After marine regression. (d) Accentuated weathering beneath former marine platform leads to the development of an etchsurface. (e) Differential uplift in the Pleistocene leads to dissection and the removal of much regolith with the remainder greatly disturbed and mixed by periglaciation. *Source*: After Jones (1999). Reproduced by permission of the Geological Society, London, and David Jones

5 The extent to which Pliocene marine transgressions invaded the lower and flatter portions of the Late Neogene land surface remains controversial. There exists evidence for a Red Crag (Pre-Ludhamian) incursion that affected the London Basin and eastern parts of the Weald, but growing uncertainty as to the validity of an earlier Lenham Beds incursion. The Red Crag incursion appears the consequence of relatively localized downwarping, and is considered to have caused minimal erosion, but sufficient to disrupt the surface duricrusts.
6 The subsequent marine regression revealed a gently inclined marine plain that suffered lowering through etchplanation that fashioned the prominent 'platforms' exposed on the flanks of the London Basin.
7 At some time after 2 million years ago, further uplift and warping occurred, possibly as discontinuous pulses through much of the Pleistocene epoch. While eastern East Anglia suffered subsidence, the remainder of southern England experienced differential uplift to a maximum of at least 250 m (Jones 1999) and possibly up to 400 m (Preece *et al.* 1990). This uplift, in conjunction with oscillating sea levels and climatic fluctuations, resulted in episodic erosion that increased in scale as relative relief developed through the Pleistocene.

Jones (1999) owns that this evolutionary model needs substantiating in a number of important regards. Uncertainty surrounds the true nature of the structural foundations of the area and its tectonic evolution, which fuels the continuing controversy over the temporal and spatial dimensions of uplift and the relative importance of Mid-Tertiary (Miocene) tectonic activity. Likewise, the recent suggestions of Pleistocene uplift and warping need confirming, elaborating, and accurately dating. Moreover, the nature and palaeoenvironmental significance of residual soils, including the varied types of silcrete, and the number, age, and geographical extent of Neogene marine incursions, most especially the baffling Lenham Beds incursion, still demand much investigation. Only after they complete this further work will geomorphologists be able to establish a fully detailed evolutionary geomorphology for this well-known region.

during the Palaeogene period, and is not a peneplain formed during the Miocene and Pliocene periods (Box 15.1).

Exhumed surfaces

Exhumed landscapes and **landforms** are common, preserved for long periods beneath sediments and then uncovered by erosion. They are common on all continents (e.g. Lidmar-Bergström 1989, 1993, 1995, 1996; Twidale 1994; Thomas 1995). **Exhumed erosion surfaces** are quite common. The geological column is packed with unconformities, which are marked by surfaces dividing older, often folded rocks from overlying, often flat-lying rocks. Some unconformities seem to be old plains, either peneplains formed by coastal erosion during a marine transgression or by fluvial erosion, or else etchplains formed by the processes of etchplanation. The overlying rocks can be marine, commonly a conglomerate laid down during a transgression, or terrestrial. The unconformity is revealed as an exhumed erosion surface when the overlying softer rocks are removed by erosion. It is debatable how the exhumed erosion surface relates to landscape evolution. If a thin cover has been stripped, then the old erosion surface plays a large part in the modern topography, but where hundreds or thousands of metres of overlying strata have been removed the exhumed erosion surface is all but a chance component part of the modern landscape, much like any other structural surface (Ollier

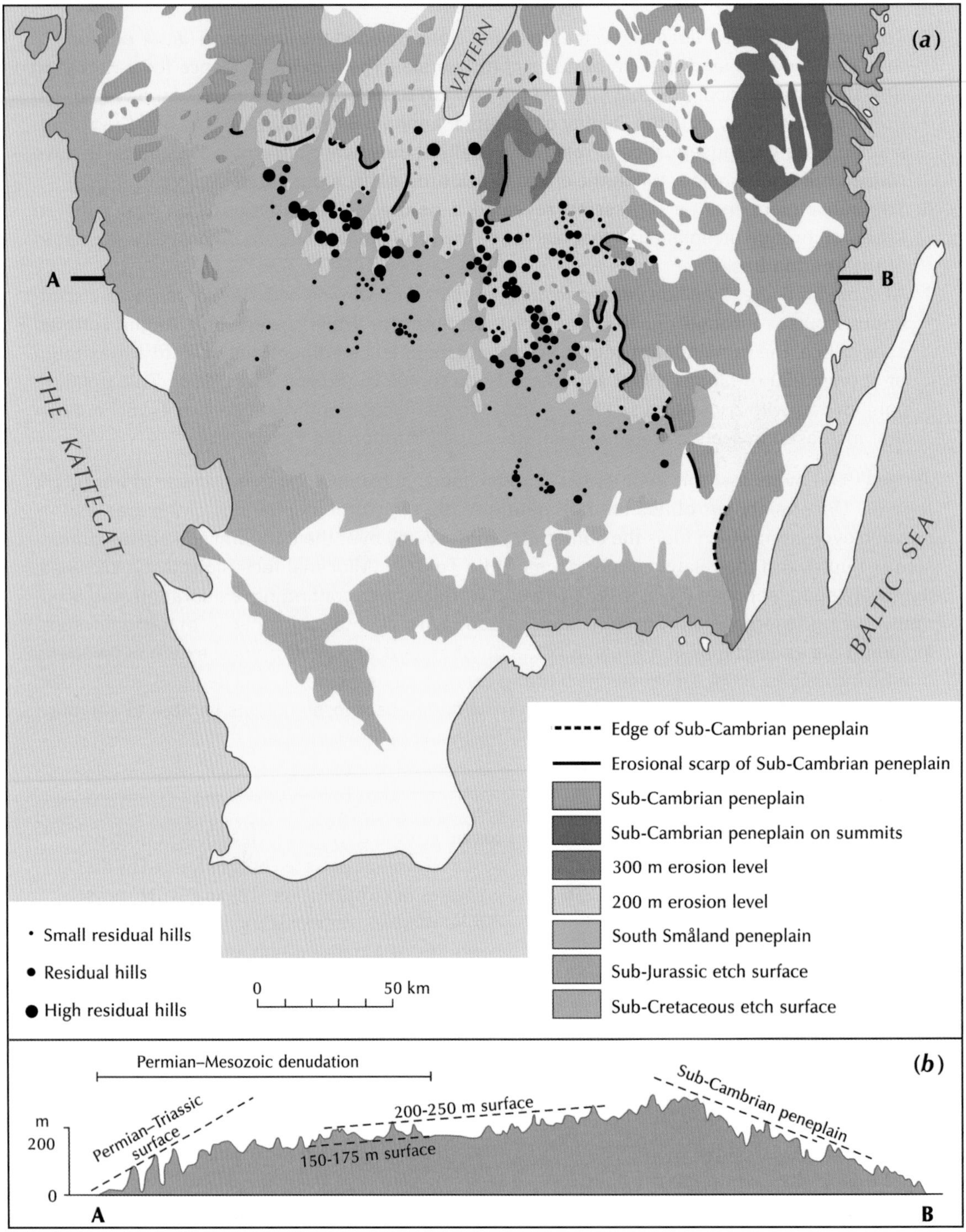

Figure 15.10 Denudation surfaces and tectonics of southern Sweden. (a) Mapped features. (b) West–east profile across the dome-like uplift of the southern Baltic Shield. Note the exhumed sub-Cretaceous hilly relief evolved from the Permo-Triassic surface. *Source*: Adapted from Lidmar-Bergström (1993, 1996)

1991, 97). The Kimberley Plateau of Western Australia bears an erosion surface carrying striations produced by the Sturtian glaciation some 700 million years ago and then covered by a glacial till. The thin till was later stripped to reveal the Kimberley surface, the modern topography of which closely matches the Precambrian topography and displays the exhumed striations (Ollier 1991, 24).

The relief differentiation on the Baltic Shield, once thought to result primarily from glacial erosion, is considered now to depend on basement-surface exposure time during the Phanerozoic aeon (Figure 15.10; Plate 15.2a–d). Three basic relief types occur on the Fennoscandian Shield (Lidmar-Bergström 1999). The first is the exhumed and extremely flat sub-Cambrian peneplain, which with sub-Vendian

Plate 15.2 (a) Exhumed sub-Cambrian peneplain 4 km west of Cambrian cover in south-east Sweden. A glaciofluvial deposit has been removed. Glacial striations are seen on the rock surface. The flat rock surface is often exposed or covered by a mainly thin layer of Quaternary deposits over large areas. (b) A regional view over the exhumed sub-Cambrian peneplain in south-east Sweden. The peneplain is seen towards the east from Aboda klint, and continues 30 km west of the border with the Cambrian cover. The peneplain is here 100 m above sea level and descends to the coast, where it disappears under Cambrian cover. The island of Jungfrun, 50 km away, can be seen from here. It is a residual hill on the sub-Cambrian peneplain that protrudes through the Cambrian cover on the sea bottom in Kalmarsund. (c) Exhumed sub-Cretaceous granite hill, Ivöklack, in north-east Scania, southern Sweden. The hill rises about 130 m above the lake level. (d) Exhumed sub-Cretaceous hilly relief along the west coast in Halland, south-west Sweden. (*All four photographs by Karna Lidmar-Bergström*)

and sub-Ordovician facets has been the starting surface for all relief upon the shield (Figure 15.11a). The second is the exhumed sub-Mesozoic etchplains, which possess an undulating and hilly relief and vestiges of a kaolinitic saprolite and Mesozoic cover rocks (Figure 15.11b, c). The third is a set of plains with residual hills that are the end result of surface denudation during the Tertiary period (Figure 15.11d). Figure 15.12 depicts the likely evolution of bedrock relief in southern Sweden. In the late Precambrian era, denudation reduced the surface of the Precambrian bedrock to an extremely flat surface, the sub-Cambrian peneplain, with residual hills only occurring as exceptions (Figure 15.12a). Starting in the Cambrian period, the sea transgressed the peneplain and Cambrian rocks were deposited on the flat surface and were succeeded by Ordovician-Carboniferous strata (Figure 15.12b). These cover rocks protected the Precambrian basement in southern Sweden from further erosion for a long time. In the Kattegat area, a thick Palaeozoic cover accumulated in the Caledonian foreland basin. In the Permian period, inversion of the Caledonian foreland basin removed the Palaeozoic cover rocks in parts of the Kattegat area and probably in most of south-western Sweden (Figure 15.12c). Through the Triassic to the early Cretaceous periods, uplift and erosion continued and the climate became humid. Kaolinitic weathering penetrated deep into the basement along fracture zones (etching), producing thick kaolinitic weathering mantles (saprolites). By alternating etching and erosion of the saprolites (stripping), the landscape developed an undulating hilly relief with Palaeozoic remnants still occurring locally on down-faulted blocks (Figure 15.12d). From the late Cretaceous to the Mid-Miocene periods, the sea transgressed across Denmark and large parts of southern Sweden, covering the area with sediments (Figure 15.12e).

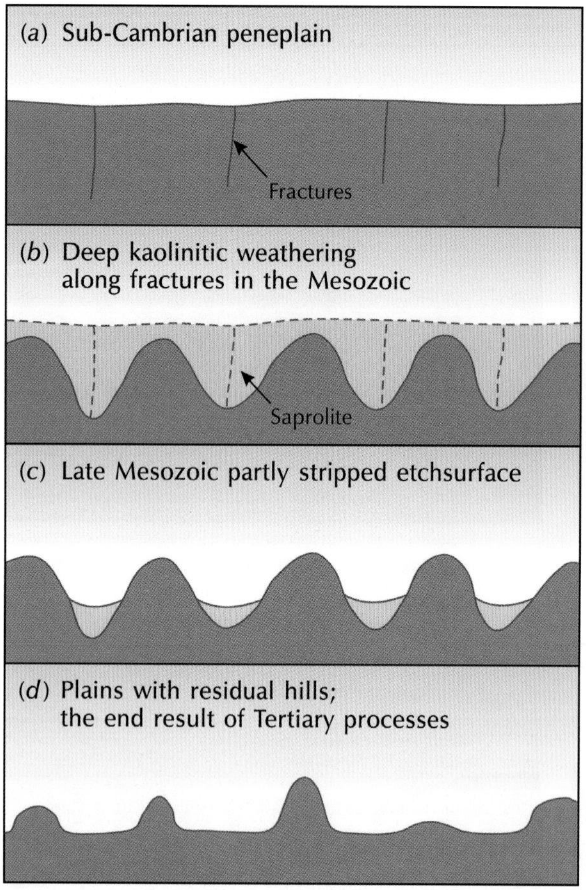

Figure 15.11 The three basic relief-types within the Fennoscandian Shield. (a) Sub-Cambrian peneplain. This exhumed and extremely flat palaeoplain, together with sub-Vendian and sub-Ordovician facets, is the starting point for all relief on the Fennoscandian shield. (b) Deep kaolinitic weathering along fractures in the Mesozoic. This is not a basic relief type, but led to (c) Late Mesozoic partly stripped etchplains, with a characteristic undulating hilly terrain and remnants of kaolinitic saprolite and Mesozoic cover rocks. (d) Late Tertiary plains with residual hills, which are the product of Tertiary surface denudation. *Source*: Adapted from Lidmar-Bergström (1999)

After a Neogene rise followed by erosion of Upper Cretaceous–Mid-Miocene cover rocks, the South Småland Peneplain developed, probably under Late Miocene dry climates, as a gently inclined flat rock surface with few residual hills (a pediplain) (Figure 15.12f). South- and west-facing escarpments formed along the elevated rim of the sub-Cambrian peneplain. Finally, after a last Neogene uplift episode, the sub-Cretaceous hilly relief was re-exposed along the coasts towards

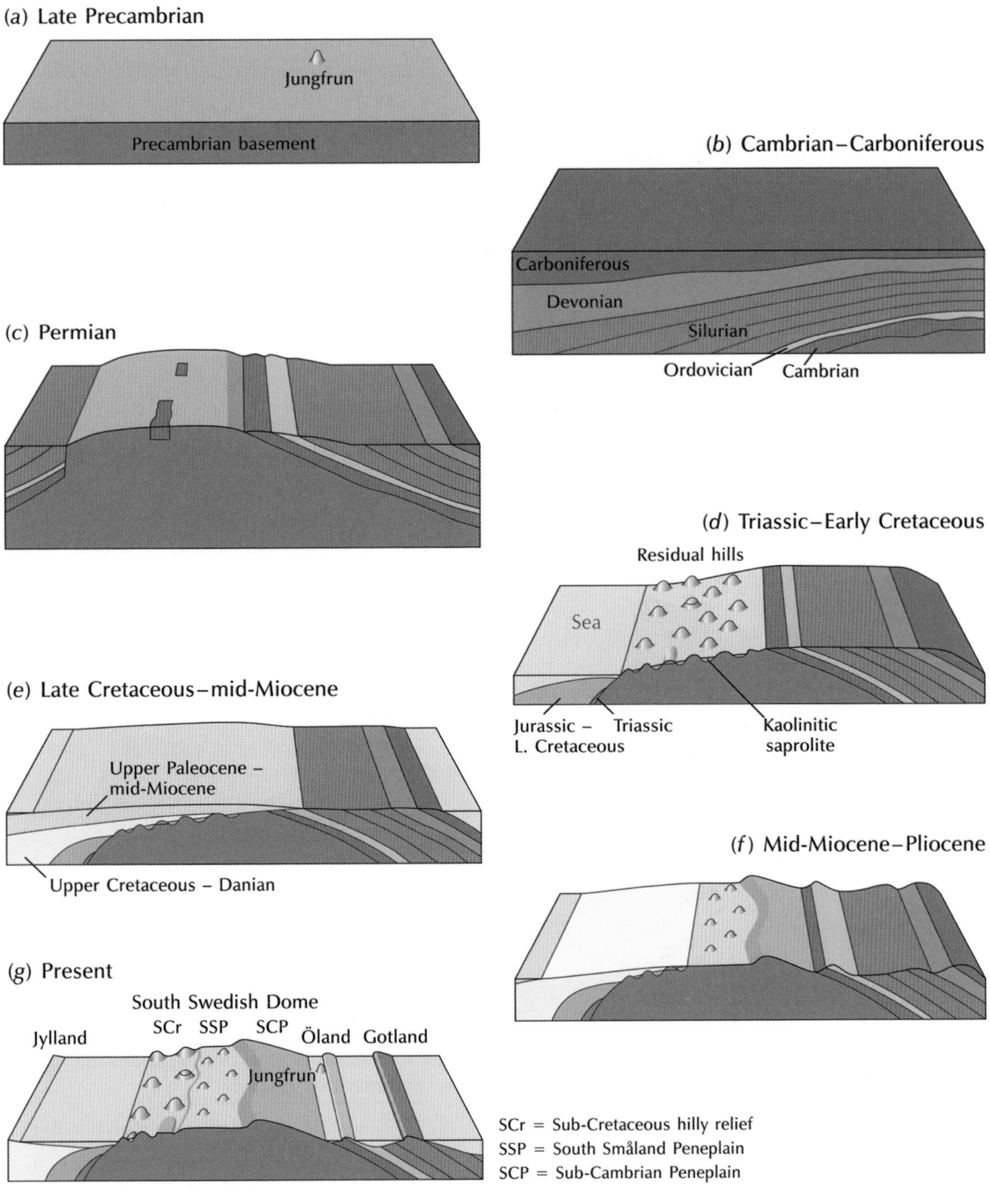

Figure 15.12 Block diagrams illustrating the development of bedrock relief in southern Sweden. (a) Late Precambrian. (b) Cambrian–Carboniferous. (c) Permian. (d) Triassic–Early Cretaceous. (e) Late Cretaceous–Mid-Miocene. (f) Mid-Miocene–Pliocene. (g) Present. *Source*: After Japsen *et al.* (2002) (see also Lidmar-Bergström 1994). Reproduced by permission of the Geological Society, London, and Peter Japsen

the south-east and the west, while the flat sub-Cambrian peneplain re-appeared at the surface in the northern and eastern part of south Sweden. Locally, sub-Cambrian facets reappeared in western Sweden (Figure 15.12g).

The Armorican Massif in northwest France is an extensive outcrop of Variscan (late Devonian to Carboniferous) basement. Its large-scale landforms consist of several planation surfaces of debatable origin and age (Figure 15.13). Paul Bessin and his colleagues (2015) studied the surfaces using detailed mapping, relative chronology, and relationships with dated outcrops of sediments and weathering products (laterite, grus, and silcretes). They propose an evolution model for these landforms and their deformation based on detailed mapping of planation surfaces, their relative chronology, and their relationships with dated outcrops of sediments and weathering products (Figure 15.14). In summary, they recognized

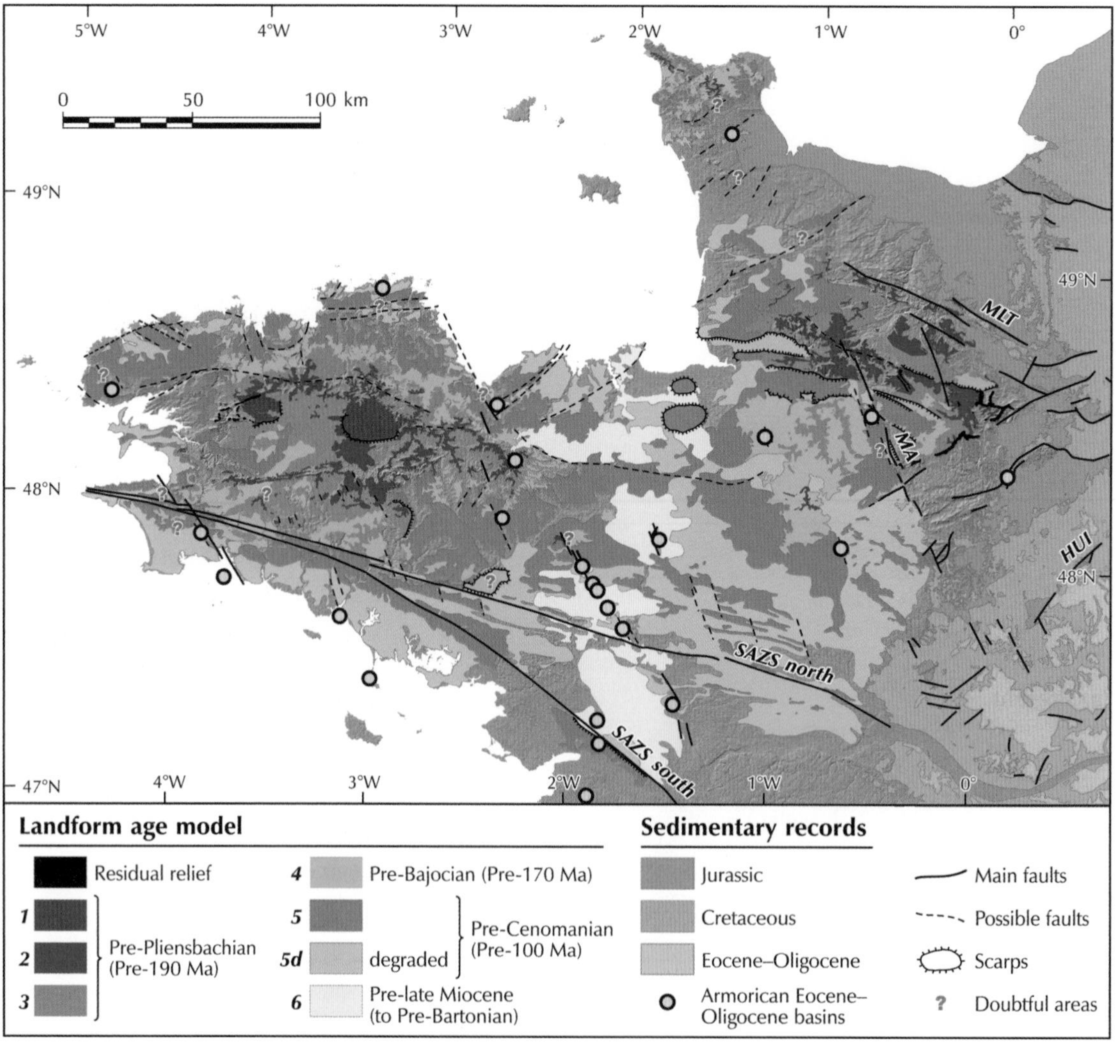

Figure 15.13 The Armorican Massif planation surfaces and their ages. *Source*: After Bessin *et al.* (2015)

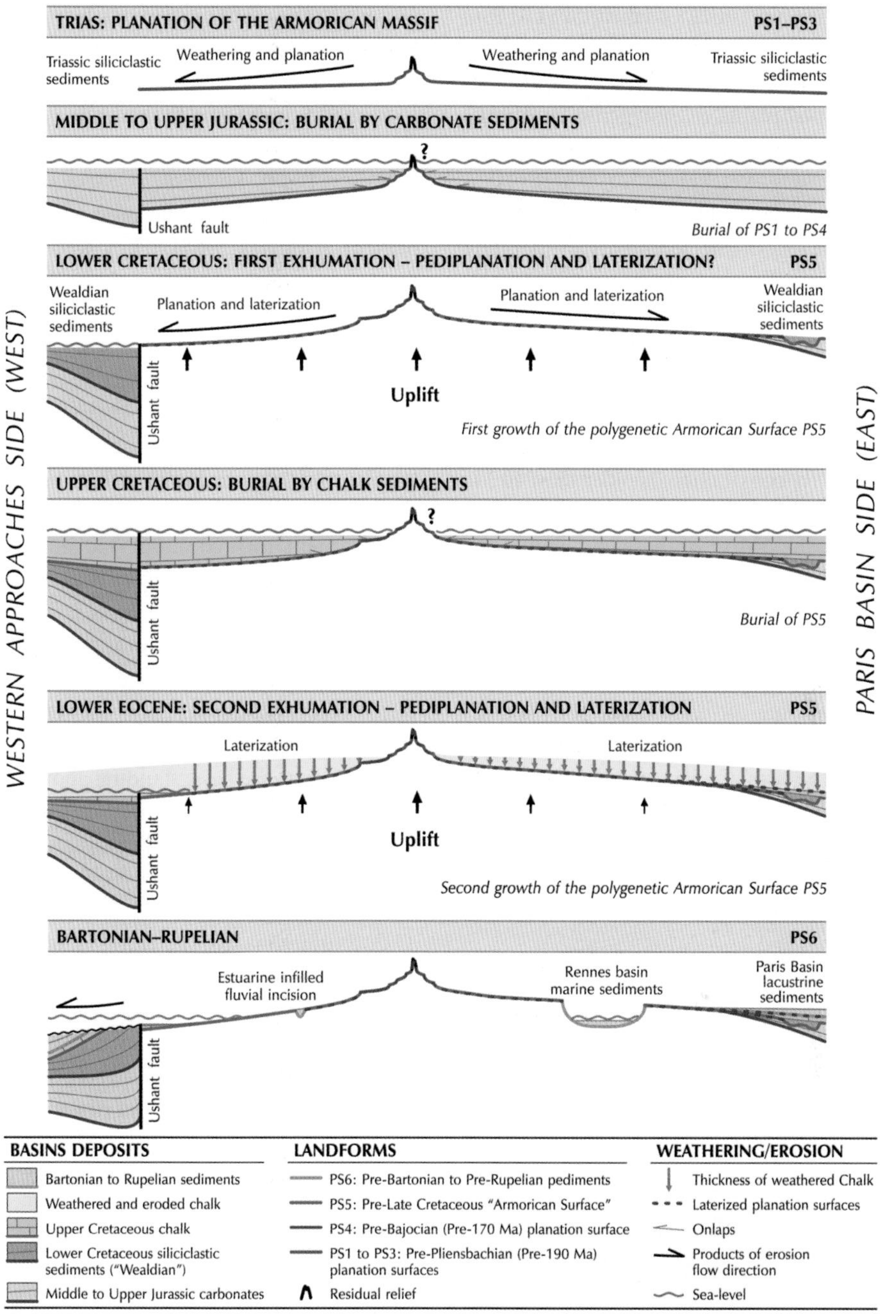

Figure 15.14 Long-term landscape evolution of the Armorican Massif from the Triassic to Oligocene periods, depicting the chief phases of burial and exhumation. The highly synthetic cross-section showing just one plateau (Lower Normandy/Western Brittany), with the Paris border on the left-hand side and the Western Approaches on the right hand-side. The pre-Bajocian period of denudation (pediments PS4) is not depicted. *Source*: After Bessin *et al.* (2015)

six stepped planation surfaces (labelled PS1 to PS6) with two successive river networks incising them later. They interpreted these landforms as pediments and pediplains or polygenic landforms (Armorican Planation Surface – PS5) resulting from two periods of etchplanation. These planation surfaces are mostly pre-late Cretaceous in age, based on the age of the sediments overlapping these pediments. The three older ones (PS1 to PS3) are pre-Pliensbachian (191–183Ma); PS4 is pre-Bajocian (170–168 Ma); PS5 (the Armorican Planation Surface) is polygenic and ranges from the base of the early Cretaceous to the base of the Bartonian (140–40 Ma); and the youngest (PS6), which is poorly constrained, is older than 40 Ma (base of the Bartonian) or 15 Ma (base of the middle Miocene). Most of these 'old' landforms are exhumed, that is they were buried by sediments and later re-exposed by denudation. At least two phases of burial and exhumation have been identified: (1) burial in Jurassic time followed by denudation during the early Cretaceous and (2) burial in late Cretaceous time followed by denudation during the latest Cretaceous to early Eocene. The depth of burial is unknown but is probably low due to the small amount of coeval siliciclastic sediments in the surrounding basins. The two periods of exhumation correspond to critical periods in the plate movements between Africa, Iberia, and Eurasia. The first is probably related to the initiation and breakup of the rift between Iberia and Eurasia, and the second to the convergence between these two plates.

Plate 15.3 Cawden Hill, an Albian (approximately 335-million-year old) reef knoll near Malham, England. *(Photograph by Richard Huggett)*

In northern England, a variety of active, exhumed, and buried limestone landforms are present (Douglas 1987). They were originally created by sedimentation early in the Carboniferous period (late Tournaisian and early Viséan ages). Subsequent tectonic changes associated with a tilt-block basement structure have effected a complex sequence of landform changes (Figure 15.15). The Waulsortian knolls (named after Waulsort in Belgium, home of the type-section of such deposits) are exhumed mounds of carbonate sediment formed about 350 million years ago. Shales and later chalk covered them, and then exhumation during the Tertiary period produced **reef knoll hills**, which are features of the present landscape (Plate 15.3; see Figure 14.12). In the Clitheroe region, they form a series of isolated hills, up to 60 m high and 100–800 m in diameter at the base, standing above the floor of the Ribble valley. The limestone fringing reefs formed in the Asbian and Brigantian ages today form prominent reef knoll hills close to Cracoe, Malham, and Settle.

Carboniferous sedimentation in the southernmost section of the Gaspé Peninsula in eastern Quebec, Canada, has fossilized a palaeosurface – the Saint Elzéar surface – that erosion is now gradually exhuming (Jutras and Schroeder 1999). Part of the surface is a nearly perfect planation surface, cut between 290 and 200 million years ago, a time spanning the Permian and Jurassic periods. The planation surface, which cuts horizontally across all geological structures, has suffered little dissection (Figure 15.16). The exhumation of the surface must also have begun by the Jurassic period following the *en bloc* uplift of the evolving Atlantic Ocean's passive margins. Some geomorphic features on the exhumed palaeosurface are guides to Carboniferous palaeoenvironments and tectonics in the area. The Saint

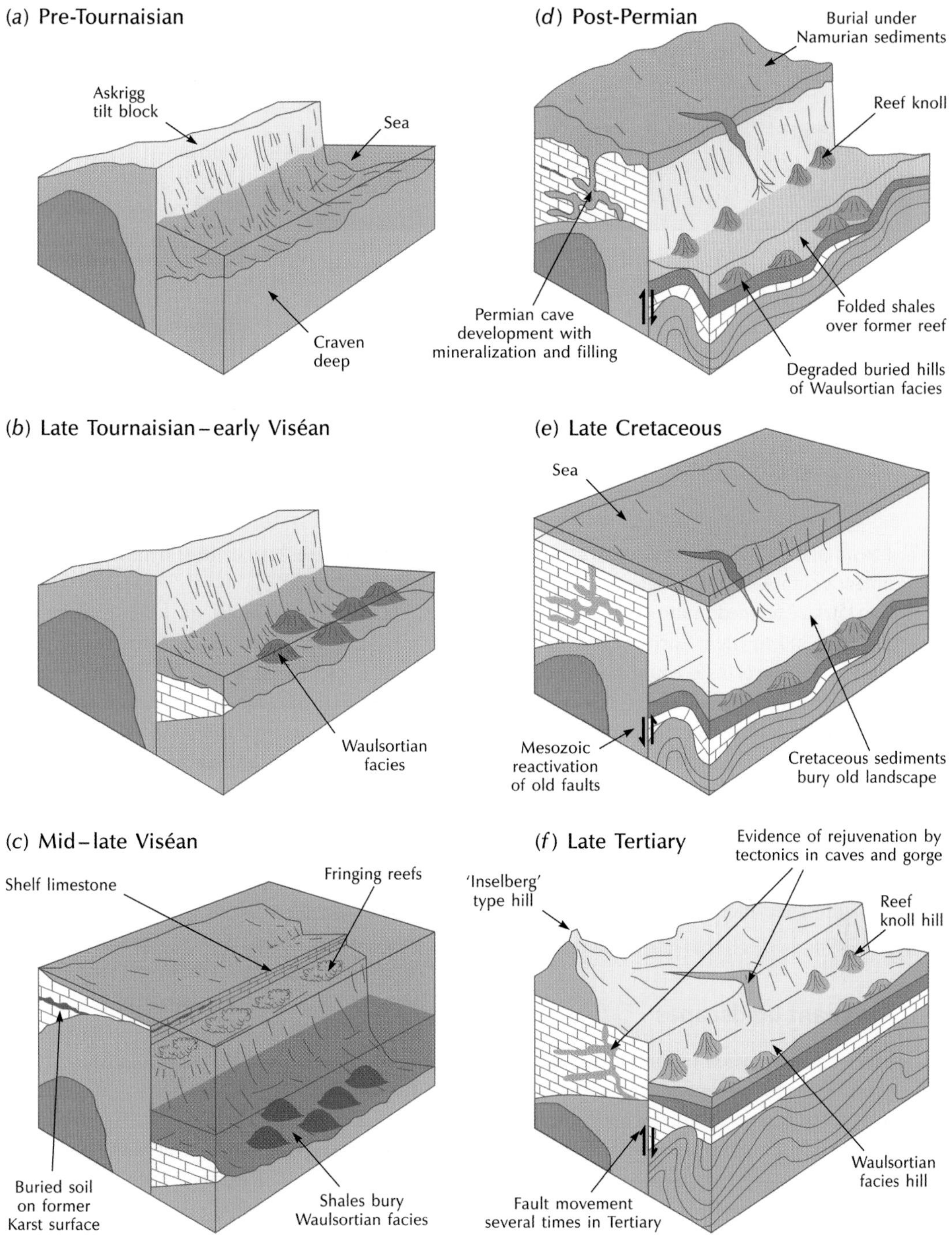

Figure 15.15 Schematic diagram showing the evolution of limestone landscapes in northern England. The Craven Fault, a major east–west fault across the northern Pennines, separates the Askrigg tilt-block from the Craven Deep. *Source*: Adapted from Douglas (1987)

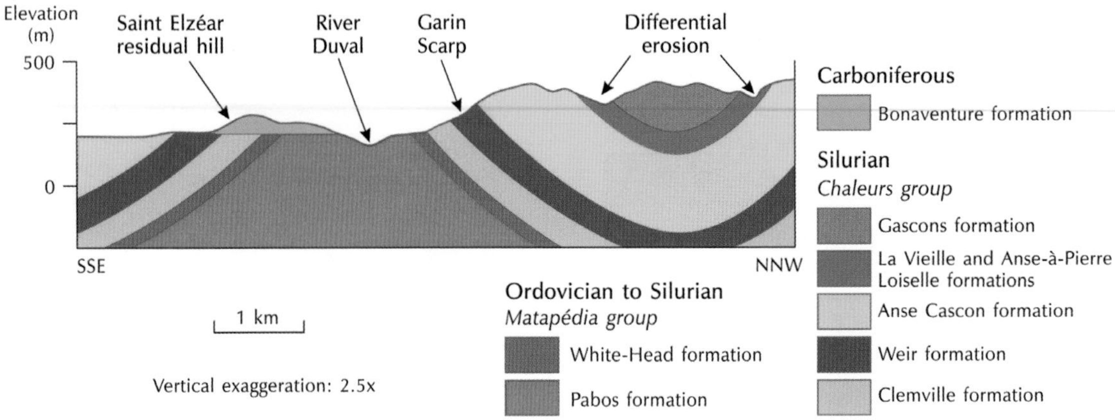

Figure 15.16 Cross-section across the Saint Elzéar region, eastern Quebec, Canada. The palaeosurface, which has been exhumed by erosion, can be seen to the south-south-east of the Garin Scarp cutting across Ordovician and Silurian rocks. *Source*: Adapted from Jutras and Schroeder (1999)

Elzéar planation surface is separated from the uplands of the Gaspesian Plateau – a higher planation surface formed in the same formations – by the 200–300-m-high Garin Scarp. So far as is known, four processes could have produced an erosion surface bounded by a scarp: faulting, etching and double planation, rock pedimentation controlled by differential erosion, and coastal erosion by a transgressive sea. Pierre Jutras and Jacques Schroeder (1999) favour the latter process and interpret the erosion surface as a wide wave-cut platform produced by the Windsor transgression. They interpret the Garin Scarp as an old sea-cliff.

Stagnant landscapes

Just what proportion of the Earth's land surface predates the Pleistocene epoch has yet to be ascertained, but it looks to be a not insignificant figure. In Australia, Gondwanan land surfaces constitute 10–20 per cent of the contemporary cratonic landscape (Twidale 1994). In south-central Brazil, on the Second Paraná Plateau, lateritic caprock on plateau interfluves has suffered very limited denudation since the Oligocene, while rivers have cut deeply into adjacent valleys in response to regional base-level change over the same period (Riffel *et al.* 2015).

Other recent work on long-term landscape evolution in varied parts of the world paints a similar picture of spatial heterogeneity in landform origin and age. As Yanni Gunnell (2015, 3) put it, 'Relatively unresponsive land surfaces coexist with hotspots of geomorphic activity at time scales compatible with the duration of the Cenozoic era'.

In the light of these findings, Gunnell (2015) suggested a **Goldilocks rule of landscape evolution** in which composite landscapes (land surfaces consisting of an odd mosaic of landscape units and landforms differing up to several tens of millions of years in age) coincide with areas where the scale of crustal uplift above marine baselevel has been intermediate, that is, neither too great nor too small. This 'rule' applies to the generations Mesozoic pediment systems, Cenozoic valley systems, and Quaternary valleys and marine abrasion platforms of western France (Bessin *et al.* 2015; see p. 471), whose evolution has depended upon the long-term presence of the Armorican Massif within the range of Mesozoic and Cenozoic cycles of eustatic fluctuation. At times, this has allowed relatively thin marine cover rocks (usually soft rocks such as Cretaceous chalk) to lap onto the basement and to protect the older land surfaces. At other times, these ancient surfaces have been re-exposed by the stripping of the cover rocks, especially during intervals of regional tectonic instability or intense tropical weathering of the chalk, or both.

An important implication of all this work is that some landforms and their associated soils can survive through various climatic changes when tectonic conditions permit. A problem arises in accounting for the survival of these **palaeoforms**. Most modern geomorphological theory would dictate that denudational processes should have destroyed them long ago. It is possible that they have survived under the exceptional circumstance of a very long-lasting arid climate, under which the erosional cycle takes a vast stretch of time to run its course (Twidale 1976, 1998, 1999). A controversial explanation is that much of the Earth's surface is, in geomorphic terms, rather inactive: the ancient landscape of south-eastern Australia, rather than being an exceptional case, may be typical of Africa and, to a lesser extent, Eurasia and the Americas (e.g. Young 1983; Twidale 1998).

Two related mechanisms might explain stagnant parts of landscapes (cf. Twidale 1999). The first mechanism is **unequal erosion**. Some parts of landscapes are more susceptible to erosion than are other parts (cf. Brunsden and Thornes 1979). Mobile, fast-responding parts (rivers, some soils, and beaches) erode readily. They quickly adopt new configurations and act as focal points for landscape change. Relatively immobile, slowly responding parts (plateaux and interfluves, some soils and weathering features) lie far from susceptible parts. This differential susceptibility of landscapes to erosion would permit fast-changing 'soft spots' to exist alongside stagnant areas. But it does not explain why some areas are stagnant. Weathering should construct regolith, and erosional processes should destroy it on all exposed surfaces, though the balance between constructive and destructive forces would vary in different environments. The second mechanism helps to explain the occurrence of stagnant areas. This is **the persistence and dominating influence of rivers** (Twidale 1997). Rivers are self-reinforcing systems: once established and dominant, they tend to sustain and augment their dominance. Thus, major rivers tend to persist in a landscape. In Australia, some modern rivers are 60 million years old and have been continuously active since their initiation in the Eocene epoch. Other equally old or even older rivers, but with slightly chequered chronologies, also persist in the landscape (see Ollier 1991, 99–103). Likewise, some landscapes reveal the ghosts of other very old rivers. Rivers of similar antiquity occur in other Gondwanan landscapes. Such long-running persistence of rivers means that parts of landscapes remote from river courses – interfluves and summits for example – may remain virtually untouched by erosive processes for vast spans of time and they are, in geomorphic terms, stagnant areas. A third possible mechanism for landscape stagnation comes from theoretical work. It was found that landscape stability depends upon timelags between soil processes, which act at right-angles to hillslopes, and geomorphic processes, which act tangentially to hillslopes (Phillips 1995). When there is no lag between debris production and its availability for removal, regolith thickness at a point along a hillside displays chaotic dynamics. On the other hand, when a time-lag is present, regolith thickness is stable and non-chaotic. The emergence of landscape stability at broad scales may therefore result from time-lags in different processes. Where regolith production is slow, and erosion even slower, stagnation might occur. Even so, the conditions necessary for the first two mechanisms to produce landscape stagnation would surely be required for a landscape to maintain stability for hundreds of millions of years.

If substantial portions of landscapes are indeed stagnant and hundreds of millions of years old, the implications for process geomorphology are not much short of sensational. It would mean that cherished views on rates of denudation and on the relation between denudation rates and tectonics would require a radical revision, and the connections between climate and landforms would be even more difficult to establish.

EVOLVING LANDSCAPES

Landscape cycles

Several geomorphologists believe that landscape history has been **cyclical** or **episodic**. The Davisian system of landscape evolution combined periods dominated by the gradual and gentle action of geomorphic processes interrupted by brief episodes of sudden and violent tectonic activity. A land mass would suffer repeated 'cycles of erosion' involving an initial rapid uplift followed by a slow wearing down. The Kingian model of repeated pediplanation envisaged long-term cycles, too. Remnants of erosion surfaces can be identified globally (King 1983). They correspond to pediplanation during the Jurassic, early to middle Cretaceous, late Cretaceous, Miocene, Pliocene, and Quaternary times (Table 15.2). However, King's views are not widely accepted, and have been challenged (e.g. Summerfield 1984; Ollier 1991, 93).

A popular theme, with several variations, is that the landscape alternates between stages of relative stability and stages of relative instability. An early version of this idea, which still has considerable currency, is the theory of biostasy and rhexistasy (Erhart 1938, 1956; cf. Butler 1959, 1967). According to this model, landscape change involves long periods of **biostasy** (biological equilibrium), associated with stability and soil development, broken in upon by short periods of **rhexistasy** (disequilibrium), marked by instability and soil erosion. During biostasy, which is the 'normal' state, streams carry small loads of suspended sediments but large loads of dissolved materials: silica and calcium are removed to the oceans, where they form limestones and chert, leaving deep ferrallitic soils and weathering profiles on the continents. Rhexistatic conditions are triggered by bouts of tectonic uplift and lead to the stripping of the ferrallitic soil cover, the headward erosion of streams, and the flushing out of residual quartz during entrenchment. Intervening plateaux become desiccated owing to a falling water table, and duricrusts form. In the oceans, red beds and quartz sands are deposited.

Later reincarnations of the stability–instability model take account of regolith, tectonics, sedi-

Table 15.2 Lester King's global planation cycles and their recognition

Cycle	Old name	New name	Recognition
I	Gondwana	Gondwana planation	Jurassic, only rarely preserved
II	Post-Gondwana	'Kretacic' planation	Early mid-Cretaceous
III	African	Moorland planation	Late Cretaceous to mid Cenozoic. Planed uplands with no trees and poor soils
IV		Rolling land-surface	Mostly Miocene. Undulating country above incised valleys
V	Post-African	Widespread landscape	Pliocene. The most widespread global planation cycle. Found mainly in basins, lowlands, and coastal plains and not in uplifted mountain regions
VI	Congo	Youngest cycle	Quaternary. Represented in deep valleys and gorges of the main rivers

Source: Adapted from Ollier (1991, 92)

mentation, and sea-level change. A **cratonic regime model**, based on studies carried out on the stable craton of Western Australia, envisaged alternating planation and transgression occurring without major disturbance for periods of up to a billion years (Fairbridge and Finkl 1980). During this long time, a **thalassocratic regime** (corresponding to Erhart's biostasy and associated with high sea levels) is interrupted by short intervals dominated by an **epeirocratic regime** (corresponding to Erhart's rhexistasy and associated with low sea levels). The alternations between thalassocratic and epeirocratic regimes may occur every 10–100 million years. However, more frequent alternations have been reported. A careful study of the Koidu etchplain in Sierra Leone has shown that interruptions mirror environmental changes and occur approximately every 1,000–10,000 years (Thomas and Thorp 1985).

A variant of the cratonic regime model explains the evolution of many Australian landscape features (Twidale 1991, 1994). An Early Cretaceous marine transgression flooded large depressed basins on the Australian land mass (Figure 15.17). The transgression covered about 45 per cent of the present continent. The new submarine basins subsided under the weight of water and sediment. Huge tracts of the Gondwanan landscape were preserved beneath the unconformity. Hinge lines (or fulcra) would have formed near shorelines. Adjacent land areas would have been uplifted, raising the Gondwanan palaeoplain, and basin margins warped and faulted. Parts of this plain were preserved on

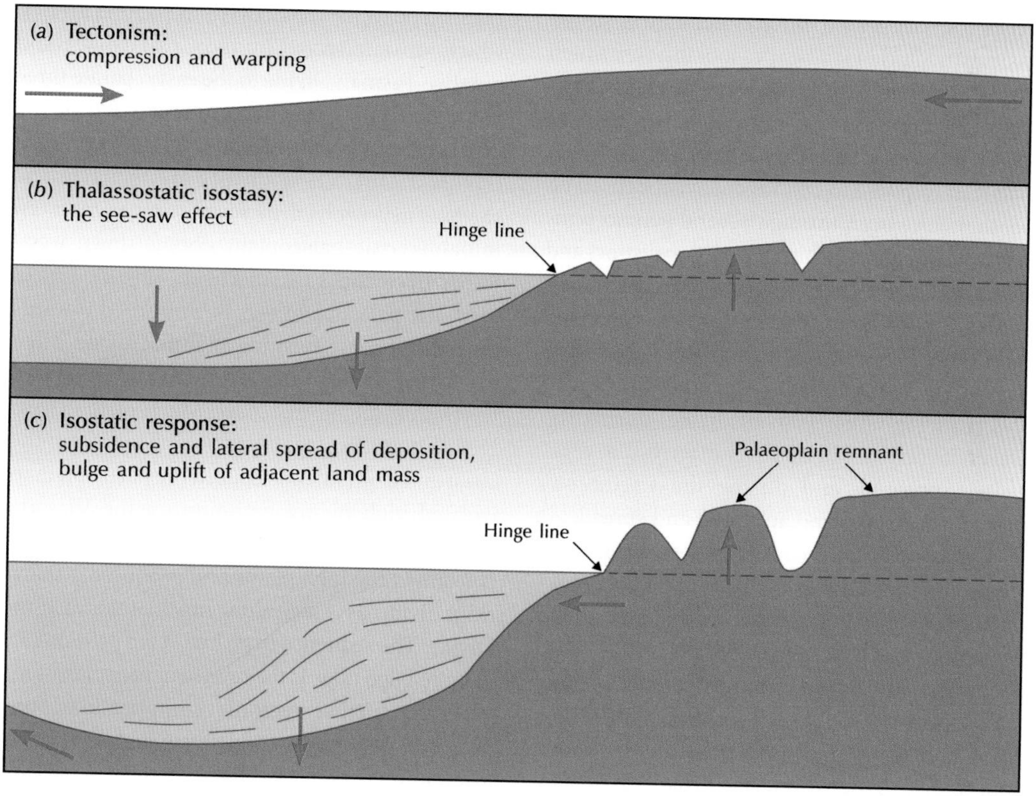

Figure 15.17 Sequence of events following a marine incursion into an Australian cratonic basin, and consequent uplift of adjacent land. *Source*: Adapted from Twidale (1994)

divides as palaeoplain remnants. Other parts were dissected and reduced to low relief by rivers graded to Cretaceous shorelines. Subsequent erosion of the Cretaceous marine sequence margins has exhumed parts of the Gondwanan surface, which is an integral part of the present Australian landscape.

Evolutionary geomorphology

The non-actualistic system of land-surface history known as **evolutionary geomorphology** (Ollier 1981, 1992) makes explicit directional change in landscape development. The argument runs that the land surface has changed in a definite direction through time, and has not suffered the 'endless' progression of erosion cycles first suggested by James Hutton and implicit in Davis's geographical cycle. An endless repetition of erosion cycles would simply maintain a steady state with Silurian landscapes looking very much like Cretaceous landscapes and modern landscapes. Evolutionary geomorphologists contend that the Earth's landscapes have evolved as a whole in response to contingent events. In doing so, they have been through several geomorphological 'revolutions', which have led to distinct and essentially irreversible changes of process regimes, so that the nature of erosion cycles has changed with time. These revolutions probably occurred during the Archaean aeon, when the atmosphere was reducing rather than oxidizing, during the Devonian period, when a cover of terrestrial vegetation appeared, and during the Cretaceous period, when grassland appeared and spread.

The breakup and coalescence of continents would also alter landscape evolution. The geomorphology of Pangaea was, in several respects, unlike present geomorphology (Ollier 1991, 212). Vast inland areas lay at great distances from the oceans, many rivers were longer by far than any present river, and terrestrial sedimentation was more widespread. When Pangaea broke up, rivers became shorter, new continental edges were rejuvenated and eroded, and continental margins warped tectonically. Once split from the supercontinent, each Pangaean fragment followed its own history. Each experienced its own unique events. These included the creation of new plate edges and changes of latitude and climate. It also involved substantial changes in drainage systems (e.g. Potter 1997; Beard 2003; Goudie 2005). The landscape evolution of each continental fragment must be viewed in this very long-term perspective. In this evolutionary context, the current fads and fashions of geomorphology – process studies, dynamic equilibrium, and cyclical theories – have limited application (Ollier 1991, 212).

Tectonic and landscape evolution in southeast Australia afford a good example of evolutionary geomorphology, with contingency playing a large role (Ollier and Pain 1994; Ollier 1995). Morphotectonic evolution in this area appears to represent a response to unique, non-cyclical events. Today, the Canobolas and Victoria divides, which are intersected by the Great Divide and putative Tasman Divide to the east separate three major basins – the Great Artesian Basin, the Murray Basin, and the Gibbsland–Otway basins (Figure 15.18). These divides are major watersheds. They evolved in several stages from an initial Triassic palaeoplain sloping down westwards from the Tasman Divide (Figure 15.19). First, the palaeoplain was downwarped towards the present coast, forming an initial divide. Then the Great Escarpment formed and retreated westwards, facing the coast. Much of the Great Divide is at this stage. Retreat of slopes from the coast and from inland reduced the palaeoplain to isolated high plains, common on the Victoria Divide. Continued retreat of the escarpment consumed the high plains and produced a sharp ridge divide, as is seen along much of the Victoria Divide. The sequence from low-relief palaeoplain to knife-edge ridge is the reverse of peneplanation. With no further tectonic complications, the present topography would presumably end up as a new lower-level plain. However, the first palaeoplain is Triassic in age, and the 'erosion cycle' is unlikely to end given continuing tectonic changes

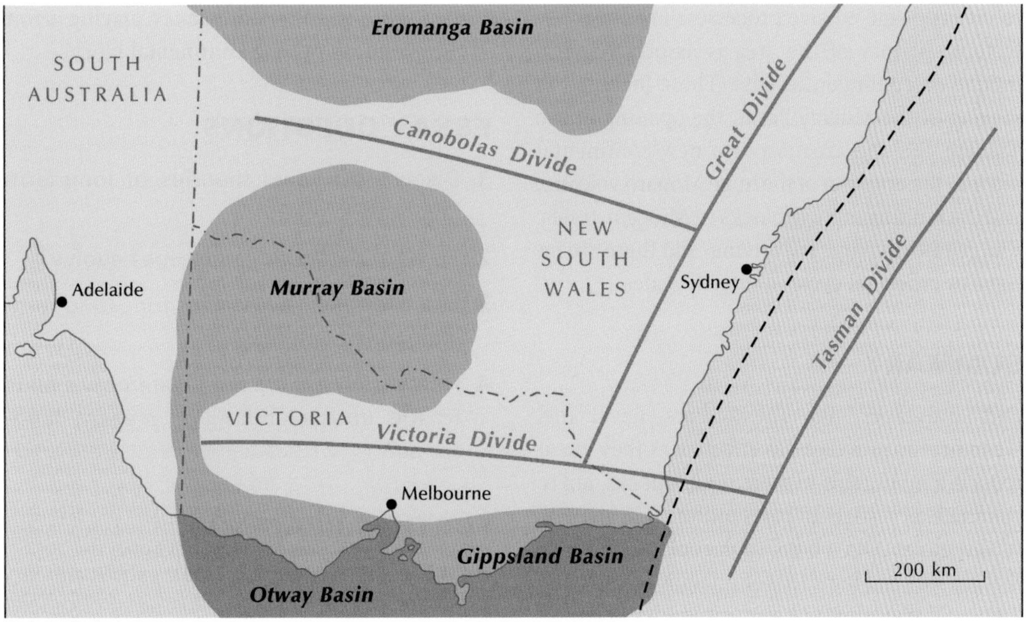

Figure 15.18 Major basins and divides of south-east Australia. The Eromanga Basin is part of the Great Artesian Basin. *Source*: Adapted from Ollier and Pain (1994)

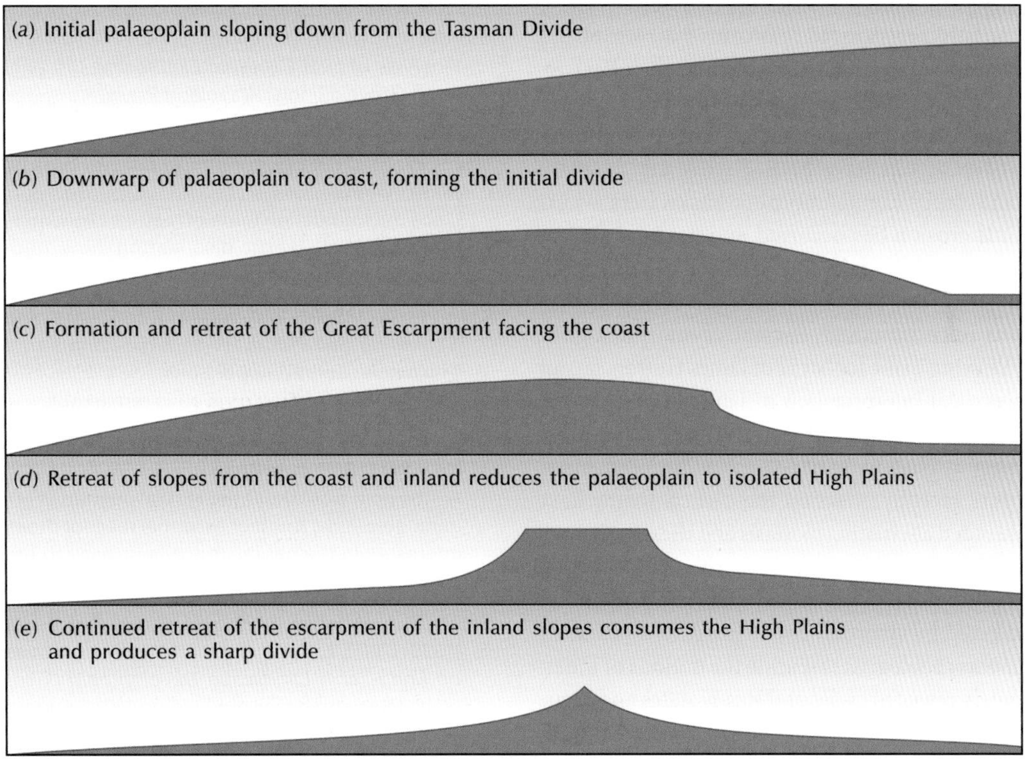

Figure 15.19 Evolution of the south-east Australian drainage divides. *Source*: Adapted from Ollier (1995)

to interrupt the erosive processes. The morphotectonic history of the area is associated with unique or contingent events. These include the sagging of the Murray Basin, the opening of the Tasman Sea and creation of a new continental margin, the eruption of the huge Monaro volcano, and the faulting of huge blocks in Miocene times. The geomorphology is evolving, and there are no signs of erosional cycles or steady states.

SUMMARY

Old landscapes, like old soldiers, never die. Geomorphic processes, as effective as they are at reducing mountains to mere monadnocks, fail to eliminate all vestiges of past landforms in all parts of the globe. Old plains (palaeoplains) survive that are tens and hundreds of millions of years old. These old plains may be various kinds of erosion surface, peneplains formed by fluvial action, pediplains and panplains formed by scarp retreat and lateral planation by rivers respectively, etchplains, or exhumed surfaces. Exhumed surfaces and landforms are old landforms that were buried beneath a cover of sediments and then later re-exposed as the cover rocks were eroded. Several exhumed palaeoplains and such other landforms as reef knolls have been discovered. Stagnant landscapes are geomorphic backwaters where little erosion has occurred and the land surface has been little altered for millions of years or far longer. They appear to be more common than was once supposed. Several geomorphologists, following in the footsteps of James Hutton, favour a cyclical interpretation of land-surface history. William Morris Davis and Lester King were doughty champions of cyclical landscape changes. More recently, ideas on the cyclical theme have included alternating biostasis and rhexistasis, and, linking geomorphic processes with plate tectonics, a cratonic regime model. All landscapes are affected by environmental change. Evolutionary geomorphologists cast aside the notions of indefinitely repeated cycles and steady states and argue for non-actualistic, directional change in land-surface history, with contingency playing a role in the evolution of each continental block.

ESSAY QUESTIONS

1 Discuss the chief theories of long-term landform evolution.

2 To what extent do landscapes evolve?

3 Discuss the evidence for long-term changes of landforms.

4 How significant are pre-Quaternary events to the understanding of present landforms?

FURTHER READING

Bell, M. and Walker, M. J. C. (2003) *Late Quaternary Environmental Change: Physical and Human Perspectives*, 2nd edn. Harlow, Essex: Prentice Hall.
This has a Quaternary science and archaeological emphasis, but is undoubtedly valuable for geomorphologists, too.

Bloom, A. L. (2004) *Geomorphology: A Systematic Analysis of Late Cenozoic Landforms*, 3rd edn. Long Grove, Illinois: Waveland Press.
Worth perusing.

Davis, W. M. (1909) *Geographical Essays*. Boston, Mass.: Ginn.
Once the foundation tome of historical geomorphology. Well worth discovering how geomorphologists used to think.

King, L. C. (1983) *Wandering Continents and Spreading Sea Floors on an Expanding Earth*. Chichester: John Wiley & Sons.
The grandeur of King's vision is remarkable.

Ollier, C. D. (1991) *Ancient Landforms*. London and New York: Belhaven Press.
A little gem from the Ollier stable, penned in his inimitable style. Another essential read for those interested in the neo-historical approach to the discipline.

Smith, B. J., Whalley, W. B., and Warke, P. A. (eds) (1999) *Uplift, Erosion and Stability: Perspectives on Long-Term Landscape Development* (Geological Society special publication 162). London: Geological Society of London.

A mixed collection of papers that should be consulted.

Twidale, C. R. (1999) Landforms ancient and recent: the paradox. *Geografiska Annaler* 81A, 431–41.
The only paper in this Further Reading section. Please read it.

Williams, M. A. J., Dunkerley, D., de Deckker, P., Kershaw, P., and Chappell, J. (1998) *Quaternary Environments*, 2nd edn. London: Arnold.
An excellent account of global fluctuations during the Quaternary

APPENDIX

THE GEOLOGICAL TIMESCALE

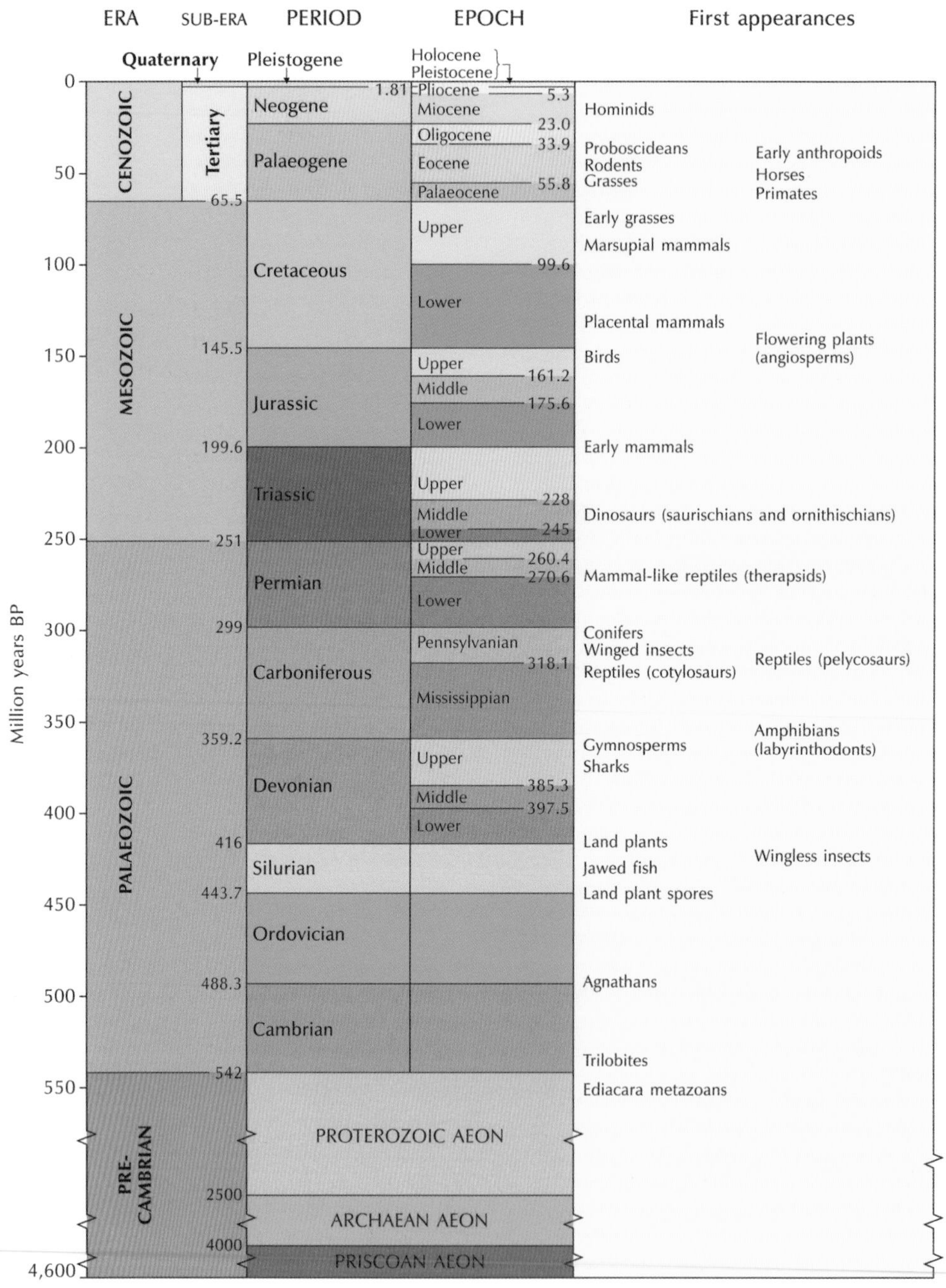

GLOSSARY

Most geomorphological terms are defined when used in the text. This glossary provides thumbnail definitions of terms that may be unfamiliar to students.

active margin The margin of a continent that corresponds with a tectonically active plate boundary.

aeolian Of, or referring to, the wind.

aeolianite Rock produced by the lithification of aeolian sediments.

aggradation A building up of the Earth's surface by the accumulation of sediment deposited by geomorphic agencies such as wind, wave, and flowing water.

alcrete A duricrust rich in aluminium, commonly in the form of hardened **bauxite**.

allitization The loss of silica and concentration of sesquioxides in the soil, with the formation of **gibbsite**, and with or without the formation of **laterite**; more or less synonymous with soluviation, ferrallitization, laterization, and latosolization.

alumina Aluminium oxide, Al_2O_3; occurs in various forms, for example as the chief constituent of **bauxite**.

alluvial Of, or pertaining to, **alluvium**.

alluvial fill The deposit of sediment laid down by flowing water in river channels.

alluvial terrace A river terrace composed of alluvium and created by renewed downcutting of the floodplain or valley floor (which leaves alluvial deposits stranded on the valley sides), or by the covering of an old river terrace with **alluvium**.

alluvium An unconsolidated, stratified deposit laid down by running water, sometimes applied only to fine sediment (**silt** and **clay**), but more generally used to include **sands** and **gravels**, too.

amphibole A group of minerals, most of which are mainly dark-coloured hydrous ferromagnesian silicates. Common in intermediate rocks and some **metamorphic** rocks.

anaerobic Depending on, or characterized by, the absence of oxygen.

andesite A grey to black, fine-grained, extrusive igneous rock that contains the minerals **plagioclase** and **hornblende** or augite. The extrusive equivalent of **diorite**.

anhydrite Anhydrous calcium sulphate, $CaSO_4$, occurring as a white mineral. Generally found in association with **gypsum**, **rock salt**, and other evaporite minerals.

anion A negatively charged **ion**.

aquifer A rock mass that readily stores and conveys **groundwater** and acts as a water supply.

aragonite A form of calcium carbonate, dimorphous with **calcite**.

arête A sharp, knifelike divide or crest.

argillaceous Referring to, containing, or composed of **clay**. Used to describe sedimentary rocks containing clay-sized material and clay minerals.

arkose A coarse-grained **sandstone** made of at least 25 per cent **feldspar** as well as **quartz**.

arroyo In the south-west USA, a small, deep, flat-floored gully cut by an ephemeral or an intermittent stream.

aspect The compass orientation of sloping ground or any other landscape feature. May be measured as an azimuth angle from North.

asthenosphere The relatively weak and ductile layer of rock lying beneath the lithosphere and occupying the uppermost part of the mantle that allows continents to move. Also called the rheosphere.

atmosphere The dusty, gaseous envelope of the Earth, retained by the Earth's gravitational field.

atoll A circular or closed-loop coral reef that encloses a lagoon.

backswamp An area of low-lying, swampy ground lying between a natural levee and the valley sides on the floodplain of an **alluvial** river.

bacteria Microorganisms, usually single-celled, that exist as free-living decomposers or parasites.

badlands A rugged terrain of steep slopes that looks like miniature mountains. Formed in weak **clay** rocks or clay-rich regolith by rapid fluvial **erosion**.

barrier reef A coral reef that is separated from the mainland shoreline by a lagoon.

basalt A hard, but easily weathered, fine-grained, dark-grey igneous rock. The commonest rock produced by a volcano, it consists mainly of calcic-**plagioclase feldspar**, augite or other **pyroxenes**, and, in some basalts, **olivine**. The fine-grained equivalent of **gabbro**.

batholith A large and deep-seated mass of igneous rock, usually with a surface exposure of more than 100 km^2.

bauxite A pale-coloured earthy mix of several hydrated aluminous ($Al_2O_3n.H_2O$) minerals. The chief ore of aluminium.

bedrock Fresh, solid rock in place, largely unaffected by weathering and unaffected by geomorphic processes.

Bernoulli effect The reduction of internal pressure with increased stream velocity in a fluid.

biogeochemical cycles The cycling of minerals or organic chemical constituents through the **biosphere**; for example, the sulphur cycle.

biosphere The totality of all living things.

biota All the animals and plants living in an area.

bluffs The steep slopes that often mark the edge of a floodplain.

breccia A bedded, rudaceous rock consisting of angular fragments of other rocks larger than 2 mm in diameter cemented in a fine matrix.

calcareous Any soil, sediment, or rock rich in calcium carbonate.

calcite A crystalline form of calcium carbonate ($CaCO_3$). The chief ingredient of **limestone**, **marble**, and **chalk**. A natural cement in many **sandstones**.

cation A positively charged **ion**.

cavitation A highly corrosive process in which water velocities over a solid surface are so high that the vapour pressure of water is exceeded and bubbles form.

chalk A soft, white, pure, fine-grained **limestone**. Made of very fine **calcite** grains with the remains of microscopic calcareous fossils.

chamosite A hydrous iron silicate.

chert A cryptocrystalline form of **silica**, a variety of chalcedony, often occurring as nodules and layers in **limestones**.

chitons Marine molluscs.

chronosequence A time sequence of landforms constructed by using sites of different ages.

clastic sediment Sediment composed of particles broken off a parent rock.

clasts Rock fragments broken off a parent rock.

clay A name commonly used to describe fine-grained sedimentary rock, plastic when wet, that consists of grains smaller than 0.002 mm, sometimes with a small portion of **silt**- and **sand**-sized particles. The grains are largely made of clay minerals but also of **calcite**, iron pyrite, altered **feldspars**, muscovite flakes, iron oxides, and organic material.

clay minerals A group of related hydrous aluminosilicates. The chief ingredients of **clay** and **mudstone**.

claystone A sedimentary rock composed of clay-sized particles (<0.002 mm in diameter) that does not split into flakes or scales. Consolidated **clay**. Claystone is a member of the **mudstone** group and is common in Palaeozoic deposits. Most Precambrian claystones have been metamorphosed to **slates** and **schists**.

cohesion In soils, the ability of **clay** particles to stick together owing to physical and chemical forces.

colloids Fine **clay**-sized particles (0.001–0.01 mm in size), usually formed in fluid suspensions.

colluvium An unconsolidated mass of rock debris at the base of a cliff or a slope, deposited by surface wash.

comminution The breaking up or grinding of sediments to form finer particles.

conglomerate A bedded, rudaceous sedimentary rock comprising rounded granules, pebbles, cobbles, or boulders of other rocks lodged in a fine-grained matrix, generally of **sand**.

connate water Meteoric water trapped in hydrous minerals and the pore spaces of sediments during deposition and out of contact with the atmosphere for perhaps millions of years.

corestone A spheroidal boulder of fresh (unweathered) rock, originally surrounded by saprolite, and formed by subsurface weathering of a joint block.

craton An old and stable part of the continental **lithosphere**, relatively undisturbed since the Precambrian era.

cryosphere All the frozen waters of the Earth (snow and ice).

cryostatic pressure The pressure caused by ice on water-saturated material sandwiched between an advancing seasonal layer of frozen ground and an impermeable layer such as permafrost.

cyanobacteria A group of unicellular and multicellular organisms, formerly called blue-green algae, that photosynthesize.

cyclic Recurring at regular intervals; for example, lunar cycles which occur twice daily, fortnightly, and so on.

dacite A fine-grained rock, the extrusive equivalent of **granodiorite**, with the same general composition as **andesite**, though with less calcic **plagioclase** and more **quartz**. Also called quartz andesite.

degradation A running down or loss of sediment.

dendrochronology The study of annual growth rings of trees. Used as a means of dating events over the last millennium or so.

denudation The sum of the processes – weathering, erosion, mass wasting, and transport – that wear away the Earth's surface.

diapir A dome or anticlinal fold produced by an uprising plume of plastic core material. The rising plume ruptures the rocks as it is squeezed up.

diatom A unicellular organism (Kingdom Protista) with a **silica** shell.

diorite A group of plutonic rocks with a composition intermediate between acidic and basic. Usually composed of dark-coloured **amphibole**, acid **plagioclase**, **pyroxenes**, and sometimes a little **quartz**.

dolerite A dark-coloured, medium-grained, hypabyssal igneous rock forming dykes and sills. Consists of **pyroxene** and **plagioclase** in equal proportions or more pyroxene than **plagioclase**, as well as a little **olivine**. An intrusive version of **gabbro** and **basalt**.

dolomite A mineral that is the double carbonate of calcium and magnesium, having the chemical formula $(CaMg)CO_3$. The chief component of dolomitic **limestones**.

eclogite A coarse-to-medium-grained igneous rock made mainly of garnet and sodic **pyroxene**.

ecosphere The global ecosystem – all life plus its life support system (air, water, and soil).

endogenic Of, or pertaining to, the Earth's interior (cf. **exogenic**).

entropy The amount of disorder in a system; a measure of the amount of energy in a system that is no longer free, in the sense of being able to perform work.

episodic Events that have a tendency to occur at discrete times.

erosion The weathering (decomposition and disintegration), solution, corrosion, corrasion, and transport of rock and rock debris.

erosion surface A more or less flat plain created by erosion; a planation surface.

eustatic Referring to a true change of sea level, in contrast to a local change caused by the upward or downward movement of the land.

exogenic Of, or pertaining to, the surface (or near the surface) of the Earth (cf. **endogenic**).

exsudation A type of salt weathering by which rock surfaces are scaled off, owing to the growth of salt and **gypsum** crystals from water raised by capillary action.

feldspar A group of minerals including **orthoclase** and microline, both of which are potassium alumino-silicates ($KAlSi_3O_8$), and the **plagioclase feldspars** (such as albite and anorthite). Albite contains more than 90 per cent sodium alumino-silicate ($NaAlSi_3O_8$); anorthite contains more than 90 per cent calcium alumino-silicate ($CaAlSi_3O_8$). Calcic feldspars are rich in anorthite. Alkali feldspars are rich in potash and soda feldspars, contain relatively large amounts of **silica** and are characteristic minerals in acid igneous rocks.

flag (flagstone) A hard, fine-grained **sandstone**, usually containing **mica**, especially along the bedding planes. Occurs in extensive thin beds with **shale** partings.

flood A short-lived but large discharge of water coursing down, and sometimes overflowing, a watercourse.

flood basalts Basalt erupted over a large area.

fulvic acid An organic acid formed from humus.

gabbro A group of dark-coloured plutonic rocks, roughly the intrusive equivalent of **basalt**, composed chiefly of **pyroxene** and **plagioclase**, with or without **olivine** and orthopyroxene.

gastropod Any mollusc of the class Gastropoda. Typically has a distinct head with eyes and tentacles, and, in most cases, a calcareous shell closed at the apex.

gibbsite A form of **alumina** and a component of **bauxite**.

gley A grey, clayey soil, sometimes mottled, formed where soil drainage is restricted.

gneiss A coarse-grained, banded, crystalline metamorphic rock with a similar mineralogical composition to **granite** (**feldspars**, **micas**, and **quartz**).

goethite A brown-coloured, hydrated oxide of iron; the main ingredient of rust.

gorge A steep-sided, narrow-floored valley cut into bedrock.

granite A coarse-grained, usually pale-coloured, acid, plutonic rock made of **quartz**, **feldspar**, and **mica**. The quartz constitutes 10–50 per cent of felsic compounds, and the ratio of alkali feldspar to total feldspar lies in the range of 65–90 per cent. Biotite and muscovite are accessories. The feldspar crystals are sometimes large, making the rock particularly attractive as a monumental stone. In the stone trade, many hard and durable rocks are called granite, though many of them are not granites according to geological definitions of the word.

granodiorite A class of coarse-grained plutonic rocks made of **quartz**, **plagioclase**, and potassium **feldspars** with biotite, **hornblende** or, more rarely, **pyroxene**.

gravel A loose, unconsolidated accumulation of rounded rock fragments, often pebbles or cobbles. Most gravels also contain **sand** and fines (**silt** and **clay**).

greywacke A dark-grey, firmly indurated, coarse-grained sandstone.

grit (gritstone) A name used, often loosely, to describe a coarse **sandstone**, especially one with angular **quartz** grains that is rough to the touch.

groundwater (phreatic water) Water lying within the saturation zone of rock and soil. Moves under the influence of gravity.

grus A saprolite on **granite** consisting of **quartz** in a **clay** matrix.

gypsum A white or colourless mineral. Hydrated calcium sulphate, $CaSO_4.2H_2O$.

halite (rock salt, common salt) Sodium chloride, NaCl, although calcium chloride and magnesium chloride are usually present, and sometimes also magnesium sulphate.

halloysite A **clay** mineral, similar to **kaolinite**, formed where aluminium and silicon are present in roughly equal amounts, providing the hydronium concentration is high enough and the concentration of bases is low.

hematite A blackish-red to brick-red or even steely-grey oxide of iron (Fe_2O_3), occurring as earthy masses or in various crystalline forms. The commonest and most important iron ore.

hillslope A slope normally produced by weathering, erosion, and deposition.

hornblende A mineral of the **amphibole** group.

humic acid An organic acid formed from humus.

hydraulic conductivity The flow rate of water through soil or rock under a unit hydraulic gradient. Commonly measured as metres per day.

hydrosphere All the waters of the Earth.

hydrostatic pressure The pressure exerted by the water at a given point in a body of water at rest. In general, the weight of water at higher elevations within the saturated zone.

hydrothermal Associated with hot water.

hydrous minerals Minerals containing water, especially water of crystallization or hydration.

hydroxyl A radical (a compound that acts as a single atom when combining with other elements to form minerals) made of oxygen and hydrogen with the formula (OH).

hypabyssal Said of rocks that solidify mainly as minor intrusions (e.g. dykes or sills) before reaching the Earth's surface.

ice age A time when ice forms broad sheets in middle and high latitudes, often in conjunction with the widespread occurrence of sea ice and permafrost, and mountain glaciers form at all latitudes.

Ice Age An old term for the full Quaternary glacial–interglacial sequence.

illite Any of three-layered, **mica**-like clay minerals.

infiltration The penetration of a fluid (such as water) into a solid substance (such as rock and soil) through pores and cracks.

inselberg A large residual hill within an eroded plain; an 'island mountain'.

intermittent stream A stream that, in the main, flows through a wet season but not through a dry season.

ion An atom or group of atoms that is electrically charged owing to the gain or loss of electrons.

ionic load The cargo of **ions** carried by a river.

island arc A curved line of volcanic islands linked to a subduction zone.

isostasy The idea of balance on the Earth's crust, in which lighter, rigid blocks of crustal material 'float' on the denser, more plastic material of the mantle. The redistribution of mass at the Earth's surface by erosion and deposition or by the growth and decay of ice upsets the balance causing the crustal blocks to float higher or lower in the mantle until a new balance is achieved.

joint block A unit of bedrock created by fractures within a rock mass.

jökulhlaup A glacier burst – the sudden release of vast volumes of water melted by volcanic activity under a glacier and held in place by the weight of ice until the glacier eventually floats.

juvenile water Primary or new water that is known not to have entered the water cycle before. It may be derived directly from magma, from volcanoes, or from cosmic sources (e.g. comets).

kaolinite A 1:1 clay mineral, essentially a hydrated aluminium silicate formed under conditions of high hydronium (hydrated hydrogen ion, H_3O) concentration and an absence of bases. Its ideal structural formula is $Al_2Si_2O_5(OH)_4$.

knickpoint An interruption or break of slope, especially a break of slope in the long profile of a river.

landslide A general term for the *en masse* movement of material down slopes.

laterite A red, iron-rich, residual material with a rich variety of definitions.

lava Molten rock.

leaching The washing-out of water-soluble materials from a soil body, usually the entire solum (the genetic soil created by the soil-forming processes), by the downwards or lateral movement of water.

limestone A bedded sedimentary rock composed largely of the mineral **calcite**.

limonite A hydrated iron oxide, $FeO(OH).nH_2O$; not a true mineral as it consists of several similar hydrated iron oxide minerals, and especially **goethite**.

lithified The state of being changed to rock, as when loose sediments are consolidated or indurated to form rocks.

lithology The physical character of a rock.

lithosphere The relatively rigid and cold top 50–200 km of the solid Earth.

mafic minerals Minerals, chiefly silicates, rich in magnesium and iron, dark-coloured and relatively dense.

magma Liquid rock coming from the mantle and occurring in the Earth's crust. Once solidified, magma produces igneous rocks.

magnesian limestone A **limestone** containing an appreciable amount of magnesium.

mantle The portion of the solid Earth lying between the core and the **lithosphere**.

marble Limestones that have been crystallized by heat or pressure, though some special types of non-metamorphosed limestones are called marble. Pure marble is white.

marcasite White iron pyrites, an iron sulphide with the identical composition to **pyrite**.

marl A soft, mainly unconsolidated rock made of **clay** or **silt** and fine-grained **aragonite** or **calcite** mud. The clay or silt fraction must lie in the range 30–70 per cent. Marls are friable when dry and plastic when wet.

marlstone Consolidated **marl**.

mesosphere A transition zone between the **asthenosphere** and the lower mantle.

metamorphism The processes by which rocks are transformed by recrystallization owing to increased heat or intense pressure or both.

meteoric water Water that is derived from precipitation and cycled through the atmosphere and the hydrosphere.

mica A group of minerals, all hydrous aluminosilicates of potassium, most members of which may be cleaved into exceptionally thin, flexible, elastic sheets. Muscovite (or white **mica**) and biotite (or dark mica) are common in **granites**.

mineral A naturally occurring inorganic substance, normally with a definite chemical composition and typical atomic structure.

monadnock An isolated mountain or large hill rising prominently from a surrounding peneplain and formed of a more resistant rock than the plain itself.

mud A moist or wet loose mixture of **silt**- and **clay**-sized particles. Clay is a mud in which clay-sized particles predominate, and silt is a mud in which silt-sized particles predominate.

mudstone A sedimentary rock, consisting mainly of clay-sized and silt-sized particles, with a massive or blocky structure and derived from mud. If **clay**-sized particles are dominant, the rock is a **claystone**; if silt-sized particles are dominant, the rock is **siltstone**. Mudstone, claystone, and siltstone are all members of the argillaceous group of clastic sedimentary rocks.

muskeg In Canada, a swamp or bog composed of accumulated bog moss (*Sphagnum*).

nappe A large body or sheet of rock that has been moved 2 km or more from its original position by folding or faulting. It may be the hanging wall of a low-angle thrust fault or a large recumbent fold.

Old Red Sandstone A thick sequence of Devonian rocks formed on land in north-west Europe some 408 to 360 million years ago.

olivine A magnesium iron silicate mineral $(Mg,Fe)_2SiO_4$, with no aluminium, usually olive-green. It is one of the commonest minerals on Earth.

orogeny The creation of mountains, especially by folding and uplift.

orthoclase Potassium aluminium silicate, an essential constituent of more acid igneous rocks, such as **granite** and **rhyolite**.

Pangaea The Triassic supercontinent comprising Laurasia to the north and Gondwana to the south.

passive margin The margin of a continent that is not associated with the active boundary of a tectonic plate and, therefore, lies within a plate.

pedosphere The shell or layer of the Earth in which soil-forming processes occur. The totality of soils on the Earth.

perennial stream A stream that flows above the surface year-round.

peridotite A coarse-grained, ultrabasic, plutonic rock, mainly made of **olivine** with or without mafic minerals.

plagioclase (or **plagioclase feldspar**) A mineral formed of aluminium silicates with calcium and sodium.

plan curvature Contour curvature, taken as negative (convex) over spurs and positive (concave) in hollows.

plinthite A hardpan or soil crust, normally rich in iron.

podzolization A suite of processes involving the chemical migration of aluminium and iron (and sometimes organic matter) from an eluvial (leached) horizon in preference to **silica**.

pore water pressure The force that builds up owing to the action of gravity on water in the pore spaces in soils and sediments.

pores Small voids within rocks, unconsolidated sediments, and soils.

porosity The amount of pore space or voids in a rock, unconsolidated sediment, or soil body. Usually expressed as the percentage of the total volume of the mass occupied by voids.

porphyry Any igneous rock that contains conspicuous phenocrysts (relatively large crystals) in a fine-grained groundmass.

pressure melting point The temperature at which ice can melt at a given pressure. The greater the pressure, the lower the pressure melting point.

profile curvature The curvature (rate of change of slope) at a point along a slope profile.

province In geology and geomorphology, a geographical entity with common geological or geomorphic attributes. It may include a single dominant structural element, as in the Snake River Plain Province in the USA, or a number of adjoining related elements, as in the Basin and Range Province in the USA.

pyrite (pyrites, iron pyrites) Iron sulphide, FeS_2; a mineral.

pyroxene A group of minerals, most of which are generally dark-coloured anhydrous ferromagnesian silicates. Characteristically occur in ultrabasic and basic rocks as the mineral augite.

quartz A widely distributed mineral with a range of forms, all made of **silica**.

quartzite **Sandstone** that has been converted into solid quartz rock, either by the precipitation of **silica** from interstitial waters (orthoquartzite) or by heat and pressure (metaquartzite). Quartzite lacks the pores of **sandstone**.

quartzose Containing **quartz** as a chief constituent.

radiolarian A unicellular organism (Kingdom Protozoa), usually with a **silica** skeleton that possesses a beautiful and intricate geometric form.

rectilinear Characterized by a straight line or lines.

regolith An accumulation of weathered and unweathered inorganic and organic material (e.g. peat) lying above fresh bedrock.

rhyolite A fine-grained, extrusive, acid igneous rock composed mainly of **quartz** and **feldspar** and commonly mica, a mineralogical equivalent of **granite**.

rock salt Halite (sodium chloride, NaCl).

sand A loose, unconsolidated sediment made of particles of any composition with diameters in the sand-sized range (0.625 to 2 mm in diameter). Most sands have a preponderance of **quartz** grains, but **calcite** grains derived from shells preponderate in some sands.

sandstone A medium-grained, bedded, clastic sedimentary rock made of abundant rounded or angular, **sand**-sized fragments in a fine-grained (**silt** or **clay**) matrix. Consolidated sand. Arkoses are sandstones rich in **feldspar**. Greywackes are sandstones containing rock fragments and clay minerals. Flags or flagstones are sandstones with flakes of **mica** occurring along the bedding planes.

saprolite Weathered or partially weathered bedrock that is *in situ* (i.e. that has not been moved).

schist A strongly foliated, crystalline rock formed by dynamic metamorphism. Readily split into thin flakes or slabs.

sediment yield The total mass of sedimentary particles reaching the outlet of a drainage basin. Usually expressed as tonnes/year, or as a specific sediment yield in tonnes/km²/year.

serpentine Any of a group of hydrous magnesium-rich silicate minerals.

shale A group of fine-grained, laminated sedimentary rocks made of **silt**- and **clay**-sized particles. Some 70 per cent of all sedimentary rocks are shales.

siderite Iron carbonate, $FeCO_3$, usually with a little manganese, magnesium, and calcium present; a mineral.

silica Chemically, silicon dioxide, SiO_2, but there are many different forms, each with their own names. For example, **quartz** is a crystalline mineral form. Chalcedony is a cryptocrystalline form, of which flint is a variety.

siliceous ooze A deep-sea pelagic sediment containing at least 30 per cent siliceous (largely silica) skeletal remains.

silicic Pertaining to, resembling, or derived from **silica** or silicon.

silt A loose, unconsolidated sediment of any composition with diameters in the silt-sized range (0.004 to 0.0625 mm in diameter). The chief component of loess.

siltstone A consolidated sedimentary rock composed chiefly of **silt**-sized particles that usually occur as thin layers and seldom qualify as formations. Consolidated silt.

slate A fine-grained, clayey metamorphic rock. It readily splits into thin slabs used in roofing.

slope An inclined surface of any part of the Earth's surface.

smectite A name for the montmorillonite group of **clay** minerals.

soil pipes Subsurface channels up to several metres in diameter, created by the dispersal of **clay** particles in fine-grained, highly permeable soil.

soil wetness The moisture content of soil.

subaerial Occurring at the land surface.

suffossion The digging or undermining of soil or rock by throughflow.

talus Rock fragments of any shape and size derived from, or lying at, the base of a cliff or steep rocky slope.

talus slope A slope made of talus.

tectosphere A name for the continental lithosphere.

terracette A small terrace; several often occur together to form a series of steps on a hillside.

toposphere The totality of the Earth's surface features, natural and human-made.

vadose water Subsurface water lying between the ground surface and the water table.

Van der Waals bonds The weak attraction that all molecules bear for one another that results from the electrostatic attraction of the nuclei of one molecule for the electrons of another.

vermiculite A **clay** mineral of the hydrous **mica** group.

viscosity The resistance of a fluid (liquid or gas) to a change of shape. It indicates an opposition to flow. Its reciprocal is fluidity.

vivianite Hydrated iron phosphate, $Fe_3P_2O_8.H_2O$; a mineral.

voids The open spaces between solid material in a porous medium.

vortex A fast-spinning or swirling mass of air or water.

vugs Pore spaces, sometimes called vugular pore spaces, formed by solution eating out small cavities.

water table The surface defined by the height of freestanding water in fissures and **pores** of saturated rock and soil.

weathering front The boundary between unaltered or fresh bedrock and saprolite. It is often very sharp.

weathering The chemical, mechanical, and biological breakdown of rocks on exposure to the **atmosphere**, **hydrosphere**, and **biosphere**.

zone A latitudinal belt.

BIBLIOGRAPHY

Abrahams, A. D., Li, G., and Atkinson, J. F. (1995) Step-pool streams: adjustment to maximum flow resistance. *Water Resources Research* 31, 2593–602.

Acreman, M. (2000) *The Hydrology of the UK: A Study of Change*. London: Routledge.

Adams, G. F. (ed.) (1975) *Planation Surfaces: Peneplains, Pediplains, and Etchplains* (Benchmark Papers in Geology 22). Stroudsburg, PA: Dowden, Hutchinson and Ross.

Affleck, J. (1970) Definition of regional structures by magnetics. In H. Johnson and B. L. Smith (eds) *The Megatectonics of Continents and Oceans*, pp. 3–11. New Brunswick, N.J.: Rutgers University Press.

Ahnert, F. (1998) *Introduction to Geomorphology*. London: Arnold.

Albritton, C. C., Brooks, J. E., Issawi, B., and Swedan, A. (1990) Origin of the Qattara Depression, Egypt. *Bulletin of the Geological Society of America* 102, 952–60.

Algeo, T. J. and Seslavinsky, K. B. (1995) The Paleozoic world: continental flooding, hypsometry, and sea-level. *American Journal of Science* 295, 787–822.

Allen, P. A. (1997) *Earth Surface Processes*. Oxford: Blackwell Science.

Allsop, D. G. (1992) *Visitor's Guide to Poole's Cavern*. Buxton, Derbyshire: Buxton and District Civic Association.

An Foras Forbartha (1981) *Areas of Scientific Interest in Ireland: National Heritage Inventory*. Dublin: An Foras Forbartha.

Andersen, B. G. and Borns, H. W., Jr (1994) *The Ice Age World*. Oslo: Scandinavian University Press.

Anderson, M. G. and Brooks, S. M. (eds) (1996) *Advances in Hillslope Processes*, 2 vols. Chichester: John Wiley & Sons.

Anderson, R. S. (1987) A theoretical model for aeolian impact ripples. *Sedimentology* 34, 943–88.

Anderson, R. S. and Anderson, S. P. (2010) *Geomorphology: the Mechanics and Chemistry of Landscapes*. Cambridge: Cambridge University Press.

Anderson, R. S. and Bunas, K. L. (1993) Grain size aggregation and stratigraphy in aeolian ripples modelled with a cellular automaton. *Nature* 365, 740–3.

Anon (1991) First International Symposium on the Conservation of our Geological Heritage, Digne, France, 11–16 June 1991: Terra Abstracts Supplement 2 to *Terra Nova*, Volume 3, 1991, 17.

Antevs, E. (1951) Arroyo cutting and filling. *Journal of Geology* 60, 375–85.

Anthony, E. and Orford, J. D. (2002) Between wave- and tide-dominated coasts: the middle ground revisited. Special issue, *Journal of Coastal Research* 36, 8–15.

Ashmore, P. (2015) Sociogeomorphology of rivers. *Geomorphology* 215, 149–56.

Attewell, P. B. and Taylor, D. (1988) Time-dependent atmospheric degradation of building stone in a polluting environment. In P. G. Marinos and G. C. Koukis (eds) *Engineering Geology of Ancient Works, Monuments and Historical Sites*, pp. 739–53. Rotterdam: Balkema.

Aubréville, A. (1949) *Climats, Forêts et Désertification de l'Afrique tropicale*. Paris: Société d'Éditions Géographiques, Maritimes et Coloniales.

Augustinus, P. G. E. F. (1989) Cheniers and chenier plains: a general introduction. *Marine Geology* 90, 219–29.

Baas, A. C. W. (2002) Chaos, fractals and self-organization in coastal geomorphology: simulating dune landscapes in vegetated environments. *Geomorphology* 48, 309–28.

Bagnold, R. A. (1941) *The Physics of Blown Sand and Desert Dunes*. London: Chapman & Hall.

Baker, A. and Genty, D. (1998) Environmental pressures on conserving cave speleothems: effects of changing surface land use and increased cave tourism. *Journal of Environmental Management* 53, 165–75.

Baker, V. R. (1977) Stream channel response to floods with examples from central Texas. *Bulletin of the Geological Society of America* 88, 1057–71.

Baker, V. R. (1981) The geomorphology of Mars. *Progress in Physical Geography* 5, 473–513.

Baker, V. R. (1983) Large-scale fluvial palaeohydrology. In K. J. Gregory (ed.) *Background to Palaeohydrology*, pp. 453–78. Chichester: John Wiley & Sons.

Baker, V. R. (2008) Paleoflood hydrology: origin, progress, prospects. *Geomorphology* 101, 1–13.

Baker, V. R. (2009) Megafloods and global paleoenvironmental change on Mars and Earth. In Mary G. Chapman and Laszlo P. Keszthelyi (eds) *Preservation of Random Megascale Events on Mars and Earth: Influence on Geologic History*, pp. 25–36. The Geological Society of America Special Paper 453. Boulder, Colorado: The Geological Society of America.

Baker, V. R. (2015) Planetary geomorphology: some historical/analytical perspectives. *Geomorphology* 240, 8–17.

Baker, V. R. and Twidale, C. R. (1991) The reenchantment of geomorphology. *Geomorphology* 4, 73–100.

Baker, V. R., Hamilton, C. W., Burr, D. M., Gulick, V., Komatsu, G., Luo, W., Rice Jr., J. W., and Rodriguez, J. A. P. (2015) Fluvial geomorphology on Earth-like planetary surfaces: a review. *Geomorphology* 245, 149–82.

Ballantyne, C. K. (2002) Paraglacial geomorphology. *Quaternary Science Reviews* 21, 1935–2017.

Ballantyne, C. K. (2007) Trotternish Escarpment, Isle of Skye. In R. G. Cooper (ed) *Mass Movements in Great Britain* (Geological Conservation Review Series, No. 33), pp. 196–204. Peterborough: Joint Nature Conservation Committee.

Ballantyne, C. K. and Harris, C. (1994) *The Periglaciation of Great Britain*. Cambridge: Cambridge University Press.

Ballantyne, C. K., Wilson, P., Gheorghiu, D., and Rodés, À. (2014) Enhanced rock-slope failure following ice-sheet deglaciation: timing and causes. *Earth Surface Processes and Landforms* 39, 900–13.

Bamber, J. L., Riva, R. E. M., Vermeersen, B. L. A., and LeBrocq, A. M. (2009) Reassessment of the potential sea-level rise from a collapse of the West Antarctic Ice Sheet. *Science* 324, 901–3.

Barry, R. G. and Ives, J. D. (1974) Introduction. In J. D. Ives and R. G. Barry (eds) *Arctic and Alpine Environments*, pp. 1–13. London: Methuen.

Barsch, D. (1996) Permafrost creep and rock glaciers. *Permafrost and Periglacial Processes* 3, 175–88.

Battiau-Queney, Y. (1984) The pre-glacial evolution of Wales. *Earth Surface Processes and Landforms* 9, 229–52.

Battiau-Queney, Y. (1987) Tertiary inheritance in the present landscape of the British Isles. In V. Gardiner (ed.) *International Geomorphology 1986, Part II* (Proceedings of the First International Conference on Geomorphology), pp. 979–89. Chichester: John Wiley & Sons.

Battiau-Queney, Y. (1991) Les marges passives. *Bulletin de l'Association de Géographes Français* 1991–2, 91–100.

Battiau-Queney, Y. (1996) A tentative classification of paleoweathering formations based on geomorphological criteria. *Geomorphology* 16, 87–102.

Baynes, E. R. C., Attal, M., Niedermann, S., Kirstein, L. A., Dugmore, A. J., and Naylor, M. (2015a) Erosion during extreme flood events dominates Holocene canyon evolution in northeast Iceland. *Proceedings of the National Academy of Sciences of the United States of America* 112, 2355–60.

Baynes, E. R. C., Attal, M., Dugmore, A. J., Kirstein, L. A., and Whaler, K. A. (2015b) Catastrophic impact of extreme flood events on the morphology and evolution of the lower Jökulsá á Fjöllum (northeast Iceland) during the Holocene. *Geomorphology* 250, 422–36.

Beard, J. S. (2003) Paleodrainage and the geomorphic evolution of passive margins in southwestern Australia. *Zeitschrift für Geomorphologie* NF 47, 273–88.

Beasley, D. B., Huggins, L. F., and Monke, E. J. (1980) ANSWERS: A model for watershed planning. *Transactions of the American Society of Agricultural Engineers* 23, 938–44.

Beaumont, C., Kooi, H., and Willett, S. (2000) Coupled tectonic-surface process models with applications to rifted margins and collisional orogens. In M. A. Summerfield (ed.) *Geomorphology and Global Tectonics*, pp. 29–55. Chichester: John Wiley & Sons.

Becker, J. J., Sandwell, D. T., Smith, W. H. F., Braud, J., Binder, B., Depner, J., Fabre, D., Factor, J., Ingalls, S., Kim, S. H., Ladner, R., Marks, K., Nelson, S., Pharaoh, A., Trimmer, R., Von Rosenberg, J., Wallace, G., and Weatherall, P. (2009) Global bathymetry and elevation data at 30 arc seconds resolution: SRTM30_PLUS. *Marine Geodesy* 32, 355–71.

Beggs, J. P. (2013) Climatediversity: a new paradigm for climate science. *Atmospheric Environment* 68, 112–13.

Bell, M. and Walker, M. J. C. (2003) *Late Quaternary Environmental Change: Physical and Human Perspectives*, 2nd edn. Harlow, Essex: Prentice Hall.

Benedict, P. C., Bondurant, D. C., McKee, J. E., Piest, R. F., Smallshaw, J., and Vanoni, V. A. (1971) Sediment transportation mechanics: Q. Genetic classification of valley sediment deposits. *Journal of the Hydraulics Division, Proceedings of the American Society of Civil Engineers* 97, 43–53.

Benn, D. I. and Evans, D. J. A. (2010) *Glaciers and Glaciation*. London and New York: Routledge.

Bennett, M. R. and Glasser, N. F. (1996) *Glacial Geology: Ice Sheets and Landforms*. Chichester: John Wiley & Sons.

Bennett, M. R. and Glasser, N. F. (2009) *Glacial Geology: Ice Sheets and Landforms*, 2nd edn. Chichester: Wiley-Blackwell.

Bennett, S. J., Ashmore, P., McKenna Neuman, C. (2015) Transformative geomorphic research using laboratory experimentation. *Geomorphology* 244, 1–8.
Berner, E. K. and Berner, R. A. (1987) *The Global Water Cycle: Geochemistry and Environment*, 1st edn. Englewood Cliffs, New Jersey: Prentice Hall.
Berner, E. K. and Berner, R. A. (2012) *Global Environment: Water, Air, and Geochemical Cycles*, 2nd edn. Princeton, N.J.: Princeton University Press.
Bessin, P., Guillocheau, F., Robin, C., Schroëtter, J.-M., and Bauer, H. (2015) Planation surfaces of the Armorican Massif (western France): denudation chronology of a Mesozoic land surface twice exhumed in response to relative crustal movements between Iberia and Eurasia. *Geomorphology* 233, 75–91.
Best, J. L. and Bristow, C. S. (eds) (1993) *Braided Rivers* (Geological Society special publication 75). London: Geological Society Publishing House.
Bintliff, J. (1976) The plain of Western Macedonia and the Neolithic site of Mea Nikomedeia. *Proceedings of the Prehistoric Society* 42, 241–62.
Bintliff, J. (2002) Time, process and catastrophism in the study of Mediterranean alluvial history: a review. *World Archaeology* 33, 417–35.
Bird, E. C. F. (1996) *Beach Management*. Chichester: John Wiley & Sons.
Bird, E. C. F. (2000) *Coastal Geomorphology: An Introduction*. Chichester: John Wiley & Sons.
Bird, E. C. F. (2008) *Coastal Geomorphology: An Introduction*, 2nd edn. Chichester: John Wiley & Sons.
Birkeland, P. W. (1999) *Soils and Geomorphology*, 3rd edn. Oxford: Oxford University Press.
Bishop, P. (2007) Long-term landscape evolution: linking tectonics and surface processes. *Earth Surface Processes and Landforms* 32, 329–65.
Bishop, P., Young, R. W., and McDougall, I. (1985) Stream profile change and long term landscape evolution: early Miocene and modern rivers of the east Australian highland crest, central New South Wales, Australia. *Journal of Geology* 93, 455–74.
Blake, S. (1989) Viscoplastic models of lava domes. *In IAVCEI (International Association of Volcanology and Chemistry of the Earth's Interior) Proceedings in Volcanology*, vol. 2. *Lava Flows and Domes*, pp. 88–126. Heidelberg: Springer.
Bliss, L. C. (1990) Arctic ecosystems: patterns of change in response to disturbance. In G. M. Woodwell (ed.) *The Earth in Transition: Patterns and Process of Biotic Impoverishment*, pp. 347–66. Cambridge: Cambridge University Press.
Bloom, A. L. (2002) Teaching about relict, no-analog landscapes. *Geomorphology* 47, 303–11.
Bloom, A. L. (2004) *Geomorphology: A Systematic Analysis of Late Cenozoic Landforms*, 3rd edn. Long Grove, Illinois: Waveland Press.
Bloom, A. L. and Yonekura, N. (1990) Graphic analysis of dislocated Quaternary shorelines. In Geophysics Study Commission (eds) *Sea-Level Change*, pp. 104–154. Washington, D.C.: National Academy Press.
Bögli, A. (1960) Kalklösung und Karrenbildung. *Zeitschrift für Geomorphologie* NF, *Supplementband* 2, 7–71.
Bonnet, S. and Crave, A. (2003) Landscape response to climate change: insights from experimental modeling and implications for tectonic versus climatic uplift of topography. *Geology* 31, 123–6.
Bosák, P., Ford, D. C., and Głazek, J. (1989) Terminology. In P. Bosák, D. C. Ford, J. Głazek, and I. Horáček (eds) *Paleokarst: A Systematic and Regional Review* (Developments in Earth Surface Processes 1), pp. 25–32. Amsterdam: Elsevier.
Bosák, P., Ford, D. C., Głazek, J., and Horáček, I. (eds) (1989) *Paleokarst: A Systematic and Regional* Review (Developments in Earth Surface Processes 1). Amsterdam: Elsevier.
Bowen, D. Q. (1973) The Quaternary deposits of the Gower. *Proceedings of the Geologists' Association* 84, 249–72.
Bowler, J. M. and Harford, L. B. (1966) Quaternary tectonics and the evolution of the riverine plain near Echuca, Victoria. *Journal of the Geological Society of Australia* 13, 339–54.
Breed, C. S., McCauley, J. F., Whitney, M. I., Tchakarian, V. D., and Laity, J. E. (1997) Wind erosion in drylands. In D. S. G. Thomas (ed.) *Arid Zone Geomorphology: Process, Form and Change in Drylands*, 2nd edn, pp. 437–64. Chichester: John Wiley & Sons.
Bremer, H. (1988) *Allgemeine Geomorphologie: Methodik – Grundvorstellungen – Ausblick auf den Landschaftshaushalt*. Berlin: Gebrüder Bornträger.
Bridge, J. S. (2003) *Rivers and Floodplains: Forms, Processes, and Sedimentary Record*. Oxford: Blackwell Science.
Brierley, G. and Hooke, J. (2015) Emerging geomorphic approaches to guide river management practices. *Geomorphology* 251, 1–5.
Brocx, M. and Semeniuk, V. (2007) Geoheritage and geoconservation – history, definition, scope and scale. *Journal of the Royal Society of Western Australia* 90, 53–87.
Broecker, W. S. (1965) Isotope geochemistry and the Pleistocene climatic record. In H. E. Wright Jr and D. G. Frey (eds) *The Quaternary of the United States*, pp. 737–53. Princeton, N.J.: Princeton University Press.
Broecker, W. S., Thurber, D. L., Goddard, J., Ku, T., Matthews, R. K., and Mesolella, K. J. (1968) Milankovitch hypothesis supported by precise dating of coral reefs and deep-sea sediments. *Science* 159, 1–4.
Brook, B. W., Ellis, E. C., Perring, M. P., Mackay, A. W., and Blomqvist, L. (2013) Does the terrestrial biosphere

have planetary tipping points? *Trends in Ecology and Evolution* 28, 396–401.

Brook, G. A. and Ford, D. C. (1978) The origin of labyrinth and tower karst and the climatic conditions necessary for their development. *Nature* 275, 493–6.

Brookes, A. J. (1985) Downstream morphological consequences of river channelisation in England and Wales. *Geographical Journal* 151, 57–65.

Brookes, A. J. and Shields, F. D. (1996) *River Channel Restoration: Guiding Principles for Sustainable Projects*. Chichester: John Wiley & Sons.

Brown, A. G., Tooth, S., Chiverrell, R. C., Rose, J., Thomas, D. S. G., Wainwright, J., Bullard, J. E., Thorndycraft, V. R., Aalto, R., and Downs, P. (2013) The Anthropocene: is there a geomorphological case? *Earth Surface Processes and Landforms* 38, 431–4.

Brown, E. H. (1980) Historical geomorphology: principles and practice. *Zeitschrift für Geomorphologie Supplementband* 36, 9–15.

Brown, H. (1956) Technological denudation. In W. L. Thomas (ed.) *Man's Role in Changing the Face of the Earth*, pp. 1023–32. Chicago, Ill.: Chicago University Press.

Brown, R. J. E. (1970) *Permafrost in Canada: Its Influence on Northern Development*. Toronto: University of Toronto Press.

Brunsden, D. and Kesel, R. H. (1973) The evolution of a Mississippi river bluff in historic time. *Journal of Geology* 81, 576–97.

Brunsden, D. and Thornes, J. B. (1979) Landscape sensitivity and change. *Transactions of the Institute of British Geographers* NS 4, 463–84.

Bryan, K. (1928) Historical evidence on changes in the channel of the Rio Puerco, a tributary of the Rio Grande in New Mexico. *Journal of Geology* 36, 265–82.

Büdel, J. (1957) Die 'Doppelten Einebnungsflächen' in den feuchten Tropen. *Zeitschrift für Geomorphologie* NF 1, 201–28.

Büdel, J. (1963) *Aufbau und Verschüttung Olympias: Mediterrane Flusstätigkeit seit der Frühantike*. Deutscher Geographentag, Heidelberg [1963], 179–83.

Büdel, J. (1977) *Klima-Geomorphologie*. Berlin, Stuttgart: Gebrüder Borntraeger.

Büdel, J. (1982) *Climatic Geomorphology*. Translated by Lenore Fischer and Detlef Busche. Princeton, N.J.: Princeton University Press.

Bull, W. B. (1992) *Geomorphic Responses to Climatic Change*. Oxford: Oxford University Press.

Buol, S. W., Southard, R. J., Graham, R. C., and McDaniel, P. A. (2003) *Soil Genesis and Classification*, 5th edn. Ames, Iowa: Iowa State Press.

Burbank, D. W. and Anderson, R. S. (2001) *Tectonic Geomorphology: A Frontier in Earth Science*. Malden, Mass.: Blackwell Science.

Burger, D. (1990) The travertine complex of Antalya/Southwest Turkey. *Zeitschrift für Geomorphologie* NF, *Supplementband* 77, 25–46.

Butler, B. E. (1959) *Periodic Phenomena in Landscapes as a Basis for Soil Studies* (CSIRO Soil Publication 14). Canberra, Australia: CSIRO.

Butler, B. E. (1967) Soil periodicity in relation to landform development in south-eastern Australia. In J. N. Jennings and J. A. Marbut (eds) *Landform Studies from Australia and New Guinea*, pp. 231–55. Cambridge: Cambridge University Press.

Butler, D. R. (1992) The grizzly bear as an erosional agent in mountainous terrain. *Zeitschrift für Geomorphologie* NF 36, 179–89.

Butler, D. R. (1995) *Zoogeomorphology: Animals as Geomorphic Agents*. Cambridge: Cambridge University Press.

Butler, D. R. and Sawyer, C. F. (2012). Introduction to the special issue – zoogeomorphology and ecosystem engineering. *Geomorphology* 157, 1–5.

Butzer, K. W. (1975) Pleistocene littoral-sedimentary cycles of the Mediterranean basin: a Mallorquin view. In K. W. Butzer and G. L. L. Isaacs (eds) *After the Australopithecines*, pp. 25–71. The Hague: Mouton.

Butzer, K. W. (1976) *Geomorphology from the Earth*. New York: Harper & Row.

Butzer, K. W. (1980) Holocene alluvial sequences: problems of dating and correlation. In R. A. Cullingford, D. A. Davidson, and J. Lewin (eds) *Timescales in Geomorphology*, pp. 131–42. Chichester: John Wiley & Sons.

Butzer, K. W. (2005) Environmental history in the Mediterranean world: cross-disciplinary investigation of cause-and-effect for degradation and soil erosion. *Journal of Archaeological Science* 32, 1773–800.

Caine, T. N. (1986) Sediment movement and storage on alpine slopes in the Colorado Rocky Mountains. In A. D. Abrahams (ed.) *Hillslope Processes*, pp. 115–37. Boston, Mass.: Allen & Unwin.

Capelotti, P. J. (2010) *The Human Archaeology of Space: Lunar, Planetary and Interstellar Relics of Exploration*. Jefferson, NC: McFarland & Co.

Carling, P., Jansen, J., and Meshkova, L. (2014) Multi-channel rivers: their definition and classification. *Earth Surface Processes and Landforms* 39, 26–37.

Carr, M. H. (2012) The fluvial history of Mars. *Philosophical Transactions for the Royal Society of London*, Series A, 370, 2193–215.

Carrivick, J. L., Tweed, F. S., Carling, P. A., Alho, P., Marren, P. M., Staines, K., Russell, A. J., Rushmer, E. L., and Duller, R. (2013) Discussion of 'Field evidence and hydraulic modelling of a large Holocene jökulhlaup at Jökulsá á Fjöllum channel, Iceland' by Douglas Howard, Sheryl Luzzadder-Beach and Timothy Beach, 2012. *Geomorphology* 201, 512–19.

Carson, M. A. and Kirkby, M. J. (2009) *Hillslope Form and Process*. Cambridge: Cambridge University Press.

Carter, R. W. G. (1992) *Coastal Environments*. London: Edward Arnold.

Carter, R. W. G., Nordstrom, K. F., and Psuty, N. P. (1990) The study of coastal dunes. In K. F. Nordstrom, N. P. Psuty, and R. W. G. Carter (eds) *Coastal Dunes: Form and Process*, pp. 1–16. Chichester: John Wiley & Sons.

Catt, J. A. (1987) Effects of the Devensian cold stage on soil characteristics and distribution in eastern England. In J. Boardman (ed.) *Periglacial Processes and Landforms in Britain and Ireland*, pp. 145–52. Cambridge: Cambridge University Press.

Chapman, P. (1993) *Caves and Cave Life* (New Naturalist Series no. 79). London: HarperCollins.

Chapman, V. J. (1977) Introduction. In V. J. Chapman (ed.) *Wet Coastal Ecosystems*, pp. 1–29. Amsterdam: Elsevier.

Chapuis, R. P. (1992) Similarity of internal stability criteria for granular soils. *Canadian Geotechnical Journal* 29, 711–13.

Charlton, R. (2008) *Fundamentals of Fluvial Geomorphology*. Abingdon: Routledge.

Chemenda, A. I., Bois, T., Bouissou, S., and Tric, E. (2009) Numerical modelling of the gravity-induced destabilization of a slope: the example of the La Clapière landslide, southern France. *Geomorphology* 109, 86–93.

Chen, A., Darbon, J., and Morel, J.-M. (2014) Landscape evolution models: a review of their fundamental equations. *Geomorphology* 219, 68–86.

Chernyakhovsky, A. G., Gradusov, B. P., and Chizhikova, N. P. (1976) Types of recent weathering crusts and their global distribution. *Geoderma* 16, 235–55.

Chin, A. (2002) The periodic nature of step–pool mountain streams. *American Journal of Science* 302, 144–67.

Chorley, R. J. (1965) A re-evaluation of the geomorphic system of W. M. Davis. In R. J. Chorley and P. Haggett (eds) *Frontiers in Geographical Teaching*, pp. 21–38. London: Methuen.

Chorley, R. J. (1978) Bases for theory in geomorphology. In C. Embleton, D. Brunsden, and D. K. C. Jones (eds) *Geomorphology: Present Problems and Future Prospects*, pp. 1–13. Oxford: Oxford University Press.

Chorley, R. J. and Kennedy, B. A. (1971) *Physical Geography: A Systems Approach*. London: Prentice Hall.

Chorley, R. J., Dunn, A. J., Beckinsale, R. P. (1964) *A History of the Study of Landforms or the Development of Geomorphology*, vol. 1, *Geomorphology Before Davis*. London: Methuen Wiley.

Chorley, R. J., Schumm, S. A., and Sugden, D. E. (1984) *Geomorphology*. London: Methuen.

Church, J. A., Clark, P. U., Cazenave, A., Gregory, J. M., Jevrejeva, S., Levermann, A., Merrifield, M. A., Milne, G. A., Nerem, R. S., Nunn, P. D., Payne, A. J., Pfeffer, W. T., Stammer, D., and Unnikrishnan, A. S. (2013) Sea level change. In *Climate Change 2013: The Physical Science Basis* (Contribution of Working Group I to the Fifth Assessment Report of the Intergovernmental Panel on Climate Change; T. F., Stocker, D. Qin, G.-K. Plattner, M. Tignor, S. K. Allen, J. Boschung, A. Nauels, Y. Xia, V. Bex, and P. M. Midgley (eds)). Cambridge, United Kingdom and New York, NY, USA: Cambridge University Press.

Church, M. (2010) The trajectory of geomorphology. *Progress in Physical Geography* 34, 265–86.

Church, M. and Ryder, J. M. (1972) Paraglacial sedimentation: a consideration of fluvial processes conditioned by glaciation. *Bulletin of the Geological Society of America* 83, 3059–71.

Church, M. and Ryder, J. M. (1989) Sedimentology and clast fabric of subaerial debris flow facies in a glacially influenced alluvial fan: a discussion. *Sedimentary Geology* 65, 195–6.

Clark, D. H., Steig, E. J., Potter, N., Jr, Gillespie, A. R. (1998) Genetic variability of rock glaciers. *Geografiska Annaler* 80, 175–82.

Clark, P. U., Dyke, A. S., Shakun, J. D., Carlson, A. E., Clark, J., Wohlfarth, B., Mitrovica, J. X., Hostetler, S. W., and McCabe, A. M. (2009) The last glacial maximum. *Science* 325, 710–14.

Clark, R. (1994) Tors, rock platforms and debris slopes at Stiperstones, Shropshire, England. *Field Studies* 8, 451–72.

Clarke, G. K. C. (2005) Subglacial processes. *Annual Review of Earth and Planetary Sciences* 33, 247–76.

Clarke, L. E. (2014) The use of live vegetation in geomorphological experiments: how to create optimal growing conditions. *Earth Surface Processes and Landforms* 39, 705–10.

Clarke, L. E. (2015) Experimental alluvial fans: advances in understanding of fan dynamics and processes. *Geomorphology* 244, 135–45.

Clayton, K. M. (1989) Sediment input from the Norfolk cliffs, eastern England: a century of coast protection and its effects. *Journal of Coastal Research* 5, 433–42.

Cole, J. W., Milner, D. M., and Spinks, K. D. (2005) Calderas and caldera structures: a review. *Earth-Science Reviews* 69, 1–26.

Collier, J. S., Oggioni, F., Gupta. S., Garcia-Mareno, D., Trentesaux, A., and De Batist, M. (2015) Streamlined islands and the English Channel megaflood hypothesis. *Global and Planetary Change* 135, 190–206.

Colls, A. E., Stokes, S., Blum, M. D., and Straffin, E. (2001) Age limits on the Late Quaternary evolution of the upper Loire River. *Quaternary Science Reviews* 20, 743–50.

Colman, S. M. and Pierce, K. L. (2000) Classification of Quaternary geochronologic methods. In J. S. Noller, J. M. Sowers, and W. R. Lettis (eds) *Quaternary Geochronology: Methods and Applications* (AGU

Reference Shelf 4), 2–5. Washington, D.C.: American Geophysical Union.

Condie, K. C. (1989) Origin of the Earth's crust. *Palaeogeography, Palaeoclimatology, Palaeoecology* (Global and Planetary Change Section) 75, 57–81.

Coney, P. J., Jones, D. L., and Monger, J. W. H. (1980) Cordilleran suspect terranes. *Nature* 288, 329–33.

Constantine, J. A., Dunne, T., Ahmed, J., Legleiter, C., and Lazarus, E. D. (2014) Sediment supply as a driver of river meandering and floodplain evolution in the Amazon Basin. *Nature Geoscience* 7, 899–903.

Cooke, R. U. (1990) *Geomorphology in Environmental Management: A New Introduction.* Oxford: Clarendon Press.

Cooke, R. U. (1994) Salt weathering and the urban water table in deserts. In D. A. Robinson and R. B. G. Williams (eds) *Rock Weathering and Landform Evolution*, pp. 193–205. Chichester: John Wiley & Sons.

Cooke, R. U., Warren, A., and Goudie, A. S. (1993) *Desert Geomorphology.* London: UCL Press.

Coombes, M. A. and Viles, H. A. (2015) Population-level zoogeomorphology: the case of the Eurasian badger (*Meles meles* L.). *Physical Geography* 36, 215–38.

Corenblit, D., Baas, A. C. W., Bornette, G., Darrozes, J., Delmotte, S., Francis, R. A., Gurnell, A. M., Julien, F., Naiman, R. J., and Steiger, J. (2011) Feedbacks between geomorphology and biota controlling Earth surface processes and landforms: a review of foundation concepts and current understandings. *Earth-Science Reviews* 106, 307–31.

Corenblit, D., Baas, A., Balke, T., Bouma, T., Fromard, F., Garófano-Gómez, V., González, E., Gurnell, A. M., Hortobágyi, B., Julien, F., Kim, D., Lambs, L., Stallins, A. J., Steiger, J., Tabacchi, E., and Walcker, R. (2015) Engineer pioneer plants respond to and affect geomorphic constraints similarly along water-terrestrial interfaces worldwide. *Global Ecology and Biogeography* 24, 1363–76.

Corenblit, D., Gurnell, A. M., Steiger, J., and Tabacchi, E. (2008) Reciprocal adjustments between landforms and living organisms: extended geomorphic evolutionary insights. *Catena* 73, 261–73.

Coulthard, T. J. and Van de Wiel, M. J. (2012) Modelling river history and evolution. *Philosophical Transactions for the Royal Society of London*, Series A, 370, 2123–42.

Cox, K. G. (1989) The role of mantle plumes in the development of continental drainage patterns. *Nature* 342, 873–6.

Coxon, P. and O'Callaghan, P. (1987) The distribution and age of pingo remnants in Ireland. In J. Boardman (ed.) *Periglacial Processes and Landforms in Britain and Ireland*, pp. 195–202. Cambridge: Cambridge University Press.

Crickmay, C. H. (1933) The later stages of the cycle of erosion. *Geological Magazine* 70, 337–47.

Crickmay, C. H. (1975) The hypothesis of unequal activity. In W. N. Melhorn and R. C. Flemal (eds) *Theories of Landform Development*, pp. 103–9. London: George Allen & Unwin.

Crofts, R. and Gordon, J. E. (2015) Geoconservation in protected areas. In G. L. Worboys, M. Lockwood, A. Kothari, S. Feary, and I. Pulsford (eds) *Protected Area Governance and Management*, pp. 531–68. Canberra, Australia: ANU Press, The Australian National University.

Crutzen, P. J. (2002) Geology of mankind. *Nature* 415, 23.

Crutzen, P. J. and Stoermer, E. F. (2000) The 'Anthropocene'. *Global Change Newsletter* 41, 17–18.

Culling, W. E. H. (1960) Analytical theory of erosion. *Journal of Geology* 68, 336–44.

Culling, W. E. H. (1963) Soil creep and the development of hillside slopes. *Journal of Geology* 71, 127–61.

Culling, W. E. H. (1965) Theory of erosion on soil-covered slopes. *Journal of Geology* 73, 230–54.

Curi, N. and Franzmeier, D. P. (1984) Toposequence of Oxisols from the central plateau of Brazil. *Soil Science Society of America Journal* 48, 341–6.

Dackombe, R. V. and Gardiner, V. (1983) *Geomorphological Field Manual.* London: Allen & Unwin.

Dalrymple, J. B., Blong, R. J., and Conacher, A. J. (1968) A hypothetical nine-unit landsurface model. *Zeitschrift für Geomorphologie* NF 12, 60–76.

Daniel, H. (2001) Replenishment versus retreat: the cost of maintaining Delaware's beaches. *Ocean and Coastal Management* 44, 87–104.

Daniels, R. B. and Hammer, R. D. (1992) *Soil Geomorphology.* New York: John Wiley & Sons.

Darwin, C. R. (1881) *The Formation of Vegetable Mould, Through the Action of Worms, with Observations on Their Habits.* London: John Murray.

Davies, G. F. (1977) Whole mantle convection and plate tectonics. *Geophysical Journal of the Royal Astronomical Society* 49, 459–86.

Davies, G. F. (1992) On the emergence of plate tectonics. *Geology* 20, 963–6.

Davies, J. L. (1980) *Geographical Variation in Coastal Development*, 2nd edn. Harlow, Essex: Longman.

Davis, R. A. and Hayes, M. O. (1984) What is a wave-dominated coast? *Marine Geology* 60, 313–29.

Davis, W. M. (1889) The rivers and valleys of Pennsylvania. *National Geographical Magazine* 1, 183–253. (Also in *Geographical Essays)*

Davis, W. M. (1898) *Physical Geography.* (Assisted by W. H. Snyder). Boston, Mass.: Ginn.

Davis, W. M. (1899) The geographical cycle. *Geographical Journal* 14, 481–504. (Also in *Geographical Essays*)

Davis, W. M. (1902) Base level, grade, and peneplain. *Journal of Geology* 10, 77–111.

Davis, W. M. (1909) *Geographical Essays.* Boston, Mass.: Ginn.

Day, M. and Urich, P. (2000) An assessment of protected karst landscapes in Southeast Asia. *Cave and Karst Science* 27, 61–70.

Day, M. J. and Chenoweth, M. S. (2004) The karstlands of Trinidad and Tobago, their land use and conservation. *Geographical Journal* 170, 256–66.

De Blasio, F. V. and Sæter, M.-B. (2009) Small-scale experimental simulation of talus evolution. *Earth Surface Process and Landforms* 34, 1685–92.

de Margerie, E. (1886) Géologie. *Polybiblion Revue Bibliographique Universelle, Partie littéraire* 2, 24, 310–30.

De Roo, A. P. J., Wesseling, C. G., Jetten, V. G., and Ritsema, C. J. (1996) LISEM: a physically-based hydrological and soil erosion model incorporated in a GIS. In *HydroGIS 96: Applications of Geographic Information Systems in Hydrology and Water Resources Management (proceedings of the Vienna Conference, April 1996)* (IAHS Publication no. 235), pp. 395–403. Rozendaalselaan, Netherlands: International Association of Hydrological Sciences.

De Waele, J., Plan, L., and Audra, P. (2009) Recent developments in surface and subsurface karst geomorphology: an introduction. *Geomorphology* 106, 1–8.

Dearing, J. A. (2008) Landscape change and resilience theory: a palaeoenvironmental assessment from Yunnan, SW China. *The Holocene* 18, 117–27.

Desrochers, A. and James, M. P. (1988) Early Palaeozoic surface and subsurface paleokarst: Middle Ordovician carbonates, Mingan Islands, Quebec. In N. P. James and P. W. Choquette (eds) *Paleokarst*, pp. 183–210. New York: Springer.

Dietrich, W. E. and Perron, J. T. (2006) The search for a topographic signature of life. *Nature* 439, 411–18: doI:10.1038/nature04452.

Dietrich, W. E. and Smith, J. D. (1983) Influence of the point bar on flow through curved channels. *Water Resources Research* 9, 1173–92.

Dietrich, W. E., Bellugi, D. G., Sklar, L. S., Stock, J. D., Heimsath, A. M., and Roering, J. J. (2003) Geomorphic transport laws for predicting landscape form and dynamics. In P. R. Wilcock and R. M. Iverson (eds) *Prediction in Geomorphology* (Geophysical Monograph 135), pp. 103–32. Washington, D.C.: American Geophysical Union.

Doerr, S. H. (1999) Karst-like landforms and hydrology in quartzites of the Venezuelan Guyana shield: pseudokarst or 'real' karst? *Zeitschrift für Geomorphologie* NF 43, 1–17.

Doerr, S. H. (2000) Morphology and genesis of some unusual weathering features developed in quartzitic sandstone, north-central Thailand. *Swansea Geographer* 35, 1–8.

Dotterweich, M. (2013) The history of human-induced soil erosion: geomorphic legacies, early descriptions and research, and the development of soil conservation – a global synopsis. *Geomorphology* 201, 1–34.

Douglas, I. (1971) Dynamic equilibrium in applied geomorphology: two case studies. *Earth Science Journal* 5, 29–35.

Douglas, I. (1987) Plate tectonics, palaeoenvironments and limestone geomorphology in west-central Britain. *Earth Surface Processes and Landforms* 12, 481–95.

Douglas, I. (1990) Sediment transfer and siltation. In B. L. Turner II, W. C. Clark, R. W. Kates, J. F. Richards, J. T. Mathews, and W. B. Meyer (eds) *The Earth as Transformed by Human Action: Global and Regional Changes in the Biosphere over the Past 300 Years*, pp. 215–34. Cambridge: Cambridge University Press with Clark University.

Douglas, I. (2000) Fluvial geomorphology and river management. *Australian Geographical Studies* 38, 253–62.

Douglas, I. and Lawson, N. (2001) The human dimensions of geomorphological work in Britain. *Journal of Industrial Ecology* 4, 9–33.

Downs, P. W. and Booth, D. B. (2011) Geomorphology in environmental management. In K. J. Gregory and A. S. Goudie (eds) *The SAGE Handbook of Geomorphology*, pp. 78–104. London: Sage Publications.

Duchaufour, P. (1982) *Pedology: Pedogenesis and Classification*. Translated by T. R. Paton. London: George Allen & Unwin.

Dunne, T. (1980) Formation and controls of channel networks. *Progress in Physical Geography* 4, 211–39.

Dunne, T. and Leopold, L. B. (1978) *Water in Environmental Planning*. San Francisco, Calif.: W. H. Freeman.

Dury, G. H. (1969) Relation of morphometry to runoff frequency. In R. J. Chorley (ed.) *Water, Earth, and Man: A Synthesis of Hydrology, Geomorphology, and Socio-Economic Geography*, pp. 419–30. London: Methuen.

Ebert, K. (2009a) Terminology of long-term geomorphology: a Scandinavian perspective. *Progress in Physical Geography* 33, 1–20.

Ebert, K. (2009b) *Cenozoic Landscape Evolution in Northern Sweden: Geomorphological Interpretation within a GIS-Framework*. PhD dissertation no. 19, Department of Physical Geography and Quaternary Geology, Stockholm University, Sweden.

Ebert, K. (2015) GIS analyses of ice-sheet erosional impacts on the exposed shield of Baffin Island, eastern Canadian Arctic. *Canadian Journal of Earth Sciences* 52, 966–79.

Edgeworth, M., deB Richter, D., Waters, C., Haff, P., Neal, C., and Price, S. J. (2015) Diachronous beginnings of the Anthropocene: the lower bounding surface of anthropogenic deposits. *The Anthropocene Review* 2, 33–58.

Ehlen, J. (2005) Above the weathering front: contrasting approaches to the study and classification of the weathered mantle. *Geomorphology* 67, 7–21.

Ellis, E. C. (2015) Ecology in an anthropogenic biosphere. *Ecological Monographs* 85, 287–331.

Elverfeldt, K. von, Embleton-Hamann, C. and Slaymaker, O. (2015) Self-organizing change? On drivers, causes and global environmental change. *Geomorphology* 252, 48–58.

Embleton, C. and King, C. A. M. (1975a) *Glacial Geomorphology*. London: Edward Arnold.

Embleton, C. and King, C. A. M. (1975b) *Periglacial Geomorphology*. London: Edward Arnold.

Embleton, C. and Thornes, J. (eds) (1979) *Process in Geomorphology*. London: Edward Arnold.

Emery, K. O. and Kuhn, G. G. (1982) Sea cliffs: their processes, profiles, and classification. *Bulletin of the Geological Society of America* 93, 644–54.

Engebretson, D. C., Kelley, K. P., Cashman, H. J., and Richards, M. A. (1992) 180 million years of subduction. *GSA–Today* 2, 93–5, and 100.

England, J. (1986) Glacial erosion of a High Arctic valley. *Journal of Glaciology* 32, 60–4.

Erhart, H. (1938) *Traité de Pédologie*, 2 vols. Strasbourg: Institut Pédologique.

Erhart, H. (1956) *La Genèse des Sols en tant que Phénomène géologique*. Paris: Masson.

Evans, D. J. A., Phillips, E. R., Hiemstra, J. F., and Auton, C. A. (2006) Subglacial till: formation, sedimentary characteristics and classification. *Earth-Science Reviews* 78, 115–76.

Evans, I. S. (1980) An integrated system of terrain analysis and slope mapping. *Zeitschrift für Geomorphologie* NF 36, 274–95.

Evans, I. S. (1994) Cartographic techniques in geomorphology. In A. S. Goudie (ed.) *Geomorphological Techniques*, 2nd edn, pp. 97–108. London and New York: Routledge.

Evans, I. S. (2012) Geomorphometry and landform mapping: what is a landform? *Geomorphology* 137, 94–106.

Eyles, N. and Scheidegger, A. E. (1995) Environmental significance of bedrock jointing in Southern Ontario, Canada. *Environmental Geology* 26, 269–77.

Eyles, N., Arnaud, E., Scheidegger, A. E., and Eyles, C. H. (1997) Bedrock jointing and geomorphology in southwestern Ontario, Canada: an example of tectonic predesign. *Geomorphology* 19, 17–34.

Fairbridge, R. W. (1980) The estuary: its definition and geodynamics cycle. In E. Olausson and I. Cato (eds) *Chemistry and Biogeochemistry of Estuaries*, pp. 1–35. Chichester: John Wiley & Sons.

Fairbridge, R. W. (2004) Classification of coasts. *Journal of Coastal Research* 20, 155–65.

Fairbridge, R. W. and Finkl, C. W., Jr (1980) Cratonic erosional unconformities and peneplains. *Journal of Geology* 88, 69–86.

Fang, X.-M., Li, J.-J., and Van der Voo, R. (1999a) Rock-magnetic and grain size evidence for intensified Asian atmospheric circulation since 800,000 years B.P. related to Tibetan uplift. *Earth and Planetary Science Letters* 165, 129–44.

Fang, X.-M., Ono, Y., Fukusawa, H., Bao-Tian, P., Li, J.-J., Dong-Hong, G., Oi, K., Tsukamoto, S., Torii, M., and Mishima, T. (1999b) Asian summer monsoon instability during the past 60,000 years: magnetic susceptibility and pedogenic evidence from the western Chinese Loess Plateau. *Earth and Planetary Science Letters* 168, 219–32.

Fei, S., Phillips, J. D., and Shouse, M. (2014) Biogeomorphic impacts of invasive species. *Annual Review of Ecology, Evolution, and Systematics* 45, 69–87.

Flint, R. F. (1971) *Glacial and Quaternary Geology*. New York: John Wiley & Sons.

Forbes, B. C. (1999) Land use and climate change on the Yamal Peninsula of north-west Siberia: some ecological and socio-economic implications. *Polar Research* 18, 367–73.

Ford, D. C. and Ewers, R. O. (1978) The development of limestone cave systems in the dimensions of length and depth. *Canadian Journal of Earth Sciences* 15, 1783–98.

Ford, D. C. and Williams, P. W. (1989) *Karst Geomorphology and Hydrology*. London: Chapman & Hall.

Ford, D. C. and Williams, P. W. (2007) *Karst Geomorphology and Hydrology*, rev. edn. Chichester: John Wiley & Sons.

Ford, T. D. (1989) Paleokarst of Britain. In P. Bosák, D. C. Ford, J. Głazek, and I. Horáček (eds) *Paleokarst: A Systematic and Regional Review* (Developments in Earth Surface Processes 1), pp. 51–70. Amsterdam: Elsevier.

Foulger, G. R., Natland, J. H., Presnall, D. C., and Anderson, D. L. (eds) (2005) *Plates, Plumes, and Paradigms* (Geological Society of America Special Paper 388). Boulder, Colo.: Geological Society of America.

Fournier, F. (1960) *Climat et érosion: La Relation entre l'érosion du sol par l'eau et les précipitations atmosphériques*. Paris: Presses Universitaires de France.

Francis, P. (1993) *Volcanoes: A Planetary Perspective*. Oxford: Clarendon Press.

Fraser, G. S. (1989) *Clastic Depositional Sequences: Processes of Evolution and Principles of Interpretation*. Englewood Cliffs, N.J.: Prentice Hall.

French, H. M. (1976) *The Periglacial Environment*, 1st edn. London and New York: Longman.

French, H. M. (1996) *The Periglacial Environment*, 2nd edn. Harlow, Essex: Addison-Wesley Longman.

French, H. M. (2007) *The Periglacial Environment*, 3rd edn. Chichester: John Wiley & Sons.

French, H. M. (ed.) (2004) *Periglacial Geomorphology* (Geomorphology: Critical Concepts in Geography, vol. V). London: Routledge.

Frumkin, A. (2013) New developments of karst geomorphology concepts. In John F. Shroder (Editor in chief), A. Frumkin (Volume Editor) *Treatise on Geomorphology, Volume 6, Karst Geomorphology*, pp. 1–13. San Diego: Academic Press.

Fryirs, K. A. and Brierley, G. J. (2013) *Geomorphic Analysis of River Systems: An Approach to Reading the Landscape*. Chichester: Wiley–Blackwell.

Fryirs, K., Brierley, G. J., and Erskine, W. D. (2012) Use of ergodic reasoning to reconstruct the historical range of variability and evolutionary trajectory of rivers. *Earth Surface Processes and Landforms* 37, 763–73.

Fukao, Y., Maruyama, S., Obayashi, M., and Inoue, H. (1994) Geological implication of the whole mantle P-wave tomography. *Journal of the Geological Society of Japan* 100, 4–23.

Furlani, S. and Ninfo, A. (2015) Is the present the key to the future. *Earth-Science Reviews* 142, 38–46.

Gabet, E. J. (2003) Sediment transport by dry ravel. *Journal of Geophysical Research* 108 (B1), 2049: doi: 10:1029/2001JB001686.

Gabet, E. J. and Dunne, T. (2003) Sediment detachment by rain power. *Water Resources Research* 39, 1002: doi: 10.1029/2001WR000656.

Gabet, E. J., Reichman, O. J., and Seabloom, E. W. (2003) The effects of bioturbation soil process and sediment transport. *Annual Review of Earth and Planetary Sciences* 31, 249–73.

Gallup, C. D., Edwards, R. L., and Johnson, R. G. (1994) The timing of high sea levels over the past 200,000 years. *Science* 263, 796–800.

Gares, P. A., Sherman, D. J., and Nordstrom, K. F. (1994) Geomorphology and natural hazards. *Geomorphology* 10, 1–18.

Geikie, A. (1905) *The Founders of Geology*, 2nd edn. London and New York: Macmillan & Co.

Genty, D., Baker, A., and Vokal, B. (2001) Intra- and inter-annual growth rate of modern stalagmites. *Chemical Geology* 176, 191–212.

Gerasimov, I. P. (1979) Anthropogene and its major problem. *Boreas* 8, 23–30.

Gerrard, J. (1992) *Soil Geomorphology: An Integration of Pedology and Geomorphology*. London: Chapman & Hall.

Gerson, R. and Grossman, S. (1987) Geomorphic activity on escarpments and associated fluvial systems in hot deserts. In M. R. Rampino, J. E. Sanders, W. S. Newman, and L. K. Königsson (eds) *Climate: History, Periodicity, and Predictability*, pp. 300–22. New York: Van Nostrand Reinhold.

Gibbs, R. J. (1970) Mechanisms controlling world water chemistry. *Science* 170, 1088–90.

Gibbs, R. J. (1973) Mechanisms controlling world water chemistry: evaporation–crystallization processes. *Science* 172, 871–2.

Gilbert, G. K. (1877) *Geology of the Henry Mountains (Utah)* (United States Geographical and Geological Survey of the Rocky Mountains Region). Washington, D.C.: United States Government Printing Office.

Gilbert, G. K. (1900) Rhythms in geologic time. *Proceedings of the American Association for the Advancement of Science* 49, 1–19.

Gilbert, G. K. (1914) *The Transportation of Debris by Running Water* (United States Geological Survey Professional Paper 86). Washington, D.C.: United States Government Printing Office.

Gilchrist, A. R., Kooi, H., and Beaumont, C. (1994) Post-Gondwana geomorphic evolution of south-western Africa: implications for the controls of landscape development from observation and numerical experiments. *Journal of Geophysical Research* 99B, 12, 211–28.

Giles, P. T. (1998) Geomorphological signatures: classification of aggregated slope unit objects from digital elevation and remote sensing data. *Earth Surface Processes and Landforms* 23, 581–94.

Giles, P. T. and Franklin, S. E. (1998) An automated approach to the classification of the slope units using digital data. *Geomorphology* 21, 251–64.

Gillieson, D. (1996) *Caves: Processes, Development and Management*. Oxford: Blackwell.

Glasser, N. F. and Barber, G. (1995) Cave conservation plans: the role of English Nature. *Cave and Karst Science* 21, 33–6.

Glasser, N. F. and Bennett, M. R. (2004) Glacial erosional landforms: origins and significance for palaeoglaciology. *Progress in Physical Geography* 28, 43–75.

Glasser, N. F., Jansson, K. M., Harrison, S., and Rivera, A. (2005) Geomorphological evidence for variations of the North Patagonian Icefield during the Holocene. *Geomorphology* 71, 263–77.

Godard, A., Lagasquie, J.-J., and Lageat, Y. (2001) *Basement Regions*. Translated by Yanni Gunnell. Heidelberg: Springer.

Golomb, B. and Eder, H. M. (1964) Landforms made by man. *Landscape* 14, 4–7.

González, F. I. (1999) Tsunami! *Scientific American* 280 (May), 44–55.

Gordon, J. E., Barron, H. F., Hansom, J. D., and Thomas, M. F. (2012) Engaging with geodiversity – why it matters. *Proceedings of the Geologists' Association* 123, 1–6.

Gossman, H. (1970) *Theorien zur Hangentwicklung in verschiedenen Klimazonen: mathematische Hangmodelle und ihre Beziehung zu den Abtragungsvorgängen* (Würzburger Geographische Arbeiten, Heft 31). Würzburg: Geographischen Instituts der Universität Würzburg.

Goudie, A. and Viles, H. (2010) *Landscapes and Geomorphology: A Very Short Introduction*. Oxford: Oxford University Press.

Goudie, A. S. (ed.) (1994) *Geomorphological Techniques*, 2nd edn. London and New York: Routledge.

Goudie, A. S. (1995) *The Changing Earth: Rates of Geomorphological Processes*. Oxford: Blackwell Science.

Goudie, A. S. (1999) Wind erosional landforms: yardangs and pans. In A. S. Goudie, I. Livingstone, and S. Stokes (eds) *Aeolian Environments, Sediments and Landforms* (British Geomorphological Research Group Symposia Series, Papers from the Fourth International Conference on Aeolian Research (ICAR 4), held in School of Geography and St Catherine's College, University of Oxford, July 1998), pp. 167–80. Chichester: John Wiley & Sons.

Goudie, A. S. (2005) The drainage of Africa since the Cretaceous. *Geomorphology* 67, 437–56.

Goudie, A. S. and Wells, G. L. (1995) The nature, distribution and formation of pans in arid zones. *Earth-Science Reviews* 38, 1–69.

Goudie, A. S., Livingstone, I., and Stokes, S. (eds) (1999) *Aeolian Environments, Sediments and Landforms*. Chichester: John Wiley & Sons.

Grab, S. W. (1994) Thufur in the Mohlesi Valley, Lesotho, southern Africa. *Permafrost and Periglacial Processes* 5, 111–18.

Grab, S. W. (2005) Aspects of the geomorphology, genesis and environmental significance of earth hummocks (thúfur, pounus): miniature cryogenic mounds. *Progress in Physical Geography* 29, 139–55.

Gray, J. M. (1982) Un-weathered, glaciated bedrock on an exposed lake bed in Wales. *Journal of Glaciology* 28, 483–97.

Gray, J. M. (1993) Quaternary geology and waste disposal in south Norfolk, England. *Quaternary Science Reviews* 12, 899–912.

Gray, J. M. (2013) *Geodiversity: Valuing and Conserving Abiotic Nature*, 2nd edn. Chichester: John Wiley & Sons.

Gray, M., Gordon, J. E., and Brown, E. J. (2013) Geodiversity and the ecosystem approach: the contribution of geoscience in delivering integrated environmental management. *Proceedings of the Geologists' Association* 124, 659–73.

Green, C. P. (1985) Pre-Quaternary weathering residues, sediment and landform development: examples from southern Britain. In K. S. Richards, R. R. Arnett, and S. Ellis (eds) *Geomorphology and Soils*, pp. 58–77. London: Allen & Unwin.

Gregory, K. J. and Goudie, A. S. (eds) (2011) *The Sage Handbook of Geomorphology*. London: Sage.

Gregory, K. J. and Lewin, J. (2014) *The Basics of Geomorphology: Key Concepts*. London: Sage.

Gregory, K. J., Lane, S. N., Lewin, J., Ashworth, P. J., Downs, P. W., Kirkby, M. J., and Viles, H. A. (2014) Communicating geomorphology: global challenges for the twenty-first century. *Earth Surface Processes and Landforms* 39, 476–86.

Grosswald, M. G. (1998) New approach to the Ice Age paleohydrology of northern Eurasia. In G. Benito, V. R. Baker, and K. J. Gregory (eds) *Palaeohydrology and Environmental Change*, pp. 199–214. Chichester: John Wiley & Sons.

Grove, A. T. (1969) Landforms and climatic change in the Kalahari and Ngamiland. *The Geographical Journal* 135, 191–212.

Grove, A. T. and Rackham, O. (2001) *The Nature of Mediterranean Europe: An Ecological History*. New Haven, Conn., and London: Yale University Press.

Gudmundsson, M. T., Sigmundsson, E., Björnsson, H. (1997) Ice-volcano interaction of the 1996 Gjálp subglacial eruption, Vatnajökull, Iceland. *Nature* 389, 954–7.

Gunnell, Y. (2015) Ancient landforms in dynamic landscapes: inheritance, transience and congruence in Earth-surface systems. *Geomorphology* 233, 1–4.

Gunnell, Y. and Fleitout, L. (2000) Morphotectonic evolution of Western Ghats, India. In M. A. Summerfield (ed.) *Geomorphology and Global Tectonics*, pp. 321–38. Chichester: John Wiley & Sons.

Gupta, A. (1983) High magnitude floods and stream channel response. In J. D. Collinson and J. Lewin (eds) *Modern and Ancient Fluvial Systems*, pp. 219–27. Oxford: Blackwell.

Gupta, S., Collier, J. S., Palmer-Felgate, A., and Potter, G. (2007) Catastrophic flooding origin of shelf valley systems in the English Channel. *Nature* 448, 342–5.

Guthrie, R. H. (2013) Mass-movement style, activity state, and distribution. In John F. Shroder (Editor in chief), R. A. Marsden and M. Stoffel (Volume Editors) *Treatise on Geomorphology, Volume 7, Mountain and Hillslope Geomorphology*, pp. 230–8. San Diego: Academic Press.

Haasis, F. W. (1935) Anthropo-ecology of China. *Ecology* 16, 655–6.

Hack, J. T. (1960) Interpretation of erosional topography in humid temperate regions. *American Journal of Science* (Bradley Volume) 258-A, 80–97.

Haff, P. K (2002) Neogeomorphology. *Eos, Transactions of the American Geophysical Union* 83 (29), 310 and 317.

Haff, P. K. (2010) Hillslopes, rivers, plows, and trucks: mass transport on Earth's surface by natural and technological processes. *Earth Surface Processes and Landforms* 35, 1157–66.

Halimov, M. and Fezer, F. (1989) Eight yardang types in Central Asia. *Zeitschrift für Geomorphologie* NF 33, 205–17.

Hall, A. M. (1991) Pre-Quaternary landscape evolution in the Scottish Highlands. *Transactions of the Royal Society of Edinburgh: Earth Sciences* 82, 1–26.

Hall, K., Thorn, C. E., Matsuoka, N., and Prick, A. (2002) Weathering in cold regions: some thoughts and perspectives. *Progress in Physical Geography* 26, 577–603.

Hambrey, M. J. (1994) *Glacial Environments*. London: UCL Press.

Hamilton, C. (2015) Getting the Anthropocene so wrong. *The Anthropocene Review* 2, 102–7.

Hamilton, C. and Grinevald, J. (2015) Was the Anthropocene anticipated? *The Anthropocene Review* 2, 59–72.

Hammond, E. H. (1954) Small-scale continental landform maps. *Annals of the Association of American Geographers* 44, 33–42.

Hansom, J. (2012) Geodiversity in a changing environment. *Scottish Geographical Journal* 128, 173–6.

Hantke, R. and Scheidegger, A. E. (1999) Tectonic predesign in geomorphology. In S. Hergarten and H. J. Neugebauer (eds) *Process Modelling and Landform Evolution* (Lecture Notes in Earth Sciences, 78), pp. 251–66. Berlin: Springer.

Haq, B. U., Hardenbol, J., and Vail, P. R. (1987) Chronology of fluctuating sea levels since the Triassic. *Science* 235, 1156–67.

Harden, C. P. (2014) The human–landscape system: challenges for geomorphologists, *Physical Geography* 35, 76–89.

Harris, P. T., Macmillan-Lawler, M., Rupp, J., and Baker, E. K. (2014) Geomorphology of the oceans. *Marine Geology* 352, 4–24.

Harvey, A. (2012) *Introducing Geomorphology: A Guide to Landforms and Processes*. Edinburgh: Dunedin Academic Press.

Harvey, G. L., Henshaw, A. J., Moorhouse, T. P., Clifford, N. J., Holah, H., Grey, J., and Macdonald, D. W. (2013) Invasive crayfish as drivers of fine sediment dynamics in rivers: field and laboratory evidence. *Earth Surface Processes and Landforms* 39, 259–71.

Hayes, M. O. (1967) Relationship between coastal climate and bottom sediment type on the inner continental shelf. *Marine Geology* 5, 11–132.

Haynes, G. (2012) Elephants (and extinct relatives) as earth-movers and ecosystem engineers. *Geomorphology* 157, 99–107.

Heginbottom, J. A. (2002) Permafrost mapping: a review. *Progress in Physical Geography* 26, 623–42.

Hengl, T. and Reuter H. I. (eds) (2009) *Geomorphometry: Concepts, Software, Applications* (*Developments in Soil Science*, vol. 33). Amsterdam: Elsevier.

Herget, J. (1998) Anthropogenic influence on the development of the Holocene terraces of the River Lippe, Germany. In G. Benito, V. R. Baker, and K. J. Gregory (eds) *Palaeohydrology and Environmental Change*, pp. 167–79. Chichester: John Wiley & Sons.

Higgins, C. G. (1975) Theories of landscape development: a perspective. In W. N. Melhorn and R. C. Flemal (eds) *Theories of Landform Development*, pp. 1–28 London: George Allen & Unwin.

Hill, S. M., Ollier, C. D., and Joyce, E. B. (1995) Mesozoic deep weathering and erosion: an example from Wilsons Promontory, Australia. *Zeitschrift für Geomorphologie* NF 39, 331–9.

Hjort, J., Gordon, J. E., Gray, M., and Hunter, M. L., Jr (2015) Why geodiversity matters in valuing Nature's stage. *Conservation Biology* 29, 630–9.

Hjulström, F. (1935) Studies of the morphological activity of rivers as illustrated by the River Fyris. *Bulletin of the Geological Institute of Uppsala* 25, 221–527.

Hodges, K. V. (2006) Climate and the evolution of mountains. *Scientific American* 295, 54–61.

Holliday, V. T. (1988) Mt Blanco revisited: soil–geomorphic implications for the ages of the upper Cenozoic Blanco and Blackwater Draw Formations. *Geology* 16, 505–8.

Holliday, V. T. (1989) The Blackwater Draw Formation (Quaternary): a 1.4-plus-m.y. record of eolian sedimentation and soil formation in the Southern High Plains. *Bulletin of the Geological Society of America* 101, 598–607.

Holmes, A. (1965) *Principles of Physical Geology*, rev. edn. London and Edinburgh: Thomas Nelson & Sons.

Hooke, J. M. (ed.) (1988) *Geomorphology in Environmental Planning*. Chichester: John Wiley & Sons.

Hooke, R. LeB. (2000) On the history of humans as geomorphic agents. *Geology* 28, 843–6.

Horton, R. E. (1945) Erosional development of streams and their drainage basins: hydrophysical approach to quantitative morphology. *Bulletin of the Geological Society of America* 56, 275–370.

Hovius, N. (1998) Controls on sediment supply by large rivers. In K. W. Shanley and P. J. McCabe (eds) *Relative Roles of Eustasy, Climate, and Tectonism in Continental Rocks* (Society for Economic Paleontology and Mineralogy, special publication 59), pp. 3–16. Tulsa, Okla.: Society for Economic Paleontology and Mineralogy.

Howard, A. D. (1965) Geomorphological systems: equilibrium and dynamics. *American Journal of Science* 263, 302–12.

Howard, A. D. (1978) Origin of the stepped topography of the Martian poles. *Icarus* 34, 581–99.

Howard, A. D. (1988) Equilibrium models in geomorphology. In M. G. Anderson (ed.) *Modelling Geomorphological Systems*, pp. 49–72. Chichester: John Wiley & Sons.

Howard, A. D., Kochel, R. C., and Holt, H. E. (eds) (1988) *Sapping Features of the Colorado Plateau: A Comparative Planetary Geology Field Guide* (Scientific and Technical Information Division, National Aeronautics and Space Administration, NASA SP–491). Washington, D.C.: United States Government Printing Office.

Howard, J. A. and Mitchell, C. W. (1985) *Phytogeomorphology*. New York and Chichester: John Wiley & Sons.

Hoyt, J. H. (1967) Barrier island formation. *Bulletin of the Geological Society of America* 78, 1125–36.

Hoyt, J. H. (1969) Chenier versus barrier, genetic and stratigraphic distinctions. *American Association of Petroleum Geologists Bulletin* 53, 299–306.

Hubbard, B. and Glasser, N. F. (2005) *Field Techniques in Glaciology and Glacial Geomorphology*. Chichester: John Wiley & Sons.

Huggett, R. J. (1975) Soil landscape systems: a model of soil genesis. *Geoderma* 13, 1–22.

Huggett, R. J. (1976) Lateral translocation of soil plasma through a small valley basin in the Northaw Great Wood, Hertfordshire. *Earth Surface Processes* 1, 99–109.

Huggett, R. J. (1985) *Earth Surface Systems* (Springer Series in Physical Environment 1). Heidelberg: Springer.

Huggett, R. J. (1989) *Cataclysms and Earth History: The Development of Diluvialism*. Oxford: Clarendon Press.

Huggett, R. J. (1991) *Climate, Earth Processes and Earth History*. Heidelberg: Springer.

Huggett, R. J. (1993) *Modelling the Human Impact on Nature: Systems Analysis of Environmental Problems*. Oxford: Oxford University Press.

Huggett, R. J. (1995) *Geoecology: An Evolutionary Approach*. London: Routledge.

Huggett, R. J. (1997a) *Catastrophism: Comets, Asteroids and Other Dynamic Events in Earth History*. London: Verso.

Huggett, R. J. (1997b) *Environmental Change: The Evolving Ecosphere*. London: Routledge.

Huggett, R. J. (2006) *The Natural History of the Earth: Debating Long-term Change in the Geosphere and Biosphere*. London: Routledge.

Huggett, R. J. (2007) A history of the systems approach in geomorphology. *Géomorphologie: Relief, Processus, Environnement* 2007(2), 145–58.

Huggett, R. J. and Cheesman, J. E. (2002) *Topography and the Environment*. Harlow, Essex: Prentice Hall.

Hughes, T. J., Denton, G. H., and Fastook, J. L. (1985) The Antarctic ice sheet: an analog for Northern Hemisphere paleo-ice sheets? In M. J. Woldenberg (ed.) *Models in Geomorphology*, pp. 25–72. Boston, Mass.: Allen & Unwin.

Hungr, O., Leroueil, S., and Picarelli, L. (2014) The Varnes classification of landslide types, an update. *Landslides* 11, 167–94.

Huntington, E. (1914) *The Climatic Factor as Illustrated in Arid America*. With contributions by Charles Schuchert, A. E. Douglas, and C. J. Kullmer (Carnegie Institution of Washington, Publication 192). Washington, D.C.: Carnegie Institution.

Hupp, C. R. and Osterkamp, W. R. (2013) Vegetation ecogeomorphology, dynamic equilibrium, and disturbance. In John F. Shroder (Editor in chief), D. R. Butler, and C. R. Hupp (Volume Eds) *Treatise on Geomorphology, Volume 12, Ecogeomorphology*, pp. 94–106. San Diego: Academic Press.

Hupy, J. P. and Koehler, T. (2012) Modern warfare as a significant form of zoogeomorphic disturbance upon the landscape. *Geomorphology* 157, 169–82.

Iaffaldano, G., Husson, L., and Bunge, H.-P. (2011) Monsoon speeds up Indian plate motion. *Earth and Planetary Science Letters* 304, 503–10.

Ibáñez, J. J., Vargas, R. J., and Vázquez-Hoehne, A. (2013) Pedodiversity state of the art and future challenges. In J. J. Ibáñez and J. G. Bockheim (eds) *Pedodiversity*, pp. 1–28. Baton Rouge, London, New York: CRC Press.

IGBP (1990) Past Global Changes Project. In *The IGBP: A Study of Global Change. The Initial Core Projects*, Chapter 7. Stockholm: International Geosphere–Biosphere Programme Secretariat.

Imaizumi, F., Hattanji, T., and Hayakawa, Y. S. (2009) Channel initiation by surface and subsurface flows in a steep catchment of the Akaishi Mountains, Japan. *Geomorphology* 115, 32–42.

Inkpen, R. and Turner, D. (2012) The topography of historical contingency. *Journal of the Philosophy of History* 6, 1–19.

Inkpen, R. J., Cooke, R. U., and Viles, H. A. (1994) Processes and rates of urban limestone weathering. In D. A. Robinson and R. B. G. Williams (eds) *Rock Weathering and Landform Evolution*, pp. 119–30. Chichester: John Wiley & Sons.

James, L. A. and Marcus, W. A. (2006) The human role in changing fluvial systems: retrospect, inventory and prospect. *Geomorphology* 79, 152–71.

Jansson, M. B. (1988) *Land Erosion by Water in Different Climates*. (Uppsala Universitet Naturgeografiska Institutionen Rapport no. 57). Uppsala: Department of Physical Geography, University of Uppsala.

Japsen, P., Bidstrup, T., and Lidmar-Bergström, K. (2002) Neogene uplift and erosion of southern Scandinavia induced by the rise of the South Swedish Dome. In A. G. Doré, J. A. Cartwright, M. S. Stoker, J. P. Turner, and N. White (eds) *Exhumation of the North Atlantic Margin: Timing, Mechanisms and Implications for Petroleum Exploration* (Geological Society, special publication 196), pp. 183–207. London: Geological Society.

Jarman, D., Wilson, P., and Harrison, S. (2013) Are there any relict rock glaciers in the British mountains? *Journal of Quaternary Science* 28, 131–43.

Jennings, J. N. (1967) Further remarks on the Big Hole, near Braidwood, New South Wales. *Helictite* 6, 3–9.

Jennings, J. N. (1971) *Karst* (An Introduction to Systematic Geomorphology, vol. 7). Cambridge, Mass. and London: MIT Press.

Jennings, J. N. (1985) *Karst Geomorphology*. Oxford and New York: Blackwell.

Jenny, H. (1980) *The Soil Resource: Origin and Behaviour* (Ecological Studies, vol. 37). New York: Springer.

Jessen, O. (1936) *Reisen und Forschungen in Angola*. Berlin: D. Reimer.

Joel, A. H. (1937) *Soil Conservation Reconnaissance Survey of the Southern Great Plains Wind-Erosion Area* (United States Department of Agriculture, Technical Bulletin no. 556). Washington, D.C.: US Government Printing Office.

Johns, J. S., Barreto, P., and Uhl, C. (1996) Logging damage during planned and unplanned logging operations in the eastern Amazon. *Forest Ecology and Management* 89, 59–77.

Johnson, J. M., Moore, L. J., Ells, K., Murray, A. B., Adams, P. N., MacKenzie III, R. A., and Jaeger, J. M. (2015) Recent shifts in coastline change and shoreline stabilization linked to storm climate change. *Earth Surface Processes and Landforms* 40, 569–85.

Johnson, R. H. (1980) Hillslope stability and landslide hazard: a case study from Longdendale, north Derbyshire, England. *Proceedings of the Geologists' Association* 91, 315–25.

Johnson, W. D. (1901) The High Plains and their utilization. *US Geological Survey, Twenty-first Annual Report*, part IV, pp. 601–741. Washington, D.C.: US Geological Survey.

Jones, C. G. (2012). Ecosystem engineers and geomorphological signatures in landscapes. *Geomorphology* 157, 75–87.

Jones, D. K. C. (1981) *Southeast and Southern England* (The Geomorphology of the British Isles). London and New York: Methuen.

Jones, D. K. C. (1999) Evolving models of the Tertiary evolutionary geomorphology of southern England, with special reference to the Chalklands. In B. J. Smith, W. B. Whalley, and P. A. Warke (eds) *Uplift, Erosion and Stability: Perspectives on Long-term Landscape Development* (Geological Society special publication 162), pp. 1–23. London: Geological Society.

Jones, J. A. A. (1997) *Global Hydrology: Process, Resources and Environmental Management*. Harlow, Essex: Longman.

Joughlin, I., Tulaczy, S., Bindschadler, R. A., and Price, S. (2002) Changes in West Antarctic ice stream velocities: observations and analysis. *Journal of Geophysical Research* 107, 2289.

Jutras, P. and Schroeder, J. (1999) Geomorphology of an exhumed Carboniferous paleosurface in the southern Gaspé Peninsula, Québec: paleoenvironmental and tectonic implications. *Géographie physique et Quaternaire* 53, 249–63.

Kamb, B. (1964) Glacier geophysics. *Science* 146, 353–65.

Kearey, P. and Vine, F. J. (1990) *Global Tectonics*. Oxford: Blackwell.

Keller, E. A. and Pinter, N. (1996) *Active Tectonics*. Upper Saddle River, N.J.: Prentice Hall.

Kennedy, B. A. (2005) *Inventing the Earth: Ideas on Landscape Development since 1740*. Oxford: Blackwell.

Kent, R. (1991) Lithospheric uplift in eastern Gondwana: evidence for a long-lived mantle plume system? *Geology* 19, 19–23.

Kerski, J. J. (2015) Geo-awareness, geo-enablement, geotechnologies, citizen science, and storytelling: geography on the world stage. *Geography Compass* 9, 14–26.

Kiernan, K. (2015) Landforms as sacred places: implications for geodiversity and geoheritage. *Geoheritage* 7, 177–93.

King, C. A. M. (1972) *Beaches and Coasts*, 2nd edn. London: Edward Arnold.

King, L. C. (1942) *South African Scenery: A Textbook of Geomorphology*. Edinburgh: Oliver & Boyd.

King, L. C. (1953) Canons of landscape evolution. *Bulletin of the Geological Society of America* 64, 721–52.

King, L. C. (1967) *The Morphology of the Earth*, 2nd edn. Edinburgh: Oliver & Boyd.

King, L. C. (1983) *Wandering Continents and Spreading Sea Floors on an Expanding Earth*. Chichester: John Wiley & Sons.

Kingma, J. T. (1958) Possible origin of piercement structures, local unconformities, and secondary basins in the eastern geosyncline, New Zealand. *New Zealand Journal of Geology and Geophysics* 1, 269–74.

Kirkby, M. J. (1971) Hillslope process-response models based on the continuity equation. In D. Brunsden (ed.) *Slope: Form and Process* (Institute of British Geographers special publication 3), pp. 15–30. London: Institute of British Geographers.

Kirkby, M. J. (1984) Modelling cliff development in South Wales: Savigear re-viewed. *Zeitschrift für Geomorphologie* NF 28, 405–26.

Kirkby, M. J. (1985) A model for the evolution of regolith-mantled slope. In M. D. Woldenberg (ed.) *Models i n Geomorphology*, pp. 213–37. London: George Allen & Unwin.

Kleman, J. (1994) Preservation of landforms under ice sheets and ice caps. *Geomorphology* 9, 19–32.

Knight, J. and Harrison, S. (eds) (2009) *Periglacial and Paraglacial Processes and Environments* (Geological Society special publication 320). London: Geological Society.

Knight, J. and Harrison, S. (eds) (2009) Periglacial and paraglacial environments: a view from the past into the future. In J. Knight and S. Harrison (eds) *Periglacial and Paraglacial Processes and Environments* (Geological Society special publication 320), pp. 1–4. London: Geological Society.

Knight, J. and Harrison, S. (2014) Limitations of uniformitarianism in the Anthropocene. *Anthropocene* 5, 71–5.

Knighton, A. D. (1998) *Fluvial Forms and Processes: A New Perspective*, 2nd edn. London: Arnold.

Knighton, A. D. and Nanson, G. C. (1993) Anastomosis and the continuum of channel pattern. *Earth Surface Processes and Landforms* 18, 613–25.

Knisel, W. G. (ed.) (1980) *CREAMS: A Field Scale Model for Chemicals, Runoff and Erosion from Agricultural Management Systems* (USDA, Conservation Research Report no. 26). Washington, D.C.: United States Department of Agriculture.

Knox, J. C. (1984) Responses of river systems to Holocene climates. In H. E. Wright Jr (ed.) *Late-Quaternary Environments of the United States*, vol. 2, *The Holocene*, pp. 21–68. London: Longman.

Kocurek, G. (1998) Aeolian system response to external forcing factors: a sequence stratigraphic view of the Saharan region. In A. S. Alsharhan, K. W. Glennie, G. L. Whittle, and C. G. St C. Kendall (eds) *Quaternary Deserts and Climatic Change*, pp. 327–37. Rotterdam: Balkema.

Kocurek, G. (1999) The aeolian rock record (Yes, Virginia, it exists, but it really is rather special to create one). In A. S. Goudie, I. Livingstone, and S. Stokes (eds) *Aeolian Environments, Sediments and Landforms* (British Geomorphological Research Group Symposia Series, Papers from the Fourth International Conference on Aeolian Research (ICAR 4), held in School of Geography and St Catherine's College, University of Oxford, July 1998), pp. 239–59. Chichester: John Wiley & Sons.

Kocurek, G., Havholm, K. G., Deynoux, M., and Blakey, R. C. (1991) Amalgamated accumulations resulting from climatic and eustatic changes, Akchar Erg, Mauritania. *Sedimentology* 38, 751–72.

Komar, P. D. (1998) *Beach Processes and Sedimentation*, 2nd edn. Upper Saddle River, N.J.: Prentice Hall.

Kondolf, M. and Pigay, H. (2002) *Methods in Fluvial Geomorphology*. New York: John Wiley & Sons.

Korup, O. (2006) Rock-slope failure and the river long profile. *Geology* 34, 45–8.

Kovanen, D. and Slaymaker, O. (2014) The paraglacial geomorphology of the Fraser Lowland, southwest British Columbia and northwest Washington. *Geomorphology* 232, 78–93.

Kronberg, B. I. and Nesbitt, H. W. (1981) Quantification of weathering, soil geochemistry and soil fertility. *Journal of Soil Science* 32, 453–9.

Kueny, J. A. and Day, M. J. (1998) An assessment of protected karst landscapes in the Caribbean. *Caribbean Geography* 9, 87–100.

Lancaster, N. (1988) Development of linear dunes in the southwestern Kalahari, southern Africa. *Journal of Arid Environments* 14, 233–44.

Lancaster, N. (1995) *Geomorphology of Desert Dunes*. London: Routledge.

Lancaster, N. (1999) Geomorphology of desert sand seas. In A. S. Goudie, I. Livingstone, and S. Stokes (eds) *Aeolian Environments, Sediments and Landforms* (British Geomorphological Research Group Symposia Series, Papers from the Fourth International Conference on Aeolian Research (ICAR 4), held in School of Geography and St Catherine's College, University of Oxford, July 1998), pp. 49–69. Chichester: John Wiley & Sons.

Lane, S. N. and Richards, K. S. (1997) Linking river channel form and process: time, space and causality revisited. *Earth Surface Processes and Landforms* 22, 249–60.

Larson, R. L. (1991) Latest pulse of the Earth: evidence for a mid-Cretaceous superplume. *Geology* 19, 547–50.

Lawrence, D. M., Slater, A. G., Tomas, R. A., Holland, M. M., and Deser, C. (2008) Accelerated Arctic land warming and permafrost degradation during rapid sea ice loss. *Geophysical Research Letters* 35, L11506; doi: 10.1029/2008GL033985.

Laycock, A. H. (1987) The amount of Canadian water and its distribution. In M. C. Healey and R. R. Wallace (eds) *Canadian Aquatic Resources* (Canadian Bulletin of Fisheries and Aquatic Sciences 215), 13–42. Ottawa: Department of Fisheries and Oceans.

Lazarus, E. D., Ellis, M. A, Murray, A. B., and Hall, D. M. (2016) An evolving research agenda for human–coastal systems. *Geomorphology* 256, 81–90.

Lehre, A. K. (1987) Rates of soil creep on colluvium-mantled hillslopes in north-central California. In R. L. Beschta, T. Blinn, G. E. Grant, F. J. Swanson, and G. G. Ice (eds) *Erosion and Sedimentation in the Pacific Rim* (International Association of Hydrological Sciences (IAHS) Publication no. 165), pp. 91–100.Wallingford, UK: IAHS Press.

Leighton, M. W. and Pendexter, C. (1962) Carbonate rock types. *American Association of Petroleum Geologists Memoir* 1, 33–61.

Lemke, P., Ren, J., Alley, R. B., Allison, I., Carrasco, J., Flato, G., Fujii, Y., Kaser, G., Mote, P., Thomas, R. H., and Zhang, T. (2007) Observations: changes in snow, ice and frozen ground. In S. Solomon, D. Qin, M. Manning, Z. Chen, M. Marquis, K. B. Averyt, M. Tignor, and H. L. Miller (eds) *Climate Change 2007: The Physical Science Basis. Contribution of Working Group I to the Fourth Assessment Report of the Intergovernmental Panel on Climate Change*, pp. 337–83. Cambridge and New York: Cambridge University Press.

Lenton, T. M. and Williams, H. T. P. (2013) On the origin of planetary-scale tipping points. *Trends in Ecology and Evolution* 28, 380–2.

Leopold, L. B. and Maddock, T., Jr (1953) *The Hydraulic Geometry of Stream Channels and Some Physiographic Implications* (United States Geological Survey Professional Paper 252). Washington, D.C.: US Government Printing Office.

Leopold, L. B., Wolman, M. G., and Miller, J. P. (1964) *Fluvial Processes in Geomorphology*. San Francisco, Calif., and London: W. H. Freeman.

Lewandowski, J. P. and McLaughlin, S. P. (1995) Managing visitors' environmental impacts in a system of sites: a network approach. *Pennsylvania Geographer* 33, 43–58.

Lewin, J. and Ashworth, P. J. (2014) The negative relief of large river floodplains. *Earth-Science Reviews* 129, 1–23.

Lewin, J., Macklin, M. G., and Woodward, J. C. (eds) (1996) *Mediterranean Quaternary River Environments*. Rotterdam: Balkema.

Lewis, S. L. and Maslin, M. A. (2015a) Defining the Anthropocene. *Nature* 519, 171–80.

Lewis, S. L. and Maslin, M. A. (2015b) A transparent framework for defining the Anthropocene Epoch. *The Anthropocene Review* 2, 128–46.

Lidmar-Bergström, K. (1988) Denudation surfaces of a shield area in south Sweden. *Geografiska Annaler* 70A, 337–50.

Lidmar-Bergström, K. (1989) Exhumed Cretaceous landforms in south Sweden. *Zeitschrift für Geomorphologie, Supplementband* 72, 21–40.

Lidmar-Bergström, K. (1993) Denudation surfaces and tectonics in the southernmost part of the Baltic Shield. *Precambrian Research* 64, 337–45.

Lidmar-Bergström, K. (1994) Morphology of the bedrock surface. In C. Fredén (ed.) *National Atlas of Sweden, Geology*, pp. 44–54. Stockholm: Almqvist & Wiksell International (in Swedish).

Lidmar-Bergström, K. (1995) Relief and saprolites through time on the Baltic Shield. *Geomorphology* 12, 45–61.

Lidmar-Bergström, K. (1996) Long term morphotectonic evolution in Sweden. *Geomorphology* 16, 33–59.

Lidmar-Bergström, K. (1999) Uplift histories revealed by landforms of the Scandinavian domes. In B. J. Smith, W. B. Whalley, and P. A. Warke (eds) *Uplift, Erosion and Stability: Perspectives on Long-term Landscape Development* (Geological Society, London, special publication 162), pp. 85–91. London: Geological Society of London.

Lidmar-Bergström, K., Ollier, C. D., and Sulebak, J. R. (2000) Landforms and uplift history of southern Norway. *Global and Planetary Change* 24, 211–31.

Limber, P. W. and Murray, A. B. (2015) Sea stack formation and the role of abrasion on beach-mantled headlands. *Earth Surface Processes and Landforms* 40, 559–68.

Linton, D. L. (1951) The delimitation of morphological regions. In L. D. Stamp and S. W. Wooldridge (eds) *London Essays in Geography*, pp. 199–218. London: Longman.

Linton, D. L. (1955) The problem of tors. *Geographical Journal* 121, 289–91.

Liu, T. and Ding, Z. (1998) Chinese loess and the paleomonsoon. *Annual Review of Earth and Planetary Sciences* 26, 111–45.

Livingstone, D. A. (1963) *Chemical Composition of Rivers and Lakes* (United States Geological Survey Professional Paper 440G). Washington, D.C.: US Government Printing Office.

Livingstone, I. (2013) Aeolian geomorphology of the Namib Sand Sea. *Journal of Arid Environments* 93, 30–9.

Livingstone, I. and Warren, A. (1996) *Aeolian Geomorphology: An Introduction*. Harlow, Essex: Longman.

Lóczy, D, and Sütő, L. (2011) Human activity and geomorphology. In K. J. Gregory and A. S. Goudie (eds) *The SAGE Handbook of Geomorphology*, pp. 260–78. London: Sage Publications.

Lorenz, E. N. (1963) Deterministic nonperiodic flow. *Journal of Atmospheric Sciences* 20, 130–41.

Lotka, A. J. (1924) *Elements of Physical Biology*. Baltimore: Williams & Wilkins.

Lowe, D. J. (1992) A historical review of concepts of speleogenesis. *Cave Science* 19, 63–90.

Löye-Pilot, M. D. and Martin, J. M. (1996) Saharan dust input to the western Mediterranean: an eleven years' record in Corsica. In S. Guerzoni and R. Chester (eds) *The Impact of Desert Dust across the Mediterranean*, pp. 191–9. Dordrecht: Kluwer Academic.

Lu, H. and Shao, Y. (2001) Toward quantitative prediction of dust storms: an integrated wind erosion modelling system and its applications. *Environmental Modelling and Software* 16, 233–49.

Ludwig, K. R., Muhs, D. R., Simmons, K. R., Halley, R. B., and Shinn, E. A. (1996) Sea-level records at ~80 ka from tectonically stable platforms: Florida and Bermuda. *Geology* 24, 211–14.

Lunardini, V. J. (1996) Climatic warming and the degradation of warm permafrost. *Permafrost and Periglacial Processes* 7, 311–20.

Lundberg, J. and Ford, D. C. (1994) Canadian landform examples: 28 dissolution pavements. *Canadian Geographer* 38, 271–5.

MacDonald, G. A. (1972) *Volcanoes*. Englewood Cliffs, N.J.: Prentice Hall.

Mackay, J. R. (1998) Pingo growth and collapse, Tuktoyaktuk Peninsula area, western Arctic coast, Canada: a long-term field study. *Géographie physique et Quaternaire* 52, 1–53.

Mackin, J. H. (1948) The concept of the graded river. *Bulletin of the Geological Society of America* 59, 463–512.

Mackintosh, D. (1869) *The Scenery of England and Wales, Its Character and Origin: Being an Attempt to Trace the Nature of the Geological Causes, especially Denudation, by which the Physical Features of the Country Have Been Produced*. London: Longmans, Green.

Macklin, M. G. and Woodward, J. C. (2009) River systems and environmental change, In J. C. Woodward (ed.) *The Physical Geography of the Mediterranean*, pp. 319–52. Oxford: Oxford University Press.

Manley, G. (1959) The late-glacial climate of North-West England. *Liverpool and Manchester Geological Journal* 2, 188–215.

Marcus, M. G. (1969) The hydrology of snow and ice. In R. J. Chorley (ed.) *Water, Earth, and Man: A Synthesis of Hydrology, Geomorphology, and Socio-Economic Geography*, pp. 359–67. London: Methuen.

Mark, D. M. and Smith, B. (2004) A science of topography: from qualitative ontology to digital representations. In M. P. Bishop and J. F. Shroder (eds) *Geographic Information Science and Mountain Geomorphology*, pp. 75–97. Berlin: Springer Verlag.

Marriner, N., Morhange, C., and Skrimshire, S. (2010) Geoscience meets the four horsemen? Tracking the rise of neocatastrophism. *Global and Planetary Change* 74, 43–8.

Martin, J.-M. and Meybeck, M. (1979) Elemental mass balance of material carried by major world rivers. *Marine Chemistry* 7, 173–206.

Martin, Y. and Church, M. (2004) Numerical modelling of landscape evolution: geomorphological perspectives. *Progress in Physical Geography* 28, 317–39.

Martini, I. P., Brookfield, M. E., and Sadura, S. (2001) *Principles of Glacial Geomorphology and Geology*. Upper Saddle River, N.J.: Prentice Hall.

Martini, J. E. J. (1981) Early Proterozoic palaeokarst of the Transvaal, South Africa. In B. F. Beck (ed.) *Proceedings of the Eighth International Congress on Speleology*, vol. 1, pp. 37–59. Huntsville, Ala.: National Speleological Society.

Maruyama, S. (1994) Plume tectonics. *Journal of the Geological Society of Japan* 100, 24–49.

Masselink, G. and Hughes, M. G. (2003) *An Introduction to Coastal Processes and Geomorphology*. London: Arnold.

Matthews, M. C., Clayton, C. R. I., and Rigby-Jones, J. (2000) Locating dissolution features in the Chalk. *Quarterly Journal of Engineering Geology* 33, 125–40.

McBratney, A. B. (1992) On variation, uncertainty and informatics in environmental soil management. *Australian Journal of Soil Research* 30, 913–35.

McCarroll, D. (1991) Relative-age dating of Inorganic deposits: the need for a more critical approach. *The Holocene* 1, 174–80.

McGee, W. J. (1888) The geology of the head of Chesapeake Bay. *Annual Report of the United States Geological Survey* 7, 537–646.

McGregor, D. M. and Thompson, D. A. (1995) *Geomorphology and Land Management in a Changing Environment*. Chichester: John Wiley & Sons.

McKee, E. D. (ed.) (1979) *A Study of Global Sand Seas* (United States Geological Survey Professional Paper 1052). Reston, Va.: United States Geological Survey.

McLean, R. F. (1967) Measurements of beachrock erosion by some tropical marine gastropods. *Bulletin of Marine Science* 17, 551–61.

Meigs, P. (1953) World distribution of arid and semiarid homoclimates. In *Reviews on Research on Arid Zone Hydrology* (UNESCO Arid Zone Programme 1), pp. 203–9. Paris: UNESCO.

Menzies, J. (1989) Subglacial hydraulic conditions and their possible impact upon subglacial bed formation. *Sedimentary Geology* 62, 125–50.

Mercier, D. (2008) Paraglacial and paraperiglacial landsystems: concepts, temporal scales and spatial distribution. *Géomorphologie: Relief, Processus, Environnement* 2008(4), 223–33.

Meybeck, M. (1979) Concentrations des eaux fluviales en éléments majeurs et apports en solution aux océans. *Revue de Géologie Dynamique et de Géographie Physique* 21, 215–46.

Meybeck, M. (1987) Les transports fluviaux en solution dans les sciences de la terre. *Bulletin de l'Institut géologique du Bassin d'Aquitaine* 41, 401–28.

Micallef, A., Ribó, M., Canals, M., Puig, P., Lastras, G., and Tubau, X. (2014) Space-for-time substitution and the evolution of a submarine canyon–channel system in a passive progradational margin. *Geomorphology* 221, 34–50.

Middleton, N. (1997) Desert dust. In D. S. G. Thomas (ed.) *Arid Zone Geomorphology: Process, Form and Change in Drylands*, pp. 413–36. Chichester: John Wiley & Sons.

Middleton, N. J. and Thomas, D. S. G. (eds) (1997) *World Atlas of Desertification*, 2nd edn. London: Arnold.

Migoń, P. (2014) The significance of landforms – the contribution of geomorphology to the World Heritage Programme of UNESCO. *Earth Surface Processes and Landforms* 39, 836–43.

Milliman, J. D. (1980) Transfer of river-borne particulate material to the oceans. In J.-M. Martin, J. D. Burton, and D. Eisma (eds) *River Inputs to Ocean Systems* (SCOR/UNEP/UNESCO Review and Workshop), pp. 5–12. Rome: FAO.

Milliman, J. D. and Meade, R. H. (1983) World-wide delivery of sediment to the oceans. *Journal of Geology* 91, 1–21.

Millington, A. C. and Townshend, J. R. G. (1987) The potential of satellite remote sensing for geomorphological investigations: an overview. In Gardiner, V. (ed.) *International Geomorphology*, pp. 331–42. Chichester: John Wiley & Sons.

Mills, H. H. (1990) Thickness and character of regolith on mountain slopes in the vicinity of Mountain Lake, Virginia, as indicated by seismic refraction, and implications for hillslope evolution. *Geomorphology* 3, 143–57.

Minár, J. and Evans, I. S. (2008) Elementary forms for land surface segmentation: the theoretical basis of terrain analysis and geomorphological mapping. *Geomorphology* 95, 236–59.

Mohr, E. J. C. and van Baren, F. A. (1954) *Tropical Soils: A Critical Study of Soil Genesis as Related to Climate, Rock and Vegetation.* New York: Wiley Interscience.

Molnar, P. and England, P. (1990) Late Cenozoic uplift of mountain ranges and global climate change: chicken or egg? *Nature* 346, 29–34.

Montgomery, D. R. and Brandon, M. T. (2002) Topographic controls on erosion rates in tectonically active mountain ranges. *Earth and Planetary Science Letters* 201, 481–9.

Montgomery, D. R. and Buffington, J. M. (1997) Channel-reach morphology in mountain drainage basins. *GSA Bulletin* 5, 596–611.

Montgomery, D. R. and Dietrich, W. E. (1988) Where do channels begin? *Nature* 336, 232–4.

Montgomery, D. R. and Dietrich, W. E. (1989) Source areas, drainage density, and channel initiation. *Water Resources Research* 25, 1907–18.

Montgomery, D. R., Balco, G., and Willett, S. D. (2001) Climate, tectonics, and the morphology of the Andes. *Geology* 29, 579–82.

Moore, I. D., Grayson, R. B., and Ladson, A. R. (1991) Digital terrain modelling: a review of hydrological, geomorphological, and biological applications. *Hydrological Processes* 5, 3–30.

Moore, I. D., Turner, A. K., Wilson, J. P., Jenson, S., and Band, L. (1993) GIS and land-surface–subsurface process modeling. In M. F. Goodchlld, B. O. Parks, and L. T. Steyaert (eds) *Environmental Modeling with GIS*, pp. 196–230. New York: Oxford University Press.

Moore, J. G. (1985) Structure and eruptive mechanism at Surtsey volcano, Iceland. *Geological Magazine* 122, 649–61.

Moore, J. G., Jakobsson, S., and Holmjarn, J. (1992) Subsidence of Surtsey volcano, 1967–1991. *Bulletin of Volcanology* 55, 17–24.

Morgan, R. P. C. (1994) The European Soil Erosion Model: an up-date on its structure and research base. In R. J. Rickson (ed.) *Conserving Soil Resources: European Perspectives*, pp. 286–99. Wallingford, UK: CAB International.

Morgan, R. P. C. (1995) *Soil Erosion and Conservation*, 2nd edn. Harlow, Essex: Longman.

Morgan, R. P. C. (2005) *Soil Erosion and Conservation*, 3rd edn. Oxford: Blackwell.

Morisawa, M. (1985) *Rivers.* Harlow, Essex: Longman.

Mörner, N.-A. (1980)The northwest European 'sea-level laboratory' and regional Holocene eustasy. *Palaeogeography, Palaeoclimatology, Palaeoecology* 29, 281–300.

Mörner, N.-A. (1987) Models of global sea-level change. In M. J. Tooley and I. Shennan (eds) *Sea-Level Changes*, pp. 332–55. Oxford: Basil Blackwell.

Mörner, N.-A. (1994) Internal response to orbital forcing and external cyclic sedimentary sequences. In P. L. De Boer and D. G. Smith (eds) *Orbital Forcing and Cyclic Sequences* (special publication no. 19 of the International Association of Sedimentologists), pp. 25–33. Oxford: Blackwell Scientific.

Mosley, M. P. (1973) Rainsplash and the convexity of badland divides. *Zeitschrift für Geomorphologie, Supplementband* 19, 10–25.

Mosley, M. P. and Zimpfer, G. L. (1978) Hardware models in geomorphology. *Progress in Physical Geography* 2, 438–61.

Mottershead, D. N. (1994) Spatial variations in intensity of alveolar weathering of a dated sandstone structure in a coastal environment, Weston-super-Mare, UK. In D. A. Robinson and R. B. G. Williams (eds) *Rock Weathering and Landform Evolution*, pp. 151–74. Chichester: John Wiley & Sons.

Moucha, R. and Forte, A. M. (2011) Changes in African topography driven by mantle convection. *Nature Geoscience* 4, 707–12.

Moulin, C. and Chiapello, I. (2006) Impact of human-induced desertification on the intensification of Sahel dust emission and export over the last decades. *Geophysical Research* Letters 33, L18808, doi: 10.1029/2006GL025923.

Murphy, P. and Cordingley, J. (1999) Some observations on the occurrence of channel karren-like features in flooded karst conduits in the Yorkshire Dales, UK. *Cave and Karst Science* 26, 129–30.

Murphy, P. J. (2000) The karstification of the Permian strata east of Leeds. *Proceedings of the Yorkshire Geological Society* 53, 25–30.

Murphy, P. J., Hall, A. M., and Cordingley, J. N. (2000) Anomalous scallop distributions in Joint Hole, Chapel-le-Dale, North Yorkshire, UK. *Cave and Karst Science* 27, 29–32.

Murray, A. B., Coco, G., Goldstein, E. B. (2014) Cause and effect in geomorphic systems: complex systems perspectives. *Geomorphology* 214, 1–9.

Murray, A. B., Knaapen, M. A. F., Tal, M., and Kirwan, M. L. (2008) Biomorphodynamics: physical–biological feedbacks that shape landscapes. *Water Resources Research* 44, W11301.

Murray, A. B., Lazarus, E., Ashton, A., Bass, A., Coco, G., Coulthard, T., Fonstad, M., Haff, P., McNamara, D., Paola, C., Pelletier, J., and Reinhardt, L. (2009) Geomorphology, complexity, and the emerging science of the Earth's surface. *Geomorphology* 103, 496–505.

Murton, J. B., Peterson, R., and Ozouf, J.-C. (2006) Bedrock fracture by ice segregation in cold regions. *Science* 314, 1127–9.

Nanson, G. C. and Croke, J. C. (1992) A genetic classification of floodplains. *Geomorphology* 4, 459–86.

Nanson, G. C. and Knighton, A. D. (1996) Anabranching rivers: their cause, character and classification. *Earth Surface Processes and Landforms* 21, 217–39.

Napieralski, J., Harbor, J., and Li, Y. (2007) Glacial geomorphology and geographic information systems. *Earth-Science Reviews* 85, 1–22.

Nash, D. B. (1981) Fault: a FORTRAN program for modeling the degradation of active normal fault scarps. *Computers & Geosciences* 7, 249–66.

Naylor, L. A., Viles, H. A., and Carter, N. E. A. (2002) Biogeomorphology revisited: looking towards the future. *Geomorphology* 47, 3–14.

Nearing, M. A., Foster, G. R., Lane, L. J., and Finkner, S. C. (1989) A process-based soil erosion model for USDA-Water Erosion Prediction Project technology. *Transactions of the American Society of Agricultural Engineers* 32, 1587–93.

Nelson, F. E. (1998) Cryoplanation terrace orientation in Alaska. *Geografiska Annaler, Series A: Physical Geography* 80, 135–51.

Nicholas, A. P. and Quine, T. A. (2007) Modelling alluvial landform change in the absence of external environmental forcing. *Geology* 35, 527–30.

Nicholas, A. P., Clarke, L., and Quine, T. A. (2009) A numerical modelling and experimental study of flow width dynamics on alluvial fans. *Earth Surface Processes and Landforms* 34, 1985–93.

Nickling, W. G. and McKenna Neuman, C. (1999) Recent investigations of airflow and sediment transport over desert dunes. In A. S. Goudie, I. Livingstone, and S. Stokes (eds) *Aeolian Environments, Sediments and Landforms* (British Geomorphological Research Group Symposia Series), Papers from the Fourth International Conference on Aeolian Research (ICAR 4), held in School of Geography and St Catherine's College, University of Oxford, July 1998), pp. 15–47. Chichester: John Wiley & Sons.

Nicod, J. (1998) Les grottes: retrospective historique et insertion des grottes aménagées dans l'espace géographique. *Annales de Géographie* 603, 508–30.

Nilsson, T. (1983) *The Pleistocene: Geology and Life in the Quaternary Ice Age.* Dordrecht, Boston, Lancaster: D. Reidel Publishing Co.

Noller, J. S., Sowers, J. M., and Lettis, W. R. (eds) (2000) *Quaternary Geochronology: Methods and Applications* (AGU Reference Shelf 4). Washington, D.C.: American Geophysical Union.

Nordstrom, K. F. (1994) Beaches and dunes of human-altered coasts. *Progress in Physical Geography* 18, 497–516.

Nunn, P. D. (1987) Small islands and geomorphology: review and prospect in the context of historical geomorphology. *Transactions of the Institute of British Geographers*, New Series 12, 227–39.

Nur, A. and Ben-Avraham, Z. (1982) Displaced terranes and mountain building. In K. J. Hsü (ed.) *Mountain Building Processes*, pp. 73–84. London and New York: Academic Press.

Ó'Cofaigh, C. (1996) Tunnel valley genesis. *Progress in Physical Geography* 20, 1–19.

O'Halloran, D., Green, C., Harley, M., Stanley, M., and Knil, J. (eds) (1994) *Geological and Landscape Conservation. Proceedings of the Malvern International Conference 1993.* London: The Geological Society, London.

Ollier, C. D. (1959) A two-cycle theory of tropical pedology. *Journal of Soil Science* 10: 137–48.

Ollier, C. D. (1960) The inselbergs of Uganda. *Zeitschrift für Geomorphologie* NF 4, 470–87.

Ollier, C. D. (1967) Landform description without stage names. *Australian Geographical Studies* 5, 73–80.

Ollier, C. D. (1968) Open systems and dynamic equilibrium in geomorphology. *Australian Geographical Studies* 6, 167–70.

Ollier, C. D. (1969) *Volcanoes* (*An Introduction to Systematic Geomorphology*, vol. 6). Cambridge, Mass. and London: MIT Press.

Ollier, C. D. (1978) Inselbergs of the Namib Desert: process and history. *Zeitschrift für Geomorphologie, Supplementband* 31, 161–76.

Ollier, C. D. (1981) *Tectonics and Landforms* (Geomorphology Texts 6). London and New York: Longman.

Ollier, C. D. (1988) Deep weathering, groundwater and climate. *Geografiska Annaler* 70A, 285–90.

Ollier, C. D. (1991) *Ancient Landforms*. London and New York: Belhaven Press.

Ollier, C. D. (1992) Global change and long-term geomorphology. *Terra Nova* 4, 312–19.

Ollier, C. D. (1995) Tectonics and landscape evolution in southeast Australia. *Geomorphology* 12, 37–44.

Ollier, C. D. (1996) Planet Earth. In I. Douglas, R. J. Huggett, and M. E. Robinson (eds) *Companion Encyclopedia of Geography*, pp. 15–43. London: Routledge.

Ollier, C. D. (2003) The origin of mountains on an expanding Earth, and other hypotheses. In G. Scalera and K.-H. Jacob (eds) *Why Expanding Earth? A Book in Honour of Ott Christoph Hilgenberg*, pp. 129–60. Rome: INGV (Instituto Nazionale di Geofísica e Vulcanologia).

Ollier, C. D. (2004a) Mountain building and climate: mechanisms and timing. *Geografía Fisica e Dinamica Quaternaria* 27, 139–49.

Ollier, C. D. (2004b) The evolution of mountains on passive continental margins. In P. N. Owens and O. Slaymaker (eds) *Mountain Geomorphology*, pp. 59–88. London: Arnold.

Ollier, C. D. (2006) Mountain uplift, climate and isostacy. *New Concepts on Global Tectonics Newsletter* 40, 14–17.

Ollier, C. D. (2010) Glaciers: science and nonsense. *Geoscientist* 20, 16–21.

Ollier, C. D. and Pain, C. (1994) Landscape evolution and tectonics in southeastern Australia. *AGSO Journal of Australian Geology and Geophysics* 15, 335–45.

Ollier, C. D. and Pain, C. (1996) *Regolith, Soils and Landforms*. Chichester: John Wiley & Sons.

Ollier, C. D. and Pain, C. F. (1981) Active gneiss domes in Papua New Guinea: new tectonic landforms. *Zeitschrift für Geomorphologie* NF 25, 133–45.

Ollier, C. D. and Pain, C. F. (1997) Equating the basal unconformity with the palaeoplain: a model for passive margins. *Geomorphology* 19, 1–15.

Ollier, C. D. and Pain, C. F. (2000) *The Origin of Mountains*. London: Routledge.

Ollier, C. D. and Tuddenham, W. G. (1962) Slope development at Coober Pedy, South Australia. *Journal of the Geological Society of Australia* 9, 91–105.

Osborne, R. A. L. and Branagan, D. F. (1988) Karst landscapes of New South Wales. *Earth-Science Reviews* 25, 467–80.

Osterkamp, T. E., Viereck, L., Shur, Y., Jorgenson, M. T., Racine, C., Doyle, A., and Boone, R. D. (2000) Observations of thermokarst and its impact on boreal forests in Alaska, USA. *Arctic, Antarctic, and Alpine Research* 32, 303–15.

Osterkamp, W. R. and Costa, J. E. (1987) Changes accompanying an extraordinary flood on a sand-bed stream. In L. Mayer and D. Nash (eds) *Catastrophic Flooding*, pp. 201–24. Boston, Mass.: Allen & Unwin.

Osterkamp, W. R. and Hupp, C. R. (2010) Fluvial processes and vegetation – glimpses of the past, present, and future. *Geomorphology* 116, 274–85.

Otto-Bliesner, B. (1995) Continental drift, runoff, and weathering feedbacks: implications from climate model experiments. *Journal of Geophysical Research* 100, 11, 537–48.

Owens, P. N. and Slaymaker, O. (2004) (eds) *Mountain Geomorphology*, pp. 59–88.London: Arnold

.Pain, C. F. and Ollier, C. D. (1995) Inversion of relief: a component of landscape evolution. *Geomorphology* 12, 151–65.

Panizza, M. (2001) Geomorphosites: concepts, methods and examples of geomorphological survey. *Chinese Science Bulletin* 46 (Issue 1 Supplement), 4–5.

Panizza, M. (2009) The geomorphodiversity of the Dolomites (Italy): a key of geoheritage assessment. *Geoheritage* 1, 33–42.

Park, R. G. (1988) *Geological Structures and Moving Plates*. London: Blackie Academic and Professional.

Parsons, G. R. and Powell, M. (2001) Measuring the cost of beach retreat. *Coastal Management* 29, 91–103.

Partridge, J. and Baker, V. R. (1987) Palaeoflood hydrology of the Salt River, Arizona. *Earth Surface Processes and Landforms* 12, 109–25.

Partridge, T. C. (1998) Of diamonds, dinosaurs and diastrophism: 150 million years of landscape evolution in southern Africa. *South African Journal of Geology* 101, 167–84.

Pearce, F. (2007) *With Speed and Violence: Why Scientists Fear Tipping Points in Climate Change*. Boston, Mass.: Beacon Press Books.

Pedro, G. (1979) Caractérisation générale des processus de l'altération hydrolitique. *Science du Sol* 2, 93–105.

Peltier, L. (1950) The geographic cycle in periglacial regions as it is related to climatic geomorphology. *Annals of the Association of American Geographers* 40, 214–36.

Peltier, W. R. (1998) Postglacial variations in the level of the sea: implications for climate dynamics and solid earth geophysics. *Reviews of Geophysics* 36, 603–89.

Peltier, W. R. (1999) Global sea level rise and glacial isostatic adjustment. *Global and Planetary Change* 20, 93–123.

Penck, A. and Brückner, A. (1901–9) *Die Alpen im Eiszeitalter*, 3 vols. Leipzig: Tauchnitz.

Penck, W. (1924) *Die morphologische Analyse, ein Kapitel der physikalischen Geologie*. Stuttgart: Engelhorn.

Penck, W. (1953) *Morphological Analysis of Landforms*. Translated and edited by H. Czech and K. C. Boswell. London: Macmillan.

Peppoloni S. and Di Capua G. (eds) (2012) *Geoethics and Geological Culture: Reflections from the Geoitalia Conference 2011*. *Annals of Geophysics* 55, No. 3.

Peppoloni, S. and Di Capua, G. (eds) (2015) *Geoethics: the Role and Responsibility of Geoscientists*. Geological Society, London, Special Publications 419, 1–4.

Petts, G. F. and Gurnell, A. M. (2005) Dams and geomorphology: research progress and future directions. *Geomorphology* 71, 27–47.

Péwé, T. L. (1991) Permafrost. In G. A. Kiersch (ed.) *The Heritage of Engineering Geology: The First Hundred Years* (Geological Society of America centennial special vol. 3), pp. 277–98. Boulder, Colo.: Geological Society of America.

Phillips, J. D. (1990) Relative importance of factors influencing fluvial loss at the global scale. *American Journal of Science* 290, 547–68.

Phillips, J. D. (1995) Time lags and emergent stability in morphogenic/pedogenic system models. *Ecological Modelling* 78, 267–76.

Phillips, J. D. (1999) *Earth Surface Systems: Complexity, Order and Scale*. Oxford: Blackwell.

Phillips, J. D. (2006) Deterministic chaos and historical geomorphology: a review and look forward. *Geomorphology* 76, 109–21.

Phillips, J. D. (2007) The perfect landscape. *Geomorphology* 84, 159–69.

Phillips, J. D. (2009) Changes, perturbations, and responses in geomorphic systems. *Progress in Physical Geography* 33, 17–30.

Phillips, J. D. (2012) Storytelling in Earth sciences: the eight basic plots. *Earth-Science Reviews* 115, 153–62.

Phillips, J. D. (2013) Networks of historical contingency in Earth surface systems. *Journal of Geology* 121, 1–16.

Phillips, J. D. (2014) Thresholds, mode switching, and emergent equilibrium in geomorphic systems. *Earth Surface Processes and Landforms* 39, 71–9.

Phillips, J. D. (2015) Badass geomorphology. *Earth Surface Processes and Landforms* 40, 22–33.

Pike, R. J. (1995) Geomorphometry: progress, practice, and prospect. *Zeitschrift für Geomorphologie Supplementband* 101, 221–38.

Pike, R. J. (1999) *A Bibliography of Geomorphometry, the Quantitative Representation of Topography: Supplements*.3.0. United States Geological Survey, Open File Report 99–140. Menlo Park, Calif.: US Department of the Interior, United States Geological Survey.

Pike, R. J., Evans I. S., and Hengl, T. (2009) Geomorphometry: a brief guide. In T. Hengl and H. I. Reuter (eds) *Geomorphometry Concepts, Software, Applications* (Developments in Soil Science, Volume 33), pp. 3–30. Amsterdam: Elsevier.

Pinchemel, P. (1954) *Les plaines de Craie du nord-ouest du Bassin Parisien et du sud-est du Bassin de Londres et leurs bordures*. Paris: Armand Colin.

Pinter, N. and Brandon, M. T. (1997) How erosion builds mountains. *Scientific American* 276 (April), 60–5.

Playfair, J. (1802) *Illustrations of the Huttonian Theory of the Earth*. London: Cadell & Davies: Edinburgh: William Creech. (A facsimile reprint was published with an introduction by George W. White in 1964 by Dover Books, New York.)

Pomerol, C. (1989) Stratigraphy of the Palaeogene: hiatuses and transitions. *Proceedings of the Geologists' Association* 100, 313–24.

Pope, G. A. (2013a) Weathering and sediment genesis. In J. Shroder, (Editor in Chief), G. A. Pope (ed.) *Treatise on Geomorphology, Vol. 4, Weathering and Soils Geomorphology*, pp. 284–93. San Diego, Cal.: Academic Press.

Pope, G. A. (2013b) Weathering in the tropics, and related extratropical processes. In: J. Shroder (Editor in Chief), G. A. Pope (ed.) *Treatise on Geomorphology, Vol. 4, Weathering and Soils Geomorphology*, pp. 179–96. San Diego, Cal.: Academic Press.

Potter, P. E. (1978) Petrology and chemistry of modern big-river sands. *Journal of Geology* 86, 423–49.

Potter, P. E. (1997) The Mesozoic and Cenozoic paleodralnage of South America: a natural history. *Journal of South American Earth Science* 10, 331–44.

Preece, R. C., Scourse, J. D., Houghton, S. D., Knudsen, K. L., and Penny, D. N. (1990) The Pleistocene sea-level and neotectonic history of the eastern Solent, southern England. *Philosophical Transactions of the Royal Society of London*, Series B, 328, 425–47.

Price, R. J. (1973) *Glacial and Fluvioglacial Landforms*. Edinburgh: Oliver & Boyd.

Pye, K. (1990) Physical and human influences on coastal dune development between the Ribble and Mersey estuaries, northwest England. In K. F. Nordstrom, N. P. Psuty, and R. W. G. Carter (eds) *Coastal Dunes: Form and Process*, pp. 339–59. Chichester: John Wiley & Sons.

Pye, K. and Sherwin, D. (1999) Loess. In A. S. Goudie, I. Livingstone, and S. Stokes (eds) *Aeolian Environments, Sediments and Landforms* (British Geomorphological Research Group Symposia Series, Papers from the Fourth International Conference on Aeolian Research (ICAR 4), held in School of Geography and St Catherine's College, University of Oxford, July 1998), pp. 213–38. Chichester: John Wiley & Sons.

Qin, S. Q., Jiao, J. J., and Wang, S. (2001) A cusp catastrophe model of instability of slip-buckling slope. *Rock Mechanics and Rock Engineering* 34, 119–34.

Rapp, A. (1960) Recent development of mountain slopes in Karkevagge and surroundings, northern Scandinavia. *Geografiska Annaler* 42, 73–200.

Rapp, A. (1986) Slope processes in high latitude mountains. *Progress in Physical Geography* 4, 531–48.

Raupach, M. R., McTainsh, G. H., and Leys, J. F. (1994) Estimates of dust mass in recent major Australian dust storms. *Australian Journal of Soil and Water Conservation* 7, 20–4.

Raymo, M. E. and Ruddiman, W. F. (1992) Tectonic forcing of late Cenozoic climate. *Nature* 359, 117–22.

Rea, B. R., Whalley, W. B., Rainey, M. M., and Gordon, J. E. (1996) Blockfields, old or new? Evidence and implications from some plateaus in northern Norway. *Geomorphology* 15, 109–21.

Renwick, W. H. (1992) Equilibrium, disequilibrium, and non-equilibrium landforms in the landscape. *Geomorphology* 5, 265–76.

Retallack, G. J. (1999) Carboniferous fossil plants and soils of an early tundra ecosystem. *Palaios* 14, 324–36.

Retallack, G. J. (2003) Soils and global change in the carbon cycle over geological time. In J. I. Drever (ed.) and H. D. Holland and K. K. Turekian (executive eds) *Treatise on Geochemistry*, vol. 5, pp. 581–605. Amsterdam: Elsevier.

Revkin, A. (1992) *Global Warming: Understanding the Forecast*. New York: Abbeville Press.

Riffel, S. B., Vasconcelos, P. M., Carmo, I. O., and Farley, K. A. (2015) Combined $^{40}Ar/^{39}Ar$ and (U–Th)/He geochronological constraints on long-term landscape evolution of the Second Paraná Plateau and its ruiniform surface features, Paraná, Brazil. *Geomorphology* 233, 52–63.

Riksen, M. J. P. M. and De Graaff, J. (2001) On-site and off-site effects of wind erosion on European light soils. *Land Degradation and Development* 12, 1–11.

Ritter, D. F., Kochel, R. C., and Miller, J. R. (1995) *Process Geomorphology*, 5th edn. Long Grove, Ill.: Waveland Press.

Rivas, V., Cendrero, A., Hurtado, M., Cabral, M., Giménez, J., Forte, L., del Río, L., Cantú, M., and Becker, A. (2006) Geomorphic consequences of urban development and mining activities: an analysis of study areas in Spain and Argentina. *Geomorphology* 73, 185–206.

Robert, A. (2003) *River Processes: An Introduction to Fluvial Dynamics*. London: Arnold.

Robinson, D. A. and Williams, R. G. B. (1992) Sandstone weathering in the High Atlas, Morocco. *Zeitschrift für Geomorphologie* NF 36, 413–29.

Robinson, D. A. and Williams, R. G. B. (1994) Sandstone weathering and landforms in Britain and Europe. In D. A. Robinson and R. G. B. Williams (eds) *Rock Weathering and Landform Evolution*, pp. 371–91. Chichester: John Wiley & Sons.

Robinson, E. (1989) European geological conservation. *Terra Nova* 1, 113–18.

Roering, J. J., Perron, J. T. and Kirchner, J. W. (2007) Functional relationships between denudation and hillslope form and relief. *Earth and Planetary Science Letters* 264, 245–58.

Rose, A. W., Hawkes, H. E., and Webb, J. S. (1979) *Geochemistry in Mineral Exploration*. London: Academic Press.

Rose, J., Moorlock, B. S. P., and Hamblin, R. J. O. (2001) Pre-Anglian fluvial and coastal deposits in Eastern England: lithostratigraphy and palaeoenvironments. *Quaternary International* 79, 5–22.

Rowley, D. B. and Currie, B. S. (2006) Palaeo-altimetry of the late Eocene to Miocene Lunpola basin, central Tibet. *Nature* 439, 677–81.

Ruddiman, W. F. (ed.) (1997) *Tectonic Uplift and Climatic Change*. New York: Plenum Press.

Rudoy, A. (1998) Mountain ice-dammed lakes of southern Siberia and their influence on the development and regime of the intracontlnental runoff systems of North Asia in the Late Pleistocene. In G. Benito, V. R. Baker, and K. J. Gregory (eds) *Palaeohydrology and Environmental Change*, pp. 215–34. Chichester: John Wiley & Sons.

Ruhe, R. V. (1960) Elements of the soil landscape. *Transactions of the Seventh International Congress of Soil Science, Madison* 4, 165–70.

Ruhe, R. V. (1975) Climatic geomorphology and fully developed slopes. *Catena* 2, 309–20.

Ryder, J. M. (1971a) The stratigraphy and morphology of paraglacial alluvial fans in south-central British Columbia. *Canadian Journal of Earth Science* 8, 279–98.

Ryder, J. M. (1971b) Some aspects of the morphometry of paraglacial alluvial fans in south-central British Columbia. *Canadian Journal of Earth Science* 8, 1252–64.

Salles, C., Poesen, J., and Govers, G. (2000) Statistical and physical analysis of soil detachment by raindrop impact: rain erosivity indices and threshold energy. *Water Resources Research* 36, 2721–9.

Sanders, N. (1966) Port Hueneme, California: a study in coastal anthropogeomorphology. *Yearbook of the Association of Pacific Coast Geographers* 28, 119–34.

Sanderson, E. W., Segan, D. B., and Watson, J. E. M. (2015) Global status of and prospects for protection of terrestrial geophysical diversity. *Conservation Biology* 29, 649–56.

Sarnthein, M. (1978) Sand deserts during glacial maximum and climatic optimum. *Nature* 272, 43–6.

Saunders, I. and Young, A. (1983) Rates of surface process on slopes, slope retreat and denudation. *Earth Surface Processes and Landforms* 8, 473–501.

Savigear, R. A. G. (1952) Some observations on slope development in South Wales. *Transactions of the Institute of British Geographers* 18, 31–52.

Savigear, R. A. G. (1956) Technique and terminology in the investigations of slope forms. In *Premier Rapport de la Commission pour l'Étude des Versants*, pp. 66–75. Amsterdam: Union Géographique Internationale.

Savigear, R. A. G. (1965) A technique of morphological mapping. *Annals of the Association of American Geographers* 55, 514–38.

Schaetzl, R. J. and Anderson, S. (2005) *Soils: Genesis and Geomorphology*. Cambridge: Cambridge University Press.

Scheidegger, A. E. (1979) The principle of antagonism in the Earth's evolution. *Tectonophysics* 55, T7–T10.

Scheidegger, A. E. (1983) Instability principle in geomorphic equilibrium. *Zeitschrift für Geomorphologie* NF TI, 1–19.

Scheidegger, A. E. (1986) The catena principle in geomorphology. *Zeitschrift für Geomorphologie* NF 30, 257–73.

Scheidegger, A. E. (1991) *Theoretical Geomorphology*, 3rd completely rev. edn. Berlin: Springer.

Scheidegger, A. E. (1994) Hazards: singularities in geomorphic systems. *Geomorphology* 10, 19–25.

Scheidegger, A. E. (1999) Morphotectonics of eastern Nepal. *Indian Journal of Landscape Systems and Ecological Studies* 22, 1–9.

Scheidegger, A. E. (2004) *Morphotectonics*. Berlin: Springer.

Scheidegger, A. E. and Hantke, R. (1994) On the genesis of river gorges. *Transactions of the Japanese Geomorphological Union* 15, 91–110.

Schumm, S. A. (1956) The evolution of drainage basin systems and slopes in badlands at Perth Amboy, New Jersey. *Bulletin of the Geological Society of America* 67, 597–646.

Schumm, S. A. (1973) Geomorphic thresholds and complex response of drainage systems. In M. Morisawa (ed.) *Fluvial Geomorphology*, pp. 299–310. Binghamton, New York: State University of New York Publications in Geomorphology.

Schumm, S. A. (1977) *The Fluvial System*. New York: John Wiley & Sons.

Schumm, S. A. (1979) Geomorphic thresholds: the concept and its applications. *Transactions of the Institute of British Geographers*, New Series 4, 485–515.

Schumm, S. A. (1981) Evolution and response of the fluvial system, sedimentologic implications. *Society of Economic Paleontologists and Mineralogists, special publication* 31, 19–29.

Schumm, S. A. (1985a) Explanation and extrapolation in geomorphology: seven reasons for geologic uncertainty. *Transactions of the Japanese Geomorphological Union* 6, 1–18.

Schumm, S. A. (1985b) Patterns of alluvial rivers. *Annual Review of Earth and Planetary Sciences* 13, 5–27.

Schumm, S. A. (1991) *To Interpret the Earth: Ten Ways to Be Wrong*. Cambridge: Cambridge University Press.

Schumm, S. A. and Lichty, R. W. (1965) Time, space and causality in geomorphology. *American Journal of Science* 263, 110–19.

Schumm, S. A. and Parker, R. S. (1977) Implications of complex response of drainage systems for Quaternary alluvial stratigraphy. *Nature* 243, 99–100.

Selby, M. J. (1980) A rock mass strength classification for geomorphic purposes: with tests from Antarctica and New Zealand. *Zeitschrift für Geomorphologie* NF 24, 31–51.

Selby, M. J. (1982) *Hillslope Materials and Processes*, 1st edn. Oxford: Oxford University Press.

Selby, M. J. (1993) *Hillslope Materials and Processes*, 2nd edn. With a contribution by A. P. W. Hodder. Oxford: Oxford University Press.

Serreze, M. C., Walsh, J. E., Chapin III, F. S., Osterkamp, T., Dyurgerov, M., Romanovsky, V., Oechel, W. C., Morison, J., Zhang, T., and Barry, R. G. (2000) Observational evidence of recent change in the northern high-latitude environment. *Climatic Change* 46, 159–207.

Sewall, J. O., Riihimaki, C. A., and Kadegis, J. (2015) Orbital control, climate seasonality, and landscape evolution in the Quaternary Rocky Mountains. *Geomorphology* 250, 89–94.

Sharp, A. D., Trudgill, S. T., Cooke, R. U., Price, C. A., Crabtree, R. W., Pickle, A. M., and Smith, D. I. (1982) Weathering of the balustrade on St Paul's Cathedral, London. *Earth Surface Processes and Landforms* 7, 387–9.

Sharples, C. (1993) *A Methodology for the Identification of Significant Landforms and Geological Sites for Geoconservation Purposes* (A Report to The Forestry Commission, Tasmania). Hobart, Tasmania: Forestry Commission.

Sharples, C. (1995) Geoconservation in forest management – principles and practice. *Tasforests* 7, 37–50.

Sharples, C. (2002) *Concepts and Principles of Geoconservation*. Hobart: Published electronically on the Tasmanian Parks and Wildlife Service website September 2002 (Version 3).

Sharples, C. (2003) *A Review of the Geoconservation Values of the Tasmanian Wilderness World Heritage Area* (Nature Conservation Report 03/06). Hobart, Tasmania: Department of Primary Industries, Water and Environment.

Shaw, J. (1994) A qualitative review of sub-ice-sheet landscape evolution. *Progress in Physical Geography* 18, 159–84.

Shaw, J., Kvill, D., and Rains, B. (1989) Drumlins and catastrophic subglacial floods. *Sedimentary Geology* 62, 177–202.

Sherlock, R. L. (1922) *Man as a Geological Agent: An Account of His Action on Inanimate Nature*. With a foreword by A. S. Woodward. London: H. F. & G. Witherby.

Shiklomanov, I. A. (1993) World fresh water resources. In P. H. Gleick (ed.) *Water in Crisis: A Guide to the World's Fresh Water Resources*, pp. 13–24. New York: Oxford University Press.

Short, A. D. (2006) Australian beach systems: nature and distribution. *Journal of Coastal Research* 22, 11–27.

Short, A. D. (2010) Beach models. Available at www.ozcoasts.org.au/conceptual_mods/beaches/beach_intro.jsp [Last accessed 3 September 2010].

Short, A. D. and Woodroffe, C. D. (2009) *The Coast of Australia*. Cambridge: Cambridge University Press.

Shreve, R. L. (1975) The probabillistic-topologic approach to drainage basin geomorphology. *Geology* 3, 527–9.

Shroder, J. F. (Editor in chief) (2013) *Treatise on Geomorphology*, 14 Volumes. San Diego: Academic Press.

Sidle, R. C. and Onda, Y. (2004) Hydrogeomorphology: overview of an emerging science. *Hydrological Processes* 18, 597–602.

Simas, T., Nunes, J. P., and Ferreira, J. G. (2001) Effects of global climate change on coastal salt marshes. *Ecological Modelling* 139, 1–15.

Simons, D. B. (1969) Open channel flow. In R. J. Chorley (ed.) *Water, Earth, and Man: A Synthesis of Hydrology, Geomorphology, and Socio-Economic Geography*, pp. 297–318. London: Methuen.

Simons, D. B. and Richardson, E. V. (1963) Forms of bed roughness in alluvial channels. *Transactions of the American Society of Civil Engineers* 128, 284–302.

Simons, M. (1962) The morphological analysis of landforms: a new view of the work of Walther Penck (1888–1923). *Transactions of the Institute of British Geographers* 31, 1–14.

Simonson, R. W. (1995) Airborne dust and its significance to soils. *Geoderma* 65, 1–43.

Simpson, G. G. (1963) Historical science. In C. C. Albritton Jr (ed.) *The Fabric of Geology*, pp. 24–48. Reading, Mass.: Addison-Wesley.

Singer, A. (1980) The paleoclimatic interpretation of clay minerals in soils and weathering profiles. *Earth-Science Reviews* 15, 303–26.

Singh, V. P. (2003) On the theories of hydraulic geometry. *International Journal of Sediment Research* 18, 196–218.

Slaymaker, O. (ed.) (2000a) *Geomorphology, Human Activity and Global Environmental Change*. Chichester: John Wiley & Sons.

Slaymaker, O. (2000b) Global environmental change: the global agenda. In O. Slaymaker (ed.) *Geomorphology, Human Activity and Global Environmental Change*, pp. 3–20. Chichester: John Wiley & Sons.

Slaymaker, O. (2009) The future of geomorphology. *Geography Compass* 3, 329–49.

Slaymaker O. (2011) Criteria to distinguish between periglacial, proglacial and paraglacial environments. *Quaestiones Geographicae* 30, 85–94.

Slaymaker, O., Spencer, T., and Embleton-Hamann, C. (2009) *Geomorphology and Global Environmental Change*. Cambridge: Cambridge University Press.

Sloss, C. R., Westaway, K. E., Hua, Q., and Murray-Wallace, C. V. (2013) An introduction to dating techniques: a guide for geomorphologists. In J. F. Shroder (Editor in Chief), A. D. Switzer and D. M. Kennedy (eds) *Treatise on Geomorphology, Vol. 14, Methods in Geomorphology*, pp. 346–69. San Diego, Cal.: Academic Press.

Small, E. E. and Anderson, B. S. (1995) Geomorphically driven late Cenozoic rock uplift in the Sierra Nevada, California. *Science* 270, 277–80.

Smart, J. S. (1978) The analysis of drainage network composition. *Earth Surface Processes* 3, 129–70.

Smith, A. G., Smith, D. G., and Funnell, B. M. (1994) *Atlas of Mesozoic and Cenozoic Coastlines*. Cambridge: Cambridge University Press.

Smith, B. J., Warke, P. A., and Moses, C. A. (2000) Limestone weathering in contemporary arid environments: a case study from southern Tunisia. *Earth Surface Processes and Landforms* 25, 1343–54.

Smith, B. J., Whalley, W. B., and Warke, P. A. (eds) (1999) *Uplift, Erosion and Stability: Perspectives on Long-Term Landscape Development* (Geological Society special publication 162). London: Geological Society of London.

Smith, M. J. and Pain, C. F. (2009) Applications of remote sensing in geomorphology. *Progress in Physical Geography* 33, 568–82

Smith, S. V., Renwick, W. H., Buddemeier, R. W., and Crossland, C. J. (2001) Budgets of soil erosion and deposition for sediments and sedimentary organic carbon across the conterminous United States. *Global Biogeochemical Cycles* 15, 697–707.

Smith, T. R. and Bretherton, F. P. (1972) Stability and the conservation of mass in drainage basin evolution. *Water Resources Research* 8, 1506–29.

Soil Survey Staff (1999) *Soil Taxonomy: A Basic System of Soil Classification for Making and Interpreting Soil Surveys*, 2nd edn (US Department of Agriculture, Natural Resources Conservation Service, Agricultural Handbook 436). Washington, D.C.: US Government Printing Office.

Sowers, J. M., Noller, J. S., and Lettis, W. R. (2000) Methods for dating Quaternary surficial materials. In J. S. Noller, J. M. Sowers, and W. R. Lettis (eds) *Quaternary Geochronology: Methods and Applications* (AGU Reference Shelf 4), 567. Washington, D.C.: American Geophysical Union.

Sparks, B. W. (1960) *Geomorphology*. London: Longmans, Green.

Sparks, B. W. (1971) *Rocks and Relief*. London: Longman.

Speight, J. G. (1974) A parametric approach to landform regions. In E. H. Brown and R. S. Waters (eds) *Progress in Geomorphology: Papers in Honour of David L. Linton* (Institute of British Geographers special publication 7), pp. 213–30. London: Institute of British Geographers.

Spencer, T. (1988) Coastal biogeomorphology. In H. A. Viles (ed.) *Biogeomorphology*, pp. 255–318. Oxford: Basil Blackwell.

Spencer, T. (1995) Potentialities, uncertainties and complexities in the response of coral reefs to future sea-level rise. *Earth Surface Processes and Landforms* 20, 49–64.

Springer, J. S. (1983) Ontario Precambrian dolomite as refractory raw material. *Ontario Geological Survey Miscellaneous Papers* 116, 303–12.

Stallard, R. F. (1995) Tectonic, environmental, and human aspects of weathering and erosion: a global review using a steady-state perspective. *Annual Review of Earth and Planetary Sciences* 23, 11–39.

Stallard, R. F. (1998) Terrestrial sedimentation and the carbon cycle: coupling weathering and erosion to carbon burial. *Global Biogeochemical Cycles* 12, 231–57.

Stallins, J. A. (2006) Geomorphology and ecology: unifying themes for complex systems in biogeomorphology. *Geomorphology* 77, 207–16.

Stamps, D. S., Iaffaldano, G., and Calais, E. (2015) Role of mantle flow in Nubia–Somalia plate divergence. *Geophysical Research Letters* 42, 290–6.

Starkel, L. (1987) Long-term and short-term rhythmicity in terrestrial landforms and deposits. In M. R. Rampino, J. E. Sanders, W. S. Newman, and L. K. Königsson (eds) *Climate: History, Periodicity, and Predictability*, pp. 323–32. New York: Van Nostrand Reinhold.

Statham, I. (1977) *Earth Surface Sediment Transport*. Oxford: Clarendon Press.

Steers, J. A. (ed.) (1971) *Applied Coastal Geomorphology*. London: Macmillan.

Steffen, W., Broadgate, W., Deutsch, L., Gaffney, O., and Ludwig, C. (2015) The trajectory of the Anthropocene: the Great Acceleration. *The Anthropocene Review* 2, 81–98.

Steiger, J. and Corenblit, D. (2012) The emergence of an 'evolutionary geomorphology'? *Central European Journal of Geosciences* 4, 376–82.

Steno, N. (1916) *The Prodomus of Nicolaus Steno's Dissertation concerning a Solid Body enclosed by Process of Nature within a Solid.* An English version with an introduction and explanatory notes by John Garrett Winter, With a foreword by William H. Hobbs (University of Michigan Humanistic Series, vol. XI, Contributions to the History of Science, Part II). New York and London: Macmillan.

Stewart, I. (1997) *Does God Play Dice? The New Mathematics of Chaos*, new edn. Harmondsworth, Middlesex: Penguin Books.

Stoddart, D. R. (ed.) (1997) *Process and Form in Geomorphology.* London: Routledge.

Stokes, S., Thomas, D. G. S., and Washington, R., (1997) Multiple episodes of aridity in southern Africa since the last interglacial period. *Nature* 388, 154–8.

Stoppani, A. (1873) *Corso di Geologia.* Miliano: G. Bernardoni, E. G. Brigola, Editori.

Strahler, A. H. and Strahler, A. N. (1994) *Introducing Physical Geography.* New York: John Wiley & Sons.

Strahler, A. H. and Strahler, A. N. (2006) *Introducing Physical Geography*, 4th edn. New York: John Wiley & Sons.

Strahler, A. N. (1952) Dynamic basis of geomorphology. *Bulletin of the Geological Society of America* 63, 923–38.

Strahler, A. N. (1980) Systems theory in physical geography. *Physical Geography* 1, 1–27.

Strakhov, N. M. (1967) *Principles of Lithogenesis*, vol. 1. Translated by J. P. Fitzsimmons, S. I. Tomkieff, and J. E. Hemingway. Edinburgh: Oliver & Boyd.

Struthers, W. A. K. (1997) From Manchester Docks to Salford Quays: ten years of environmental improvements in the Mersey Basin Campaign. *Journal of the Chartered Institution of Water and Environmental Management* 11, 1–7.

Sugden, D. E. and John, B. S. (1976) *Glaciers and Landscape: A Geomorphological Approach.* London: Edward Arnold.

Summerfield, M. A. (1984) Plate tectonics and landscape development on the African continent. In M. Morisawa and J. T. Hack (eds) *Tectonic Geomorphology* (The Symposia on Geomorphology, International Series no. 15), pp. 27–51. Boston, Mass.: George Allen & Unwin.

Summerfield, M. A. (1991) *Global Geomorphology: An Introduction to the Study of Landforms.* Harlow, Essex: Longman.

Summerfield, M. A. (ed.) (2000) *Geomorphology and Global Tectonics.* Chichester: John Wiley & Sons.

Summerfield, M. A. (2005) A tale of two scales, or the two geomorphologies. *Transactions of the Institute of British Geographers*, New Series 30, 402–15.

Summerfield, M. A. (2007) Internal-external Interactions in the Earth system. In I. Douglas, R. J. Huggett, and C. R. Perkins (eds) *Companion Encyclopedia of Geography: From Local to Global*, 2nd edn, pp. 93–107. London: Routledge.

Summerfield, M. A. and Hulton, N. J. (1994) Natural controls of fluvial denudation in major world drainage basins. *Journal of Geophysical Research* 99, 13, 871–84.

Summerfield, M. A. and Thomas, M. F. (1987) Long-term landform development: editorial comment. In V. Gardiner (ed.) *International Geomorphology 1986, Part II* (Proceedings of the First International Conference on Geomorphology), pp. 927–33. Chichester: John Wiley & Sons.

Sunamura, T. (1992) *Geomorphology of Rocky Coasts.* Chichester: John Wiley & Sons.

Surian, N. (1999) Channel changes due to river regulation: the case of the Piave River, Italy. *Earth Surface Processes and Landforms* 24, 1135–51.

Svendsen, J. I., Alexanderson, H., Astakhov, V. I., Demidov, I., Dowdeswell, J. A., Funder, S., Gataullin, V., Henriksen, M., Hjort, C., Houmark-Nielsen, M., Hubberten, H. W., Ingólfsson, Ó. Jakobsson, M., Kjæri, K. H., Larsen, E., Lokrantz, H., Lunkka, J. P., Lysâ, A., Mangerud, J., Matiouchkov, A., Murray, A., Möller, P., Niessen, F., Nikolskaya, O., Polyak, L., Saarnisto, M., Siegert, C., Siegert, M. J., Spielhagen, R. F., and Stein, R. (2004) Late Quaternary ice sheet history of northern Eurasia. *Quaternary Science Reviews* 23, 1229–71.

Sweeting, M. M. (1950) Erosion cycles and limestone caverns in the Ingleborough District of Yorkshire. *Geographical Journal* 115, 63–78.

Syvitski, J. P. M., Vörösmarty, C. J., Kettner, A. J., and Green, P. (2005) Impact of humans on the flux of terrestrial sediment to the global coastal ocean. *Science* 308, 376–80.

Tansley, A. G. (1935) The use and abuse of vegetational concepts and terms. *Ecology* 16, 284–307.

Tarboton, D. G., Bras, R. L., and Rodriguez-Iturbe, I. (1992) A physical basis for drainage density. *Geomorphology* 5, 59–76.

Tarolli, P. (2014) High-resolution topography for understanding Earth surface processes: opportunities and challenges. *Geomorphology* 216, 295–312.

Tarolli, P., Arrowsmith, J. R. and Vivoni, E. R. (2009) Understanding Earth surface processes from remotely sensed digital terrain models. *Geomorphology* 113, 1–3.

Tatsumi, Y. (2005) The subduction factory: how it operates in the evolving Earth. *GSA Today* 15(7), 4–10.

Taylor, G. and Eggleton, R. A. (2001) *Regolith Geology and Geomorphology.* Chichester: John Wiley & Sons.

Tchakerian, V. P. (1999) Dune palaeoenvironments. In A. S. Goudie, I. Livingstone, and S. Stokes (eds)

Aeolian Environments, Sediments and Landforms (British Geomorphological Research Group Symposia Series, Papers from the Fourth International Conference on Aeolian Research (ICAR 4), held in School of Geography and St Catherine's College, University of Oxford, July 1998), pp. 261–92. Chichester: John Wiley & Sons.

Terry, J. P. (2005) Karst distribution and controls on Yoron-Jima, an emerged reef island in subtropical Japan. *Journal of Cave and Karst Studies* 67, 48–54.

Thom, R. (1975) *Structural Stability and Morphogenesis: Outline of a General Theory of Models.* Reading, Mass.: Benjamin.

Thomas, D. S. G. (1989) The nature of arid environments. In D. S. G. Thomas (ed.) *Arid Zone Geomorphology: Process, Form and Change in Drylands*, pp. 1–8. London: Belhaven Press.

Thomas, D. S. G. (ed.) (2011) *Arid Zone Geomorphology: Process, Form and Change in Drylands*, 3rd edn. Chichester: John Wiley & Sons.

Thomas, D. S. G. and Shaw, P. A. (1991) 'Relict' desert dune systems: interpretations and problems. *Journal of Arid Environments* 20, 1–14.

Thomas, M. F. (1965) Some aspects of the geomorphology of tors and domes in Nigeria. *Zeitschrift für Geomorphologie* NF 9, 63–81.

Thomas, M. F. (1974) *Tropical Geomorphology: A Study of Weathering and Landform Development in Warm Climates.* London and Basingstoke: Macmillan.

Thomas, M. F. (1989a) The role of etch processes in landform development. I. Etching concepts and their application. *Zeitschrift für Geomorphologie* NF 33, 129–42.

Thomas, M. F. (1989b) The role of etch processes in landform development. II. Etching and the formation of relief. *Zeitschrift für Geomorphologie* NF 33, 257–74.

Thomas, M. F. (1994) *Geomorphology in the Tropics: A Study of Weathering and Denudation in Low Latitudes.* Chichester: John Wiley & Sons.

Thomas, M. F. (1995) Models for landform development on passive margins: some implications for relief development in glaciated areas. *Geomorphology* 12, 3–15.

Thomas, M. F. and Thorp, M. B. (1985) Environmental change and episodic etchplanation in the humid tropics of Sierra Leone: the Koidu etchplain. In I. Douglas and T. Spencer (eds) *Environmental Change and Tropical Geomorphology*, pp. 239–67. London: George Allen & Unwin.

Thompson, W. F. (1990) Climate related landscapes in world mountains: criteria and maps. *Zeitschrift für Geomorphologie Supplementband* 78.

Thorarinsson, S. (1964) *Surtsey, the New Island in the North Atlantic.* New York: Viking Press.

Thorn, C. E. (1988) *An Introduction to Theoretical Geomorphology.* Boston, Mass.: Unwin Hyman.

Thorn, C. E. (1992) Periglacial geomorphology: what, where, when? In J. C. Dixon and A. D. Abrahams (eds) *Periglacial Geomorphology*, pp. 3–31. Chichester: John Wiley & Sons.

Thorn, C. E. and Hall, K. (2002) Nivation and cryoplanation: the case for scrutiny and integration. *Progress in Physical Geography* 26, 533–50.

Thorn, C. E. and Welford, M. R. (1994) The equilibrium concept in geomorphology. *Annals of the Association of American Geographers* 84, 666–96.

Thornbury, W. D. (1954) *Principles of Geomorphology*, 1st edn. New York: John Wiley & Sons.

Thorne, C. R., Hey, R. D., and Newson, M. D. (1997) *Applied Fluvial Geomorphology for River Engineering and Management.* Chichester: John Wiley & Sons.

Thornes, J. B. (1983) Evolutionary geomorphology. *Geography* 68, 225–35.

Thornes, J. B. (ed.) (1990) *Vegetation and Erosion: Processes and Environments.* Chichester: John Wiley & Sons.

Thornthwaite, C. W. (1948) An approach towards a rational classification of climate. *Geographical Review* 38, 55–94.

Thornthwalte, C. W., Sharpe, C. F. S., and Dosch, E. F. (1942) *Climate and Accelerated Erosion in the Arid and Semi-arid Southwest with Special Reference to the Polacca Wash Drainage Basin, Arizona* (US Department of Agriculture, Technical Bulletin 808). Washington, D.C.: US Government Printing Office.

Tîrlă, L. and Vijulie, I. (2013) Structural–tectonic controls and geomorphology of the karst corridors in alpine limestone ridges: Southern Carpathians, Romania. *Geomorphology* 197, 123–36.

Tómasson, H. (1996) The jökulhlaup from Katla in 1918. *Annals of Glaciology* 22, 249–54.

Tooth, S. (2009) Invisible geomorphology? *Earth Surface Processes and Landforms* 34, 752–4.

Tooth, S. and Nanson, G. C. (1999) Anabranching rivers of the Northern Plains of arid central Australia. *Geomorphology* 29, 211–33.

Trenhaile, A. S. (1987) *The Geomorphology of Rock Coasts.* Oxford: Oxford University Press.

Trenhaile, A. S. (1997) *Coastal Dynamics and Landforms.* Oxford: Clarendon Press.

Trenhaile, A. S. (1998) *Geomorphology: A Canadian Perspective.* Toronto: Oxford University Press.

Tricart, J. (1985) Evidence of Upper Pleistocene dry climates in northern South America. In I. Douglas and T. Spencer (eds) *Environmental Change and Tropical Geomorphology*, pp. 197–217. London: George Allen & Unwin.

Tricart, J. and Cailleux, A. (1972) *Introduction to Climatic Geomorphology.* Translated from the French by Conrad J. Kiewiet de Jonge. London: Longman.

Trimble, S. W. (1983) A sediment budget for Coon Creek basin in the driftless area, Wisconsin. *American Journal of Science* 283, 454–74.

Trimble, S. W. (1995) Catchment sediment budgets and change. In A. Gurnell and G. Petts (eds) *Changing River Channels*, pp. 201–15. Chichester: John Wiley & Sons.

Trimble, S. W. (1999) Rates of soil erosion. *Science* 286, 1477–8.

Trudgill, S. (1985) *Limestone Geomorphology*. Harlow: Longman.

Tsou, C.-Y., Chigira, M., Matshushi, Y., and Chen, S.-C. (2015) Deep-seated gravitational deformation of mountain slopes caused by river incision in the Central Range, Taiwan: spatial distribution and geological characteristics. *Engineering Geology* 196, 126–38.

Tucker, G. E. and Hancock, G. R. (2010) Modelling landscape evolution. *Earth Surface Processes and Landforms* 35, 28–50.

Twidale, C. R. (1971) *Structural Landforms: Landforms Associated with Granitic Rocks, Faults, and Folded Strata* (*An Introduction to Systematic Geomorphology*, vol. 5). Cambridge, Mass. and London: MIT Press.

Twidale, C. R. (1976) On the survival of paleoforms. *American Journal of Science* 276, 77–95.

Twidale, C. R. (1991) A model of landscape evolution involving increased and increasing relief amplitude. *Zeitschrift für Geomorphologie* NF 35, 85–109.

Twidale, C. R. (1994) Gondwanan (Late Jurassic and Cretaceous) palaeosurfaces of the Australian craton. *Palaeogeography, Palaeoclimatology, Palaeoecology* 112, 157–86.

Twidale, C. R. (1997) Persistent and ancient rivers: some Australian examples. *Physical Geography* 18, 291–317.

Twidale, C. R. (1998) Antiquity of landforms: an 'extremely unlikely' concept vindicated. *Australian Journal of Earth Sciences* 45, 657–68.

Twidale, C. R. (1999) Landforms ancient and recent: the paradox. *Geografiska Annaler* 81A, 431–41.

Twidale, C. R. (2002) The two-stage concept of landform and landscape development involving etching: origin, development and implications of an idea. *Earth-Science Reviews* 57, 37–74.

Twidale, C. R. (2004) River patterns and their meaning. *Earth-Science Reviews* 67, 159–218.

Twidale, C. R. and Campbell, E. M. (1993) *Australian Landforms: Structure, Process and Time*. Adelaide: Gleneagles Publishing.

Twidale, C. R. and Campbell, E. M. (1995) Pre-Quaternary landforms in the low latitude context: the example of Australia. *Geomorphology* 12, 17–35.

Twidale, C. R. and Campbell, E. M. (2005) *Australian Landforms: Understanding a Low, Flat, Arid and Old Landscape*. Kenthurst, New South Wales: Rosenberg Publishing.

Twidale, C. R. and Lageat, Y. (1994) Climatic geomorphology: a critique. *Progress in Physical Geography* 18, 319–34.

Twidale, C. R., Bourne, J. A., and Smith, D. M. (1974) Reinforcement and stabilisation mechanisms in landform development. *Révue de Géomorphologie Dynamique* 28, 81–95.

Twidale, C. R., Bourne, J. A., and Vidal-Romani J. R. (1999) Bornhardt inselbergs in the Salt River Valley, south of Kellerberrin, Western Australia (with notes on a tessellated pavement in granite and pinnacles in laterite). *Journal of the Royal Society of Western Australia* 82, 33–49.

Vail, P. R., Audemard, F., Bowman, S. A., Eisner, P. N., and Periz-Cruz, G. (1991) The stratigraphic signatures of tectonics, eustasy and sedimentology: an overview. In G. Einsele, W. Ricken, and A. Seilacher (eds) *Cycles and Events in Stratigraphy*, pp. 617–59. Berlin: Springer.

Vail, P. R., Mitchum, R. M., Jr. and Thompson III, S. (1977) Seismic stratigraphy and global changes of sea level. Part 4: Global cycles of relative changes of sea levels. In C. E. Payton (ed.) *Seismic Stratigraphy: Applications to Hydrocarbon Exploration* (American Association of Petroleum Geologists memoir 26), pp. 83–97. Tulsa, Okla.: American Association of Petroleum Geologists.

van Andel, T. H., Runnels, C. N., and Pope, K. O. (1986) Five thousand years of land use and abuse in the Southern Argolid, Greece. *Hesperia* 55, 103–28.

Van Dyke, C. (2013) Channels in the making – an appraisal of channel evolution models. *Geography Compass* 7, 759–77.

van Loon, A. J. (2004) From speculation to model: the challenge of launching new ideas in the Earth sciences. *Earth-Science Reviews* 65, 305–13.

van Maanen, B., Nicholls, R. J., French, J. R., Barkwith, A., Bonaldo, D., Burningham, H., Murray, A. B., Payo, A., Sutherland, J., Thornhill, G., Townend, I. H., van der Wegen, M., and Walkden, M. J. A. (2015) Simulating mesoscale coastal evolution for decadal coastal management: a new framework integrating multiple, complementary modelling approaches. *Geomorphology* In press.

van't Hoff, J. H. (1884) *Études de Dynamique chemique*. Amsterdam: Friederik Muller.

Varnes, D. J. (1978) Slope movement and types and processes. In R. L. Schuster and R. J. Krizek (eds) *Landslides: Analysis and Control* (Transportation Research Board special report 176), pp. 11–33. Washington, D.C.: National Academy of Sciences.

Viles, H. A. (ed.) (1988) *Biogeomorphology*. Oxford: Basil Blackwell.

Viles, H. and Spencer, T. (1996) *Coastal Problems: Geomorphology, Ecology and Society at the Coast*. London: Arnold.

Vita-Finzi, C. (1969) *The Mediterranean Valleys: Geological Changes in Historical Times*. Cambridge: Cambridge University Press.

Vita-Finzi, C. (1973) *Recent Earth History*. London and Basingstoke: Macmillan.

Vörösmarty, C. J., Meybeck, M., Fekete, B., Sharma, K., Green, P., and Syvitski, J. P. M. (2003) Anthropogenic sediment retention: major global impact from registered river impoundments. *Global and Planetary Change* 39, 169–90.

Walker, D. A., Webber, P. J., Binnian, E. F., Everett, K. R., Lederer, N. D., Nordstrand, E. A., and Walker, M. D. (1987) Cumulative impacts of oil fields on northern Alaskan landscapes. *Science* 238, 757–61.

Walker, J. (1999) The application of geomorphology to the management of river-bank erosion. *Journal of the Chartered Institution of Water and Environmental Management* 13, 297–300.

Walker, M. J. C. (2005) *Quaternary Dating Methods*. Chichester: John Wiley & Sons.

Wallace, R. E. (1977) Profiles and ages of young fault scarps, north-central Nevada. *Bulletin of the Geological Society of America* 88, 1267–81.

Wallin, T. R. and Harden, C. P. (1996) Estimating trail-related soil erosion in the humid tropics: Jatun Sacha, Ecuador, and La Selva, Costa Rica. *Ambio* 25, 517–22.

Walling, D. E. and Fang, D. (2003) Recent trends in the suspended sediment load of the world's rivers. *Global and Planetary Change* 39, 111–26.

Walling, D. E. and Webb, B. W. (1986) The dissolved load of rivers: a global overview. In K. J. Gregory (ed.) *Dissolved Load of Rivers and Surface Water Quantity/Quality Relationships* (IAHS publication 141), pp. 3–20. Rozendaalselaan, Netherlands: International Association of Hydrological Sciences.

Walther, J. (1900) *Das Gesetz der Wüstenbildung in Gegenwart und Vorzeit*. Berlin: Dietrich Reimer.

Waltz, J. P. (1969) Ground water. In R. J. Chorley (ed.) *Water, Earth, and Man: A Synthesis of Hydrology, Geomorphology, and Socio-Economic Geography*, pp. 259–67. London: Methuen.

Warburton, J. and Caine, N. (1999) Sorted patterned ground in the English Lake District. *Permafrost and Periglacial Processes* 10, 193–7.

Warburton, J. and Danks, M. (1998) Historical and contemporary channel change, Swinhope Burn. In J. Warburton (ed.) *Geomorphological Studies in the North Pennines: Field Guide*, pp. 77–91. Durham: Department of Geography, University of Durham, British Geomorphological Research Group.

Warburton, J., Danks, M., and Wishart, D. (2002) Stability of an upland gravel-bed stream, Swinhope Burn, Northern England. *Catena* 49, 309–29.

Ward, S. D. and Evans, D. F. (1976) Conservation assessment of British limestone pavements based on floristic criteria. *Biological Conservation* 9, 217–33.

Warwick, G. T. (1953) *The Geomorphology of the Dove–Manifold Region*. Unpublished PhD thesis, University of Birmingham, England.

Warwick, G. T. (1964) Dry valleys in the southern Pennines. *Erdkunde* 18, 116–23.

Washburn, A. L. (1979) *Geocryology: A Survey of Periglacial Processes and Environments*. London: Edward Arnold.

Waters, C. N., Zalasiewicz, J., Summerhayes, C., Barnosky, A. D., Poirier, C., Galuszka, A., Cearreta, A., Edgeworth, M., Ellis, E. C., Ellis, M., Jeandel, C., Leinfelder, R., McNeill, J. R., Richter, D. deB., Steffen, W., Syvitski, J., Vidas, D., Wagreich, M., Williams, M., Zhisheng, A., Grinevald, J., Odada, E., Oreskes, N., and Wolfe, A. P. (2016) The Anthropocene is functionally and stratigraphically distinct from the Holocene. *Science* 351 (6269), 137–47.

Weaver, T. and Dale, D. (1978) Trampling effects of hikers, motorcycles and horses in meadows and forests. *Journal of Applied Ecology* 15, 451–7.

Weertman, J. (1957) On the sliding of glaciers. *Journal of Glaciology* 3, 33–8.

Wellman, H. W. and Wilson, A. T. (1965) Salt weathering, a neglected geological erosion agent in coastal and arid environments. *Nature* 205, 1097–8.

Wells, G. (1989) Observing earth's environment from space. In L. Friday and R. Laskey (eds) *The Fragile Environment: The Darwin College Lectures*, pp. 148–92. Cambridge: Cambridge University Press.

Werner, B. T. and McNamara, D. E. (2007) Dynamics of coupled human–landscape systems. *Geomorphology* 91, 393–407.

Westbroek, P. (1991) *Life as a Geological Force: Dynamics of the Earth*. New York: W. W. Norton.

Wheaton, J. M., Gibbins, C., Wainwright, J., Larsen, L., and McElroy, B. (2011) Preface: multiscale feedbacks in ecogeomorphology. *Geomorphology* 126, 265–8.

Whipple, K. X. (2004) Bedrock rivers and the geomorphology of active orogens. *Annual Review of Earth and Planetary Sciences* 32, 151–85.

White, J. F. (1966) Concave–convex landslopes: a geometrical study. *Ohio Journal of Science* 66, 592–608.

White, W. B. (1976) Cave minerals and speleothems. In T. D. Ford and C. H. D. Cullingford (eds) *The Science of Speleology*, pp. 267–327. London: Academic Press.

White, W. D., Jefferson, G. L., and Hama, J. F. (1966) Quartzite karst in southeastern Venezuela. *International Journal of Speleology* 2, 309–14.

Whitesides, C. J. (2015) The bioturbation of Olympic marmots (*Marmota olympus*) and their impacts on soil properties. *Physical Geography* 36, 202–14.

Whitney, J. W. and Harrington, C. D. (1993) Relict colluvial boulder deposits as paleoclimatic indicators in the Yucca Mountain region, southern Nevada. *Bulletin of the Geological Society of America* 105, 1008–18.

Widdowson, M. (ed.) (1997) *Palaeosurfaces: Recognition, Reconstruction and Palaeoenvironmental Interpretation* (special publication 120). London: Geological Society.

Wilkinson, B. H. (2005) Humans as geological agents: a deep-time perspective. *Geology* 33, 161–4.

Willett, S. D. (1999) Orography and orogeny: the effects of erosion on the structure of mountain belts. *Journal of Geophysical Research* 104, 28, 957–81.

Williams, M. A. J., Dunkerley, D., de Deckker, P., Kershaw, P., and Chappell, J. (1998) *Quaternary Environments*, 2nd edn. London: Arnold.

Williams, M., Zalasiewicz, J., Haff, P. K., Schwägerl, C., Barnosky, A. D., and Ellis, E. C. (2015) The Anthropocene biosphere. *The Anthropocene Review* 2015, 1–24.

Williams, P. J. and Smith, M. W. (1989) *The Frozen Earth: Fundamentals of Geocryology*. Cambridge: Cambridge University Press.

Williams, P. W. (1969) The geomorphic effects of groundwater. In R. J. Chorley (ed.) *Water, Earth, and Man: A Synthesis of Hydrology, Geomorphology and Socio-Economic Geography*, pp. 269–84. London: Methuen.

Williams, R. B. G. and Robinson, D. A. (1989) Origin and distribution of polygonal cracking of rock surfaces. *Geografiska Annaler* 71A, 145–59.

Wilson, I. (1971) Desert sandflow basins and a model for the development of ergs. *Geographical Journal* 137, 180–99.

Wilson, J. P. (2012) Digital terrain modeling. *Geomorphology* 137, 107–21.

Wilson, J. P. and Gallant J. C. (eds) (2000) *Terrain Analysis: Principles and Applications*. New York: John Wiley & Sons.

Wilson, J. P. and Seney, J. P. (1994) Erosional impacts of hikers, horses, motorcycles and off-road bicycles on mountain trails in Montana. *Mountain Research and Development* 14, 77–88.

Wilson, J. T. (1968) Static or mobile Earth: the current scientific revolution. *Proceedings of the American Philosophical Society* 112, 309–20.

Wilson, P. (2011) Lake District hillslopes. *Geology Today* 27, 149–53.

Wirthmann, A. (2000) *Geomorphology of the Tropics*. Translated by Detlef Busche. Berlin, Heidelberg: Springer.

Wischmeier, W. H. and Smith, D. D. (1978) *Predicting Rainfall Erosion Losses: A Guide to Conservation Planning* (USDA Agricultural Handbook 537). Washington, D.C.: United States Department of Agriculture, Science and Education Administration.

Wohl, E. (2015) Particle dynamics: the continuum of bedrock to alluvial river segments. *Geomorphology* 241, 192–208.

Wohletz, K. H. and Sheridan, M. F. (1983) Hydrovolcanic explosions. II. Evolution of basaltic tuff rings and tuff cones. *American Journal of Science* 283, 385–413.

Wolman, M. G. and Miller J. P. (1960) Magnitude and frequency of forces in geomorphic processes. *Journal of Geology* 68, 54–74.

Womack, W. R. and Schumm, S. A. (1977) Terraces of Douglas Creek, northwestern Colorado: an example of episodic erosion. *Geology* 5, 72–6.

Wood, A. (1942) The development of hillside slopes. *Proceedings of the Geologists' Association* 53, 128–39.

Woodroffe, C. D. (1990) The impact of sea-level rise on mangrove shorelines. *Progress in Physical Geography* 14, 483–520.

Woodroffe, C. D. (2002) *Coasts: Form, Process and Evolution*. Cambridge: Cambridge University Press.

Woodward, H. P. (1936) Natural Bridge and Natural Tunnel, Virginia. *Journal of Geology* 44, 604–16.

Wooldridge, S. W. (1932) The cycle of erosion and the representation of relief. *Scottish Geographical Magazine* 48, 30–6.

Wooldridge, S. W. and Linton, D. L. (1939) *Structure, Surface and Drainage in South-East England* (Institute of British Geographers special publication 10). London: Institute of British Geographers.

Wooldridge, S. W. and Linton, D. L. (1955) *Structure, Surface and Drainage in South-East England*. London: George Philip.

Worster, D. (1979) *Dust Bowl: The Southern Plains in the 1930s*. New York: Oxford University Press.

Wright, J. V., Smith, A. L., and Self, S. (1980) A working terminology of pyroclastic deposits. *Journal of Volcanology and Geothermal Research* 8, 315–36.

Wright, L. D. (1985) River deltas. In R. A. Davies (ed.) *Coastal Sedimentary Environments*, pp. 1–76. New York: Springer.

Wright, T. J., Ebinger, C., Biggs, J., Ayele, A., Yirgu, G., Keir, D., and Stork, A. (2006) Magma-maintained rift segmentation at continental rupture in the 2005 Afar dyking episode. *Nature* 442, 291–4.

Yatsu, E. (1988) *The Nature of Weathering: An Introduction*. Tokyo: Sozosha.

Young, A. (1971) Slope profile analysis: the systems of best units. In D. Brunsden (compiler) *Slopes Form and Process* (Institute of British Geographers special publication 3), pp. 1–13. London: Institute of British Geographers.

Young, A. (1974) The rate of slope retreat. In E. H. Brown and R. S. Waters (eds) *Progress in Geomorphology: Papers in Honour of David L. Linton* (Institute of British Geographers special publication 7), pp. 65–78. London: Institute of British Geographers.

Young, A. and Saunders, I. (1986) Rates of surface processes and denudation. In A. D. Abrahams (ed.) *Hillslope Processes*, pp. 1–27. Boston, Mass.: Allen & Unwin.

Young, R. W. (1983) The tempo of geomorphological change: evidence from southeastern Australia. *Journal of Geology* 91, 221–30.

Zalasiewicz, J., Waters, C. N., Williams, M., Barnosky, A. D., Cearreta, A., Crutzen, P., Ellis, E., Ellis, M. A., Fairchild, I. J., Grinevald, J., Haff, P. K., Hajdas, I., Leinfelder, R., McNeill, J., Odada, E. O., Poirier, C., Richter, D., Steffen, W., Summerhayes, C., Syvitski, J. P. M., Vidas, D., Wagreich, M., Wing, S. L., Wolfe, A. P., An, Z., and Oreskes, N. (2015) When did the Anthropocene begin? A mid-twentieth century boundary level is stratigraphically optimal. *Quaternary International* 383, 196–203.

Zalasiewicz, J., Williams, M., Smith, A., Barry, T. L., Coe, A. L., Bown, P. R., Brenchley, P., Cantrill, D., Gale, A., Gibbard, P., Gregory, F. J., Hounslow, M. W., Kerr, A. C., Pearson, P., Knox, R., Powell, J., Waters, C., Marshall, J., Oates, M., Rawson, P., and Stone, P. (2008) Are we now living in the Anthropocene? *GSA Today* 18, 4–8.

Zemp, M., Hoelzle, M., and Haeberli, W. (2009) Six decades of glacier mass-balance observations: a review of the worldwide monitoring network. *Annals of Glaciology* 50, 101–11.

Zemp, M., Roer, I., Kääb, A., Hoelzle, M., Paul, F., and Haeberli, W. (eds) (2008) *Global Glacier Changes: Facts and Figures*. Zurich, Switzerland: UNEP (United Nations Environment Programme), World Glacier Monitoring Service.

Zobeck, T. M., Parker, N. C., Haskell, S., and Guoding, K. (2000) Scaling up from field to region for wind erosion prediction using a field-scale wind erosion model and GIS. *Agriculture, Ecosystems and Environment* 2, 247–59.

INDEX

Page numbers in **bold** represent figures.
Page numbers in *italics* represent tables.